how things work

THE PHYSICS OF EVERYDAY LIFE

Louis A. Bloomfield

The University of Virginia

3rd Edition

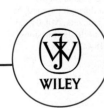

John Wiley & Sons, Inc.

*To Karen for your friendship, insight, and support all these years,
to Elana and Aaron for sharing your wonderful life adventures with us,
to Sadie for being endlessly entertaining, and to the students of the
University of Virginia for inspiring this book with your curiosity,
enthusiasm, and humanity.*

SENIOR ACQUISITIONS EDITOR	Stuart Johnson
PROJECT EDITOR	Geraldine Osnato
EXECUTIVE PUBLISHER	Kaye Pace
EXECUTIVE MARKETING MANAGER	Amanda Wygal
EDITORIAL ASSISTANT	Alyson Rentrop/Krista Jarmas
PRODUCTION MANAGER	Pamela Kennedy
SENIOR PRODUCTION EDITOR	Sarah Wolfman-Robichaud
SENIOR CREATIVE DIRECTOR/DESIGNER	Harry Nolan
SENIOR ILLUSTRATION EDITOR	Sandra Rigby
PHOTO EDITOR	Tara Sanford
SENIOR MEDIA EDITOR	Tom Kulesa
ASSISTANT MEDIA EDITOR	Gabriel Dillon
COPYEDITOR	Connie Parks
PROJECT MANAGER	Jane Shifflet/GTS Companies: PA/York Campus
COVER PHOTO	Image Source Limited/Index Stock

This book was set in 10/12 Times Roman by GTS Companies and printed and bound by Quebecor World Perú. The cover was printed by Quebecor World Perú.

This book is printed on acid free paper. ∞

To order books or for customer service please, call 1-800-CALL WILEY (225-5945).

ISBN-13 978- 0-471-46886-8
ISBN-10 0-471-46886-X

10 9 8 7 6 5 4 3 2

Foreword

In today's world we are surrounded by science and by the technology that has grown out of that science. For most of us, this is making the world increasingly mysterious and somewhat ominous as technology becomes ever more powerful. For instance, we are confronted by many global environmental questions such as the dangers of greenhouse gases and the best choices of energy sources. These are questions that are fundamentally technical in nature and there is a bewildering variety of claims and counterclaims as to what is "the truth" on these and similar important scientific issues. For many people, the reaction is to throw up their hands in hopeless frustration and accept that the modern world is impossible to understand and one can only huddle in helpless ignorance at the mercy of its mysterious and inexplicable behavior.

In fact, much of the world around us and the technology of our every day lives is governed by a few basic physics principles, and once these principles are understood, the world and the vast array of technology in our lives become understandable and predictable. How does your microwave oven heat up food? Why is your radio reception bad in some places and not others? And why can birds happily land on a high voltage electrical wire? The answers to questions like these are obvious once you know the relevant physics. Unfortunately, you are not likely to learn that from a standard physics course or physics textbook. There is a large body of research showing that instead of providing this improved understanding of every day life, most introductory physics courses are doing quite the opposite. In spite of the best intentions of the teachers, most students are "learning" that physics is abstract, uninteresting, and unrelated to the world around them.

How Things Work is a dramatic step towards changing that by presenting physics in a new way. Instead of starting out with abstract principles that leave the reader with the idea that physics is about artificial and uninteresting ideas, Lou Bloomfield starts out talking about real objects and devices that we encounter in our everyday lives. He then shows how these seemingly magical devices can be understood in terms of the basic physics principles that govern their behavior. This is much the way that most physics was discovered in the first place; people asked why the world around them behaved as it did and as a result discovered the principles that explained and predicted what they observed.

I have been using this book in my classes for several years and I continue to be impressed with how Lou can take seemingly highly complex devices and strip away the complexity to show how at their heart are simple physics ideas. Once these ideas are understood, they can be used to understand the behavior of many devices we encounter in our daily lives, and often even fix things that before had seemed impossibly complex. In the process of teaching from this book, I have increased my own understanding of the physics behind much of the world around me. In fact, after consulting *How Things Work,* I have had the confidence to confront both plumbers and air-conditioner repairmen to tell them (correctly as it turned out) that their diagnosis did not make sense and they needed to do something different to solve my plumbing and AC problems. Now I am regularly amused at the misconceptions some trained physicists have about some of the physics they encounter in their daily lives, such as how a microwave oven works and why it can be made out of metal walls, but putting aluminum foil in it is bad. It has convinced me that we need to take the approach used in this book in far more of our science texts.

Of course, the most important impact is on the students in my classes that use this book. These are typically nonscience students majoring in fields such as film studies, classics, English, business, etc. They often come to physics with considerable trepidation. It is inspiring to see many of them discover to their surprise that physics is very different from what they thought— that physics can actually be interesting and useful and makes the world a much less mysterious and more understandable place. I remember many examples of seeing this in action: the student who, after learning how both speakers and TVs work, was suddenly able to understand that it was not magic that putting his large speaker next to the TV distorted the picture but in fact it was just physics, and now he knew just how to fix it; the young woman scuba diver who, after learning about light and color, suddenly interrupted class to announce that now she understood why it was that you could tell how deep you were by seeing what color lobsters appeared; or the students who announced that suddenly it made sense that the showers on the first floor of the dorm worked better than those on the second floor. In addition, of course everyone is excited to learn how a microwave oven works and why there are these strange rules as to what you can and cannot put in it. These examples are particularly inspiring to a teacher, because they tell you that the students are not just learning the material presented in class, but they are then able to apply that understanding to new situations in a useful way, something that happens far too seldom in science courses.

Whether a curious layperson, a trained physicist, or a beginning physics student, most everyone will find this book an interesting and enlightening read and will go away comforted in that the world is not so strange and inexplicable after all.

Carl Wieman
Nobel Laureate in Physics 2001
CASE/Carnegie US University Professor of the Year 2004

Contents

Preface

This book is an unconventional introduction to physics and science that starts with whole objects and looks inside them to see what makes them work. It's written for students who seek a connection between science and the world in which they live. Though reluctant to study science as an academic exercise, these individuals show remarkable enthusiasm for it when it's presented in context. Many of the 7500 students I've taught during the past fifteen years have been surprised at their own interest in the course, have looked forward to classes, have asked insightful questions, have experimented on their own, and have found themselves explaining to friends and family how things in their world work.

How Things Work brings science to the reader rather than the reverse. Like the course in which it developed, this book has always been for nonscientists and is written with their interests in mind. Nonetheless, it has attracted students from the sciences, engineering, architecture, and other technical fields who wish to put scientific concepts into context.

Most physics texts develop the principles of physics first and present real-life examples of these principles reluctantly if at all. That teaching method is abstract and inaccessible, providing few conceptual footholds for students as they struggle to understand unfamiliar principles. After all, the comforts of experience and intuition lie in the examples, not in the principles. While a methodical and logical development of scientific principles can be satisfying to the seasoned scientist, it's alien to someone who doesn't even recognize the language being used.

This book is written in English and organized in a case-study fashion. It conveys an understanding and appreciation for physics by finding physics concepts and principles within the familiar objects of everyday experience. Because its structure is defined by real-life examples, this book necessarily discusses concepts as they're needed and then revisits them later on when they reappear in other objects. What better way is there to show the universality of the natural laws?

Changes in the Third Edition

Content Changes

- **Reorganization of chapters and sections.** While it's possible to study all of physics in a single object, most objects exemplify a particular area of physics especially well. Rather than looking for too many disparate physical issues, I now concentrate the exploration of each object on the physics it portrays best. The result is a smoother, more coherent discussion with fewer digressions or jarring transitions. Achieving such steady focus required that I rearrange the objects slightly and separate them into 16 chapters rather than the 14 chapters of the second edition.

- **More even treatment of formulaic physics.** Although this book is primarily about concepts, it contains just enough formulaic physics to show that physics can make precise, quantitative predictions about the world around us. Even a nonscientist may find those formulae useful from time to time. While the second edition was inconsistent in offering formulae for some of the physics concepts it introduced, the third edition is more uniform in that regard. Readers with no interest in quantitative physics, however, can safely disregard most of those formulae.

- **Wider coverage of physics concepts.** No one book can or should cover all of physics, but the second edition had some significant omissions. This edition provides new or improved coverage of some important physics issues, including material phases and phase transitions, mechanical waves, quantum physics, physics of solids, atomic structure, vision, radioactivity and half-life, and relativity.

- **Rewritten chapters on fluids and motion, electricity, and magnetism and electrodynamics.** Both students and instructors struggled with these three chapters in the second edition. In response to much constructive criticism, I rewrote these chapters almost from scratch. The new versions are much more approachable, employ better objects, and develop the physics of those objects more gradually and more carefully.

- **Keeping up with new technology.** While physics at the introductory level changes fairly slowly, the objects in which that physics appears change almost daily. This edition reflects many of those changes, particularly in the sections on audio players, lasers and LEDs, and cameras. I have omitted a section on television from this edition because television is in such a state of flux that whatever I write now will be obsolete or irrelevant in a year or two.

- **Inclusion of both SI and English units.** Scientists may wish otherwise, but the vast majority of people in the United States will continue to think in English units for the foreseeable future. In recognition of that reality, this edition provides both SI (metric) and English units whenever appropriate.

- **Print book and websites are separate.** The second edition tried to straddle the print and electronic worlds, an idea with more promise than practicality. The third edition returns entirely to print, with reliable websites that supplement the book instead of trying to be part of it.

Feature Changes

- **Common misconceptions and intuition alert boxes.** What people have heard or think that they've experienced is not always how our world actually works. There are a number of common misconceptions and misleading intuitions that can easily confuse students and shake their confidence in what they are learning. In this edition, I explicitly discuss the conflicts between what students naively expect of the world and what physics has to say about it. For students to succeed in placing physics in their everyday lives, they need to resolve the contradictions between what they are taught and what they think they already know. These boxed features will help students achieve that resolution.

- **Page references to chapter summaries and answers to questions.** A reader seeking a more detailed introduction to a chapter is now guided directly to the chapter summary. And when looking for the answer and explanation to a Check Your Understanding or Check Your Figures question, they'll know the page on which to look.

- **Better visual cues for key points.** Most key physics concepts and laws now appear in boxed statements. This highlighting calls attention to important concepts and makes the material easier to review.

The Goals of This Book

As they read this book, students should:

1. Begin to see science in everyday life. Science is everywhere; we need only open our eyes to see it. We're surrounded by things that can be understood in terms of science, much of which is within a student's reach. Seeing science doesn't mean that when viewing an oil painting they should note only the selective reflection of incident light waves by organic and inorganic molecules. Rather, they should realize that there's a beauty to science that complements aesthetic beauty. They can learn to look at a glorious red sunset and appreciate both its appearance and why it exists.

2. Learn that science isn't frightening. The increasing technological complexity of our world has instilled within most people a significant fear of science. As the gulf widens between those who create technology and those who use it, their ability to understand one another and communicate diminishes. The average person no longer tinkers with anything and many modern devices are simply disposable, being too complicated to modify or repair. To combat the anxiety that accompanies unfamiliarity, this book shows students that most objects can be examined and understood, and that the science behind them isn't scary after all. The more we understand how others think, the better off we'll all be.

3. Learn to think logically in order to solve problems. Because the universe obeys a system of well-defined rules, it permits a logical understanding of its behaviors. Like mathematics and computer science, physics is a field of study where logic reigns supreme. Having learned a handful of simple rules, students can combine them logically to obtain more complicated rules and be certain that those new rules are true. So the study of physical systems is a good place to practice logical thinking.

4. Develop and expand their physical intuition. When you're exiting from a highway, you don't have to consider velocity, acceleration, and inertia to know that you should brake gradually—you already have physical intuition that tells you the consequences of doing otherwise. Such physical intuition is essential in everyday life, but it ordinarily takes time and experience to acquire. This book aims to broaden a student's physical intuition to situations they normally avoid or have yet to encounter. That is, after all, one of the purposes of reading and scholarship: to learn from other people's experiences.

5. Learn how things work. As this book explores the objects of everyday life, it gradually uncovers most of the physical laws that govern the universe. It reveals those laws as they were originally discovered: while trying to understand real objects. As they read this book and learn these laws, students should begin to see the similarities between objects, shared mechanisms, and recurring themes that are reused by nature or by people. This book reminds students of these connections and is ordered so that later objects build on their understanding of concepts encountered earlier.

6. Begin to understand that the universe is predictable rather than magical. One of the foundations of science is that effects have causes and don't simply occur willy-nilly. Whatever happens, we can look backward in time to find what caused it. We can also predict the future to some extent, based on insight acquired from the past and on knowledge of the present. And where predictability is limited, we can understand those limitations. What distinguishes the physical sciences and mathematics from other fields is that there are often absolute answers, free from inconsistency, contraindication, or paradox. Once students understand how the physical laws govern the universe, they can start to appreciate that perhaps the most magical aspect of our universe is that it is not magic; that it is orderly, structured, and understandable.

7. Obtain a perspective on the history of science and technology. None of the objects that this book examines appeared suddenly and spontaneously in the workshop of a single individual who was oblivious to what had been done before. These objects were developed in the context of history by people who were generally aware of what they were doing and usually familiar with any similar objects that already existed. Nearly everything is discovered or developed when related activities make their discoveries or developments inevitable and timely. To establish that historical context, this book describes some of the history behind the objects it discusses.

Student Website

This book is supported by an array of web-based student supplements. Starting from the website on this book's cover, a student has free access to resources that include:

- **Review questions and physics concept summaries**

- **Additional chapters, sections, and subsections**

- **Additional exercises, problems, and cases**

Instructor Website and Instructor's Manual

The instructor's website, accessible from the same URL, provides everything described above, plus access to:

- **Additional homework questions and solutions**
- **Test questions and solutions**

Wiley PLUS

The third edition of *How Things Work* is available in *Wiley PLUS. Wiley PLUS* is an on-line tool that allows for easy course administration. Selected end of chapter problems as well as the entire Test Bank have been coded for on-line homework assignments. *Wiley PLUS* also features a multimedia version of the text that can be purchased at a significantly reduced price.

- **Organizational ideas for designing a course**
- **Demonstration ideas for each section**
- **Lecture slides for each section**
- **Artwork to use in presentations**
- **Resource lists**

Acknowledgments

I didn't write this book in a vacuum, thank goodness. Its preparation required the help of a great many wonderful people to whom I am enormously indebted. First among them is my editor, Stuart Johnson, whose enthusiasm for this project continues unabated after nearly a decade. It has also been a pleasure working almost daily with Geraldine Osnato during the book's development process and Sarah Wolfman-Robichaud during its production; I couldn't ask for two more friendly, encouraging collaborators. Harry Nolan has designed a beautiful format for the book, Tara Sanford has located great photographs, Sandra Rigby and Julie Horan have done a stunning job with the line art, and Connie Parks has been eagle-eyed in finding my clumsy mistakes before they get into print. Thanks to Dana Kasowitz and Matthew Breitwisch for helping me learn what instructors and students want from this book, to Amanda Wygal for helping to plan the book and coordinate its dissemination, and to Krista Jarmas and Aly Rentrop for able assistance along the way. Thanks also to my family, Karen, Elana, and Aaron Bloomfield, for helping me with everything from editing text, to discussing standing waves, to photographing whirling wineglasses.

I continue to enjoy tremendous assistance from colleagues here and elsewhere who have supported the *How Things Work* concept, discussed it with me, and often taught the course themselves. These people include Bascom Deaver, Michael Fowler, Tom Gallagher, Bob Jones, Richard Lindgren, Despina Louca, Rick Marshall, and Rob Watkins at the University of Virginia and J. Robert Anderson, Nora Berrah, Katy Disney, Ursula Gibson, Laura Green, Robert Hubel, Larry Hunter, Edwin Jones, Julian Krolik, John Krupczak, Laura Lising, Alan Nathan, Mike Noel, David Ollis, Promod Pratap, Chuck Stone, Richard Superfine, Kristin Wedding, Bob Welsh, and Carl Wieman at other institutions. I am particularly grateful to Carl Wieman for writing the foreword to this book and expressing therein his vision for a physics education that is accessible and valuable to everyone.

Behind the scenes in my class has always been a talented lecture-demonstration and computer staff. Without the creativity and energy of Mike Timmins, John Malone, Roger Staton, Atsushi Yoshida, and Bryan Wright, both my teaching and this book's photographs would be far less interesting.

Nothing informs science teaching like active science research and I am exceedingly grateful to Wendy Fuller-Mora, Denise Caldwell, Hollis Wickman, and Uma Venkateswaran of the National Science Foundation for their steadfast support of my scientific studies.

The best way to discover how students learn science, however, is to teach it. I am ever so grateful to the students of the University of Virginia for being such eager, enthusiastic, and interactive participants in this long educational experiment. It has been a delight and a privilege to get to know so many of them as individuals and their influence on this enterprise has been immeasurable.

Lastly, this book has benefited more than most from the constructive criticism of a number of talented reviewers. In addition to getting a better sense of how to present the material in this book, I have learned a great deal of physics from their reviews. My deepest thanks to all of these fine people for reviewing the third edition:

Timothy Bolton
Kansas State University

Dennis Duke
CSIT, Florida State University

Wayne Garver
University of Missouri- St. Louis

Laura Greene
University of Illinois, Urbana-
Champaign

Frank Hartranft
University of Nebraska at Omaha

John L. Hubisz
North Carolina State University

Edwin R. Jones
University of South Carolina

Lois Breur Krause
Clemson University

John Krupczak
Hope College

Lloyd Makarowitz
SUNY, Farmingdale

Promod Pratap
University of North Carolina,
Greensboro

Chuck Stone
North Carolina A&T State
University

John Tanis
Western Michigan University

Ju Xin
Bloomsburg University

I have always felt that the real test of this book, and of any course taught from it, is its impact on students' lives long after their classroom days are done. It is my sincere hope that many of these students will find themselves looking at things in the world around them years later with understanding and insight that they would not have had were it not for their encounter with this book.

Louis A. Bloomfield
Charlottesville, Virginia
bloomfield@virginia.edu

The Laws of Motion Part 1 ↓

The purpose of this book is to broaden your perspectives on familiar objects and situations by helping you understand the physical processes that make them work. While science is part of our daily existence—not some special activity we do only occasionally, if at all—most of us ignore it or take it for granted. In this book we'll counter that tendency by seeking out science in the world around us, in the objects we encounter every day. We'll see that seemingly "magical" objects and effects are really very straightforward once we know a few of the physical concepts that make them possible. In short, we'll learn about *physics*—the study of the material world and the rules that govern its behavior.

To help us get started, this first pair of chapters will do two main things: introduce the language of physics, which we'll be using throughout the book, and present the basic laws of motion on which everything else will rest. In later chapters, we'll explore objects that are interesting and important, both in their own right and because of the scientific issues they raise. Most of these objects, as we'll see, involve many different aspects of physics and thus bring variety to each section and chapter. But these first two chapters are special because they must provide an orderly introduction to the discipline of physics itself.

EXPERIMENT: Removing a Tablecloth from a Table

One famous "magical" effect is the feat of removing a tablecloth from a set table without breaking the dishes on top. The person performing this stunt pulls the tablecloth out from under the place settings in one lightning-swift motion. With a little luck—not

Courtesy Lou Bloomfield

to mention a smooth, slippery tablecloth—the covering slides off the table suddenly, leaving the dishes behind and virtually unaffected.

With a little practice, you too can do this stunt. Choose a slick, unhemmed tablecloth, one with no flaws that might catch on the dishes. A soft, flexible material such as silk helps because you can then pull the cloth downward and over the edge of the table. When you finally get up the nerve to try this stunt—with unbreakable dishes, of course—make sure you pull as abruptly as possible, keeping the time you spend moving the cloth out from under the dishes to an absolute minimum. It helps to hold the cloth with your palms downward and to let the cloth hang loosely between each hand and the table so that you can get your hands moving before the cloth snaps taut and begins to slide off the table. Don't make the mistake of starting slowly or you'll be picking up pieces.

Before you perform this stunt, try to *predict* what will happen when you pull the cloth. How far will the dishes move—or will they not move at all? How important is the speed with which you pull the cloth? How does the weight of each dish affect its movement or lack of movement? Is the surface texture of each dish important? How would rubbing each dish with wax paper alter the results?

Now give the tablecloth a pull and *observe* what happens. Hopefully, the table will remain set. If it doesn't, try again, but this time vary the tablecloth's speed or the types of dishes or the way you pull the cloth. See if you can *measure* the effects of these variations on the dishes. Does everything work as you expected? Do the results *verify* your predictions, or were those predictions wrong?

If you don't have a suitable tablecloth, or any dishes you care to risk, there are many similar experiments you can try. Put several coins on a sheet of paper and whisk that sheet out from under them. Or stack several books on a table and use a stiff ruler to knock out the bottom one. Most impressive of all is to balance a short eraser-less pencil on top of a wooden needlepoint ring that is itself balanced on the open mouth of a glass bottle. If you yank the ring away quickly enough, the pencil will be left behind and will drop right into the bottle.

We'll return to the tablecloth stunt at the end of this chapter. In the meantime, we'll explore some of the physics concepts that help explain why your stunt worked—or, if it didn't, why the floor is now covered with dishes.

Chapter Itinerary

To examine these concepts, we'll look carefully at three kinds of everyday activities and objects: (1) *skating,* (2) *falling balls,* and (3) *ramps.* In *skating,* we'll see how objects move when nothing pushes on them. In *falling balls,* we'll see how that movement can be influenced by gravity. In *ramps,* we'll explore mechanical advantage and how gradual inclines make it possible to lift heavy objects without pushing very hard. For a more complete preview of what we'll examine in this chapter, flip ahead to the Chapter Summary on p. 33.

These activities may seem mundane, but understanding them in terms of basic physical laws requires considerable thought. These two introductory chapters will be like climbing up the edge of a high plateau: the ascent won't be easy, and our destination will be hidden from view. But once we arrive at the top, with the language and basic concepts of physics in place, we'll be able to explain a broad variety of objects with only a small amount of additional effort. And so we begin the ascent.

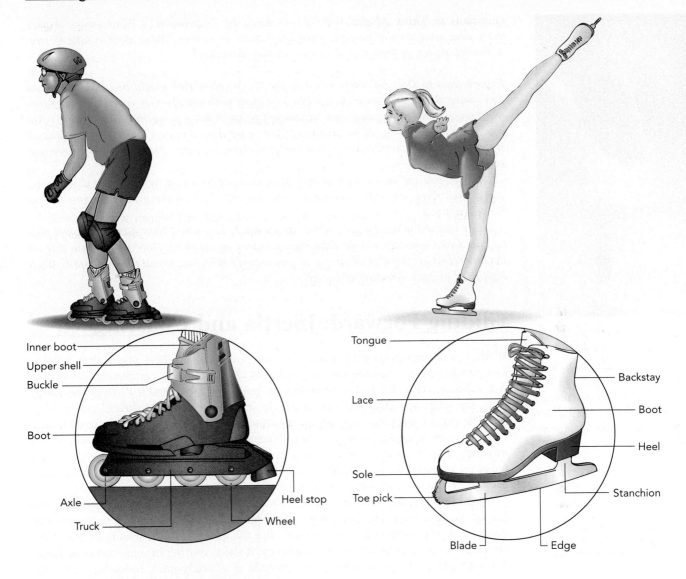

Inner boot
Upper shell
Buckle
Boot
Axle
Truck
Heel stop
Wheel

Tongue
Lace
Sole
Toe pick
Blade
Edge
Backstay
Boot
Heel
Stanchion

SECTION 1.1 **Skating**

Like many sports, skating is trickier than it appears. As a first-time skater, you're likely to find yourself getting up repeatedly from the ground or ice, and it takes some practice before you can glide smoothly forward or come gracefully to a stop. But whether you're wearing ice skates or Rollerblades, the physics of your motion is surprisingly simple. When you're on a level surface with your skates pointing forward, you coast!

Coasting is one of the most basic concepts in physics and our starting point in this book. Joining it in this section will be starting, stopping, and turning, which together will help us understand the first few laws of motion. We'll leave sloping surfaces for the section on ramps and won't have time to teach you how to do spins or win a race. Nonetheless, our exploration of skating will get us well on the road to an understanding of the fundamental principles that govern all motion.

Questions to Think About: *What do we mean by "movement"? What makes skaters move and, once they're moving, what keeps them in motion? What does it take to stop a moving skater or turn that skater in another direction?*

Experiments to Do: *An hour or two on the ice or roller rink would be ideal, but if you don't have skates then try a skateboard or a chair with wheels. Get yourself moving forward on a level surface and then let yourself coast. What's propelling you forward? Are you being pushed forward by anything? Does your direction ever reverse as you coast? How would you describe where you are at a given moment? How would you measure your speed?*

Before you run into a wall or tree, slow yourself to a stop. What was it that slowed you down? Were you still coasting as you stopped? Did anything push on you as you slowed yourself?

Get yourself moving again. What caused you to speed up? How quickly can you pick up speed and what do you do differently to speed up quickly? Now turn to one side or the other. Did anything push on you as you turned? What happened to your speed? What happened to your direction of travel?

Gliding Forward: Inertia and Coasting

While you're putting on your skates, let's take a moment to think about what happens to a person who has nothing pushing on her at all. When she's completely free of outside influences (Fig. 1.1.1), free of pushes and pulls, does she stand still? Does she move? Does she speed up? Does she slow down? In short, what does she do?

The correct answer to that apparently simple question eluded people for thousands of years; even Aristotle, perhaps the most learned philosopher of the classical world, was mistaken about it (see ❏). What makes this question so tricky is that objects on earth are never truly free of outside influences; instead, they all push on, rub against, or interact with one another in some way or other.

As a result, it took the remarkable Italian astronomer, mathematician, and physicist Galileo Galilei many years of careful observation and logical analysis to answer that question ❏. The solution he came up with, like the question itself, sounds simple: if the person is stationary, she will remain stationary; if she is moving in some particular direction, she will continue moving in that direction at a steady pace, following a straight-line path. This property of steady motion in the absence of any outside influence is called **inertia.**

> ### Inertia
> A body in motion tends to remain in motion; a body at rest tends to remain at rest.

The main reason why Aristotle failed to discover inertia, and why we often overlook inertia ourselves, is friction. When you slide across the floor in your shoes, friction quickly slows you to a stop and masks your inertia. To make inertia more obvious, we have to get rid of friction. That's why you're wearing skates.

Chris Trotman/Duomo Photography, Inc

Fig. 1.1.1 Skater Michelle Kwan glides without any horizontal influences. If she's stationary, she'll tend to remain stationary; if she's moving, she'll tend to continue moving.

❏ Aristotle (Greek philosopher, 384–322 B.C.) theorized that objects' velocities were proportional to the forces exerted on them. While this theory correctly predicted the behavior of a sliding object, it incorrectly predicted that heavier objects should fall faster than lighter objects. Nonetheless, Aristotle's theory was retained for a long time, in part because finding the simpler and more complete theory was hard and in part because the scientific method of relating theory and observation took time to develop.

Skates almost completely eliminate friction, at least in one direction, so that you can glide effortlessly across the ice or roller rink and experience your own inertia. For simplicity, let's imagine that your skates are perfect and experience no friction at all as you glide. Also, for this and the next couple of sections, we'll forget not only about friction but also about air resistance. As long as the air is calm and you're not moving too fast, air resistance isn't all that important to skating anyway.

Now that you're ready to skate, we'll begin to examine five important physical quantities relating to motion and look at their relationships to one another. These quantities are position, velocity, mass, acceleration, and force.

Let's begin by describing where you are. At any particular moment, you're located at a **position,** that is, at a specific point in space. Whenever we report your position, it's always as a **distance** and **direction** from some reference point: how many meters north of the refreshment stand or how many kilometers west of Cleveland.

Position is an example of a vector quantity. A **vector quantity** consists of both a magnitude and a direction; the **magnitude** tells you how much of the quantity there is, while the direction tells you which way the quantity is pointing. Vector quantities are common in nature. When you encounter one, pay attention to the direction part; if you're looking for buried treasure 30 paces from the old tree but forget that it's due east of that tree, you'll have a lot of digging ahead of you.

You're on your feet and beginning to skate. If you're moving, then your position is changing. In other words, you have a velocity. **Velocity** measures how quickly your position changes; it's our second vector quantity and consists of the speed at which you're moving and the direction in which you're heading. Your **speed** is the distance you travel in a certain amount of time,

$$\text{speed} = \frac{\text{distance}}{\text{time}},$$

and the direction you're heading might be east, north, or down—if you're taking a spill.

But when you're gliding freely, with nothing pushing you horizontally, your velocity is particularly easy to describe. Since you travel at a steady pace along a straight-line path, your velocity never changes—it is constant. For example, if you're heading west at a speed of 10 meters-per-second (33 feet-per-second), you will have that same velocity indefinitely. A speed of 10 meters-per-second means that, if you travel for 1 second at your present speed, you'll cover a distance of 10 meters. Since your velocity is constant, you'll travel 100 meters in 10 seconds, 1000 meters in 100 seconds, and so on. Furthermore, the path you'll take is a straight line. In a word, you coast.

Thanks to your skates, we can now restate the previous description of inertia in terms of velocity: an object that is not subject to any outside influences moves at a constant velocity, covering equal distances in equal times along a straight-line path. This statement is frequently identified as **Newton's first law of motion,** after its discoverer, the English mathematician and physicist Sir Isaac Newton ❏. The outside influences referred to in this law are called **forces,** a technical term for pushes and pulls.

Newton's First Law of Motion
An object that is not subject to any outside forces moves at a constant velocity, covering equal distances in equal times along a straight-line path.

❏ While a professor in Pisa, Galileo Galilei (Italian scientist, 1564–1642) was obliged to teach the natural philosophy of Aristotle. Troubled with the conflict between Aristotle's theory and observations of the world around him, Galileo devised experiments that measured the speeds at which objects fall and determined that all falling objects fall at the same rate.

❏ In 1664, while Sir Isaac Newton (English scientist and mathematician, 1642–1727) was a student at Cambridge University, the university was forced to close for 18 months because of the plague. Newton retreated to the country and discovered the laws of motion and gravitation and invented the mathematical basis of calculus. These discoveries, along with his observation that celestial objects such as the moon obey the same simple physical laws as terrestrial objects such as an apple (a new idea for the time), are recorded in his *Philosophiæ Naturalis Principia Mathematica,* first published in 1687. That book is perhaps the most important and influential scientific and mathematical work of all time.

Intuition Alert: Coasting

Intuition says that when nothing pushes on an object, that object slows to a stop; you must push it to keep it going.

Physics says that when nothing pushes on an object, that object coasts at constant velocity.

Resolution: Objects usually experience hidden forces, such as friction or air resistance, that tend to slow them down. Eliminating those hidden forces is difficult, so that you rarely see the full coasting behavior of force-free objects.

▶ Check Your Understanding #1: A Puck on Ice

for answers, see page **34**

Why does a moving hockey puck continue to slide across an ice rink even though no one is pushing on it?

The Alternative to Coasting: Acceleration

As you glide forward with nothing pushing you horizontally, what prevents your speed and direction from changing? The answer is your mass. **Mass** is the measure of your inertia, your resistance to changes in velocity. Almost everything in the universe has mass. Because you have mass, your velocity will change only if something pushes on you—that is, only if you experience a force. You'll keep moving steadily in a straight path until something pushes on you with a force to stop you or send you in another direction. Force is our third vector quantity, having both a magnitude and a direction. After all, a push to the right is different from a push to the left.

When something pushes on you, your velocity changes; in other words, you accelerate. **Acceleration,** our fourth vector quantity, measures how quickly your velocity changes. *Any* change in your velocity is acceleration, whether you're speeding up, slowing down, or even turning. If either your speed or direction of travel is changing, you're accelerating!

Like any vector quantity, acceleration has a magnitude and a direction. To see how these two parts of acceleration work, imagine that you're at the starting line of a race, waiting for it to begin. The starting buzzer sounds and you're off! You dig your skates into the surface beneath you and begin to accelerate—your speed increases and you cover ground more and more quickly. The magnitude of your acceleration depends on how hard the skating surface pushes you forward. If it's a long race and you're not in a hurry, you obtain a modest push from the surface and the magnitude of your acceleration is small. Your velocity changes slowly. But if the race is a sprint and you haven't a moment to spare, you spring forward hard and the surface exerts an enormous forward force on you. The magnitude of your acceleration is large and your velocity changes rapidly. In this case, you can actually feel your inertia opposing your efforts to pick up speed.

But acceleration has more than just a *magnitude*. When you start the race, you also select a *direction* for your acceleration—the direction toward which your velocity is shifting with time. This acceleration is in the same direction as the force causing it. If you obtain a forward force from the surface, you'll accelerate forward—your velocity will shift more and more forward. If you obtain a sideways force from the surface, well. . . . The other racers will have to jump out of your way as you career into the wall. They'll laugh all the way to the finish line at your failure to recognize the importance of direction in the definitions of force and acceleration.

Once you're going fast enough, you can stop fighting inertia and begin to glide. You coast forward at a constant velocity. Now inertia is helping you; it keeps you moving steadily along even though nothing is pushing you forward. (Recall that we're neglecting friction and air resistance. In reality, those effects push you backward and gradually slow you down as you glide. However, since we're ignoring them in this section, your motion is smooth and steady.)

But even when you're not trying to speed up or slow down, you can still accelerate. As you steer your skates or go over a bump, you experience sideways or up–down forces that change your *direction of travel* and thus cause you to accelerate.

Finally the race is over and you skid to a stop. You're accelerating again. This time you're accelerating backward, in the direction opposite your forward velocity. While we often call this process *deceleration,* it's just a special type of acceleration. Your forward velocity gradually diminishes until you come to rest.

To help you recognize acceleration, here are some accelerating objects:

1. A runner who's leaping forward at the start of a race—the runner's velocity is changing from zero to forward so the runner is accelerating *forward.*
2. A bicycle that's stopping at a crosswalk—its velocity is changing from forward to zero so it's accelerating *backward* (it's decelerating).
3. An elevator that's just starting upward from the first floor to the fifth floor—its velocity is changing from zero to upward so it's accelerating *upward.*
4. An elevator that's stopping at the fifth floor after coming from the first floor—its velocity is changing from upward to zero so it's accelerating *downward.*
5. A car that's beginning to shift left to pass another car—its velocity is changing from forward to left-forward so it's accelerating mostly *leftward.*
6. An airplane that's just beginning its descent—its velocity is changing from level-forward to descending-forward so it's accelerating mostly *downward.*
7. Children riding a carousel around in a circle—while their speeds are constant, their directions of travel are always changing. We'll discuss the directions in which they're accelerating in Section 3.3.

Here are some objects that are *not* accelerating:

1. A parked car—its velocity is always zero.
2. A car traveling straight forward on a level road at a steady speed—no change in its speed or direction of travel.
3. A bicycle that's climbing up a smooth, straight hill at a steady speed—no change in its speed or direction of travel.
4. An elevator that's moving straight upward at a steady pace, halfway between the first floor and the fifth floor—no change in its speed or direction of travel.

Seeing acceleration isn't as easy as seeing velocity. You must watch skaters closely for some time to see whether or not they're accelerating. If their paths aren't straight or if their speeds aren't steady, then they're accelerating.

→ Check Your Understanding #2: Changing Trains

for answers, see page 34

Trains spend much of their time coasting along at constant velocity. When does a train accelerate forward? backward? leftward? downward?

How Forces Affect Skaters

Now that we've learned what acceleration is, let's see how you accelerate in response to a particular force. First, your acceleration depends on the strength of that force: the stronger the force, the more you accelerate. But your acceleration also depends on your mass: the more massive you are, the less you accelerate. For example, it's easier to change your velocity before you eat Thanksgiving dinner than afterward.

There is a simple relationship between the force exerted on you, your mass, and your acceleration. Your acceleration is equal to the force exerted on you divided by your mass or, as a word equation,

$$\text{acceleration} = \frac{\text{force}}{\text{mass}}. \tag{1.1.1}$$

Your acceleration, as we've seen, is in the same direction as the force exerted on you.

This relationship was deduced by Newton from his observations of motion and is referred to as **Newton's second law of motion.** Structuring the relationship this way sensibly distinguishes the causes (force and mass) from their effect (acceleration). However, it has become customary to rearrange this equation to eliminate the division. The relationship then takes its traditional form, which can be written in a word equation:

$$\text{force} = \text{mass} \cdot \text{acceleration}, \tag{1.1.2}$$

in symbols:

$$\mathbf{F} = m \cdot \mathbf{a,}$$

and in everyday language:

Throwing a baseball is much easier than throwing a bowling ball (Fig. 1.1.2).

Remember that in Eq. 1.1.2 the direction of the acceleration is the same as the direction of the force.

Jeff Christensen/Liaison Agency, Inc./Getty Images

Steven E. Sutton/Duomo Photography, Inc

Fig. 1.1.2 A baseball accelerates easily because of its small mass. A bowling ball has a large mass and is harder to accelerate.

Newton's Second Law of Motion
The force exerted on an object is equal to the product of that object's mass times its acceleration. The acceleration is in the same direction as the force.

Because it's an equation, the two sides of Eq. 1.1.1 are equal. Your acceleration equals the force on you divided by your mass. Since your mass can't change without a visit to the snack bar, Eq. 1.1.1 indicates that an increase in the force on you is accompanied by a similar increase in your acceleration. That way, as the right side of the equation increases, the left side increases to keep the two sides equal. Thus the harder something pushes on you, the more rapidly your velocity changes.

We can also compare the effects of equal forces on two different masses, for example, you and the former sumo wrestler to your left. Equation 1.1.1 indicates that an increase in mass must be accompanied by a corresponding decrease in acceleration. Sure enough, your velocity changes more rapidly than the velocity of the sumo wrestler when the two of you are subjected to identical forces (Fig. 1.1.3).

So far we've explored five principles:

1. Your position indicates exactly where you're located.
2. Your velocity measures how quickly your position changes.
3. Your acceleration measures how quickly your velocity changes.
4. In order for you to accelerate, something must exert a force on you.
5. The more mass you have, the less acceleration you experience for a given force.

We've also encountered five important physical quantities—mass, force, acceleration, velocity, and position—as well as some of the rules that relate them to one another. Much of the groundwork of physics rests on these five quantities and on their interrelationships.

Skating certainly depends on these quantities. We can now see that, in the absence of any horizontal forces, you either remain stationary or coast along at a constant velocity. To start, stop, or turn, something must push you horizontally and that something is the ice or pavement. We haven't talked about how you obtain horizontal forces from the ice or pavement and we'll leave that problem for later sections. But as you skate, you should be aware of these forces and notice how they change your speed, direction of travel, or both. Learn to watch yourself accelerate.

Fig. 1.1.3 If you give these two skaters identical pushes, the boy will accelerate more quickly than the girl. That's because the girl has more mass than the boy.

Check Your Understanding #3: Hard to Stop

for answers, see page 35

It's much easier to stop a bicycle traveling toward you at 5 kilometers-per-hour (3 miles-per-hour) than an automobile traveling toward you at the same velocity. What accounts for this difference?

Check Your Figures #1: At the Bowling Alley

for answers, see page 36

Bowling balls come in various masses. Suppose that you try bowling two different balls, one with twice the mass of the other. If you push on them with equal forces, which one will accelerate faster and how much faster?

Several Skaters: Frames of Reference

While skating alone is peaceful, it's usually more fun with other skaters around. That way, you have people to talk to and an audience for your athleticism and artistry.

However, with several skaters coasting on the ice at once, there's a question of perspective. As you glide steadily past a friend, the two of you see the world somewhat differently. From your perspective, you are motionless and your friend is moving. But from your friend's perspective, your friend is motionless and you are moving. Who is right?

It turns out that you're both right and that physics has a way of accommodating this apparent paradox. Each of you is observing the world from a different **inertial frame of reference,** the viewpoint of an **inertial** object—an object that is not accelerating and that moves according to Newton's first law. One of the remarkable discoveries of Galileo and Newton is that the laws of physics work perfectly in any inertial frame of reference. From an inertial frame, everything you see in the world around you obeys the laws of motion that we're in the process of learning.

Since both you and your friend are coasting, each of you views the world from an inertial frame of reference and sees the surrounding objects moving in perfect accordance with the laws of motion. Some objects travel at constant velocity while others accelerate in response to forces. But because the two of you are observing those objects from different inertial frames, you will disagree on the particular values of some of the physical quantities you might measure.

In the present case, you see yourself as motionless because you view the world from your own inertial frame. In that frame, your friend is coasting westward at 2 m/s (6.6 ft/s). However, your friend sees things differently. In your friend's inertial frame, your friend is motionless and you yourself are coasting eastward at 2 m/s. As long as the two of you don't try to compare the positions or velocities of objects you observe, or certain physical quantities derived from those values, there will be no disagreements and no inconsistencies. But if you forget to watch where you're going and crash into a wall, don't expect your friend to sympathize when you claim that you were motionless and that the moving wall ran into you. That's not how your friend saw it.

Each time we examine an object in this book, we'll pick a specific inertial frame of reference from which to view that object. We'll normally select an inertial frame that makes the object and its motions appear as simple as possible and then stick with that frame consistently. The best choice of inertial frame will usually be so obvious that we'll adopt it without even a moment's thought. But on occasion, we'll have to pick the frame carefully and deliberately. Finally, while there are formal methods for working with two or more inertial frames at once, we'll leave that for another book.

> ### Check Your Understanding #4: Two Views

for answers, see page 35

You are standing on the sidewalk, watching a train coast eastward at constant velocity. Your friend is riding in that train. In her inertial frame of reference, the sweater in her lap is motionless. Describe the sweater's motion in your inertial frame of reference.

Measure for Measure:
The Importance of Units

If you went to the grocery store and asked for "6 sugar," the clerk wouldn't know how much sugar to give you. The number *6* wouldn't be enough information; you need to specify which units—cups, pounds, cubes, or tons—you have in mind. This need to specify units applies to almost all physical quantities—velocity, force, mass, and so on—and has led our society to develop units that everyone agrees on, also known as **standard units.**

For example, when you say that a skater's speed is 20 miles-per-hour, you have chosen "miles-per-hour" as the standard unit of speed and you're asserting that the skater is moving 20 times that fast. You can report the skater's speed as a multiple of any standard unit of speed—feet-per-second, yards-per-day, and inches-per-century, to name only a few—and you can always find a simple relationship to convert from one unit of speed to another. For example, to convert from miles-per-hour to kilometers-per-hour, you multiply by 1.609 kilometers/mile. Using that technique, you'll find that the skater's speed is 32.2 kilometers-per-hour.

Many of the common units in the United States come from the old **English system of units,** which most of the world has abandoned in favor of the **SI units** (Systéme International d'Unités). The continued use of English units in the United States often makes life difficult. If you have to triple a cake recipe that calls for ¾ cup of milk, you must work hard to calculate that you need 2¼ cups. Then you go to buy 2¼ cups of milk, which is slightly more than half a quart, but end up buying two pints instead. You now have 14 ounces of milk more than you need. But is that 14 fluid ounces or 14 ounces of weight? And so it goes.

The SI system has two important characteristics that distinguish it from the English system and make it easier to use:

1. Different units for the same physical quantity are related by factors of 10.
2. Most of the units are constructed out of a few basic units: the meter, the kilogram, and the second.

Let's start with the first characteristic: different units for the same physical quantity are related by factors of 10. When measuring volume, 1000 milliliters is exactly 1 liter and 1000 liters is exactly 1 meter^3. When measuring mass, 1000 grams is exactly 1 kilogram and 1000 kilograms is exactly 1 metric ton. Because of this consistent relationship, enlarging a recipe that's based on the SI system is as simple as multiplying a few numbers. You never have to think about converting pints into quarts, teaspoons into tablespoons, or ounces into pounds. Instead, if you want to triple a recipe that calls for 500 milliliters of sugar, you just multiply the recipe by 3 to obtain 1500 milliliters of sugar. Since 1000 milliliters is 1 liter, you'll need 1.5 liters of sugar. Converting milliliters to liters is as simple as multiplying by 0.001 liter/milliliter. (See Appendix B for more conversion factors.)

SI units remain somewhat mysterious to many U.S. residents, even though some of the basic units are slowly appearing on our grocery shelves and highways. As a result, while the SI system really is more sensible than the old English system, developing a feel for some SI units is still difficult. How many of us know our heights in meters (the SI unit of length) or our masses in kilograms (the SI unit of mass)? If your car is traveling 200 kilometers-per-hour and you pass a police car, are you in trouble? Yes, because 200 kilometers-per-hour is about 125 miles-per-hour. Actually, the hour is not an SI unit—the SI unit of time is the second—but the hour remains customary for describing long periods of time. Thus the kilometer-per-hour is a unit that is half SI (the kilometer part) and half customary (the hour part).

The second characteristic of the SI system is its relatively small number of basic units. So far, we've noted the SI units of mass (the **kilogram,** abbreviated kg), length (the **meter,** abbreviated m), and time (the **second,** abbreviated s). One kilogram is about the mass of a liter of water; one meter is about the length of a long stride; one second is about the time it takes to say "one banana." From these three basic units, we can create several others, such as the SI units of velocity (the **meter-per-second,** abbreviated m/s) and acceleration (the **meter-per-second^2,** abbreviated m/s^2). One meter-per-second is a healthy walking speed; one meter-per-second^2 is about the acceleration of an elevator after the door closes and it begins to head upward. This conviction that many units are best constructed out of other, more basic units dramatically simplifies the SI system; the English system doesn't usually suffer from such sensibility.

The SI unit of force is also constructed out of the basic units of mass, length, and time. If we choose a 1-kilogram object and ask just how much force is needed to make that object accelerate at 1 meter-per-second^2 we define a specific amount of force. Since 1 kilogram is the SI unit of mass and 1 meter-per-second^2 is the SI unit of acceleration, it's only reasonable to let the force that causes this acceleration be the SI unit of force: the **kilogram-meter-per-second^2.** Since this composite unit sounds unwieldy but is very important, it has been given its own name: the **newton** (abbreviated N)—after, of course, Sir Isaac, whose second law defines the relationship among mass, length, and time that the unit expresses. One newton is about the weight of 18 U.S. quarter dollars; if you hold 18 quarters steady in your hand, you'll feel a downward force of about 1 newton.

Because a complete transition to the SI system will take generations, this book uses both unit systems whenever possible. Although it will emphasize the SI system, English and customary units may give you a better intuitive feel for a particular physical quantity. A bullet train traveling "67 meters-per-second" doesn't mean much to most of us, while one moving "150 miles-per-hour" (150 mph) or "240 kilometers-per-hour" (240 km/h) should elicit our well-deserved respect.

Quantity	SI Unit	English Unit	SI → English	English → SI
Position	meter (m)	foot (ft)	1 m = 3.2808 ft	1 ft = 0.30480 m
Velocity	meter-per-second (m/s)	foot-per-second (ft/s)	1 m/s = 3.2808 ft/s	1 ft/s = 0.30480 m/s
Acceleration	meter-per-second^2 (m/s^2)	foot-per-second^2 (ft/s^2)	1 m/s^2 = 3.2808 ft/s^2	1 ft/s^2 = 0.30480 m/s^2
Force	newton (N)	pound-force (lbf)*	1 N = 0.22481 lbf	1 lbf = 4.4482 N
Mass	kilogram (kg)	pound-mass (lbm)*	1 kg = 2.2046 lbm	1 lbm = 0.45359 kg

*The English units of force and mass are both called the pound. To distinguish these two units, it has become standard practice to identify them explicitly as pound-force and pound-mass.

Check Your Understanding #5: Going for a Walk

for answers, see page 35

If you're walking at a pace of 1 meter-per-second, how many miles will you travel in an hour?

SECTION 1.2 Falling Balls

We've all dropped balls from our hands or seen them arc gracefully through the air after being thrown. These motions are simplicity itself and, not surprisingly, they're governed by only a few universal rules. We encountered several of those rules in the previous section, but we're about to examine our first important type of force: gravity. Like Newton, who reportedly began his investigations after seeing an apple fall from a tree, we'll start simply by exploring gravity and its effects on motion in falling objects.

Questions to Think About: What do we mean by "falling," and why do balls fall? Which falls faster: a heavy ball or a light ball? Can a ball that's heading upward still be falling? How does gravity affect a ball that's thrown sideways?

Experiments to Do: A few seconds with a baseball will help you see some of the behaviors that we'll be exploring. Toss the ball into the air to various heights, catching it in your hand as it returns. Have a friend time the flight of the ball. As you toss the ball higher, how much more time does it spend in the air? How does it feel coming back to your hands? Is there any difference in the impact it makes? Which takes the ball longer: rising from your hand to its peak height or returning from its peak height back to your hand?

*   Now drop two different balls—a baseball, say, and a golf ball. If you drop them simultaneously, without pushing either one up or down, does one ball strike the ground first, or do they arrive together? Now throw one ball horizontally while dropping the second. If they both leave your hands at the same time and the first one's initial motion is truly horizontal, which one reaches the ground first?*

Weight and Gravity

Like everything else around us, a ball has a weight. For example, a golf ball weighs about 0.45 N (0.10 lbf). But what is weight? Evidently it's a force, since both the newton (N) and the pound-force (lbf) are units of force. But to understand what weight really is—and, in particular, where it comes from—we need to look at gravity.

Gravity is a physical phenomenon that produces an attractive force between every pair of objects in the universe. In our daily lives, however, the only object massive enough and near enough to have an obvious gravitational effect on us is our planet, the earth. Gravity weakens with distance and the moon and sun are so far away that we notice their gravities only through such subtle effects as the ocean tides.

The earth's gravity exerts a downward force on any object near its surface. That object is attracted directly toward the center of the earth with a force we call the object's **weight** (Fig. 1.2.1). Remarkably enough, this weight is exactly proportional to the object's mass—if one ball has twice the mass of another ball, it also has twice the weight. Such a relationship between weight and mass is astonishing because weight and mass are very different attributes: weight is how hard gravity pulls on a ball, and mass is how difficult that ball is to accelerate. Because of this proportionality, a ball that's heavy is also hard to shake!

An object's weight is also proportional to the local strength of gravity, which is measured by a downward vector called the **acceleration due to gravity.** We'll explain its odd name shortly, but at the surface of the earth the acceleration due to gravity is about 9.8 N/kg (1.0 lbf/lbm). That value means that a mass of 1 kilogram has a weight of 9.8 newtons, and that a mass of 1 pound-mass has a weight of 1 pound-force.

More generally, an object's weight is equal to the product of its mass times the acceleration due to gravity, which can be written as a word equation:

$$\text{weight} = \text{mass} \cdot \text{acceleration due to gravity},\qquad\qquad \textbf{(1.2.1)}$$

in symbols:

$$\mathbf{w} = m \cdot \mathbf{g},$$

and in everyday language:

> *You can lose weight either by reducing your mass or by going someplace, like a small planet, where the gravity is weaker.*

But why *acceleration* due to gravity? What acceleration do we mean? To answer that question, let's consider what happens to a ball when you drop it.

If the only force on the ball is its weight, the ball accelerates downward; in other words, it falls. While a ball moving through the earth's atmosphere encounters additional forces due to air resistance, let's ignore those forces for the time being. Doing so costs us only a little in terms of accuracy—the effects of air resistance are negligible as long as the ball is dense and its speed relatively small—and allows us to focus exclusively on the effects of gravity.

How much does the falling ball accelerate? According to Eq. 1.1.1, the ball's acceleration is equal to the force exerted on it divided by its mass. But because the ball is *falling,* the only force on it is its own weight. That weight, according to Eq. 1.2.1, is

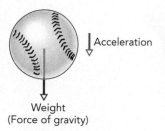

Acceleration

Weight
(Force of gravity)

Fig. 1.2.1 A ball experiencing the force of gravity. It accelerates downward.

equal to the ball's mass times the acceleration due to gravity. Using word algebra, we get

$$\text{falling ball's acceleration} = \frac{\text{ball's weight}}{\text{ball's mass}}$$

$$= \frac{\text{ball's mass} \cdot \text{acceleration due to gravity}}{\text{ball's mass}}$$

$$= \text{acceleration due to gravity}.$$

As you can see, the falling ball's acceleration is equal to the acceleration due to gravity. So acceleration due to gravity really is an acceleration after all: it's the acceleration of a freely falling object. Moreover, the units of acceleration due to gravity can be transformed easily from those relating weight to mass, 9.8 N/kg (1.0 lbf/lbm), into those describing the acceleration of free fall, 9.8 m/s^2 (32 ft/s^2).

Thus a ball falling near the earth's surface experiences a downward acceleration of 9.8 m/s^2 (32 ft/s^2), regardless of its mass. This downward acceleration is substantially more than that of an elevator starting its descent. When you drop a ball, it picks up speed very quickly in the downward direction.

Because all falling objects at the earth's surface accelerate downward at exactly the same rate, a billiard ball and a bowling ball dropped simultaneously from the same height will reach the ground together. (Remember that we're not considering air resistance yet.) Although the bowling ball weighs more than the billiard ball, it also has more mass; so while the bowling ball experiences a larger downward force, its larger mass ensures that its downward acceleration is equal to that of the lighter and less massive billiard ball.

Check Your Understanding #1: Weight and Mass

for answers, see page **35**

Out in deep space, far from any celestial object that exerts significant gravity, would an astronaut weigh anything? Would that astronaut have a mass?

Check Your Figures #1: Weighing In on the Moon

for answers, see page **36**

You're in your spacecraft on the surface of the moon. Before getting into your suit, you weigh yourself and find that your moon weight is almost exactly one-sixth your earth weight. What is the moon's acceleration due to gravity?

The Velocity of a Falling Ball

We're now ready to examine the motion of a falling ball near the earth's surface. A falling ball is one that has only the force of gravity acting on it and gravity, as we've seen, causes any falling object to accelerate downward at a constant rate. But we're usually less interested in the falling object's acceleration than we are in its position and velocity. Where will the object be in 3 seconds, and what will its velocity be then? When you're trying to summon up the courage to jump off the high dive, you want to know how long it'll take you to reach the water and how fast you'll be going when you hit.

The first step in answering these questions is to look at how a ball's velocity is related to the time you've been watching it fall. To do that, you'll need to know the ball's *initial velocity,* that is, its speed and direction at the moment you start watching it. If you drop the ball from rest, its initial velocity is zero.

You can then describe the ball's present velocity in terms of its initial velocity, its acceleration, and the time that has passed since you started watching it. Because a constant acceleration causes the ball's velocity to change by the same amount each second, the ball's present velocity differs from its initial velocity by the product of the acceleration times the time over which you've been watching it. We can relate these quantities as a word equation:

$$\text{present velocity} = \text{initial velocity} + \text{acceleration} \cdot \text{time}, \qquad (1.2.2)$$

in symbols:

$$\mathbf{v} = \mathbf{v_o} + \mathbf{a} \cdot t,$$

and in everyday language:

A stone dropped from rest descends faster with each passing second, but you can give it a boost by throwing it downward instead of just letting go.

For a ball falling from rest, the initial velocity is zero, the acceleration is downward at 9.8 m/s^2 (32 ft/s^2), and the time you've been watching it is simply the time since it started to drop (Fig. 1.2.2). After one second, the ball has a downward velocity of 9.8 m/s (32 ft/s). After two seconds, the ball has a downward velocity of 19.6 m/s (64 ft/s). After three seconds, its downward velocity is 29.4 m/s (96 ft/s), and so on. Since the ball's motion is strictly vertical, we often put a negative sign in front of the acceleration to indicate the direction. By convention, we say that a negative sign means "down."

Check Your Understanding #2: Half a Fall

for answers, see page 35

You drop a marble from rest and after 1 s, its velocity is 9.8 m/s (32 ft/s) in the downward direction. What was its velocity after only 0.5 s of falling?

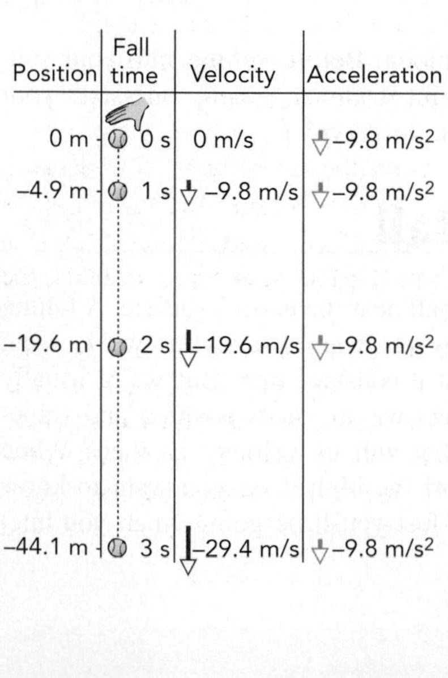

Position	Fall time	Velocity	Acceleration
0 m	0 s	0 m/s	−9.8 m/s^2
−4.9 m	1 s	−9.8 m/s	−9.8 m/s^2
−19.6 m	2 s	−19.6 m/s	−9.8 m/s^2
−44.1 m	3 s	−29.4 m/s	−9.8 m/s^2

Fig. 1.2.2 The moment you let go of a ball that was resting in your hand, it begins to fall. Its weight causes it to accelerate downward. After 1 second, it has fallen 4.9 m and has a velocity of 9.8 m/s downward. After 2 seconds, it has fallen 19.6 m and has a velocity of 19.6 m/s downward, and so on. As the ball continues to accelerate downward, its velocity continues to increase downward. Negative values for the position and velocity are meant to indicate downward movement, caused by a negative or downward acceleration.

► Check Your Figures #2: The High Dive

for answers, see page 36

If it takes you about 1.4 s to reach the water from the 10-meter (32-foot) diving platform, how fast are you going just before you enter the water?

The Position of a Falling Ball

The ball's velocity continues to increase as it falls, but where exactly is the ball located? To answer that question, you need to know the ball's *initial position,* that is, where it was when you started to watch it fall. If you dropped the ball from rest, the initial position was your hand and you can define that spot as 0.

You can then describe the ball's present position in terms of its initial position, its initial velocity, its acceleration, and the time that has passed since you started watching it. However, because the ball's velocity is changing, you can't simply multiply its present velocity by the time that it's been falling to determine how much the ball's present position differs from its initial position. Instead, you must use the ball's average velocity during the whole period you've been watching it. Since the ball's velocity has been changing uniformly from its initial velocity to its present velocity, the ball's average velocity is exactly halfway in between the two:

$$\text{average velocity} = \text{initial velocity} + \tfrac{1}{2} \cdot \text{acceleration} \cdot \text{time}.$$

The ball's present position differs from its initial position by the product of this average velocity times the time over which you've been watching it. We can relate these quantities as a word equation:

$$\text{present position} = \text{initial position} + \text{initial velocity} \cdot \text{time} + \tfrac{1}{2} \cdot \text{acceleration} \cdot \text{time}^2, \quad \textbf{(1.2.3)}$$

in symbols:

$$\mathbf{x} = \mathbf{x_0} + \mathbf{v_0} \cdot t + \tfrac{1}{2} \cdot \mathbf{a} \cdot t^2,$$

and in everyday language:

> *The longer a stone has been falling, the more its height diminishes with each*
> *passing second. However, it won't overtake a stone that was dropped next to*
> *it at an earlier time or dropped from beneath it at the same time.*

For a ball falling from rest, the initial velocity is zero, the acceleration is downward at 9.8 m/s^2 (32 ft/s^2), and the time you've been watching it is simply the time since it started to drop (Fig. 1.2.2). After one second, the ball has fallen 4.9 m (16 ft). After two seconds, the ball has fallen downward a total of 19.6 m (64 ft). After three seconds, the ball has fallen a total of 44.1 m (145 ft), and so on.

Equations 1.2.2 and 1.2.3 depend on the definition of acceleration as the measure of how quickly *velocity* changes and the definition of velocity as the measure of how quickly *position* changes. Because the acceleration of a falling ball doesn't change with time, the two equations can be derived using algebra. But in more complicated situations, where an object's acceleration changes with time, predicting its position and velocity usually requires the use of calculus. *Calculus* is the mathematics of change, invented by Newton to address just these sorts of problems.

❑ In 1782, William Watts, a plumber from Bristol, England, patented a technique for forming perfectly spherical, seamless lead shot for use in guns. His idea was to pour molten lead through a sieve suspended high above a pool of water. The lead droplets cool in the air as they fall, solidifying into perfect spheres before reaching the water. Shot towers based on this idea soon appeared throughout Europe and eventually in the United States. Nowadays, iron shot has all but replaced environmentally dangerous lead shot. Iron shot is cast, rather than dropped, because the longer cooling time needed to solidify molten iron would require impractically tall shot towers.

We've been discussing what happens to a falling ball, but we could have chosen another object instead. Everything falls the same way; heavy or light, large or small, all objects take the same amount of time to fall some distance near the earth's surface, as long as they're dense enough to overcome air resistance. If there were no air, this statement would be exactly true for any object; a feather and a lead brick would plummet downward together if you dropped them simultaneously.

Now that we've explored acceleration due to gravity, we can see why a ball dropped from a tall ladder is more dangerous than the same ball dropped from a short stool. The farther the ball has to fall, the longer it takes to reach the ground and the more time it has to accelerate. During its long fall from the tall ladder, the ball acquires a large downward velocity and becomes very hard to stop. If you try to catch it, you'll have to exert a very large upward force on it to accelerate it upward and bring it to rest quickly. Exerting that large upward force may hurt your hand.

The same notion holds if you're the falling object. If you leap off a tall ladder, a substantial amount of time will pass before you reach the ground. By the time you hit the ground, you will have acquired considerable downward velocity. The ground will then accelerate you upward and bring you to rest with a very large and unpleasant upward force. (For an interesting and less painful application of long falls, see ❑.)

Check Your Understanding #3: Half a Fall Again

for answers, see page 35

You drop a marble from rest and after 1 s, it has fallen downward a distance of 4.9 m (16 ft). How far had it fallen after only 0.5 s?

Check Your Figures #3: Extreme Physics

for answers, see page 36

You're planning to construct a bungee-jumping amusement at the local shopping center. If you want your customers to have a 5-second free-fall experience, how tall will you need to build the tower from which they'll jump? (Don't worry about the extra height needed to stop people after the bungee pulls taut.)

How a Thrown Ball Moves: Projectile Motion

If the only force acting on an object is its weight, then the object is falling. So far, we've explored this principle only as it pertains to balls dropped from rest. However a thrown ball is falling, too; once it leaves your hand, it's subject only to the downward force of gravity and it falls. It may seem odd but even though it's initially traveling upward, a ball tossed upward is accelerating downward at 9.8 m/s^2 (32 ft/s^2). As a result, the tossed ball's upward velocity diminishes, it stops rising, its velocity becomes downward, and it eventually returns to the ground.

Equation 1.2.2 still describes how the ball's velocity depends on the fall time, but now the initial velocity isn't zero; it points in the upward direction! If you toss a ball straight up in the air, it leaves your hand with a large upward velocity (Fig. 1.2.3). As soon as you let go of it, it begins to accelerate downward. If the ball's initial upward velocity is 29.4 m/s (96 ft/s), then after one second its upward velocity is 19.6 m/s (64 ft/s). After another second, its upward velocity is only 9.8 m/s (32 ft/s).

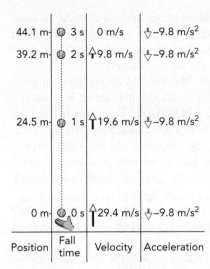

Position	Fall time	Velocity	Acceleration
44.1 m	⦿ 3 s	0 m/s	↧ −9.8 m/s²
39.2 m	⦿ 2 s	↥9.8 m/s	↧ −9.8 m/s²
24.5 m	⦿ 1 s	↥19.6 m/s	↧ −9.8 m/s²
0 m	⦿ 0 s	↥29.4 m/s	↧ −9.8 m/s²

Fig. 1.2.3 The moment you let go of a ball thrown straight upward, it begins to accelerate downward at 9.8 m/s². The ball rises but its upward velocity diminishes steadily until it momentarily comes to a stop. It then descends with its downward velocity increasing steadily. In this example, the ball rises for 3 s and comes to rest. It then descends for 3 s before returning to your hand in a very symmetrical flight.

After a third second, the ball momentarily comes to a complete stop with a velocity of zero. It then descends from this peak height, falling just as it did when you dropped it from rest.

The ball's flight before and after its peak is symmetrical. It travels upward quickly at first, since it has a large upward velocity. As its upward velocity diminishes, it travels more and more slowly until it comes to a stop. It then begins to descend, slowly at first and then faster and faster as it continues its constant downward acceleration. The time the ball takes to rise from its initial position in your hand to its peak height is exactly equal to the time it takes to descend back down from that peak to your hand. Equation 1.2.3 indicates how the position of the ball depends on the fall time, with the initial velocity being the upward velocity of the ball as it leaves your hand.

The larger the initial upward velocity of the ball, the longer it rises and the higher it goes before its velocity is reduced to zero. It then descends for the same amount of time it spent rising. The higher the ball goes before it begins to descend, the longer it takes to return to the ground and the faster it's traveling when it arrives. That's why catching a high fly ball with your bare hands stings so much: the ball is traveling very, very fast when it hits your hands, and a large force is required to bring the ball to rest quickly.

What happens if you don't toss the ball exactly straight up? Suppose you throw the ball upward at some angle. The ball still rises to a peak height and then begins to descend; but as it rises and descends, it also travels away from you horizontally so that it strikes the ground at some distance from your feet. How much does this horizontal travel complicate the motion of a falling ball?

The answer is not very much. One of the beautiful simplifications of physics is that we can often treat an object's vertical motion independently of its horizontal motion. This technique involves separating the vector quantities—acceleration, velocity, and position—into **components,** those portions of the quantities that lie along specific directions (Fig. 1.2.4). For example, the vertical component of an object's position is that object's altitude.

If you know a ball's altitude, you know only part of its position; you still need to know where it is to your left or right and to your front or back. In fact, you can completely specify its position (or any other vector quantity) in terms of three components along three directions that are perpendicular (or at right angles) to one another. This

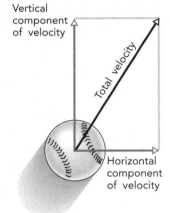

Fig. 1.2.4 Even if the ball has a velocity that is neither purely vertical nor purely horizontal, its velocity may nonetheless be viewed as having a vertical component and a horizontal component. Part of its total velocity acts to move this ball upward, and part of its total velocity acts to move this ball in a horizontal direction.

means that you can completely specify the ball's position by its vertical altitude, its horizontal distance to your left or right, and its horizontal distance in front or in back of you. For example, the ball might be 10 m above you, 3 m to your left, and 2 m behind you. These three distances indicate precisely where the ball is located.

Up until now, we've actually been examining only the vertical components of position, velocity, and acceleration. But now we're going to let the ball move horizontally so that we also have to consider what happens to the two horizontal components of each vector quantity. Keeping track of all three components of each quantity is difficult. Since we're interested in the motion of a tossed ball, we can eliminate the left–right component by throwing the ball forward; that way, the ball will move only in a vertical plane that extends directly in front of us. We can then specify the ball's position as its altitude and its horizontal distance in front of us.

Because the falling ball's acceleration is constant and independent of where the ball is near the earth's surface, the ball's horizontal motion is independent of its vertical motion. We already know about the falling ball's vertical motion, but what is its horizontal motion? It appears we need some new relationships to describe how the horizontal components of position, velocity, and acceleration change with time. Fortunately, we can reuse Eqs. 1.2.2 and 1.2.3.

Although Eqs. 1.2.2 and 1.2.3 were introduced to describe a ball's vertical motion, they're actually more general. They relate three vector quantities—position, velocity, and constant acceleration—to one another. They describe how an object's position and velocity change with time when the object undergoes constant acceleration, regardless of the direction of that constant acceleration.

These equations also apply to the components of position, velocity, and constant acceleration. If you add the words "vertical component of" in front of each vector quantity in Eqs. 1.2.2 and 1.2.3, the equations correctly describe the object's vertical motion. The same can be done with the object's horizontal motion. Our previous look at vertical motion implicitly inserted the words "vertical component of" into the equations. Now we'll try inserting the words "horizontal component of" to understand the tossed ball's horizontal motion.

As soon as the ball leaves your hand, its motion can be broken into two parts: a vertical motion and a horizontal motion (Fig. 1.2.4). Part of the ball's initial velocity is in the upward direction, and that vertical velocity component determines the object's ascent and descent. Part of the ball's initial velocity is in the horizontal or downfield direction, and that horizontal velocity component determines the ball's drift downfield.

Because gravity is the only force on the ball and it acts only in the downward vertical direction, there is no horizontal component of acceleration. The horizontal velocity component therefore remains constant, and the ball travels downfield at a steady rate throughout its flight (Fig. 1.2.5). Overall, the upward vertical component of the ball's initial velocity determines how high the ball goes and how long it stays aloft before striking the ground, while the horizontal component of the initial velocity determines how quickly the ball travels downfield during its time aloft (Fig. 1.2.6).

When the ball hits the ground it still has its original horizontal velocity component, but its vertical velocity component is now in the downward direction. The total velocity of the ball is composed of these two components. The ball starts with its velocity up and forward and ends with its velocity down and forward.

If you want a ball or shot put to hit the ground as far from your feet as possible, you should keep it aloft for a long time *and* give it a sizable horizontal component of velocity; in other words, you must achieve a good balance between the vertical

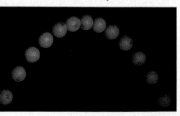

Courtesy Lou Bloomfield

Fig. 1.2.5 This golf ball drifts steadily to the right after being thrown because gravity affects only the ball's vertical component of velocity.

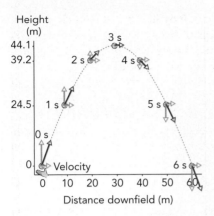

Fig. 1.2.6 If you throw a ball upward, at an angle, part of the initial velocity will be in the vertical direction and part will be in the horizontal direction. The vertical and horizontal motions will take place independently of one another. The ball will rise and fall just as it did in Figs. 1.2.2 and 1.2.3; at the same time, however, it will move downfield. Because there is no horizontal acceleration (gravity acts only in the vertical direction), the horizontal velocity remains constant, as indicated by the velocity arrows. In this example, the ball travels 10 m downfield each second and strikes the ground 60 m from your feet after a 6-s flight through the air.

and horizontal velocity components (Fig. 1.2.7). These components of velocity together determine the ball's flight path, its **trajectory.** If you throw the ball straight up, it will stay aloft for a long time but will not travel downfield at all (and you will need to wear a helmet). If you throw the ball directly downfield, it will have a large initial horizontal velocity component but will hit the ground almost immediately.

Neglecting air resistance and the altitude difference between your throwing arm and the ground that the ball will eventually hit, your best choice is to throw the ball at an angle of 45° above horizontal. At that angle, the initial upward velocity component will be the same as the initial downfield velocity component. The ball will stay aloft for a reasonably long time and will make good use of that time to move downfield. Other angles won't make such good use of the initial speed to move the ball downfield. (A discussion of how to determine the horizontal and vertical components of a velocity appears in Appendix A.)

These same ideas apply to two baseballs, one dropped from a cliff and the other thrown horizontally from that same cliff. If both leave your hands at the same time, they will both hit the ground below at the same time (Fig. 1.2.8). The fact that the second ball has an initial horizontal velocity doesn't affect the time it takes to descend to the ground, because the horizontal and vertical motions are independent. Of course, the ball thrown horizontally will strike the ground far from the base of the cliff, while the dropped ball will land directly below your hand.

Richard Megna/Fundamental Photographs

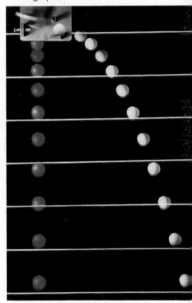

> Check Your Understanding #4: Aim High

for answers, see page 35

Why must a sharpshooter or an archer aim somewhat above her target? Why can't she simply aim directly at the bull's-eye to hit it?

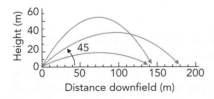

Fig. 1.2.7 If you want the ball to hit the ground as far from your feet as possible, given a certain initial speed, throw the ball at 45°. Halfway between horizontal and vertical, such a throw gives the ball equal initial vertical and horizontal components of velocity. The ball then stays aloft for a relatively long time and makes good use of that flight time to travel downfield.

Fig. 1.2.8 When these two balls were dropped, they accelerated downward at the constant rate of 9.8 m/s^2 and their velocities increased steadily in the downward direction. Even though one of them was initially moving to the right, they descended together.

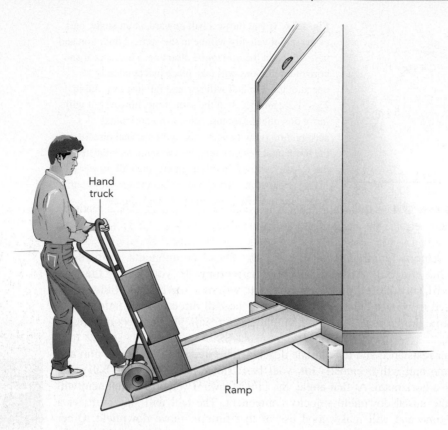

Hand
truck

Ramp

SECTION 1.3 Ramps

In the previous section, we looked at what happens to an object experiencing only a single force: the downward force of gravity. But what happens to objects that experience two or more forces at the same time? Imagine, for example, an object resting on the floor. That object experiences both the downward force of gravity and an upward force from the floor. If the floor is level, the object doesn't accelerate; but if the floor is tilted, so that it forms a ramp, the object accelerates downhill. In this section, we'll examine the motion of objects traveling along ramps. Also called inclined planes, ramps are common tools that make things easier to move.

Questions to Think About: *How does a ramp make it possible for one person to lift a very heavy object? Why is lowering a heavy object so much easier than raising it? What changes about that object as you raise it? Why is a steep hill so much scarier to ski or sled down than a more gradual slope? Why is the steeper hill so much harder to bicycle up?*

Experiments to Do: *Place a book on a slippery but level table or board. Hold the book steady for a second and then let go of it without pushing on it. Why does the book remain motionless?*

Now, equip yourself with a pencil. Have a friend tilt the surface of the table or board slightly so that the book begins to slide downhill. Can you stop the book by pushing on it with the pencil? Place the book back on the table and have your friend tilt the table more sharply. Does the tilt of the table affect your ability to stop the book from sliding? Now try to push the book uphill with the pencil (a) when the table is slightly tilted and (b) when it is more sharply tilted. Which task requires more force? Why?

Evidently a gentle push is all that may be needed to raise a relatively heavy object if you use a ramp to help you. To understand why this feat is possible, we need to explore a handful of physical concepts and a few basic laws of motion.

A Piano on the Sidewalk

Imagine that you have a friend who's a talented but undiscovered pianist. She's renting a new apartment, and because she can't afford professional movers (Fig. 1.3.1), she's asked you to help her move her baby grand. Fortunately, her new apartment is only on the second floor. But the two of you still face a difficult challenge: how do you get that heavy piano up there? More importantly, how do you keep it from falling on you during the move?

The problem is that you can't push upward hard enough to lift the entire piano at once. One solution to this problem, of course, would be to break the piano into pieces and carry them up one by one. But this method has obvious drawbacks: your friend isn't expecting a firewood delivery. A better solution would be to find something else to help you push upward, and one of your best choices would be the simple machine known as a ramp.

Throughout the ages, **ramps,** also known as inclined planes, have made tasks like piano-moving possible. Because ramps can exert the enormous upward forces needed to lift stone and steel, they've been essential building equipment since the days of the pyramids. To see how ramps provide these lifting forces, we'll continue to explore the example of the piano, looking first at the force that the piano experiences when it touches a surface. For the time being, we'll continue to ignore friction and air resistance; they'd needlessly complicate our discussion. Besides, as long as the piano is on wheels, friction is negligible.

Courtesy Jerry Ohlinger's Movie Material Store

Fig. 1.3.1 A ramp would make this move much easier.

> Check Your Understanding #1: Brick Work
>
> for answers, see page 35
>
> Which requires larger forces: lifting a pile of bricks one at a time or lifting them all together?

The Sidewalk Pushes Back: Newton's Third Law

With the piano resting on the sidewalk outside the apartment you make a startling discovery: the piano is *not* falling. Has gravity disappeared? The answer to that question would be painfully obvious if your foot were underneath one of the piano's wheels. No, the piano's weight is still all there. But something is happening at the surface of the sidewalk to keep the piano from falling. Let's take a careful look at the situation.

To begin with, the piano is clearly pushing down hard on the sidewalk. That's why you're keeping your toes out of the way. But the presence of a new downward force *on the sidewalk* doesn't explain why *the piano* isn't falling. Instead, we must look at the sidewalk's response to the piano's downward push: the sidewalk pushes upward on the piano! You can feel this response by leaning over and pushing down on the sidewalk with your hand—the sidewalk will push back. Those two forces, your downward push on the sidewalk and its upward push on your hand, are exactly equal in magnitude but opposite in direction.

This observation, that two things exert equal but opposite forces on one another, isn't unique to sidewalks, pianos, or hands; in fact, it's always true. If you push on any object, that object will push back on you with an equal amount of force in exactly the opposite direction. This rule—often expressed as "for every action, there is an equal but opposite reaction"—is known as **Newton's third law of motion,** the last of his three laws.

Newton's Third Law of Motion

For every force that one object exerts on a second object, there is an equal but oppositely directed force that the second object exerts on the first object.

The universality of this law is astounding. Whether an object is large or small, hard or soft, stationary or faster than a rocket, if you can push on it, it *will* push back on you with an equal but oppositely directed force.

In the present case, the sidewalk and piano push on one another with equal but oppositely directed forces. Of this pair of equal-but-opposite forces, only one force acts *on the piano*: the sidewalk's upward push. This upward push on the piano is what keeps the piano from falling. We've solved the mystery.

Intuition Alert: Action and Reaction

Intuition says that when you push an object that's moving away from you, it pushes back more gently on you than you push on it and when you push an object that's moving toward you, it pushes back harder on you than you push on it.

Physics says that when you push on an object, it always pushes back on you exactly as hard as you push on it.

Resolution: It's difficult to push on an object that's moving away from you, so you naturally push on it more gently than you expect. The gentle force it exerts on you is simply an equal but oppositely directed response to your gentle force on it. In contrast, it's difficult not to push strongly on an object that's moving toward you, so you naturally push on it harder than you expect. The strong force it exerts on you is again an equal but oppositely direct response to your strong force on it.

Summary of Newton's Laws of Motion

1. An object that is not subject to any outside forces moves at a constant velocity, covering equal distances in equal times along a straight-line path.
2. The force exerted on an object is equal to the product of that object's mass times its acceleration. The acceleration is in the same direction as the force.
3. For every force that one object exerts on a second object, there is an equal but oppositely directed force that the second object exerts on the first object.

► Check Your Understanding #2: Swinging

for answers, see page **35**

You are pushing a child on a playground swing. If you exert a 50-N (11-lbf) force on him as he is swinging away from you, how much force will he exert back on you?

Looking for Support and Adding Up the Forces

Although we've figured out why the piano isn't falling, we still don't know what type of force the sidewalk is using to hold it up or why that upward force so perfectly balances the piano's downward weight.

Let's begin with the type of force: since two objects can't occupy the same space at the same time, their surfaces push apart whenever they're in contact. They exert **support forces** on one another, each pushing the other directly away from its surface—the direction *normal* or *perpendicular* to its surface (see ❑). Since the sidewalk is horizontal, the support force it exerts on the piano is vertical—straight up (Fig. 1.3.2).

But how large is that support force? To answer this question, suppose the sidewalk's support force was strong enough to make the piano accelerate upward. The piano would soon lift off the sidewalk and, as their surfaces stopped touching, the sidewalk's support force on the piano would grow weaker. Alternatively, suppose the sidewalk's support force was weak enough to let the piano accelerate downward. The piano would soon drop into the sidewalk and, as their surfaces overlapped more, the sidewalk's support force on the piano would grow stronger.

Because of these behaviors, the sidewalk's upward support force on the piano adjusts automatically until it exactly balances the piano's downward weight and the piano accelerates neither up nor down. And when you sit on the piano during a break, the sidewalk's upward support force quickly readjusts to balance your weight as well.

Another way to state that the upward support force on the piano exactly balances the piano's downward weight is to say that the **net force** on the piano is zero, meaning that the sum of all the forces on the piano is zero. Objects often experience more than one force at a time, and it's the net force, together with the object's mass, that determines how it accelerates. When you and your friend push the piano in the same direction, your forces add up, assisting one another so that the piano accelerates in that direction (Fig. 1.3.3*a*). When the two of you push the piano in opposite directions, your forces oppose and at least partially cancel one another (Fig. 1.3.3*b*).

And when the two of you push the piano at an angle with respect to one another, the net force points somewhere in between. For example, if you push the piano eastward while your friend pushes it northward, the net force will point to the northeast and the piano will accelerate in that direction (Fig. 1.3.3*c*). The precise angle of the net force and the piano's subsequent acceleration depends on exactly how hard each person pushes. For most of the following discussion, we'll need only a rough estimate of the net force's magnitude and direction, and we'll obtain that estimate using common sense.

Apart from direction, there's one crucial difference between the force gravity exerts on the piano and the support force the sidewalk exerts on it. While the piano's weight is dispersed throughout the piano, the sidewalk's upward support force acts only on the piano's wheels. Even when the net force on the piano is zero, having individual forces

❑ Forces that are directed exactly away from surfaces are called **normal forces,** since the term **normal** is used by mathematicians to describe something that points exactly away from a surface—at right angles or perpendicular to that surface.

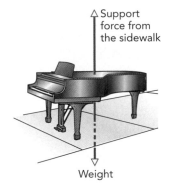

Fig. 1.3.2 A piano resting on the sidewalk. The sidewalk exerts an upward support force that exactly balances the piano's downward weight. The net force on the piano is zero, so the piano doesn't accelerate.

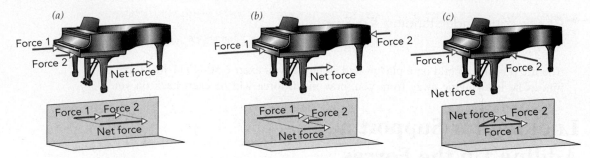

Fig. 1.3.3 When several forces act simultaneously on an object, the object responds to the sum of the forces. This sum is called the net force, and it has both a magnitude and a direction. Here, as elsewhere in this book, the length of each force arrow indicates its magnitude. You can sum the forces graphically by arranging the force arrows in sequence, head to tail. The net force arrow then points from the tail of the first arrow to the head of the last arrow. Some of the force arrows are shown displaced slightly for clarity, a shift that doesn't affect the summing process.

act on it at different locations can lead to considerable stress within the piano. If it weren't built so well, the piano might lose a leg or two during the move.

> ### Common Misconceptions: Newton's Third Law and Balanced Forces
>
> **Misconception:** When you push on an object and it pushes back on you with an equal but opposite force, those two forces exactly balance one another and nothing accelerates!
>
> **Resolution:** While the object pushes back on you just as hard as you push on it, those two forces act on *two different objects* and cannot balance one another. The object you are pushing experiences only your force on it and accelerates.

Check Your Understanding #3: Riding the Elevator

for answers, see page 35

As you ride upward in an elevator at a constant velocity, what two forces act on your body and what is the net force on you?

What the Piano Needs: Energy

As you approach the task of lifting your friend's piano into her apartment, you might begin to worry about safety. There is clearly a difference between the piano resting on the sidewalk and the piano suspended on a board just outside the second floor apartment. After all, which one would you rather be sitting beneath? The elevated piano has something that the piano on the sidewalk doesn't have: the ability to produce motion and structural rearrangement (i.e., damage) in itself and the things beneath it. This capacity to make things happen is called *energy,* and the process of making them happen is called *work.*

Energy and work are both important physical *quantities,* meaning that both are measurable. For example, you can measure the amount of energy in the suspended piano, and the amount of work the piano does when the board breaks and it falls to the sidewalk. As you may suspect, the physical definitions of energy and work are somewhat different from those of common English. Physical **energy** isn't the exuberance of a child at the amusement park or the contents of a large cup of coffee; instead, it's defined as the capacity to do work. Similarly, physical **work** doesn't refer to activities at the office or in the yard; instead, it refers to the process of transferring energy.

Energy is what's transferred, and work does the transferring. The most important characteristic of energy is that it's conserved. In physics, a **conserved quantity** is one that can't be created or destroyed but can be transferred between objects or, in the case of energy, converted from one form to another. Conserved quantities are very special in physics; there are only a few of them. An object that has energy can't simply make that energy disappear; it can only get rid of energy by giving it to another object, and it makes this transfer by doing work on that object.

The relationship between energy and work is analogous to the relationship between money and spending: money is what is transferred, and spending does the transferring. Sensible, law-abiding citizens don't create or destroy money; instead they transfer it among themselves through spending. Just as the most interesting aspect of money is spending it, so the most interesting aspect of energy is doing work with it. We can define money as the capacity to spend, just as we define energy as the capacity to do work.

So far we've been using a circular definition: work is the transfer of energy and energy is the capacity to do work. But what is involved in doing work on an object? You do work on an object by exerting a force on it as it moves in the direction of that force. As you throw a ball, exerting a forward force on the ball and moving the ball forward, you do work on the ball; as you lift a rock, pushing upward and moving the rock upward, you do work on the rock. In both cases, you transfer energy from yourself to an object by doing work on it.

This transferred energy is often apparent in the object. When you throw a ball, it picks up speed and undergoes an increase in **kinetic energy**—energy of motion that allows the ball to do work on whatever it hits. And when you lift a rock, it shifts farther from the earth and undergoes an increase in **gravitational potential energy**—energy stored in the gravitational forces between the rock and the earth that allows the rock to do work on whatever it falls on. In general, **potential energy** is energy stored in the forces between or within objects.

Returning to the task at hand, it's now apparent that raising the piano to the second floor apartment is going to increase the piano's gravitational potential energy. Since energy is a conserved quantity, this additional energy must come from something else. Unfortunately, that something is you! To deliver the piano, you are going to have to provide it with the gravitational potential energy it needs by doing exactly that amount of work on it. And as we'll see, you can do that work the hard way by carrying it up a ladder or the easy way by pushing it up a ramp.

▶ Check Your Understanding #4: No Shortage of Energy

for answers, see page **35**

Do any of these objects have energy they can spare: a compressed spring, an inflated toy balloon, a stick of dynamite, and a falling ball?

Lifting the Piano: Doing Work

To do work on an object, you must push on it while it moves in the direction of your push. The work you do on it is the product of the force you exert on it times the distance it travels along the direction of your force. We can express this relationship as a word equation:

$$work = force \cdot distance, \qquad (1.3.1)$$

in symbols:

$$W = \mathbf{F} \cdot \mathbf{d},$$

and in everyday language:

> *If you're not pushing or it's not moving, then you're not working.*

Calculating the work you do is simple if the object moves exactly in the direction of your push: you simply multiply your force times the distance it travels. But if the object doesn't move in the direction of your push, you must multiply your force times the *component* of the object's motion along the direction of your force.

As long as the angle between your force and the object's motion is small, you can often ignore this complication. But as the angle becomes larger, the work you do on the object decreases. When the object moves at right angles to your force, the work you do on it drops to zero—it's not moving along the direction of your force at all. And for angles larger than 90°, the object moves *opposite* your force and the work you do on it actually becomes negative!

Recalling that forces always come in equal but oppositely directed pairs, we can now explain why energy is conserved: whenever you do work on an object, that object simultaneously does an equal amount of negative work on you! After all, if you push an object and it moves along the direction of your force, then it pushes back on you and you move along the direction opposite its force. You do positive work on it and it does negative work on you.

For example, when you lift the piano to judge its weight, you push it up as it moves up and thus do work on it. At the same time, the piano pushes your hand down but your hand moves up, so it does negative work on your hand. Overall, the piano's energy increases by exactly the same amount that your energy decreases—a perfect transfer! The energy that you're losing is mostly food energy, a form of chemical potential energy, and the energy the piano is gaining is mostly gravitational potential energy.

When you lower the piano after realizing that it's too heavy to carry up a ladder, the process is reversed and the piano transfers energy back to you. Now the piano does work on you and you do an equal amount of negative work on the piano. The piano is losing mostly gravitational potential energy and you are gaining mostly thermal energy— a disordered form of energy that we'll examine in Section 2.2. Your body just isn't good at storing work done on it, so it simply gets hotter. Nonetheless, it's usually easier to have work done on you than to do work on something else. That's why it's easier to lower objects than to lift them.

Finally, when you hold the piano motionless above the pavement, while waiting for your friend to reinstall the wheel that fell off, you and the piano do no work on one another. You are simply converting chemical potential energy from your last meal into thermal energy in your muscles and get overheated, probably in more ways than one.

> **Conserved Quantity: Energy** **Transferred By: Work**
>
> *Energy:* The capacity to do work. Energy has no direction. It can be hidden as potential energy.
>
> *Kinetic Energy:* The form of energy contained in an object's motion.
>
> *Potential Energy:* The form of energy stored in the forces between or within objects.
>
> *Work:* The mechanical means for transferring energy; work = force · distance.

Check Your Understanding #5: Pitching

for answers, see page **35**

When you throw a baseball horizontally, you're not pushing against gravity. Are you doing any work on the baseball?

Check Your Figures #1: Light Work, Heavy Work

for answers, see page **36**

You are moving books to a new shelf, 1.20 m (3.94 ft) above the old shelf. The books weigh 10.0 N (2.25 lbf) each and you have 10 of them to move. How much work must you do on them as you move them? Does it matter how many you move at once?

Gravitational Potential Energy

How much work would you do on the piano while lifting it straight up a ladder to the apartment? Apart from a little extra shove to get the piano moving upward, lifting it would entail supporting the piano's weight while it coasted upward at constant velocity from the sidewalk to the second floor. Since you would be pushing upward on the piano with a force equal in amount to its weight, the work you would do on it would be the product of its weight times the distance you lifted it.

As the piano rises in this scenario, its gravitational potential energy increases by an amount equal to the work you do on it. If we agree that the piano has zero gravitational potential energy when it rests on the sidewalk, then the suspended piano's gravitational potential energy is simply its weight times its height above the sidewalk. Since the piano's weight is equal to its mass times the acceleration due to gravity, its gravitational potential energy is its mass times the acceleration due to gravity times its height above the sidewalk.

These ideas are not limited to pianos. You can determine the gravitational potential energy of any object by multiplying its mass times the acceleration due to gravity times its height above the level at which its gravitational potential energy is zero. This relationship can be expressed as a word equation:

gravitational potential energy = mass · acceleration due to gravity · height, **(1.3.2)**

in symbols:

$$U = m \cdot g \cdot h,$$

and in common language:

The higher they were, the harder they hit.

Of course, if you know the object's weight, you can use it in place of the object's mass times the acceleration due to gravity.

So what is the piano's gravitational potential energy when it reaches the second floor? If it weighs 2000 N (450 lbf) and the second floor is 5 m (16 ft) above the sidewalk, you will have done 10,000 N·m (about 7200 ft·lbf) of work in lifting it up there, and the piano's gravitational potential energy will thus be 10,000 N·m. The **newton-meter** is the SI unit of energy and work; it's so important that it has its own name, the **joule** (abbreviated J). At the second floor, the piano's gravitational potential energy is 10,000 J.

A few everyday examples should give you a feeling for how much energy a joule is. Lifting a liter bottle of water 10 centimeters (4 inches) upward requires about 1 J of work. A 100-watt lightbulb needs 100 J every second to operate. Your body is able to extract about 2,000,000 J from a slice of cherry pie. When you're bicycling or rowing hard, your body can do about 1000 J of work each second. A typical flashlight battery has about 10,000 J of stored energy.

Quantity	SI Unit	English Unit	SI → English	English → SI
Energy	joule (J) (1 J = 1 N·m)	foot-pound (ft·lbf)	1 J = 0.73757 ft·lbf	1 ft·lbf = 1.3558 J

▶ Check Your Understanding #6: Mountain Biking

for answers, see page 35

Bicycling to the top of a mountain is much harder than rolling back down to the bottom. At which place do you have the most gravitational potential energy?

▶ Check Your Figures #2: Watch Out Below

for answers, see page 36

If you carry a U.S. penny (0.0025 kg) to the top of the Empire State Building (443.2 m or 1453.7 ft), how much gravitational potential energy will it have?

Lifting the Piano with a Ramp

Unfortunately, you probably can't carry a grand piano up a ladder by yourself. You need a ramp, both to help you support the piano and to make it easier for you to raise the piano to the second floor.

Like the sidewalk, a ramp exerts a support force on the piano to prevent the piano from passing through its surface. However, since the ramp isn't exactly horizontal, that support force isn't exactly vertical (Fig. 1.3.4). The piano's weight still points straight down, but since the ramp's support force doesn't point straight up, the two forces can't balance one another. There is a nonzero net force on the piano.

This net force can't point into or out of the ramp. If it did, the piano would accelerate into or out of the ramp and the two objects would soon either lose contact or travel through one another. Instead, the net force points exactly along the surface of the ramp—a direction *tangent* or *parallel* to the surface. More specifically, it points directly downhill so the piano accelerates down the ramp!

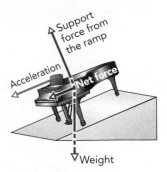

But because this net force is much smaller than the piano's weight, the piano's acceleration down the ramp is slower than if it were falling freely. This effect is familiar to anyone who has bicycled downhill or watched a cup slip slowly off a tilted table. While gravity is still responsible, these objects accelerate more slowly than falling and in the direction of the downward slope.

Therein lies the beauty of the ramp. By putting the piano on a ramp, you let the ramp support most of the piano's weight. The piano only experiences a small residual net force pushing it downhill along the ramp. If you now push uphill on the piano with a force that exactly balances that downhill force, the net force on the piano drops to zero and the piano stops accelerating. If you push uphill a little harder, the piano will accelerate up the ramp!

Fig. 1.3.4 A piano sliding on a frictionless ramp while experiencing a force due to gravity. It accelerates down the ramp more slowly than it would if it were falling freely.

How does the ramp change the job of moving the piano? Suppose that you build a ramp 50 m long that extends from the sidewalk to the apartment's balcony, 5 m above the pavement. This ramp is sloped so that traveling 10 m uphill along its surface lifts the piano only 1 m upward (Fig. 1.3.5). Because of the ramp's 10 to 1 grade, you can push the 2000-N (450-lbf) piano up it at constant velocity with a force of only 200 N (45 lbf). Most people can push that hard, so the moving job is now realistic. To reach the apartment, you must push the piano 50 m along this ramp with a force of 200 N so that you will do a total of 10,000 J of work.

By pushing the piano up the ramp, you've used physical principles to help you perform a task that would otherwise have been nearly impossible. But you didn't get something for nothing. The ramp is much longer than the ladder, and you have had to push the piano for a longer distance in order to raise it to the second floor. Of course, you have had to push with less force.

Remarkably, the amount of work you do in either case is 10,000 J. In carrying the piano up the ladder, the force you exert is large but the distance the piano travels in the direction of that force is small. In pushing the piano up the ramp, the force is small but the distance is large. Either way, the final result is the same: the piano ends up on the second floor with an additional 10,000 J of gravitational potential energy and you have done 10,000 J of work. Expressed graphically in an equation, this relationship would appear as follows:

$$\text{work} = \text{large force} \cdot \text{small distance} = \text{small force} \cdot \text{large distance}.$$

In the absence of friction, the amount of work you do on the piano to get it to the second floor doesn't depend on how you raise it. No matter how you move that piano

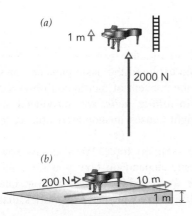

Fig. 1.3.5 To lift a piano weighing 2000 N, you can either (*a*) push it straight up or (*b*) push it along a ramp. To keep the piano moving at a constant velocity, you must make sure it experiences a net force of zero. If you lift it straight up the ladder in (*a*), you must exert an upward lifting force of 2000 N to balance the piano's downward weight. If you push it up the ramp shown in (*b*), you will only have to push the piano uphill with a force of 200 N in order to give the piano a net force of zero.

up to the second floor, its gravitational potential energy will increase by 10,000 J so you'll have to do 10,000 J of work on it. Even if you disassemble the piano into parts, carry them individually up the stairs, and reconstruct the piano in your friend's living room, you will have done 10,000 J of work lifting the piano.

Unless you're an experienced piano tuner, you'll probably be better off sticking with the ramp. It offers an easy method for one person to lift a baby grand piano. The ramp provides **mechanical advantage**—the process whereby a mechanical device redistributes the amounts of force and distance that go into performing a specific amount of mechanical work. In moving the piano with the help of the ramp, you've performed a task that would normally require a large force over a small distance by supplying a small force over a large distance. You might wonder whether the ramp itself does any work on the piano; it doesn't. Although the ramp exerts a support force on the piano and the piano moves along the ramp's surface, this force and the distance traveled are at right angles to one another. The ramp does no work on the piano.

Mechanical advantage occurs in many situations involving ramps. For example, it appears when you ride a bicycle up a hill. Climbing a gradual hill takes far less uphill force than climbing a steep hill of the same height. Since your pedaling ultimately provides the uphill force, it's much easier to climb the gradual hill than the steep one. Of course, you must travel a longer distance along the road as you climb the gradual hill than you do on the steep hill, so the work you do is the same in either case.

Ramps and inclined planes show up in many devices, where they usually reduce the forces needed to perform otherwise difficult tasks. They also change the character of certain activities. Skiing wouldn't be very much fun if the only slopes available were horizontal or vertical. By choosing ski slopes of various grades, you can select the net forces that set you in motion. Gentle slopes leave only small net forces and small accelerations; steep slopes produce large net forces and large accelerations.

Finally, our observation about mechanical advantage is this: mechanical advantage allows you to do the same work, but you must make a trade-off—you must choose whether you want a large force or a large distance. The product of the two parts, force times distance, remains the same.

> ➤ Check Your Understanding #7: Access Ramps
>
> for answers, see page 35

Ramps for handicap entrances to buildings are often quite long and may even involve several sharp turns. A shorter, straighter ramp would seem much more convenient. What consideration leads the engineers designing these ramps to make them so long?

Epilogue for Chapter 1

In this chapter we examined three everyday things and explored the basic physical laws that govern their behaviors. In *skating,* we looked at the concept of inertia and observed that objects accelerate only in response to forces. In *falling balls,* we introduced an important type of force—weight—and saw how weight causes unsupported objects to fall with equal downward accelerations.

In *ramps,* we encountered another type of force—support force. We also saw how the work done in changing an object's altitude doesn't depend on how you raise the object, since work is actually the mechanical means for transferring energy from one object to another. Energy, we noted, is one of the conserved physical quantities that govern the motion of objects in our universe. The particular type of energy we studied in *ramps* was gravitational potential energy—the potential energy associated with the force of gravity.

Explanation: Removing a Tablecloth from a Table

The dishes remain in place because of their inertia. As we've seen, an object in motion tends to remain in motion, while an object at rest tends to remain at rest.

Before you pull on the tablecloth, the dishes sit motionless on its surface and they tend to remain that way. By sliding the tablecloth off the table as quickly and smoothly as possible, you ensure that whatever force the tablecloth exerts on each dish occurs only for a very short time. As a result, the dishes undergo only the tiniest changes in velocity and remain essentially stationary on the tabletop.

Chapter Summary

How Skating Works: When you're gliding forward on frictionless skates, you experience no horizontal forces and move at a constant velocity. Inertia alone carries you forward. To change your velocity—to accelerate—something must exert a horizontal force on you. A forward force speeds you up while a backward force slows you down. Since sideways forces change your direction of travel, they, too, make you accelerate.

How Falling Balls Work: Any ball that's subject only to the force of gravity is a falling ball. It accelerates downward at a steady rate. Gravity affects only the ball's vertical motion, causing the ball's vertical component of velocity to increase steadily in the downward direction. If the ball were initially moving horizontally, it would continue that horizontal motion and drift steadily downfield as it falls.

A falling ball that's initially rising soon stops rising and begins to descend. The larger its initial upward component of velocity, the longer the ball rises and the greater its peak height. When the ball then begins to descend, the peak height determines how long it takes for the ball to reach the ground.

When you throw a ball, the vertical component of initial velocity determines how long the ball remains aloft. The horizontal component of initial velocity determines how quickly the ball moves downfield. A thrower intuitively chooses an initial speed and direction for a ball so that it moves just the right distance downfield by the time it descends to the desired height.

How Ramps Work: An object at rest on level ground experiences two forces: its downward weight and an upward support force from the ground that exactly balances that weight. The net force on the object is thus zero. But if the ground is replaced with a ramp, the support force is no longer directly upward and the net force on the object isn't zero. Instead, the net force points downhill along the ramp and is equal to the weight of the object multiplied by the ratio of the ramp's rise to the ramp's length. If a 10-m-long ramp rises 1 m in height, then this ratio is 1 m divided by 10 m or 0.10. The net force downhill along this ramp is thus only 10% of the object's weight.

To stop an object from accelerating down a ramp, you must balance the downhill force by pushing equally hard uphill. In fact, if you exert more force up the ramp than it experiences down the ramp, the object will begin to accelerate up the ramp.

It takes less force to push an object up a ramp than to lift it directly upward, but you must push that object a longer distance along the ramp. Overall, the work you do in raising the object from one height to another is the same, whether or not you use the ramp. However, the ramp gives you mechanical advantage, allowing you to do work that would require an unrealistically large force by instead exerting a much smaller force for a much longer distance.

Important Laws and Equations

1. Newton's First Law of Motion: An object that is free from all outside forces travels at a constant velocity, covering equal distances in equal times along a straight-line path.

2. Newton's Second Law of Motion: An object's acceleration is equal to the force exerted on that object divided by the object's mass, or

$$\text{force} = \text{mass} \cdot \text{acceleration}. \qquad (1.1.2)$$

3. Relationship between Mass and Weight: An object's weight is equal to its mass times the acceleration due to gravity, or

$$\text{weight} = \text{mass} \cdot \text{acceleration due to gravity}. \qquad (1.2.1)$$

4. The Velocity of an Object Experiencing Constant Acceleration: The object's present velocity differs from its initial velocity by the product of its acceleration times the time since it was at that initial velocity, or

$$\text{present velocity} = \text{initial velocity} + \text{acceleration} \cdot \text{time}. \qquad (1.2.2)$$

5. The Position of an Object Experiencing Constant Acceleration: The object's present position differs from its initial position by the product of its average velocity since it was at that initial position times the time since it was at that initial position, or

$$\text{present position} = \text{initial position} + \text{initial velocity} \cdot \text{time} + \tfrac{1}{2} \cdot \text{acceleration} \cdot \text{time}^2. \qquad (1.2.3)$$

6. Newton's Third Law of Motion: For every force that one object exerts on a second object, there is an equal but oppositely directed force that the second object exerts on the first object.

7. The Definition of Work: The work done on an object is equal to the product of the force exerted on that object times the distance that object travels in the direction of the force, or

$$\text{work} = \text{force} \cdot \text{distance}. \qquad (1.3.1)$$

8. Gravitational Potential Energy: An object's gravitational potential energy is its mass times the acceleration due to gravity times its height above a zero level, or

$$\text{gravitational potential energy} = \text{mass} \cdot \text{acceleration due to gravity} \cdot \text{height}. (1.3.2)$$

Check Your Understanding—Answers

Section 1.1 SKATING

1. The puck coasts across the ice because it has inertia.
Why: A hockey puck resting on the surface of wet ice is almost completely free of horizontal influences. If someone pushes on the puck, so that it begins to travel with a horizontal velocity across the ice, inertia will ensure that the puck continues to slide at constant velocity.

2. The train accelerates forward when starting out from a station, backward when it arrives at the next station, to the left when it turns left, and downward when it begins its descent out of the mountains.
Why: Whenever the train changes its speed or its direction of travel, it is accelerating. When it speeds up on leaving a station, it is accelerating forward (more forward-directed speed). When it slows down at the next station, it is accelerating backward (more backward-directed speed or, equivalently, less forward-directed speed). When it turns left, it is accelerating to the left (more leftward-directed speed). When it begins to descend, it is accelerating downward (more downward-directed speed).

3. An automobile has a much greater mass than a bicycle.
Why: To stop a moving vehicle, you must exert a force on it in the direction opposite its velocity. The vehicle will then accelerate backward so that it eventually comes to rest. If the vehicle is heading toward you, you must push it away from you. The more mass the vehicle has, the less it will accelerate in response to a certain force and the longer you will have to push on it to stop it completely. While it's easy to stop a bicycle by hand, stopping even a slowly moving automobile by hand requires a large force exerted for a substantial amount of time.

4. The sweater is coasting eastward at constant velocity.
Why: While both of you agree that the sweater is not accelerating and that it is moving according to Newton's first law, you disagree on its specific velocity. She sees the sweater at rest while you see it coasting eastward at constant velocity. Your viewpoints are equally valid.

5. About 2.24 miles.
Why: There are many different units in this example, so we must do some converting. First, an hour is 3600 seconds, so that in an hour of walking at 1 meter-per-second, you will have walked 3600 meters. Second, a mile is about 1609 meters so that each time you travel 1609 meters, you have traveled a mile. By walking 3600 meters, you have completed 2 miles and are about one-quarter of the way into your third mile.

Section 1.2 FALLING BALLS

1. The astronaut would have zero weight but would still have a normal mass.
Why: Weight is a measure of the force exerted on the astronaut by gravity. Far from the earth or any other large object, the astronaut would experience virtually no gravitational force and would have zero weight. But mass is a measure of inertia and doesn't depend at all on gravity. Even in deep space, it would be much harder to accelerate a school bus than to accelerate a baseball because the school bus has more mass than the baseball.

2. 4.9 m/s (16 ft/s) in the downward direction.
Why: A freely falling object accelerates downward at a steady rate. Its velocity changes by 9.8 m/s (16 ft/s) in the downward direction each and every second. In half a second, the marble's velocity changes by only half that amount or 4.9 m/s (16 ft/s).

3. About 1.2 m (4 ft).
Why: While a freely falling object's velocity changes steadily in the downward direction, its height is more complicated. When you drop the marble from rest, it starts its descent slowly but picks up speed and covers the downward distance faster and faster. In the first 0.5 s, it travels only a quarter of the distance it travels in the first 1 s, or about 1.2 m (4 ft).

4. The bullet or arrow will fall in flight so she must compensate for its loss of height.
Why: To hit the bull's-eye, the sharpshooter or archer must aim above the bull's-eye because the projectile will fall in flight. The

longer the bullet or arrow is in flight, the more it will fall and the higher she must aim. As the distance to the target increases, the flight time increases and her aim must move upward.

Section 1.3 RAMPS

1. Lifting the bricks all together.
Why: To lift a brick upward steadily, you must support it with an upward force equal in amount to its downward weight. If you lift several bricks at once, you will have to support all of them together and that will involve a larger upward force.

2. 50 N (11 lbf).
Why: Whenever you exert a 50-N force on an object, whether it's moving or stationary, it will exert a 50-N force back on you. There are no exceptions. If that object is a friend, it doesn't matter whether she is stationary or moving or wearing roller skates or even sound asleep: she will push back with 50 N of force. She has no choice in the matter. Similarly, if someone pushes on you, you will feel yourself pushing back. That's how Newton's third law works.

3. The two forces are the downward force of gravity (your weight) and an upward support force from the floor. They balance, so that the net force on you is zero.
Why: Whenever anything is moving with constant velocity, it's not accelerating and thus has zero net force on it. Although the elevator is moving upward, the fact that you are not accelerating means that the car must exert an upward support force on you that exactly balances your weight. You experience zero net force.

4. Yes, all of them.
Why: Each of these four objects can easily do work on you and thereby give you some of its spare energy. It does this work by pushing on you as you move in the direction of that push.

5. Yes.
Why: Any time you exert a force on an object and the object moves in the direction of that force, you are doing work on the object. Since gravity doesn't affect horizontal motion, the work you do on the baseball as you throw it ends up in the baseball as kinetic energy (energy of motion). As anyone who has been hit by a pitch can attest, a moving baseball has more energy than a stationary baseball.

6. At the top of the mountain.
Why: Bicycling up the mountain is hard because you must do work against the force of gravity. This work is stored as an increasing gravitational potential energy on your way uphill. Gravity then does work on you as you roll downhill and your gravitational potential energy decreases.

7. The engineers must limit the amount of force needed to propel a wheelchair steadily up the ramp. The steeper the ramp, the more force is required.
Why: A person traveling in a wheelchair on a level surface experiences little horizontal force and can move at constant velocity with very little effort. But climbing a ramp at constant

velocity requires a substantial uphill force equal in magnitude to the downhill force from gravity. The steeper the ramp, the more uphill force is needed to maintain constant velocity. A 12 to 1 grade (12 meters of ramp surface for each meter of rise in height) is the accepted limit to how steep such a long ramp can be.

Check Your Figures—Answers

Section 1.1 SKATING

1. The less massive ball will accelerate twice as rapidly.
Why: Equation 1.1.1 shows that an object's acceleration is inversely proportional to its mass:

$$\text{acceleration} = \frac{\text{force}}{\text{mass}}.$$

If you push on both bowling balls with equal forces, then their accelerations will only depend on their masses. Doubling the mass on the right side of this equation halves the acceleration on the left side. That means that the more massive ball will accelerate only half as quickly as the other ball.

Section 1.2 FALLING BALLS

1. About 1.6 m/s^2 (5.3 ft/s^2).
Why: You can rearrange Eq. 1.2.1 to show that the acceleration due to gravity is proportional to an object's weight:

$$\text{acceleration due to gravity} = \frac{\text{weight}}{\text{mass}}.$$

Your mass doesn't change in going to the moon, so any change in your weight must be due to a change in the acceleration due to gravity. Since your moon weight is one-sixth of your earth weight, the moon's acceleration due to gravity must be one-sixth that of the earth or about 1.6 m/s^2.

2. About 14 m/s (45 ft/s, 50 km/h, or 31 mph).
Why: The downward acceleration due to gravity is 9.8 m/s^2 (32 ft/s^2). You fall for 1.4 s, during which time your velocity increases steadily in the downward direction. Since you start with zero velocity, Eq. 1.2.2 gives a final velocity of

$$\text{final velocity} = 9.8 \text{ m/s}^2 \cdot 1.4 \text{ s} = 13.72 \text{ m/s}.$$

Since the time of the fall is only given to two digits of accuracy (1.4 s could really be 1.403 s or 1.385 s), we shouldn't claim that our calculated final velocity is accurate to four digits. We should round the value to 14 m/s (45 ft/s).

3. About 122 meters (402 feet or a 40-story building).

Why: As they fall, the jumpers will travel downward at ever increasing speeds. Since the jumpers start from rest and fall downward for 5 seconds, we can use Eq. 1.2.3 to determine how far they will fall:

$$\text{final height} = \text{initial height} - \tfrac{1}{2} \cdot 9.8 \text{ m/s}^2 \cdot (5 \text{ s})^2$$

$$= \text{initial height} - 122.5 \text{ m}.$$

The downward acceleration is indicated here by the negative change in height. At the end of 5 seconds, the jumpers will have fallen more than 122 m (402 ft) and will be traveling downward at about 50 m/s (160 ft/s). The tower will need additional height to slow the jumpers down and begin bouncing them back upward. Clearly, a 5-second free fall is pretty unrealistic. Try for a 2- or 3-second free fall instead.

Section 1.3 RAMPS

1. It takes 120 N · m (88.6 ft · lbf), no matter how many you lift at once.
Why: To keep each book from accelerating downward, you must support its weight with an upward force of 10.0 N. You must then move it upward 1.20 m. The work you do pushing upward on the book as it moves upward is given by Eq. 1.3.1:

$$\text{work} = \text{force} \cdot \text{distance} = 10.0 \text{ N} \cdot 1.20 \text{ m} = 12.0 \text{ N} \cdot \text{m}.$$

It takes 12.0 N · m of work to lift each book, whether you lift it together with other books or all by itself. The total work you must do on all 10 books is 120 N · m.

2. About 11 J.
Why: The penny's gravitational potential energy is given by Eq. 1.3.2:

$$\text{gravitational potential energy}$$
$$= 0.0025 \text{ kg} \cdot 9.8 \text{ N/kg} \cdot 443.2 \text{ m}$$
$$= 11 \text{ N} \cdot \text{m} = 11 \text{ J}.$$

This 11-J increase in energy would be quite evident if you were to drop the penny. In principle, the penny could accelerate to very high speed (up to 340 km/h or 210 mph) and do lots of damage when it hit the ground. Fortunately, a falling penny actually tumbles and is slowed so much by the air that it isn't particularly dangerous.

Exercises

1. A dolphin can leap several meters above the ocean's surface. Why doesn't gravity stop the dolphin from leaving the water?

2. As you jump across a small stream, does a horizontal force keep you moving forward? If so, what is that force?

3. Why does stamping your feet clean the snow off them?

4. Why does tapping your toothbrush on the sink dry it off?

5. The back of your car seat has a headrest to protect your neck during a collision. What type of collision causes your head to press against the headrest?

6. An unseatbelted driver can be injured by the steering wheel during a head-on collision. Why does the driver hit the steering wheel when the car suddenly comes to a stop?

7. Why do loose objects on the dashboard slide to the right when the car turns suddenly to the left?

8. Why is your velocity continuously changing as you ride on a carousel?

9. When you apply the brakes on your bicycle, which way do you accelerate?

10. One type of home coffee grinder has a small blade that rotates very rapidly and cuts the coffee beans into powder. Nothing prevents the coffee beans from moving so why don't they get out of the way when the blade begins to push on them?

11. A blacksmith usually hammers hot metal on the surface of a massive steel anvil. Why is this more effective than hammering the hot metal on the surface of a thin steel plate?

12. A sprinter who is running a 200-m race travels the second 100 m in much less time than the first 100 m. Why?

13. If you pull slowly on the top sheet of a pad of paper, the whole pad will move. But if you yank suddenly on that sheet, it will tear away from the pad. What causes these different behaviors?

14. A ball falls from rest for 5 seconds. Neglecting air resistance, during which of the 5 seconds does the ball's speed increase most?

15. If you drop a ball from a height of 4.9 m, it will hit the ground 1 s later. If you fire a bullet exactly horizontally from a height of 4.9 m, it will also hit the ground 1 s later. Explain.

16. An acorn falls from a branch located 9.8 m above the ground. After 1 second of falling, the acorn's velocity will be 9.8 m/s downward. Why hasn't the acorn hit the ground?

17. A diver leaps from a 50-m cliff into the water below. The cliff is not perfectly vertical so the diver must travel forward several meters in order to avoid the rocks beneath him. In fact, he leaps directly forward rather than upward. Explain why a forward leap allows him to miss the rocks.

18. The kicker in a sporting event isn't always concerned with how far downfield the ball travels. Sometimes the ball's flight time is more important. If he wants to keep the ball in the air as long as possible, which way should he kick it?

19. The heads of different golf clubs are angled to give the golf ball different initial velocities. The golf ball's speed remains almost constant, but the angle changes with the different clubs. Neglecting any air effects, how does changing the initial angle of the ball affect the distance the ball travels?

20. A speedboat is pulling a water-skier with a rope, exerting a large forward force on her. The skier is traveling forward in a straight line at constant speed. What is the net force she experiences?

21. Your suitcase weighs 50 N. As you ride up an escalator toward the second floor, carrying that suitcase, you are traveling at a constant velocity. How much upward force must you exert on the suitcase to keep it moving with you?

22. What is the net force on (a) the first car, (b) the middle car, and (c) the last car of a metro train traveling at constant velocity?

23. Two teams are having a tug-of-war with a sturdy rope. It has been an even match so far, with neither team moving. What is the net force on the left team?

24. When you kick a soccer ball, which pushes on the other harder: your foot or the soccer ball?

25. The earth exerts a downward force of 850 N on a veteran astronaut as he works outside the space shuttle. What force (if any) does the astronaut exert on the earth?

26. A car passes by, heading to your left, and you reach out and push it toward the left with a force of 50 N. Does this moving car push on you and, if so, with what force?

27. Which is larger: the force the earth exerts on you or the force you exert on the earth?

28. Comic book superheroes often catch a falling person only a hairsbreadth from the ground. Why would this rescue actually be just as fatal for the victim as hitting the ground itself?

29. You accidentally miss the doorway and run into the wall. You suddenly experience a backward force that is several times larger than your weight. What's the origin of this force?

30. When you fly a kite, there is a time when you must do (positive) work on the kite. Is that time when you let the kite out or when you pull it in?

31. Which does more work in lifting a grain of rice over its head: an ant or a person? Use this result to explain how insects can perform seemingly incredible feats of lifting and jumping.

32. Are you doing work while kneading bread? If so, when?

33. While hanging a picture, you accidentally dent the wall with a hammer. Did the hammer do work on the wall?

34. You're cutting wood with a handsaw. You have to push the saw away from you as it moves away from you and pull the saw toward you as it moves toward you. When are you doing work on the saw?

35. The steel ball in a pinball game rolls around a flat, tilted surface. If you flick the ball straight uphill, it gradually slows to a stop and then begins to roll downhill. Which way is the ball accelerating as it rolls uphill? downhill?

36. Why do less snow and other debris accumulate on a steep roof than on a flatter roof?

37. You roll a marble down a playground slide that starts level, then curves downward, and finally curves very gradually upward so that it's level again at the end. Where along its travel does the marble experience its greatest acceleration? its greatest speed?

38. When you're roller skating on level pavement, you can maintain your speed for a long time. But as soon as you start up a gradual hill, you begin to slow down. What slows you?

39. When the brakes on his truck fail, the driver steers it up a runaway truck ramp. As the truck rolls up the ramp, it slows to a stop. What happens to the truck's kinetic energy, its energy of motion?

Problems

1. If your car has a mass of 800 kg, how much force is required to accelerate it forward at 4 m/s2?

2. If your car accelerates from rest at a steady rate of 4 m/s2, how soon will it reach 88.5 km/h (55.0 mph or 24.6 m/s)?

3. On Mars, the acceleration due to gravity is 3.71 m/s2. What would a rock's velocity be 3 s after you dropped it on Mars?

4. How far would a rock fall in 3 s if you dropped it on Mars? (See Problem 3.)

5. How would your Mars weight compare to your earth weight? (See Problem 3.)

6. A basketball player can leap upward 0.5 m. What is his initial velocity at the start of the leap?

7. How long does the basketball player in Problem 6 remain in the air?

8. A sprinter can reach a speed of 10 m/s in 1 s. If the sprinter's acceleration is constant during that time, what is the sprinter's acceleration?

9. If a sprinter's mass is 60 kg, how much forward force must be exerted on the sprinter to make the sprinter accelerate at 0.8 m/s2?

10. How much does a 60-kg person weigh on earth?

11. If you jump upward with a speed of 2 m/s, how long will it take before you stop rising?

12. How high will you be when you stop rising in Problem 11?

13. How much force must a locomotive exert on a 12,000-kg boxcar to make it accelerate forward at 0.4 m/s2?

14. How long will it take the boxcar in Problem 13 to reach its cruising speed of 100 km/h (62 mph or 28 m/s)?

15. The builders of the pyramids used a long ramp to lift 20,000-kg (22-ton) blocks. If a block rose 1 m in height while traveling 20 m along the ramp's surface, how much uphill force was needed to push it up the ramp at constant velocity?

16. How much work was done in raising one of the blocks in Problem 15 to a height of 50 m?

17. What is the gravitational potential energy of one of the blocks in Problem 15 if it's now 75 m above the ground?

18. As water descends from the top of a tall hydroelectric dam, its gravitational potential energy is converted to electric energy. How much gravitational potential energy is released when 1000 kg of water descends 200 m to the generators?

19. The tire of your bicycle needs air so you attach a bicycle pump to it and begin to push down on the pump's handle. If you exert a downward force of 25 N on the handle and the handle moves downward 0.5 m, how much work do you do?

20. You're using a wedge to split a log. You are hitting the wedge with a large hammer to drive it into the log. It takes a force of 2000 N to push the wedge into the wood. If the wedge moves 0.2 m into the log, how much work have you done on the wedge?

21. The wedge in Problem 20 acts like a ramp, slowly splitting the wood apart as it enters the log. The work you do on the wedge, pushing it into the log, is the work it does on the wood, separating its two halves. If the two halves of the log only separate by a distance of 0.05 m while the wedge travels 0.2 m into the log, how much force is the wedge exerting on the two halves of the log to rip them apart?

22. You're sanding a table. You must exert a force of 30 N on the sandpaper to keep it moving steadily across the table's surface. You slide the paper back and forth for 20 minutes, during which time you move it 1000 m. How much work have you done?

The Laws of Motion Part 2

In the first chapter, we learned about how things move from place to place and began to develop an understanding of energy, an important conserved quantity. But motion doesn't always involve a change of position, and energy isn't the only conserved quantity that exists in nature. In this chapter, we'll take a look at a second type of motion—rotation—and at two other conserved quantities—momentum and angular momentum. Spinning objects are quite common, and we'll find it useful to understand their laws of motion before proceeding much further. Once we get those additional concepts under our belts, we'll be ready to explore the physics behind a broad assortment of mechanical objects.

EXPERIMENT: A Spinning Pie Dish

Spinning a dish on the top of a narrow post seems like a simple activity. But don't let its uncomplicated appearance deceive you: there are lots of physics involved in keeping the dish turning, in gradually slowing it down, and in preventing it from falling off the post.

An easy way to experiment with a spinning dish is to tape a pencil vertically to the edge of a table or chair so that its eraser projects several inches upward into the air. To avoid wobbling problems, the tape should hold the pencil rigidly in place, and the table or chair should be sturdy and stable.

Now prepare to balance a metal pie dish on the eraser before giving it a twist. If you don't have a pie dish, you can use a Frisbee, a deep plastic plate, or a shallow plastic bowl instead. Be creative. You probably have something that will work; just don't use Grandma's heirloom porcelain unless you're willing to face the possible consequences.

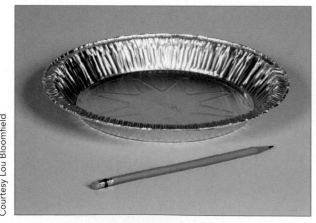
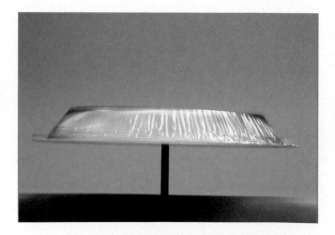

Courtesy Lou Bloomfield

The first thing you'll need to do is to balance the dish on the eraser. Try to *predict* whether this task will be easy or hard. Will it matter whether the dish opens up or down? *Observe* what happens when you set the dish on the eraser. Did you *verify* your predictions?

Find an arrangement in which the dish balances well and then give the dish a gentle spin. What sort of influence do you have to exert on the dish to start it rotating? If the dish doesn't wobble and the pencil remains stationary, the dish should spin for a while before coming to a stop. Once you're no longer touching the dish, what keeps it turning? On the other hand, why doesn't it keep turning forever?

Now flip the pencil over so that its sharp point projects upward. What will happen when you place the dish on that point? Will the dish still balance? When you spin the dish, will it turn for a longer or shorter time than on the bare eraser? Try to *measure* how changing the pivot on which the dish spins affects the duration of that spin. If the dish is soft and the point digs into it, protect the dish's bottom by taping a coin to it. How does this improved pivot affect the dish's rotation? Can you prolong the spin by taping weights around the outer edge of the dish? Is there a way to get the dish spinning just by blowing on it? Can you relate this motion to that of a spinning skater or a bicycle wheel?

Chapter Itinerary

We're going to explore the laws of rotational motion and two new conserved quantities in the context of three everyday objects: (1) *seesaws,* (2) *wheels,* and (3) *bumper cars.* In *seesaws,* we'll look at twists and turns, and see how two children manage to rock a seesaw back and forth. In *wheels,* we'll examine how friction affects motion and learn how wheels make a vehicle more mobile. In *bumper cars,* we'll learn the physics behind collisions and uncover some of the simple rules that govern what initially appear to be complicated motions. For a more complete preview of what we'll examine in this chapter, flip ahead to the Chapter Summary on p. 74.

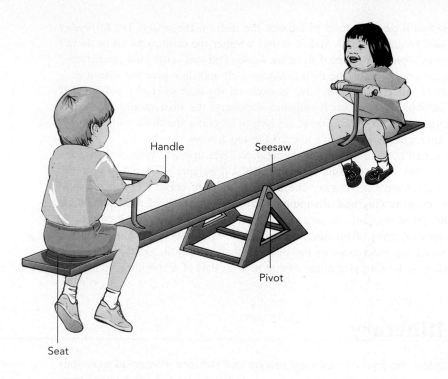

Handle   Seesaw

Pivot

Seat

SECTION 2.1 **Seesaws**

The ramp that we examined in Section 1.3 is only one tool that provides mechanical advantage. In this section, we'll look at another such device: the type of lever known as a seesaw. As we discuss seesaws, we'll revisit many of the laws of motion that we encountered in the previous chapter. However, we'll see these laws in a new context: rotational motion.

Questions to Think About: *A playground seesaw only balances when the children riding it are properly situated. What do we mean by a balanced seesaw? Why does it matter just where the children sit on the seesaw? What are they doing to make the balanced seesaw rock back and forth? Who is doing work on whom as they rock?*

Experiments to Do: *To get a feel for how levers work, find a rigid ruler with a hole in its center—the kind that can be clipped into a three-ring binder. If you support the ruler by putting the tip of an upright pencil into the central hole, you'll find that the ruler balances; that is, it either remains stationary, at whatever orientation you choose, or rotates steadily about the central hole. (Eventually, the ruler comes to rest because of friction, a detail that we'll continue to ignore for now.) Now, push on one end of the ruler. What happens? Try pushing the ruler's end toward its central hole. What happens then? What is the most effective way to make the ruler spin?*

Now lay the pencil on a table and place the ruler flat on top of it so that the pencil and the ruler are at right angles, or perpendicular, to each other. If you center the ruler on the pencil, the ruler will balance. Load the two ends of the ruler with coins or other small weights, trying as you do to keep the ruler balanced. Try placing the coins at different positions relative to the pencil. Is there any way you can balance a light weight on one end with a heavy weight on the other end?

The Seesaw

Any child who has played on a seesaw with friends of different sizes knows that the toy works best for two children of roughly the same weight (Fig. 2.1.1*a*). Evenly matched riders balance each other, and this balance allows them to rock back and forth easily. In contrast, when a light child tries to play seesaw with a heavy child, the heavy child's side of the seesaw drops rapidly and hits the ground with a thud (Fig. 2.1.1*b*). The light child is tossed into the air.

There are several solutions to the heavy child/light child problem. Of course, two light children could try to balance one heavy child. But most children eventually figure out that if the heavy child sits closer to the seesaw's pivot, the seesaw will balance (Fig. 2.1.1*c*). The children can then make the seesaw tip back and forth easily, just as it does when two evenly matched children ride at its ends. This is a pretty useful trick, and we'll explore it later in this section. First, though, we'll need to look carefully at the nature of rotational motion.

For simplicity, let's ignore the mass and weight of the seesaw itself. There are then only three forces acting on the seesaw shown in Fig. 2.1.1: two downward forces (the weights of the two children) and one upward force (the support force of the central pivot). Seeing those three forces, we may immediately think about net forces and begin to look for some overall acceleration of this toy and its riders. But we know that the seesaw remains where it is in the playground and isn't likely to head off for Kalamazoo or the center of the earth anytime soon. Because the seesaw's fixed pivot always provides just enough upward or sideways force to keep the seesaw from accelerating as a whole, the seesaw always experiences zero net force and never leaves the playground. Overall movement of an object from one place to another is called **translational motion.** While the seesaw never experiences this kind of motion, it can turn around the pivot, and thus it experiences a different kind of motion. Motion around a fixed point (which prevents translation) is called **rotational motion.** The hands of a clock experience rotational motion as they go around in a circle.

Rotational motion is what makes a seesaw interesting. The whole point of a seesaw is that it can rotate so that one child rises and the other descends. (You may not think of going up and down as rotating, but if the ground weren't there, the seesaw would be able to rotate in a big circle.) What causes the seesaw to rotate, and what observations can we make about the process of rotation?

To answer those questions, we'll need to examine several new physical quantities associated with rotation and explore the laws of rotational motion that relate them to one another. We'll do these things both by studying the workings of seesaws and other rotating objects and by looking for analogies between translational motion and rotational motion.

Imagine holding onto the seesaw in Fig. 2.1.1*a* to keep it level for a moment while the child on the left climbs off the seesaw. Now imagine letting go of the seesaw. As soon as you let go, the seesaw will begin to rotate, and the child on the right will descend toward the ground. The seesaw's motion will be fairly slow at first, but it will move more and more quickly until that child strikes the ground with a teeth-rattling thump.

If we focus only on the rotation itself, we might describe the motion of the seesaw in the following way:

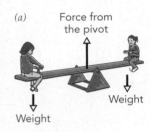

(a)

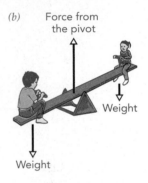

(b)

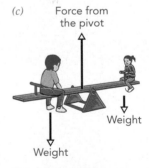

(c)

Fig. 2.1.1 (*a*) When two children of equal weight sit at opposite ends of a seesaw, it balances. (*b*) When their weights are not equal, the heavy child descends. (*c*) If the heavy child moves closer to the pivot, the seesaw can balance.

"The seesaw starts out not rotating at all. When we release the seesaw, it begins to rotate clockwise. The seesaw's rate of rotation increases steadily in the clockwise direction until the moment the seesaw strikes the ground."

This description sounds a lot like the description of a falling ball released from rest:

"The ball starts out not moving at all. When we release the ball, it begins to move downward. The ball's rate of translation increases steadily in the downward direction until the moment the ball strikes the ground."

The statement about the seesaw involves rotational motion, while the statement about the ball involves translational motion. Their similarity isn't a coincidence; the concepts and laws of rotational motion have many analogies in the concepts and laws of translational motion. The familiarity that we've acquired with translational motion will help us examine rotational motion.

> ➤ Check Your Understanding #1: Wheel of Fortune Cookies
>
> *for answers, see page* **75**

The guests at a large table in a Chinese restaurant use a revolving tray, a lazy Susan, to share the food dishes. How does the lazy Susan's motion differ from that of the passing dessert cart?

The Motion of a Dangling Seesaw

In the previous chapter we looked at the concept of translational inertia, which holds that a body in motion tends to stay in motion and a body at rest tends to stay at rest. This concept led us to Newton's first law of translational motion. Inserting the word "translational" here is a useful revision because we're about to encounter analogous concepts associated with rotational motion. We'll begin that encounter by observing a seesaw that's free of outside rotational influences. We'll then examine how the seesaw responds to outside influences such as its pivot or a handful of young riders. Because of the similarities between rotational and translational motions, this section will closely parallel our earlier examination of skating and falling balls.

Let's suppose that your local playground is installing a new seesaw and that this seesaw is presently dangling from a rope. The rope is attached to the middle of the seesaw in such a way that it supports the seesaw's weight but exerts no other influences on the seesaw. Most importantly, let's suppose that the dangling seesaw can spin and pivot with complete freedom—nothing pushes on it or twists it—and that the rope doesn't get tangled or in the way (Fig. 2.1.2). This dangling seesaw is free to turn in any direction, even completely upside down. You, the observer, are standing motionless near the seesaw. When you look over at the seesaw, what does it do?

If the seesaw is stationary, then it will remain stationary. However, if it's rotating, it will continue rotating at a steady pace, about a fixed line in space. What keeps the seesaw rotating? Its **rotational inertia.** A body that's rotating tends to remain rotating; a body that's not rotating tends to remain not rotating. That's how our universe works.

To describe the seesaw's rotational inertia and rotational motion more accurately, we'll need to identify several physical quantities associated with rotational motion. The first is the seesaw's orientation. At any particular moment, the seesaw is oriented in a certain way—that is, it has an **angular position**. Angular position describes the seesaw's

Fig. 2.1.2 A seesaw that's dangling from a rope at its middle. Since nothing twists it, the seesaw rotates steadily about a fixed line in space.

orientation relative to some reference orientation; it can be specified by determining how far the seesaw has rotated away from its reference orientation and the axis or line about which that rotation has occurred. The seesaw's angular position is a vector quantity of relatively minor importance, pointing along the rotation axis with a magnitude equal to the rotation angle (Fig. 2.1.3).

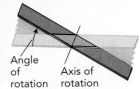

Fig. 2.1.3 You can specify this seesaw's angular position, relative to its horizontal reference orientation, as the axis about which it was rotated to reach its new orientation and the angle through which it was rotated.

The SI unit of angular position is the **radian**, the natural unit for angles. It's a natural unit because it follows directly from geometry, not from an arbitrary human choice or convention the way most units do. Geometry tells us that a circle of radius 1 has a circumference of 2π. By letting arc lengths around that circle's circumference specify angles, we are using radians. For example, there are 2π radians (or 360°) in a full circle and $\pi/2$ radians (or 90°) in a right angle. Since the radian is a natural unit, it is often omitted from calculations and derived units.

If the seesaw is rotating, then its angular position is changing; in other words, it has an angular velocity. **Angular velocity** is our first important vector quantity of rotational motion and measures how quickly the seesaw's angular position changes; it consists of the angular speed at which the seesaw is rotating and the axis about which that rotation proceeds. The seesaw's **angular speed** is its change in angle divided by the time required for that change:

$$\text{angular speed} = \frac{\text{change in angle}}{\text{time}}.$$

The SI unit of angular velocity is the **radian-per-second** (abbreviated 1/s).

The seesaw's **axis of rotation** is the line in space about which the seesaw is rotating. But just knowing that line isn't quite enough: is the seesaw rotating clockwise or counterclockwise?

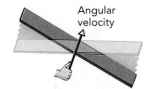

Angular velocity

Fig. 2.1.4 This seesaw is spinning about the rotation axis shown. The direction of the seesaw's angular velocity is defined by the right-hand rule.

To resolve this ambiguity, we take advantage of the fact that any line has two directions to it. Once we have identified the line about which the seesaw is rotating, we can look down that line at the seesaw from either direction. From one direction, the seesaw appears to be rotating clockwise; from the other direction, counterclockwise. By convention, we choose the direction in which the seesaw appears to be rotating clockwise and say that the seesaw's rotation axis points away from our eye toward the seesaw. This convention is called the **right-hand rule** because if the fingers of your right hand are curling around the axis in the way the seesaw is rotating, then your thumb is pointing along the seesaw's rotation axis (Fig. 2.1.4).

Remembering this convention isn't as important as understanding why we must specify the direction about which rotation occurs when describing a rotating object's angular velocity. Just as translational velocity consists of a translational speed and a direction in which the translational motion occurs, so angular velocity consists of a rotational speed and a direction about which the rotational motion occurs.

We're now prepared to describe the rotational motion of the dangling seesaw. Because of its freedom from outside influences and its rotational inertia, its angular velocity is constant. The dangling seesaw just keeps on turning and turning, always at the same angular speed, always about the same axis of rotation.

As you might suspect, this observation isn't unique to seesaws. It is **Newton's first law of rotational motion**, which states that a rigid object that is not wobbling and is not subject to any outside influences rotates at a constant angular velocity, turning equal amounts in equal times about a fixed axis of rotation. The outside influences referred to in this law are called **torques**—a technical term for twists and spins. When you twist off the lid of a jar or spin a top with your fingers, you're exerting a torque.

This law excludes objects that wobble or can change shape as they rotate because those objects have more complicated motions. They are covered instead by a more general principle—the conservation of angular momentum—that we'll learn later on.

> **Newton's First Law of Rotational Motion**
> A rigid object that is not wobbling and is not subject to any outside torques rotates at a constant angular velocity, turning equal amounts in equal times about a fixed axis of rotation.

> **Check Your Understanding #2: Going for a Spin**
> *for answers, see page* 75

A rubber basketball floats in a swimming pool. It experiences zero torque, no matter which end of it is up. If you spin the basketball and then let go, how will it move?

The Seesaw's Center of Mass

Even without visiting the playground, you can find many objects that are nearly free from torques: a baton thrown overhead by a baton twirler, for example, or a juggler's club whirling through the air between two clowns. These motions, however, are complicated because those freely moving objects rotate and translate at the same time. The spinning baton travels up and down, the turning club arcs through the air, and, if the rope breaks, our seesaw will fall as it spins. How can we distinguish their translational motions from their rotational motions?

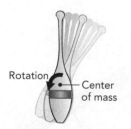

Fig. 2.1.5 This club spins about its center of mass, which remains stationary.

Once again, we can make use of a wonderful simplification of physics. There's a special point in or near a free object about which all its mass is evenly balanced and about which it naturally spins—its **center of mass.** The axis of rotation passes right through this point so that, as the free object rotates, the center of mass doesn't move unless the object has an overall translational velocity. The center of mass of a typical ball is at its geometrical center, while the center of mass of a less symmetrical object depends on how the mass of that object is distributed. You can begin to find a small object's center of mass by spinning it on a smooth table and looking for the fixed point about which it spins (Fig. 2.1.5).

Center of mass allows us to separate an object's translational motion from its rotational motion. As a juggler's club arcs through space, its center of mass follows the simple path we discussed in Section 1.2 on falling balls (Fig. 2.1.6). At the same time, the club's rotational motion about its center of mass is that of an object that's free of outside torques: if it's not wobbling, it rotates with a constant angular velocity.

In the course of this book we'll encounter many objects that translate and rotate simultaneously, and it's worth remembering that we can often separate these two motions by paying attention to an object's center of mass. For example, the workers installing our seesaw will locate its pivot strategically at or very near the seesaw's center of mass. As a result, the pivot will prevent any translational motion of the seesaw while permitting nearly free rotational motion of the seesaw about its center of mass, at least about one axis.

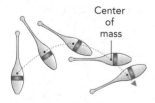

Fig. 2.1.6 A juggler's club that is traveling through space rotates about its center of mass as its center of mass travels in the simple arc associated with a falling object.

> **Check Your Understanding #3: Tracking the High Dive**
> *for answers, see page* 75

When a diver does a rigid, open somersault off a high diving board, his motion appears quite complicated. Can this motion be described simply? How?

How the Seesaw Responds to Torques

The workers are eating lunch, so the seesaw is still hanging from the rope. Why can't this dangling seesaw change its rotational speed or axis of rotation? Because it has rotational mass ❏. **Rotational mass** is the measure of an object's *rotational* inertia, its resistance to changes in its *angular* velocity. An object's rotational mass depends both on its ordinary mass and on how that mass is distributed within the object. The SI unit of rotational mass is the **kilogram-meter^2** (abbreviated kg·m^2). Because the seesaw has rotational mass, its angular velocity will change only if something twists it or spins it. In other words, it must experience a torque.

Torque—our second important vector quantity of rotational motion—has both a magnitude and a direction. The more torque you exert on the seesaw, the more rapidly its angular velocity changes. Depending on the direction of the torque, you can make the seesaw turn more rapidly or less rapidly or even rotate about a different axis. But how do you determine the direction of a particular torque? One way is to imagine exerting this torque on a stationary ball floating in water (Fig. 2.1.7a,b). The ball will begin to rotate, acquiring a nonzero angular velocity (Fig. 2.1.7c). The direction of this angular velocity is that of the torque. The SI unit of torque is the **newton-meter** (abbreviated N·m).

The larger an object's rotational mass, the more slowly its angular velocity changes in response to a specific torque (Fig. 2.1.8). You can easily spin a basketball with the tips of

❏ For clarity and simplicity, this book refers to the measure of an object's rotational inertia as "rotational mass." However, this quantity is known more formally as **"moment of inertia."**

Chris Harvey/Stone/Getty Images

Fig. 2.1.8 Spinning a merry-go-round is difficult because of its large rotational mass. Despite the large torque exerted by this boy, the merry-go-round's angular velocity increases slowly.

(a)

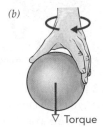

(b)

▽ Torque

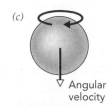

(c)

▽ Angular velocity

Fig. 2.1.7 If you start with a ball that's not spinning (*a*) and twist it with a torque (*b*), the ball will acquire an angular velocity (*c*) that's in the same direction as that torque.

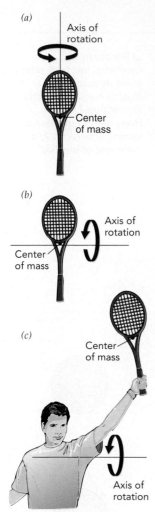

(a)

Axis of rotation

Center of mass

(b)

Axis of rotation

Center of mass

(c)

Center of mass

Axis of rotation

Fig. 2.1.9 A tennis racket's rotational mass depends on the axis about which it rotates. Its rotational mass is small (*a*) when it rotates about its handle and large (*b*) when it rotates head-over-handle. (*c*) If you make it rotate about your shoulder, its rotational mass becomes even larger.

your fingers, but it's much harder to spin a bowling ball. The bowling ball's larger rotational mass comes about primarily because it has a greater ordinary mass than the basketball.

But rotational mass also depends on an object's shape, particularly on how far each portion of its ordinary mass is from the axis of rotation. The farther a portion of mass is from that axis, the more rapidly it must accelerate as the entire object undergoes angular acceleration and the more leverage it has with which to oppose that acceleration. We'll examine levers shortly, but the consequence of these two effects of distance from the rotation axis is that each portion of mass contributes to the object's rotational mass in proportion to the square of its distance from that axis. That's why an object that has most of its mass located near the axis of rotation will have a much smaller rotational mass than an object of the same mass that has most of its mass located far from that axis. Thus a ball of pizza dough has a smaller rotational mass than the finished pizza. And the bigger the pizza gets, the harder it is to start or stop spinning.

Because an object's rotational mass depends on how far its mass is from the axis of rotation, changes in the axis of rotation are likely to change its rotational mass. For example, less torque is required to spin a tennis racket about its handle (Fig. 2.1.9*a*) than to flip the racket head-over-handle (Fig. 2.1.9*b*). When you spin the tennis racket about its handle, the axis of rotation runs right through the handle so that most of the racket's mass is fairly close to the axis and the rotational mass is small. When you flip the tennis racket head-over-handle, the axis of rotation runs across the handle so that both the head and the handle are far away from the axis and the rotational mass is large. The tennis racket's rotational mass becomes even larger when you hold it in your hand and make it rotate about your shoulder rather than its center of mass (Fig. 2.1.9*c*).

When something exerts a torque on the dangling seesaw, its angular velocity changes; in other words, it undergoes angular acceleration, our third important vector quantity of rotational motion. **Angular acceleration** measures how quickly the seesaw's *angular* velocity changes. It's analogous to acceleration, which measures how quickly an object's *translational* velocity changes. Just as with acceleration, angular acceleration involves both a magnitude and a direction. An object undergoes angular acceleration when its angular speed increases or decreases or when its angular velocity changes directions. The SI unit of angular acceleration is the **radian-per-second^2** (abbreviated $1/s^2$).

There is a simple relationship between the torque exerted on the seesaw, its rotational mass, and its angular acceleration. The seesaw's angular acceleration is equal to the torque exerted on it divided by its rotational mass or, as a word equation,

$$\text{angular acceleration} = \frac{\text{torque}}{\text{rotational mass}}. \qquad (2.1.1)$$

The seesaw's angular acceleration, as we've seen, is in the same direction as the torque exerted on it.

This relationship is **Newton's second law of rotational motion.** Structuring the relationship this way distinguishes the causes (torque and rotational mass) from their effect (angular acceleration). Nonetheless, it has become customary to rearrange the relationship to eliminate the division. In its traditional form, the relationship can be written in a word equation:

$$\text{torque} = \text{rotational mass} \cdot \text{angular acceleration}, \qquad (2.1.2)$$

in symbols:

$$\tau = I \cdot \alpha,$$

and in everyday language:

Spinning a marble is much easier than spinning a merry-go-round.

It resembles Newton's second law of translational motion (force = mass · acceleration), except that torque has replaced force, rotational mass has replaced mass, and angular acceleration has replaced acceleration. This new law doesn't apply to wobbling objects, however, because they're being affected by more than one rotational mass simultaneously (see the discussion of tennis rackets above) and follow a much more complicated law.

> ## Newton's Second Law of Rotational Motion
> The torque exerted on an object that is not wobbling is equal to the product of that object's rotational mass times its angular acceleration. The angular acceleration points in the same direction as the torque.

Because it's an equation, the two sides of Eq. 2.1.1 are equal. Any change in the torque you exert on the seesaw must be accompanied by a proportional change in its angular acceleration. As a result, the harder you twist or spin the seesaw, the more rapidly its angular velocity changes.

We can also compare the effects of a specific torque on two different rotational masses. Equation 2.1.1 indicates that a decrease in rotational mass must be accompanied by a corresponding increase in angular acceleration. If we replace the playground seesaw with one from a dollhouse, the rotational mass will decrease and the angular acceleration will increase. The angular velocity of a doll's seesaw thus changes more rapidly than the angular velocity of a playground seesaw when the two experience identical torques.

In summary:

1. Your angular position indicates exactly how you're oriented.
2. Your angular velocity measures how quickly your angular position changes.
3. Your angular acceleration measures how quickly your angular velocity changes.
4. In order for you to undergo angular acceleration, something must exert a torque on you.
5. The more rotational mass you have, the less angular acceleration you experience for a given torque.

Quantity	SI Unit	English Unit	SI → English	English → SI
Angular position	radian (1)	radian (1)		
Angular velocity	radian-per-second (1/s)	radian-per-second (1/s)		
Angular acceleration	radian-per-second^2 (1/s^2)	radian-per-second^2 (1/s^2)		
Torque	newton-meter (N·m)	foot-pound (ft·lbf)	1 N·m = 0.73757 ft·lbf	1 ft·lbf = 1.3558 N·m
Rotational mass	kilogram-meter^2 (kg·m^2)	pound-foot^2 (lbm·ft^2)	1 kg·m^2 = 23.730 lbm·ft^2	1 lbm·ft^2 = 0.042140 kg·m^2

This summary of the physical quantities of rotational motion is analogous to the one for translational motion on p. 43. Take a moment to compare the two.

Check Your Understanding #4: The Merry-Go-Round

for answers, see page 76

The merry-go-round is a popular playground toy (see Fig. 2.1.8). Already challenging to spin empty, a merry-go-round is even harder to start or stop when there are lots of children on it. Why is it so difficult to change a full merry-go-round's angular velocity?

Check Your Figures #1: Hard to Turn

for answers, see page 77

Automobile tires are normally hollow and filled with air. If they were made of solid rubber, their rotational masses would be about 10 times as large. With the wheel lifted off the ground, how much more torque would an automobile have to exert on a solid tire to make it undergo the same angular acceleration as a hollow tire?

Forces, Torques, and Seesaws

The workers have finally installed the seesaw. They have mounted it on a pivot that passes directly through its center of mass, so that the pivot coincides with a natural rotation axis of the seesaw. The pivot thus supports the seesaw's weight while leaving it free to obey Newton's first law of rotational motion. That is, the unoccupied seesaw rotates with constant angular velocity about its pivot.

The unoccupied seesaw is balanced, meaning that it has zero torque on it. As a result, it experiences no angular acceleration. You might think that a balanced seesaw always remains horizontal, but that isn't necessarily so. What is certain is that its angular velocity is constant. If it's rotating, then it continues to rotate steadily about the pivot; if it's stationary, then it remains stationary at its current tilt, whether horizontal or not.

To change the seesaw's angular velocity, you must exert a torque on it. But how do you actually exert a torque? You put your hand on one end of the seesaw and push that end down (Fig. 2.1.10a). The seesaw begins to rotate, and your end soon hits the ground. You have exerted a torque on the seesaw.

But you started by exerting a *force* on the seesaw—you pushed on it—so forces and torques must be related somehow. Sure enough, a force can produce a torque and a torque can produce a force. To help us explore that relationship, let's think of all the ways not to exert a torque on a seesaw.

What happens if you push on the seesaw right where the pivot passes through it (Fig. 2.1.10b)? Nothing—no angular acceleration. If you move a little away from the pivot, you can get the seesaw rotating but you have to push hard. You do much better to push on the end of the seesaw, where even a small force can start the seesaw rotating. The distance from the pivot to the place where you push on the seesaw is called the **lever arm;** in general, the longer the lever arm, the less force it takes to cause a particular angular acceleration. Our first observation about producing a torque with a force is this: you obtain more torque by exerting that force farther from the pivot or axis of rotation. In other words, the torque is proportional to the lever arm.

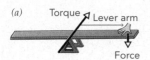

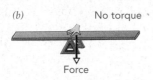

Fig. 2.1.10 (a) When you push down on the seesaw, far from its pivot, you exert a torque on it. But when you (b) exert your force at the pivot or (c) exert your force toward the pivot, you exert no torque.

Another ineffective way to start the seesaw rotating is to push its end directly toward or away from the pivot (Fig. 2.1.10c). A force exerted toward or away from the axis of rotation doesn't produce any torque about that axis. At least a component of the force you exert must be perpendicular to the lever arm, which is actually a vector pointing along the seesaw's surface from the pivot to the place where you push on the seesaw. Our second observation about producing a torque with a force is that you must exert at least a component of that force perpendicular to the lever arm. Only that component of force contributes to the torque. To produce the most torque, push perpendicular to the lever arm.

We can summarize these two observations as follows: the torque produced by a force is equal to the product of the lever arm times that force, where we include only the component of the force that is perpendicular to the lever arm. This relationship can be written as a word equation:

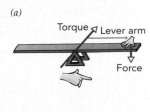

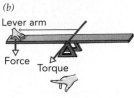

$$\text{torque} = \text{lever arm} \cdot \text{force perpendicular to lever arm}, \qquad (2.1.3)$$

in symbols:

$$\tau = r \cdot F_\perp,$$

and in everyday language:

Fig. 2.1.11 The torque on a seesaw obeys a right-hand rule: if your index finger points along the lever arm and your middle finger points along the force, your thumb points along the torque.

When twisting an unyielding object, it helps to use a long stick.

The directions of the force and lever arm also determine the direction of the torque. The three directions follow another right-hand rule (Fig. 2.1.11). If you point your right index finger in the direction of the lever arm and your bent middle figure in the direction of the force, then your thumb will point in the direction of the torque. Thus in Fig. 2.1.11a, the lever arm points to the right, the force points downward, and the resulting torque points into the page so that the seesaw undergoes angular acceleration in the clockwise direction. In Fig. 2.1.11b, the lever arm has reversed directions and so has the torque.

What happens if you and a friend push down simultaneously on both seats at once? Then you produce two torques on the seesaw about its pivot, and these torques have opposite directions. The seesaw responds to the **net torque** it experiences, the sum of all the individual torques on the seesaw. Since your two torques oppose one another, they at least partially cancel. If you carefully exert identical downward forces at identical distances from the pivot, the magnitudes of the two torques will be exactly equal and the torques will sum to zero. The seesaw will experience zero net torque, and it will be balanced.

This observation explains the need for careful seating of the children on the seesaw. Each child's weight exerts a downward force on the seesaw and by properly distributing those weights on both sides of the pivot, the torques that they produce can be made to sum to zero. With zero net torque about its pivot, the seesaw balances.

In fact, the weight of the seesaw itself is balanced in this manner. While each end's weight exerts a torque on the seesaw, those two torques sum to zero and have no overall effect on the seesaw's rotation.

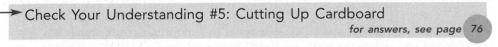

► Check Your Understanding #5: Cutting Up Cardboard

for answers, see page 76

When you cut cardboard with a pair of scissors, it's best to move the cardboard as close as possible to the scissors' pivot. Explain.

▶ Check Your Figures #2: A Few Loose Screws

for answers, see page 77

You're trying to remove some rusty screws from your refrigerator, using an adjustable wrench with a 0.2-meter (20-centimeter) handle. Although you push as hard as you can on the handle, you can't produce enough torque to loosen one of the screws. You have a 1.0-meter (100-cm) pipe that you can slip over the handle of the wrench to make the wrench effectively 1 meter long. How much more torque will you then be able to exert on the screw?

Net Torque and Mechanical Advantage

The amount of torque that a child's weight produces on a seesaw depends on that child's distance from the pivot. If the child sits on the pivot, the lever arm is zero and she produces no torque; but if she sits at the extreme end of the seesaw, the lever arm is long and she produces a large torque. She can adjust her torque by moving along the seesaw because the seesaw provides her with mechanical advantage. As we saw in Section 1.3, mechanical advantage appears when a device redistributes the amounts of force and distance used to produce a particular amount of work. The seesaw allows a small force exerted at its end to do the same work as a large force exerted near its pivot.

To see how mechanical advantage appears in a seesaw, think of what happens when two children sit on its ends. If two 5 year olds, each weighing 200 N (45 lbf), sit at opposite ends of the seesaw, 2 m (6.6 ft) from the pivot (Fig. 2.1.12a), each one exerts a torque of 400 N·m (300 ft·lbf) on the seesaw about its pivot (200 N·2 m = 400 N·m). But because these torques are oppositely directed, they add to zero. The net torque on the seesaw is zero and the seesaw balances.

If you replace one of the 5 year olds with a 400-N (90-lbf) teenager, the teenager must sit at half the distance from the pivot (Fig. 2.1.12b). Doubling the force while halving the lever arm leaves the torque unchanged at 400 N·m. The two children again produce equal but oppositely directed torques about the pivot, so that the net torque on the seesaw is zero and the seesaw balances. This effect explains how a small child at the end of the seesaw can balance a large child nearer the pivot.

With the seesaw balanced, nothing is accelerating. Each child experiences zero net force; the seesaw pushes up on the small child with a force of 200 N (45 lbf) and on the large child with a force of 400 N (90 lbf). Ultimately, it's the small child's 200-N (45-lbf) weight that gives rise to the 400-N (90-lbf) supporting force experienced by the large child. The seesaw's mechanical advantage allows the small child to support and lift the much heavier child. This effect, where a small force on one part of a rotating system produces a large force elsewhere in that system, is an example of the mechanical advantage associated with levers.

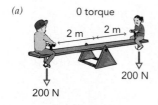

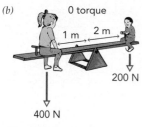

Fig. 2.1.12 (a) When two children of equal weight sit at equal distances from the pivot, they produce equal but oppositely directed torques about the pivot. These torques sum to zero so that the seesaw experiences zero net torque. (b) When one child weighs twice as much as the other, the seesaw balances when the heavy child sits at half the distance from the pivot.

▶ Check Your Understanding #6: Pulling Nails

for answers, see page 76

Some hammers have a special claw designed to remove nails from wood. When you slide the claw under the nail's head and rotate the hammer by pulling on its handle, the claw pulls the nail out of the wood. The hammer's head contacts the wood to form a pivot that's about 10 times closer to the nail than to the handle. The torque you exert on the hammer twists it in one direction, while the torque that the nail exerts on the hammer twists it in the opposite direction. The hammer isn't undergoing any significant angular acceleration, so the torques must nearly balance. If you're exerting a force of 100 N (22 lbf) on the hammer's handle, how much force is the nail exerting on the hammer's claw?

Riding a Seesaw

Each seesaw in Fig. 2.1.12 is balanced, meaning that the net torque on it is zero. Although each child's weight exerts a torque on the seesaw, the two torques sum to zero. Since the seesaw experiences zero net torque and no angular acceleration, it continues rotating at constant angular velocity.

However, as it presently stands, a balanced seesaw should either remain motionless forever or else rotate endlessly in the same direction. Children are unlikely to wait motionless forever, and endless rotation implies that the children will be upside-down periodically. We've obviously neglected a few details.

What do the children do when the seesaw is motionless? To start the seesaw moving, they have to unbalance the seesaw. One of the children must change the torque she exerts on it. She can either change the downward force she exerts on the seesaw or change the distance between that force and the pivot. Actually, children change both the force and the lever arm frequently without even thinking about it. If a child leans inward, toward the pivot, the lever arm decreases and the child exerts less torque on the seesaw; as a result, the seesaw begins to rotate and the child rises. If the child pushes on the ground with his feet, the ground exerts an upward force on him, reducing the force and torque he exerts on the seesaw; again, the seesaw begins to rotate and the child rises.

So either by leaning or by pushing on the ground, the children can start an initially motionless, balanced seesaw rotating. Similarly, when one end of the seesaw hits the ground, the ground exerts a strong, upward support force on it. Located far from the pivot and almost perpendicular to that lever arm, this force produces a huge torque on the seesaw and abruptly stops it from rotating. The angular acceleration is so uncomfortably large that most children push on the ground with their feet to cushion the impact. The child on the ground can continue to push down with her feet until the seesaw rotates in the opposite direction. That child begins to rise and the other child descends. When the other end of the seesaw reaches the ground, this cycle begins again.

As they play on a seesaw, the two children frequently change the torques they exert on it so that it tips back and forth. During the moments when a child is pushing on the ground or leaning inward or outward to get a stationary seesaw moving, the seesaw is no longer balanced. A balanced seesaw has zero angular acceleration; it's only by unbalancing the seesaw that the children can change the angular velocity of the seesaw.

► Check Your Understanding #7: Rocking the Boat

for answers, see page 76

Loading a large container ship requires some care in balancing the cargo and fastening it down firmly. The effective pivot about which the ship can rotate in the water is located roughly along the centerline of the ship, from its bow to its stern. Why is improperly fastened-down cargo so dangerous on such a ship, possibly causing it to capsize during a storm?

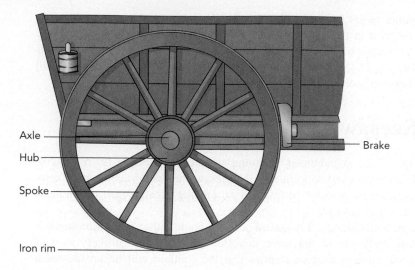

Axle
Hub
Spoke
Iron rim
Brake

Wheels

Like ramps and levers, wheels are simple tools that make our lives easier. But a wheel's main purpose isn't mechanical advantage, it's overcoming friction. Up until now, we've ignored friction, looking at the laws of motion as they apply only in idealized situations. But our real world does have friction, and an object in motion tends to slow down and stop because of it. One of our first tasks in this section will therefore be to understand friction—though, for the time being, we'll continue to neglect air resistance.

Questions to Think About: If objects in motion tend to stay in motion, why is it so hard to drag a heavy box across the floor? If objects should accelerate downhill on a ramp, why won't a plate slide off a slightly tilted table? What makes the wheels of a cart turn as you pull the cart forward? How does spinning its wheels propel a car forward?

Experiments to Do: To observe the importance of wheels in eliminating friction, try sliding a book along a flat table. Give the book a push and see how quickly it slows down and stops. Which way is friction pushing on the book? Does the force that friction exerts on the book depend on how fast the book is moving? Let the book come to a stop. Is friction still pushing on the book when it isn't moving? If you push gently on the stationary book, what force does friction exert on it?

Lay down three or four round pencils, parallel to one another and a few inches apart. Rest the book on top of the pencils and give the book a push in the direction that the pencils can roll. Describe how the book now moves. What do you think has caused the difference?

Moving a File Cabinet: Friction

When we imagined moving your friend's piano into a new apartment back in Section 1.3, we neglected a familiar force—friction. Luckily for us, your friend's piano had wheels on its legs, and wheels facilitate motion by reducing the effects of friction. We'll focus on wheels in this section. But first, to help us understand the relationship between wheels and friction, we'll look at another item that needs to be moved—your friend's file cabinet.

The file cabinet is resting on a smooth and level hardwood floor; it's full of sheet music and weighs about 1000 N (225 lbf). Despite its large mass, you know that it should accelerate in response to a horizontal force, so you give it a gentle push toward the door. Nothing happens. Something else must be pushing on the file cabinet in just the right way to cancel your force and keep it from accelerating. Undaunted, you push harder and harder until finally, with a tremendous shove, you manage to get the file cabinet sliding across the floor. But the cabinet moves slowly even though you continue to push on it. Something else is pushing on the file cabinet, trying to stop it from moving.

That something else is **friction,** a force that opposes the relative motion of two surfaces in contact with one another. Two surfaces that are in **relative motion** are traveling with different velocities so that a person standing still on one surface would observe the other surface as moving. In opposing relative motion, friction exerts forces on both surfaces in directions that tend to bring them to a single velocity.

For example, when the file cabinet slides by itself toward the left, the floor exerts a rightward frictional force on it (Fig. 2.2.1). The frictional force exerted on the file cabinet, *toward the right,* is in the direction opposite the file cabinet's velocity, *toward the left.* Since the file cabinet's acceleration is in the direction opposite its velocity, the file cabinet slows down and eventually comes to a stop.

According to Newton's third law of motion, an equal but oppositely directed force must be exerted by the file cabinet on the floor. Sure enough, the file cabinet does exert a leftward frictional force on the floor. However, the floor is rigidly attached to the earth, so it accelerates very little. The file cabinet does almost all the accelerating, and soon the two objects are traveling at the same velocity.

Frictional forces always oppose relative motion, but they vary in strength according to (1) how tightly the two surfaces are pressed against one another, (2) how slippery the surfaces are, and (3) whether or not the surfaces are actually moving relative to one another. First, the harder you press two surfaces together, the larger the frictional forces they experience. For example, an empty file cabinet slides more easily than a full one. Second, roughening the surfaces generally increases friction, while smoothing or lubricating them generally reduces it. Riding a toboggan down the driveway is much more interesting when the driveway is covered with snow or ice than when the driveway is bare asphalt. We'll examine the third issue later on.

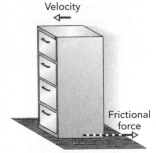

Fig. 2.2.1 A file cabinet sliding to the left across the floor. The file cabinet experiences a frictional force toward the right that gradually brings it to a stop.

Check Your Understanding #1: The One That Got Away

for answers, see page 76

Your table at the restaurant isn't level, and your water glass begins to slide slowly downhill toward the edge. Which way is friction exerting a force on it?

A Microscopic View of Friction

As the file cabinet slides by itself across the floor, it experiences a horizontal frictional force that gradually brings it to a stop. But from where does this frictional force come? The obvious forces on the file cabinet are both vertical, not horizontal: the cabinet's weight is downward and the support force from the floor is upward. How can the floor exert a horizontal force on the file cabinet?

The answer lies in the fact that neither the bottom of the file cabinet nor the top of the floor is perfectly smooth. They both have microscopic hills and valleys of various sizes.

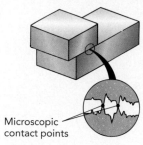

Microscopic
contact points

Fig. 2.2.2 Two surfaces
that are pressed against
one another actually touch
only at specific contact
points. When the surfaces
slide across one another,
these contact points
collide, producing sliding
friction and wear.

The file cabinet is actually supported by thousands of tiny contact points, where the file cabinet directly touches the floor (Fig. 2.2.2). As the file cabinet slides, the microscopic projections on the bottom of the file cabinet pass through similar projections on the top of the floor. Each time two projections collide, they experience horizontal forces. These tiny forces oppose the relative motion and give rise to the overall frictional forces experienced by the file cabinet and floor. Because even an apparently smooth surface still has some microscopic surface structure, all surfaces experience friction as they rub across one another.

Increasing the size or number of these microscopic projections by roughening the surfaces generally leads to more friction. If you put sandpaper on the bottom of the file cabinet, it will experience larger frictional forces as it slides across the floor. On the other hand, a microscopically smoother "nonstick" surface, like that used in modern cookware, would let the file cabinet slide easily.

Increasing the number of contact points by squeezing the two surfaces more tightly together also leads to more friction. The microscopic projections simply collide more often. That's why adding more sheet music to the file cabinet would make it harder to slide. Doubling the file cabinet's weight would roughly double the number of contact points and make it about twice as hard to move across the floor. A useful rule of thumb is that the frictional forces between two surfaces are proportional to the forces pressing those two surfaces together.

Friction also causes wear when the colliding contact points break one another off. With time, this wear can remove large amounts of material so that even seemingly indestructible stone steps are gradually worn away by foot traffic. The best way to reduce wear between two surfaces (other than to insert a lubricant between them) is to polish them so that they are extremely smooth. The smooth surfaces will still touch at contact points and experience friction as they slide across one another, but their contact points will be broad and round and will rarely break one another off during a collision.

▶ Check Your Understanding #2: Weight and Friction

for answers, see page **76**

How much harder is it to slide a stack of two identical books across a table than it is to slide just one of those books?

Static Friction, Sliding Friction, and Traction

There are really two kinds of friction—sliding and static. When two surfaces are moving across one another, **sliding friction** acts to stop them from sliding. But even when those surfaces have the same velocity, **static friction** may act to keep them from starting to slide across one another in the first place.

You find it particularly hard to start the file cabinet sliding across the floor. Contact points between the cabinet and floor have settled into one another, so a small push does nothing. Static friction is always exerting a frictional force that exactly balances your push. Since the net force on the file cabinet is zero, it doesn't accelerate.

However, the force that static friction can exert is limited. To get the file cabinet moving, you need to give it a mighty shove and thereby exert more horizontal force on it than static friction can exert in the other direction. The net force on the file cabinet is then no longer zero and it accelerates.

Once the file cabinet is moving, static friction is replaced by sliding friction. Because sliding friction acts to bring the file cabinet back to rest, you must push on the cabinet to keep it moving. With the file cabinet sliding across the floor, however, the contact points between the surfaces no longer have time to settle into one another, and they consequently experience weaker horizontal forces. That's why the force of sliding friction is generally weaker than that of static friction and why it's easier to keep the file cabinet moving than it is to get it started.

Both forms of friction are incorporated in the concept of **traction**—the largest amount of frictional force that the file cabinet can obtain from the floor at any given moment. When the cabinet is stationary, its traction is equal to the maximum amount of force that static friction can exert on it. But once it begins to slide across the floor, its traction reduces to the amount of force that sliding friction exerts.

While the file cabinet's traction is a nuisance that you must overcome, the traction of your shoes on the floor is crucial. Unless you can push against the wall, your shoes are going to need enough traction to provide the horizontal force required to move the file cabinet. Let's hope you're wearing your Doc Martens!

➤ Check Your Understanding #3: Skidding to a Stop

for answers, see page **76**

Antilock brakes keep an automobile's wheels from locking and skidding during a sudden stop. Apart from issues of steering, what is the advantage of preventing the wheels from skidding (sliding) on the pavement?

Work, Energy, and Power

There is another difference between static and sliding friction: sliding friction wastes energy. It can't make that energy disappear altogether because energy, as we've seen, is a conserved quantity: it can't be created or destroyed. But energy can be transferred between objects or converted from one form to another. What sliding friction does is convert useful, **ordered energy**—energy that can easily be used to do work—into relatively useless, disordered energy. This disordered energy is called **thermal energy** and is the energy we associate with temperature. It's sometimes called *internal energy* or *heat*. Sliding friction makes things hotter by turning work into thermal energy.

As we saw in Section 1.3, energy is the capacity to do work and is transferred between objects by doing that work. Energy can also change forms, appearing as either kinetic energy in the motions of objects or as potential energy in the forces between or within those objects. With practice, you can "watch" energy flow through a system just as an accountant watches money flow through a company.

The most obvious form of energy is kinetic energy, the energy of motion. It's easy to see when kinetic energy is transferred into or out of an object. As kinetic energy leaves an object, the object slows down; thus moving water slows down as it turns a gristmill, and a bowling ball slows down as it knocks over bowling pins. Conversely, as kinetic energy enters an object, the object speeds up. A baseball moves faster as you do work on it during a pitch; you're transferring energy from your body into the baseball, where the energy becomes kinetic energy in the baseball's motion.

Potential energy is stored in the forces between or within objects, and usually isn't as visible as kinetic energy. It can take many different forms, some of which appear in

| Table 2.2.1 | Several Forms and Examples of Potential Energy |

FORM OF POTENTIAL ENERGY	EXAMPLE
Gravitational potential energy	A bowling ball at the top of a hill
Elastic potential energy	A wound clock spring
Electrostatic potential energy	A cloud in a thunderstorm
Chemical potential energy	A firecracker
Nuclear potential energy	Uranium

Table 2.2.1. In each case nothing is moving; but because the objects still have a great potential to do work, they contain potential energy.

We measure energy in many common units: joules (J), calories, food Calories (also called kilocalories), and kilowatt-hours, to name only a few. All of these units measure the same thing, and they differ from one another only by numerical conversion factors, some of which can be found in Appendix B. For example, 1 food Calorie is equal to 1000 calories or 4187 J. Thus a jelly donut with about 250 food Calories contains about 1,000,000 J of energy. Since a joule is the same as a newton-meter, 1,000,000 J is the energy you'd use to lift your friend's file cabinet into the second-floor apartment 200 times (1000 N times 5 m upward is 5000 J of work per trip). No wonder eating donuts is hard on your physique!

Of course, you can eventually use up the energy in a jelly donut; it just takes time. You can only do so much work each second. The measure of how quickly you do work is **power**—the amount of work you do in a certain amount of time, or

$$\text{power} = \frac{\text{work}}{\text{time}}.$$

The SI unit of power is the **joule-per-second,** also called the **watt** (abbreviated W). Other units of power include Calories-per-hour and horsepower; like the units for energy, these units differ only by numerical factors, which are again listed in Appendix B. For example, 1 horsepower is equal to 745.7 W. Since a 1-horsepower motor does 745.7 J of work each second, and since it takes 5000 J of work to move the file cabinet to the second floor, that motor has enough power to do the job in about 6.7 s.

► Check Your Understanding #4: Apple Overtures

for answers, see page 76

Trace the flow of energy as an archer shoots an apple off the head of her assistant with an arrow.

Friction and Thermal Energy

But what about the *thermal* energy produced by sliding friction? Is thermal energy a new kind of potential energy or an alternative to kinetic energy?

In truth, it's neither. Thermal energy is actually a mixture of ordinary kinetic and potential energies. But unlike the kinetic energy in a moving ball or the potential energy in an elevated piano, the kinetic and potential energies in thermal energy are disordered at the atomic and molecular level. Thermal energy makes every microscopic particle in an object jiggle independently; at any moment, each particle has its own tiny supply of potential and kinetic energies, and this dispersed energy is collectively referred to as thermal energy.

As you push the file cabinet across the floor, you do work on it, but it doesn't pick up speed. Instead, sliding friction converts your work into thermal energy, so that the cabinet becomes hotter as the energy you transfer to it disperses among its particles. But while sliding friction easily turns work into thermal energy, there's no easy way to turn thermal energy back into work. Disorder not only makes things harder to use, but it is also difficult to undo. When you drop your favorite coffee mug on the floor and it shatters into a thousand pieces, the cup is still all there, but it's disordered and thus much less useful. Just as dropping the pieces on the floor a second time isn't likely to reassemble your cup, energy converted into thermal energy can't easily be reassembled into useful, ordered energy.

Sliding friction always converts at least some work into thermal energy. Since two surfaces sliding across one another experience frictional forces that oppose their relative motion, sliding friction does negative work on them; it extracts energy from a sliding object and converts that energy into thermal energy. Thus while you do work on the file cabinet by pushing it across the floor, sliding friction does negative work on it. The file cabinet's kinetic energy doesn't change very much, but its thermal energy continues to increase.

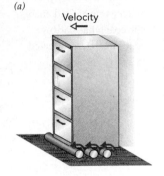

In contrast, static friction doesn't convert work into thermal energy. Since two surfaces experiencing static friction don't move relative to one another, there is no distance traveled and thus no work done. You can push against the stationary file cabinet all day without doing any work on it. Even if you lift the file cabinet upward with your hands (no easy task), static friction between your hands and the file cabinet's sides merely assists you in doing work on the file cabinet itself. As you lift the file cabinet upward, all of your work goes into increasing the file cabinet's gravitational potential energy.

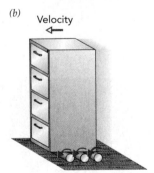

> ## Check Your Understanding #5: Burning Rubber
>
> *for answers, see page* 76
>
> If you push too hard on your car's accelerator pedal when the traffic light turns green, your wheels will slip and you'll leave a black trail of rubber behind. Such a "jackrabbit start" can cause as much wear on your tires as 50 km (31 miles) of normal driving. Why is skidding so much more damaging to the tires than normal driving?

Wheels

You've wrestled your friend's file cabinet out the door of the old apartment and are now dragging it along the sidewalk. You're doing work against sliding friction the whole way, producing large amounts of thermal energy in both the bottom of the cabinet and the surface of the sidewalk. You're also damaging both objects, since sliding friction is wearing out their surfaces. The four-drawer file cabinet may be down to three drawers by the time you arrive at the new apartment.

Fortunately, there are mechanical systems that can help you move one object across another without sliding or sliding friction. The classic example is a roller (Fig. 2.2.3). If you place the file cabinet on rollers, those rollers will rotate as the file cabinet moves so that their surfaces never slide across the bottom of the cabinet or the top of the sidewalk. To see how the rollers work, make a fist with one hand and roll it across the palm of your other hand. The skin of one hand doesn't slide across the skin of the other hand; since this silent motion doesn't convert work into thermal energy, your skin remains

Fig. 2.2.3 (*a*) A file cabinet that's supported on turning rollers experiences only static friction. (*b*) Since the top surface of a roller moves forward with the file cabinet, while its bottom surface stays behind with the sidewalk, the roller's center of mass moves only half as fast as the file cabinet. As a result, the rollers are soon left behind.

Bettman/Corbis Images

Fig. 2.2.4 As this stagecoach rolls forward, sliding friction between its axles and hubs converts some of its kinetic energy into thermal energy. To reduce this wasted energy, the coach has narrow axles that are lubricated with axle grease.

cool. Now slide your two open palms across one another; this time, sliding friction warms your skin.

Although the rollers don't experience sliding friction, they do experience static friction. The top of each roller is touching the bottom of the cabinet, and the two surfaces move along together because of static friction; they grip one another tightly until the roller's rotation pulls them apart. A similar process takes place between the rollers and the top of the sidewalk; static friction exerts torques on the rollers and hence is what makes them rotate in the first place. Again, you can illustrate this behavior with your hands. Try to drag your fist across your open palm. Just before your fist begins to slide, you'll feel a torque on it. Static friction between the skins of your two hands, acting to prevent sliding, causes your fist to begin rotating just like a roller.

Once you get the file cabinet moving on rollers, you can keep it rolling along the level sidewalk indefinitely. Without any sliding friction, the cabinet doesn't lose kinetic energy, so it continues at constant velocity without your having to push it. However, the rollers move out from under the file cabinet as it travels, and you frequently have to move a roller from the back of the cabinet to the front. In fact, you need at least three rollers to ensure that the file cabinet never falls to the ground when a roller pops out the back. Although the rollers have eliminated sliding friction, they've created another headache—one that makes the prospect of a long trip unappealing. Is there another device that can reduce sliding friction without requiring constant attention?

One alternative would be a four-wheeled cart. The simplest cart rests on fixed poles or axles that pass through central holes or hubs in the four wheels (Fig. 2.2.4). The ground exerts upward support forces on the wheels, the wheels exert upward support forces on the axles, and the axles support the cart and its contents. As the cart moves forward, its wheels turn so that their bottom surfaces don't slide or skid across the ground; instead, each wheel lowers a portion of its surface onto the sidewalk, leaves it there briefly to experience static friction, and then raises it back off the sidewalk, with a new portion of wheel surface taking its place. Thus there is only static friction between the cart's wheels and the ground.

Unfortunately, as each wheel rotates, its hub slides across the stationary axle at its center (Fig. 2.2.5). This sliding friction wastes energy and causes wear to both hub and axle. However, the narrow hub moves relatively slowly across the axle so that the work and wear done each second are small. Still, this sliding friction is undesirable and can be reduced significantly by lubricating the hub and axle with "axle grease."

A better solution is to insert rollers between the hub and axle (Fig. 2.2.6). The result is a roller bearing—a mechanical device that minimizes sliding friction between a hub and an axle. A complete bearing consists of two rings separated by rollers that keep those rings from rubbing against one another. In this case, the bearing's inner ring is attached to the stationary axle while its outer ring is attached to the spinning wheel hub. The nondriven wheels of an automobile are supported by such bearings on essentially stationary axles. A nondriven bicycle wheel is similarly supported on a stationary axle, but its bearings use balls instead of rollers—ball bearings. When either vehicle starts forward, static friction from the ground exerts torques on its free wheels and they begin to turn.

A car's driven wheels are also supported by roller bearings, but these bearings act somewhat differently. Because the engine must be able to exert a torque on each driven wheel, those wheels are rigidly connected to their axles. As the engine spins one of these axles, the axle spins its wheel. A bearing prevents the spinning axle from rubbing against the car's frame. This bearing's outer ring is attached to the stationary car frame while its inner ring is attached to the spinning axle.

As the driven wheel begins to spin, it experiences static friction with the ground and the ground pushes horizontally on the wheel's bottom to keep it from skidding. Since that is the only horizontal force on the automobile, the automobile accelerates forward.

Velocity

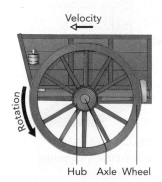

Rotation

Hub   Axle   Wheel

Fig. 2.2.5 A wheel rotates as it travels so that no sliding friction occurs between it and the ground. The load is imparted to the wheel through an axle that enters the central hole of the wheel, the hub. As the wheel turns, the hub rubs against the axle, causing some sliding friction and wear.

Recognizing a good idea when you think of it, you load the file cabinet into the back of your Jaguar XK8 convertible and climb into the driver's seat. The car isn't quite as responsive as usual because of the added mass, but it's still able to accelerate respectably. In a few seconds, you're cruising down the road toward the new apartment and a very grateful friend.

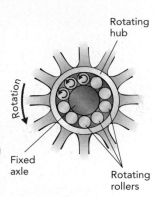

Check Your Understanding #6: Jewel Movements

for answers, see page 76

Many antique mechanical watches and clocks proudly proclaim that they have "jewel movements." Gears in these timepieces turn on axles that are pointed at either end and are supported at those ends by very hard, polished gemstones. What is the advantage of having needlelike ends on an axle and supporting those needles with smooth, hard jewels?

Kinetic Energy

As you near your destination, you begin thinking about the car's brakes. They're designed to stop the car by turning its kinetic energy into thermal energy. They'll perform their task by rubbing stationary brake pads against spinning metal discs, so that sliding friction can transform the energy. Although you're confident that those brakes are up to the task, just how much kinetic energy are they going to have to convert into thermal energy?

One way is to determine the car's kinetic energy is to calculate the work its engine did on it while bringing it from rest to its current speed. The result of that calculation is that the moving car's kinetic energy is equal to one-half of its mass times the square of its speed. This relationship can be written as a word equation:

$$\text{kinetic energy} = \tfrac{1}{2} \cdot \text{mass} \cdot \text{speed}^2, \qquad (2.2.1)$$

in symbols:

$$K = \tfrac{1}{2} \cdot m \cdot v^2,$$

and in everyday language:

> *Racing around at twice the speed takes four times the energy.*

With you and the file cabinet on board, the XK8 has a mass of about 2000 kg (4400 lbm). At a speed of 100 km/h (62 mph), it has over 770,000 J of kinetic energy. That enormous energy is four times what it would be at 50 km/h (31 mph), so put down your cell phone and drive carefully. The dramatic increase in kinetic energy that results from a modest increase in speed explains why high-speed crashes are far deadlier than those at lower speeds and why that police officer is checking out your car with a radar gun. Red cars get all the attention.

You're traveling safely within the speed limit and exchange a polite wave with the officer. However, you soon pass another car that has been stopped for a ticket. The light on the nearby police car spins round and round, and rotating objects have kinetic energy, too. Like the kinetic energy of translational motion, the kinetic energy of rotational

Fig. 2.2.6 In a roller bearing, the hub of the wheel doesn't touch the axle directly. Instead, the two are separated by a set of rollers that turn with the hub. The bottom few rollers bear most of the load since the hub pushes up on them and they push up on the axle. As the wheel turns, the rollers recirculate, traveling up to the right and over the top of the axle before returning down to the left to bear the load once again. The rollers, wheel, and axle experience only static friction, not sliding friction. In a ball bearing, the cylindrical rollers are replaced by spherical balls.

motion depends on the light's inertia and speed. But for a spinning light, it's the rotational inertia and rotational speed that matter. The light's kinetic energy is equal to one-half of its rotational mass times the square of its angular speed. This relationship can be written as a word equation:

$$\text{kinetic energy} = \tfrac{1}{2} \cdot \text{rotational mass} \cdot \text{angular speed}^2, \qquad \textbf{(2.2.2)}$$

in symbols:

$$K = \tfrac{1}{2} \cdot I \cdot \omega^2,$$

and in everyday language:

It takes a very energetic person to spin his wheels twice as fast.

With the ticket complete, the police car pulls out into traffic with its light still spinning. The light's total kinetic energy is now the sum of two parts: translational kinetic energy and rotational kinetic energy. Its translational kinetic energy depends on the speed of the light's center of mass, which is equal to the police car's speed through traffic. And its rotational kinetic energy depends on the angular speed at which the light turns about its center of mass.

As the police car disappears in the distance, it occurs to you that the spinning wheels of your car also have rotational kinetic energy that adds to the car's substantial translational kinetic energy. Still, you trust your brakes. In a few minutes, you arrive at your destination and brake to a stop. Although you're aware of the added mass as the car decelerates less quickly than usual, the brakes successfully transform the car's kinetic energy into thermal energy. You've reached your goal safely and are now a hero.

Check Your Understanding #7: Throwing a Fastball

for answers, see page 76

A typical grade-school pitcher can throw a baseball at 80 km/h (50 mph), but only a few professional athletes have the extraordinary strength needed to throw a baseball at twice that speed. Why is it so much harder to throw the baseball only twice as fast?

Check Your Figures #1: Blowing in the Wind

for answers, see page 77

The air in a hurricane travels at 200 km/h (124 mph). How much more kinetic energy does 1 kg of this air have than 1 kg of air moving at only 20 km/h?

Check Your Figures #2: Playing Around at the Playground

for answers, see page 77

When children climb onto a playground merry-go-round, they increase its rotational inertia. If the children triple the merry-go-round's rotational mass, how will they alter the kinetic energy it has when it spins at a certain angular speed?

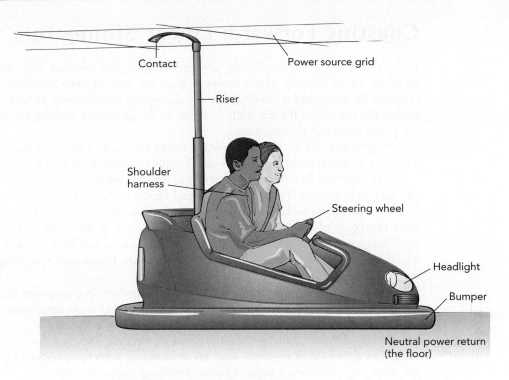

Contact

Power source grid

Riser

Shoulder
harness

Steering wheel

Headlight

Bumper

Neutral power return
(the floor)

SECTION 2.3 Bumper Cars

While car crashes normally aren't much fun outside of movies or television, there is one delightful exception: bumper cars. For a few minutes, drivers in this amusement park ride race madly about an oval track, deliberately crashing their vehicles into one another and laughing hysterically at the violent impacts. Jolts, jerks, and spins are half the fun, and it's a wonder that no one gets whiplash. But hidden in the excitement are several important physics concepts that influence everything from tennis to billiards.

Questions to Think About: *Why does a stationary car begin rolling forward after being struck by a moving car? What aspects of motion are passed between cars as they collide? Why does your car jolt more when the car that hits it contains two big adults rather than one small child? What would happen if the bumper cars had hard steel bumpers rather than soft rubber ones? Why is your car often set spinning by collisions, and what keeps it spinning?*

Experiments to Do: *Place a coin on a smooth table and flick a second, identical coin so that it slides along the table and strikes the stationary coin squarely. What happens? Try this experiment again, but now use two coins with different masses. How is the collision different? Does it matter which coin you crash into the other?*

Now line up several identical coins so that they touch and slide another coin into one end of this line. How does the collision affect the coin that was originally moving? How does it affect the line of coins? What was transferred among the coins by the collision?

Now stand a coin on its edge and flick it so that it spins rapidly. Did you give it something that keeps it spinning? Why does the coin eventually stop spinning?

Coasting Forward: Linear Momentum

Bumper cars are small, electrically powered vehicles that can turn on a dime and are protected on all sides by rubber bumpers. Each car has only two controls: a pedal that activates its motor and a steering wheel that controls the direction in which the motor pushes the car. Since the car itself is so small, its occupants account for much of the car's total mass and rotational mass.

Imagine that you have just sat down in one of these cars and put on your safety strap. The other people also climb into their cars, usually one person per car, and the ride begins.

With your car free to move or turn, you quickly become aware of its translational and rotational inertias. The car's translational inertia makes it hard to start or stop, and its rotational inertia makes it difficult to spin or stop from spinning. While we've seen these two types of inertia before, let's take another look at them and at how they affect your bumper car. This time, we'll see that they're associated with two new conserved quantities—linear momentum and angular momentum. It turns out that energy isn't the only conserved quantity in nature!

When fast-moving bumper cars crash into one another, they exchange more than just energy. Energy is directionless—it's not a vector quantity—yet these cars seem to be exchanging some quantity of motion that has a direction associated with it. For example, if your car is hit squarely by a rightward moving car, then your car's motion shifts somewhat rightward in response. Your car is receiving a vector quantity of motion from the other car, a conserved vector quantity known as linear momentum.

Linear momentum, usually just called **momentum,** is the measure of an object's translational motion—its tendency to continue moving in a particular direction. Roughly speaking, your car's momentum indicates which way it's heading and just how difficult it was to get the car moving with its current velocity.

The car's momentum is its mass times its velocity and can be written as a word equation:

$$\text{momentum} = \text{mass} \cdot \text{velocity}, \tag{2.3.1}$$

in symbols:

$$\mathbf{p} = m \cdot \mathbf{v},$$

and in everyday language:

It's hard to stop a fast-moving truck.

Note that momentum is a vector quantity and that it has the same direction as the velocity. As we might expect, the faster your car is moving or the more mass it has, the more momentum it has in the direction of its motion. The SI unit of momentum is the **kilogram-meter-per-second** (abbreviated kg·m/s).

To physicists, conserved quantities are rare treasures that make it easier to understand otherwise complicated motions. Like all conserved quantities, momentum can't be created or destroyed. It can only be transferred between objects. Momentum plays a very basic role in bumper cars: the whole point of crashing them into one another is to enjoy the momentum transfers. During each collision, momentum shifts from one car to the other so that they abruptly change their speeds or directions or both. As long as these momentum transfers aren't too jarring, everyone has a good time.

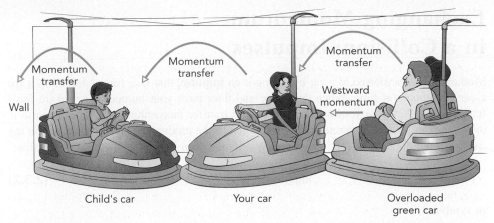

Fig. 2.3.1 Your car is hit by a fast-moving, massive car with westward momentum. Much of that westward momentum is transferred to your car. You crash into a child's car, transferring the westward momentum to it. It then crashes into the wall, transferring the westward momentum to the wall.

You've stopped your car, so it has zero velocity and zero momentum. To begin moving again, something must transfer momentum to your car. While you could press the pedal and let the motor gradually transfer momentum from the ground to your car, that's not much fun. Instead, you let two grinning couch potatoes in an overloaded green car slam into you at breakneck speed (Fig. 2.3.1).

The green car was heading westward, and in a few moments your car is moving westward, too, while the green car has slowed significantly. Before you recover from the jolt, your car pounds a child's car westward and your car slows down abruptly. Finally, its impact with a wall stops the child's car. Despite disapproving looks from the child's parents, there's no harm done. Overall, westward momentum has flowed from the spudmobile to your car, to the child's car, and into the wall. No momentum has been created or destroyed; you've all simply enjoyed passing it along from car to car.

Check Your Understanding #1: Stuck on the Ice

for answers, see page 77

Suppose you're stuck in the middle of a frozen lake, with a surface so slippery that you can't get any traction. You take off a shoe and throw it toward the southern shore. You find yourself coasting toward the northern shore and soon escape from the lake. Why did this scheme work?

Check Your Figures #1: Follow That Train!

for answers, see page 78

The bad guys are getting away in a four-car train and you're trying to catch them. The train has a mass of 20,000 kg and it's rolling forward at 22 m/s (80 km/h or 50 mph). What is the train's momentum?

Exchanging Momentum
in a Collision: Impulses

Momentum is transferred to a car by giving it an **impulse,** that is, a force exerted on it for a certain amount of time. When the motor and floor push your bumper car forward for a few seconds, they give your car an impulse and transfer momentum to it. This impulse is the change in your car's momentum and is equal to the product of the force exerted on the car times the duration of that force. This relationship can be written as a word equation:

$$\text{impulse} = \text{force} \cdot \text{time}, \qquad (2.3.2)$$

in symbols:

$$\Delta \mathbf{p} = \mathbf{F} \cdot t,$$

and in everyday language:

The harder and longer you push a bobsled forward at the start of a race, the more momentum it will have when it starts down the hill.

The more force or the longer that force is exerted, the larger the impulse and the more your car's momentum changes. Remember that an impulse, like momentum itself, is a vector quantity and has a direction. If your aim is off and the misdirected impulse you obtain from the floor sends you crashing into the wall, don't say you hadn't been warned!

Different forces exerted for different amounts of time can transfer the same momentum to a car:

$$\text{impulse} = \text{large force} \cdot \text{short time}$$
$$= \text{small force} \cdot \text{long time}. \qquad (2.3.3)$$

Thus you can get your car moving with a certain forward momentum either by letting the motor and floor push on it with a small forward force of long duration or by letting the colliding green car push on it with a large forward force of short duration.

We can now explain why bumper cars have soft rubber bumpers. If the bumpers were hard steel, the collision between the green car and your car would last only an instant and would involve an enormous forward force. You'd be in need of a neck brace and the services of a personal injury lawyer. However, amusement parks don't like lawsuits and sensibly limit the car impact forces. To do this, they use rubber bumpers and rather slow-moving cars.

Nonetheless, you can get a pretty good jolt when you collide head-on with another car. Your two cars then start with oppositely directed momenta, and the collision roughly exchanges those momenta between cars. In almost no time, you go from heading forward to heading backward. The impulse that causes this reversal of motion is especially large because it not only stops your forward motion, it also causes you to begin heading backward.

Why should momentum be a conserved quantity? It's conserved because of Newton's third law of motion. When one car exerts a force on a second car for a certain amount of time, the second car exerts an equal but oppositely directed force on the first car for exactly the same time. Because of the equal but oppositely directed nature of the two forces, cars that push on one another receive impulses that are equal in amount but opposite in direction. Since the momentum gained by one car is exactly equal to the momentum lost by the other car, we say that momentum is transferred from one car to the other.

The more mass a car has, the less its velocity changes as a consequence of a momentum transfer. That's why the green car doesn't stop completely when it crashes into your car, while your car speeds up dramatically. The green car has so much forward momentum that transferring a fraction of it to your car causes a large change in your car's velo-city. Like a bug being hit by an automobile windshield, your bumper car does most of the accelerating.

Conserved Quantity: Momentum   **Transferred By: Impulse**

Momentum: The measure of an object's translational motion—its tendency to continue moving in a particular direction. Momentum is a vector quantity, meaning that it has a direction. It has no potential form and therefore cannot be hidden; momentum = mass · velocity.

Impulse: The mechanical means for transferring momentum; impulse = force · time.

Common Misconceptions: Momentum and Force

Misconception: A massive, moving object carries a force with it—the "force of its momentum."

Resolution: Although the impulses that transfer momentum involve forces, momentum itself does not. A moving object carries only momentum, not force. Most importantly, a coasting object is free of any net force. But when that object hits an obstacle, the two will exchange momentum via impulses and those impulses will involve forces.

► Check Your Understanding #2: Bowling Them Over

for answers, see page 77

When a beanbag hits the wall, it transfers all of its forward momentum to the wall and comes to a stop. When a rubber ball hits the wall, it transfers all of its forward momentum, comes to a stop, and then rebounds. During the rebound it transfers still more forward momentum to the wall. If you wanted to knock over a weighted bowling pin at the county fair, which would be the more effective projectile: the rubber ball or the beanbag, assuming they have identical masses and you throw them with identical velocities?

► Check Your Figures #2: Stop That Train!

for answers, see page 78

The engine of the train you're chasing (see Check Your Figures #1) has broken down, but it's still rolling forward. To stop it, you grab onto the last car and begin to drag your boot heels on the ground. The backward force on the train is 200 N. How long will it take you to stop the train?

Spinning in Circles: Angular Momentum

When bumper cars are set spinning during crashes, they are exchanging yet another conserved quantity. Like momentum, it's a conserved vector quantity, but now it's associated with the angular speed and direction of rotational motion around a specific pivot. For example, when your car receives a glancing blow from a car that is circling you clockwise, your car's rotational motion or spin shifts somewhat clockwise in response. Your car is receiving a vector quantity of motion from the other car, a conserved vector quantity known as angular momentum.

Angular momentum is the measure of an object's rotational motion—its tendency to continue spinning about a particular axis. Simply put, your car's angular momentum indicates the direction of its rotation and just how difficult it was to get it spinning with its current angular velocity. The car's angular momentum is its rotational mass times its angular velocity and can be written as a word equation:

$$\text{angular momentum} = \text{rotational mass} \cdot \text{angular velocity}, \qquad (2.3.4)$$

in symbols:

$$\mathbf{L} = I \cdot \boldsymbol{\omega},$$

and in everyday language:

It's hard to stop a spinning carousel.

Note that angular momentum is a vector quantity and that it has the same direction as the angular velocity. The faster your car is spinning or the larger its rotational mass, the more angular momentum it has in the direction of its angular velocity. The SI unit of angular momentum is the **kilogram-meter^2-per-second** (abbreviated kg·m^2/s).

Angular momentum is another conserved quantity, so it can't be created or destroyed; it can only be transferred between objects. For your car to begin spinning, something must transfer angular momentum to it, and your car will then continue to spin until it transfers this angular momentum elsewhere. But to study angular momentum properly, we must pick the pivot about which all the spinning will occur. In the present situation, a good choice for this pivot is your car's initial center of mass.

Your car is stationary again, so it has zero angular velo-city and zero angular momentum. Suddenly, a purple car sweeps by and strikes your car a glancing blow (Fig. 2.3.2). Because the purple car was circling your car counterclockwise, it had counterclockwise angular momentum about the pivot. Its impact transfers some of this angular momentum to your car, which begins spinning counter-clockwise itself. Since it has given up some of its angular momentum, the purple car circles your car less rapidly. Your car gradually stops spinning as its wheels and friction transfer the angular momentum to the ground and earth. Overall, no angular momentum was created or

Fig. 2.3.2 Since the purple car is circling your car counterclockwise, it has counterclockwise angular momentum. When it hits your car, it transfers some of that angular momentum to your car. Because of this transfer, the purple car stops circling quickly as your car begins to spin counterclockwise.

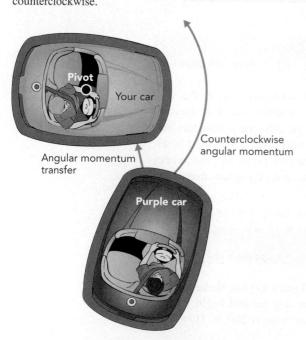

Pivot

Your car

Counterclockwise angular momentum

Angular momentum transfer

Purple car

destroyed during the collision. Instead, it was transferred from the purple car to your car to the earth.

►Check Your Understanding #3: Many Happy Re-Turns
for answers, see page 77

Satellites are often set spinning during launch in order to give them added stability. When astronauts visit these satellites years later, they find them still spinning. Why don't the satellites stop spinning?

►Check Your Figures #3: Want to Go for a Spin?
for answers, see page 78

Spinning satellites are particularly stable. Suppose that the astronauts launching a particular satellite decide to increase its angular velocity by a factor of 5. How will that change affect the satellite's angular momentum?

Glancing Blows: Angular Impulses

Angular momentum is transferred to a car by giving it an **angular impulse,** that is, a torque exerted on it for a certain amount of time. When the purple car hits your car and exerts a torque on it briefly, it gives your car an angular impulse and transfers angular momentum to it. This angular impulse is the change in your car's angular momentum and is equal to the product of the torque exerted on your car times the duration of that torque. This relationship can be written as a word equation:

$$\text{angular impulse} = \text{torque} \cdot \text{time}, \qquad \textbf{(2.3.5)}$$

in symbols:

$$\Delta \mathbf{L} = \boldsymbol{\tau} \cdot t,$$

and in everyday language:

> *To get a merry-go-round spinning rapidly, you must twist it hard and for a long time.*

The more torque or the longer that torque is exerted, the larger the angular impulse and the more your car's angular momentum changes. Once again, an angular impulse is a vector quantity and has a direction. Had the purple car been circling your car clockwise when it struck the glancing blow, its angular impulse would have been in the opposite direction and you'd be spinning the other way.

Different torques exerted for different amounts of time can transfer the same angular momentum to a car:

$$\text{angular impulse} = \text{large torque} \cdot \text{short time}$$
$$= \text{small torque} \cdot \text{long time.} \qquad \textbf{(2.3.6)}$$

Thus you can get your car spinning with a certain angular momentum either by letting the motor and floor twist it with a small torque of long duration or by letting the colliding purple car twist it with a large torque of short duration. As with linear momentum, sudden transfers of angular momentum can break things, so the cars are designed to limit their impact torques to reasonable levels. Even so, you may find yourself reaching for the motion sickness bag after a few spinning collisions.

Why should angular momentum be a conserved quantity? Like linear momentum, angular momentum is conserved because of Newton's third law of motion. In this case, we're referring to **Newton's third law of rotational motion:** if one object exerts a torque on a second object, then the second object will exert an equal but oppositely directed torque on the first object.

> ### Newton's Third Law of Rotational Motion
> For every torque that one object exerts on a second object, there is an equal but oppositely directed torque that the second object exerts on the first object.

Fig. 2.3.3 When Kristi Yamaguchi pulls in her arms, she reduces her rotational mass. Since she is experiencing zero net torque, her angular momentum must remain constant and she begins to spin more rapidly.

When one car exerts a torque on a second car for a certain amount of time, the second car exerts an equal but oppositely directed torque on the first car for exactly the same amount of time. Because of the equal but oppositely directed nature of the two torques, cars that exert torques on one another receive angular impulses that are equal in amount but opposite in direction. Since the angular momentum gained by one car is exactly equal to the angular momentum lost by the other car, we say that angular momentum is transferred from one car to the other.

Because a car's angular momentum depends on its rotational mass, two different cars may end up rotating at different angular velocities even though they have identical angular momenta. For example, when the purple car hits the overloaded green car and transfers angular momentum to it, the green car's enormous rotational mass makes it spin relatively slowly. The same sort of behavior occurs with linear momentum, where a car's mass affects how fast it travels when it's given a certain amount of linear momentum. But while a bumper car can't change its mass, it can change its rotational mass. If it does so while it's spinning, its angular *momentum* won't change, but its angular *velocity* will!

To see this change in angular velocity, consider the overloaded green car. Its two large occupants are disappointed with the ride because their huge mass and rotational mass prevent them from experiencing the intense jolts and spins that you've been enjoying. Suddenly they get a wonderful idea. As their car slowly spins, one of them climbs into the other's lap and the two sit very close to the car's center of mass. By rearranging the car's mass this way, they have reduced the car's overall rotational mass and the car actually begins to spin faster than before.

Since the green car's mass has been redistributed, it's no longer a freely turning rigid object covered by Newton's first law of rotational motion. However, it is freely turning and thus covered by a more general and equally useful rule: an object that is not subject to any outside torques has constant angular momentum. Since the car's rotational mass has become smaller, its angular velocity must increase in order to keep its angular momentum constant. That's just what happens. This effect of changing one's rotational mass explains how an ice skater can achieve an enormous angular velocity by pulling herself into a thin, spinning object on ice (Fig. 2.3.3).

Conserved Quantity: Angular Momentum   **Transferred By: Angular Impulse**

Angular Momentum: The measure of an object's rotational motion—its tendency to continue spinning about a particular axis. Angular momentum is a vector quantity, meaning that it has a direction. It has no potential form and therefore cannot be hidden; angular momentum = rotational mass · angular velocity.

Angular Impulse: The mechanical means for transferring angular momentum; angular impulse = torque · time.

Summary of Newton's Laws of Rotational Motion

1. A rigid object that is not wobbling and is not subject to any outside torques rotates at a constant angular velocity, turning equal amounts in equal times about a fixed axis of rotation.

2. The torque exerted on an object that is not wobbling is equal to the product of that object's rotational mass times its angular acceleration. The angular acceleration points in the same direction as the torque.

3. For every torque that one object exerts on a second object, there is an equal but oppositely directed torque that the second object exerts on the first object.

Note: These laws are the rotational analogs of the translational laws on p. 28.

➤ Check Your Understanding #4: Spinning the Merry-Go-Round

for answers, see page 77

A person who is initially motionless starts a merry-go-round spinning and then returns to being motionless. If angular momentum is truly conserved, what is the source of the angular momentum that the spinning merry-go-round now has?

➤ Check Your Figures #4: Spin Away!

for answers, see page 78

How much longer will it take the astronauts launching the satellite in Check Your Figures #3 to bring it to the faster angular velocity, if they use the initially planned torque?

The Three Conserved Quantities

As you drive your bumper car around the oval, its motion is governed in large part by three conserved quantities: energy, linear momentum, and angular momentum (Table 2.3.1). While you can exchange those quantities with the earth and the power company by steering your car or switching on its motor, most of the interesting exchanges involve collisions.

Each time your car shoves another car forward, your car does work on that other car and transfers energy to it. Each time your car pushes another car northward briefly, your car gives a northward impulse to that other car and transfers northward momentum to it. And each time your car twists another car clockwise about its center of mass, your

| Table 2.3.1 | The Three Conserved Quantities of Motion and Their Transfer Machanisms |

CONSERVED QUANTITY	TRANSFER MECHANISM
Energy	Work
Linear momentum	Impulse
Angular momentum	Angular impulse

car gives a clockwise angular impulse to the other car and transfers clockwise angular momentum to it. These exchanges of energy, momentum, and angular momentum are fast and furious and make for an exciting ride.

Quantity	SI Unit	English Unit	SI → English	English → SI
Momentum	kilogram-meter-per-second (kg·m/s)	pound-foot-per-second (lbm·ft/s)	1 kg·m/s = 7.2329 lbm·ft/s	1 lbm·ft/s = 0.13826 kg·m/s
Angular momentum	kilogram-meter^2-per-second (kg·m^2/s)	pound-foot^2-per-second (lbm·ft^2/s)	1 kg·m^2/s = 23.730 lbm·ft^2/s	1 lbm·ft^2/s = 0.042140 kg·m^2/s

▶ Check Your Understanding #5: Hitting the Wall

for answers, see page 77

You're backing out of a parking space and accidentally hit a concrete wall. The wall doesn't move and your car sustains some damage. Did your car transfer any energy or momentum to the wall?

Potential Energy and Acceleration

Shortly before the ride stops, you notice that there is a low point in the floor. After years of use, its metal surface has dented into a bowl-shaped depression and you observe that cars naturally tend to roll into this bowl and accelerate toward its bottom. We've seen this tendency to accelerate downhill before with ramps, but now let's look at it in terms of energy: a car always accelerates in the direction that reduces its total potential energy as quickly as possible. Since a lone car's only potential energy is gravitational potential energy, it accelerates in such a way as to reduce its gravitational potential energy as quickly as possible: down the steepest route to the bottom of the bowl.

This behavior of accelerating in the direction that reduces total potential energy as quickly as possible is universal. Potential energy and forces are related to one another, so this rule is really just a way to determine the direction of the net force on an object or its parts. An object accelerates in the direction of the net force on it, which is also the direction that will reduce its total potential energy as quickly as possible. This rule is a useful way to determine how motion will proceed: which way a spring will leap, a chair will tip, or a bumper car will roll. We'll use it frequently in this book.

Potential Energy and Acceleration
An object accelerates in the direction that reduces its total potential energy as quickly as possible.

➤ Check Your Understanding #6: Heading Down

for answers, see page 77

When you pull a child back on a playground swing and let go, which way does that child accelerate?

Epilogue for Chapter 2

In this chapter we looked at rotating and colliding objects and studied the physical laws that describe their motions. In *seesaws,* we examined rotational inertia and saw how torques cause angular accelerations. We also noticed how useful it can be to separate an object's rotational motion from its translational motion. In *wheels,* we discussed another important type of force, friction, as well as a new type of energy, thermal energy—the energy associated with heat and temperature. In *bumper cars,* we introduced two more conserved physical quantities: momentum and angular momentum. As we'll see, following the flows of energy, momentum, and angular momentum between objects often helps in understanding how those objects work.

Explanation: Spinning a Pie Dish

Because the balanced dish has rotational inertia, torques are required both to start it spinning and to stop it from doing so. When you twist the dish with your hand, the torque you exert gives the dish an angular impulse and sets it spinning with a certain amount of angular momentum. If the pivot were truly frictionless and there were no air resis-tance, the dish would spin indefinitely because it would be unable to get rid of its angular momentum. However, friction in the pivot exerts a small but significant torque that opposes the dish's motion and gradually slows it down. As this frictional torque transfers angular momentum out of the dish and into the pencil, chair, and earth, the dish turns more and more slowly until it finally comes to a stop. The sharper the pivot and the smaller the contact area between point and dish, the less frictional torque the dish experiences and the longer it spins.

Balancing the dish is easy as long as it's upside-down. With its edge drooping downward, the dish has relatively little gravitational potential energy and is surprisingly stable. If it begins to tip to one side, the dish's average height rises and so does its gravitational potential energy. Since objects naturally accelerate in whatever direction lowers their potential energy as quickly as possible, the upside-down dish quickly tips back toward level after being disturbed. In contrast, an upright dish is virtually impossible to balance on a point because any tip will lower its gravitational potential energy and lead quickly to catastrophe. We'll look at these stabilizing/destabilizing effects more carefully later on in this book.

Chapter Summary

How Seesaws Work: A seesaw is a rotating toy that works best when it's almost perfectly balanced, meaning that it experiences zero net torque. The seesaw's pivot usually passes through its center of mass so that the seesaw balances when it's not occupied. The riders arrange themselves so that the torques they exert on the seesaw cancel one another completely. The seesaw then experiences zero net torque and zero angular acceleration, and it rotates with constant angular velocity. It either remains motionless or turns steadily in one direction or the other.

To make the seesaw tip back and forth, the riders subtly adjust the torques they exert on the seesaw. They do this either by leaning, thus varying their distances from the pivot, or by pushing against the ground with their feet, thus varying the forces they exert on the seesaw. In either case, they unbalance the seesaw, and it experiences both a net torque and an angular acceleration. By rhythmically changing the net torque on the seesaw, the riders cause it to rotate back and forth.

How Wheels Work: Wheels facilitate motion by eliminating or reducing sliding friction between an object and a surface. The wheels convey the support forces needed to hold the object up but allow the object to move without sliding. As a cart with freely turning wheels moves along the surface, static friction between each wheel and the surface exerts a torque on that wheel and causes it to turn. However, rubbing may occur between the wheel's hub and the axle, where sliding friction can waste energy and cause wear. To eliminate this sliding friction, roller or ball bearings are often used.

The torque that causes a powered wheel on a vehicle to turn comes from an engine by way of an axle. In this case, static friction between the outside of the wheel and the ground exerts a torque on the wheel that opposes the torque from the engine. This static frictional force also contributes to the net force on the vehicle and causes it to accelerate.

Once supported on wheels and bearings, objects can move freely and can retain linear momentum, angular momentum, and energy for long periods of time. By eliminating sliding friction, wheels can also keep objects from converting ordered energy into thermal energy. Wheels allow vehicles to hold onto these conserved quantities for extended periods and make transportation far more practical.

How Bumper Cars Work: Since they start from rest, bumper cars must obtain their initial momenta and angular momenta from the ground and their initial kinetic energies from the power company. They do this with the help of motors and wheels, which gradually transfer energy, momentum, and angular momentum into the cars.

Once the cars are moving, they can begin to exchange those conserved quantities by way of collisions. Each impact usually changes the cars' speeds and directions of travel in a manner that may seem rather complicated. However, following the exchanges of momentum, angular momentum, and energy often makes it easier to understand these collisions.

Cars containing massive riders respond weakly when collisions transfer momentum and angular momentum to them. That's because their large masses and rotational masses minimize their changes in velocity and angular velocity. Because of their small masses, children experience the wildest rides.

Important Laws and Equations

1. Newton's First Law of Rotational Motion: A rigid object that is not wobbling and is not subject to any outside torques rotates at a constant angular velocity, turning equal amounts in equal times about a fixed axis of rotation.

2. Newton's Second Law of Rotational Motion: The torque exerted on an object is equal to the product of that object's rotational mass times its angular acceleration, or

$$\text{torque} = \text{rotational mass} \cdot \text{angular acceleration.} \quad (2.1.2)$$

The angular acceleration points in the same direction as the torque. This law doesn't apply to objects that are wobbling.

3. Relationship between Force and Torque: The torque produced by a force is equal to the product of the lever arm times the component of that force perpendicular to the lever arm, or

$$\text{torque} = \text{lever arm} \cdot \text{force perpendicular to lever arm.} \quad (2.1.3)$$

4. Kinetic Energy: An object's translational kinetic energy is one-half of its mass times the square of its speed, or

$$\text{kinetic energy} = \tfrac{1}{2} \cdot \text{mass} \cdot \text{speed}^2. \quad (2.2.1)$$

An object's rotational kinetic energy is one-half of its rotational mass times the square of its angular speed, or

$$\text{kinetic energy} = \tfrac{1}{2} \cdot \text{rotational mass} \cdot \text{angular speed}^2. \quad (2.2.2)$$

5. Linear Momentum: An object's linear momentum is its mass times its velocity, or

$$\text{linear momentum} = \text{mass} \cdot \text{velocity.} \quad (2.3.1)$$

6. The Definition of Impulse: The impulse given to an object is equal to the product of the force exerted on that object times the length of time that force is exerted, or

$$\text{impulse} = \text{force} \cdot \text{time.} \quad (2.3.2)$$

7. Angular Momentum: An object's angular momentum is its rotational mass times its angular velocity, or

$$\text{angular momentum} = \text{rotational mass} \cdot \text{angular velocity.} \quad (2.3.4)$$

8. The Definition of Angular Impulse: The angular impulse given to an object is equal to the product of the torque exerted on that object times the length of time that torque is exerted, or

$$\text{angular impulse} = \text{torque} \cdot \text{time.} \quad (2.3.5)$$

9. Newton's Third Law of Rotational Motion: For every torque that one object exerts on a second object, there is an equal but oppositely directed torque that the second object exerts on the first object.

10. Potential Energy and Acceleration: An object accelerates in the direction that reduces its total potential energy as quickly as possible.

Check Your Understanding—Answers

Section 2.1 SEESAWS

1. The lazy Susan undergoes rotational motion while the dessert cart undergoes translational motion.
Why: The lazy Susan has a fixed pivot at its center. This pivot never goes anywhere, no matter how you rotate the lazy Susan. In contrast, the dessert cart moves about the room and has no fixed point. The server can rotate the dessert cart when necessary, but its principal motion is translational.

2. It will continue to spin at a steady pace about a fixed rotational axis (although friction with the water will gradually slow the ball's rotation).
Why: Because the basketball is free of torques, the outside influences that affect rotational motion, it has a constant angular velocity. If you spin the basketball, it will continue to spin about whatever axis you chose. If you don't spin the basketball, its angular velocity will be zero and it will remain stationary.

3. Yes. His center of mass falls smoothly, obeying the rules governing falling objects. As he falls, his body rotates at constant angular velocity about his center of mass.
Why: Like a thrown football or tossed baton, the diver is a rigid, rotating object. His motion can be separated into translational motion of his center of mass (it falls) and rotational motion about his center of mass (he rotates about it at constant angular velocity). While the diver may never think of his motion in these terms, he is aware intuitively of the need to handle both his rotational and translational motions carefully. Hitting the water with his chest because he mishandled his rotation isn't much more fun than hitting the board because he mishandled his translation.

4. The full merry-go-round has a huge rotational mass.

Why: Starting or stopping a merry-go-round involves angular acceleration. As the pusher, you exert a torque on the merry-go-round and it undergoes angular acceleration. But this angular acceleration depends on the merry-go-round's rotational mass, which in turn depends on how much mass it has and how far that mass is from the axis of rotation. With many children adding to the merry-go-round's rotational mass, its angular acceleration tends to be small.

5. The closer the cardboard is to the pivot, the more force it must exert on the scissors to produce enough torque to keep the scissors from rotating closed. When the cardboard is unable to produce enough torque, the scissors cut through it.

Why: When you place paper close to the pivot of a pair of scissors, you are requiring that paper to exert enormous forces on the scissors to keep them from rotating closed. Rotations are started and stopped by torques, and forces exerted close to the pivot exert relatively small torques.

6. About 1000 N (220 lbf).

Why: Since the nail is 10 times closer to the pivot, the nail must exert 10 times the force on the hammer to create the same magnitude of torque as you do pulling on the handle. As the nail pulls on the hammer, the hammer pulls on the nail. Although the wood exerts frictional forces on the nail to keep it from moving, the extracting force overwhelms this friction and the nail slides slowly out of the wood.

7. A substantial shift in the cargo's position during a storm can unbalance the ship, creating a net torque on the ship, and cause it to begin rotating about the effective pivot. The ship may then capsize.

Why: Although most boats can compensate for some amount of cargo imbalance, shifting cargo can easily flip even a fairly stable boat. It happens frequently in real life, often with fatal consequences. Some boats, particularly canoes and racing shells, are notoriously sensitive to unbalanced loading and are easily flipped by careless or moving occupants.

Section 2.2 WHEELS

1. Friction is pushing the glass uphill.

Why: The glass is sliding downhill across the top of the stationary table. Since friction always opposes relative motion, it pushes the glass uphill, in the direction opposite its motion.

2. About twice as hard.

Why: The frictional forces between the table and books are roughly proportional to the forces pressing them together. The book's weight is what pushes them together. When you effectively double the book's weight, by stacking a second book on top of it, you double the frictional forces between the table and the books.

3. If the wheels continue to turn, they experience static friction with the pavement. If they lock and begin to skid, they experi-

ence sliding friction. Since the traction provided by static friction is greater than that provided by sliding friction, the car will decelerate faster if the wheels don't skid.

Why: For a rapid stop, the car needs the maximum possible force in the direction opposite its velocity. The most effective way to obtain that stopping force from the road is with static friction between the turning wheels and the pavement. Sliding friction, the result of skidding tires, is much less effective at stopping the car, wears out the tires, and diminishes the driver's ability to steer the vehicle.

4. The archer does work on the string and bow as she draws the arrow back. (Chemical energy from her body is transferred to the bow, where it is stored as elastic potential energy.) As she releases the arrow, the string and bow do work on the arrow. (Elastic potential energy is transferred to the arrow, where it becomes kinetic energy.) Finally, the arrow does work on the apple, knocking it off the head of the assistant. (Kinetic energy is transferred from the arrow to the apple.)

Why: Because energy is conserved, we could in principle follow it back to the origins of the universe. Whatever energy we see around us now was somewhere in our universe yesterday, last week, and a million years ago, although its form may have changed. It will still be in our universe next year, too, but maybe not in so useful a form.

5. Normal driving involves mostly static friction because the surfaces of the tires don't slide across the pavement. Skidding involves sliding friction as the tire surfaces move independently of the pavement. Because it involves sliding friction, skidding creates thermal energy and damages the tires.

Why: The expression "burn rubber" is an appropriate name for skidding during a jackrabbit start. Substantial thermal energy is produced, and a trail of hot rubber is left on the pavement behind the car. At drag races, the frictional heating that results from skidding at the start can be so severe that the tires actually catch on fire.

6. Because all the supporting forces are very close to the axis of rotation, the jewels exert almost zero torque on the axle. The axle turns remarkably freely.

Why: Mechanical timepieces need almost ideal motion to keep accurate time. One of the best ways to allow a rotating object free movement is to support it exactly on the axis of rotation, where the support can't exert torque on the object.

7. Doubling the speed of the baseball requires quadrupling the energy transferred to it by the pitcher.

Why: To throw a 160-km/h (100-mph) fastball, a major league pitcher must put four times as much kinetic energy into both the ball and his arm as when pitching an 80-km/h slow ball. He also pitches the fastball in half the time needed to pitch the slow ball. Overall, he must do four times as much work throwing a fastball and he must do that work in one-half the time. That means the pitcher produces eight times as much power while throwing a fastball as when throwing a slow ball. No wonder amateurs have trouble duplicating that feat.

Section 2.3 BUMPER CARS

1. By transferring southward momentum to the shoe, you would obtain northward momentum.

Why: Initially, both you and your shoe have zero momentum. But when you throw the shoe southward, you give it southward momentum. Since the only source of that southward momentum is you, you must have lost southward momentum. A negative amount of southward momentum is actually northward momentum, and thus you coast northward. Interestingly enough, the total momentum of you and the shoe hasn't changed. It's still zero, as it must be because momentum is conserved. It has simply been redistributed.

2. The bouncy rubber ball would be more effective.

Why: Either projectile will transfer all of its original momentum to the bowling pin while coming to a stop. But then the bouncy rubber ball will bounce back and continue to exert a force on the bowling pin. The impulse (force · time) delivered by the rubber ball will be greater than that delivered by the beanbag because the ball will exert its forward force for a longer time (during stopping *and* rebounding). The ball will rebound with its momentum reversed, having transferred roughly twice its original momentum to the pin.

3. The satellites are unable to get rid of their angular momentum.

Why: Because of their extreme isolation, orbiting satellites have nothing with which to exchange angular momentum. The angular momentum given to them at launch stays with them indefinitely, so they continue to spin for decades.

4. It came from the entire earth.

Why: Because the person stood on the earth as he started the merry-go-round spinning, he transferred angular momentum from the earth to the merry-go-round. The merry-go-round spins in one direction, and the earth's rotation changes ever so slightly in the other direction. Because the earth is so huge and has such an enormous rotational mass, its slight change in rotation is undetectable.

5. Your car transferred momentum but no energy.

Why: To transfer momentum to the wall, your car must give it an impulse: it must push on the wall for an amount of time. It did so and thus transferred all its backward momentum to the wall. But to transfer energy to the wall, your car must do work on it: it must push on the wall as the wall moves in the direction of that push. But the wall is immobile, so your car couldn't do work on it. Instead, the car's energy stayed in the car, where it caused damage.

6. The child accelerates forward, in the direction that will reduce the child's potential energy as quickly as possible.

Why: The child has only one form of potential energy—gravitational potential energy. This gravitational potential energy is lowest when the child is directly below the swing's supporting bar. The child accelerates forward because that will put the child below the support as quickly as possible.

Check Your Figures—Answers

Section 2.1 SEESAWS

1. About 10 times as much torque.

Why: To keep the angular acceleration in Eq. 2.1.1 unchanged while increasing the rotational mass by a factor of 10, the torque must also increase by a factor of 10. Solid tires are extremely difficult to spin or to stop from spinning, which is why automobiles use hollow tires.

2. Five times as much torque as before.

Why: The pipe increases the wrench's lever arm by a factor of 5, from 0.2 meter to 1.0 meter. According to Eq. 2.1.3, the same force exerted five times as far from the pivot will produce five times as much torque about that pivot. Extending the handle of a lever-like tool is a common technique to increase the available torque, although it can be hazardous for both the tool and its user. Some tools that are designed for such extreme use come with removable handle extensions.

Section 2.2 WHEELS

1. One hundred times as much kinetic energy.

Why: Because kinetic energy is proportional to speed squared, the kilogram of air in the hurricane moves 10 times as fast but has 100 times as much kinetic energy as the slower moving air. This enormous increase in energy is what makes a hurricane's wind dangerous. The air's terrific speed also brings large quantities of it to you quickly, so that the wind power arriving each second is overwhelming.

2. The children will triple the merry-go-round's kinetic energy.

Why: The kinetic energy of a spinning object is proportional to its rotational mass. Since the children triple the merry-go-round's rotational mass, they triple its kinetic energy.

Section 2.3 BUMPER CARS

1. 440,000 kg·m/s, in the forward direction.
Why: You can use Eq. 2.3.1 to calculate the train's momentum from its mass and velocity:

$$\text{linear momentum} = 20,000 \text{ kg} \cdot 22 \text{ m/s}$$
$$= 440,000 \text{ kg} \cdot \text{m/s}.$$

That momentum is in the same direction the train is moving, the forward direction.

2. 2200 s, so kiss your boot heels goodbye!
Why: To stop the train, you must give it a backward impulse that completely cancels its forward momentum. Since its forward momentum is 440,000 kg·m/s, the backward impulse must be 440,000 kg·m/s. Since 200 N can also be

written as 200 kg·m/s², we can use Eq. 2.3.2 to find the time:

$$\text{time} = \frac{440,000 \text{ kg} \cdot \text{m/s}}{200 \text{ kg} \cdot \text{m/s}^2}$$
$$= 2200 \text{ s}.$$

3. The angular momentum will increase by a factor of 5.
Why: Because the satellite's angular momentum is proportional to its angular velocity, spinning it five times faster will increase its angular momentum by that same factor.

4. It will take them five times as long.
Why: To reach the new, faster angular velocity, the astronauts will need an angular impulse that's five times as large as originally planned. Since they will be using the same torque, they will have to exert that torque for five times as long.

Exercises

1. The chairs in an auditorium aren't all facing the same direction. How could you describe their angular positions in terms of a reference orientation and a rotation?

2. When an airplane starts its propellers, they spin slowly at first and gradually pick up speed. Why does it take so long for them to reach their full rotational speed?

3. A mechanic balances the wheels of your car to make sure that their centers of mass are located exactly at their geometrical centers. Neglecting friction and air resistance, how would an improperly balanced wheel behave if it were rotating all by itself?

4. An object's center of mass isn't always inside the object, as you can see by spinning it. Where is the center of mass of a boomerang or a horseshoe?

5. Why is it hard to start the wheel of a roulette table spinning, and what keeps it spinning once it's started?

6. Why can't you open a door by pushing its doorknob directly toward or away from its hinges?

7. Why can't you open a door by pushing on its hinged side?

8. It's much easier to carry a weight in your hand when your arm is at your side than it is when your arm is pointing straight out in front of you. Use the concept of torque to explain this effect.

9. A gristmill is powered by falling water, which pours into buckets on the outer edge of a giant wheel. The weight of the water turns the wheel. Why is it important that those buckets be on the wheel's outer edge?

10. How does the string of a yo-yo get the yo-yo spinning?

11. One way to crack open a walnut is to put it in the hinged side of a door and then begin to close the door. Why does a small force on the door produce a large force on the shell?

12. A common pair of pliers has a place for cutting wires, bolts, or nails. Why is it so important that this cutter be located very near the pliers' pivot?

13. You can do push-ups with either your toes or your knees acting as the pivot about which your body rotates. When you pivot about your knees, your feet actually help you to lift your head and chest. Explain.

14. Tightrope walkers often use long poles for balance. Although the poles don't weigh much, they can exert substantial torques on the walkers to keep them from tipping and falling off the ropes. Why are the poles so long?

15. Some racing cars are designed so that their massive engines are near their geometrical centers. Why does this design make it easier for these cars to turn quickly?

16. How does a bottle opener use mechanical advantage to pry the top off a soda bottle?

by pushing far from pivot you exert more [handwritten]

17. A jar-opening tool grabs onto a jar's lid and then provides a long handle for you to turn. Why does this handle's length help you to open the jar?

18. When you climb out on a thin tree limb, there's a chance that the limb will break off near the trunk. Why is this disaster most likely to occur when you're as far out on the limb as possible?

19. How does a crowbar make it easier to lift the edge of a heavy box a few centimeters off the ground?

20. The basket of a wheelbarrow is located in between its wheel and its handles. How does this arrangement make it relatively easy for you to lift a heavy load in the basket?

21. Skiers often stop by turning their skis sideways and skidding them across the snow. How does this trick remove energy from a skier, and what happens to that energy?

friction [handwritten]

22. A horse does work on a cart it's pulling along a straight, level road at a constant speed. The horse is transferring energy to the cart, so why doesn't the cart go faster and faster? Where is the energy going?

23. Explain why a rolling pin flattens a piecrust without encountering very much sliding friction as it moves.

24. Professional sprinters wear spikes on their shoes to prevent them from sliding on the track at the start of a race. Why is energy wasted whenever a sprinter's foot slides backward along the track?

25. A yo-yo is a spool-shaped toy that spins on a string. In a sophisticated yo-yo, the end of the string forms a loop around the yo-yo's central rod so that the yo-yo can spin almost freely at the end of the string. Why does the yo-yo spin longest if the central rod is very thin and very slippery?

26. As you begin pedaling your bicycle and it accelerates forward, what is exerting the forward force that the bicycle needs to accelerate?

27. When you begin to walk forward, what exerts the force that allows you to accelerate?

Static frictional force from pavement [handwritten]

28. If you are pulling a sled along a level field at constant velocity, how does the force you are exerting on the sled compare to the force of sliding friction on its runners?

29. Why does putting sand in the trunk of a car help to keep the rear wheels from skidding on an icy road?

30. When you're driving on a level road and there's ice on the pavement, you hardly notice that ice while you're heading straight at a constant speed. Why is it that you only notice how slippery the road is when you try to turn left or right, or to speed up or slow down?

31. Describe the process of writing with chalk on a blackboard in terms of friction and wear.

32. Falling into a leaf pile is much more comfortable than falling onto the bare ground. In both cases you come to a complete stop, so why does the leaf pile feel so much better?

33. In countless movie and television scenes, the hero punches a brawny villain who doesn't even flinch at the impact. Why is the immovable villain a Hollywood fantasy?

34. Why can't an acrobat stop himself from spinning while he is in midair?

35. While a gymnast is in the air during a leap, which of the following quantities must remain constant for her: velocity, momentum, angular velocity, or angular momentum?

36. If you sit in a good swivel chair with your feet off the floor, the chair will turn slightly as you move about but will immediately stop moving when you do. Why can't you make the chair spin without touching something?

37. When a star runs out of nuclear fuel, gravity may crush it into a neutron star about 20 km (12 miles) in diameter. While the star may have taken a year or so to rotate once before its collapse, the neutron star rotates several times a second. Explain this dramatic increase in angular velocity.

38. A toy top spins for a very long time on its sharp point. Why does it take so long for friction to slow the top's rotation?

39. It's easier to injure your knees and legs while hiking downhill than while hiking uphill. Use the concept of energy to explain this observation.

40. When you first let go of a bowling ball, it's not rotating. But as it slides down the alley, it begins to rotate. Use the concept of energy to explain why the ball's forward speed decreases as it begins to spin.

41. Firefighters slide down a pole to get to their trucks quickly. What happens to their gravitational potential energy, and how does it depend on the slipperiness of the pole?

Problems

1. When you ride a bicycle, your foot pushes down on a pedal that's 17.5 cm (0.175 m) from the axis of rotation. Your force produces a torque on the crank attached to the pedal. Suppose that you weigh 700 N. If you put all your weight on the pedal while it's directly in front of the crank's axis of rotation, what torque do you exert on the crank?

2. An antique carousel that's powered by a large electric motor undergoes constant angular acceleration from rest to full rotational speed in 5 seconds. When the ride ends, a brake causes it to decelerate steadily from full rotational speed to rest in 10 seconds. Compare the torque that starts the carousel to the torque that stops it.

3. When you start your computer, the hard disk begins to spin. It takes 6 seconds of constant angular acceleration to reach full speed, at which time the computer can begin to access it. If you wanted the disk drive to reach full speed in only 2 seconds, how much more torque would the disk drive's motor have to exert on it during the starting process?

4. An electric saw uses a circular spinning blade to slice through wood. When you start the saw, the motor needs 2 seconds of constant angular acceleration to bring the blade to its full angular velocity. If you change the blade so that the rotating portion of the saw now has three times its original rotational mass, how long will the motor need to bring the blade to its full angular velocity?

5. When the saw in Problem 4 slices wood, the wood exerts a 100-N force on the blade, 0.125 m from the blade's axis of rotation. If that force is at right angles to the lever arm, how much torque does the wood exert on the blade? Does this torque make the blade turn faster or slower?

6. When you push down on the handle of a doll-like wooden nutcracker, its jaw pivots upward and cracks a nut. If the point at which you push down on the handle is five times as far from the pivot as the point at which the jaw pushes on the nut, how

much force will the jaw exert on the nut if you exert a force of 20 N on the handle? (Assume all forces are at right angles to the lever arms involved.)

7. Some special vehicles have spinning disks (flywheels) to store energy while they roll downhill. They use that stored energy to lift themselves uphill later on. Their flywheels have relatively small rotational masses but spin at enormous angular speeds. How would a flywheel's kinetic energy change if its rotational mass were five times larger but its angular speed were five times smaller?

8. What is the momentum of a fly if it's traveling 1 m/s and has a mass of 0.0001 kg?

9. Your car is broken, so you're pushing it. If your car has a mass of 800 kg, how much momentum does it have when it's moving forward at 3 m/s (11 km/h)?

10. You begin pushing the car forward from rest (see Problem 9). Neglecting friction, how long will it take you to push your car up to a speed of 3 m/s on a level surface if you exert a constant force of 200 N on it?

11. When your car is moving at 3 m/s (see Problems 9 and 10), how much translational kinetic energy does it have?

12. No one is driving your car (see Problems 9, 10, and 11) and it crashes into a parked car at 3 m/s. Your car comes to a stop in just 0.1 s. What force did the parked car exert on it to stop it that quickly?

13. You're at the roller-skating rink with a friend who weighs twice as much as you do. The two of you stand motionless in the middle of the rink so that your combined momentum is zero. You then push on one another and begin to roll apart. If your momentum is then 450 kg·m/s to the left, what is your friend's momentum?

Mechanical Objects Part 1

Now that we've surveyed the laws of motion, we can begin using those laws to explain the behaviors of everyday objects. But while we can already understand some of the central features at work in a toy wagon, a weight machine, or a ski lift, we're still missing a number of mechanical concepts that are important in the world around us. In this chapter, we'll look at some of those additional concepts.

One of the most important will be acceleration. If we treat acceleration passively, it can be fairly uninteresting: we push on the cart and the cart accelerates. But if we think of the concept more actively—for example, if we envision ourselves on a roller coaster as it plummets down that first big hill—then acceleration becomes much more intriguing. In fact, we might even need to hold on to our hats.

EXPERIMENT: Swinging Water Overhead

To examine some of the novel effects of acceleration, try experimenting with a bucket of water. If you're careful, you can swing this bucket over your head and upside

Courtesy Lou Bloomfield

down without spilling a drop. In the process, you'll be demonstrating a number of important physical concepts.

To do this experiment, you'll need a bucket with a handle. (You might substitute some equivalent container; even a plastic cup will do in a pinch.) Fill the bucket partway full of water and then hold it by the handle so that it hangs down by your side.

Now swing the bucket backward about an eighth of a turn and bring it forward rapidly. In one smooth, fluid motion, swing it forward, up, and over your head. Continue this motion all the way around behind you and then bring the bucket forward again. You'll need to swing the bucket quickly to avoid getting wet. As you swing the bucket around and around, you'll notice that the water stays in it even when it travels over your head and is upside-down. Why doesn't the water fall out?

You can carry this experiment a step further by swinging the bucket at various speeds—that is, if you don't mind getting wet. First, try to *predict* what will happen if you swing the bucket less rapidly or more rapidly. Now do the experiments and *observe* what happens. Did the experiments *verify* your predictions?

As you swing the bucket over your head, try to *measure* how strongly the bucket pulls on your hand. Is the pull stronger or weaker when you swing less rapidly? more rapidly? Is there any relationship between the upward pull you feel from the inverted bucket and the water's tendency to remain inside?

You might vary this experiment in several ways. Try swinging a plastic cup held in your fingers, or try placing a full wine glass in the bucket and swinging the two objects together. In the latter case, you'll find that the wine will stay in the glass and the glass will stay at the bottom of the bucket, even when the bucket is upside down.

By the way, the hardest part of all these tricks is stopping. To avoid a catastrophe, you'll need to do the same thing you did to get started, only in reverse. Come to a smooth, gradual stop about an eighth of a turn in front of you, and then let the bucket loosely drop back to your side. If you stop moving the bucket too abruptly, the water, wine, or glass will spill or smash. Why do you suppose that happens?

Chapter Itinerary

In this chapter, we'll examine three types of everyday objects: (1) *spring scales,* (2) *bouncing balls,* and (3) *carousels and roller coasters.* In *spring scales,* we'll review the relationship between mass and weight and explore how the distortion of a spring can be used to measure an object's weight. In *bouncing balls,* we'll study how balls store and return energy and how their bouncing depends both on their own characteristics and on those of the objects they hit. And in *carousels and roller coasters,* we'll look at how acceleration gives rise to gravity-like apparent forces that can make us scream with delight at the amusement park. For a more complete preview of what we'll examine in this chapter, skip ahead to the Chapter Summary on p. 106.

The concepts these objects illustrate can explain other phenomena as well. Almost any solid object, from a mattress to a diving board to a tire, behaves like a spring scale's spring when you push on it. Bouncing balls offer a view of collisions that will help you comprehend what happens when two cars crash or when a hammer hits a nail. And the sensations associated with acceleration that you experience on a roller coaster are also present when you ride in airplanes, on subways, or on swing sets. When it comes to the physics of everyday objects, there really is nothing new under the sun.

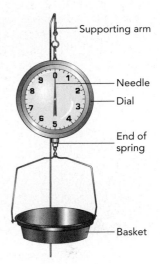

Supporting arm

Needle

Dial

End of
spring

Basket

SECTION 3.1 Spring Scales

*How much of you is there? From day to day, depending on how much you eat, the amount
of you stays approximately the same. But how can you tell how much that is? The best
measure of quantity is mass: kilograms of gold, beef, grain, or you. Mass is the measure of an object's inertia and, as we saw in Section 1.1, doesn't depend on the object's
environment or on gravity. A kilogram box of cookies always has a mass of 1 kilogram,
no matter where in the universe you take it.*

*But mass is difficult to measure directly. Moreover, the very concept of mass is only
about 300 years old. Consequently, people began quantifying the material in an object
by measuring its weight. Spring scales eventually became one of the simplest and most
practical tools for accomplishing this task, and they are still found in bathrooms and
grocery stores today. They really do contain springs, although these are normally hidden from view.*

*Questions to Think About: How is your weight related to your mass? If the earth's grav-
ity became twice as strong, how would your mass be affected? What about your weight?
Does jumping up and down change either your mass or your weight? If you stand on a
strong spring, how does your weight affect the shape of the spring? Why should there
be a relationship between your weight and how much the spring bends?*

*Experiments to Do: Find a hanging spring scale of the type used in the produce section
of a grocery store and watch the scale's basket and weight indicator as you put objects
in the basket. What happens to the basket as you fill it up? Can you find a relationship
between the basket's height and the weight reported by the scale? If you drop something
into the basket, instead of lowering it gently, how does the scale respond? Why does the
weight indicator bounce back and forth rhythmically? What happens to the gravitational
potential energy of the dropped item?*

*Now find a spring bathroom scale—the short, flat kind with a rotating dial. Stand
on it. Why does it read your correct weight only when you are standing still? If you jump
upward, how does the scale's reading change? What about if you let yourself drop?
Bounce up and down gently. How does the scale's average reading compare with your
normal weight? Does bouncing really change your weight? You can also change the
scale's reading by pushing on the floor, wall, or other nearby objects. Which way must
you push to increase the scale's reading? to decrease it? When you change the reading
in these ways, are you actually changing your weight?*

Why You Must Stand Still on a Scale

Whenever you stand on a scale in your bathroom or place a melon on a scale at the grocery store, you are measuring weight. An object's weight is the force exerted on it by gravity, usually the earth's gravity. When you stand on a bathroom scale, the scale measures
just how much upward force it must exert on you in order to keep you from moving downward toward the earth's center. As in most scales you'll encounter, the bathroom scale
uses a spring to provide this upward support. If you're stationary, you're not accelerating,
so your downward weight and the upward force from the spring must cancel one another;
that is, they must be equal in magnitude but opposite in direction so that they sum to zero

net force. Consequently, although the scale actually displays how much upward force it's exerting on you, that amount is also an accurate measure of your weight.

This subtle difference between your actual weight and what the scale is reporting is important. While an object's weight depends only on its gravitational environment, not on its motion, the weighing process itself is extremely sensitive to motion. If anything accelerates during the weighing process, the scale may not report the object's true weight. For example, if you jump up and down while you're standing on a scale, the scale's reading will vary wildly. You're accelerating, so your downward weight and the upward force from the scale no longer cancel. If you want an accurate measurement of your weight, therefore, you have to stand still.

But even if you stand still, weighing is not a perfect way to quantify the amount of material in your body. That's because your weight depends on your environment. If you always weigh yourself in the same place, the readings will be pretty consistent, as long as you don't routinely eat a dozen jelly doughnuts for lunch. But if you moved to the moon, where gravity is weaker, you'd weigh only about one-sixth as much as on the earth. Even a move elsewhere on the earth will affect your weight: the earth bulges outward slightly at the equator, and gravity there is about 0.5% weaker than at the poles. That change, together with a small acceleration effect due to the earth's rotation, means that a scale will read 1.0% less when you move from the north pole to the equator. Obviously, moving south is not a useful weight-loss plan.

> ### Check Your Understanding #1: Space Merchants
> for answers, see page 107
>
> You're opening a company that will export gourmet food from the earth to the moon. You want the package labels to be accurate at either location. How should you label the amount of food in each package—by mass or by weight?

Stretching a Spring

So you know now that when you put a melon in the basket of a scale at the grocery store and read its weight from the scale's dial, the scale is actually reporting just how much upward force its spring is exerting on the melon. While your shopping cart could support the melon equally well, there's no simple way to determine just how much upward force the cart exerts on the melon. Therein lies the beauty, and the utility, of a spring: a simple relationship exists between its length and the forces it's exerting on its ends. The spring scale can therefore determine how much force it's exerting on the melon by measuring the length of its spring.

The springs shown in Fig. 3.1.1 consist of a wire coil that pulls inward on its ends when it's stretched and pushes outward on them when it's compressed. If a coil spring is neither stretched nor compressed, it exerts no forces on its ends.

The top spring (Fig. 3.1.1a) is neither stretched nor compressed, so that when it lies on a table to keep it from falling, its ends remain motionless. Those ends are in **equilibrium**—experiencing zero net force. As the phrase "zero net force" suggests, equilibrium occurs whenever the forces acting on an object sum perfectly to zero so that the object doesn't accelerate. When you sit still in a chair, for example, you are in equilibrium.

The spring in Fig. 3.1.1a is also at its equilibrium length, its natural length when you leave it alone. No matter how you distort this spring, it tries to return to this equilibrium length. If you stretch it so that it's longer than its equilibrium length, it will pull inward on its end. If you compress it so that it's shorter than its equilibrium length, it will push outward on its ends.

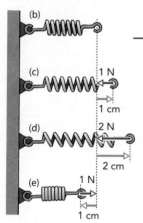

Fig. 3.1.1 Five identical springs. The ends of spring (*a*) are free so that it can adopt its equilibrium length. The left ends of the other springs are fixed so that only their right ends can move. When the free end of a spring (*b*) is moved away from its original equilibrium position (*c, d,* and *e*), the spring exerts a restoring force on that end that is proportional to the distance between its new position and the original equilibrium position.

Let's attach the left end of our spring to a post (Fig. 3.1.1*b*) and look at the behavior of its free right end. With nothing pushing or pulling on the spring, this free end will be in equilibrium at a particular location—its **equilibrium position.** Since the spring's end naturally returns to this equilibrium position if we stretch or compress it and then let go, the end is in a **stable equilibrium.**

But what happens if we pull the free end to the right and don't let go? The spring now exerts a steady inward force on that end, trying to return it to its original equilibrium position. The more we stretch the spring, the more inward force it exerts on the end. This inward force is exactly proportional to how far we stretch the end away from its original equilibrium position. Since the end of the spring in Fig. 3.1.1*c* has been pulled 1 cm to the right of its original equilibrium position, the spring now pulls this end to the left with a force of 1 N; if the end is instead pulled 2 cm to the right, as it has been in Fig. 3.1.1*d*, the spring pulls it to the left with a force of 2 N. This proportionality continues to work even when we compress the spring: in Fig. 3.1.1*e*, the end has been pushed 1 cm to the left, and the spring is pushing it to the right with a force of 1 N.

The force exerted by a coil spring thus has two interesting properties. First, this force is always directed so as to return the spring to its equilibrium length. We call this kind of force a **restoring force** because it acts to restore the spring to equilibrium. Second, the spring's restoring force is proportional to how far it has been distorted (stretched or compressed) from its equilibrium length.

These two observations are expressed in **Hooke's law,** named after the Englishman Robert Hooke, who first demonstrated it in the late seventeenth century. This law can be written in a word equation:

$$\text{restoring force} = -\text{spring constant} \cdot \text{distortion}, \tag{3.1.1}$$

in symbols:

$$\mathbf{F} = -k \cdot \mathbf{x},$$

and in everyday language:

The farther you squeeze a roll of paper towels, the harder it pushes back.

Here the **spring constant,** k, is a measure of the spring's stiffness. The larger the spring constant—that is, the firmer the spring—the larger the restoring force the spring exerts for a given distortion. The negative signs in these equations indicate that a restoring force always points in the direction opposite the distortion.

> **Hooke's Law**
> The restoring force exerted by an elastic object is proportional to how far it has been distorted from its equilibrium shape.

Springs are distinguished by their stiffness, as measured by their spring constants. Some springs are **soft** and have small spring constants—for example, the one that pops the toast out of your toaster, which you can easily compress with your hand. Others, like the large springs that suspend an automobile chassis above the wheels, are **firm** and have large spring constants. But no matter the stiffness, all springs obey Hooke's law.

Hooke's law is remarkably general and isn't limited to the behavior of coil springs. Almost anything you distort will pull or push back with a force that's proportional to how far you've distorted it away from its equilibrium length—or, in the case of a

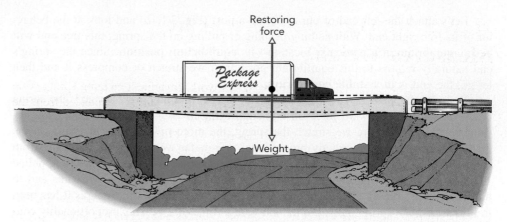

Fig. 3.1.2 A steel bridge sags under the weight of a truck. The bridge bends downward until the upward restoring force it exerts on the truck exactly balances the truck's weight.

complicated object, its equilibrium shape. Equilibrium shape is the shape an object adopts when it's not subject to any outside forces. If you bend a tree branch, it will push back with a force proportional to how far it has been bent. If you pull on a rubber band, it will pull back with a force proportional to how far it has been stretched, up to a point. If you squeeze a ball, it will push outward with a force proportional to how far it has been compressed. If a heavy truck bends a bridge downward, the bridge will push upward with a force proportional to how far it has been bent (Fig. 3.1.2).

There is a limit to Hooke's law, however. If you distort an object too far, it will usually begin to exert less force than Hooke's law demands. This is because you will have exceeded the **elastic limit** of the object and will probably have permanently deformed it in the process. If you pull on a spring too hard, you'll stretch it forever; if you push on a branch too hard, you'll break it. But as long as you stay within the elastic limit, almost everything obeys Hooke's law—a rope, a ruler, an orange, and a trampoline.

Distorting a spring requires work. When you stretch a spring with your hand, pulling its end outward, you transfer some of your energy to the spring. The spring stores this energy as **elastic potential energy.** If you reverse the motion, the spring returns most of this energy to your hand, while a small amount is converted to thermal energy by frictional effects inside the spring itself. Work is also required to compress, bend, or twist a spring. In short, a spring that is distorted away from its equilibrium shape always contains elastic potential energy.

➤ Check Your Understanding #2: Going Down Anyone?

for answers, see page 107

As you watch people walk off the diving board at a pool, you notice that it bends downward by an amount proportional to each diver's weight. Explain.

➤ Check Your Figures #1: A Sinking Sensation

for answers, see page 108

You're hosting a party in your third-floor apartment. When the first 10 guests begin standing in your living room, you notice that the floor has sagged 1 centimeter in the middle. How far will the floor sag when 20 guests are standing on it? when 100 guests are standing on it?

How a Hanging Grocery Scale Measures Weight

We're now ready to understand how spring scales work. Imagine a hanging spring scale of the kind used to weigh produce. Inside this scale is a coil spring, suspended from the ceiling by its upper end (Fig. 3.1.3). Hanging from its lower end is a basket. For the sake of simplicity, imagine that this basket has little or no weight. With no force pulling down on it, the scale's spring adopts its equilibrium length, and the basket, experiencing zero net force, is in a position of stable equilibrium. If you shift the basket up or down and then let go of it, the spring will push it back to this position.

When you place a melon in the basket, the melon's weight pushes it downward. The basket starts descending and as it does, the spring stretches and begins to exert an upward force on the basket. The more the spring stretches, the greater this upward force so that eventually the spring is stretched just enough so that its upward force exactly supports the melon's weight. The basket is now in a new stable equilibrium position—again experiencing zero net force.

But how does the scale determine the melon's weight? It uses Hooke's law. Once the basket has adopted its new equilibrium position, where the melon's weight is exactly balanced by the upward force of the spring, the amount the spring has stretched is an accurate measure of the melon's weight.

The scales in Fig. 3.1.3 differ only in the way they measure how far the spring has stretched beyond its equilibrium length. The top scale uses a pointer attached to the end of the spring, while the bottom scale uses a rack and pinion gear system that converts the small linear motion of the stretching spring into a much more visible rotary motion of the dial needle. The rack is the series of evenly spaced teeth attached to the lower end of the spring; the pinion is the toothed wheel attached to the dial needle. As the spring stretches, the rack moves downward, and its teeth cause the pinion to rotate. The farther the rack moves, the more the pinion turns, and the higher the weight reported by the needle.

Each of these spring scales reports a number for the weight of the melon you put in the basket. In order for that number to mean something, the scale has to be calibrated. **Calibration** is the process of comparing a local device or reference to a generally accepted standard to ensure accuracy. To calibrate a spring scale, the device or its reference components must be compared against standard weights. Someone must put a standard weight in the basket and measure just how far the spring stretches. Each spring is different, although spring manufacturers try to make all their springs as identical as possible.

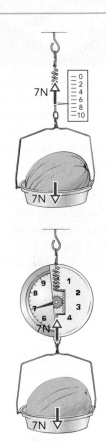

Fig. 3.1.3 Two spring scales weighing melons. Each scale balances the melon's downward weight with the upward force of a spring. The heavier the melon, the more the spring will stretch before it exerts enough upward force to balance the melon's weight. The top scale has a pointer to indicate how far the spring has stretched and thus how much the melon weighs. The bottom scale has a rack and pinion gear that turns a needle on a dial. As the comblike rack moves up and down, it turns the toothed pinion gear.

► Check Your Understanding #3: Scaling Down

for answers, see page 107

If you pull the basket of a hanging grocery store scale downward 1 centimeter, it reports a weight of 5 N (about 1.1 lbf) for the contents of its basket. If you pull the basket downward 3 centimeters, what weight will it report?

Bouncing Bathroom Scales

As we noted earlier, the most common type of bathroom scale is also a spring scale (Fig. 3.1.4). When you step on this kind of scale, you depress its surface and levers inside it pull on a hidden spring. That spring stretches until it exerts, through the levers, an upward force on you that is equal to your weight. At the same time, a rack and pinion mechanism inside

Courtesy Lou Bloomfield

Fig. 3.1.4 When you step on this bathroom scale, its surface moves downward slightly and compresses a stiff spring. The extent of this compression is proportional to your weight, which is reported by the dial on the left. Levers inside the scale make it insensitive to exactly where you stand.

the scale turns a wheel with numbers printed on it. When the wheel stops moving, you can read one of these numbers through a window in the scale. Because which number you see depends on how far the spring has stretched, this number indicates your weight.

However, the wheel rocks briefly back and forth around your actual weight before it settles down. The wheel moves because you're bouncing up and down as the scale gradually gets rid of excess energy. When you first step on the scale's surface, its spring is not stretched and it isn't pushing up on you at all. You begin to fall. As you descend, the spring stretches and the scale begins to push up on your feet. But by the time you reach the equilibrium height, where the scale is exactly supporting your weight, you are traveling downward quickly and coast right past that equilibrium. The scale begins to read more than your weight.

The scale now accelerates you upward. Your descent slows, and you soon begin to rise back toward equilibrium. Again you coast past the proper height, but now the scale begins to read less than your weight. You are bouncing up and down because you have excess energy that is shifting back and forth between gravitational potential energy, kinetic energy, and elastic potential energy. This bouncing continues until sliding friction in the scale has converted it all into thermal energy. Only then does the bouncing stop and the scale read your correct weight.

The bouncing that you experience about this stable equilibrium is a remarkable motion, one that we'll study in detail when we examine clocks in Chapter 9. You are effectively a mass supported by a spring and your rhythmic rise and fall is that of a *harmonic oscillator*. Harmonic oscillators are so common and important in nature that Chapter 9 is devoted to them. The details can wait, but there are two features of your present situation that we'll examine now.

First, your total potential energy is at its minimum when you're at the stable equilibrium. Even though both gravitational and elastic potentials are involved, their sum increases as you shift away from the equilibrium. Since an object always accelerates so as to reduce its total potential energy as quickly as possible, you always accelerate toward the stable equilibrium.

Second, your kinetic energy reaches a peak as you pass through the stable equilibrium. Having accelerated toward that equilibrium until the moment of arrival, you're moving fast and coast right through it. But once you leave it, you begin to accelerate toward it again. That acceleration is backward, opposite your velocity, so you are decelerating. Therefore, you reached your peak speed and kinetic energy at the moment you passed through equilibrium. As you bounce up and down, waiting for the scale to waste your excess energy, that excess transforms back and forth rhythmically between kinetic and potential forms.

Common Misconceptions: Equilibrium and Motionlessness

Misconception: An object at equilibrium is motionless.

Resolution: An object at equilibrium is not accelerating, but its velocity may not be zero. If it was moving when it reached equilibrium it will coast at constant velocity.

▶ Check Your Understanding #4: Weighed Down

for answers, see page 107

When you step on the surface of a spring bathroom scale, you can feel it move downward slightly. How is the distance that the scale's surface moves downward when you step on it related to the weight it reports?

Using Several Scales at Once

One scale is enough for you, but how can you weigh an upright piano? It's too heavy and awkward for a single scale, but two scales will do the trick. If you put one scale under each side of the piano, the scales will work together to support its weight. Each scale will report just how much upward force it's exerting on the piano, so the sum of the two measurements will equal the piano's overall weight (Fig. 3.1.5).

The specific readings of the two scales will depend on the position of the piano's center of gravity. Its **center of gravity** is the effective location of its weight and coincides with its center of mass. Because the piano's longest and heaviest strings are on its left side, the piano's center of gravity is to the left of its middle. As a result, the left scale must support more of the piano's weight and it reads higher than the right scale.

We can explain the different readings by considering rotational motion. Like the see-saw in Section 2.1, the piano can rotate about its center of mass and will undergo angular acceleration in response to a net torque. To avoid angular acceleration so that it can rest motionless on the scales, the piano must be in **rotational equilibrium,** that is, it must experience zero net torque.

Because the piano's weight effectively acts at the piano's center of gravity, it has no lever arm and exerts no torque on the piano about its center of mass. However, the scales do exert torques on the piano about its center of mass. The left scale pushes up on the piano's left side, thereby producing a clockwise torque on the piano. From Eq. 2.1.3, the amount of this torque is the product of the left horizontal lever arm times the left upward force. The right scale similarly produces a counterclockwise torque on the piano and the amount of that torque is the product of the right horizontal lever arm times the right upward force.

For the piano to be in rotational equilibrium, these two torques must cancel; they must be equal in amount, but opposite in direction. Their amounts will be equal when:

$$\text{left lever arm} \cdot \text{left force} = \text{right lever arm} \cdot \text{right force}.$$

Since the left lever arm is shorter than the right lever arm, the left force must be proportionately larger than the right force. That's why the left scale reads higher than the right scale.

This effect, where the scale that is more nearly beneath an object's center of gravity must support more of the object's weight, is familiar to anyone who has moved heavy objects. If you and a friend try to carry the piano in Fig. 3.1.5, the person carrying the piano's left side will bear more of the burden. And if an object is tipped so that its center of gravity is almost directly above one of the movers, as it is in Fig. 1.3.1, that mover will support almost the entire weight of the object.

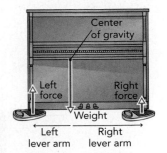

Fig. 3.1.5 You can weigh an upright piano by placing a spring scale under each end. Each scale exerts an upward force on the piano to support it, and the piano's weight is equal to the sum of those two forces, as measured by the scales.

Check Your Understanding #5: No Need to Unpack

for answers, see page 107

Somewhere in your enormous suitcase are several extremely heavy books. How can you locate those books among the much lighter clothes without unpacking the entire suitcase?

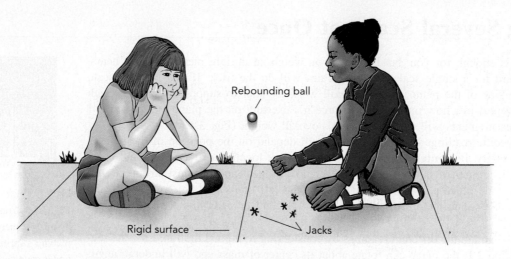

Rebounding ball

Rigid surface ———   Jacks

SECTION 3.2 **Bouncing Balls**

If you visit a toy or sporting goods store, you'll find many different balls—almost a unique ball for every sport or ball game. These balls differ in more than just size and weight. Some are very hard, others very soft; some are smooth, others rough or ridged.

In this section we'll focus primarily on another difference: the ability to bounce. A superball, for example, bounces extraordinarily well, while a foam rubber ball hardly bounces at all. Even balls that appear identical can be very different; a new tennis ball bounces much better than an old one. We'll begin this section by exploring these differences.

Questions to Think About: Is it possible for a ball to bounce higher than the height from which it was dropped? Where does a ball's kinetic energy go as it bounces, and what happens to the energy that doesn't reappear after the bounce? What happens when a ball bounces off a moving object, such as a baseball bat? What role does the baseball bat's structure have in the bouncing process? Does it matter which part of a baseball bat hits the ball?

Experiments to Do: Drop a ball on a hard surface and watch it bounce. What happens to the ball's shape during the bounce? Hold the ball in your hands and push its surface inward with your fingers. What is the relationship between the force it exerts on your fingers and how far inward you dent it? Denting the ball takes work. Why? How does the ball's energy change as it dents? What happens to the ball's energy as its shape returns to normal (to equilibrium)?

Drop the ball from various heights and see if you can find a simple relationship between the ball's initial height and the height to which it rebounds. Now drop the ball on a soft surface, such as a pillow, or on a lively surface, such as an inflated balloon. Why does the surface it hits change the way the ball bounces?

The Way the Ball Bounces: Balls as Springs

In many ways, balls are perfect objects. What would most sports be like without them? How would industrial machines function without ball bearings to keep them from grinding to a halt? Their simple shapes, uncomplicated motions, and ability to bounce make balls both fascinating and useful. Most balls are spherical, meaning that when no outside forces act on them they adopt spherical equilibrium shapes. But some balls, such as those used in U.S. football and rugby, have equilibrium shapes that are not spherical but oblong.

The term "equilibrium shape," of course, is one we've seen already: the previous section used it to describe springs. This is no coincidence, for a spherical ball behaves like a spherical spring. In fact, everything we associate with springs has some place in the behavior of balls. For example, when you push a ball's surface inward, it exerts an outward restoring force on you. When you do work on the ball as you distort its surface, it stores some of this work as elastic potential energy and when you let the ball return to its equilibrium shape, it releases that stored energy.

This springlike behavior is evident when a ball collides with the floor or with a bat. The ball's surface distorts during the collision, giving it elastic potential energy that is released when the ball rebounds. If the ball is moving, then some of this stored energy comes from the ball's kinetic energy. If the ball hits a moving surface, then some of this stored energy comes from the surface's kinetic energy. Much of the stored energy reappears as kinetic energy in the ball and surface as the ball rebounds.

Some balls bounce better than others. We frequently call a very bouncy ball "lively" and a ball that doesn't bounce well "dead." One way to look at a ball's liveliness is to compare kinetic energies before and after the bounce. We can do that by dividing the bounce into two halves: the collision and the rebound (Fig. 3.2.1). During the collision, the ball and surface convert some of their overall kinetic energy into elastic potential energy and thermal energy. The amount of kinetic energy transformed at impact is called the **collision energy.** A lively ball does a good job of converting the collision energy into elastic potential energy, while a dead ball converts most of it into thermal energy (Fig. 3.2.2).

During the rebound, the ball and surface push away from one another, converting elastic potential energy back into kinetic energy. The total amount of kinetic energy released as the surface and ball push apart is the **rebound energy.** Collision energy that doesn't reappear as rebound energy has been transformed into thermal energy.

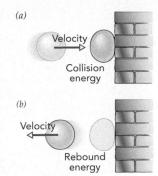

(a) Velocity · Collision energy

(b) Velocity · Rebound energy

Fig. 3.2.1 A bounce from a wall has two halves: (*a*) the collision and (*b*) the rebound. During the collision between the ball and the wall, some of their kinetic energy is transformed into other forms—an amount called the collision energy. During the rebound, some stored energy is released as kinetic energy—an amount called the rebound energy. The rebound energy is always less than the collision energy because some energy is lost as thermal energy. However, a lively ball wastes less energy than a dead one.

Courtesy Lou Bloomfield

Fig. 3.2.2 When a tennis ball hits the floor, it dents inward to store energy and then rebounds somewhat more slowly than it arrived. These images show the ball's position at twelve equally spaced times. Is the ball bouncing to the left or the right? How can you tell?

Table 3.2.1 Approximate Energy Ratios and Coefficients of Restitution for a Variety of Balls

TYPE OF BALL	REBOUND ENERGY COLLISION ENERGY	COEFFICIENT OF RESTITUTION
Superball	0.81	0.90
Racquet ball	0.72	0.85
Golf ball	0.67	0.82
Tennis ball	0.56	0.75
Steel ball bearing	0.42	0.65
Baseball	0.30	0.55
Foam rubber ball	0.09	0.30
"Unhappy" ball	0.01	0.10
Beanbag	0.002	0.04

The ratio of rebound energy to collision energy (Table 3.2.1) determines how high a ball will bounce when you drop it from rest onto a hard floor (Fig. 3.2.3). An ideally elastic ball would have a ratio of 1.00 and would rebound to its initial height. But a real ball wastes some of the collision energy and rebounds to a lesser height. The height from which you drop the ball is proportional to its initial gravitational potential energy and therefore to its collision energy. The height to which it rebounds is similarly proportional to its rebound energy. The ratio of these two heights is therefore a good measure of the ratio of the rebound energy to the collision energy. The smaller this ratio, the less kinetic energy the ball receives during the rebound and the weaker the bounce.

While that energy ratio is often useful, a ball is traditionally characterized by its **coefficient of restitution**—the ratio of its rebound *speed* to its collision *speed* when it bounces off a hard, immobile surface:

$$\text{coefficient} = \frac{\text{outgoing speed of ball}}{\text{incoming speed of ball}} \tag{3.2.1}$$

Scientists have found that, for most balls, this speed ratio remains constant over a wide range of collision speeds. A ball that rebounds with the same speed that it had when it collided with the surface has a coefficient of restitution of 1.00. A superball is almost this lively, with a coefficient of restitution of about 0.90. Thus, when a superball traveling at 10 km/h collides with a concrete wall, it rebounds at about 9 km/h. In contrast, a foam rubber ball's coefficient of restitution is about 0.30, while that of a beanbag is almost zero.

(a)

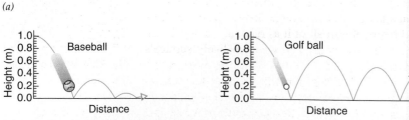

(b)

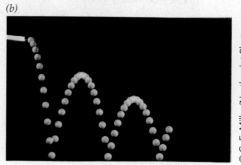

Fig. 3.2.3 (*a*) A baseball wastes 70% of the collision energy as thermal energy and bounces weakly. In contrast, a golf ball wastes only 30% of the collision energy and bounces well. (*b*) The photo at right of a golf ball illustrates this in stop-motion.

Actually, the energy ratio and the coefficient of restitution are directly related. Remember, a ball's kinetic energy equals half its mass times its velocity squared, or $\frac{1}{2}\,mv^2$. Even if we don't know the masses of the balls, we know that the energy ratio is equal to the square of the speed ratio. Thus if a foam rubber ball rebounds at only 0.30 times its collision speed, it retains only 0.30^2—0.09 times, or 9%—of its original kinetic energy, and the remaining 91% is converted to thermal energy in the rubber and air that make up the ball. A superball, in contrast, retains about 81% of its original kinetic energy after a bounce.

Balls bounce best when they store energy through compression rather than through surface bending. That's because most ball materials, such as leather or leather-like plastics, experience lots of wasteful internal friction during bending. Since solid balls involve compression, they usually bounce well, whether they're made of rubber, wood, plastic, or metal. But air-filled balls bounce well only when properly inflated. A normal basketball, which stores most of its energy in its compressed air, has a high coefficient of restitution. In contrast, an underinflated basketball, which experiences lots of surface bending during a collision, barely bounces at all. Similarly, a tennis ball bounces best when new; after a while, the compressed air inside leaks out and the ball's coefficient of restitution drops.

▶ Check Your Understanding #1: A Game of Marbles

for answers, see page **107**

As you head into the park to play a game of marbles with your friends, several of the glass marbles fall through a hole in your marble bag and bounce nicely on the granite walkway. How can a marble bounce?

How the Surface Affects the Bounce

If the surface on which a ball bounces isn't perfectly hard, that surface will contribute to the bouncing process. It will distort and store energy when the ball hits it and will return some of this stored energy to the rebounding ball. Overall, the collision energy is shared between the ball and the surface, both of which behave as springs, and each provides part of the rebound energy.

Just how the collision energy is distributed between the surface and ball depends on how stiff each one is. During the bounce, they push on one another with equal but oppositely directed forces. Since the forces denting them inward are equal, the work done in distorting each object is proportional to how far inward it dents. Whichever object dents the farthest receives the most collision energy.

Since the ball usually distorts more than the surface it hits, most of the collision energy normally goes into the ball. As a result, you might expect the ball to provide most of the rebound energy, too. However, that's not always true. Some lively elastic surfaces store collision energy very efficiently and return almost all of it as rebound energy. Since a relatively dead ball wastes most of the collision energy it receives, a lively surface's contribution to the rebound energy can be very important to the bounce. A lively racket is critical to the game of tennis because the racket's strings provide much of the rebound energy as the ball bounces off the racket (Fig. 3.2.4). Trampolines and springboards are even more extreme examples, with surfaces so lively that they can even make people bounce. People, like beanbags, have coefficients of restitution near zero; when you land on a trampoline, it receives and stores most of the collision energy and then provides most of the rebound energy.

Courtesy Louis Bloomfield

Fig. 3.2.4 When a tennis ball bounces from a moving racket, both the ball and racket dent inward. The ball and racket share the collision energy almost evenly.

The stiffnesses of the ball and surface also determine how much force each object exerts on the other and thus how quickly the collision proceeds. Hard objects resist denting much more strongly than soft objects. When both objects are very hard, the forces involved are large and the acceleration is rapid. Thus a steel ball rebounds very quickly from a concrete floor because the two exert enormous forces on one another. If the ball and/or surface are relatively soft, the forces are weaker and the acceleration is slower.

What if the surface that a ball hits isn't very massive? In that case, the surface may do part or all of the "bouncing." During the bounce, the ball and the surface accelerate in opposite directions and share the rebound energy. Massive surfaces, such as floors and walls, accelerate little and receive almost none of the rebound energy. But when the surface a ball hits is not very massive, you may see it accelerate. When a ball hits a lamp on the coffee table, the ball will do most of the accelerating, but the lamp is likely to fall over, too.

Similarly, when a baseball strikes a baseball bat, the ball and bat accelerate in opposite directions. The more massive the bat, the less it accelerates. To ensure that most of the rebound energy went to the baseball, the legendary hitters of the early twentieth century used massive bats. Such bats are no longer in vogue because they're too difficult to swing. But in the early days of baseball, when pitchers were less skillful, massive bats drove many long home runs.

▶ Check Your Understanding #2: The Game Begins

for answers, see page 107

You are playing the game of marbles on a soft dirt field. The goal is to knock glass marbles out of a circle by hitting them with other marbles. You initially drop several marbles onto the ground inside the circle and they hardly bounce at all. What prevents them from bouncing well here?

How a Moving Surface Affects the Bounce

The last paragraph describes the act of hitting a baseball with a moving baseball bat as though it were a case of "bouncing." That might sound a little strange. When a baseball hits a stationary bat, the ball bounces. But if a moving bat hits a stationary baseball, is it proper to say that the ball bounces?

The answer is yes. In fact, which object is moving and which is stationary depends on your point of view—your inertial frame of reference. A fly resting on the baseball will claim that the baseball is stationary and that it's about to be struck by a moving bat. Another fly resting on the baseball bat will claim that the bat is stationary and that it's about to be struck by a moving baseball. Which fly has the correct inertial frame of reference?

As we noted in Section 1.1, both frames of reference are equally valid. An inertial frame of reference is one that's not accelerating and is thus either stationary or moving at a constant velocity. As long as you view the world around you from an inertial frame of reference, the laws of motion will accurately describe what you see, and energy, momentum, and angular momentum will all be conserved.

But frames of reference aren't the first things you think of during a baseball game. When you swing a bat and drive the pitch toward center field, your main concern is how fast the outgoing ball is traveling toward the bleachers. A speedy ball will be a home run, while a slower ball will be an out. What determines the ball's outgoing speed?

With both bat and ball moving relative to the playing field, there are several useful inertial frames of reference from which to study the collision. However, we'll find it

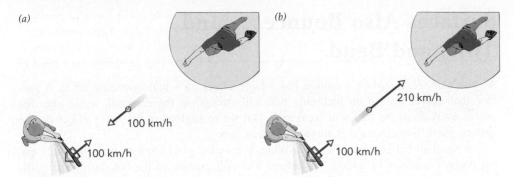

(a)

(b)

210 km/h

100 km/h

100 km/h   100 km/h

Fig. 3.2.5 (a) Before they collide, the bat and ball are approaching one another at an overall speed of 200 km/h. (b) After the collision, the two are separating from one another at a speed of 110 km/h. However, because the bat is moving toward the pitcher at 100 km/h, the outgoing ball is traveling at 210 km/h in that same direction.

easier to focus on how quickly the bat and ball move toward or away from one another. This relative motion is what matters most in a collision. After all, whether a rock hits a bottle or a bottle hits a rock, it's going to be bad for the bottle.

When a ball bounces off a *moving* surface that's rigid and massive, the ball's coefficient of restitution still applies. But now we must use a more general form of that speed ratio. This improved version divides the speed at which the ball and surface separate after the bounce by the speed at which they approach before the bounce:

$$\text{coefficient of restitution} = \frac{\text{speed of separation}}{\text{speed of approach}}. \qquad \textbf{(3.2.2)}$$

When the surface is stationary, Eq. 3.2.2 is equivalent to Eq. 3.2.1.

To see how this generalization allows us to explain why the baseball you hit is now traveling over the center fielder's head, let's examine the collision between bat and ball. Let's suppose that, just before the collision (Fig. 3.2.5a), the pitched baseball was approaching home plate at 100 km/h (62 mph) and that, as you swung to meet the ball, your bat was moving toward the pitcher at 100 km/h. Since each object is moving toward the other, their speed of approach is the sum of their individual speeds, or 200 km/h (124 mph).

The baseball's coefficient of restitution is 0.55, so after the collision (Fig. 3.2.5b) the speed of separation will be only 0.55 times the speed of approach, or 110 km/h. The outgoing ball and the swinging bat separate from one another at 110 km/h. Since the bat is still moving toward the pitcher at 100 km/h, the ball must be traveling toward the pitcher even faster: at 100 km/h plus 110 km/h or a total speed of 210 km/h (130 mph)! That's why it flies past everyone and into the stands.

Check Your Understanding #3: Marble Frames of Reference

for answers, see page 107

Two of you flick your marbles into the circle simultaneously from opposite sides of the circle and they collide head on. Each marble was traveling forward at 1 m/s (3.3 ft/s). From the inertial reference frame of your marble, what was the velocity of the other person's marble just before the two marbles hit?

Surfaces Also Bounce . . . and Twist and Bend

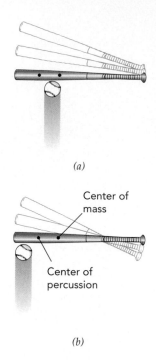

(a)

Center of mass

Center of percussion

(b)

(c)

Fig. 3.2.6 When a ball hits a bat, the bat experiences both acceleration and angular acceleration. (*a*) If the ball hits near the bat's center, the angular acceleration is small and the bat's handle accelerates backward. (*b*) If the ball hits near the bat's end, the angular acceleration is large and the bat's handle accelerates forward. (*c*) But if the ball hits the bat's center of percussion, the angular acceleration is just right to keep the handle from accelerating.

As you can see, a surface's motion has a large effect on a ball bouncing off it. A surface that moves toward an incoming ball will strengthen the rebound, while one that moves away from the ball will weaken it. But we're neglecting the ball's effect on the surface itself. Sometimes that surface bounces, too.

A baseball bat is a case in point. When you swing your bat into a pitched ball, the bat doesn't continue on exactly as before. The ball pushes on the bat during the collision and the bat responds in a number of interesting ways.

First, as we noted before, the bat rebounds from the ball. The bat decelerates slightly during the collision so that its speed after the impact is a little less than before it. Since the ball's final speed depends on the bat's final speed, a slower bat means a slower ball. Thus we've slightly overestimated the ball's speed as it heads toward the bleachers.

Second, the ball's impact sets the bat spinning. When the ball pushes on the bat and makes it accelerate backward, it also exerts a torque on the bat about its center of mass and makes it undergo angular acceleration (Fig. 3.2.6). While these two types of acceleration might seem inconsequential, your hands notice their effects. The bat's acceleration tends to yank its handle toward the catcher, while its angular acceleration tends to twist its handle toward the pitcher. The extent of these two motions depends on just where the ball hits the bat. If the ball hits a point known as the **center of percussion,** the handle experiences no overall acceleration (Fig. 3.2.6*c*). The smooth feel of such a collision explains why the center of percussion, located about 7 inches from the end of the bat, is known as a "sweet spot."

Finally, the collision often causes the bat to vibrate. Like a xylophone bar struck by a mallet (Fig. 3.2.7*a*), the bat bends back and forth rapidly with its ends and center moving in opposite directions (Fig. 3.2.7*b*). These vibrations sting your hands and can even break a wooden bat. However, near each end of the bat, there's a point that doesn't move when the bat vibrates—a **vibrational node.** When the ball hits the bat at its node, no vibration occurs. Instead, the bat emits a crisp, clear "crack" and the ball travels farther. Fortunately, the bat's vibrational node and its center of percussion almost coincide so you can hit the ball with both sweet spots at once.

As these handle motions and bending vibrations illustrate, the science and engineering of bats is surprisingly complicated. That's why bat manufacturers are forever developing better, more potent ones. Like the makers of golf clubs, tennis rackets, bowling equipment, and billiard tables, they dream of bounces so perfect that all of the collision energy is stored and returned as rebound energy. Known as **elastic collisions,** such perfect bounces are common among the tiny atoms in a gas, but unattainable for ordinary objects; there are just too many ways for any large item to divert or dissipate energy, including as thermal energy, sound, vibration, and light.

Although equipment manufacturers must resign themselves to **inelastic collisions**— collisions that fail to return some of the collision energy as rebound energy—they get closer to perfection every year. They also push right up to the limits of regulation, and occasionally beyond, in their quests for maximum performance.

Multiwalled aluminum, titanium, and composite bats are a case in point. Each of these hollow bats is soft enough to dent considerably during its impact with the ball so that it receives much of the collision energy. Its circular barrel flattens into an oval, storing the collision energy beautifully, and then kicks back to circular over a time that is

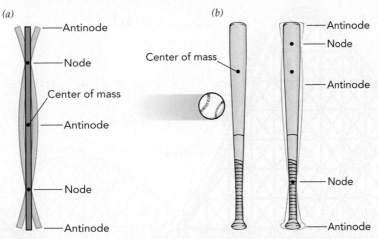

Fig. 3.2.7 (*a*) When struck by a mallet, a xylophone bar vibrates with its middle and ends moving back and forth in opposite directions. The parts that move farthest are antinodes, while the points that don't move at all are nodes. (*b*) When struck by a ball, a baseball bat vibrates in a similar fashion. However, an impact at one of the bat's nodes causes no vibration.

well matched to the timing of the bounce itself. Unlike the hard surface of a wooden bat, which barely dents and thus barely participates in the energy storage process, one of these hi-tech bats acts as a trampoline—it stores and returns so much of the collision energy that it substantially increases the outgoing speed of the batted ball. You can find equally hi-tech equipment at the golf shop or tennis store. We're in an era when scientific analysis and design are radically altering sports.

Check Your Understanding #4: Mass and Marbles

for answers, see page 108

The marbles you're playing with are not all the same size and mass. You notice that larger marbles are particularly effective at knocking other marbles out of the circle. You decide to use a 10-cm-diameter glass ball as a marble, expecting to clean out the entire circle. But when you flick it with your thumb, your thumb merely bounces off. Why doesn't the glass ball move forward quickly?

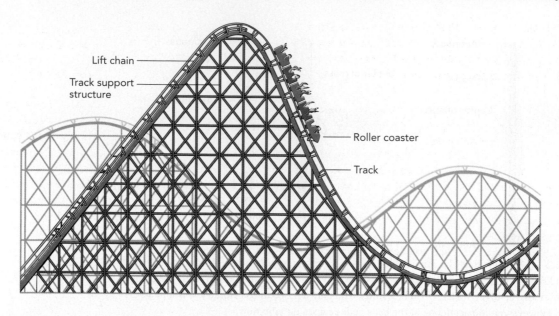

Lift chain

Track support
structure

Roller coaster

Track

SECTION 3.3 Carousels and Roller Coasters

As your sports car leaps forward at a green light, you're pressed firmly back against your seat. It's as though gravity were somehow pulling you down and backward at the same time. But it's not gravity that pulls back on you; it's your own inertia preventing you from accelerating forward with the car.

When this happens, you're experiencing the feeling of acceleration. We encounter this feeling many times each day, whether through turning in an automobile or riding up several floors in a fast elevator. But nowhere are the feelings of acceleration more acute than at the amusement park. We accelerate up, down, and around on the carousel, back and forth in the bumper cars, and left and right in the scrambler. The ultimate ride, of course, is the roller coaster, which is one big, wild experience of acceleration. When you close your eyes on a straight stretch of highway, you can hardly tell the automobile is moving. But when you close your eyes on a roller coaster, you have no trouble feeling every last turn in the track. It's not the speed you feel, but the acceleration. What is often called motion sickness should really be called acceleration sickness.

Questions to Think About: *How does your body feel its own weight? When you swing a bucket full of water around in a circle, as we asked you to do in the chapter opening, why does the bucket pull outward on you? Why can you swing that bucket completely over your head without spilling the water inside it? What keeps you from falling out of a roller coaster as it goes over the top of a loop-the-loop? Which car of a roller coaster should you sit in to experience the best ride?*

Experiments to Do: *To begin associating the familiar sensations of motion with the physics of acceleration, travel as a passenger in a vehicle that makes lots of turns and stops. Close your eyes and see whether you can tell which way the vehicle is turning and when it's starting or stopping. Which way do you feel pulled when the vehicle turns left? turns right? starts? stops? How is this sensation related to the direction of the* *vehicle's acceleration? Now find a time when the vehicle is traveling at constant veloc-ity on a level path and see if you feel any sensations that tell you which way it's head-ing. Try to convince yourself that it's heading backward or sideways rather than forward. Which is easier to feel: your acceleration or your velocity?*

The Experience of Acceleration

Nothing is more central to the laws of motion than the relationship between force and acceleration. Up until now, we've looked at forces and noticed that they can produce accelerations; in this section we'll take the opposite perspective, looking at accelerations and noticing that they require forces. For you to accelerate, something must push or pull on you. Just where and how that force is exerted on you determine what you feel when you accelerate.

The backward sensation you feel as your car accelerates forward is caused by your body's inertia, its tendency not to accelerate (Fig. 3.3.1). The car and your seat are accel-erating forward, and since the seat acts to keep you from traveling through its surface, it exerts a forward support force on you that causes you to accelerate forward. But the seat can't exert a force uniformly throughout your body. Instead, it pushes only on your back, and your back then pushes on your bones, tissues, and internal organs to make them accel-erate forward. Each piece of tissue or bone is responsible for the forward force needed to accelerate forward the tissue in front of it. A whole chain of forces, starting from your back and working forward toward your front, cause your entire body to accelerate forward.

Let's compare this situation with what happens when you're standing motionless on the floor. Since gravity exerts a downward force on you that's distributed uniformly throughout your body, each part of your body has its own independent weight; these individual weights, taken together, add up to your total weight. The floor, for its part, is exerting an upward support force on you that keeps you from accelerating downward through its surface. But the floor can't exert a force uniformly throughout your body. Instead, it pushes only on your feet, and your feet then push on your bones, tissues, and internal organs to keep them from accelerating downward. Each piece of tissue or bone is responsible for the upward force needed to keep the tissue above it from accelerating downward. A whole chain of forces, starting from your feet and working upward toward your head, keep your entire body from accelerating downward.

As you probably noticed, the two previous paragraphs are very similar. But so are the sensations of gravity and acceleration. When the ground is preventing you from falling, you feel "heavy"; your body senses all the internal forces needed to support its pieces so that they don't accelerate, and you interpret these sensations as weight. When the car seat is causing you to accelerate forward, you also feel "heavy"; your body senses all the inter-nal forces needed to accelerate its pieces forward, and you interpret these sensations as weight. This time you experience the weightlike sensation toward the back of the car.

Try as you may, you can't distinguish the weightlike sensation that you experience as you accelerate from the true force of gravity. And you're not the only one fooled by acceleration. Even the most sophisticated laboratory instruments can't determine directly whether they are experiencing gravity or are accelerating. However, despite the con-vincing sensations, the backward heavy feeling in your gut as you accelerate forward is the result of inertia and is not due to a real backward force. We'll call this experience a **feeling of acceleration** ❑. It always points in the direction opposite the acceleration that causes it, and its strength is proportional to that acceleration.

If you accelerate forward quickly, the backward feeling of acceleration you experi-ence can be quite large. However, you don't experience this feeling of acceleration all by

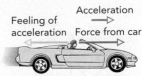

Fig. 3.3.1 As you accelerate forward in a car, you feel a gravity-like feeling of acceleration in the direction opposite to the acceleration. This feeling of acceleration is really the mass of your body resisting acceleration.

❑ The experience of acceleration that this book calls a feeling of acceleration is known elsewhere as a fictitious or apparent force. Since it's not a real force, our term aims to reduce the confusion.

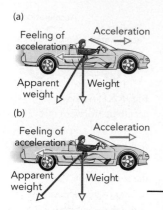

Fig. 3.3.2 (*a*) When you accelerate forward gently, the backward feeling of acceleration is small and your apparent weight is mostly downward. (*b*) When you accelerate forward quickly, you experience a strong backward feeling of acceleration and your apparent weight is backward and down.

itself; you also experience your downward weight, and together these effects feel like an especially strong weight at an angle somewhere between straight down and the back of the car. We'll call your combined experience of weight and feeling of acceleration your **apparent weight.** The faster you accelerate, the stronger the backward feeling of acceleration and the more your apparent weight points toward the back of the car (Fig. 3.3.2).

Feelings of acceleration don't have to point backward. They can also point forward or even to the side. When you turn your car to the left, you're accelerating leftward and experience a strong rightward feeling of acceleration. When you're not driving and can safely close your eyes, you should be able to feel acceleration in any direction.

> Check Your Understanding #1: The Feel of a Tight Turn
>
> *for answers, see page* 108
>
> You're sitting in the passenger seat of a racing car that is moving rapidly along a level track. The track takes a sharp turn to the left and you find yourself thrown against the door to your right. What horizontal forces are acting on you and what feelings of acceleration do you experience?

Carousels

When you ride on a carousel, you travel in a circle around a central pivot. That's an unusual motion. If you were experiencing zero net force, you would travel in a straight line at a steady pace in accordance with Newton's first law. But since your path is circular instead of straight, you must be experiencing a nonzero net force and you must be accelerating.

Which way are you accelerating? Remarkably, you are always accelerating toward the center of the circle. To see why that's so, let's look down on a simple carousel that's turning counterclockwise at a steady pace (Fig. 3.3.3). At first, the boy riding the carousel is directly east of its central pivot and is moving northward (Fig. 3.3.3*a*). If nothing were pulling on the boy, he would continue northward and fly off the carousel. Instead, he follows a circular path by accelerating toward the pivot, that is, toward the west. As a result, his velocity turns toward the northwest and he heads in that direction. To keep from flying off the carousel, he must continue to accelerate toward the pivot, which is now southwest of him (Fig. 3.3.3*b*). His velocity turns toward the west, and he follows the circle in that direction. And so it goes (Fig. 3.3.3*c*).

The boy's body is always trying to go in a straight line, but the carousel keeps pulling him inward so that he accelerates toward the central pivot. The boy is experiencing **uniform circular motion.** "Uniform" means that the boy is always moving at the same speed, although his direction keeps changing. "Circular" describes the path the boy follows as he moves, his trajectory.

Like any object undergoing uniform circular motion, the boy is always accelerating toward the center of the circle. An acceleration of this type, toward the center of a circle, is called a **centripetal acceleration** and is caused by a centrally directed force, a **centripetal force.** A centripetal force is not a new, independent type of force like gravity, but the net result of whatever forces act on the object. Centripetal means "center-seeking" and a centripetal force pushes the object toward that center. The carousel uses support forces and friction to exert a centripetal force on the boy, and he experiences a centripetal acceleration. Amusement park rides often involve centripetal acceleration (Fig. 3.3.4).

The amount of acceleration the boy experiences depends on his speed and the radius of the carousel. The faster the boy is moving and the smaller the radius of his circular trajectory, the more he accelerates. His acceleration is equal to the square of his speed divided by the radius of his path.

But we can also determine the boy's acceleration from the carousel's angular speed and its radius. The faster the carousel turns and the larger the radius of the boy's circular trajectory, the more he accelerates. His acceleration is equal to the square of the carousel's angular speed times the radius of his path. We can express these two relationships as word equations:

$$\text{acceleration} = \frac{\text{speed}^2}{\text{radius}} = \text{angular speed}^2 \cdot \text{radius,} \qquad \textbf{(3.3.1)}$$

in symbols:

$$a = \frac{v^2}{r} = \omega^2 \cdot r,$$

and in everyday language:

Making a tight, high-speed turn involves lots of acceleration.

Since the boy is accelerating inward, toward the center of the circle, he experiences a feeling of acceleration outward, away from the center of the circle. The boy feels as though his weight is pulling him outward as well as down and he clings tightly to the carousel to keep from falling off.

Weight and feeling of acceleration can differ significantly in strength. While your weight is the specific gravitational force you experience near the earth's surface, you can experience a feeling of acceleration that has any strength or direction. If you are moving very rapidly around a small circle, you can easily experience a feeling of acceleration that is stronger than your weight.

Your feeling of acceleration can be measured relative to your weight. When a feeling of acceleration matches the feeling of your weight, it is said to be 1 gravity or 1g, for short. To experience a 1g feeling of acceleration, you must accelerate in the opposite direction at 9.8 m/s² (32 ft/s²), the acceleration due to gravity. If you accelerate 5 times that quickly, on the scrambler or on an airplane maneuvering sharply, you'll experience a 5g feeling of acceleration.

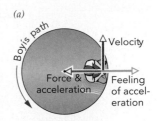
(a)

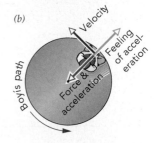
(b)

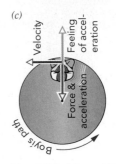
(c)

Fig. 3.3.3 A boy riding on a turning carousel is always accelerating toward the central pivot. His velocity vector shows that he is moving in a circle but his acceleration vector points toward the pivot. When he is heading north (*a*), he is accelerating toward the west. His velocity gradually changes direction until he is heading northwest (*b*), at which time he is accelerating toward the southwest. He turns further until he is heading west (*c*) and is then accelerating toward the south. (North is upward.)

Stuart Dee/The Image Bank/Getty Images

Fig. 3.3.4 The people on this ride travel in a circle, pulled inward by cords so that they accelerate toward the center of the circle.

→ Check Your Understanding #2: Banking on a Curve

for answers, see page 108

Your racing car comes to a banked, left-hand turn, which it completes easily. Why is it essential that a racetrack turn be banked so that it slopes downhill toward the center of the turn?

→ Check Your Figures #1: Going for a Spin

for answers, see page 108

Some children are riding on a playground carousel with a radius of 1.5 m (4.9 ft). The carousel turns once every two seconds. How quickly are the children accelerating?

Roller Coaster Acceleration

While roller coasters offer interesting visual effects, such as narrowly missing obstacles, and strange orientations, such as upside down, the real thrill of roller coasters comes from their accelerations. Plenty of other amusement park rides suspend you sideways or upside down, so that you feel ordinary gravity pulling at you from unusual angles. But why pay for these when you can stand on your head for free? For a *real* thrill, you need acceleration to give you the weightless feeling you experience as a roller coaster dives over its first big hill or the several-g sensation you feel as you go around a sharp corner. A change in the amount of "gravity" you feel is much more exciting than a change in its direction. We're now prepared to look at a roller coaster and understand what you feel as you go over hills (Fig. 3.3.5) and loop-the-loops (Fig. 3.3.6).

Every time the roller coaster accelerates, you experience a feeling of acceleration in the direction opposite your acceleration. That feeling of acceleration gives you an apparent weight that's different from your real weight. As we saw with a car, rapid forward acceleration tips your apparent weight backward toward the rear of the vehicle, while rapid deceleration tips it toward the front. But a roller coaster can do something a car can't: it can accelerate downward rapidly! In that case, the feeling of acceleration you experience is upward and opposes your downward weight so that they partially cancel. As a result, your apparent weight is less than your real weight and points downward or perhaps, if the downward acceleration is fast enough, points upward.

Consider that last possibility: if you accelerate downward at just the right rate, the upward feeling of acceleration will exactly cancel your downward weight. You will feel perfectly weightless, as though gravity didn't exist at all. The rate of downward acceleration that causes this perfect cancellation is that of a freely falling object. Your weight won't have changed, but the roller coaster will no longer be supporting you and you will be accelerating downward at 9.8 m/s^2 (32 ft/s^2). You will experience the same sensations as a diver who has just stepped off the high dive.

Since freely falling objects are subject only to the forces of gravity, they don't have to push on one another to keep their relative positions. As you fall,

Fig. 3.3.5 As this roller coaster plunges down the first hill, its last car is pulled over the edge by the cars in front of it and its riders feel almost weightless.

Harry DiOrio/The Image Works

your hat and sunglasses will fall with you and won't require any support forces from your head. Even if your sunglasses come off, they will hover in front of you as the two of you accelerate downward together. Similarly, your internal organs don't need to support one another, and the absence of internal support forces gives rise to the exhilarating sensation of free fall.

However, a roller coaster is attached to a track, and its rate of downward acceleration can actually exceed that of a freely falling object. In those special situations, the track will be assisting gravity in pushing the roller coaster downward. As a rider, you will feel less than weightless. The upward feeling of acceleration will be so large that your apparent weight will be in the upward direction, as though the world had turned upside down!

Fig. 3.3.6 As it goes over the loop-the-loop, this roller coaster is accelerating rapidly toward the center of the circle. When it reaches the top of the loop, the track pushes the roller coaster downward and the riders feel pressed into their seats. If they close their eyes, they won't even be able to tell that they're upside-down.

Check Your Understanding #3: Drop Tower Panic

for answers, see page 108

A drop tower is a terrifying amusement park ride in which you are strapped into a seat, lifted high in the air, and then dropped. While you're in free fall, what is your apparent weight?

Roller Coaster Loop-the-Loops

Figure 3.3.7 shows a single-car roller coaster at various points along a simple track with one hill and one loop-the-loop. Weight, feeling of acceleration, and apparent weight are all vector quantities, as are the car's velocity and acceleration. These quantities are indicated with arrows of varying lengths that show each vector's direction and magnitude. The longer the arrow, the greater the magnitude of the quantity it represents.

At the top of the first hill of Fig. 3.3.7, the single-car roller coaster is almost stationary (1). You, the rider, feel only your weight, straight down—nothing exciting yet. But as soon as the car begins its descent, accelerating down the track, a feeling of acceleration appears pointing up the track (2). The combination of your weight and the feeling of acceleration gives you an apparent weight that is very small and points down and into the track. Most people find this sudden reduction in apparent weight terrifying. Our bodies are very sensitive to partial weightlessness, and this falling sensation is half the fun of a roller coaster. An astronaut, falling freely in space, has this disquieting weightless feeling for days on end. No wonder astronauts have such frequent troubles with motion (or rather acceleration) sickness.

On the earth, the weightless feeling can't last. It occurs only during downward acceleration and disappears as your car levels off near the bottom of the hill. By the time the car begins its rise into the loop-the-loop, it is traveling at maximum speed and has begun to accelerate upward (3). This upward acceleration creates a downward feeling of acceleration so that your apparent weight is huge and downward. You feel pressed into your seat as you experience 2 or 3g's.

The trip through the loop-the-loop resembles uniform circular motion. It's roughly like taking a single turn around a vertical carousel. However, as the car rises into the

Fig. 3.3.7 A single-car roller coaster going over the first hill and a loop-the-loop. At each point along the track, the car experiences its weight, a feeling of acceleration due to its current acceleration, and an apparent weight that is the sum of those two. The apparent weight always points toward the track and the car doesn't fall off it.

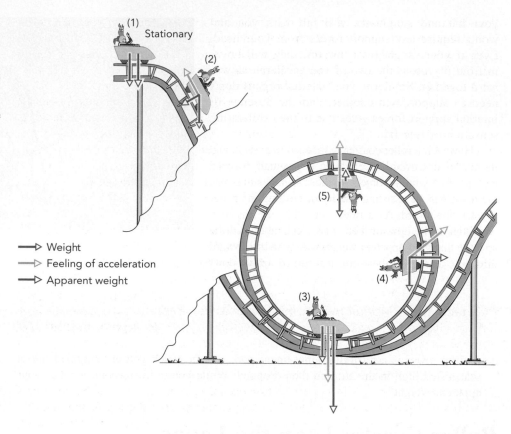

(1)
Stationary
(2)

(5)

(4)

(3)

→ Weight
→ Feeling of acceleration
→ Apparent weight

Fig. 3.3.8 The roller coaster pushes these riders inward as it travels around the loop-the-loop. The riders experience strong feelings of acceleration outward and can hardly tell that they're turning upside down.

Chad Slattery/Stone/Getty Images

loop-the-loop, some of its kinetic energy becomes gravitational potential energy and the car slows down. As the car descends out of the loop-the-loop, this gravitational potential energy returns to kinetic form and the car speeds up. As a result of these speed changes, your acceleration is not exactly inward toward the center of the loop-the-loop and the feeling of acceleration you experience is not exactly outward. Still, an inward acceleration and an outward feeling of acceleration are good approximations for what occurs.

Halfway up the loop-the-loop, your true acceleration is inward and downward, so the feeling of acceleration you experience is outward and upward (4). Your apparent weight is still much more than your weight and is directed outward. You feel pressed into your seat, and the car itself is pressed against the track (Fig. 3.3.8).

Finally, you reach the top of the loop-the-loop (5). The car has slowed somewhat as the result of its climb against the force of gravity. But it's still accelerating toward the center of the circle, and you still experience a feeling of acceleration outward and, in this case, upward. Your weight is downward, but the upward feeling of acceleration exceeds your weight. Your apparent weight is upward (Fig. 3.3.6)!

Not only does the inverted car stay on its track, but you feel a weak weightlike sensation pressing you into your seat. Actually, the car is pushing on you to help gravity accelerate you around the loop.

If your hat were to come off at the top of the loop, it would land in your seat, even though that seems to involve some sort of upward movement. In fact, the car is accelerating almost directly downward at a rate that's faster than that of your freely falling hat. Gravity and the track together push the car downward so fast that it overtakes the hat—your hat is really falling but the car is plummeting even faster.

In truth, a typical loop-the-loop isn't perfectly circular; it's more sharply curved on top than on its sides or bottom. This varying radius curve, known as a clothoid, is chosen for safety and comfort. By sharpening the curve only on top, the clothoid track maximizes the downward acceleration there while reducing the acceleration elsewhere on the track. High acceleration is important only when the roller coaster is upside-down. Everywhere else, it simply makes the riders feel heavy and uncomfortable, particularly at the bottom of the loop where the coaster is tra-veling fastest and accelerating upward rapidly.

Many roller coaster tracks are designed so that acceleration leaves the riders pressed into their seats even when the cars go upside down. In principle, roller coasters that travel on such tracks don't need seat belts to prevent riders from falling out (although seat belts are comforting to the passengers and insurance companies). But some roller coaster tracks have special cars and seat belts that permit them to have apparent weights away from the track. These roller coasters can and do go upside down without enough downward acceleration to keep the riders in their seats. In such roller coasters, the riders feel like they are hanging and if one of them were to lose a hat, it would fall to the ground rather than into the car.

What about a roller coaster with more than one car? For the most part, the same rules apply. However, new forces now act on each car: forces exerted by the other cars in the train. The effects of these cars are most pronounced at the top of the first and biggest hill. As the train disconnects from the lift chain and approaches the descent, it is rolling forward slowly and the first cars are well over the crest of the hill before they pick up much speed (Fig. 3.3.9a). They're pulling hard on the cars behind them, and those cars are pulling back, slowing their descent. By the time the train is moving fast, the first car is well down the hill. By then, the track is beginning to turn upward and the first car's riders experience mostly upward acceleration and downward feeling of acceleration. That's why riders in the first few cars of a roller coaster don't feel much weightlessness.

In contrast, the last car is moving at high speed early in its descent. It undergoes a dramatic downward acceleration as it's yanked over the crest of the first hill by the cars in front of it (Fig. 3.3.9b). As a result, its riders feel large upward feelings of acceleration and quite extreme weightlessness. In fact, the designers of the track must be careful not to make the downward acceleration too rapid, or the roller coaster will flick the riders in the last car right out of their seats.

Obviously, it does matter where you sit on a roller coaster. The first seat offers the most exciting view, but it provides less than spectacular weightless feelings. The last car almost always offers the best weightless feelings. Probably the dullest seat in the roller coaster is the second; it offers a relatively tame ride and an unchanging view of the people in the front seat.

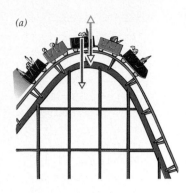

(a)

→ Weight
⇾ Feeling of acceleration
➡ Apparent weight

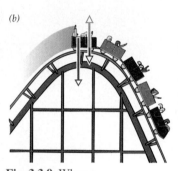

(b)

Fig. 3.3.9 When a multicar roller coaster descends the first hill, the ride experienced in the first cars is different from that in the last cars. (a) The first cars travel over the crest of the hill slowly and reach high speed only well down the hill. The cars behind them slow their descent. (b) The last cars are whipped over the top and are traveling very rapidly early on. The last cars accelerate downward dramatically as they go over the first hill and their riders experience a strong feeling of weightlessness.

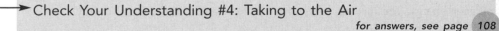

► Check Your Understanding #4: Taking to the Air

for answers, see page **108**

Your racing car travels over a bump in the track and suddenly becomes airborne. What keeps you in the air? Has gravity disappeared?

Epilogue for Chapter 3

In this chapter we looked at the physical concepts involved in four types of simple machines. In *spring scales,* we explored the relationship between the force acting on a spring and its distortion, and we examined how this distortion can be used to measure an object's weight. In *bouncing balls,* we examined the process of storing and releasing kinetic energy during a collision. As we saw, both the ball and the object it hits contribute to the bounce.

In *carousels and roller coasters,* we explored both the sensations we feel as we accelerate and why those sensations occur. We saw that circular motion involves a centrally directed acceleration that gives rise to an outwardly directed feeling of acceleration.

Explanation: Swinging Water Overhead

As you swing the bucket over your head, you are pulling downward on it and causing it to accelerate downward very rapidly. The water remains in the inverted bucket because the bucket is accelerating downward faster than gravity alone can accelerate the water. Although the water is free to fall, the bucket overtakes the falling water. As a result, the water is pressed into the bottom of the bucket. The same effect occurs when you push a book downward rapidly with your open palm. Although the book is free to fall, it remains pressed into your palm because your hand is accelerating it downward faster than gravity. Finally, if you stop swinging the bucket too abruptly, the bucket's contents will not decelerate with it. Instead, they will spill or smash.

Chapter Summary

How Spring Scales Work: A spring scale measures an object's weight by supporting that object with a spring. When the object is at rest, the spring's upward force exactly balances the object's downward weight; the scale then reports the upward restoring force its spring is exerting on the object. As Hooke's law describes, that restoring force is proportional to the spring's distortion, so the scale can determine it by measuring how far the spring has bent. This measurement is often done mechanically and is reported with a needle or dial.

How Bouncing Balls Work: A ball behaves like a spherical or oblong spring. It stores elastic potential energy as it distorts away from its equilibrium shape and releases some of that energy as it returns to normal. When a ball strikes a surface, some kinetic energy is removed from the ball and the surface and is either stored within those objects as elastic potential energy or lost as thermal energy. As the objects rebound, some of the stored energy becomes kinetic energy again. The kinetic energy returned—the rebound energy—is always less than the kinetic energy initially removed from the objects—the collision energy. The missing energy has been converted into thermal energy.

How Carousels and Roller Coasters Work: A carousel uses centripetal acceleration to give each rider an outward feeling of acceleration. Combined with weight, this gravity-like sensation gives the rider an apparent weight that points downward and outward.

A roller coaster also uses rapid acceleration to create unusual apparent weights for its riders. Each time the coaster accelerates on a hill or turn, the rider experiences a feeling of acceleration in the direction opposite the acceleration. This feeling of acceleration, combined with the rider's weight, creates an apparent weight that varies dramatically in amount and direction throughout the ride. It's this fluctuating apparent weight, particularly its near approaches to zero, that make riding a roller coaster so exciting.

Important Laws and Equations

1. Hooke's Law: The restoring force exerted by an elastic object is proportional to how far it is from its equilibrium shape, or

$$\text{restoring force} = -\text{spring constant} \cdot \text{distortion}. \quad (3.1.1)$$

2. Acceleration of an Object in Uniform Circular Motion: An object in uniform circular motion has a centripetal acceleration equal to the square of its speed divided by the radius of its circular trajectory, which is equal to the square of its angular speed times the radius of its circular trajectory, or

$$\text{acceleration} = \frac{\text{speed}^2}{\text{radius}} = \text{angular speed}^2 \cdot \text{radius}. \quad (3.3.1)$$

Check Your Understanding—Answers

Section 3.1 SPRING SCALES

1. You should sell by mass, for example, by kilogram or pound-mass.
Why: If you label your product by weight, you are specifying the force that gravity exerts on it near the earth's surface. When it's exported to the moon, such a product will weigh just ⅙ as much, and your company may be fined for selling underweight groceries. If you label the packages according to their masses, that labeling will remain correct no matter where you ship the packages. Mass is the measure of inertia and depends only on the object, not on its environment.

2. The diving board is behaving as a spring, bending downward in proportion to the weight of each diver.
Why: The heavier the diver, the more the board bends downward before exerting enough upward force on the diver to balance the diver's weight.

3. 15 N (about 3.3 lbf).
Why: The scale's dial is simply reporting the position of its basket. The dial is calibrated so that a 1-centimeter drop in the basket indicates that the spring is pulling up on it with a force of 5 N. Since the spring's restoring force is described by Hooke's law, a 3-centimeter drop in the basket means that the spring is exerting an upward force of 15 N on the basket.

4. The distance the scale's surface moves downward is proportional to the weight it reports.
Why: The scale's spring is connected to its surface by levers so that as the surface moves downward, the spring distorts by a proportional amount. The spring's distortion is reported on the dial. Thus the dial's reading is proportional to the surface's downward movement.

5. Lift each corner of the suitcase and see which one requires the most upward force. The books should be near that corner.
Why: By lifting each corner of the suitcase, you are determining the location of the suitcase's center of gravity. The more upward force required to support a corner, the closer it is horizontally to the suitcase's center of gravity and the books themselves.

Section 3.2 BOUNCING BALLS

1. When the marble collides with the hard granite surface, it dents inward and converts most of its kinetic energy into elastic potential energy. It then rebounds, converting this stored energy back into kinetic energy as it bounces from the surface.
Why: An elastic marble has a very high coefficient of restitution and bounces well from a hard surface.

2. The soft dirt distorts more than the hard marble and receives most of the collision energy. It converts most of that energy into thermal energy so the marble rebounds weakly.
Why: While a marble bounces nicely on a hard surface, the dirt is soft and receives virtually all of the collision energy when the hard marble hits it. The dirt distorts during the impact but stores little energy because it's not very elastic. The marble doesn't rebound much.

3. 2 m/s (6.6 ft/s) toward you.
Why: The velocities reported in the question are those observed by people sitting still with respect to the circle. From the inertial reference frame of your marble, the circle itself is heading in your direction at 1 m/s. Since the other person's marble is moving in your direction at 1 m/s faster than the circle, its total velocity according to your marble is 2 m/s in your direction.

4. The glass ball is so much more massive than your thumb that your thumb receives almost all the rebound energy. Your thumb bounces, not the glass ball.

Why: In any collision, it's the least massive object that experiences the greatest acceleration and that receives the largest share of the rebound energy. The effect is similar to what would happen if you swung a light aluminum baseball bat at a pitched bowling ball. The bat would rebound wildly but the bowling ball would continue to travel toward the catcher. This same effect is true in automobile collisions, where a massive sedan is much less disturbed than the tiny, subcompact with which it collides.

Section 3.3 CAROUSELS AND ROLLER COASTERS

1. The car seat and door are exerting a leftward force on you, causing you to accelerate leftward with the car. You also experience a rightward feeling of acceleration, as your inertia acts to keep you from accelerating leftward.

Why: As the car turns left, it accelerates toward the left. The car seat and right door together exert a leftward force on you, to prevent the car from driving out from under you. Because it has mass, your body resists this leftward acceleration. You feel your body trying to go in a straight line, which would carry it out the right door of the car as the car accelerates toward the left.

2. A banked turn is needed so that the support force exerted by the racetrack on the car's wheels can provide at least some of the centripetal force needed to accelerate the car around the turn.

Why: As a car and driver travel around a circular turn, they are accelerating toward the center of the track and require a huge centripetal force inward. On a level track, the only horizontal force available is static friction between the ground and the car's tires. If static friction is unable to provide enough inward force, the car will skid off the track, following a straight-line path. This type of accident is typical of a highway curve on an icy day and is why designers bank the curves. The banks are ramps sloping down toward the center so that the horizontal component of the support force exerted by the track on the car's wheels provides an additional, inward, centripetal force to help that car accelerate around the curve.

3. Zero.

Why: As you fall freely, you are accelerating downward at the full acceleration due to gravity. The upward feeling of acceleration you experience exactly matches your downward weight, so that your apparent weight is zero. You feel perfectly weightless.

4. Gravity is still present, but your inertia prevents you from following a rapid downturn in the surface you are traveling along.

Why: If the road you are traveling along suddenly turns downward, you must accelerate downward to stay in contact with its surface. The steeper and more abrupt the descent, the more downward acceleration you need. The only downward force you experience is your weight, which can cause a downward acceleration of no more than 9.8 m/s^2 (32 ft/s^2). If the surface drops out from under you faster than that, you will become airborne. You will then be falling freely and accelerating downward as fast as gravity will permit. Eventually, you will fall to the surface. In many sports, including skiing, motorcycle racing, and skateboarding, a person traveling along an uneven surface becomes airborne after passing over a bump.

Check Your Figures—Answers

Section 3.1 SPRING SCALES

1. 2 centimeters and 10 centimeters (assuming that the floor doesn't break).

Why: A floor, like most suspended surfaces, behaves like a spring. Your floor distorts 1 centimeter before it exerts an upward restoring force equal to the weight of 10 guests. It will thus distort 2 centimeters before supporting 20 guests and 10 centimeters before supporting 100 guests. While this distortion should be within the elastic limit of the floor beams, it may cause the plaster and paint to crack. If the beams break, the floor will collapse.

Section 3.3 CAROUSELS AND ROLLER COASTERS

1. About 15 m/s^2 (50 ft/s^2).

Why: Because the children are in uniform circular motion, their acceleration is given by Eq. 3.3.1. The carousel turns once every 2 seconds, so its angular velocity is 2π radians divided by 2 seconds. Omitting "radians" because they're the natural unit of angle, the carousel's angular velocity is π 1/s. Since its radius is 1.5 meters, the children's acceleration is

$$\text{children's acceleration} = (\pi \text{ 1/s})^2 \cdot 1.5 \text{ m} = 14.8 \text{ m/s}^2.$$

Since our measurements of the carousel's radius and its turning time are only accurate to about 10%, our calculation of the children's acceleration is only accurate to about 10%. We report it as 15 m/s^2 (50 ft/s^2).

Exercises

1. In what way does the string of a bow and arrow behave like a spring?

2. As you wind the mainspring of a mechanical watch or clock, why does the knob get harder and harder to turn?

3. Curly hair behaves like a weak spring that can stretch under its own weight. Why is a hanging curl straighter at the top than at the bottom?

4. When you lie on a spring mattress, it pushes most strongly on the parts of you that stick into it the farthest. Why doesn't it push up evenly on your entire body?

5. If you pull down on the basket of a hanging grocery store scale so that it reads 15 N, how much downward force are you exerting on the basket?

6. While you're weighing yourself on a bathroom scale, you reach out and push downward on a nearby table. Is the weight reported by the scale high, low, or correct?

7. There's a bathroom scale on your kitchen table and your friend climbs up to weigh himself on it. One of the table's legs is weak and you're afraid that he'll break it, so you hold up that corner of the table. The table remains level as you push upward on the corner with a force of 100 N. Is the weight reported by the scale high, low, or correct?

8. If you put your bathroom scale on a ramp and stand on it, will the weight it reports be high, low, or correct?

9. When you step on a scale, it reads your weight plus the weight of your clothes. Only your shoes are touching the scale, so how does the weight of the rest of your clothes contribute to the weight reported by the scale?

10. To weigh an infant you can step on a scale once with the infant and then again without the infant. Why is the difference between the scale's two readings equal to the weight of the infant?

11. An elastic ball that wastes 30% of the collision energy as heat when it bounces on a hard floor will rebound to 70% of the height from which it was dropped. Explain the 30% loss in height.

12. The best running tracks have firm but elastic rubber surfaces. How does a lively surface assist a runner?

13. Why is it so exhausting to run on soft sand?

14. Steep mountain roads often have emergency ramps for trucks with failed brakes. Why are these ramps most effective when they are covered with deep, soft sand?

15. There have been baseball seasons in which so many home runs were hit that people began to suspect that something was wrong with the baseballs. What change in the baseballs would account for them traveling farther than normal?

16. During rehabilitation after hand surgery, patients are often asked to squeeze and knead putty to strengthen their muscles. How does the energy transfer in squeezing putty differ from that in squeezing a rubber ball?

17. Your car is on a crowded highway with everyone heading south at about 100 km/h (62 mph). The car ahead of you slows down slightly and your car bumps into it gently. Why is the impact so gentle?

18. Bumper cars are an amusement park ride in which people drive small electric vehicles around a rink and intentionally bump them into one another. All of the cars travel at about the same speed. Why are head-on collisions more jarring than other types of collisions?

19. When two trains are traveling side by side at breakneck speed, it's still possible for people to jump from one train to the other. Explain why this can be done safely.

20. If you drop a steel marble on a wooden floor, why does the floor receive most of the collision energy and contribute most of the rebound energy?

21. A RIF (reduced injury factor) baseball has the same coefficient of restitution as a normal baseball except that it deforms more severely during a collision. Why does this increased deformability lessen the forces exerted by the ball during a bounce and reduce the chances of its causing injury?

22. Padded soles in running shoes soften the blow of hitting the pavement. Why does padding reduce the forces involved in bringing your foot to rest?

23. Some athletic shoes have inflatable air pockets inside them. These air pockets act like springs that become stiffer as you pump up the air pressure. High pressure also makes you bounce back up off the floor sooner. Why does high pressure shorten the bounce time?

24. Why does it hurt less to land on a soft foam pad than on bare concrete after completing a high jump?

25. Why must the surface of a hammer be very hard and stiff for it to drive a nail into wood?

26. Some amusement park rides move you back and forth in a horizontal direction. Why is this motion so much more disturbing to your body than cruising at a high speed in a jet airplane?

27. You are traveling in a subway along a straight, level track at a constant velocity. If you close your eyes, you can't tell which way you're heading. Why not?

28. Moving a can of spray paint rapidly in one direction will not mix it nearly as well as shaking it back and forth. Why is it so important to change directions as you mix the paint?

29. Why does a baby's rattle only make noise when the baby moves it back and forth and not when the baby moves it steadily in one direction?

30. In some roller coasters, the cars travel through a smooth tube that bends left and right in a series of complicated turns. Why does the car always roll up the right-hand wall of the tube during a sharp left-hand turn?

31. Railroad tracks must make only gradual curves to prevent trains from derailing at high speeds. Why is a train likely to derail if it encounters a sharp turn while it's traveling fast?

32. Police sometimes use metal battering rams to knock down doors. They hold the ram in their hands and swing it into a door from about 1 m away. How does the battering ram increase the amount of force the police can exert on the door?

33. When a moving hammer hits a nail, it exerts the enormous force needed to push the nail into wood. This force is far greater than the hammer's weight. How is it produced?

34. A hammer's weight is downward, so how can a hammer push a nail upward into the ceiling?

35. As you swing back and forth on a playground swing, your apparent weight changes. At what point do you feel the heaviest?

36. Some stores have coin-operated toy cars that jiggle back and forth on a fixed base. Why can't these cars give you the feeling of actually driving in a drag race?

37. A salad spinner is a rotating basket that dries salad after washing. How does the spinner extract the water?

38. People falling from a high diving board feel weightless. Has gravity stopped exerting a force on them? If not, why don't they feel it?

39. When your car travels rapidly over a bump in the road, you suddenly feel weightless. Explain.

40. Astronauts learn to tolerate weightlessness by riding in an airplane (nicknamed the "vomit comet") that follows an unusual trajectory. How does the pilot direct the plane in order to make its occupants feel weightless?

41. You board an elevator with a large briefcase in your hand. Why does that briefcase suddenly feel particularly heavy when the elevator begins to move upward?

42. As your car reaches the top in a smoothly turning Ferris wheel, which way are you accelerating?

Problems

1. Your new designer chair has an S-shaped tubular metal frame that behaves just like a spring. When your friend, who weighs 600 N, sits on the chair, it bends downward 4 cm. What is the spring constant for this chair?

2. You have another friend who weighs 1000 N. When this friend sits on the chair from Problem 1, how far does it bend?

3. You're squeezing a springy rubber ball in your hand. If you push inward on it with a force of 1 N, it dents inward 2 mm. How far must you dent it before it pushes outward with a force of 5 N?

4. When you stand on a particular trampoline, its springy surface shifts downward 0.12 m. If you bounce on it so that its surface shifts downward 0.30 m, how hard is it pushing up on you?

5. Engineers are trying to create artificial "gravity" in a ring-shaped space station by spinning it like a centrifuge. The ring is

100 m in radius. How quickly must the space station turn in order to give the astronauts inside it apparent weights equal to their real weights at the earth's surface?

6. A satellite is orbiting the earth just above its surface. The centripetal force making the satellite follow a circular trajectory is just its weight, so its centripetal acceleration is about 9.8 m/s^2 (the acceleration due to gravity near the earth's surface). If the earth's radius is about 6375 km, how fast must the satellite be moving? How long will it take for the satellite to complete one trip around the earth?

7. When you put water in a kitchen blender, it begins to travel in a 5-cm-radius circle at a speed of 1 m/s. How quickly is the water accelerating?

8. In Problem 7, how hard must the sides of the blender push inward on 0.001 kg of the spinning water?

CHAPTER 4

Mechanical Objects Part 2

No matter how sophisticated the modern machines around us appear, most of them are based in large part on the simple principles that we have already encountered. In this chapter, we'll take a look at two more fascinating machines and see what makes them tick. As we do, we'll find ourselves revisiting familiar issues and exploring a few new ones all the way to the frontiers of science and the cosmos.

EXPERIMENT: High Flying Balls

Among the physical issues that we'll discuss in this chapter are the reaction effects that push rockets forward. These reaction effects appear in an interesting way in a simple experiment that involves two different sized balls, a basketball and a tennis ball.

If you can't find a basketball and a tennis ball, any two lively balls of very different masses will do. Drop the balls separately and see how high they bounce. You'll immediately notice that an individual ball can't bounce higher than the point from which you dropped it. Such a rebound would give it more energy than it had originally. In fact, some of the ball's energy will be lost to thermal energy so that it will not even reach its original height.

Courtesy Lou Bloomfield

But what will happen when you stack the smaller ball on top of the larger ball and drop the two balls together? Think about the sequence of events that will occur and try to *predict* what will happen. Now give it a try. Make sure that the small ball remains directly above the large ball as the two balls fall to the floor. *Observe* what happens and see whether the outcome *verifies* your prediction. *Measure* the heights of the rebounds and try to explain the effect. You could also try the same experiment with balls of various sizes to gather more insight into what is happening.

Chapter Itinerary

The objects in this chapter combine many of the concepts we have discussed and add a few new ones of their own. These objects are (1) *bicycles* and (2) *rockets*. In *bicycles,* we'll see how motion and clever design can make an apparently unstable vehicle stable enough to ride easily, even without hands on the handlebars. In *rockets,* we'll look at the principle of action and reaction, and see how this basic idea makes it possible for spacecraft to leave the earth's surface and even travel toward the stars. For a more complete preview of this chapter, turn ahead to the Chapter Summary on p. 134.

This chapter revisits many ideas from the previous chapters but it also introduces a variety of new concepts we can use. In *bicycles,* we'll learn about dynamic stability and explore issues that waiters, water-skiers, and warehouse clerks must deal with all the time. We'll also encounter gyroscopes and their peculiar responses to torques. In *rockets,* we'll see how gravity causes objects to orbit one another, giving rise to the intricate motions of the planets, moons, and comets. We'll also look at new physics that appears near the extremes of speed and gravity.

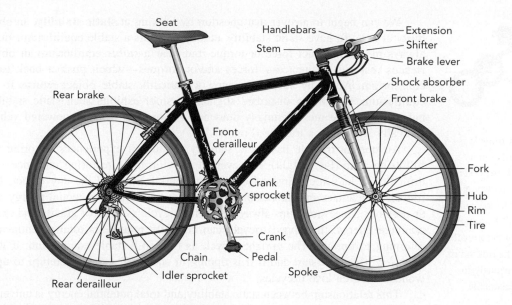

Seat

Handlebars — Extension
Stem — Shifter
— Brake lever

Shock absorber
Rear brake — Front brake

Front
derailleur

Fork

Crank
sprocket
Hub
Rim
Tire

Crank

Chain — Pedal

Idler sprocket — Spoke

Rear derailleur

SECTION 4.1 Bicycles

A bicycle is a wonderfully energy efficient, human-powered vehicle. Its wheels allow its rider to coast forward easily on level surfaces and accelerate effortlessly down hills. Compare the easy motion of a bicycle to that of walking, which requires effort every step of the way. Bicycles are very simple machines, and most of their moving parts are quite visible: the pedals, sprockets, brakes, and steering mechanisms, to name a few. Their simplicity and visibility make bicycles relatively easy to fix, even for a novice.

Questions to Think About: Why is a two-wheeled bicycle preferable to the apparently more stable three-wheeled tricycle? Why do you lean a bicycle in the direction that you are turning? How is it possible to ride a bicycle without hands on the handlebar? How does pedaling a bicycle make it move forward? Why does a bicycle have such a complicated drive system between its pedals and its rear wheel?

Experiments to Do: If you know how to ride a bicycle, pay attention to its stability as you ride it. Notice how you lean the bicycle in the direction that you are turning and how the bicycle naturally steers into that turn as you begin to lean. This automatic steering is part of the bicycle's self-stabilizing behavior. As you ride, observe how fast the pedals are turning and how hard you must push on the pedals to keep them turning at that rate. Go up and down some hills in various gears and see how each gear-choice affects the pedaling rate and the forces you must exert on the pedals. Why do you choose a certain gear for a certain situation? What pedaling rate do you find most comfortable? What pedaling force feels best?

Tricycles and Static Stability

Bicycles have a stability problem. With only two wheels to support them, stationary bicycles tip over easily. Why then do we use two-wheeled bicycles for transportation?

Fig. 4.1.1 A tricycle is very stable when it's standing still. However, it tips over easily during a high-speed turn because the rider can't lean in the direction that the tricycle is turning. The rider must also pedal furiously to move at a reasonable speed.

We can begin to answer that question by looking at **static stability,** an object's stability at rest. To have static stability, an object needs a stable equilibrium. Equilibrium always means zero net force or torque, but near a *stable* equilibrium an object experiences restoring influences—forces and/or torques—which push it back toward that equilibrium. Aided by such influences, a statically stable object returns to its stable equilibrium after being disturbed slightly. A stool exhibits such static stability, but a balancing bicycle most definitely does not. If you want a pedal-powered vehicle that's statically stable, try a tricycle (Fig. 4.1.1).

An upright tricycle has static stability because it experiences restoring influences following a tip (Fig. 4.1.2a). However, identifying those restoring influences directly is difficult, so let's focus instead on total potential energy. Like any object, the tipped tricycle accelerates in whichever direction reduces its total potential energy as quickly as possible. Since small tips always raise the tricycle's center of gravity and gravitational potential energy, the tipped tricycle can certainly reduce its total potential energy by returning to upright. The upright tricycle is thus in a stable equilibrium; if it isn't tipping yet, it doesn't start and if it is tipping, it accelerates so as to return to upright. No wonder children love tricycles!

This relationship between static stability and total potential energy is universal. There is no need to look directly for the restoring influences, which may be complicated anyway. When an object is situated so that its total potential energy rises if it shifts in any

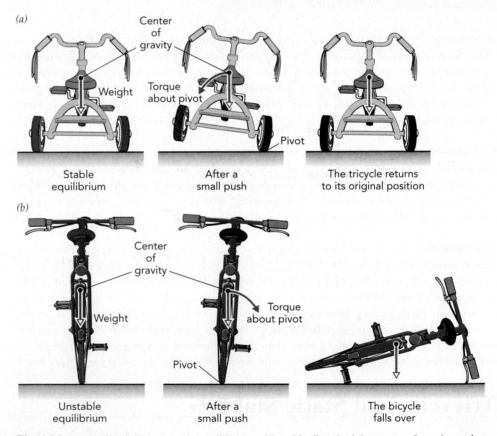

Fig. 4.1.2 (*a*) A tricycle is in a stable equilibrium. When it's disturbed, its center of gravity and gravitational potential energy rise, and it experiences restoring forces that return it to upright. (*b*) A bicycle is in an unstable equilibrium. Any disturbance causes it to fall.

direction, it's in a stable equilibrium and will tend to remain there. That useful rule applies not only to tricycles, but to everything from canoes to mobiles. If you want an object to be stable at rest, ensure that any small shift increases its total potential energy.

> **Stable Equilibrium and Potential Energy**
> An object is in a stable equilibrium when any small shift increases its total potential energy.

That general observation underlies a simple rule of thumb for the behavior of an independent object like a tricycle resting on a surface: that object will be in a stable equilibrium so long as its center of gravity is positioned over its **base of support**—the polygon defined by its contact points with the ground. This rule follows from geometry: if the object's center of gravity is positioned over its base of support, then tipping it raises its center of gravity and increases its gravitational potential energy. Assuming no other potential energies are significant, the tipped object will accelerate back toward that equilibrium and the equilibrium will be stable.

Geometry also sets the limits of the object's stability: if its center of gravity moves outside the original base of support, it will no longer experience restoring influences and can tip over completely. That's why leaning back too far while on a tricycle is a recipe for disaster! The tricycle's three wheels define a triangular base of support, so it will be statically stable only so long as its rider keeps the overall center of gravity above that triangle.

But with just two wheels touching the ground, a bicycle has no base of support (Fig. 4.1.2b) and therefore no static stability. Although it's at equilibrium when perfectly upright, this equilibrium isn't stable. If the bicycle tips to one side, its center of gravity will descend and its gravitational potential energy will decrease. The bicycle will actually accelerate *away* from equilibrium! The upright bicycle is thus in an **unstable equilibrium;** instead of experiencing restoring forces or torques, the disturbed bicycle will tip farther and faster until it hits the ground.

> **Unstable Equilibrium and Potential Energy**
> An object's equilibrium is unstable when a small shift can decrease its total potential energy.

Check Your Understanding #1: Keeping Coffee in Its Place

for answers, see page 135

Some travel mugs taper outward at the bottom so that they have very wide bases. Why does this shape make such a mug particularly stable and keep it from flipping over during the morning commute?

Bicycles and Dynamic Stability

Static stability matters most to people who have difficulty balancing. That's why children learn to ride tricycles first (Fig. 4.1.1). But when a tricycle is moving, static stability doesn't guarantee safety. If a child rolls down a steep hill and then makes a sudden sharp turn, he or she will probably flip over. What has gone wrong?

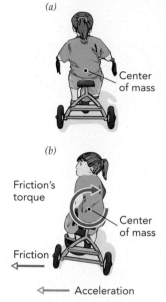

(a)

Center of mass

(b)

Friction's torque

Center of mass

Friction

Acceleration

Fig. 4.1.3 (*a*) A tricycle that is heading straight is stable because any tip causes its center of gravity to move upward. (*b*) During a fast left turn, however, the tricycle accelerates left and friction exerts a large leftward force on the wheels. This frictional force produces a torque about the tricycle and rider's combined center of mass and can cause the tricycle to tip over.

A moving tricycle stays upright only if the girl riding it avoids sudden accelerations, such as a turn at high speed. To make such a turn, she steers the tricycle's wheels so that friction with the pavement pushes the tricycle to one side (Fig. 4.1.3). Frictional forces on the wheel accelerate the tricycle to that side, redirecting its speed so that it turns. Of course, the girl needs to turn, too, so the tricycle pushes her along with it. As long as the turn is slow, a gentle push is all that's required and the tricycle and girl turn together safely.

But if the turn is too abrupt, the girl doesn't complete the turn along with the tricycle. Instead, her body goes straight as the tricycle twists out from under her. Crash. As you can see, a tricycle has good static stability but poor **dynamic stability**—stability in motion.

The tricycle flips because it can't handle the enormous torque that friction exerts on it during a sharp turn. Because the horizontal frictional force that turns the tricycle is exerted well below the tricycle and rider's combined center of mass (Fig. 4.1.3), it produces a torque about that center of mass. If this torque is small, the tricycle's static stability will provide a restoring torque in the opposite direction that prevents any angular acceleration. But if the turn is too sharp, the huge frictional torque will overwhelm the limited restoring torque and the tricycle and rider will flip over. During high-speed turns, the tricycle is dynamically *unstable.*

Since the goal of a wheeled vehicle is to go somewhere, dynamic stability is ultimately more important than static stability. And while a bicycle lacks static stability, a moving bicycle is remarkably stable. Its dynamic stability is so good that it's almost hard to tip over and can even be ridden without any hands on the handlebars. This feat is a popular daredevil stunt among children who haven't yet realized how easy it is.

As British physicist David Jones discovered (Fig. 4.1.4), the bicycle's incredible dynamic stability results from its tendency to steer automatically in whatever direction it's leaning. For example, if the bicycle begins leaning to the left, the front wheel will automatically steer toward the left so as to return the bicycle to an upright position. While a stationary bicycle falls over when it's disturbed from its unstable equilibrium, a forward-moving bicycle naturally drives under the combined center of mass and returns to that unstable equilibrium.

There are two physical mechanisms acting together to produce this automatic steering effect: one involving rotation and one involving potential energy. The first mechanism is based on the wheels alone: they behave as *gyroscopes*. Because each wheel is spinning, it has angular momentum and tends to continue spinning at a constant angular

Courtesy Dr. David Jones

Fig. 4.1.4 In 1970, British physicist David Jones investigated the origins of a bicycle's dynamic stability. He built a series of "unridable" bicycles, including one that had a front wheel so small that it often skidded and became almost red hot as the bicycle was ridden. Jones found that gyroscopic effects alone did not account for a bicycle's stability. He discovered that this stability also comes from the shape of the front fork.

speed about a fixed axis in space. Since a wheel's angular momentum can be changed only by a torque, it naturally tends to keep its upright orientation.

But angular momentum alone doesn't prevent the bicycle from tipping over, any more than it prevents a tricycle from doing so. Instead it prompts the bicycle to steer automatically by way of gyroscopic **precession**—the pivoting of a gyroscope's rotational axis caused by a torque exerted perpendicular to its angular momentum. When a bicycle is upright, the pavement exerts no perpendicular torque on the front wheel. But when the bicycle leans to the left, the pavement's upward support force no longer points at the wheel's center of mass and it produces a perpendicular torque on the wheel. This torque is what makes that wheel precess; its axis of rotation pivots toward the left and thereby steers the bicycle to safety!

Fig. 4.1.5 A bicycle is stable when moving forward in part because its front wheel touches the ground behind the steering axis. As a result, the front wheel naturally steers in the direction that the bicycle is leaning and returns the bicycle to an upright position.

Assisting gyroscopic precession in this automatic steering process is a second effect due to potential energy. Because of the shape and angle of the fork supporting its front wheel, a leaning bicycle can lower its center of gravity and total potential energy by steering its front wheel in the direction that the bicycle is leaning. When the bicycle leans to the left, its front wheel accelerates toward the left—the direction that lowers the bicycle's total potential energy as quickly as possible. Once again the bicycle automatically steers in the direction that it's leaning and avoids falling over. These self-correcting effects explain why a riderless bicycle stays up so long when you roll it forward or down a hill.

There is little a bicycle designer can do to change the gyroscopic effect, but the potential energy effect depends on fork shape and angle. To be stable, the front wheel must touch the ground behind the steering axis (Fig. 4.1.5). If the fork is flawed, so that the wheel touches the ground ahead of the steering axis, the bicycle will steer the wrong way when it leans and be virtually unridable.

The front fork of a typical adult's bicycle arcs forward so that the wheel touches the ground just behind the steering axis. This situation leaves the bicycle dynamically stable enough to ride yet highly maneuverable. In contrast, the front fork of a typical child's bicycle is relatively straight so that the wheel touches the ground far behind the steering axis. The child's bicycle is therefore more dynamically stable than the adult's bicycle but also less easy to turn. That trade-off between dynamic stability and maneuverability is universal, appearing not only in pedal-powered vehicles, but in cars, boats, and aircraft as well.

▶ Check Your Understanding #2: Like a Rolling Coin

for answers, see page 135

Standing a coin on edge is difficult, but a rolling coin stays upright easily. What effect is stabilizing the moving coin?

Leaning While Turning

So why does a bicycle rider lean during a turn? The answer is that leaning can balance out the torque that friction exerts on him during the turn, the same frictional torque that flipped our unfortunate tricycle rider. With the proper lean, the bicyclist can safely complete even the sharpest turn.

As he rides along, the bicyclist tries to keep himself and the bicycle in rotational equilibrium, that is, experiencing zero net torque about their combined center of mass. A net torque would cause them to undergo angular acceleration and flip. Since they experience a frictional torque each time they turn, the bicyclist balances that torque by leaning himself and/or the bicycle toward the inside of the turn. The pavement's upward

(a)

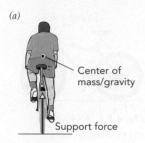

Center of
mass/gravity

Support force

(b)

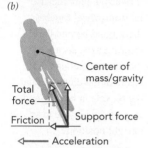

Center of
mass/gravity

Total
force

Friction

Support force

◄——— Acceleration

Fig. 4.1.6 (*a*) A bicycle
that is heading straight is in
rotational equilibrium when
it's perfectly upright. The
support force from the road
produces no torque about
its center of mass. (*b*) A
bicycle that is turning left
is in rotational equilibrium
when it's tilted to the left.
Together, the support and
frictional forces from the
road produce no torque
about its center of mass.

Wesley Hitt/Liaison Agency,
Inc./Getty Images

Fig. 4.1.7 As they round
a turn during a race, these
cyclists lean toward the
inside of the turn. This
leaning prevents the
pavement from producing
torques on them and
tipping them over.

support force on the wheels then produces a torque on them about their combined center of mass and that new torque opposes the frictional torque. When those two torques sum to zero, the bicyclist and bicycle are safely in rotational equilibrium.

Since both of those opposing torques are produced by pavement forces on the wheels, we can combine them into a single torque; we can first add the friction and support forces and then see how much torque that overall force produces on the bicyclist. What we'll find in doing this addition is that the torque drops to zero when the pavement pushes the wheels directly toward the bicycle and rider's combined center of mass. As we saw in Section 2.1, a force exerted directly toward a pivot produces zero torque about that pivot. To stay in rotational equilibrium as he rides, the bicyclist always places the combined center of mass directly in line with the pavement's force on the wheels.

For example, when the bicycle is heading straight, the rider can avoid a net torque by keeping the bicycle upright. The pavement is then pushing the wheels straight upward, directly toward the combined center of mass (Fig. 4.1.6*a*). But when the bicycle is turning left, the rider must lean himself and/or the bicycle to the left to stay in rotational equilibrium. That's because each wheel of the turning bicycle is experiencing not only an upward support force, but also the leftward frictional force it needs to accelerate through the turn (Fig. 4.1.6*b*). The road's total force on the turning bicycle therefore points up and to the left. By leaning leftward just the right amount, the rider ensures that this total force points directly toward the combined center of mass and exerts no torque about it (Fig. 4.1.7).

After a while, leaning the bike while turning becomes so automatic, so habitual, that you don't even think about it. You simply can't ride a bicycle without leaning as you turn. Moreover, even if you turn a bicycle or motorcycle so sharply that it skids, leaning can still keep it safely in rotational equilibrium.

But only a statically unstable vehicle can lean, so a statically stable one must rely on restoring torques to keep it near its rotational equilibrium. As we saw with tricycles, restoring torques are limited and a vehicle can tip out of its rotational equilibrium during rapid accelerations. A car, truck, or SUV will flip if it turns too sharply, and some are particularly prone to such catastrophes. The higher a vehicle's center of mass and the narrower its base of support, the more limited its restoring torques and the more easily it flips during turns. SUVs are surprisingly vulnerable to such rollover accidents, and small trucks that have been boosted up to resemble tractor-trailer cabs are truly hazardous. Even some production cars have been found unsafe in this regard.

► **Check Your Understanding #3: Cutting Corners**

for answers, see page **135**

Why does a skier lean in the direction of a turn during a downhill run?

Pedaling Bicycles

So far, we have only discussed stability. But once we've settled on a bicycle as the most likely configuration for a useful person-powered vehicle, we need to figure out how to person-power it. The rider could push his feet on the ground, but that would be pretty inconvenient and even dangerous at high speeds. Instead, we use foot pedals to produce a torque on one of the wheels. But which wheel, and how do we produce that torque?

The original answer was to power the front wheel, using cranks attached directly to its axle. A crank is simply a lever that projects from the axle and produces a torque on that axle as you push its free end around in a circle. Each bicycle crank has a pedal installed on its free end so that you can use your foot to push on it. With its axle suspended in bearings, the front wheel turns as you pedal it. Friction between the turning wheel and the ground then pushes the bicycle forward.

Of course, you can't get something for nothing: ground friction also pushes back on the wheel, opposing its rotation. That's why you have to keep pedaling: your pedaling twists the wheel forward, while ground friction twists it backward. When those two torques balance, you'll move along at a steady pace.

This pedaling method is still used in children's tricycles, but it has three drawbacks. First, pedaling the front wheel interferes with its other responsibility: steering. Second, you can't take a break from pedaling; if the vehicle is moving, so are the pedals. Third, you can produce more than enough torque on the front wheel, but you often have trouble moving your legs quickly enough to keep up with the pedals. When you ride on level ground, you find yourself pedaling furiously and feeling very little resistance from the pedals.

The frantic pedaling problem goes to the heart of all pedal-powered vehicles, vehicles that draw power from you, the rider, to overcome inertia, air resistance, and perhaps an uphill slope. You provide that power by doing work on the pedals—a certain amount of work each second. Since work is the product of force times distance, you can do the same work on the pedals each second—that is, you can provide the same power to the vehicle—by pushing the pedals down hard as they turn slowly or by pushing them down gently as they turn quickly. However, for reasons having to do with physiology more than physics, your legs are best at providing power when you are pushing the pedals down medium hard as they turn medium fast.

Unfortunately, the pedals of an ordinary tricycle turn too quickly and too easily to make good use of your pedaling power. When the tricycle is moving fast on level ground, you can barely keep up with its pedals, let alone do work on them. The only time a tricycle makes good use of your capacity to supply power is when it's going uphill at moderate speed; only then do the pedals move medium fast and require that you push on them medium hard.

An early solution to the frantic pedaling problem was to use a gigantic front wheel. In such a configuration, one turn of the wheel would take you a considerable distance, so that you no longer had to pedal furiously while traveling at high speed along a level road. At the same time, ground friction exerted a larger backward torque on the huge wheel, so that you had to push harder on the pedals to keep the wheel turning steadily. At last you could reach your peak performance on level ground, pushing medium hard on the pedals as they turned medium fast. The pennyfarthing of the mid-nineteenth century was this sort of bicycle (Fig. 4.1.8). But pedaling still interfered with steering and you couldn't stop pedaling while the bicycle was moving. Furthermore, this bicycle had a new problem: you couldn't push hard enough with your feet to keep its front wheel turning steadily on uphill stretches.

These problems were solved by removing the cranks from the front wheel's axle and using an indirect drive scheme to convey power to the rear wheel. Employing toothed sprockets and a chain loop, that indirect drive allows the pedals and the wheels to turn at different rates. This change lets you to use mechanical advantage to choose how you supply power to the bicycle: whether you exert large forces on slowly moving pedals or small forces on rapidly moving pedals or, ideally, medium forces on medium-fast moving pedals. Whether you're zooming along a level road or grinding slowly up a steep hill, you can always find a drive setting or "gear" that lets you comfortably supply your maximum power.

Lastly, the nonstop pedaling problem is solved by incorporating a one-way drive or freewheel in the hub of the rear wheel. This freewheel (Fig. 4.1.9) allows the rear wheel

Fig. 4.1.8 The pennyfarthing used a large, directly driven front wheel to permit the rider to travel at a reasonable speed without having to pedal very rapidly. Its name came from its resemblance to two coins, the large English penny and the smaller farthing.

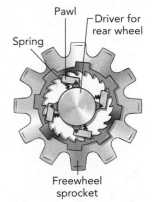

Fig. 4.1.9 The ratchet in a bicycle freewheel. If the relative rotation of the inner and outer parts is in the correct direction, the pawls transmit torque from the outer part to the inner part. If the relative rotation direction is reversed, the pawls compress the springs and skip along the teeth on the inside of the outer part. No torque is transmitted.

Fig. 4.1.10 A modern bicycle. The rear wheel is driven by a chain that allows the rider to vary the mechanical advantage between the pedals and the rear wheel. A freewheel in the hub of the rear wheel lets that wheel turn freely in one direction. This free motion allows the bicycle to coast forward, even when the pedals are stationary.

to turn freely in one direction so that you can stop pedaling as you coast forward. A modern bicycle with these improvements appears in Fig. 4.1.10.

Check Your Understanding #4: Bigger Isn't Always Better

for answers, see page 135

Why is it so difficult to climb a hill on a pennyfarthing?

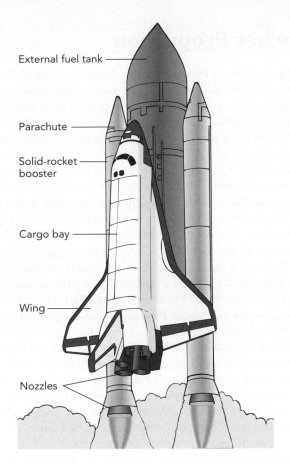

External fuel tank

Parachute

Solid-rocket booster

Cargo bay

Wing

Nozzles

SECTION 4.2 Rockets and Space Travel

Despite the complexity of modern spacecraft, the rocket is one of the simplest of all machines. It makes use of the very basic principle that every action has a reaction. A rocket is propelled forward by pushing material out its tail. But despite its simplicity, people have been developing better and better rockets for more than 700 years. Rockets are now used for such pursuits as space exploration, weaponry, rescue operations, and amusement.

Questions to Think About: *What pushes a rocket forward? How can a rocket work in space, where there seems to be nothing to push against? Why do modern rockets have such fancy exhaust nozzles? What is the fastest speed that a rocket can reach? Why do satellites travel endlessly around the earth? What sort of path does a rocket take as it travels from planet to planet?*

Experiments to Do: *While an afternoon spent launching model rockets (available at hobby and toy stores) would be the best introduction to this section, a toy balloon will do just fine. Blow up the toy balloon and let it go. It will sail around the room. What pushes the balloon forward? Are the room's air and walls involved in the propulsion or is the balloon propelled by the very act of ejecting gas through its opening? How could you verify your answer to the previous question?*

Rocket Propulsion

Among a rocket's most impressive features are its ability to propel itself forward even in the complete isolation of space and the astonishing speeds that it can achieve with that propulsion. It somehow manages to push itself forward without any outside help and to use that forward push to accelerate seemingly without limits.

Of course, a rocket can't really push itself forward, any more than you can lift yourself up by your boots, and it can't accelerate forever. In reality, it obtains a forward force, a **thrust** force, by pushing against its own limited store of fuel, and when that fuel runs out, it stops accelerating. To understand how a rocket obtains thrust from its fuel supply, let's look at how Newton's third law, the one describing action and reaction, applies to rockets.

Imagine that you're sitting in the middle of a frozen pond with zero velocity and no momentum. It's a warm day and the wet ice is remarkably slippery. Try as you like, you can't seem to get moving at all. How can you get off the ice?

Because of your inertia, the only way you can start moving is if something pushes on you. Sure, you could order a pizza and then push against the delivery truck when it arrives. But instead you follow the ideas we discussed on p. 66. You remove a shoe and throw it as hard as you can toward the east side of the pond (Fig. 4.2.1). As you throw the shoe, you exert a force on it with your hand. The shoe accelerates and heads off across the ice.

What happens to you? You head off toward the west side of the pond! You're moving because when you pushed the shoe toward the east side of the pond, it pushed you equally hard toward the west side of the pond. In the process, you transferred momentum to the shoe and it transferred momentum in the opposite direction to you. Momentum isn't being created or destroyed, it's only being redistributed. Even after you let go of the shoe, your combined momentum remains at zero. The shoe has as much momentum in one direction as you have in the other.

Of course, you are much more massive than the shoe, so you end up traveling slower than it does. Momentum is the product of mass times velocity, so the more massive the object, the less velocity it needs for the same amount of momentum. Still, you've achieved what you set out to do: you're sliding slowly toward the west side of the pond.

Your final speed is limited because you only managed to transfer a small amount of momentum to the shoe and thus received only a small amount of opposite momentum in return. If you'd been able to throw the shoe faster or if you'd thrown a whole box of shoes, you'd have transferred more momentum and would be going faster.

Instead of throwing shoes, you'd have done better to throw very fast-moving gas molecules. Even at room temperature, the molecules in air are traveling about 1800 km/h (1100 mph). When gas molecules are heated to roughly 2800 °C (5000 °F), as they are in a liquid-fuel rocket engine, they move about three times that fast. If you throw something in one direction at that kind of speed, you receive quite a lot of momentum in the other direction.

That's what a conventional rocket engine does (Fig. 4.2.2). It uses a chemical reaction to create very hot exhaust gas from fuels contained entirely within the rocket itself. What started as potential energy in the stored chemical fuels becomes thermal energy in the hot, burned exhaust gas. This thermal energy is mostly kinetic energy, hidden in the random motion of the tiny molecules themselves. The rocket engine's nozzle steers most of this random motion in one direction, and the engine obtains thrust in the opposite direction.

If you've ever watched the launch of a large rocket, you've probably noticed the bell-shaped nozzles through which the exhaust flows (Fig. 4.2.3). Each nozzle allows the rocket to obtain as much forward momentum as possible from its exhaust by directing that exhaust backward and accelerating it to the greatest possible speed. As we'll see in

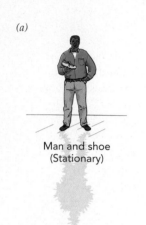

(a)

Man and shoe
(Stationary)

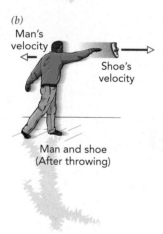

(b)

Man's
velocity

Shoe's
velocity

Man and shoe
(After throwing)

Fig. 4.2.1 A man who is holding a shoe while standing still on ice has zero momentum. Once he has thrown the shoe to the right, the shoe has a momentum to the right and the man has a momentum to the left. The total momentum of the man and shoe is still zero. Because the man is much more massive than the shoe, the shoe moves much faster than the man.

Chapter 6, nozzles allow gases to convert their various internal energies into kinetic energy and are ideally suited for directing and accelerating gases. In the case of rocket exhaust, the most effective nozzle shape is a converging–diverging one, called a de Laval nozzle after its Swedish inventor, Carl Gustaf de Laval ❑.

To understand fully why this complicated nozzle structure works so well for rocket engines, we'd need to examine the physics of flowing gases up to and beyond the speed of sound. We'll encounter some of these issues later in this book, but for now a brief summary will have to suffice.

Inside the rocket and before the de Laval nozzle, the hot exhaust gas is tightly packed and its pressure is enormous. Like the gas in a spray bottle, this exhaust gas accelerates rapidly through the nozzle toward the lower pressure environment outside. The narrowing throat of that nozzle aids its acceleration, up to a point. When the gas reaches the narrowest part of the nozzle, it's traveling at the speed of sound and its characteristics begin to change dramatically. To coax the supersonic exhaust gas to accelerate still further, the nozzle stops narrowing and begins to widen. The tightly packed exhaust gas expands in volume as it flows through that widening bell and thereby prepares to enter the more open environment outside the nozzle.

Just how wide the diverging end of the de Laval nozzle must be to obtain the maximum thrust from its exhaust gas depends the nozzle's surroundings. Near sea level, the exhaust gas flows into ordinary air outside the nozzle and a relatively narrow de Laval nozzle works best. At high altitude or in space, the exhaust gas enters thinner air or nothing at all, so a wider de Laval nozzle is more ideal. Rockets typically make a compromise in their nozzle shapes so as to operate reasonably well in all environments.

By the time the gas reaches the end of the de Laval nozzle, it has converted most of its original energy into kinetic energy, with its velocity directed away from the nozzle. In fact, because the gas actually continues to burn even as it flows through the nozzle, its kinetic energy and speed keep rising until they reach fantastic levels. With the help of the de Laval nozzle, exhaust gas leaves the rocket's engine at an **exhaust velocity** or backward-directed flow speed of between 10,000 and 16,000 km/h (6,000 and 10,000 mph).

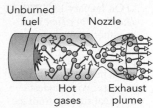

Unburned fuel Nozzle

Hot gases Exhaust plume

Fig. 4.2.2 A molecular picture of what happens in a chemical rocket engine. The engine burns its fuel in a confined chamber, and the exhaust gas flows out of a nozzle. The nozzle converts the random, thermal motions of the exhaust gas molecules into directed motion away from the rocket engine.

❑ Swedish inventor and engineer Carl Gustaf de Laval's (1845–1913) invention of the converging–diverging nozzle predates the modern development of rockets by several decades. He invented this nozzle as a way to make steam turbines more efficient and is credited with laying the foundation for all future turbine technology. De Laval is also known for his invention of the cream separator for milk.

Courtesy NASA

Fig. 4.2.3 The space shuttle's nozzles are designed to push its rocket exhaust downward as long glowing plumes. The gas pushes back, lifting the shuttle upward into space.

❏ On January 13, 1920, *The New York Times* ran an editorial attacking Robert Goddard for proposing that rockets could be used for travel in space. With modest financial support from the Smithsonian Institution, Goddard was pioneering the development of liquid fuel rockets. The editorial began: "That Professor Goddard, with his 'chair' in Clark College and the countenancing of the Smithsonian Institution, does not know the relation of action to reaction, and of the need to have something better than a vacuum against which to react—to say that would be absurd. Of course he only seems to lack the knowledge ladled out daily in high schools."

As it creates this plume of exhaust, the rocket pushes the gas backward and gives it backward momentum. The gas completes the momentum transfer by pushing the rocket forward. The very act of ejecting the exhaust is all that's required to obtain forward thrust; the rocket doesn't need anything external to push "against" and will operate perfectly well in empty space (see ❏). When it pushes hard enough on its exhaust, the rocket can not only support its own weight, it can even accelerate upward. The space shuttle weighs about 20,000,000 N (4,500,000 lbf) at launch but its thrust is about 30,000,000 N (6,750,000 lbf). That means that the space shuttle can accelerate upward at about half the acceleration due to gravity! As the shuttle consumes its fuel, so that its weight and mass diminish, it can accelerate upward even more rapidly.

Common Misconceptions: Action and Reaction in Rockets

Misconception: A rocket needs some external object to react against in order to push itself forward.

Resolution: While rocket propulsion does involve a pair of equal but opposite forces, action and reaction, the rocket is pushing its exhaust backward (action) and the exhaust is pushing the rocket forward (reaction). What this exhaust plume hits, if anything, makes no difference to the propulsion effect.

➤ Check Your Understanding #1: A Rocket with a Head Start

for answers, see page 135

When a missile is launched from beneath the wing of a fighter aircraft, what does it push against in order to accelerate forward?

The Ultimate Speed of a Spacecraft

At rest on the launch pad, a rocket consists principally of a spacecraft and a supply of fuel. Once the rocket's engine begins to fire, exhaust from the burned fuel accelerates backward and the spacecraft accelerates forward. The fuel is gradually consumed until eventually it runs out and the spacecraft coasts along on its own. Although weight and air resistance influence this story, let's neglect both for now to see what determines the spacecraft's eventual speed.

It might seem that the ultimate speed of the spacecraft is limited to the rocket's exhaust speed. Remarkably, there is no such limit. As long as the rocket keeps pushing exhaust backward, it will continue to accelerate forward. However, for the spacecraft to reach extremely high speed, the rocket must push the vast majority of its initial mass backward as exhaust. For example, if the rocket's initial mass is 90% fuel (i.e., it starts with 9 parts fuel and 1 part spacecraft), then we might expect the spacecraft to end up heading forward at about 9 times the speed of the exhaust gas. After all, that arrangement seems to satisfy the conservation of momentum.

Unfortunately, that analysis overestimates the spacecraft's speed. Because the rocket still has fuel on board as it accelerates forward, some of the forward momentum it obtains from its exhaust goes into its remaining fuel rather than into the spacecraft. Since that fuel will soon be ejected from the rocket as exhaust, giving it precious forward momentum is wasteful and the spacecraft therefore ends up with less forward momentum. While the total momentum of the spacecraft and fuel still always sum to zero, this inopportune momentum transfer from the exhaust to the fuel reduces both the spacecraft's speed forward and the exhaust's average speed backward.

Despite this problem, a spacecraft can still travel faster than the speed of its rocket exhaust; it just needs more fuel. Neglecting air resistance and weight, the spacecraft's final speed is given by the rocket equation:

$$\text{spacecraft speed} = \text{exhaust speed} \cdot \log_e\left(\frac{\text{mass}_{\text{spacecraft}} + \text{mass}_{\text{fuel}}}{\text{mass}_{\text{spacecraft}}}\right). \quad \textbf{(4.2.1)}$$

For a rocket that is 90% fuel at launch, its spacecraft can reach 2.3 times the speed of its exhaust gas. If it can eject more than 90% of its initial mass as exhaust, it can go even faster.

But there's a problem with trying to burn up and eject a huge fraction of the rocket's original mass as exhaust. It's difficult to construct a rocket that is 99.99% fuel and 0.01% spacecraft. Instead, space-bound rockets use several separate stages, each stage much smaller than the previous stage. Once the first stage has used up all of its fuel, the whole stage is discarded and a new, smaller rocket begins to operate. In this manner, the rocket behaves as though it's ejecting almost all of its mass as rocket exhaust. With the help of stages and lots of fuel, rockets can travel substantially faster than their exhaust velocities and reach earth orbit or the solar system beyond. (For ongoing developments in single-stage rockets, see ❏.)

❏ In 1989, the U.S. government began a program to develop a reusable rocket vehicle that could achieve earth orbit with only a single stage. Nothing but fuel would be jettisoned during launch so that the vehicle could travel to and from orbit repeatedly with only refueling and minimal maintenance between flights. The challenges facing this program are formidable. Even with liquid hydrogen and oxygen as its fuels, almost 90% of this vehicle's launch weight must be fuel. Nonetheless, efforts to produce such single stage to orbit (SSTO) vehicles are proceeding.

Check Your Understanding #2: Not Everything Is Disposable

for answers, see page **135**

The space shuttle doesn't have the staged look of expendable rockets. How does it manage to eject most of its launch mass as exhaust?

Orbiting the Earth

If the spacecraft was heading straight up when it ran out of fuel, it will either fall back to the ground or leave the earth forever (more on that later). But if it was heading primarily horizontally when its engine turned off, it may find itself circling the earth endlessly. With no atmosphere to affect it, the spacecraft follows a path determined only by inertia and gravity, and since the spacecraft's weight causes it to accelerate toward the center of the earth, its trajectory can bend into a huge elliptical loop around the earth.

The spacecraft is following an orbit around the earth. An **orbit** is the path an object takes as it falls freely around a celestial object. Although the spacecraft accelerates directly toward the earth's center at every moment, its huge horizontal speed prevents it from actually hitting the earth's surface. In effect, the spacecraft perpetually misses the earth as it falls (Fig. 4.2.4). To orbit the earth just above the atmosphere, a spacecraft must travel at the enormous speed of 7.9 km/s (about 17,800 mph) and will circle the earth once every 84 minutes.

However, the farther the spacecraft's orbit is from the earth's surface, the longer its **orbital period**—the time it takes to complete one orbit. First, since the spacecraft must travel farther to complete the larger orbit, the trip takes longer. Second, the spacecraft

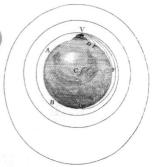

Courtesy New York Public Library

Fig. 4.2.4 Newton's drawing of a cannonball fired horizontally from the top of a tall mountain. As the cannonball's speed increases, it travels farther from the mountain before hitting the earth. Eventually, the cannonball moves so quickly that the curved earth drops away beneath it and it never hits the earth at all. The cannonball then orbits the earth.

must travel slower in order to follow a circular path around the earth because the earth's gravity becomes weaker with distance.

Chapter 1 noted that gravity attracts every object in the universe toward every other object in the universe (see ❏). In particular, objects are attracted toward the earth. Near the earth's surface, an object's weight is simply its mass times 9.8 N/kg, the acceleration due to gravity. But as the object's distance from the center of the earth increases, the acceleration due to gravity diminishes. Equation 1.2.1 is only an approximation, valid for objects near the earth's surface.

A more general formula relates the gravitational forces between two objects to their masses and the distance separating them. These forces are equal to the gravitational constant times the product of the two masses, divided by the square of the distance separating them. This relationship, discovered by Newton and called the **law of universal gravitation,** can be written as a word equation:

$$\text{force} = \frac{\text{gravitational constant} \cdot \text{mass}_1 \cdot \text{mass}_2}{(\text{distance between masses})^2}, \tag{4.2.2}$$

in symbols:

$$F = \frac{G \cdot m_1 \cdot m_2}{r^2},$$

and in common language:

The pull of gravity is strongest between massive objects but diminishes rapidly with distance.

Note that the force on mass_1 is directed toward mass_2 and the force on mass_2 is directed toward mass_1. Those two forces are equal in magnitude but oppositely directed. The **gravitational constant** is a fundamental constant of nature, with a measured value of $6.6720 \times 10^{-11} \text{ N} \cdot \text{m}^2/\text{kg}^2$.

The Law of Universal Gravitation
Every object in the universe attracts every other object in the universe with a force equal to the gravitational constant times the product of the two masses, divided by the square of the distance separating the two objects.

This relationship describes any gravitational attraction, whether it's between two planets or between the earth and you. The effective location of an object's mass is its center of mass, so the distance used in Eq. 4.2.2 is the distance separating the two centers of mass. For a spacecraft orbiting the earth just above its atmosphere, that distance is roughly the earth's radius of 6378 km (3964 miles). But for a spacecraft far above the atmosphere, the distance is larger and the force of gravity is weaker. That spacecraft experiences a smaller acceleration due to gravity. To give it the additional time it needs for its path to bend around in a circle, the high-altitude spacecraft must travel more slowly than the low-altitude spacecraft. This reduced speed explains the long orbital periods of high-altitude spacecraft.

At 35,900 km (22,300 miles) above the earth's surface, the orbital period reaches 24 hours. A satellite traveling eastward in such an orbit turns with the earth and is said to be *geosynchronous.* If a geosynchronous satellite orbits the earth around the equator, it's also *geostationary*—it always remains over the same spot on the earth's surface. Such a fixed orientation is useful for communications and weather satellites.

Not all orbits are circular. The orbits of some spacecraft are elliptical, so that their altitudes vary up and down once per orbit. At *apogee,* its greatest distance from the earth's center, a spacecraft travels relatively slowly because it has converted some of its kinetic energy into gravitational potential energy. At *perigee,* its smallest distance from the earth's center, the spacecraft travels relatively rapidly because it has converted some of its gravitational potential energy into kinetic energy. Of course, the perigee should not bring the spacecraft into the earth's atmosphere or it will crash.

The orbit of a spacecraft can also be hyperbolic. If the spacecraft is traveling too fast, the earth will be unable to bend its path into a closed loop and the spacecraft will coast off into interplanetary space. The spacecraft's path near the earth is then a hyperbola. The spacecraft only follows this hyperbolic path once and then drifts away from the earth forever.

A spacecraft usually enters a hyperbolic orbit by firing its rocket engine. It starts in an elliptical orbit around the earth and uses its rocket engine to increase its kinetic energy. The spacecraft then arcs away from the earth and its kinetic energy gradually transforms into gravitational potential energy. But the earth's gravity becomes weaker with distance and the spacecraft's gravitational potential energy slowly approaches a maximum value even as its distance from the earth becomes infinite. If the spacecraft has more than enough kinetic energy to reach this maximum gravitational potential energy, it will be able to escape completely from the earth's gravity.

The speed that a spacecraft needs in order to escape from the earth's gravity is called the **escape velocity.** This escape velocity depends on the spacecraft's altitude and is about 11.2 km/s (25,000 mph) near the earth's surface. A spacecraft traveling at more than the escape velocity follows a hyperbolic orbital path and heads off toward the other planets or beyond.

Common Misconceptions: Astronauts and "Weightlessness"

Misconception: An astronaut orbiting the earth is too far from the earth to experience gravity and is truly weightless.

Resolution: The astronaut is still so near the earth's surface that she experiences almost her full earth weight. She only feels weightless because she is in free fall.

▶Check Your Understanding #3: Speeding Up the Lunar Month

for answers, see page 135

The moon orbits the earth every 27.3 days, at a distance of 384,400 km from the earth's center of mass. For the moon to orbit in less time, how would its distance from the earth have to change?

▶Check Your Figures #1: Attractive Cars

for answers, see page 135

How much force does a 1000-kg automobile exert on an identical car located 10 m away?

Orbiting the Sun: Kepler's Laws

Once it escapes from the earth's gravity and again turns off its rocket engine, the spacecraft behaves like a tiny planet and orbits the sun. If you watch it patiently as it travels and compare its orbital motion with the motions of the planets themselves, you may begin to notice three universal features of all these solar orbits. First recognized by German astronomer Johannes Kepler (1571–1630) through his careful analysis of the extensive observational data collected by Danish astronomer Tycho Brahe, those three orbital behaviors are known as Kepler's laws.

Kepler's first law is already rather familiar to us from our examination of earth orbits. This law describes the shape of the spacecraft's looping orbit around the sun: it's an ellipse, with the sun at one focus of that ellipse (Fig. 4.2.5). An ellipse isn't an arbitrary oval; it's a planar curve with two foci and a rule stating that each point on the curve has the same sum of distances to the two foci. In this case, the sun occupies one focus and the other focus is empty. If you add the distance from the spacecraft to the sun plus the distance from the spacecraft to the empty focus, that sum will remain constant as the spacecraft orbits the sun. A circular orbit around the sun is a particularly simple elliptical one; its two foci coincide and the sun occupies them both.

Kepler recognized that every object orbiting the sun follows such an elliptical path. The planets move along nearly circular ellipses while the comets travel in highly elongated ones. Our spacecraft's orbit may be circular or elongated, depending on its position and velocity at the time its engine stopped firing. To reach another planet, the spacecraft's solar orbit and that of its destination planet must overlap and the two objects must reach that overlapping point at the same time. Traveling from planet to planet is clearly a tricky business.

Newton later recognized that these elliptical orbits are a direct consequence of the law of universal gravitation (Eq. 4.2.2) and its inverse square relationship between force and distance (force $\propto$ 1/distance^2). Any other relationship between force and distance would yield curving paths that don't close on themselves at all, let alone form elliptical loops. As you watch the spacecraft follow its elliptical orbit around the sun, you are witnessing an elegant exhibition of the law of universal gravitation.

> **Kepler's First Law: Orbits**
> All planets move in elliptical orbits, with the sun at one focus of the ellipse.

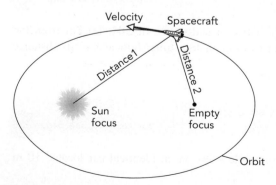

Fig. 4.2.5 A spacecraft's orbit around the sun is an ellipse, with the sun occupying one focus and the other focus empty. The sum of Distance 1 and Distance 2 is the same for all points on this ellipse.

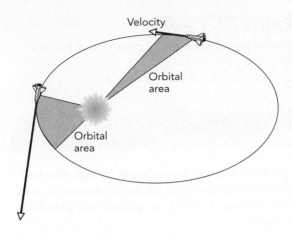

Fig. 4.2.6 The orbiting spacecraft sweeps out the same orbital area each second, despite variations in its distance from the sun. This steady sweep is a result of the spacecraft's constant angular momentum about the sun.

Kepler's second law describes the area swept out by a line stretching from the sun to the spacecraft: that line sweeps out equal areas in equal times (Fig. 4.2.6). Regardless of how circular or elongated the spacecraft's orbit is, or where the spacecraft is along that orbit, the area marked off each second by that moving line is always the same.

This observation demonstrates another physical law: conservation of angular momentum. Since the sun's gravity pulls the spacecraft directly toward the sun, it exerts no torque on the spacecraft about the sun and the spacecraft's angular momentum about the sun is constant. Remarkably, the rate at which this line sweeps out area is proportional to the spacecraft's angular momentum, so the steadiness of that sweep demonstrates the constancy of the spacecraft's angular momentum.

Kepler's Second Law: Areas
A line stretching from the sun to a planet sweeps out equal areas in equal times.

Kepler's third law describes the spacecraft's orbital period around the sun: the square of its orbital period is proportional to the cube of its mean distance from the sun, that is, the average of its *perihelion* (closest distance to the sun, Fig. 4.2.7) and *aphelion* (farthest distance to the sun). This relationship can be derived from the law of universal gravitation (Eq. 4.2.2), the equations describing centripetal acceleration (e.g., Eq. 3.3.1), and Newton's second law (Eq. 1.1.2).

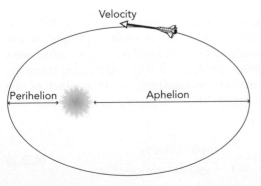

Fig. 4.2.7 The square of the spacecraft's orbital period is proportional to the cube of its mean distance from the sun—the average of its perihelion and its aphelion.

Kepler's Third Law: Periods
The square of a planet's orbital period is proportional to the cube of that planet's mean distance from the sun.

▶ Check Your Understanding #4: Out of Round

for answers, see page 135

Your spacecraft has just left the earth behind and is now orbiting the sun independently. Its elongated orbit has the same mean distance from the sun as the earth's orbit, but its aphelion is twice its perihelion. Compare its speeds at aphelion and perihelion. Will it ever meet up with the earth again?

Travel to the Stars: Special Relativity

Despite formidable challenges, it may one day be possible for manned spacecraft to venture away from the solar system and travel to the stars. The distances involved are so vast that the only way to cover them in an astronaut's lifetime would be to move at fantastic speeds, speeds comparable to that of light itself.

Should spacecraft one day be able to attain such enormous speeds, they'll find that the basic laws of motion, the Galilean and Newtonian laws that we have been learning up to this point, are incomplete. Although extremely accurate at ordinary speeds, those laws falter near the **speed of light** (exactly 299,792,458 m/s). They turn out to be low-speed approximations for the more accurate laws of motion developed by Einstein in 1905 (❏). Built on the observation that light always travels at the same speed, regardless of an observer's inertial frame of reference, these **relativistic laws of motion** are accurate at any speed. They are part of Einstein's **special theory of relativity,** the conceptual framework that describes space, time, and motion in the absence of gravity.

We saw in Section 1.1 that observers in different inertial frames of reference can disagree on an object's position and velocity. Special relativity recognizes that those observers can also disagree on the distance and time separating two events. More broadly, two inertial observers who are in relative motion perceive space and time somewhat differently. If they're moving at ordinary speeds, that difference in perceptions is negligible and the Newtonian laws of motion are nearly perfect. But if they're moving relative to one another at a substantial fraction of the speed of light, then they perceive space and time quite differently. In that case, the Newtonian approximations fail and the full laws of relativity are required.

Special relativity has many consequences for high-speed space travel, but we'll concentrate on how relativity alters two familiar conserved quantities: momentum and energy. At low speeds, our spacecraft's momentum takes its usual Newtonian value: mass times velocity (Eq. 2.3.1). But with increasing speed, a new relativistic factor enters the picture: $(1 - \text{speed}^2/\text{light speed}^2)^{-\frac{1}{2}}$. **Relativistic momentum** is the product of the object's mass times its velocity times that factor. This relationship can be written as a word equation:

❏ In 1905, while Albert Einstein (German-born Swiss then American physicist, 1879–1955) was working as a patent examiner in Bern, he published four revolutionary papers in three different areas of physics. For the 26-year-old doctoral student from Germany, it was quite a banner year. Though Einstein is often portrayed as an elderly, wild-haired gentleman, his most important contributions to science were made when he was a vibrant young man who had only just married his first wife two years earlier.

$$\text{relativistic momentum} = \frac{\text{mass} \cdot \text{velocity}}{\sqrt{1 - \text{speed}^2/\text{light speed}^2}}, \qquad (4.2.3)$$

in symbols:

$$\mathbf{p} = \frac{m \cdot \mathbf{v}}{\sqrt{1 - v^2/c^2}},$$

and in common language:

For a spacecraft to reach the speed of light, its momentum would have to be infinite.

At ordinary speeds, the relativistic factor is so nearly equal to 1 that this relativistic relationship is beautifully approximated by the Newtonian one. But as the spacecraft's speed increases to a substantial fraction of that of light itself, the relativistic factor spoils the simple proportionality between momentum and velocity. Momentum then increases more rapidly than velocity. One result of this change is that it becomes impossible to reach the speed of light, let alone exceed it. Even if the spacecraft's thrust increases its forward momentum at a steady rate, its speed will increase less and less quickly. It will approach but never reach the speed of light.

A similar change happens to the spacecraft's energy as it approaches the speed of light. At low speeds, our isolated spacecraft's kinetic energy takes its usual Newtonian value: half its mass times the square of its speed (Eq. 2.2.1). But at high speeds, we must begin using **relativistic energy.** Relativistic energy is the product of the object's mass times the square of the speed of light times the relativistic factor. This relationship can be written as a word equation:

$$\text{relativistic energy} = \frac{\text{mass} \cdot \text{light speed}^2}{\sqrt{1 - \text{velocity}^2/\text{light speed}^2}}, \qquad \textbf{(4.2.4)}$$

in symbols:

$$E = \frac{m \cdot c^2}{\sqrt{1 - v^2/c^2}},$$

and in common language:

A spacecraft starts with a rest energy and its energy grows toward infinity as its speed approaches the speed of light.

At ordinary speeds, the spacecraft's relativistic energy can be approximated as:

$$\text{relativistic energy} \simeq \text{mass} \cdot \text{light speed}^2 + \tfrac{1}{2} \cdot \text{mass} \cdot \text{speed}^2.$$

The usual Newtonian kinetic energy appears at the right in this approximation, but to its left is a new energy that we've never seen before. Called the *rest energy,* it's present even when the spacecraft is motionless. Because the rest energy is constant, it doesn't affect low-speed motion and was overlooked in the Newtonian laws. However, this energy associated with mass itself (stated symbolically as $E = mc^2$) does have consequences and is surely the most famous feature of the special theory of relativity.

The relativistic version of energy has two implications for our spacecraft. First, the spacecraft's energy increases so quickly as it nears the speed of light that it can never reach that speed. Second, the spacecraft's initial store of energy before launch is associated with its initial mass. That mass and energy are so closely related is something to which we'll return in Chapter 16.

Check Your Understanding #5: Stellar Tugboats

for answers, see page 135

A stellar tugboat is exerting a steady forward force on a spacecraft that is already traveling close to the speed of light. Although the force is constant, the spacecraft is speeding up less and less quickly. Why?

Check Your Figures #2: The Real Speed Limit

for answers, see page 136

A spacecraft passes your planetary base going half the speed of light. Using special equipment, you transfer exactly enough momentum to that spacecraft to double its forward momentum. How fast is it going now?

Check Your Figures #3: It Packs a Wallop

for answers, see page 136

The speed of a 1-kg minispacecraft is $\sqrt{3/4}$ times the speed of light. What is its relativistic energy and what fraction of that energy is rest energy?

Visiting the Stars: General Relativity

In its travels near stars and other massive celestial objects, our spacecraft is likely to encounter another surprise: the Newtonian view of gravity is also an approximation! Near extremely dense, massive objects, gravity is no longer accurately described by Newton's law of universal gravitation. Instead, understanding gravity requires a new conceptual framework that Einstein first presented in 1916, the **general theory of relativity.**

This new framework is based on the observation that you can't distinguish between downward gravity and upward acceleration. As we saw in Section 3.3, they feel exactly the same. For example, if you feel heavy as you stand inside a closed spacecraft, you can't be sure whether you're experiencing a weight due to downward gravity or a feeling of acceleration due to upward acceleration. In fact, the spacecraft's scientific instruments can't help you because they can't distinguish the effects of gravity from those of acceleration, either.

At the heart of this problem is the concept of mass. Up to this point, we have seen mass play two apparently different roles that we could refer to as gravitational mass and inertial mass. When you're experiencing a weight, your **gravitational mass** is acting together with gravity to make you feel heavy. When you're experiencing a feeling of acceleration, your **inertial mass** is acting together with acceleration to make you feel heavy. But, in spite of their different roles, these two masses seem to be related. Without exception, an object that has a large inertial mass and is therefore difficult to shake also has a large gravitational mass and is therefore hard to support against gravity. In fact, the two masses seem to be the same. That observation led Einstein to propose the **principle of equivalence:** that these two masses, gravitational and inertial, are truly identical and therefore that no experiment you perform inside your spacecraft can distinguish between free fall and the absence of gravity. The general theory of relativity is based on this principle of equivalence.

As long as your spacecraft stays in regions of weak gravity, Newtonian's law of universal gravitation will adequately describe its motion. But at the extremes of gravity, the general theory of relativity is necessary. That theory describes a universe in which massive objects distort the structure of nearby space and time, and in which extreme masses produce extreme distortions. One of the most startling predictions of this theory is the existence of objects so radical in their gravitational warping of nearby space and time that they are **black holes**—spherical or nearly spherical surfaces from which not even light can escape. A number of black holes have been discovered, including an enormous one at the center of our galaxy. You might want to avoid them.

Check Your Understanding #6: Rip-van-Twinkle

for answers, see page **135**

You awaken aboard your small spacecraft only to discover that you have been asleep for almost 20 years and the crew has vanished. The windows are closed and you have no idea where your spacecraft is located or what it is doing. You realize that you feel pulled toward the floor of the spacecraft. Are you experiencing weight or a feeling of acceleration?

Epilogue for Chapter 4

This chapter discussed the physical concepts behind two types of machines. In *bicycles,* we investigated the concept of dynamic stability, where an object that falls over when stationary becomes remarkably stable in motion. We also saw how the need for mechanical advantage between the pedals and the wheels led to the development of the modern multispeed bicycle.

In *rockets,* we studied the ways in which reaction forces propel rockets forward. We found that a rocket's ultimate speed is not limited by its exhaust velocity, allowing it to lift itself into orbit around the earth. We also learned that gravity weakens as the distance between two gravitating objects increases, making it possible for a spaceship to break free from earth's gravity and begin orbiting the sun. Finally, we took a brief look at the exotic physics a spacecraft would encounter if it approached the speed of light or passed through the extreme gravity near some celestial objects.

Explanation: High Flying Balls

The top ball doesn't bounce off the ground, it bounces off the bottom ball. The top ball actually completes its bounce as the bottom ball is heading upward. After colliding, the two balls push off one another in just the same way that a rocket pushes off its exhaust. The balls exchange momentum and energy as they push against one another and the small top ball—which has less mass and accelerates much more easily than the bottom ball—ends up with a large upward momentum and more than its fair share of energy. It flies upward as though it were hit by a massive upward-moving *bat*. In fact, it has been hit by a massive upward-moving *ball* and it rebounds at high speed.

If the small ball's mass were really negligible in comparison to that of the large ball and the balls bounced without wasting any energy, then the small ball would rebound from the big ball traveling 3 times as fast as when it arrived. Its kinetic energy would be 9 times as great as before and it would rebound to 9 times its original height. Of course, real balls aren't perfect so the tennis ball won't bounce quite so high. Still, the effect is pretty impressive.

Chapter Summary

How Bicycles Work: While a stationary bicycle tips over easily, a moving bicycle is remarkably stable. It remains upright with the help of two stabilizing effects of motion—one due to gyroscopic precession and the other due to the shape and angle of the front fork. These effects work together to steer the bicycle in the direction that it's leaning. Whenever it tips, it automatically drives under its center of gravity and returns to upright. Leaning is also an essential part of turns. By leaning himself and/or the bicycle properly during a turn, the bicyclist ensures that there is no overall torque on the bicycle and that it doesn't tip over.

The rider powers the bicycle by pushing its pedals around in a circle. Since the rider can provide the most pedaling power when the pedals are turning at a moderate rate and he is pushing on them with moderate forces, a modern multispeed bicycle allows him to adjust the relative rotation rates of the pedals and the wheels. By choosing the right gear, the bicycle allows him to provide his peak power comfortably.

How Rockets Work: A rocket obtains its thrust by ejecting gas from an engine in its tail. The rocket pushes on the gas and the gas pushes back. This gas usually comes from burning chemical fuels contained entirely inside the rocket itself and is released from the rocket's engine through a nozzle. A carefully designed nozzle permits the rocket to make efficient use of the energy stored in the fuel by ensuring that gas leaves the rocket at the maximum possible speed. The rocket's thrust is used to lift the rocket against the force of gravity and to accelerate it upward. Once in space, with its engine inactive, the rocket will orbit the earth or sun, or it may travel off into interstellar space.

Important Laws and Equations

1. The Law of Universal Gravitation: Every object in the universe attracts every other object in the universe with a force equal to the gravitational constant times the product of the two masses, divided by the square of the distance separating the two objects, or

$$\text{force} = \frac{\text{gravitational constant} \cdot \text{mass}_1 \cdot \text{mass}_2}{(\text{distance between masses})^2}. \quad (4.2.2)$$

2. Kepler's First Law: All planets move in elliptical orbits, with the sun at one focus of the ellipse.

3. Kepler's Second Law: A line stretching from the sun to a planet sweeps out equal areas in equal times.

4. Kepler's Third Law: The square of a planet's orbital period is proportional to the cube of that planet's mean distance from the sun.

5. Relativistic Momentum: An object's relativistic momentum is the product of its mass times its velocity times the relativistic factor, or

$$\text{relativistic momentum} = \frac{\text{mass} \cdot \text{velocity}}{\sqrt{1 - \text{speed}^2/\text{light speed}^2}}. \quad (4.2.3)$$

6. Relativistic Energy: An object's relativistic energy is the product of its mass times the square of the speed of light times the relativistic factor, or

$$\text{relativistic energy} = \frac{\text{mass} \cdot \text{light speed}^2}{\sqrt{1 - \text{velocity}^2/\text{light speed}^2}}. \quad (4.2.4)$$

Check Your Understanding—Answers

Section 4.1 BICYCLES

1. The mug's wide base ensures that its total potential energy will continue to increase as it tips, even if that tip becomes quite severe, so it will naturally return to its stable equilibrium. **Why:** A mug with a narrow base is stable, but only if it never tips very far. As soon as its center of gravity moves beyond its original base of support, over it will go. But a mug with a wide base has a broad base of support and will recover even from severe tips.

2. Gyroscopic precession.
Why: On edge, the coin has little or no static stability and tips over easily. But when it's rolling forward quickly, the same gyroscopic precession effect that causes a bicycle to steer under its center of gravity also dynamically stabilizes the coin's otherwise unstable equilibrium.

3. To make sure that the force on her skis is directed toward her center of mass.
Why: There are many sports and situations where a person must lean in the direction of a turn. Since a turn always involves horizontal acceleration and horizontal force, the person must shift her center of mass toward the inside of the turn. When she leans just the right amount, the net force on her feet points directly toward her center of mass and exerts no torque on her about her center of mass. She remains in rotational equilibrium and doesn't tip over.

4. Each turn of the huge wheel takes you a long distance up the hill and requires lots of work. To do this much work in a single turn of the pedals, you must push on the pedals very hard.
Why: As you pedal a bicycle, you are doing work on the pedals; you push the pedals as they move in the direction of that push. If too much work is required, as is the case when you're going uphill on a pennyfarthing, the force required becomes unbearable.

Section 4.2 ROCKETS AND SPACE TRAVEL

1. It pushes against its own exhaust, as do all rockets.
Why: While it may appear that the plume of exhaust beneath a ground-launched rocket is what lifts that rocket upward, the ground itself doesn't contribute to the thrust. The very act of pushing the gas out the nozzle propels the rocket forward.

2. It's actually staged subtly. Its two solid fuel boosters are effectively the first stage, the external liquid fuel tank is the second stage, and the orbiter itself is the third stage.
Why: Although the space shuttle is not stacked one stage on the next like a Saturn or Delta rocket, it doesn't travel from ground to space as a single object. It discards empty fuel containers as it accelerates upward. First to go are the two boosters, followed by the external fuel tank. The final mass of the orbiter itself is much less than what left the launch pad.

3. The distance between the earth and moon would have to decrease.
Why: The moon behaves just like a spacecraft orbiting the earth at a distance of 384,400 km. Such a spacecraft would also have an orbital period of 27.3 days. To reduce this orbital period, the moon would have to move closer to the earth so that the earth's gravity could bend its path more rapidly.

4. It is moving half as fast at aphelion as at perihelion and it will meet the earth again in exactly one year.
Why: In accordance with Kepler's first law, the spacecraft is traveling in an elliptical orbit. Since Kepler's second law requires it to sweep out area at a constant rate, the spacecraft must move half as fast when it is twice as far from the sun. Finally, because the spacecraft's mean distance from the sun is the same as that of the earth, the two have the same orbital periods. They'll both complete their orbits in one year and meet up at that time.

5. The tugboat is steadily increasing the spacecraft's forward momentum, but in accordance with the relativistic relationship between momentum and velocity, the spacecraft's velocity increases ever more slowly.
Why: At ordinary speeds, a steady transfer of momentum to an object will yield a steady increase in its speed. But near the speed of light, the object's increase in speed no longer keeps pace with its increase in momentum.

6. Trick question! You can't possibly determine the answer!
Why: The central principle of general relativity is that no experiment you make inside your spacecraft can distinguish between the consequences of gravity and the consequences of acceleration. You're simply going to have to look out the window to figure it out.

Check Your Figures—Answers

Section 4.2 ROCKETS AND SPACE TRAVEL

1. About 6.73×10^{-7} N.

Why: We use Eq. 4.2.2 to obtain the force:

$$\text{force} = \frac{6.6720 \times 10^{-11} \text{ N} \cdot \text{m}^2/\text{kg}^2 \cdot 1000 \text{ kg} \cdot 1000 \text{ kg}}{(10 \text{ m})^2}$$

$$= 6.6720 \times 10^{-7} \text{ N}.$$

This force is roughly equal to the weight of a grain of sand. No wonder it's hard to feel gravity from anything but the entire earth.

2. $\sqrt{4/7}$ times the speed of light.

Why: When the spacecraft is traveling at $\frac{1}{2}$ the speed of light, Eq. 4.2.3 gives its forward momentum as $\sqrt{1/3}$ times its mass times the speed of light. Doubling that momentum and solving Eq. 4.2.3 for the velocity yields $\sqrt{4/7}$ times the speed of light.

3. Its total energy is 1.8×10^{17} J, of which half is rest energy.

Why: For an object traveling at $\sqrt{3/4}$ times the speed of light, Eq. 4.2.4 gives a total energy of 2 times its mass times the square of the speed of light. For a 1-kg object, that energy is 1.8×10^{17} J. Its rest energy is simply its mass times the square of the speed of light, or 9.0×10^{16} J. Both values of energy are astonishingly large.

Exercises

1. Sprinters start their races from a crouched position with their bodies well forward of their feet. This position allows them to accelerate quickly without tipping over backward. Explain this effect in terms of torque and center of mass.

2. If the bottom of your bicycle's front wheel becomes caught in a storm drain, your bicycle may flip over so that you travel forward over the front wheel. Explain this effect in terms of rotation, torque, and center of mass.

3. When you turn while running, you must lean in the direction of a turn or risk falling over. If you lean left as you turn left, why don't you fall over to the left?

4. If a motorcycle accelerates too rapidly, its front wheel will rise up off the pavement. During this stunt the pavement is exerting a forward frictional force on the rear wheel. How does that frictional force cause the front wheel to rise?

5. As a skateboard rider performs stunts on the inside of a U-shaped surface, he often leans inward toward the middle of the U. Why does leaning keep him from falling over?

6. Most racing cars are built very low to the ground. While this design reduces air resistance, it also gives the cars better dynamic stability on turns. Why are these low cars more stable than taller cars with similar wheel spacings?

7. The crank of a hand-operated kitchen mixer connects to a large gear. This gear meshes with smaller gears attached to the mixing blades. Since each turn of the crank makes the blades spin several times, how is the force you exert on the crank handle related to the forces the mixing blades exert on the batter around them?

8. The starter motor in your car is attached to a small gear. This gear meshes with a large gear that's attached to the engine's crankshaft. The starter motor must turn many times to make the crankshaft turn just once. How does this gearing allow modest forces inside the starter motor to turn the entire crankshaft?

9. A bread-making machine uses gears to reduce the rotational speed of its mixing blade. While its motor spins about 50 times each second, the blade spins only once each second. The motor provides a certain amount of work each second, so why does this arrangement of gears allow the machine to exert enormous forces on the bread dough?

10. The chain of a motorcycle must be quite strong. Since the motorcycle has only one sprocket on its rear wheel, the top of the chain must pull forward hard to keep the rear wheel turning as the motorcycle climbs a hill. Why would replacing the motorcycle's rear sprocket for one with more teeth make it easier for the chain to keep the rear wheel turning?

11. As you clear the sidewalk with a leaf blower, the blower pushes you away from the leaves. What is pushing on the blower so that it can push on you?

12. When a sharpshooter fires a pistol at a target, the gun recoils backward very suddenly, leaping away from the target. Explain this recoil effect in terms of the transfer of momentum.

13. Which action will give you more momentum toward the north: throwing one shoe southward at 10 m/s or two shoes southward at 5 m/s?

14. Do both of the actions in Exercise 13 take the same amount of energy? If not, which one requires more energy?

15. You are propelling yourself across the surface of a frozen lake by hitting tennis balls toward the southern shore. From your perspective, each ball you hit heads southward at 160 km/h. You have a huge bag of balls with you and are already approaching the northern shore at a speed of 160 km/h (100 mph). When you hit the next ball southward, will you still accelerate northward?

16. Can you use the tennis ball scheme of Exercise 15 to propel yourself to any speed, or are you limited to the speed at which you can hit the tennis balls?

17. You and a friend are each wearing roller skates. You stand facing each other on a smooth, level surface and begin tossing a heavy ball back and forth. Why do you drift apart?

18. The time it takes a relatively small object to orbit a much larger object doesn't depend on the mass of the small object. Use an astronaut who is walking in space near the space shuttle to illustrate that point.

19. As the moon orbits the earth, which way is the moon accelerating?

20. Which object is exerting the stronger gravitational force on the other, the earth or the moon, or are the forces equal in magnitude?

21. To free an Apollo spacecraft from the earth's gravity took the efforts of a gigantic Saturn V rocket. Freeing a lunar module from the moon's gravity took only a small rocket in the lunar module's base. Why was it so much easier to escape from the moon's gravity than from that of the earth?

22. Spacecraft in low earth orbit take about 90 minutes to circle the earth. Why can't they be made to orbit the earth in half that amount of time?

23. As a comet approaches the sun, it arcs around the sun more rapidly. Explain.

24. Mars has a larger orbital radius than the earth. Compare the solar years on those two planets.

Problems

1. If an 80-kg baseball pitcher wearing frictionless roller skates picks up a 0.145-kg baseball and pitches it toward the south at 42 m/s (151 km/h or 94 mph), how fast will he begin moving toward the north?

2. How high above the earth's surface would you have to be before your weight would be only half its current value?

3. As you walked on the moon, the earth's gravity would still pull on you weakly and you would still have an earth weight. How large would that earth weight be, compared to your earth weight on the earth's surface? (Note: The earth's radius is 6378 km and the distance separating the centers of the earth and moon is 384,400 kilometers.)

4. If you and a friend 10 m away each have masses of 70 kg, how much gravitational force are you exerting on your friend?

5. The gravity of a black hole is so strong that not even light can escape from within its surface or "event horizon." Even outside that surface, enormous energies are needed to escape. Suppose that you are 10 km away from the center of a black hole that has a mass of 10^{31} kg. If your mass is 70 kg, how much do you weigh?

6. In Problem 5, how much work would something have to do on you to lift you 1 m farther away from the black hole?

7. A 1000-kg spacecraft passes you traveling forward at ½ the speed of light. What is its relativistic momentum?

8. A spacecraft passes you traveling forward with at ⅓ the speed of light. By what factor would its relativistic momentum increase if its speed doubled?

9. For a rocket traveling at 10,000 m/s, by what factor does its relativistic momentum differ from its ordinary momentum?

10. What is the rest energy of 1 kg of uranium?

11. How much mass has a rest energy of 1000 J?

12. A 1000-kg spacecraft passes you at ½ the speed of light. What is its relativistic energy?

13. A spacecraft passes you at ⅓ the speed of light. By what factor would its relativistic energy increase if its speed doubled?

14. By what factor does the relativistic energy in Problem 13 increase if you subtract out the constant rest energy?

CHAPTER 5

Fluids

So far all of the everyday objects we've examined have been solids. But since gases and liquids are also important parts of the world around us—as the air we breathe, the water we swim in, and even the blood we pump through our veins—we'll now turn to objects that, unlike solids, don't have well-defined shapes. These objects are called fluids, and the study of fluid behavior and motion is a broad field, extending across the sciences and engineering. Fluid dynamics, often called hydrodynamics, is as important to an oil-well engineer as to an animal physiologist or a stellar astrophysicist. The tools used to analyze fluids are somewhat more complicated than for solids because fluids themselves are more complicated: it's hard to exert a force directly on them, and, even if we could, they usually don't move as a single rigid object. In this chapter, we'll look at some of the concepts and tools needed to understand their complex behaviors.

EXPERIMENT: A Cartesian Diver

One of these concepts is buoyancy: an object immersed in a fluid experiences an upward force from that fluid. This buoyant force is what lifts a helium balloon into the

Courtesy Lou Bloomfield

sky and suspends a boat on the surface of water. How the object responds to buoyancy depends on the relative densities of the object and the fluid surrounding it, where density is the ratio of mass to volume. As we'll see in this chapter, an object that's more dense than the surrounding fluid sinks, while one that's less dense than that fluid floats.

To see the importance of density in determining whether an object sinks or floats, you can construct a simple toy called a Cartesian diver. This once-popular parlor gadget consists of a small air-filled vial floating in a sealed container of water. Normally, the air bubble inside the vial keeps it floating at the surface of the water, but whenever you squeeze the container, the vial sinks.

To make a Cartesian diver, you'll need only a plastic soda bottle and a small vial that's open at one end. The vial can be made of almost anything—plastic, metal, or glass—as long as it's dense enough to sink in water. Fill the plastic soda bottle full of water and float the vial in it upside down; air trapped inside the vial should keep the vial afloat. Now slowly reduce the size of the air bubble inside the vial until the vial barely floats. You can make this adjustment by tipping the vial to let some of its air escape or by removing it from the bottle and pouring water into it. Once you have the vial floating only a few millimeters out of the water at the top of the soda bottle, cap the bottle and prepare to test your diver.

Before you squeeze the soda bottle, try to *predict* what will happen when you do. How will squeezing the bottle affect the air bubble and the vial's position in the water? Now squeeze the bottle hard and *observe* the results. Did they *verify* your predictions? *Measure* the air bubble's size as you squeeze the bottle. How does the bubble's size depend on how hard you squeeze the bottle? Why should the two be related? What is the relationship between the size of the air bubble and the diver's height in the water?

As you release the pressure on the bottle, the diver will float back up to the surface. Why does the sunken diver suddenly become buoyant again? By carefully squeezing the bottle, you can even make the diver hover in the middle of the bottle. Try making the diver hover while your eyes are closed. Why is hovering so difficult to sustain? Why must you watch the diver to make it hover?

Chapter Itinerary

We'll return to the diver at the end of the chapter. But first we'll examine two things from the world around us: (1) *balloons* and (2) *water distribution*. In *balloons,* we'll explore how the concepts of pressure and buoyancy help explain how the earth's atmosphere keeps hot air and helium balloons from falling to the ground. In *water distribution,* we'll see both how pressure propels water through plumbing and the ways in which water can contain energy. For a more complete preview of this chapter, jump ahead to the Chapter Summary on p. 160.

The issues we'll be looking at crop up frequently in our everyday experiences. Pressure plays an important role in aerosol cans, steam engines, firecrackers, and even the weather; buoyancy supports ships on water and keeps oil above vinegar in a bottle of salad dressing. Just as important, these concepts will lay the groundwork for Chapter 6, where we'll examine objects in which motion affects the behavior of fluids.

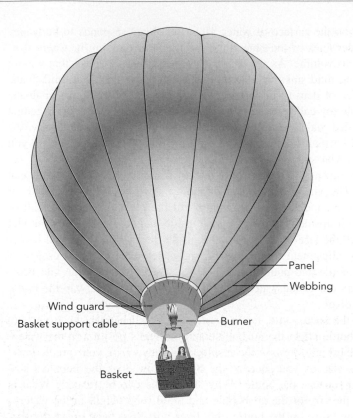

Panel

Webbing

Wind guard

Basket support cable

Burner

Basket

SECTION 5.1   Balloons

Because gravity gives every object near the earth's surface a weight proportional to its mass, objects fall when you drop them. Why then does a helium-filled balloon—which, after all, is just another object with a mass and a weight—sail upward into the sky when you let go of it? Does the balloon have a negative mass and a negative weight, or are we forgetting something?

We're forgetting air—specifically, the layer of air that sits atop the earth's surface and is held in place by gravity. Since this air is difficult to see and moves out of our way so easily, we often forget that it's there. But air sometimes makes itself noticeable. When you ride a bicycle, you feel its forces; when you blow up a beach ball, you see that it takes up space. And when you release a helium balloon, air lifts the balloon upward.

Questions to Think About: Since most objects fall to the ground through the atmosphere, why doesn't the atmosphere itself fall downward? Why is the air "thinner" in the mountains than at sea level? If you suck all of the air out of a plastic bag, what squeezes the bag into a thin sheet? Why does blowing air into the bag make it inflate? What happens to the bag's total mass when you fill it with air? with hot air? with helium? If you took a sturdy helium balloon to the moon, where there is no air, and then released it, which way would it move?

Experiments to Do: Pull on the string of a helium balloon to get a feel for how it behaves. If it pulls upward on your fingers, does that mean its weight (and mass) is negative? How would an object with a negative mass respond to a force? Shake the balloon and convince yourself that the balloon's mass is positive. Do you think there are any objects with negative masses?

Since the balloon's mass and weight are both positive, gravity must be pulling the balloon downward. But how can the stationary balloon pull upward on your finger? What other forces might be pushing the balloon upward? You can enhance these upward forces by partially submerging the balloon in a container of water. Where else do similar upward forces appear in everyday life?

Take the balloon for a ride in a car. Which way does the balloon move when you start suddenly? when you stop suddenly? Again, it seems as though the balloon's mass is negative. What is pushing on the balloon to make it move in this counterintuitive way?

Air and Air Pressure

Hot-air and helium balloons are supported by the air around them. Although these balloons have positive masses and downward weights, the surrounding air pushes upward on them hard enough to balance their weights so that they float. To understand balloons, we must start by understanding air.

Like the objects we've already studied, air has mass and weight. Unlike those objects, however, it has no fixed shape or size. You can mold 1 kg of air into any form you like and it can occupy a wide range of **volumes.** Air is **compressible,** that is, you can squeeze a certain mass of it into almost any space. For example, 1 kg of air could fill a single scuba tank or a whole basketball arena.

This flexibility of size and shape originates in the microscopic nature of air. Air is a **gas,** a substance consisting of tiny, individual particles that travel around independently. These individual particles are atoms and molecules. An **atom** is the smallest portion of an element that retains all of the chemical characteristics of that element; a **molecule,** an assembly of two or more atoms, is the smallest portion of a chemical compound that retains all of the characteristics of that compound. A molecule's atoms are held together by **chemical bonds,** linkages formed by electromagnetic forces between the atoms.

Air particles are extremely small, less than a millionth of a millimeter in diameter. Most are nitrogen and oxygen molecules, but others include carbon dioxide, water, methane, and hydrogen molecules, and argon, neon, helium, krypton, and xenon atoms. Those atoms, which don't make strong chemical bonds and rarely form molecules, are called **inert gases** because of their chemical inactivity.

Like tiny marbles, these air particles have sizes, masses, and weights. But while marbles quickly settle to the ground when you spill them from a bag, air particles don't seem to fall at all. Why don't they pile up on the earth's surface?

The answer has to do with air's thermal energy, specifically the portion of that energy contained in the motions of the air particles. This **internal kinetic energy** keeps the tiny air particles moving, spinning, and away from the earth's surface. In contrast, real marbles are too massive and heavy to be moved noticeably by thermal energy. The air's internal kinetic energy per particle is measured as its **temperature;** the greater that energy per particle, the hotter the air. While air's thermal energy also includes a portion stored in the forces between particles, this **internal potential energy** is negligible because the average forces between air particles are so weak.

An expanded view of air would reveal countless individual particles in frenetic **thermal motion** (Fig. 5.1.1*a*). At room temperature, these particles travel at bulletlike speeds of roughly 500 m/s (1100 mph), but they collide so often that they make little

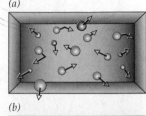

(a)

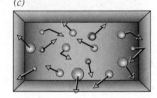

(b)

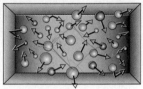

(c)

Fig. 5.1.1 (*a*) As air particles bounce off surfaces, they exert pressure on those surfaces; the amount of pressure depends on the air's temperature and on how densely its particles are packed. (*b*) Packing the air particles more densely increases the number of particles that hit the surfaces each second. (*c*) Increasing the temperature of the air increases the speed of the particles (shown by the arrows) so that they hit the surfaces harder and more frequently. Either change, in speed or collision frequency, increases the air's pressure.

progress in any particular direction. Between collisions, air particles travel in nearly straight-line paths because gravity has too little time to make them fall very far.

Let's ignore gravity for the moment and consider what happens in a box containing 1 kg of air. The air particles whiz around inside the box and each time a particle bounces off a wall of the box, it exerts a force on that wall. Although the individual forces are tiny, the number of particles is not, and together they produce a large average force. The size of this total force depends on the wall's **surface area;** the larger its surface area, the more average force it experiences. In order to characterize the air, however, we don't really need to know the wall's surface area; instead, we can refer to the average force the air exerts on each unit of surface area, a quantity called **pressure.**

Pressure is measured in units of force-per-area. Since the SI unit of surface area is the **meter^2** (abbreviated m^2 and often referred to as the square meter), the SI unit of pressure is the **newton-per-meter^2.** This unit is also called the **pascal** (abbreviated Pa), after French mathematician and physicist Blaise Pascal. One pascal is a small pressure; in contrast, the air around you has a pressure of about 100,000 Pa (2100 lbf/ft^2 or 15 lbf/inch^2), so that it exerts a force of about 100,000 N on a 1-m^2 surface. Since 100,000 N (22,500 lbf) is about the weight of a city bus, air pressure can exert enormous forces on large surfaces.

Besides pushing on the walls of our hypothetical box, air also pushes on any object immersed in it. Its particles bounce off the object's surfaces, pushing them inward. As long as the object can withstand these compressive forces, the air won't greatly affect it since the uniform air pressure ensures that the forces on all sides of the object cancel one another perfectly. A sheet of paper, for example, will experience zero net force because the forces exerted on its two sides will add to zero.

Air particles also bounce off one another, so that air pressure exerts forces on air, too. A cube of air inserted into the box experiences all of the inward forces that a cube of metal would experience. The air around the cube pushes inward on it, and the cube pushes outward on the air around it. Since the net force on the cube of air is zero, the cube doesn't accelerate.

▶ Check Your Understanding #1: Getting a Grip on Suction

for answers, see page 161

After you push a suction cup against a smooth wall, the elastic cup bends back and a small, empty space is created between the cup and the wall. What keeps the suction cup against the wall?

Pressure, Density, and Temperature

Since air pressure is produced by bouncing air particles, it depends on how often, and how hard, those particles hit a particular region of surface. The more frequent or harder the impacts, the greater is the air pressure.

To increase the rate at which air particles hit a surface, we can pack them more tightly. If we add another 1 kg of air to our hypothetical box, we double the number of air particles in the same volume, which doubles the rate at which they hit each surface and therefore doubles the pressure (Fig. 5.1.1*b*). Air's pressure is thus proportional to its **density,** that is, its mass per unit of volume. Since the SI unit of volume is the **meter^3** (abbreviated m^3 and often referred to as the cubic meter), the SI unit of density is the **kilogram-per-meter^3** (abbreviated kg/m^3). The air around you has a density of about

1.25 kg/m^3 (0.078 lbm/ft^3). Water, in contrast, has a much greater density of about 1000 kg/m^3 (62.4 lbm/ft^3).

We can also increase the rate at which air particles hit a surface by speeding them up (Fig. 5.1.1c). If we double the internal kinetic energy of the air in our box, we double the average kinetic energy of each particle. Because a particle's kinetic energy depends on the square of its speed, doubling its kinetic energy increases its speed by a factor of $\sqrt{2}$. As a result, each particle hits the surface $\sqrt{2}$ times as often and exerts $\sqrt{2}$ times as much average force when it hits. With each particle exerting $\sqrt{2} \cdot \sqrt{2}$ or two times as much average force, the pressure doubles. Air's pressure is thus proportional to the average kinetic energy of its particles—to their average internal kinetic energies.

This average kinetic energy per particle is measured by the air's temperature; the hotter the air, the larger the average kinetic energy per particle and the higher the air's pressure. But the most convenient scale for relating the temperature of air to its pressure isn't the common **Celsius** (°C) or **Fahrenheit** (°F) scale; instead, it's a special **absolute temperature scale.** The SI scale of absolute temperature is the **Kelvin** scale (K). When the air's temperature is 0 K (−273.15 °C or −459.67 °F), it contains no internal kinetic energy at all and has no pressure; this temperature is called **absolute zero.** The Kelvin scale is identical to the Celsius scale, except that it's shifted so that 0 K is equal to −273.15 °C. In addition to associating the zero of temperature with the zero of internal kinetic energy, the Kelvin scale avoids the need for negative temperatures. Room temperature is about 293 K.

Since air pressure is proportional to both the air's density and its absolute temperature, we can express the relationship among these quantities in the following way:

$$\text{pressure} \propto \text{density} \cdot \text{absolute temperature}. \qquad \textbf{(5.1.1)}$$

This proportionality is useful, since it allows us to predict what will happen if we change the temperature or density of a specific gas, such as air. But it has its limitations; in particular, it doesn't work if we compare the pressures of two different gases, such as air and helium, which differ in their chemical compositions. To make such a comparison, we'll need to improve on Eq. 5.1.1. We'll do that later when we examine helium balloons.

Even in describing a specific gas, Eq. 5.1.1 has other shortcomings. The main problem is that real gas particles aren't completely independent of one another. If the temperature drops too low, the particles begin to stick together to form a liquid and Eq. 5.1.1 becomes invalid. But despite its limitations, this simple relationship between pressure, density, and temperature will prove useful in understanding how hot-air balloons float: it will help us understand the basic structure of the earth's atmosphere, the origins of the upward force that keeps a hot-air balloon aloft, and the reason why hot air rises.

Quantity	SI Unit	English Unit	SI → English	English → SI
Area	meter^2 (m^2)	foot^2 (ft^2)	1 m^2 = 10.764 ft^2	1 ft^2 = 0.092903 m^2
Volume	meter^3 (m^3)	foot^3 (ft^3)	1 m^3 = 35.315 ft^3	1 ft^3 = 0.028317 m^3
Pressure	pascal (Pa)	pound-force-per-foot^2 (lbf/ft^2)	1 Pa = 0.020885 lbf/ft^2	1 lbf/ft^2 = 47.880 Pa
Density	kilogram-per-meter^3 (kg/m^3)	pound-mass-per-foot^3 (lbm/ft^3)	1 kg/m^3 = 0.062428 lbm/ft^3	1 lbm/ft^3 = 16.018 kg/m^3

► Check Your Understanding #2: Snacks That Go Pop in the Night
for answers, see page **161**

If you remove a partially filled container of food from the refrigerator and allow it to warm to room temperature, the lid will often bow outward and may even pop off. What has happened?

The Earth's Atmosphere

Most of the mass of the earth's atmosphere is contained in a layer less than 6 km (4 miles) thick. Since the earth is 12,700 km (7,900 miles) in diameter, this layer is relatively thin—so thin that, if the earth were the size of a basketball, it would be no thicker than a sheet of paper.

The atmosphere stays on the earth's surface because of gravity. Every air particle, as we've seen, has a weight. Just as a marble thrown upward eventually falls back to the ground, so the particles of air keep returning toward the earth's surface. Although the particles are moving too fast for gravity to affect their motions significantly over the short term, gravity works slowly to keep them relatively near the earth's surface. An air particle, like a rapidly moving marble, may appear to travel in a straight line at first, but it will arc over and begin to fall downward eventually. Only the lightest and fastest moving particles in the atmosphere, hydrogen molecules and helium atoms, occasionally manage to escape from earth's gravity and drift off into interplanetary space.

While gravity pulls the atmosphere downward, air pressure pushes the atmosphere upward. As the air particles try to fall to the earth's surface, their density increases and so does their pressure. It's this air pressure that supports the atmosphere and prevents it from collapsing into a thin pile on the ground.

To understand how gravity and air pressure structure the atmosphere, picture a 1-m-square column of the atmosphere as though it were a tall stack of 1-kg air blocks (Fig. 5.1.2). These blocks support one another with air pressure to form a stack of about 10,000 blocks. The bottom block must support the weight of all the blocks above it and is tightly compressed, with a height of about 0.8 m, a density of about 1.25 kg/m^3, and a pressure of about 100,000 Pa. A block farther up in the stack has less weight to support and is less tightly compressed. The higher in the stack you look, the lower the density of the air and the less the air pressure.

The atmosphere has essentially the same structure as this stack of blocks. The air near the ground supports the weight of several kilometers of air above it, giving it a density of about 1.25 kg/m^3 and a pressure of about 100,000 Pa; at higher altitudes, however, the air's density and pressure are reduced, since there is less atmosphere overhead and the air doesn't have to support as much weight. High-altitude air is thus "thinner" than low-altitude air. Whatever the altitude, the pressure of the surrounding air is referred to as **atmospheric pressure.**

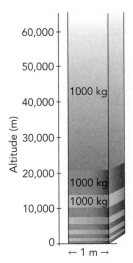

Fig. 5.1.2 The air in a 1-m^2 column of atmosphere has a mass of about 10,000 kg. The bottom 1000 kg is the most tightly compressed, because it supports the most weight above it. At higher altitudes, the air is less tightly compressed because it has less weight above it.

► Check Your Understanding #3: Mountain Travel Is a Pain in the Ears
for answers, see page **161**

As you drive up and down in the mountains, you may feel a popping in your ears as air moves to equalize the pressures inside and outside your eardrum. What causes these pressure changes?

The Lifting Force on a Balloon: Buoyancy

So far we've examined air, air pressure, and the atmosphere. While it may seem that we've avoided dealing with balloons, these topics really are involved in keeping a hot-air or helium balloon aloft. As we've seen, the air in the earth's atmosphere is a **fluid,** a shapeless substance with mass and weight. This air has a pressure and exerts forces on the surfaces it touches; that pressure is greatest near the ground and decreases with increasing altitude. Air pressure and its variation with altitude allow air to lift a hot-air or helium balloon through an effect known as buoyancy.

Buoyancy was first described more than two thousand years ago by the Greek mathematician Archimedes (287–212 B.C.). Archimedes realized that an object partially or wholly immersed in a fluid is acted on by an upward **buoyant force** equal to the weight of the fluid it displaces. **Archimedes' principle** is actually very general and applies to objects floating or submerged in any fluid, including air, water, or oil. The buoyant force originates in the forces a fluid exerts on the surfaces of an object. We've seen that such forces can be quite large but tend to cancel one another. How then can pressure create a nonzero total force on an object, and why should that force be in the upward direction?

Archimedes' Principle
An object partially or wholly immersed in a fluid is acted on by an upward buoyant force equal to the weight of the fluid it displaces.

Without gravity the forces would cancel each other perfectly because the pressure of a stationary fluid would be uniform throughout. But gravity causes a stationary fluid's pressure to decrease with altitude. For example, when nothing is moving, the air pressure beneath an object is always greater than the air pressure above it. Thus air pushes upward on the object's bottom more strongly than it pushes downward on the object's top, and the object consequently experiences an upward overall force from the air—a buoyant force.

How large is the buoyant force on this object? It's equal in magnitude to the weight of the fluid that the object displaces. To understand this remarkable result, imagine replacing the object with a similarly shaped portion of the fluid itself (Fig. 5.1.3*a*). Since the buoyant force is exerted by the surrounding fluid, not the object, it doesn't depend on the object's composition. A balloon filled with helium will experience the same buoyant force as a similar balloon filled with water or lead or even air. So replacing the object with a similarly shaped portion of fluid will leave the buoyant force on it unchanged.

But a portion of fluid suspended in more of the same fluid doesn't accelerate anywhere; it just sits there, so the net force on it is clearly zero. It has a downward weight, but that weight must be canceled by some upward force that can only come from the surrounding fluid. This upward force is the buoyant force, and it's always equal in magnitude to the weight of the object-shaped portion of fluid, the fluid displaced by the object.

This buoyant principle explains why some objects float while others sink. An object placed in a fluid experiences two forces: its downward weight and an upward buoyant force. If its weight is more than the buoyant force, it will accelerate downward

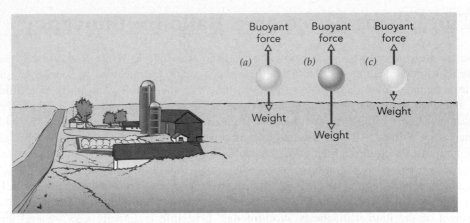

Fig. 5.1.3 (*a*) A portion of air immersed in that same air experiences an upward buoyant force equal to its weight and doesn't accelerate. (*b*) An object that is heavier than the air it displaces sinks, while (*c*) another object that is lighter than the air it displaces floats.

(Fig. 5.1.3*b*); if its weight is less than the buoyant force, it will accelerate upward (Fig. 5.1.3*c*). And if the two forces are equal, it won't accelerate at all and will maintain a constant velocity.

Whether or not an object will float in a fluid can also be viewed in terms of density. An object that has an average density greater than that of the surrounding fluid sinks, while one that has a lesser average density floats. A water-filled balloon, for example, will sink in air because water and rubber are more dense than air. If you double the volume of the balloon, you double both its weight and the buoyant force on it, so it still sinks. The total volume of an object is less important than how its density compares to that of the surrounding fluid.

> ### Check Your Understanding #4: Why People Don't Float in Air
> *for answers, see page* **161**

If a person displaces 0.08 m³ (2.8 ft³) of air, what is the buoyant force he experiences?

Hot-Air Balloons

Since air is very light, with a density of only 1.25 kg/m³ (0.078 lbm/ft³), few objects float in it. One of these rare objects is a perfectly empty balloon. Assuming that the balloon has a very thin outer shell or *envelope,* it will weigh almost nothing and have an average density near zero. Because its negligible weight is less than the upward buoyant force it experiences, the empty balloon will float upward nicely.

Unfortunately, this empty balloon won't last long. Because it's surrounded by atmospheric pressure air, each square meter of its envelope will experience an inward force of 100,000 N. With nothing inside the balloon to support its envelope against this crushing force, it will smash flat. A thick, rigid envelope might be able to withstand the pressure of the surrounding air, but then the balloon's average density would be large and it would sink. So an empty balloon won't work.

What will work is a balloon filled with something that exerts an outward pressure on the envelope equal to the inward pressure of the surrounding air. Then each portion of the envelope will experience zero net force and the balloon will not be crushed. We could fill the balloon with outside air, but that would make its average density too high. Instead, we need a gas that has the same pressure as the surrounding air but a lower density.

One gas that has a lower density at atmospheric pressure is hot air. Filling our balloon with hot air takes fewer particles than filling it with cold air, since each hot-air particle is moving faster and contributes more to the overall pressure than does a cold-air particle. A hot-air balloon contains fewer particles, has less mass, and weighs less than it would if it contained cold air. Now we have a practical balloon with an average density less than that of the surrounding air. The buoyant force it experiences is larger than its weight, and up it goes (Fig. 5.1.4).

Because the air pressure inside a hot-air balloon is the same as the air pressure outside the balloon, the air has no tendency to move in or out (an issue we will cover in the next section), and the balloon doesn't need to be sealed (Fig. 5.1.5). A large propane burner, located beneath the balloon's open end, heats the air that fills the envelope. The hotter the air in the envelope, the lower its density and the less the balloon weighs. The balloon's pilot controls the flame so that the balloon's weight is very nearly equal to the buoyant force on the balloon. If the pilot raises the air's temperature, particles leave the envelope, the balloon's weight decreases, and the balloon rises. If the pilot allows the air to cool, particles enter the envelope, the balloon's weight increases, and the balloon descends.

But even if the pilot heats the air very hot, the balloon won't rise upward forever. As the balloon ascends, the air becomes thinner and the pressure decreases both inside and outside the envelope. Although the balloon's weight decreases as the air thins out, the buoyant force on it decreases even more rapidly, and it becomes less effective at lifting its cargo. When the air becomes too thin to lift the balloon any higher, the balloon reaches a *flight ceiling* above which it can't rise, even if the pilot turns the flame on full blast. For each hot-air temperature, then, there is a cruising altitude at which the balloon will hover. When the balloon reaches that altitude, it's in a stable equilibrium. If the balloon shifts downward for some reason, the net force on it will be upward; if it shifts upward, the net force on it will be downward.

Check Your Understanding #5: Ballooning Weather

for answers, see page 161

Can a hot-air balloon lift more on a hot day or a cold day?

Helium Balloons

Although the particles in hot and cold air are similar, there are fewer of them in each cubic meter of hot air than in each cubic meter of cold air. We call the number of particles per unit of volume **particle density,** and hot air has a smaller particle density than cold air (Fig. 5.1.6). Because they contain similar particles, hot air also has a smaller density than cold air and is lifted upward by the buoyant force.

But there's another way to make one gas float in another: use a gas consisting of very light particles. Helium atoms, for example, are much lighter than air particles. When they have equal pressures and temperatures, helium gas and air also have equal particle densities. Since each helium atom weighs 14% as much as the average air particle, 1 m³ of helium weighs only 14% as much as 1 m³ of air. Thus a helium-filled balloon has

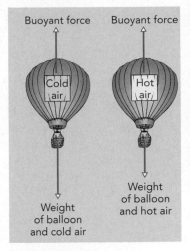

Fig. 5.1.4 A balloon filled with hot air contains fewer air particles and weighs less than a balloon filled with cold air. If the balloon's weight is small enough, the net force on the balloon will be in the upward direction and the balloon will accelerate upward.

Courtesy Lou Bloomfield

Fig. 5.1.5 The bottom of a hot-air balloon is open so that heated air can flow in and cold air can flow out. The heated air displaces more than its weight in cold air and makes the balloon lighter.

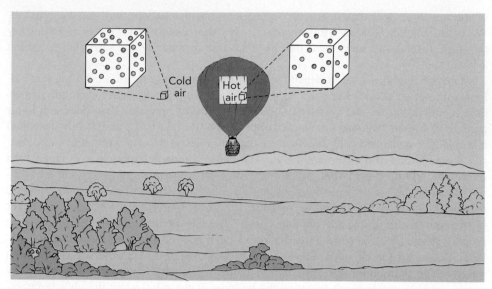

Fig. 5.1.6 A cube of hot air contains fewer air particles than a similar cube of cold air. Since it weighs less than the cold air it displaces, the hot air inside the balloon experiences an upward buoyant force that is greater than its weight.

only a fraction of the weight of the air it displaces, and the buoyant force carries it upward easily.

Why should air and helium have the same particle densities whenever their pressures and temperatures are equal? Because a gas particle's contribution to the pressure doesn't depend on its mass (or weight). At a particular temperature, each particle in a gas has the same average kinetic energy in its translational motion, regardless of its mass. Although a helium atom is much less massive than a typical air particle, the average helium atom moves much faster and bounces more often. As a result, lighter but faster-moving helium atoms are just as effective at creating pressure as heavier but slower-moving air particles.

Thus, if you allow the helium atoms inside a balloon to spread out until the pressures and temperatures inside and outside the balloon are equal, the particle densities inside and outside the balloon will also be equal (Fig. 5.1.7). Since the helium atoms inside the balloon are lighter than the air particles outside it, the balloon weighs less than the air it displaces, and it will be lifted upward by the buoyant force.

The pressure of a gas is proportional to the product of its particle density and its absolute temperature, as the following formula indicates:

$$\text{pressure} \propto \text{particle density} \cdot \text{absolute temperature.} \qquad \textbf{(5.1.2)}$$

This proportionality holds regardless of the gas's chemical composition. Our previous proportionality, Eq. 5.1.1, worked only as long as the gas's composition didn't change, so that density and particle density remained proportional to one another. But now we have a relationship with a wider applicability.

Equation 5.1.2, with an associated constant of proportionality, is called the **ideal gas law.** This law relates pressure, particle density, and absolute temperature for a gas in which the particles are perfectly independent. It's also fairly accurate for real gases in which the particles do interact somewhat. The constant of proportionality is the

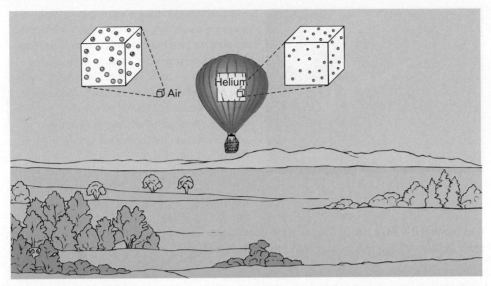

Fig. 5.1.7 A cube of helium gas contains the same number of particles as a similar cube of air, but each helium particle weighs less than the average air particle. Since it weighs less than the air it displaces, the helium inside the balloon experiences an upward buoyant force that is greater than its weight.

Boltzmann constant, with a measured value of 1.381×10^{-23} Pa · m³/(particle · K). Using the Boltzmann constant, the ideal gas law can be written as a word equation:

pressure = Boltzmann constant · particle density · absolute temperature, **(5.1.3)**

in symbols:

$$p = k \cdot \rho_{\text{particle}} \cdot T,$$

and in everyday language:

Don't incinerate a spray can. A hot, dense gas tends to burst its container.

The Ideal Gas Law
The pressure of a gas is equal to the product of the Boltzmann constant times the particle density times the absolute temperature.

Helium isn't the only "lighter-than-air" gas. Hydrogen gas, which is half as dense as helium, is also used to make balloons float. But don't expect hydrogen to lift twice as much weight as helium. A balloon's lifting capacity is the difference between the upward buoyant force it experiences and its downward weight. Although the gas in a hydrogen balloon weighs half that in a similar helium balloon, the balloons experience the same buoyant force. Thus the hydrogen balloon's lifting capacity is only slightly more than that of the helium balloon. Hydrogen's main advantage is that it's cheap and plentiful, while helium is scarce (see ❏). But because hydrogen is also dangerously

❏ Helium gas is obtained as a by-product of natural gas production from underground reservoirs in the United States, where it formed through the gradual radioactive decay of uranium and other unstable elements. While some of this gas is saved for industrial and commercial use, much is simply released into the atmosphere. The only other source of helium is the atmosphere, where helium is present at a level of 5 parts per million. Once the underground stores are consumed, helium will become a relatively rare and expensive gas.

Corbis Images

Fig. 5.1.8 Rigid airships were appropriately named because they were truly ships that floated through the air. Unfortunately, the hydrogen gas that filled most airships, making them light enough that air could lift them, is very flammable. The *Hindenburg* burned on May 6, 1937, while trying to land at Lakehurst, New Jersey. Because hydrogen is so buoyant in air, most of the combustion occurred above the airship and many passengers survived.

❏ Even helium-filled airships were easily destroyed by bad weather. The *Shenandoah,* one of two U.S. airships based on the German designs, was destroyed by air turbulence on September 3, 1925, near Ava, Ohio. Crowds from a local fair immediately poured over the wreckage, collecting souvenirs.

flammable, it's avoided in situations where safety is important (Fig. 5.1.8). However, even helium-filled airships can have problems (see ❏).

Check Your Understanding #6: What Not to Put in a Balloon

for answers, see page 161

A carbon dioxide molecule is heavier than an average air particle. If you pour carbon dioxide gas from a cup, which way does it flow in air, up or down?

Check Your Figures #1: Popped Out of the Refrigerator

for answers, see page 162

When you take an air-filled plastic container out of the refrigerator, it warms from 2 °C to 25 °C. How much does the pressure of the air inside it change?

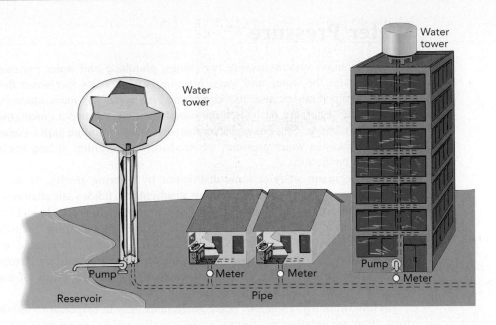

SECTION 5.2 Water Distribution

Now that we've explored the behavior of objects in fluids, let's turn to the behavior of fluids in objects. In this section, as we examine how plumbing distributes water, we'll see that pressure, density, and weight are just as important in plumbing as they are in ballooning. To keep things simple, we'll focus on the causes of water's motion through plumbing, leaving most of the complications associated with the motion itself for the next chapter. For example, we'll temporarily ignore drag and viscosity and the fascinating pressure changes that accompany fluid motion.

Questions to Think About: Why is the water pressure higher in the basement than it is in the attic? Why does a deep water well require a pump at the bottom? If a water tower is only a storage device, why is it so tall? Why do skyscrapers have complicated plumbing systems that include reservoirs at various levels in the building? What causes water to flow up through a drinking straw to your mouth?

Experiments to Do: To see the effects of pressure and weight on water, try these simple experiments with a drinking straw. First, "suck" water up the straw from a glass to your mouth. Are you exerting an attractive force on the water, or is some other force pushing it upward toward your mouth? With the straw full of water, seal the top with your finger; keeping the seal tight, remove the straw from the glass. What happens to the water inside? What happens to the water when you release the seal? Now blow into one end of a straw full of water while sealing the far end with your finger. When you release the seal on the far end, what happens to the water? What forces are responsible for this effect?

Water Pressure

Water distribution systems require two things: plumbing and water pressure. Plumbing is what delivers the water, and water pressure is what starts that water flowing. Water pressure is important because, like everything else, water has mass and accelerates only when pushed. If nothing pushed on the water when you opened a faucet, the water simply wouldn't budge. Since the pushes that send water through pipes come principally from differences in water pressure, we need to look carefully at how such pressure is created and controlled.

We'll begin our study of water distribution by ignoring gravity. As we've seen with the atmosphere, gravity creates **pressure gradients** in fluids—distributions of pressure that vary continuously with position. Pressure decreases with altitude and increases with depth, and these vertical pressure gradients complicate plumbing in hilly cities and skyscrapers. But if all of our plumbing is in a level region—for example, a single-story house in a very flat city—our job is much simpler. With no significant changes in height, we can safely ignore gravity, since no water is supporting the weight of water above it and gravity's effects are minimal.

In this simplified situation, water accelerates only in response to unbalanced pressures. Just as unbalanced forces make a solid object accelerate, so unbalanced pressures make a fluid accelerate. If the water inside a pipe is exposed to a uniform pressure throughout, then each portion of the water feels no net force and doesn't accelerate; it either remains stationary or coasts in a straight line at a steady pace (Fig. 5.2.1). But if the pressure is out of balance, the water accelerates toward the region of lowest pressure.

This acceleration doesn't mean that the water will instantly begin moving toward the lowest pressure. Because of its inertia, water changes velocity gradually: it speeds up, slows down, or turns to the side, depending on where the lowest pressure is located. A complicated arrangement of high and low pressures can steer water through an intricate maze of pipes, and that is exactly how water reaches your home from the city pumping station. Every change in its velocity during its trip through the plumbing is caused by a pressure imbalance.

You can create a pressure imbalance in water simply by squeezing parts of it. The pressure in a squeezed portion will rise and it will accelerate toward lower pressures elsewhere. Since this sort of pressure change isn't caused by the water's motion, it's a **static variation** in pressure. But the motion of water itself can also affect its pressure, and such **dynamic variations** in pressure can be complicated and fascinating. As we'll see in the next chapter, they contribute to such diverse effects as the spray of a garden hose nozzle, the lift on an airplane's wing, and the curve of a curve ball.

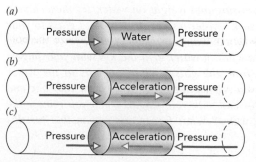

(a)

(b)

(c)

Fig. 5.2.1 (a) If the water in a horizontal pipe is exposed to a uniform pressure, then it will not accelerate. (b, c) If the pressure along the pipe is not uniform, however, the imbalance will create a net force on each portion of the water and the water will accelerate toward the side with the lower pressure.

→ Check Your Understanding #1: Under Pressure in the Garden
                                                    for answers, see page 162

With the water faucet open and the nozzle at the end of your garden hose shut tightly, the hose is full of high-pressure water. Why doesn't this water accelerate?

Creating Water Pressure with Water Pumps

To start water flowing through the plumbing in a level house or city, you need a water pump—a device that uses mechanical work to deliver pressurized water through a pipe. At its most basic level, a water pump squeezes a portion of water to raise its local pressure and keeps squeezing as that water accelerates and flows out toward regions of lower pressure elsewhere in the plumbing.

To understand how a pump works, picture a sealed plastic soda bottle full of water. When you don't squeeze the bottle, the pressure inside it is atmospheric and uniform (remember we're neglecting gravity). But when you squeeze the sides of the bottle and push inward on the water, the water responds by pushing outward on you—Newton's third law—and it does this by increasing its pressure. The harder you push, the harder the water pushes back and the greater its pressure becomes.

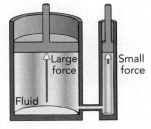

Fig. 5.2.2 The force a pressurized fluid exerts on a piston is proportional to the surface area of that piston. This fact underlies hydraulic systems, in which a small force exerted on a small piston pressurizes a trapped liquid so that it exerts a large force on a large piston.

Like all liquids, water is **incompressible**—its volume doesn't change as its pressure increases—so the bottle won't get smaller. But the rise in water pressure inside it can be substantial. It doesn't take much force, exerted with your thumb on a small area of the bottle, to increase the pressure in the bottle from atmospheric to twice that value or more.

The pressure increases uniformly throughout the water bottle, an observation known as **Pascal's principle**—a change in the pressure of an enclosed incompressible fluid is conveyed undiminished to every part of the fluid and to the surfaces of its container. This uniform pressure rise leads to a large upward force on the bottle's cap. If the cap were wider and had more surface area, the upward force on it might be large enough to blow it off the bottle. That effect is the basis for hydraulic systems and lifts, where pressure produced in an incompressible fluid by a small force exerted on a small area of the fluid's container results in a large force exerted on a large area of the fluid's container (Fig. 5.2.2). It also explains why plastic drinking bottles usually have small caps and why wide-mouth plastic bottles are better suited for candies, cookies, and nuts.

> **Pascal's Principle**
> A change in the pressure of an enclosed incompressible fluid is conveyed undiminished to every part of the fluid and to the surfaces of its container.

It's time to remove the cap from the water bottle. Now when you squeeze the bottle and increase the local water pressure, the water can move. As its pressure rises inside the bottle, the water begins to accelerate toward the lower pressure above the bottle's open top and the result is a fountain. You are pumping water!

You are also doing work: as the water flows out of the bottle, your hands move inward. Since you are pushing inward on the water and the water is moving inward, you are doing work on the water. Pumps do work when they deliver pressurized water, and pressurized water carries with it the energy associated with that work.

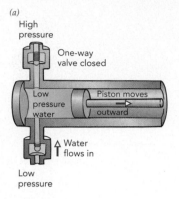

(a)
High pressure

One-way valve closed

Low pressure water

Piston moves outward

Water flows in

Low pressure

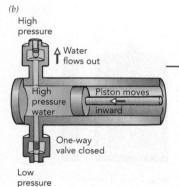

(b)
High pressure

Water flows out

High pressure water

Piston moves inward

One-way valve closed

Low pressure

Fig. 5.2.3 Water is pumped from a region of low pressure to a region of high pressure by a reciprocating piston pump. (*a*) As the piston is drawn outward, water flows into the cylinder from the low-pressure region. (*b*) As the piston is pushed inward, the inlet one-way valve closes and water is driven out of the cylinder and into the high-pressure region.

While a water bottle can act as a pump briefly, it soon runs out of water. A more practical pump appears in Fig. 5.2.3. In this pump, a piston slides back and forth in the open end of a hollow cylinder, making a watertight seal. Pushing inward on that piston squeezes any water in the cylinder and raises the local water pressure. Water begins to flow.

But in contrast to our simple bottle, this pump's cylinder is easy to refill. The cylinder actually has two openings, each with a valve that permits water to flow only in one direction. Water can leave the cylinder only through the top opening and can enter only through the bottom opening. As you push the pump's piston into the water-filled cylinder, the water pressure in the cylinder rises and water accelerates and flows out through the top valve. As you pull the pump's piston out of the water-filled cylinder, the water pressure inside the cylinder drops and water accelerates and flows in through the bottom valve. In fact, as you withdraw the piston, the pressure inside the cylinder drops below atmospheric pressure, so that even water in an open reservoir nearby will accelerate toward the partial vacuum in the cylinder and refill it.

▶ Check Your Understanding #2: Working a Water Pump

for answers, see page 162

Which normally requires more work: pulling the piston of a water pump out of the cylinder or pushing it back in?

Moving Water: Pressure and Energy

The pump of Fig. 5.2.3 can draw low-pressure water from a pond and fill a hose with high-pressure water. If the other end of the hose is open, the water will accelerate toward lower pressure at that end and will have considerable kinetic energy as it sprays out of the hose. From where does this kinetic energy come?

The energy comes from you and the pump. As you push inward on the piston, pressurizing the water and squeezing it out through the top valve, you're doing work on the water because you're exerting an inward force on the water's surface and the water is moving inward. The amount of work you do is equal to the product of the water pressure times the volume of water you pump. This simple relationship between work, pressure, and volume comes about because the inward *force* you exert on the water with the piston is equal to the water pressure times the surface area of the piston, and because the *distance* the water moves in the direction of that force is equal to the volume of water pumped divided by the surface area of the piston. *Force* times *distance* equals *work*.

As you pump the water, it sprays out of the open hose. The energy that makes the water accelerate out of the hose actually travels through the water directly from the pump to the end of the hose. Since water is incompressible, each time a liter of water leaves the pump, a liter of water also leaves the hose. While the water never really stores any energy, the pump gives each liter of water a certain amount of energy as it leaves the hose so we can imagine that this energy is associated with the water and not with the pump. We create a useful fiction: **pressure potential energy.** Water that's under pressure has a pressure potential energy equal to the product of the water's volume times its pressure.

Because pressure potential energy actually comes from the pump, it vanishes as soon as you break the link between the water and the pump: you can't save a bottle of high-pressure water and expect it to retain this potential energy. The concept of pressure potential energy is only meaningful if the water is flowing freely so that water leaving the plumbing is immediately replaced; then whatever energy leaves the plumbing as kinetic energy in the water is put back into the plumbing by the pump. Actually, the details of the pump aren't as important as the idea that any water moving through the plumbing is immediately replaced by more water with the same pressure. As long as the water is flowing steadily, you can safely use the concept of pressure potential energy, even if you don't know where the pump is or whether there actually is one.

Pressure potential energy is most meaningful in **steady-state flow**—a situation in which fluid flows continuously and steadily through a stationary environment, without starting or stopping or otherwise changing its characteristics anywhere. You can tell you're watching steady-state flow when you can't detect the passage of time in the fluid or its environment. Water spraying steadily out of a hose, wind blowing smoothly across your motionless face, and a gentle current flowing in a quiet river are all cases of steady-state flow in fluids.

Without gravity, the energy in a certain volume of water in steady-state flow is equal to the sum of its pressure potential energy and its kinetic energy. We've already seen that the pressure potential energy is the product of the water's volume times its pressure. The water's kinetic energy is given by Eq. 2.2.1 as one-half the product of its mass times the square of its speed. Since water's mass is its density times its volume, this sum is

$$\text{energy} = \text{pressure potential energy} + \text{kinetic energy}$$
$$= \text{pressure} \cdot \text{volume} + \tfrac{1}{2} \cdot \text{density} \cdot \text{volume} \cdot \text{speed}^2. \qquad \textbf{(5.2.1)}$$

If we divide both sides of this expression by the volume involved, we can obtain another useful form of this relationship:

$$\frac{\text{energy}}{\text{volume}} = \frac{\text{pressure potential energy}}{\text{volume}} + \frac{\text{kinetic energy}}{\text{volume}}$$
$$= \text{pressure} + \tfrac{1}{2} \cdot \text{density} \cdot \text{speed}^2. \qquad \textbf{(5.2.2)}$$

As each volume of water moves along with the flow, it is soon replaced by a new volume of water. Because the flow is steady-state, the energy in the new volume of water must be exactly the same as in the volume that preceded it; thus the energy in each volume of water that flows along a particular path must be identical. The particular path that a volume of water takes is called **a streamline,** and the energy-per-volume of fluid along a streamline is constant:

$$\frac{\text{energy}}{\text{volume}} = \frac{\text{pressure potential energy}}{\text{volume}} + \frac{\text{kinetic energy}}{\text{volume}}$$
$$= \text{pressure} + \tfrac{1}{2} \cdot \text{density} \cdot \text{speed}^2$$
$$= \text{constant } (\textit{along a streamline}). \qquad \textbf{(5.2.3)}$$

Equation 5.2.3 is called **Bernoulli's equation,** after Swiss mathematician Daniel Bernoulli ❏ whose work led to its development, although Swiss mathematician Leonhard Euler (1707–1783) actually completed it.

Because energy is conserved, an incompressible fluid such as water that's in steady-state flow can exchange pressure for speed or speed for pressure as it flows along a streamline. As water accelerates out of a hose with a nozzle, for example, its pressure drops but its speed increases because it's converting pressure potential energy into kinetic energy. And as that moving water sprays against the car you're washing, it slows down but its pressure increases because it's converting kinetic energy back into pressure potential energy. In both cases, the water's total energy is constant.

❏ As a professor in Basel, Daniel Bernoulli (Swiss mathematician, 1700–1782) taught not only physics, but also botany, anatomy, and physiology. He correctly proposed that the pressure a gas exerts on the walls of its container results from the countless impacts of tiny particles that make up the gas. He also derived an important relationship between the pressure, motion, and height of a fluid— Bernoulli's equation.

→ Check Your Understanding #3: How Does Your Garden Grow?
for answers, see page 162

Water in your garden hose has considerable pressure and arcs several meters through the air as you water your plants. What is the water pressure in the falling water once it leaves the end of the hose?

Gravity and Water Pressure

Gravity creates a pressure gradient in water: the deeper the water, the more weight there is overhead and the greater the pressure. Since water is much denser than air, water pressure increases rapidly with depth. In a vertical pipe that's open on top, the water's surface is at atmospheric pressure (about 100,000 Pa), but only 10 m (33 feet) below the water's surface, the pressure has already doubled to 200,000 Pa. At that modest depth, the water overhead weighs as much as the air overhead, even though the atmosphere is several kilometers thick.

The shape of the pipe doesn't affect the relationship between pressure and depth. No matter how complicated the plumbing, the pressure of stationary water inside it increases with depth by 10,000 Pa per meter or 10,000 Pa/m (Fig 5.2.4). This uniform pressure gradient creates an upward buoyant force on anything immersed in the water. In fact, that buoyant force is what supports the water itself (Fig. 5.2.5).

The dependence of water pressure on depth has a number of important implications for water distribution. First, water pressure at the bottom of a tall pipe is substantially higher than at the top of that same pipe. Consequently, if only a single pipe supplies water to a skyscraper, then the water pressure on the ground floor will be dangerously high while the pressure in the penthouse will be barely enough for a decent shower. Tall buildings must therefore handle water pressure very carefully; they can't simply supply water to every floor directly from the same pipe.

Second, pressure in a city water main does more than just accelerate water out of a showerhead; it also supports water in the pipes of multistory buildings. Lifting water to the third floor against the downward force of gravity requires a large upward force, and that force is provided by water pressure. The higher you want to lift the water, the more water pressure you need at the bottom of the plumbing. Lifting the water also requires energy, which is often provided by a water pump.

Third, as water travels up and down the streets of a hilly city, its pressure varies with height. In the valleys the pressure can be very large, and at the tops of hills the pressure can be very small. Water mains in valleys must therefore be particularly strong to keep from bursting. The large pressure in a valley is quite useful because it helps push the water back uphill on the other side of the valley (Fig. 5.2.6). Nonetheless, a hilly city must have pumping stations and other water pressure control systems located throughout in order to provide reasonable water pressures to all the buildings, regardless of their altitudes.

Apart from those pressure control systems, even a hilly city's plumbing often involves steady-state flow and can be explained using a version of Bernoulli's equation that includes gravity. But before we explore that topic, let's take a look at the non-steady-state flow that occurs when water in an isolated section of plumbing has free surfaces that can move up or down.

The simplest case is plumbing that's open on top, so that all the water's free surfaces are at atmospheric pressure. Like any object, water accelerates in the direction that

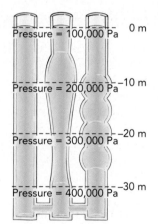

Fig. 5.2.4 The pressure of stationary water in pipes increases with depth by about 10,000 Pa per meter of depth. *The shape of the pipes doesn't matter.* For plumbing that's open on top and connected near the bottom, as shown here, water will tend to flow until its height is uniform throughout the plumbing.

Pressure = 100,000 Pa 0 m
Pressure = 200,000 Pa -10 m
Pressure = 300,000 Pa -20 m
Pressure = 400,000 Pa -30 m

Courtesy Department of Water and Power of the City of Los Angeles

Fig. 5.2.6 Los Angeles receives much of its water from Owens Valley, 300 km north. The water negotiates the mountains and valleys in between, driven by gravity alone. Giant pipes allow pressure to build during downhill stretches in order to push the water back uphill later on. Parts of the 1913 aqueduct support so much pressure that the steel pipe used in them has to be more than an inch thick.

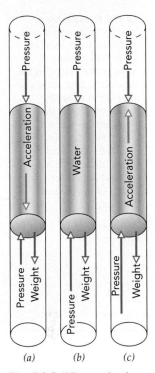

(a)   *(b)*   *(c)*

Fig. 5.2.5 When a pipe is oriented vertically, gravity affects the motion of water in the pipe. (*a*) If the water's pressure doesn't change with depth, the water will accelerate downward (fall) because of its weight. (*b*) If the water's pressure increases with depth by 10,000 Pa/m, the water won't accelerate. (*c*) If the water's pressure increases with depth by more than that amount, the water will accelerate upward.

lowers its total potential energy as quickly as possible. With neither the immovable plumbing beneath it nor the uniform air pressure above it contributing any potential energy, water's only potential energy is gravitational and it accelerates in the direction that lowers that gravitational potential as quickly as possible.

If the water's free surface is higher in one place than another, it can reduce its average height and therefore its gravitational potential energy by letting its highest water fill in its lowest valley. After some sloshing about, the water settles down in stable equilibrium with all of its free surfaces smooth and level at a single, uniform height. No matter how complicated the plumbing, water that is open to the air on top always "seeks its level." The natural flow associated with this leveling effect is often used in water delivery (see 🔲).

But if you seal off part of the isolated plumbing and reduce the pressure above one of the water's free surfaces, that surface will rise higher than all of the others. It will rise until the added pressure produced by the taller column of water replaces the missing pressure above the water's free surface. The less pressure there is above that surface, the higher the water must rise to make up for the missing pressure. This effect lifts water in a drinking straw and allows it to travel between two open containers in order to "seek its level" through an elevated pipe known as a siphon (Fig. 5.2.7).

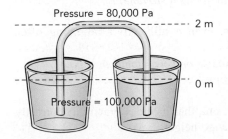

Pressure = 80,000 Pa

2 m

0 m

Pressure = 100,000 Pa

Fig. 5.2.7 The two open containers of water are connected by a siphon. This U-shaped tube permits water to flow until its level is equal in both containers. The sturdy tube permits the water pressure in the siphon to drop below atmospheric pressure.

❏ The Romans used gravity to convey water to Rome from sources up to 90 km away. A very gradual slope in the aqueducts kept the water moving in spite of frictional effects that opposed the water's progress. Poisoning from the lead pipes used in some of the aqueducts is blamed in part for the decay of the Roman Empire.

However, removing all of the air pressure above water's free surface inside a long straw or siphon will raise its height only about 10 m (33 feet) above the level of the water elsewhere in an open container. Even with no pressure above it, this 10-m column of elevated water completely replaces the absent air pressure and thus prevents water in the rest of the plumbing from lifting it any higher. It's therefore impossible to draw water from a deep well simply by lowering a pipe into that well and reducing the air pressure in the pipe: the water will rise no more than 10 m upward. Instead, a pump must be attached to the bottom of the pipe to pressurize the water and push it all the way to the top of the pipe.

➤ **Check Your Understanding #4: What's the Water Pressure?**

for answers, see page 162

If all of the water to a 400-m-tall skyscraper were delivered from a single pipe, how much higher would the water pressure be on the ground floor than on the top floor?

Moving Water Again: Gravity

As we've seen, it takes pressure and energy to lift water to the third floor of a building. We can now expand our statement of energy conservation in fluids undergoing steady-state flow to include gravity and gravitational potential energy.

Water's gravitational potential energy is equal to its weight times its height (the force required to lift it times the distance it has been lifted), and its gravitational potential energy-per-volume is its weight-per-volume times its height. Since its weight-per-volume is its density times the acceleration due to gravity, water's gravitational potential energy-per-volume is its density times the acceleration due to gravity times its height.

If we include gravitational potential energy in Eq. 5.2.2 and recognize that, for fluids in steady-state flow along a streamline, the energy-per-volume is constant, we obtain a relationship that can be written as a word equation:

$$\frac{\text{energy}}{\text{volume}} = \frac{\text{pressure potential energy}}{\text{volume}} + \frac{\text{kinetic energy}}{\text{volume}} + \frac{\text{gravitational potential energy}}{\text{volume}}$$

$$= \text{pressure} + \tfrac{1}{2} \cdot \text{density} \cdot \text{speed}^2 + \text{density} \cdot \text{acceleration due to gravity} \cdot \text{height}$$

$$= \text{constant } (along\ a\ streamline), \tag{5.2.4}$$

in symbols:

$$p + \tfrac{1}{2} \cdot \rho \cdot v^2 + \rho \cdot g \cdot h = \text{constant } (along\ a\ streamline),$$

and in everyday language:

> *When a stream of water speeds up in a nozzle or flows uphill in a pipe,*
> *its pressure drops.*

This is a revised version of Bernoulli's equation, one that includes gravity. It correctly describes steady-state flow in streamlines that change height.

> ### Bernoulli's Equation
> For an incompressible fluid in steady-state flow, the sum of its pressure potential energy, its kinetic energy, and its gravitational potential energy is constant along a streamline. Equation 5.2.4 expresses this law as a formula.

Because energy is conserved, an incompressible fluid such as water that's in steady-state flow can exchange its speed, pressure, and height for one another. Thus as water flows downward, its speed or pressure or both increase; if it falls from an open faucet, its speed increases; and if it descends steadily inside a uniform pipe, its pressure increases. The reverse happens as water flows upward. Water rising from a fountain loses speed as it ascends while water rising steadily in a uniform pipe loses pressure on its way up.

This interchangeability of height, pressure, and speed makes it possible to pressurize plumbing by connecting a tall column of water to the pipes. That's why cities, communities, and even individual buildings have water towers (Fig. 5.2.8). A water tower is built at a relatively high site within the region it serves. A pump fills the water tower with water, and then gravity maintains a constant high pressure throughout the plumbing that connects to it (Fig. 5.2.9). The water is at atmospheric pressure at the top of the water tower, but the pressure is much higher at the bottom; at the base of a 50-m-high water tower, for example, the pressure is about 600,000 Pa or six times atmospheric pressure.

In addition to providing a fairly steady pressure in the water mains, a water tower stores energy efficiently and can deliver that energy quickly. When water is drawn out of the water tower, its gravitational potential energy at the top becomes pressure potential energy at the bottom. The water tower replaces a pump, supplying a steady flow of water at an almost constant high pressure. But unlike a pump, the water tower can supply this high-pressure water at an enormous rate. As long as the water level doesn't drop too far, high-pressure water keeps flowing.

Steve Lewis/The Image Bank/Getty Images

Fig. 5.2.8 Many buildings in New York City have water towers on their roofs. These towers maintain water pressure in the plumbing and help in firefighting.

Check Your Understanding #5: Water Power

for answers, see page 162

A hydroelectric power plant extracts energy from water that has descended from an elevated reservoir in a pipe. In the reservoir, this energy takes the form of gravitational potential energy. Just before the power plant, what form does the energy take?

Check Your Figures #1: Up or Out

for answers, see page 162

If the water pressure at the entrance to a building is 1,000,000 Pa, how high can the water rise up inside the building and how fast will it flow out of a faucet right at the entrance? (A liter of water inside a pipe has a mass of 1 kg.)

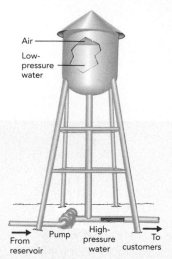

Fig. 5.2.9 A water tower uses the weight of water to create a large water pressure near the ground. The higher the tower, the greater the pressure near the ground. The water tower is able to maintain the water pressure passively and doesn't require constant pumping. Even during periods of peak water consumption, it maintains a fairly steady pressure. When the water level in the water tower drops below a certain set point, a pump refills the tower.

Epilogue for Chapter 5

In this chapter we investigated some of the basic concepts associated with fluids. In *balloons,* we explored the concept of pressure and the way in which air pressure structures and supports the earth's atmosphere. We saw that increased air pressure beneath an object produces an upward buoyant force on that object, and that this buoyant force can float objects, such as hot-air and helium balloons, that are less dense than the surrounding air.

In *water distribution,* we examined how water pressure causes water to accelerate through plumbing, from high pressure to low pressure. We then focused on ways to produce water pressure, either with pumps or with gravity. By studying the forms of energy in water, we were led to Bernoulli's equation, which describes the interconversion of a fluid's energy between pressure potential energy, kinetic energy, and gravitational potential energy. Although the most dramatic applications of Bernoulli's equation are ahead of us, we've already used it to understand the changes in pressure and speed that accompany water's movement up and down in pipes and in fountains.

Explanation: A Cartesian Diver

The diver floats because its average density, that is, the mass of the vial and its contents divided by the volume of space those two components occupy, is less than the density of water. Since the upward buoyant force on the diver exceeds its downward weight, the diver floats upward toward the top of the bottle. When the diver begins to stick out of the water, it displaces less water and more air and the buoyant force it experiences decreases. Eventually it experiences zero net force and floats without accelerating at the water's surface.

When you squeeze the soda bottle, you increase the pressure inside it. Because water is incompressible, its density doesn't change as the pressure goes up. However, the air bubble inside the vial is compressed and takes up less space inside the vial. Water flows into the vial and increases the average density of the vial and its contents. When the average density of the diver finally exceeds the density of water, the diver sinks.

To keep the diver hovering in the water, you must adjust the water pressure until the diver's average density is exactly that of water. This adjustment is impossible to make without looking at the diver. Even the slightest overpressure or underpressure will cause the diver to drift slowly down or up.

Chapter Summary

How Balloons Work: A hot-air balloon floats because its total weight (basket, envelope, and hot air) is less than that of the cooler air it displaces. By heating the air in the envelope with a flame, the pilot reduces its density. As the air warms up, fewer particles are needed to fill the envelope, the extra particles flow out of the envelope through the opening at its bottom, and the balloon becomes lighter.

A helium balloon also weighs less than the air it displaces. But its lower weight is due to the lightness of the individual helium atoms, each of which weighs much less than the air particle it replaces. By filling a balloon with helium, the balloon's weight is dramatically reduced. Since the buoyant force exerted on the balloon by the surrounding air exceeds the balloon's weight, the balloon accelerates upward.

How Water Distribution Works: Water distribution begins when a pump transfers low-pressure water from a reservoir to high-pressure plumbing. Along a level path, water accelerates toward regions of lower pressure, such as open hoses or showerheads. Pressure imbalances allow the water to negotiate bends in the pipes en route to its destination. During its travels, the water may rise or fall in height; as it does, its pressure changes, the pressure decreasing as the water moves upward and increasing as the water moves downward. In low-lying regions, where the water pressure may be too high to use directly, a pressure regulator may have to be added to the plumbing. In high-lying regions, where the water pressure may be too low to be practical, an additional pump may have to be employed to boost its pressure.

Important Laws and Equations

1. Archimedes' Principle: An object partially or wholly immersed in a fluid is acted on by an upward buoyant force equal to the weight of the fluid it displaces.

2. The Ideal Gas Law: The pressure of a gas is equal to the product of the Boltzmann constant times the particle density times the absolute temperature, or

$$\text{pressure} = \text{Boltzmann constant} \cdot \text{particle density} \cdot \text{absolute temperature.} \qquad (5.1.3)$$

3. Pascal's Principle: A change in the pressure of an enclosed incompressible fluid is conveyed undiminished to every part of the fluid and to the surfaces of its container.

4. Bernoulli's Equation: For an incompressible fluid in steady-state flow, the sum of its pressure potential energy, its kinetic energy, and its gravitational potential energy is constant along a streamline, or

$$\frac{\text{energy}}{\text{volume}} = \frac{\text{pressure potential energy}}{\text{volume}} + \frac{\text{kinetic energy}}{\text{volume}}$$
$$+ \frac{\text{gravitational potential energy}}{\text{volume}}$$
$$= \text{pressure} + \tfrac{1}{2} \cdot \text{density} \cdot \text{speed}^2 \qquad (5.2.4)$$
$$+ \text{density} \cdot \text{acceleration due to gravity} \cdot \text{height}$$
$$= \text{constant } (\textit{along a streamline}).$$

Check Your Understanding—Answers

Section 5.1 BALLOONS

1. Air pressure.
Why: Because the space between the suction cup and the wall is empty, the pressure there is zero. Pressure of the surrounding air exerts large inward forces on the outsides of both the cup and the wall, squeezing them together. As long as there is no air between them to push outward, the cup and wall remain tightly attached. Once air leaks into the suction cup, it's easily detached from the wall.

2. The pressure of the air trapped in the container increases as its temperature increases, causing the lid to bulge outward.
Why: Whenever a trapped quantity of gas changes temperature, it also changes volume or pressure or both. In this case, warming the air trapped in the container causes its pressure to increase. The unbalanced pressures inside and outside the container cause the lid to bow outward or even to pop off.

3. As you change altitude, the atmospheric pressure changes.
Why: The air inside your ears is trapped, so that its temperature, density, and pressure are normally constant. As your altitude changes, the pressure outside your ear changes and your eardrum experiences a net force. It bows inward or outward, muting the sounds you hear and causing some discomfort. The pressure imbalance is relieved during swallowing, when air can flow into or out of your eardrum.

4. About 1 N (0.22 lbf).
Why: Air exerts a buoyant force on him equal to the weight of the air he displaces. The density of air near sea level is about 1.25 kg/m^3, so 0.08 m^3 of air has a mass of about 0.1 kg (1.25 kg/m^3 times 0.08 m^3) and a weight of about 1 N. So the upward buoyant force on him is about 1 N. This buoyant force due to the air is real and reduces the weight you read when you stand on a scale by about 0.125%.

5. It can lift more on a cold day.
Why: On a cold day, the outside air is relatively dense and the buoyant force on a balloon is larger than it would be on a hot day. The hot air in the balloon will cool off more quickly on a cold day, but the balloon will be able to carry a heavier load. Airplanes also fly better on cold days.

6. It flows down.
Why: Carbon dioxide, found in carbonated beverages, dry ice, and fire extinguishers, is heavier than air because its molecules

are heavier than air particles. The carbon dioxide you pour from the cup has the same pressure and temperature as the air around it and thus the same particle density. But each carbon dioxide molecule weighs more, so the carbon dioxide is the denser gas (it has higher mass density) and flows down in air. This tendency to flow along the floor makes carbon dioxide very good at extinguishing low-lying flames by depriving them of oxygen.

Section 5.2 WATER DISTRIBUTION

1. The water pressure (force on a unit of surface area) inside the hose is uniform throughout, so the water experiences no net force and doesn't accelerate.

Why: In the absence of gravity, fluids accelerate only when they experience pressure imbalances. Since water throughout the hose is at the same pressure, there is no pressure imbalance and no acceleration. When you open the nozzle, the pressure at that end of the hose drops and the water accelerates toward it.

2. Pushing it back in usually requires more work.

Why: When you pull the piston out of the cylinder, you are moving air out of the way and creating a partial vacuum inside the cylinder. This action requires a modest amount of work because the air doesn't push terribly hard on the back of the piston and pressure from the water flowing into the cylinder assists you. But as you push the piston back into the cylinder, you are pressurizing the water. Depending on the pressure of water in the outlet hose, the water in the cylinder may exert a very large force on the piston. In that case, you must do a great deal of work on the water as you push the piston inward and drive the water out of the cylinder.

3. Atmospheric pressure.

Why: As the water accelerates out of the hose, its pressure drops. It is converting pressure potential energy into kinetic energy. The water pressure drops until it reaches the pressure of the surrounding air, atmospheric pressure.

4. About 4,000,000 Pa (40 atmospheres) higher.

Why: The weight of water inside the pipe would create an enormous excess pressure near the bottom of the building. Water spraying from an open faucet on the first floor at this enormous pressure could accelerate to 319 km/h (200 mph), as it does in some high-pressure jet washing and cutting machines.

5. Pressure potential energy (and some kinetic energy).

Why: As the water descends inside the pipe, its gravitational potential energy is converted into pressure potential energy. The water reaching the power plant is under enormous pressure, and it's this pressure that exerts the forces needed to turn the turbines that run the generators. Work is required to turn the turbines, so the water gives up much of its energy in the power plant. This energy leaves the power plant via the electric power lines.

Check Your Figures—Answers

Section 5.1 BALLOONS

1. It increases by 8.4%.

Why: To use Eq. 5.1.3 to determine the pressure change, we need temperatures measured on an absolute scale, such as the Kelvin scale. Since 0 °C is about 273 K, 2 °C is about 275 K and 25 °C is about 298 K, we can write Eq. 5.1.3 twice, once for each temperature:

$$\text{pressure}_{298\ K} = \text{Boltzmann constant} \cdot \text{particle density} \cdot 298\ K,$$

$$\text{pressure}_{275\ K} = \text{Boltzmann constant} \cdot \text{particle density} \cdot 275\ K.$$

The particle density of the air in the container can't change as it warms up because its volume is fixed. Therefore we can divide the upper equation by the lower one and cancel the Boltzmann constant and particle density on the right-hand side:

$$\frac{\text{pressure}_{298\ K}}{\text{pressure}_{275\ K}} = \frac{298\ K}{275\ K}$$

$$= 1.084.$$

The pressure in the container thus increases by a factor of almost 1.084, or about 8.4%. This elevated pressure will cause the container to emit a "pop" sound when you open it.

Section 5.2 WATER DISTRIBUTION

1. It can rise up about 100 m or emerge from the faucet at about 45 m/s (101 mph).

Why: As the water flows through the pipe (a streamline), its pressure potential energy can become gravitational potential energy or kinetic energy. At the start, the water's energy is all pressure potential energy so, from Eq. 5.2.4, the water's energy-per-volume is 1,000,000 Pa. If the water flows up the pipe, that energy will become gravitational potential energy. We can rearrange Eq. 5.2.4 to find the height it can reach:

$$\text{height} = \frac{\text{energy}}{\text{volume}} \cdot \frac{1}{\text{density} \cdot \text{acceleration due to gravity}}$$

$$= 1{,}000{,}000\ \text{Pa} \cdot \frac{1}{1000\ \text{kg/m}^3 \cdot 9.8\ \text{m/s}^2} = 102\ \text{m}.$$

If the water flows out the faucet, that energy will become kinetic energy. We can also rearrange Eq. 5.2.4 to find the speed it will obtain:

$$\text{speed} = \sqrt{\frac{\text{energy}}{\text{volume}} \cdot \frac{2}{\text{density}}}$$

$$= \sqrt{\frac{2{,}000{,}000\ \text{Pa}}{1000\ \text{kg/m}^3}} = 45\ \text{m/s}.$$

Exercises

1. A helium-filled balloon floats in air. What will happen to an air-filled balloon in helium? Why?

2. A log is much heavier than a stick, yet both of them float in water. Why doesn't the log's greater weight cause it to sink?

3. An automobile will float on water as long as it doesn't allow water to leak inside. In terms of density, why does admitting water cause the automobile to sink?

4. Many grocery stores display frozen foods in bins that are open at the top. Why doesn't the warm room air enter the bins and melt the food?

5. Some clear toys contain two colored liquids. No matter how you tilt one of those toys, one liquid remains above the other. What keeps the upper liquid above the lower liquid?

6. When the car you are riding in stops suddenly, heavy objects move toward the front of the car. Explain why a helium-filled balloon will move toward the rear of the car.

7. Water settles to the bottom of a tank of gasoline. Which takes up more space: 1 kg of water or 1 kg of gasoline?

8. Oil and vinegar salad dressing settles with the oil floating on top of the vinegar. Explain this phenomenon in terms of density.

9. When a fish is floating motionless below the surface of a lake, what is the amount and direction of the force the water is exerting on it?

10. Some fish move extremely slowly, and it's hard to tell whether they are even alive. However, if a fish is floating at a middle height in your aquarium and not at the top or bottom of the water, you can be pretty certain that it's alive. Why?

11. A barometer, which is often used to monitor the weather, is a device that measures air pressure. How could you use a barometer to measure your altitude as you climbed in the mountains?

12. If you seal a soft plastic bottle or juice container while hiking high in the mountains and then return to the valley, the container will be dented inward. What causes this compression?

13. Many jars have dimples in their lids that pop up when you open the jar. What holds the dimple down while the jar is sealed, and why does it pop up when the jar is opened?

14. If you place a hot, wet cup upside down on a smooth counter for a few seconds, you may find it difficult to lift up again. What is holding that cup down on the counter?

15. You seal a rigid container that is half full of hot food and put it in the refrigerator. Why is the container's lid bowed inward when you look at it later?

16. Why aren't there any thermometers that read temperatures down to -300 °C?

17. You use your breath to inflate a large rubber tube and then ride down a snowy hill on it. After a few minutes in the snow the tube is underinflated. What happened to the air?

18. A marshmallow is filled with air bubbles. Why does a marshmallow puff up when you toast it?

19. Wasp and hornet sprays proudly advertise just how far they can send insecticide. How does the pressure inside the spray can affect that distance, and why is the direction of the spray important (vertical vs. horizontal)?

20. Ice tea is often dispensed from a large jug with a faucet near the bottom. Why does the speed of tea flowing out of the faucet decrease as the jug empties?

21. Why must tall dams be so much thicker at their bases than at their tops?

22. Waterproof watches have a maximum depth to which they can safely be taken while swimming. Why?

23. When you stand in a pool with water up to your neck, you find that it's somewhat more difficult to breathe than when you're out of the water. Why?

24. How does pushing on the plunger of a syringe cause medicine to flow into a patient through a hollow hypodermic needle?

25. Why must the pressure inside a whistle teakettle exceed atmospheric pressure before the whistle can begin to make noise?

26. Each time you breathe in, air accelerates toward your nose and lungs. How does the pressure in your lungs compare with that in the surrounding air as you breathe in?

27. You can inflate a plastic bag by holding it up so that it catches the wind. Use Bernoulli's equation to explain this effect.

28. When someone pulls a fire alarm in a skyscraper, pumps increase the water pressure in the section of the building nearest that alarm box. How does this pressure change assist firefighters who must battle the blaze?

Problems

1. The particle density of standard atmospheric air at 273.15 K (0 °C) is 2.687×10^{25} particles/m^3. Using the ideal gas law, calculate the pressure of this air.

2. How much force is the air exerting on the front surface of this book?

3. If you fill a container with air at room temperature (300 K), seal the container, and then heat the container to 900 K, what will the pressure be inside the container?

4. An air compressor is a device that pumps air particles into a tank. A particular air compressor adds air particles to its tank until the particle density of the inside air is 30 times that of the outside air. If the temperature inside the tank is the same as that outside, how does the pressure inside the tank compare to the pressure outside?

5. If you seal a container of air at room temperature (20 °C) and then put it in the refrigerator (2 °C), how much will the pressure of the air in the container change?

6. If you submerge an 8-kg log in water and it displaces 10 kg of water, what will the net force on the log be the moment you let go of the log?

7. If your boat weighs 1200 N, how much water will it displace when it's floating motionless at the surface of a lake?

8. The density of gold is 19 times that of water. If you take a gold crown weighing 30 N and submerge it in water, what will the buoyant force on the crown be?

9. How much upward force must you exert on the submerged crown in Problem 8 to keep it from accelerating?

10. How could you use your results from Problems 8 and 9 to determine whether the crown was actually gold rather than gold-plated copper? (The density of copper is nine times that of water.)

11. Your town is installing a fountain in the main square. If the water is to rise 25 m (82 feet) above the fountain, how much pressure must the water have as it moves slowly toward the nozzle that sprays it up into the air?

12. Rather than putting a pump in the fountain (see Problem 11), the town engineer puts a water storage tank in one of the nearby high-rise office buildings. How high up in that building should the tank be for its water to rise to 25 m when spraying out of the fountain? (Neglect friction.)

13. To clean the outside of your house you rent a small high-pressure water sprayer. The sprayer's pump delivers slow-moving water at a pressure of 10,000,000 Pa (about 150 atmospheres). How fast can this water move if all of its pressure potential energy becomes kinetic energy as it flows through the nozzle of the sprayer?

14. When the water from the sprayer in Problem 13 hits the side of your house, it slows to a stop. If it hasn't lost any energy since leaving the sprayer, what is the pressure of the water at the moment it stops completely?

15. To dive far below the surface of the water, a submarine must be able to withstand enormous pressures. At a depth of 300 m, what pressure does water exert on the submarine's hull?

CHAPTER 6

Fluids and Motion

Fluids are fascinating when they move. Stationary water and air may be essential to life, but they're also fairly simple; only their pressures vary from place to place and even these are determined primarily by gravity. But rushing rivers or gusts of wind, with their wonderful variety of simple and complicated behaviors, are much more interesting. And the motion of fluids isn't just interesting; it's also important, since our world is filled with objects and machines that work in whole or part because of the behaviors of moving fluids. In this chapter, we'll look at several situations in which fluid motion contributes to the way things work.

EXPERIMENT: A Vortex Cannon

Fluids are real; they exist independent of any solids that might move through them. That notion is easy to accept in reference to water, since we can see that water doesn't

Courtesy Lou Bloomfield

wait for a boat to pass before moving in interesting ways. But air is harder to visualize in this fashion because we seldom see it by itself, apart from its effects on buildings or airplanes or our skin.

To begin seeing air as something tangible and to demonstrate the rich possibilities of motion available to it, build a vortex cannon—a device that sends rings of air sailing across a room. While a serious vortex cannon is best constructed from a large drum or crate, you can make a simple one from an empty cardboard cereal box.

Seal the rectangular box on all edges with tape, and cut a circular hole 5 centimeters (2 inches) in diameter in the center of one face. To use your cannon, just tap hard on the other face of the box. Rings of air will leap out of the hole and sail across the room at about 5 m/s (11 mph).

Although you can't see these rings, you can follow their motion by looking for their effects on objects they encounter. They can easily blow out candles or rustle light window drapes across a small room. If you have a friend blow some rings at you, you'll feel exactly where they hit on your face or shirt.

To see these rings directly, fill the cereal box with smoke, mist, or dust. Try to *predict* what the rings will look like when they emerge from the hole. Will the smoke in each ring be stationary, or will it be swirling about the ring in some manner? How will the ring's size and speed of travel depend on the size of the hole from which it emerges? Will it depend on how hard you tap the cereal box? Will the ring's size and speed change in flight?

Now tap the box and *observe* the smoke rings. What motions do you see? *Measure* the size and speed of the rings as they fly. Did you *verify* your predictions? Change the hole size and see how this change affects the rings and their motion. As you can see, the air can execute some very complicated movements all by itself. In this chapter, we will examine how moving air, water, and other fluids affect our everyday lives.

Chapter Itinerary

In particular, we'll explore (1) *garden watering*, (2) *balls and air,* and (3) *airplanes*. In *garden watering*, we'll look at how water's pressure and speed vary as it flows through a faucet, a hose, and a nozzle. In *balls and air*, we'll investigate the effects of air on the motions of balls. Lastly, in *airplanes*, we'll study the ways in which moving air supports and propels airplanes in flight. A more complete preview can be found in the Chapter Summary on p. 197.

This chapter continues to develop the concept of energy conservation along a streamline that was introduced in Chapter 5. It also brings up several new types of forces that are present when fluids move past one another or past solid objects. These ideas are present not only in the topics of this chapter, but also in many other commonplace activities, from washing windows with a hose to pumping water with a windmill.

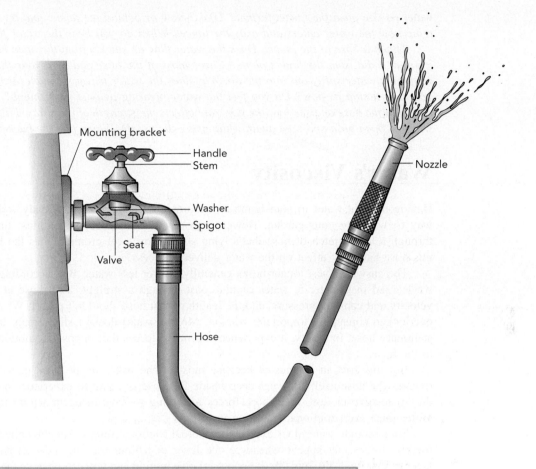

Mounting bracket
Handle
Stem
Washer
Spigot
Seat
Valve
Hose
Nozzle

SECTION 6.1 **Garden Watering**

Tending a flower garden often involves watering. While this once meant walking the garden's paths with a watering can, modern plumbing has made such effort unnecessary. With a hose and nozzle attached to a faucet, you can do your job without leaving your lawn chair. But while the tools involved—faucets, hoses, and nozzles—are simple and unsophisticated, the principles behind them are not. All three make elegant use of the laws of fluid flow, letting openings and channels control the delivery rate and speed of the water so that it arcs gracefully through the air to the farthest reaches of your garden.

Questions to Think About: *Where is the water when the faucet is closed, and why does it begin flowing through the hose when you open the faucet? What determines the rate at which water flows through the faucet or the hose? If honey ran through your hose instead of water, how would that affect the flow? Why does water make noise as you sprinkle your garden? Why does a nozzle make the water spray so fast and so far? Why do the pipes sometimes clank when you abruptly close the faucet?*

Experiments to Do: *Open a faucet gradually and watch the water begin to flow. What's pushing the water out of the faucet? What happens to the flow rate and speed of the* *water as you open the faucet further? Look below or behind the faucet and try to determine how the water enters and exits the faucet. When do you hear the water flowing?*

Attach a hose to the faucet. Does the water flow as quickly from the open end of the hose as it did from the faucet alone? Cover most of the hose end with your thumb and watch the water spray out into the air. Why does the water travel so much farther when you almost stop its flow? Do you feel the water pressing against your thumb? Attach a nozzle to the hose and see how the flow rate affects the strength of the spray. Fill a bucket with the open hose and then again while using the nozzle. Which fills the bucket fastest?

Water's Viscosity

Having brought water to your home in the previous chapter, we're already well on our way to watering your garden. However, to reach the garden, water must first travel through a long stretch of hose that's lying straight on level ground. Does the length of this hose have any effect on the water delivery process?

The answer is yes, longer hoses generally deliver less water. But, according to what we learned in Chapter 5, water should coast through a straight, level hose at constant velocity and constant pressure, and the length of that hose shouldn't matter. We evidently overlooked something important: friction. Moving water doesn't slide freely through a stationary hose. In reality, it experiences frictional forces that oppose its motion relative to the hose.

But this friction is unusual because most of the water in the hose never actually touches the hose itself. If water deep inside the hose is going to experience any forces due to relative motion, then those forces are going to have to occur *within* the water. Water must exert frictional forces on itself!

Sure enough, water does experience internal frictional forces. They're called **viscous forces**—forces that appear whenever one layer of a fluid tries to slide across another layer of that fluid. Viscous forces oppose relative motion and you can observe their effects easily when you pour honey out of a jar. The honey at the jar's surface is stuck there by chemical forces and remains stationary. But even honey that's far from the walls can't move easily; it experiences viscous forces as it tries to move relative to nearby honey. Since honey is a "thick" or *viscous* fluid, viscous forces act quite effectively to keep all the honey moving with nearly the same velocity. Since the honey at the walls can't move, viscous forces tend to prevent any of the honey from moving.

Water isn't as thick as honey (Table 6.1.1), so it's less resistant to relative motion. The measure of this resistance to relative motion within a fluid is called **viscosity,** and

Table 6.1.1 Approximate Viscosities of a Variety of Fluids

FLUID	VISCOSITY*
Helium (2 K)	$0 \text{ Pa} \cdot \text{s}$
Air (20 °C)	$0.0000183 \text{ Pa} \cdot \text{s}$
Water (20 °C)	$0.00100 \text{ Pa} \cdot \text{s}$
Olive oil (20 °C)	$0.084 \text{ Pa} \cdot \text{s}$
Shampoo (20 °C)	$100 \text{ Pa} \cdot \text{s}$
Honey (20 °C)	$1000 \text{ Pa} \cdot \text{s}$
Glass (540 °C)	$10^{12} \text{ Pa} \cdot \text{s}$

* The pascal-second (abbreviated $\text{Pa} \cdot \text{s}$ and synonymous with $\text{kg/m} \cdot \text{s}$) is the SI unit of viscosity. Only the superfluid portion of ultracold liquid helium exhibits zero viscosity.

water's viscosity is less than that of honey. In fact, if you heat the water up it will become even less viscous and thus flow more easily. Typical of most liquids, this decrease in viscosity with increasing temperature reflects the molecular origins of viscous forces: the molecules in a liquid stick to one another, forming weak chemical bonds that require energy to break. In a hot liquid, the molecules have more thermal energy, so they break these bonds more easily in order to move past one another (see ❏).

❏ Your car's engine is protected by motor oil with a carefully chosen viscosity. If that oil were too thin, it would flow out from between surfaces and wouldn't keep them from rubbing against one another. If that oil were too thick, the engine would waste power moving its parts through the oil. Years ago, you had to change your motor oil for the season. Thick 40 weight motor oil was used in summer because hot weather made it thinner; thin 10 weight oil was used in winter because cold weather made it thicker. But a modern, multigrade oil maintains a nearly constant viscosity over a wide range of temperatures and need not be changed with the seasons. This oil contains tiny molecular chains that ball up when cold but straighten out when hot. These chains thicken hot oil so that 10W-40 oil resembles 10 weight oil in winter and 40 weight oil in summer.

➤ Check Your Understanding #1: Keeping Warm on a Windy Day
for answers, see page 198

A loosely woven wool sweater has many tiny air passages between the wool fibers, yet it dramatically reduces the rate at which air flows through to your skin when you stand in a breeze. Why doesn't air flow easily through the gaps between the fibers?

Flow in a Straight Hose: The Effect of Viscosity

Viscosity slows the flow of water through your hose. Chemical forces between the hose and the outmost layer of water hold that layer of water stationary, and this motionless layer exerts viscous forces on the layer of moving water inside it. As this second layer slows, it exerts viscous forces on yet another layer. Layer by layer, viscous forces hold back the moving water until even water at the center of the hose feels viscosity's slowing effects (Fig. 6.1.1). Although water at the center of the hose moves faster than water in any other layer, it's still affected by the stationary hose.

These viscous forces impede water delivery. Instead of coasting effortlessly through the straight, level hose, real water needs a pressure gradient to push it steadily forward. Like the file cabinet sliding on the sidewalk in Section 2.2, water must be pushed through the hose if it's to maintain a continuous flow. And like that file cabinet, the water becomes hotter as the work done pushing it forward is wasted and becomes thermal energy.

However, unlike the forces of ordinary sliding friction—which don't depend on relative velocities—viscous forces become larger as the relative velocities within a fluid increase. That's because as two layers of water slide past one another faster, their molecules collide harder and more frequently. Since it experiences stronger viscous forces, fast-moving water wastes more energy per meter and needs a larger pressure gradient to keep it moving steadily through a hose than does slow-moving water.

Because of viscous forces, the amount of water flowing steadily through a hose depends on four factors:

1. It's inversely proportional to the water's viscosity. The more viscous the water, the more difficulty it has flowing through the hose.
2. It's inversely proportional to the length of the hose. The longer the hose, the more opportunity viscous forces have to slow the water down.
3. It's proportional to the pressure difference between the hose's inlet and its outlet. This pressure difference determines the water's pressure gradient and thus how hard the water is pushed forward through the hose.
4. It's proportional to the fourth power of the diameter of the hose. Tripling the hose's diameter provides the water with nine times as much room and also allows water near the hose's center to move nine times faster.

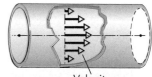

Velocity

Fig. 6.1.1 The speed of water flowing through a pipe is not constant across the pipe. The water near the walls is stationary, while the water at the center of the pipe moves the fastest. The differences in velocity are the results of viscous forces.

We can turn all these proportional relationships into an equation by adding the correct numerical constant ($\pi/128$). The final relationship is called **Poiseuille's law** and can be written as a word equation:

$$\text{volume} = \frac{\pi \cdot \text{pressure difference} \cdot \text{pipe diameter}^4}{128 \cdot \text{pipe length} \cdot \text{fluid viscosity}}, \qquad (6.1.1)$$

in symbols:

$$\frac{\Delta V}{\Delta t} = \frac{\pi \cdot \Delta p \cdot D^4}{128 \cdot L \cdot \eta},$$

and in everyday language:

It's hard to squeeze honey through a long, thin tube.

Poiseuille's Law
The volume of fluid flowing through a cylindrical pipe each second is equal to ($\pi/128$) times the pressure difference (Δp) across that pipe times the pipe's diameter to the fourth power, divided by the pipe's length times the fluid's viscosity (η).

It's hardly surprising that the flow rate depends in this manner on the pressure difference, pipe length, and viscosity; we've all observed that low water pressure or a long hose lengthens the time needed to fill a bucket with water and that viscous syrup pours slowly from a bottle. But the dependence of the flow rate on the fourth power of pipe diameter may come as a surprise. Even a small change in the diameter of a hose significantly changes the amount of water that hose delivers each second. The two most common garden hoses in the United States have diameters of 5/8 inch and 3/4 inch, and while these hoses differ by a seemingly insignificant 20% or a factor of 1.2 in diameter, the 3/4 inch hose can carry about 1.2^4 or 2 times as much water as the 5/8 inch hose (see ❏s).

We can also look at viscous forces in terms of total energy. By opposing the flow of water through a hose, viscous forces do negative work on it and reduce its total energy—the energy considered in Bernoulli's equation, which doesn't include thermal energy. Just how much total energy the water retains depends on how fast it moves inside the hose. If you allow lots of water to leave the hose, water will move through it quickly and encounter large viscous forces. In the process, most of the water's total energy will be wasted as thermal energy and the water will pour gently out of the end of the hose.

But if you partially block the hose's opening with your thumb and reduce the flow, water will travel slowly through the hose and encounter smaller viscous forces. As a result, the water will retain most of its total energy and will still be at high pressure when it reaches your thumb. This high-pressure water will then accelerate to enormous speed as it passes through the narrow opening and sprays out into the air.

We can now explain why water delivery systems normally use the widest pipes that are practical and affordable. In contrast to a narrow hose, wide pipes can carry large amounts of water while letting that water travel slowly, experience weak viscous forces, and waste little of its total energy. In such energy efficient water delivery systems, friction

❏ To deliver large amounts of water at high pressure or velocity, fire hoses must have large diameters. When filled with high-pressure water, these wide hoses become stiff and heavy, making them difficult to handle. Chemical additives that decrease water's viscosity allow firefighters to use narrower, lighter, and more flexible hoses.

❏ Very large diameter pipes are required to transport crude oil across the Alaskan wilderness. The distances are long and the fluid is viscous, although it is heated to lower its viscosity.

is insignificant and Bernoulli's equation (Eq. 5.2.4) accurately predicts water's properties throughout its trip.

➤ Check Your Understanding #2: Air Ducts

for answers, see page **198**

The long air ducts used to ventilate homes and businesses usually have very large diameters. These ducts are often visible near the ceilings of modern warehouse-style stores and restaurants as pipes roughly 0.5 m across. Why must the ducts be so large in diameter?

➤ Check Your Figures #1: Old Plumbing

for answers, see page **199**

When your friend's house was new, the kitchen faucet could deliver 0.50 liter per second (0.50 L/s). But mineral deposits have built up in the pipes over the years and reduced their effective diameters by 20%. How much water can the faucet deliver now?

Flow in a Bent Hose: Dynamic Pressure Variations

On reaching your garden, the hose bends toward the right and the flowing water bends with it. That water is accelerating as it turns and, as we observed in Chapter 5, water accelerates horizontally only in response to unbalanced pressures. Since the hose is motionless, the unbalanced pressures inside it must be caused by the water itself; the water is experiencing dynamic pressure variations.

To understand these dynamic pressure variations, let's follow the streamlines as water traverses this bend. Although we've just introduced viscous forces, we're going to ignore them here for clarity and simplicity. While viscous forces are certainly important in the long, narrow hose, the bend is so short that viscous forces have little effect on what happens to the water passing through it.

Neglecting viscous forces, the water's total energy is constant along each streamline and we can observe the interchanges of energy allowed by Bernoulli's equation (Eq. 5.2.4). However, since the hose rests on level ground, water's gravitational potential energy can't vary and the only interchanges we'll see are between pressure potential energy and kinetic energy.

Figure 6.1.2 shows the water's steady-state flow pattern near the bend. We're looking down on the hose in this calculated drawing and, as indicated by the black streamlines, water that is initially flowing straight ahead arcs rightward at the bend and eventually continues directly toward the right.

Water approaches the bend through a straight section of hose in which it travels at constant velocity and has a uniform pressure. Its velocity is constant because the straight hose directs all the streamlines forward and because the water moving along a given streamline can't change its speed; if it tried to speed up, it would leave an empty space behind it; if it tried to slow down, it would cause a "traffic jam." The water's pressure is uniform throughout this straight section because constant velocities mean no accelerations and thus no pressure differences.

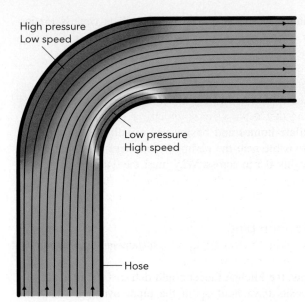

High pressure
Low speed

Low pressure
High speed

Hose

Fig. 6.1.2 Water in a bent hose experiences changes in speed and pressure. The black streamlines show the paths the water takes as it flows around the bend. The spacing between streamlines indicates flow speed (wider space is slower flow), and the background color indicates pressure (violet is higher pressure; red is lower pressure).

The water's constant velocity and uniform pressure are represented visually in Fig. 6.1.2. You can see the local water velocity by looking at the direction and spacing of the streamlines. Streamlines always point in the direction of the local water velocity and their spacing varies inversely with the local water speed. Streamlines that become more widely spaced denote decreasing speed—water that slows down spreads out sideways and its streamlines separate from one another. Streamlines that become more narrowly spaced denote increasing speed—water that speeds up stretches out along its path and its streamlines draw toward one another. Since the streamlines leading up to the bend are straight and evenly spaced, we know that water moves along each streamline at a constant velocity.

You can see the local water pressure in Fig. 6.1.2 by looking for the colors of the rainbow. Colors toward the violet end of the spectrum denote higher pressures while those toward the red end of the spectrum denote lower pressures. Since the straight section has a uniform blue-green color, the water there has a uniform pressure.

Once the water starts bending toward the right, its velocities and pressures begin to vary. Since the water is accelerating toward the inside of the bend, there must be a pressure imbalance pushing it in that direction. Sure enough, the turning stream of water develops higher local pressure (violet) near the outside of the bend and lower local pressure near the inside of the bend (red). A similar pressure imbalance accompanies any bend in a fluid's path: the pressure is always higher on the outside of the bend than it is on the inside of that bend. After all, that pressure imbalance is what causes the fluid to bend!

Bends and Pressure Imbalances
When the path of a fluid in steady-state flow bends, the pressure on the outside of the bend is always higher than the pressure on the inside of the bend.

To keep the total energy constant along a streamline, each decrease in the water's local pressure is accompanied by an increase in the water's local speed and vice versa.

Water arcing around the outside of the bend slows down (the streamline spacing widens) as its pressure rises, while the water arcing around the inside of the bend speeds up (the streamline spacing narrows) as its pressure drops.

As the hose straightens out beyond the bend, water's pressures and speeds return to what they were before the bend. Water from the outside of the bend speeds up and its pressure drops, while water from the inside of the bend slows down and its pressure rises. In the straight section following the bend, water's velocity is once again constant along each streamline and its pressure is uniform.

Odd as these pressure and speed changes may seem, they are quite real and have real consequences. If your hose were clear and you could introduce thin threads of dye into the flowing water, you'd see these dyed streamlines arc around the bend just as they do in Fig. 6.1.2. And if the hose were weak and couldn't tolerate excessive pressure, it would be most likely to burst on the outside of the bend, where the local water pressure is highest.

You might wonder which causes which: does each pressure change cause a speed change or does each speed change cause a pressure change? The answer is that they occur together and are equally entitled to be called cause and effect. Once the steady-state flow pattern has established itself, water following a particular streamline experiences rises and falls in pressure at the same time that it experiences decreases and increases in speed. The two simply go hand in hand.

> **Check Your Understanding #3: Washing Your Shirt with a Spoon**
> *for answers, see page* **198**

You're washing dishes when the stream of water falling from the kitchen faucet follows the curve of a spoon and sprays up onto your shirt. As the stream bends upward, where was its pressure greatest?

Flow through a Nozzle: From Pressure to Speed

When water finally flows through the nozzle at the end of your hose, it exchanges its remaining pressure potential energy for kinetic energy and sprays out into the garden. The nozzle's narrowing channel initiates this energy transformation so that low-speed, high-pressure water entering the nozzle becomes fast-moving, atmospheric-pressure water leaving the nozzle.

Figure 6.1.3 shows that as water passes through the nozzle, the narrowing channel herds all the streamlines together so that its local speed increases. The water following each streamline is speeding up in order to squirt through the bottleneck without causing a backup. And this increase in water's local speed is accompanied by a decrease in water's local pressure, as indicated by the color shift toward the red end of the spectrum.

By the time the water leaves the hose nozzle, its pressure has dropped all the way to atmospheric pressure and it has turned all of its available pressure potential energy into kinetic energy. It emerges as a narrow stream of fast-moving water and arcs gracefully through the air. No wonder you can reach the farthest parts of your garden with water when you use a nozzle.

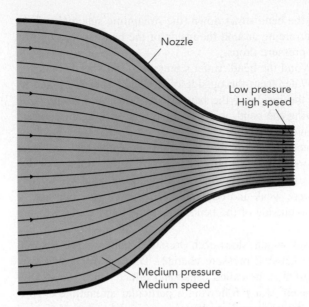

Nozzle

Low pressure
High speed

Medium pressure
Medium speed

Fig. 6.1.3 Water flowing through a nozzle speeds up and its pressure drops. The narrowing spacing between streamlines indicates that the flow speed is increasing while the color shift from violet toward red indicates that the pressure is dropping.

Common Misconceptions: Speed and Pressure in Fluids
Misconception: A fast-moving fluid always has a low pressure.
Resolution: The pressure of a specific portion of fluid depends on its circumstances and can take any value, high or low. However, if a fluid speeds up without descending as it flows along a streamline in steady-state flow, then its pressure will decrease. In that special context, the *faster* moving fluid has a *lower* pressure.

Check Your Understanding #4: Cleaning House

for answers, see page 198

As air rushes steadily into the narrow opening of a vacuum cleaner attachment, it accelerates to high speed and its pressure drops well below atmospheric pressure. From where does the air's newfound kinetic energy come?

The Onset of Turbulence

As you direct the stream of water at the plants in your garden, you notice two interesting phenomena: first, the stream pushes on any surface that slows it down and, second, it tends to break up into fragments as it flows around obstacles. The pushing effect is another Bernoulli result: when the stream encounters a surface, it slows down and spreads out sideways. As the water slows close to the surface, its pressure there rises above atmospheric pressure and it is this elevated pressure that actually pushes the surface forward.

But the breakup effect is something new. In trying to go around the obstacle, the stream of water loses its orderly structure and disintegrates into a swirling, hissing froth. Actually, the hiss you hear is familiar; you heard it as you opened the faucet to start water flowing through the hose. That faucet uses a movable stopper to control the flow of water into the hose; as you opened the faucet, you gradually removed this stopper

Courtesy Lou Bloomfield

Fig. 6.1.4 Water flows slowly past rocks in the stream on the left, and its viscosity keeps it smooth and laminar. Water flows quickly past rocks in the stream on the right, and its inertia separates it into swirling, splashing pockets of turbulence.

from the water pipe to allow water to flow more freely into the hose and the faucet hissed. Whether water encounters a plant or a faucet stopper, there is something about high speeds and obstacles that upsets the smoothly flowing fluid.

Up until now, we've discussed only **laminar flow**—smooth, silent flow that's characterized by simple streamlines. In laminar flow, adjacent regions of a fluid always remain adjacent. For example, if you place two drops of dye near one another in a smoothly flowing stream, they will remain close together indefinitely as they follow streamlines in the laminar flow (Fig. 6.1.4). Laminar flow is the orderly result of viscous forces, which tend to bring adjacent portions of fluid to the same velocity. When viscosity dominates a fluid's motion, the flow is usually laminar.

But as the stream flows swiftly past rocks and obstacles, its streamlines break up into the eddies and churning "white water" that make rafting exciting. The dye is quickly dispersed in this frenzied **turbulence.** The stream is experiencing **turbulent flow**—roiling, noisy flow in which adjacent regions of fluid soon become separated from one another as they move independently in unpredictable directions. Turbulent flow is the disorderly consequence of inertia, which tends to propel each portion of fluid independently according to its own momentum. When inertia dominates a fluid's motion, the flow is usually turbulent.

The plants and faucet stopper are evidently initiating turbulence in what had been laminar flows; flows that were dominated by viscous forces are suddenly dominated by inertia instead. Whether a flow is laminar or turbulent depends on several characteristics of the fluid and its environment:

1. The fluid's viscosity. Viscous forces tend to keep nearby regions of fluid moving together, so high viscosity favors laminar flow (Fig. 6.1.5).
2. The fluid's speed past a stationary obstacle. The faster the fluid is moving, the more quickly two nearby regions of fluid can become separated and the harder it is for viscous forces to keep them together.
3. The size of the obstacle the fluid encounters. The larger the obstacle, the more likely that it will cause turbulence because viscous forces will be unable to keep the fluid ordered over such a long distance.
4. The fluid's density. The denser the fluid, the less it responds to viscous forces and the more likely it is to become turbulent.

Rather than keeping track of all four physical quantities independently, English mathematician and engineer Osborne Reynolds (1842–1912) found that they could be

Courtesy Lou Bloomfield

Fig. 6.1.5 Honey's large viscosity keeps it flowing smoothly (laminar flow) when you pour it. Colored water's small viscosity allows it to splash about (turbulent flow).

combined into a single number that permits a comparison of seemingly different flows. The **Reynolds number** is defined as

$$\text{Reynolds number} = \frac{\text{density} \cdot \text{obstacle length} \cdot \text{flow speed}}{\text{viscosity}}. \qquad \textbf{(6.1.2)}$$

The units on the right side of Eq. 6.1.2 cancel one another so that the Reynolds number is dimensionless, that is, it's just a simple number, such as 10 or 25,000. As the Reynolds number increases, the flow goes from viscous dominated to inertia dominated and therefore from laminar to turbulent. In his experiments, Reynolds found that turbulence usually appears when the Reynolds number exceeds roughly 2300. You can observe this transition by moving a 1-cm-thick (0.4-inch-thick) stick through still water. If you move the stick slowly, about 10 cm/s (4 inch/s), the Reynolds number will be about 1000 and the flow around the stick will be laminar. But if you speed the stick up to about 50 cm/s (20 inch/s), the Reynolds number will rise to about 5000 and the flow will become turbulent.

One of the most common features of turbulent flows is the **vortex,** a whirling region of fluid that moves in a circle around a central cavity. A vortex resembles a miniature tornado, with its cavity created by inertia as the fluid spins. Vortices are easily visible behind a canoe paddle or in a mixing bowl. Once an object moves fast enough through a fluid to create turbulence, these vortices begin to form. Each vortex builds up behind the object but is soon whisked away to form a wake of *shed vortices* (Fig. 6.1.6).

While laminar flow is fully predictable, turbulent flow exhibits chaotic behavior or **chaos:** you can no longer predict exactly where any particular drop of water will go. The study of chaos is a relatively new field of science. Because a **chaotic system**—a system exhibiting chaos—is exquisitely sensitive to initial conditions, even the slightest change in those conditions may produce profound changes in its situation later on.

Even when you can't see turbulent water flow, you can usually hear it. The churning motion of turbulence converts some of the water's total energy into thermal energy

Courtesy Peter Bradshaw, Stanford University

Fig. 6.1.6 When water flows rapidly around a cylinder, its flow becomes turbulent. A pattern of swirling vortices forms to the right of this cylinder.

and sound. The turbulence near the faucet slightly reduces the water's total energy as it enters the hose and therefore its speed as it emerges from the nozzle and sprays toward your garden.

A different sound occurs when you abruptly close the nozzle and stop the flow of water. Moving water has momentum, and stopping it suddenly requires an enormous backward force. Since the slowing flow is not steady state, Bernoulli's equation doesn't apply and the pressure can surge to astronomical values near the front of the moving water. This pressure surge is what accelerates the water backward to slow it down and also what leads to the loud "thump" sound you hear as the water stops. Known as **water hammer,** the surging pressure in front of stopping water jerks the nozzle, swells the hose, and may even rattle the pipes in your home.

Check Your Understanding #5: Urban Windstorms

for answers, see page **198**

On a windy day in a city with many tall buildings, leaves and papers can be seen swirling about in the air or on the sidewalks. What causes these whirling air currents?

Check Your Figures #2: Wind on the Open Road

for answers, see page **199**

Is the flow of air around a convertible laminar or turbulent as the convertible cruises down the highway? (Air's viscosity is given in Table 6.1.1.)

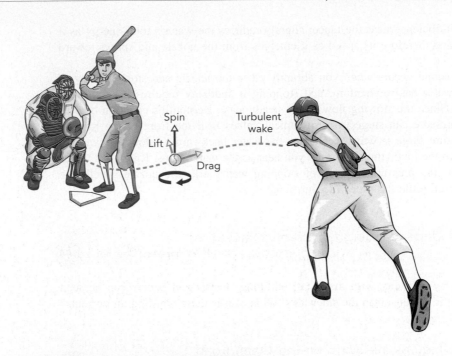

Balls and Air

Much of the subtlety and nuance in games such as baseball and golf come from the way balls interact with air. If baseball were played on the moon, which has no air, the sinking fastball would be the only interesting pitch. Moon golfers wouldn't have to worry about hooks or slices. In this section we will investigate how air affects the flight of balls and other related objects.

Questions to Think About: Why can you throw a real baseball so much farther than a hollow plastic one? Why does a long fly ball appear to drop straight down when you try to catch it in deep center field? What kind of force could make a curveball curve? What makes a well-hit golf ball hang in the air before falling to the green? How can a knuckleball or spitball jitter about in flight?

Experiments to Do: To make air's effects most apparent, you need a ball that weighs little but has lots of surface area. A beach ball is ideal, but a whiffle ball or hollow plastic ball will also do nicely. See how far you can throw it. How does it stop? Does it slow down and lose height gradually, or does it stop rapidly and fall to the ground? Now make the ball spin as you throw it. Why does the ball curve in flight? Does a fast spin make the ball curve more or less? Which way is the ball spinning, and how is the spin related to the direction of its curve? Change the direction of spin. Which way does the ball curve now?

When a Ball Moves Slowly: Laminar Airflow

One of the first things you might notice if you joined a new baseball franchise on the moon is that pitched balls reach home plate faster than back at home. Since the moon has no atmosphere, there is no air resistance to slow a ball down. In the previous section, we saw how objects affect moving fluids. Now as we study **aerodynamics,** the science of air's dynamic interactions, we'll see how fluids affect moving objects.

In air, a moving ball experiences **aerodynamic forces,** that is, forces exerted on it by the air because of their relative motion. These consist of **drag forces** that push the ball downwind and **lift forces** that push the ball to one side or the other (Fig. 6.2.1). We'll begin our study of ball aerodynamics with drag forces, commonly called "air resistance," and we'll start with a slow-moving ball. The reason for starting slow is that at low speeds, viscous forces are able to organize the air as it flows around the ball; viscosity dominates over inertia and the airflow around the slow-moving ball is laminar.

Figure 6.2.2 shows the pattern of laminar airflow around a slow-moving ball. Actually, the pattern is the same whether the ball moves slowly through the air or the air moves slowly past the ball. For simplicity, let's move along with the ball and study the airflow from the ball's inertial frame of reference. In that inertial frame, the ball appears stationary with the air flowing past it.

The slow-moving air separates neatly around the front of the ball and comes back together behind it. It produces a **wake,** an air trail behind the ball, that's smooth and free of turbulence. But the air's speed and pressure aren't uniform all the way around the ball. The airflow bends several times as it travels around the ball and, as we saw in the previous section, such bends always involve pressure imbalances. Since the air pressure far from the ball is steadfastly atmospheric, those pressure imbalances are always caused by pressure variations near the ball's surface. Whenever air bends away from the ball, so that the ball is on the outside of a bend, the pressure near the ball must be higher than atmospheric. And whenever the air bends toward the ball, so that the ball is on the inside of a bend, the pressure near the ball must be lower than atmospheric.

With that introduction, let's examine the slow-moving airflow around the ball. Air heading toward the ball's front bends away from it, so the pressure near the front of the ball must be higher than atmospheric. This rise in air pressure is accompanied by a decrease in **airspeed**—the air's speed relative to the ball. Figure 6.2.2 indicates the pressure rise by a color shift toward the violet end of the spectrum and the decrease in airspeed by the widening separation of the streamlines.

Air rounding the ball's sides bends toward it, so the pressure near the sides of the ball must be below atmospheric. This drop in air pressure is accompanied by an increase in

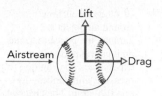

Fig. 6.2.1 The two types of aerodynamic forces exerted on objects by air are drag and lift. Drag is exerted parallel to the onrushing airstream and slows the object's motion through the air. Lift is exerted perpendicular to that airstream so that it pushes the object to one side or the other. Lift is not necessarily in the upward direction.

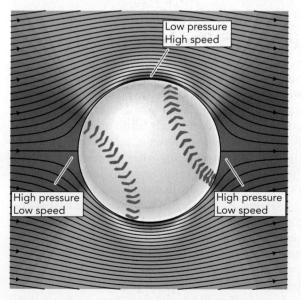

Low pressure
High speed

High pressure
Low speed

High pressure
Low speed

Fig. 6.2.2 The airflow around a slowly moving ball is laminar. Air slows down in front of and behind the ball (widely spaced streamlines), and its pressure increases (shifts toward the violet end of the spectrum). Air speeds up at the sides of the ball (narrowly spaced streamlines), and its pressure decreases (shifts toward the red end of the spectrum). However, the pressure forces on the ball balance one another perfectly, and it experiences no pressure drag. Only viscous drag is present to affect the ball.

❑ When the airflow around an object is laminar, the pressure forces on it cancel perfectly and it experiences no drag due to pressure imbalances—no *pressure* drag. The absence of pressure drag was a great puzzlement to early aerodynamicists, who knew that the airflow around dust is laminar and that it experiences a drag force. This mystery was named d'Alembert's paradox, after Jean Le Rond d'Alembert (1717–1783), the French mathematician who first recognized it. D'Alembert and his contemporaries didn't know about the viscous drag force, which is what really slows dust's motion through the air.

airspeed. Figure 6.2.2 indicates the pressure drop by a color shift toward the red end of the spectrum and the increase in airspeed by the narrowing separation of the streamlines.

The laminar airflow continues around to the back of the ball and then trails off behind it. Since the departing air again bends away from the ball, the pressure near the back of the ball must be higher than atmospheric. Figure 6.2.2 indicates the pressure rise as a shift toward violet and the accompanying decrease in airspeed as a widening separation of the streamlines.

It may seem strange that the air pressure can be different at different points on the ball, but that's what happens in a flowing stream of air. It's particularly remarkable that low-pressure air at the sides of the ball is able to flow around to the back of the ball, where the pressure is higher. This air is experiencing a pressure imbalance that pushes it backward, opposite its direction of travel. But a pressure imbalance causes acceleration, not velocity, and the low-pressure air flowing past the sides of the ball has enough energy and forward momentum to carry it all the way to the back of the ball. Although this air slows as it flows into the rising pressure, it manages to complete its journey.

The airflow around the ball is symmetric, and the forces that air pressure exerts on the ball are also symmetric. These pressure forces cancel one another perfectly so that the ball experiences no overall force due to pressure. Most importantly, the high pressure in front of the ball is balanced by the high pressure behind it. As a result of this symmetric arrangement, the only aerodynamic force acting on the ball is **viscous drag**—the downstream frictional force caused by layers of viscous air sliding across the ball's surface (see ❑).

Though we'll soon see that viscous drag is only a small fraction of the air resistance experienced by sports balls, it's the force that suspends dust in the air for hours and is an important issue for airplane wings. It's also a force that we've encountered before: viscous drag slowed water in your garden hose in the previous section!

▶Check Your Understanding #1: Smooth Flow in a Stream
for answers, see page 198

When water in a stream flows slowly past a small rock, the water in front of the rock slows down and its increased pressure lifts the water level slightly. The water level behind the rock also rises slightly. Explain.

When a Ball Moves Fast: Turbulent Airflow

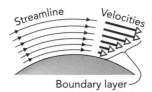

Fig. 6.2.3 As air flows past a surface, a thin layer of it is slowed by viscous drag forces. This boundary layer is laminar at low Reynolds numbers and doesn't become turbulent until the Reynolds number exceeds about 100,000.

Balls don't always experience laminar airflow. Turbulence is common, particularly in sports, and brings with it a new type of drag force. When the air flowing around a ball is turbulent, the air pressure distribution is no longer symmetric and the ball experiences **pressure drag**—the downstream force exerted by unbalanced pressures in the moving air. These unbalanced pressures exert an overall force on the ball that slows its motion through the air.

A ball can experience turbulent airflow and pressure drag when the Reynolds number exceeds about 2000. The Reynolds number, introduced in the previous section, combines the ball's size and speed with the air's density and viscosity to give an indication of whether the airflow is dominated by viscosity or inertia. At low Reynolds numbers, the air's viscosity dominates over its inertia and the airflow is laminar. But at high Reynolds numbers, air's inertia dominates over its viscosity and the airflow tends to become turbulent. This turbulence, however, won't start until something triggers it and viscosity provides that trigger.

To understand viscosity's role, we must look at the air near the ball's surface. Even in a strong wind, viscous forces slow down a thin **boundary layer** of air near the ball's surface (Fig. 6.2.3). Discovered by Ludwig Prandtl ❑ with help from Gustave Eiffel

(Fig. 6.2.4), this boundary layer moves more slowly and has less total energy than the freely flowing air farther from the surface.

As air flows toward the back of the ball, it travels through an **adverse pressure gradient**—a region of rising pressure that pushes backward on the air and causes it to decelerate. While the freely flowing airstream outside the boundary layer has enough energy and forward momentum to continue onward and reach the back of the ball on its own, air in the boundary layer does not. It needs a forward push.

At low Reynolds numbers, the entire airstream helps to push that boundary layer all the way to the back of the ball and the airflow remains laminar. But at high Reynolds numbers, viscous forces between the freely flowing airstream and the boundary layer are too weak to keep the boundary layer moving forward into the rising pressure behind the ball.

Without adequate help, the boundary layer eventually **stalls**—it comes to a stop and thereby spoils steady-state flow. More horrible still, this stalled boundary layer air is pushed backward by the adverse pressure gradient and it returns all the way to the sides of the ball. As it does, it cuts like a wedge between the ball and the freely flowing airstream. The result is an aerodynamic catastrophe: the airstream separates from the ball, leaving a huge turbulent wake or air pocket behind the ball (Fig. 6.2.5).

Because of this turbulent wake, air no longer bends smoothly away from the back of the ball and there is no rise in pressure there. Instead, the pressure behind the ball is roughly atmospheric. The absence of a high-pressure region behind the ball spoils the symmetry of pressure forces on the ball and those forces no longer cancel. The ball experiences an overall pressure force downwind—the force of pressure drag. In effect, the ball is transferring forward momentum to the air in its turbulent wake and dragging that wake along with it.

Pressure drag slows the flight of almost any ball moving faster than a snail's pace. The pressure drag force is roughly proportional to the cross-sectional area of the turbulent air pocket and to the square of the ball's speed through the air. For a ball moving at a moderate speed, the air pocket is about as wide as the ball and the ball experiences a large pressure drag force.

❑ Among Ludwig Prandtl's (German engineer, 1875–1953) many pivotal contributions to aerodynamic theory is the concept of boundary layers in fluid motion. Prandtl was so engrossed in establishing Göttingen as the world's foremost aerodynamic research facility that he did not have time to court a wife. Deciding he should be married, Prandtl wrote his former advisor's wife, asking to marry one of her two daughters but not specifying which one. The family selected the eldest daughter, and the wedding took place.

Courtesy Lou Bloomfield

Fig. 6.2.4 Early experiments in aerodynamics were performed by Gustave Eiffel (French engineer, 1832–1923), who designed the tower that bears his name. In the 1890s, Eiffel dropped objects of various sizes and shapes from his tower and measured the drag that they experienced. His work was used by Prandtl to explain the reduction in drag that accompanies the appearance of turbulent boundary layers.

Fig. 6.2.5 When a ball's speed gives it a Reynolds number between about 2000 and 100,000, its laminar boundary layer stalls in the rising pressure behind the ball. The resulting reversed flow causes the main airflow to separate from the ball's surface, leaving a large, turbulent wake. The average pressure behind the ball remains low, and the ball experiences a large pressure drag.

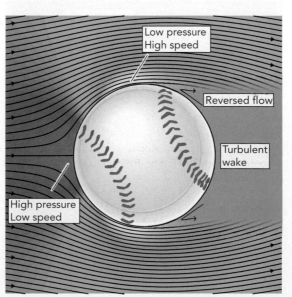

Low pressure
High speed

Reversed flow

Turbulent wake

High pressure
Low speed

→ Check Your Understanding #2: Leaving No Trace

for answers, see page **199**

When your canoe coasts extremely slowly across the water of a still lake, it leaves almost no trail in the water behind it. When you paddle it swiftly through the water, however, the canoe leaves a swirling wake. Explain this difference.

The Dimples on a Golf Ball

If this were the whole story, you would never hit a home run at a baseball game or a 250-yard drive on the golf course. But inertia has yet another card to play.

At very high Reynolds numbers the boundary layer itself becomes turbulent (Fig. 6.2.6). It loses its laminar streamlines and begins to mix rapidly within itself and with the freely flowing airstream nearby. This mixing brings additional energy and forward momentum into the boundary layer and makes it both harder to stop and more resistant to reversed flow. Although this turbulent boundary layer still stalls before reaching the back of the ball, the stalled air flows upstream only a short distance. While the freely flowing airstream still separates from the ball, that separation occurs far back on the ball and the resulting turbulent wake is relatively small (Fig. 6.2.7).

As a result of this smaller air pocket, the pressure drag is reduced from what it would be without the turbulent boundary layer. The effect of replacing the laminar boundary layer with a turbulent one is enormous; it's the difference between a golf drive of 70 yards and one of 250 yards! The effects of the Reynolds number on the airflow around a ball are summarized in Table 6.2.1.

Delaying the airflow separation behind the back of the ball is so important to distance and speed that the balls of many sports are designed to encourage a turbulent boundary layer (Fig. 6.2.8). Rather than waiting for the Reynolds number to exceed 100,000, the point near which the boundary layer spontaneously becomes turbulent, these balls "trip" the boundary layer deliberately (Fig. 6.2.9). They introduce some impediment to laminar flow, such as hair or surface irregularities, which causes the air near the

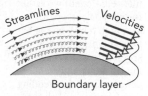

Streamlines   Velocities

Boundary layer

Fig. 6.2.6 When the Reynolds number exceeds about 100,000, the boundary layer of air flowing past a surface becomes turbulent. This whirling fluid brings in extra energy and momentum from the freely flowing airstream and can travel deep into a region of increasing pressure.

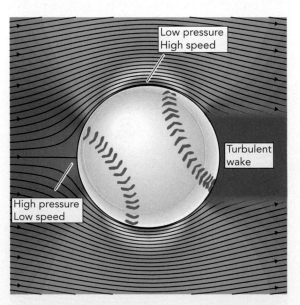

Low pressure
High speed

Turbulent
wake

High pressure
Low speed

Fig. 6.2.7 When a ball travels fast enough that its Reynolds number exceeds 100,000, its boundary layer becomes turbulent. This turbulent layer travels much of the way around the back of the ball before it separates from the surface. The freely flowing air follows it, and the two leave a relatively small turbulent wake. The ball experiences only a modest pressure drag.

Table 6.2.1 Effects of Reynolds Number on the Airflow around a Ball or Other Object

REYNOLDS NUMBER	BOUNDARY LAYER	TYPE OF WAKE	MAIN DRAG FORCE
< 2000	Laminar	Small laminar	Viscous
2000–100,000	Laminar	Large turbulent	Pressure
>100,000	Turbulent	Small turbulent	Pressure

Courtesy Rebus, Inc

ball's surface to tumble about and become turbulent. The drop in pressure drag more than makes up for the small increase in viscous drag. That's why a tennis ball has fuzz and a golf ball has dimples.

So how much does drag affect balls in various sports? For those that involve rapid movements through air or water, the answer is quite a bit. Drag forces increase dramatically with speed; as soon as a turbulent wake and pressure drag appear, the drag force increases as the square of a ball's speed. As a result, baseball pitches slow significantly during their flights to home plate, and the faster they're thrown, the more speed they lose. A 90-mph fastball loses about 8 mph en route, while a 70-mph curveball loses only about 6 mph.

A batted ball fares slightly better because it travels fast enough for the boundary layer around it to become turbulent, an effect that appears at around 160 km/h (100 mph). While the resulting reduction in drag explains why it's possible to hit a home run, the presence of air drag still shortens the distance the ball travels by as much as 50%. Without air drag, a routine fly ball would become an out-of-the-park home run. To compensate for air drag, the angle at which the ball should be hit for maximum distance isn't the theoretical 45° above horizontal discussed in Section 1.2. Because of the ball's tendency to lose downfield velocity, it should be hit at a little lower angle, about 35° above horizontal (Fig. 6.2.10).

Since the ball loses much of its horizontal component of velocity during its trip to the outfield, a long fly ball tends to drop almost straight down as you catch it. Gravity causes it to move downward, but drag almost stops its horizontal motion away from home plate. Drag also limits the downward speed of a falling ball to about 160 km/h (100 mph). That's the baseball's **terminal velocity,** the downward velocity at which the upward drag force exactly balances its downward weight and it stops accelerating. Even if you drop a baseball from an airplane, its velocity will not exceed this value.

Fig. 6.2.8 Early golf balls (*left*) were handmade of leather and stuffed with feathers. Golf became popular when cheap balls made of a hard rubber called gutta-percha became available. But new, smooth "gutties" didn't travel very far; they flew better when they were nicked and worn. Manufacturers soon began to produce balls with various patterns of grooves on them (*bottom and right*), and those balls traveled dramatically farther than smooth ones. Modern golf balls (*top*) have dimples instead of grooves.

Check Your Understanding #3: Designing a Great Sports Car

for answers, see page 199

As an automobile designer, your job is to minimize the aerodynamic drag experienced by the car on which you are working. Where should you try to locate the point at which the airflow separates from the car?

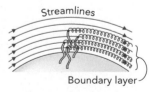

Streamlines

Boundary layer

Fig. 6.2.9 The boundary layer can be made turbulent at Reynolds numbers below 100,000 by "tripping" it with obstacles such as fuzz or dimples.

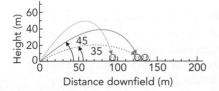

Fig. 6.2.10 Air drag slows the flight of a batted ball so that the ideal angle at which to hit it isn't the theoretical 45° of Fig. 1.2.7. An angle of roughly 35° above horizontal will achieve the maximum distance.

Curveballs and Knuckleballs

The drag forces on a ball push it downstream, parallel to the onrushing air. But in some cases, the ball may also experience lift forces—forces that are exerted perpendicular to the airflow (Fig. 6.2.1). To experience drag, the ball only has to slow the airflow down; to experience lift, the ball must deflect the airflow to one side or the other. Although its name implies an upward force, lift can also push the ball toward the side or even downward.

Curveballs and knuckleballs both use lift forces. In each of these famous baseball pitches, the ball deflects the airstream toward one side and the ball accelerates toward the other. Again we have action and reaction—the air and the ball push off one another. Getting the air to push the ball sideways is no small trick. Explaining it isn't easy either, but here we go.

A curveball is thrown by making the ball spin rapidly about an axis perpendicular to its direction of motion. The choice of this axis determines which way the ball curves. In Fig. 6.2.11, the ball is spinning clockwise, as viewed from above. With this choice of rotation axis, the ball curves to the pitcher's right because the ball experiences two lift forces to the right. One is the Magnus force, named after the German physicist H. G. Magnus (1802–1870) who discovered it. The other is a force we will call the wake deflection force.

The **Magnus force** occurs because the spinning ball carries some of the viscous air around with it (Fig. 6.2.11a). The steady-state flow pattern that forms around this ball is asymmetric: the airstream that moves with the turning surface is much longer than the airstream that moves opposite that surface. Since the longer airstream bends mostly

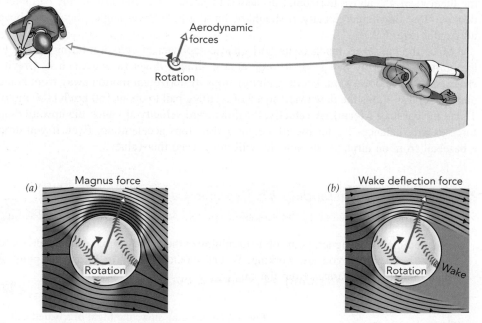

Fig. 6.2.11 A rapidly rotating baseball experiences two lift forces that cause it to curve in flight. (a) The Magnus force occurs because air flowing around the ball in the direction of its rotation bends mostly toward it, while air flowing opposite its rotation bends mostly away from it. (b) The wake deflection force occurs because air flowing around the ball in the direction of its rotation remains attached to the ball longer and the ball's wake is deflected.

toward the baseball, the average pressure on that side of the ball must be below atmospheric. The shorter airstream bends mostly away from the ball, so the average pressure on that side of the ball must be above atmospheric. Because the pressure forces on the ball's sides don't balance one another, the ball experiences the Magnus force toward the low-pressure side—the side turning toward the pitcher—and deflects in that direction. The airflow deflects in the opposite direction.

In laminar flow, the Magnus force is the only lift force acting on a spinning object. But a pitched baseball has a turbulent wake behind it and is also acted on by the **wake deflection force.** This force appears when the ball's rapid rotation deforms the wide, symmetric wake (Fig. 6.2.5) that develops behind it at high Reynolds numbers. When the ball isn't spinning, the freely flowing airstream separates from the ball approximately at its side and that separation is symmetric all the way around the ball's middle. But when the ball is spinning (Fig. 6.2.11b), the moving surface pushes on the airstream with viscous forces. As a result, airstream separation is delayed on one side of the ball and hastened on the other. The overall wake of air behind the ball is thus deflected to one side and the ball experiences the wake deflection force toward the opposite side—the side turning toward the pitcher. The wake deflection force and the Magnus force both push the ball in the same direction.

Of these two forces, the wake deflection force is probably the more important for a curveball, although the Magnus force is usually given all the credit. A skillful pitcher can make a baseball curve about 0.3 m (12 inches) during its flight from the mound to home plate—the more spin, the more curve. The pitcher counts on this change in direction to confuse the batter. The pitcher can also choose the *direction* of the curve by selecting the axis of the ball's rotation. The ball will always curve toward the side of the ball that is turning toward the pitcher. When thrown by a right-handed pitcher, a proper curveball curves down and to the left, a slider curves horizontally to the left, and a screwball curves down and to the right.

When the pitcher throws a ball with backspin, so that the top of the ball turns toward the pitcher, the ball experiences an upward lift force. In baseball this force isn't strong enough to overcome gravity, but it does make the pitch hang in the air unusually long. And in golf, where the club can give the ball enormous backspin, the ball really does lift itself upward so that it flies down the fairway like a glider.

However, there are some cases when a ball's behavior stems from its *lack* of spin. In baseball, for example, a knuckleball is thrown by giving the ball almost no rotation. Its seams are then very important. As air passes over a seam, the flow is disturbed so that the ball experiences a sideways aerodynamic force: a lift force. The ball flutters about in a remarkably erratic manner. Releasing the ball without making it spin is difficult and requires great skill. Pitchers who are unable to throw a knuckleball legally sometimes resort to lubricating their fingers so that the ball slips out of their hands without spinning. Like its legal relative, the so-called spitball dithers about and is hard to hit. The same is true for a scuffed ball.

Check Your Understanding #4: Center Court

for answers, see page **199**

One of the most difficult and effective strokes in tennis is the topspin lob, in which the top of the ball spins away from the player who hit it. Which way is the lift force on this ball directed?

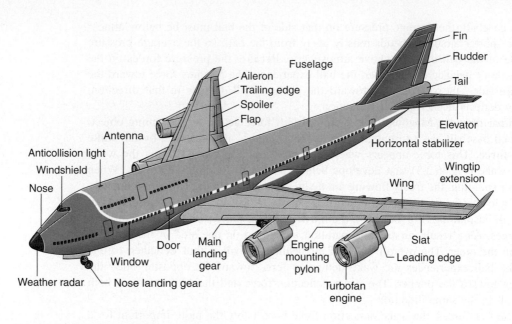

SECTION 6.3   **Airplanes**

We have now set the stage for the ultimate aerodynamic machines, airplanes. Freed from contact with the ground, airplanes are affected only by aerodynamic forces and gravity, hopefully in that order. Despite their complex appearances, airplanes employ physical principles that we have already successfully examined. But while this section revisits many familiar concepts, it also explores new territory. For example, you may have already figured out what type of aerodynamic force holds an airplane up, but what type of aerodynamic force keeps it moving forward?

Questions to Think About: *Why are the wings of small, propeller-driven aircraft relatively large and bowed in comparison to those of jets? Why does a commercial airplane extend slats and flaps from its wings during takeoffs and landings? What pushes airplanes forward in flight? How can some airplanes fly upside down? Why do most fast commercial aircraft employ jet engines and not propellers?*

Experiments to Do: *The best experiment for this section is to take a plane flight or at least to visit the airport and watch the planes.*

As you sit in the plane during takeoff, feel the plane accelerating forward. If you're on a commercial jet, notice that the slats and flaps on the airplane's wings are extended during takeoff, making the wings wider and more curved. How could this increased width and curvature help the plane take off? The pilot holds the plane on the ground until it reaches the proper speed, then quickly tips it upward into the air. An invisible vortex of air peels away from the trailing edge of the wing and the plane lifts off the ground.

Once airborne, the airplane retracts its landing gear, slats, and flaps. Watch the trailing edge of each wing as the plane turns or changes altitudes; you'll see various surfaces there move up or down. Similar motions occur on the tail. How do these surfaces control the plane's orientation?

Near its destination, the airplane prepares for landing. Again the slats and flaps are extended. Watch as spoilers on the tops of the wings pop up and down noisily. How do these surfaces affect the drag force on the plane? The plane's landing gear extends, and it touches down on the runway. The propellers or jet engines abruptly begin to slow the airplane, assisted by the spoilers on the wings. Feel the plane accelerating backward. Another invisible vortex of air peels away from the trailing edge of the wing, rotating in the opposite direction from the first vortex, and the flight is over.

Airplane Wings: Streamlining

By now you've probably realized that an airplane is supported in flight by an upward lift force on its wings and that this lift force comes from deflecting the passing airflow downward. Each wing is an **airfoil,** an aerodynamically engineered surface that's designed to obtain particular lift and drag forces from the air flowing past it. More specifically, each wing is shaped and oriented so that, during flight, the airstream flowing over the wing bends downward toward its top surface while the airstream under the wing bends downward away from its bottom surface. These bends are associated with pressure changes near the wing itself and are responsible for the upward lift force that suspends the airplane in the sky.

However, to get a more complete understanding of how the wing develops this lift, let's go for a flight. Imagine yourself in an airplane that has just begun rolling down the runway. From your perspective, air is beginning to flow past each of the airplane's wings. When this moving air encounters the wing's *leading edge*, it separates into two airstreams: one traveling over the wing and the other under it (Fig. 6.3.1). These airstreams continue onward until they leave the wing's *trailing edge*. Since the airplane's nose is still on the ground, the wing is essentially horizontal and the airflow around it is simple and symmetric.

Since the wing isn't deflecting the airflow yet, it's experiencing no lift, only drag. But while this drag pushes the airplane downwind and thus opposite its forward motion along the runway, it's unusually weak. The wing produces almost no turbulent wake and

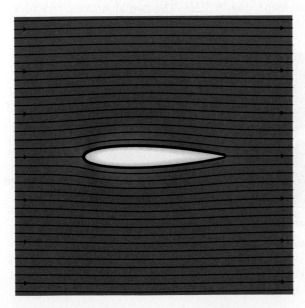

Fig. 6.3.1 An airplane wing is a streamlined airfoil and the airflow around it is laminar. This horizontal wing is symmetric, top and bottom, and the airflow splits evenly into airstreams above and below it. Since it doesn't deflect the airflow, it experiences no lift.

thus experiences almost no pressure drag. What little drag it does experience is mostly viscous drag, essentially surface friction with the passing air.

Although the wing's near lack of air resistance should surprise you, you probably take it for granted. That's because you've often observed that such "streamlined" objects cut through the air particularly well. Something about having a long, tapered tail allows the wing to avoid the flow separation and turbulent wake that occur behind an unstreamlined ball.

What makes the horizontal wing **streamlined** is the extremely gradual rise in air pressure after its widest point. While this gently rising pressure pushes the wing's boundary layer backward, opposite the direction of flow, the force it exerts is so weak that the layer doesn't stall. Driven onward by viscous forces from the freely flowing airstream, the wing's boundary layer manages to keep moving forward all the way to the wing's trailing edge and never triggers flow separation. The wing produces almost no turbulent wake and experiences almost no pressure drag.

> Check Your Understanding #1: Slicing through the Air
>
> for answers, see page 199

The fastest bicycles are those with fairings—streamlined shells that reduce air resistance dramatically. How do those fairings work?

Airplane Wings: Producing Lift

With so little air resistance, the airplane accelerates forward rapidly and soon reaches takeoff speed. The pilot then raises the airplane's nose so that its wings are no longer horizontal and they begin to experience upward lift forces. The airplane's total lift soon exceeds its weight and it begins to accelerate upward into the air. The airplane is flying!

But let's take a closer look at the moment of takeoff. If you could see the airflow and were paying close attention, you'd notice a remarkable sequence of events that begin when the wings tilt upward.

At first, the airflow around the tilted wings continues to travel horizontally on average, although it develops a peculiar shape (Fig. 6.3.2a). The two airstreams, one over

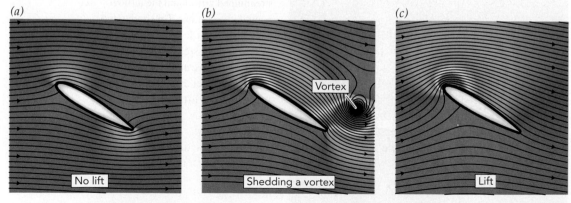

Fig. 6.3.2 (*a*) Although this wing's leading edge has been tipped upward, giving it a positive angle of attack, the airflow around it is relatively symmetric and produces no lift. (*b*) The kink at the trailing edge of the wing is unstable and is blown away or *shed* as a horizontal vortex. (*c*) The resulting airflow is deflected downward and the wing experiences an upward lift force.

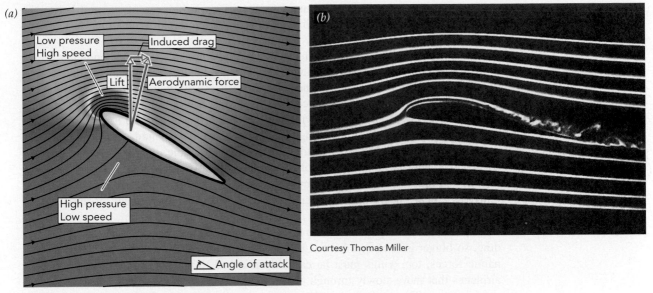

Fig. 6.3.3 (*a*) This airplane wing is shaped and oriented so that both airstreams, over it and under it, bend downward. The wing experiences a large aerodynamic force that points upward and slightly downstream. The upward component of this force is lift. The downstream component is induced drag. (*b*) Smoke trails in a wind tunnel show the airflow past a wing.

the tilted wing and one under it, each bend twice—once up and once down. As we saw while studying balls, when an airstream bends toward the wing, the pressure near the wing is below atmospheric and when an airstream bends away from the wing, the pressure near the wing is above atmospheric. Since each airstream bends equally toward and away from the wing, it experiences no overall deflection or average pressure change and provides the wing with no overall lift.

But the lower airstream is making a sharp bend around the wing's trailing edge, essentially an upward kink. Air's inertia makes such a kink unstable and it soon blows away from the wing's trailing edge as a swirling horizontal vortex of air (Fig. 6.3.2*b*). After shedding that vortex, the wing establishes a new, stable flow pattern in which both airstreams pass smoothly away from the wing's trailing edge (Fig. 6.3.2*c*), a situation named the *Kutta condition* after the German mathematician M. Wilhelm Kutta (1867–1944).

In this new pattern, the airstream flowing over the wing is longer than the airstream flowing under it and both bend downward (Fig. 6.3.3). The upper airstream bends primarily toward the wing, so the air's pressure just above the wing is below atmospheric (a shift toward red) and its speed is increased (narrowly spaced streamlines). In contrast, the lower airstream bends primarily away from the wing, so the air's pressure just below the wing is above atmospheric (a shift toward violet) and its speed is decreased (widely spaced streamlines). The air pressure is now higher under the wing than over it, so this new flow pattern produces upward lift. The air now supports your plane and up you go.

Another way to think about this lift is as a deflection of the airflow. Air approaches the wing horizontally but leaves heading somewhat downward. To cause this deflection, the wing must push the airflow downward. In reaction, the airflow pushes the wing upward and produces lift. In other words, the wing transfers downward momentum to the air and is left with upward momentum as a result. These two explanations for lift—

the Bernoullian view that lift is caused by a pressure difference above and below the wing and the Newtonian view that lift is caused by a transfer of momentum to the air—are perfectly equivalent and equally valid.

However, the overall aerodynamic force on the wing isn't quite perpendicular to the onrushing air; it tilts slightly downwind. The perpendicular component of this aerodynamic force is lift, but the downwind component is a new type of drag force—induced drag. **Induced drag** is a consequence of energy conservation: in addition to transferring momentum to the passing air, the wing also transfers some energy to it. The air extracts that energy from the wing by pushing the wing downwind with induced drag and thereby doing negative work on it. Since induced drag is undesirable, the airplane minimizes it by using as much air mass as possible to obtain its lift. A larger mass of air carries away the airplane's unwanted downward momentum while moving downward less quickly and with less kinetic energy. Since larger wings obtain their lift from larger air masses, they experience less induced drag.

Unfortunately, larger wings also have more surface area and experience more viscous drag, so bigger isn't always better. And because wing shape and airspeed affect aerodynamic forces, too, wings must be carefully matched to their airplanes. Small propeller airplanes that move slowly through the air need relatively large, highly curved wings to support them. Those wings are often asymmetric—more curved on top than on bottom to make maximum use of the limited, low-speed air they encounter each second. Commercial and military jets fly faster and encounter far more high-speed air each second, so they can get by with relatively small, moderately curved wings.

But even at constant airspeed, a wing's lift can be adjusted by varying its **angle of attack**—the angle at which it approaches the onrushing air. The larger the angle of attack, the more the two airstreams bend and the greater the wing's lift. Because the wings are rigidly attached to the plane, the pilot tips the nose of the plane upward to increase the lift and downward to reduce the lift. That's why raising the plane's nose during takeoff is what finally makes the plane leap up into the air.

Since lift depends so strongly on a wing's angle of attack, some planes can be flown upside down. As long as the inverted wing is tilted properly, it obtains upward lift and supports the plane. But this feat is easiest when a plane's wing has the same curvature, top and bottom. That's why stunt fliers who regularly fly upside down often use sport aircraft that have symmetric or nearly symmetric wings.

> ► Check Your Understanding #2: Blowing in the Wind
>
> for answers, see page 199

The sail of a small sailboat bows forward and outward so that wind traveling around the sail's outside surface bends toward the sail, while wind traveling across its inside surface bends away from the sail. How does this arrangement propel the sailboat across the water?

Lift Has Its Limits: Stalling a Wing

There's a limit to how much lift the pilot can obtain by increasing the wing's angle of attack because tilting the wing gradually transforms it from streamlined to **blunt,** that is, to having a rapid rise in air pressure after its widest point. As we saw for balls, blunt objects generally experience airflow separation and pressure drag. Indeed, beyond a certain angle of attack, the airstream over the top of a wing separates from its surface and

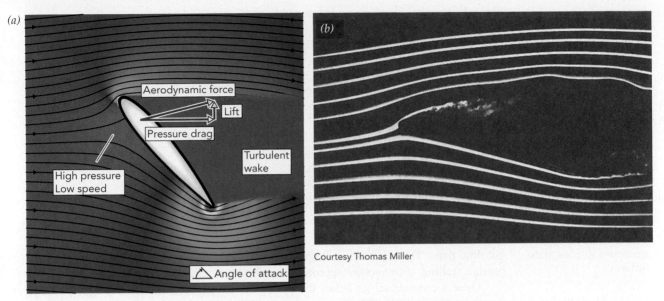

Courtesy Thomas Miller

Fig. 6.3.4 (*a*) A wing stalls when the airstream over the top of the wing separates from its surface. A turbulent air pocket forms above the wing, making it much less efficient. The wing's lift decreases because the average pressure above the thickest part of the wing becomes higher, and the drag increases because the average pressure above the trailing edge becomes lower. (*b*) Smoke trails in a wind tunnel show that the air separates from the surface and becomes turbulent as it flows over a stalled wing.

the wing stalls. This separation starts when air in the upper boundary layer is brought to standstill by the rapidly rising pressure beyond the wing's widest point. Once this boundary layer stalls, it shaves most of the airstream away from the wing's upper surface.

The separated airstream over the top of the stalled wing leaves a billowing storm of turbulence beneath it (Fig. 6.3.4). This airstream separation is an aerodynamic catastrophe for the airplane. Because the average pressure above the wing increases, the wing loses much of its lift. And the appearance of a turbulent wake heralds the arrival of severe pressure drag. The plane slows dramatically and drops like a rock.

To avoid stalling, pilots keep the angle of attack within a safe range. But the possibility of stalling also limits the minimum speed at which the airplane will fly. As the airplane slows down, the pilot must increase its angle of attack to maintain adequate lift. Below a certain speed, the airplane can't obtain that lift without tilting its wings until they stall. It can no longer fly.

To avoid stalling, a plane must never fly slower than this minimum speed, particularly during landings and takeoffs. For a small, propeller-driven plane with highly curved wings, the minimum flight speed is so low that it's rarely an issue. For a commercial jet, however, the minimum airspeed is about 220 km/h (140 mph). Airplanes taking off or landing this fast would require very long runways on which to build up or get rid of speed. Instead, commercial jets have wings that can change shape during flight. Slats move forward and down from the leading edges of the wings, and flaps move back and down from the trailing edges (Fig. 6.3.5). With both slats and flaps extended, the wing becomes larger and more strongly curved, similar to the wings of a small propeller plane, and the minimum safe airspeed drops to a reasonable 150 km/h (95 mph). Vanes near the flaps also emerge during landings to direct high-energy air from beneath the wings

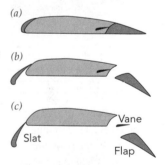

Fig. 6.3.5 At cruising speed, an airplane's wings are moderately curved airfoils (*a*). But during takeoffs (*b*) and landings (*c*), slats are extended from the leading edges and flaps from the trailing edges. The airfoils become much more highly curved, generating more lift at low speeds. During landing, a vane is also extended for boundary layer control to prevent stalling.

❑ Airplane designers can reduce the dangers of stalling by adding special boundary layer control devices to their aircraft. Narrow metal strips called *vortex generators*, which stick up from the surfaces of wings, introduce turbulence into the boundary layers over the wings. This turbulent flow allows higher energy air to mix with the boundary layers so that they can continue forward into rising pressure. This process helps keep the airstreams attached to the surfaces.

Fig. 6.3.6 This vertical wingtip keeps air from flowing around the end of the wing, a motion that would otherwise leave a powerful vortex in the air trailing behind the plane. Such wingtip vortices waste energy and are hazardous for other aircraft.

Courtesy Lou Bloomfield

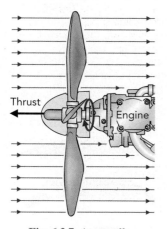

Fig. 6.3.7 A propeller behaves like a rotating wing. As the propeller turns, its blades create lift in the forward direction. This lift pushes the engine and the aircraft forward through the air, so that it's called thrust.

onto the flaps. These jets of air keep the boundary layers moving downstream and help prevent stalling. (For another approach to stall prevention, see ❑).

Once a commercial jet lands, flat panels on the top surfaces of its wings are tilted upward and cause the airflow to separate from the tops of the wings. The resulting turbulence created by these spoilers reduces the lift of the wings and increases their drag, so that the plane doesn't accidentally start flying again. Even before landing, the spoilers are sometimes used to slow the plane and help it descend rapidly toward an airport.

In flight, a wing does more than just push the passing air downward; it also twists the air near its tip. Since the air pressure below the wing is greater than the air pressure above it, air tends to flow around the wing's tip from bottom to top. The plane soon leaves this air behind, but not before the air has acquired lots of angular momentum and kinetic energy. A swirling vortex thus emerges from each wingtip and trails behind the plane for several kilometers, like an invisible tornado. A wingtip vortex from a jumbo jet can flip a small aircraft that flies through it or give passengers in a much larger plane an unexpected thrill. For safety, air traffic controllers are careful to keep planes from flying through one another's wakes and schedule them at least 90 s apart on runways. Some modern airplanes have vertical wingtip extensions that reduce these vortices, both to save energy and to diminish the hazard (Fig. 6.3.6).

> Check Your Understanding #3: Stunt Flying

for answers, see page 199

A pilot normally tips the plane's nose upward in order to gain altitude. But if the pilot tries to make the plane rise too quickly, the plane will suddenly begin to drop. What is happening?

Propellers

For a plane to obtain lift, it needs airspeed; air must flow across its wings. And since drag forces push it downwind, a plane in level flight can't maintain its airspeed unless something pushes it upwind. That's why a plane has propellers or jet engines: to push the air backward so that the air pushes the plane forward, action and reaction.

A propeller is an assembly of rotating wings. Extending from its central hub are two or more blades that together form a sophisticated fan (Fig. 6.3.7). These blades have

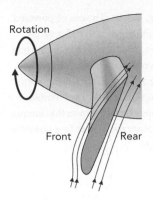

Fig. 6.3.8 As the propeller blade rotates, the flow of air around it creates a low pressure in front of it (left) and a high pressure behind it (right). The blade experiences a lift force that pushes the propeller and plane forward (toward the left). Induced drag tends to slow the rotation of the propeller.

□ One of the principal sources of noise in submarines is the turbulence created by their propellers. To reduce this turbulence, the propellers of modern nuclear submarines are designed to avoid water flow separation and stalling.

airfoil cross sections and are designed to create forward lift forces when the propeller turns and the blades move through the air.

As a propeller blade slices through the air, the airstreams bending around that blade experience pressure variations (Fig. 6.3.8). The forward airstream bends toward the blade's front surface, so the pressure in front of the blade drops below atmospheric. And the rearward airstream bends away from the blade's rear surface, so the pressure behind the blade rises above atmospheric. The resulting pressure difference exerts a forward lift or **thrust** force on the propeller.

The propeller blades have all the features, good and bad, of airplane wings. Their thrust increases with size, front-surface curvature, *pitch* (i.e., angle of attack), and airspeed; in other words, the larger the propeller, the faster it turns, and the more its blades are angled into the wind, the more thrust it produces. The blades themselves have a twisted shape to accommodate the variations in airspeed along their lengths, from hub to tip.

And like a wing, a propeller stalls when the airflow separates from the front surfaces of its blades; it suddenly becomes more of an air-mixer than a propeller. This stalled-wing behavior was the standard operating condition for air and marine propellers (see □) before the work of Wilbur Wright in 1902 (see □). The Wrights were among the first people to study aerodynamics with a wind tunnel (Fig. 6.3.9), and their

□ In addition to achieving the first self-propelled flight of an airplane in 1903, Orville (1871–1948) and Wilbur (1867–1912) Wright (American aviators) were exceptionally accomplished aerodynamicists. In 1902, Wilbur was the first person to recognize that a propeller is actually a rotating wing. Propellers up until his time were little more than rotating paddles, more effective at stirring the air than propelling the plane. Wilbur's aerodynamically redesigned propeller made flight possible and dominated aircraft design for a decade.

Courtesy Smithsonian Institution

Fig. 6.3.9 The Wright brothers were accomplished aerodynamicists, using this wind tunnel to study and perfect wings and propellers for their airplanes.

Fig. 6.3.10 The era of powered flight began at 10:35 a.m. on Dec. 17, 1903, when the Wright Flyer lifted Orville Wright into the air over Kitty Hawk, North Carolina. His brother Wilbur stands beside him in this unique photograph of that first powered flight.

Courtesy Smithsonian Institution

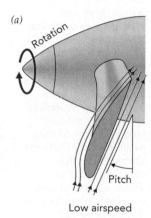

(a)

Rotation

Pitch

Low airspeed

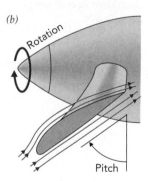

(b)

Rotation

Pitch

High airspeed

Fig. 6.3.11 At low airspeeds (*a*), the propeller blade approaches nearly stationary air as it rotates. At high airspeeds (*b*), air rushes past the propeller, so the blade must swivel forward to meet it. The blade's angle of attack is called its pitch.

methodical and scientific approach to aeronautics allowed them to achieve the first powered flight (Fig. 6.3.10).

A propeller also experiences induced drag. As the propeller's thrust pushes the plane through the air, induced drag extracts energy from the propeller. To keep the propeller turning steadily, an engine must do work on the propeller. Propellers are driven by high-performance reciprocating engines, like those found in automobiles, or the turbojet engines that we'll discuss later.

Propellers aren't perfect; they have three serious limitations. First, a propeller exerts a torque on the passing air, so that air exerts a torque on the propeller. This reaction torque can flip a small plane. To minimize torque problems, some planes use pairs of oppositely turning propellers and single-propeller planes usually locate their propellers in front, so that the spinning air can return angular momentum to them while passing over their wings.

A second problem with a propeller is that its thrust diminishes as the plane's forward speed increases. When the airplane is stationary, a propeller blade moves through motionless air (Fig. 6.3.11*a*). But when the airplane is traveling fast, the air approaches that same propeller blade from the front of the plane (Fig. 6.3.11*b*). To retain its thrust at higher airspeeds, the propeller blade must increase its pitch—it must swivel forward—to meet this onrushing air.

The third and most discouraging problem with propellers, especially in high-speed aircraft, is drag. To keep up with the onrushing air at high airspeeds, the propellers must turn at phenomenal rates. The tips of the blades travel so fast that they exceed the **speed of sound**—the fastest speed at which a fluid such as air can convey forces from one place to another. When the blade tip exceeds this speed, the air near the tip doesn't accelerate until the tip actually hits it. Instead of flowing smoothly around the tip, the air forms a **shock wave**—a narrow region of high pressure and temperature caused by the supersonic impact—and the propeller stalls. That's why propellers aren't useful on high-speed aircraft.

Check Your Understanding #4: Circulating the Air

for answers, see page 199

The only difference between a fan and a propeller is what moves, the air or the object. Which side of each fan blade experiences the lowest air pressure, the inlet or the outlet side?

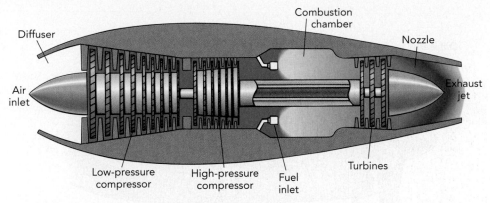

Fig. 6.3.12 The turbojet engine operates by compressing incoming air with a series of fanlike blades. Fuel is mixed with the high-pressure air and the mixture is ignited. The high-energy, high-pressure air accelerates out the rear of the jet, does work on the turbines, and leaves at a greater speed than it had when it arrived. The engine has accelerated the air backward and experiences a thrust forward.

Jet Engines

Unlike propellers, jet engines work well at high speeds. While a propeller tries to operate directly in the high-speed air approaching the plane, a jet engine first slows this air down to a manageable speed. To achieve this change in speeds, the jet engine makes wonderful use of Bernoulli's equation.

A turbojet engine is depicted in Fig. 6.3.12. During flight, air rushes into the engine's inlet duct or *diffuser* at about 800 km/h (500 mph), the speed of the plane. Once inside that diffuser, the air slows down and its pressure increases, but its total energy is unchanged. The air then passes through a series of fanlike compressor blades that push it deeper into the engine, doing work on it and increasing both its pressure and its total energy. By the time the air arrives at the combustion chamber, its pressure is many times atmospheric.

Now fuel is added to the air and the mixture is ignited. Since hot air is less dense than cold air, the hot exhaust gas takes up more space than it did before combustion. Furthermore, combustion subdivides the fuel molecules into smaller pieces that therefore take up still more volume. This hot exhaust gas pours out of the combustion chamber, traveling faster than when it entered.

The pressure of the exhaust gas is still very high as it streams through a windmill-like turbine. The air does work on that turbine and thereby spins the compressor for the incoming air. After the turbine, the high-pressure gas finally accelerates through the engine's outlet nozzle and emerges into the open sky at atmospheric pressure and extraordinarily high speed.

Overall, the engine slows the air down, adds energy to it, and then lets it accelerate back to high speed. Because the engine has added energy to the air, the air leaves the engine traveling faster than when it arrived. The air's increased backward velocity means that the jet engine has pushed it backward and the air has reacted by exerting a forward thrust force on the jet engine. In other words, the airplane has obtained forward momentum by giving the departing air backward momentum.

The turbojet is less energy efficient than it could be. Since it gives backward momentum to a relatively small mass of air, that air ends up traveling overly fast and

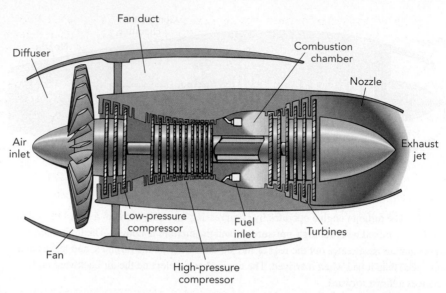

Fig. 6.3.13 The turbofan engine adds a giant fan to the shaft of a normal turbojet engine. Most of the air passing through the fan bypasses the turbojet and returns directly to the airstream around the engine. Because the fan does work on this air, it leaves the engine at a higher speed than it had when it arrived. The air has transferred forward momentum to the engine and the plane.

❏ Ramjets are jet engines that have no moving parts. Air that approaches the engine at supersonic speeds interacts with carefully tapered surfaces so that its own forward momentum compresses it to high density. The engine then adds fuel to this pressurized air, ignites the mixture, and allows the hot burned gas to expand out of a nozzle. The engine pushes this exhaust backward, and the exhaust propels the engine and airplane forward. Although the air enters the engine at supersonic speeds, it passes through the combustion chamber much more slowly. In a supersonic combustion ramjet or "scramjet," the fuel and air mixture flows through the combustion chamber at supersonic speeds. This motion makes it extremely difficult to keep the fuel burning because the flame tends to flow downstream and out of the engine. The flame can't advance through the mixture faster than the speed of sound, so it won't spread upstream fast enough to stay in the engine on its own.

with excessive kinetic energy. To make the engine more efficient, it should give backward momentum to a larger mass of air.

The turbofan engine solves this problem by using a turbojet engine to spin a huge fan (Fig. 6.3.13). Since this fan is located in the engine's inlet duct, the air's speed decreases and its pressure increases before it enters the fan. The fan then does work on the air and further increases its pressure. While about 5% of this air then enters the turbojet engine, the vast majority of it accelerates out the back of the fan duct and emerges into the open sky at atmospheric pressure and increased speed. Overall, the fan has pushed the air backward and the air has pushed the fan forward, producing forward thrust.

Like a turbojet, the turbofan slows air down, adds energy to it, and then lets it accelerate back to high speed. But because the turbofan engine moves more air than a turbojet engine, it gives that air less energy and uses less fuel. The huge fanlike engines on many jumbo jets are turbofans. (For another type of jet engine, see ❏.)

Check Your Understanding #5: Energy and a Jet Engine

for answers, see page 199

A jet engine somehow slows the air down, adds energy to it at low speed, and then returns the air to high speed. Why doesn't slowing the air down waste lots of energy?

Epilogue for Chapter 6

In this chapter, we looked at a number of objects that use moving fluids to perform their tasks. In *garden watering,* we saw how water moves through openings and channels. We looked at the effects of viscosity and the importance of Bernoulli's equation in describing the conversion of pressure potential energy into kinetic energy and vice versa. Two types of fluid flow appeared—laminar and turbulent. While laminar flow is smooth and predictable, we saw that turbulent flow involves unpredictability and chaos, a common behavior in our complicated universe.

In the section on *balls and air,* we examined the ways in which moving air exerts forces on larger objects, both in the downstream direction as drag and in a perpendicular direction as lift. We learned about two different types of drag forces and how these can be controlled or reduced by choosing the shapes or motions of the balls.

Finally, in *airplanes,* we explored the aerodynamics of those remarkable machines. We saw how they use air to support and propel themselves. We also examined the limitations of wings and saw what can go wrong if those limitations are exceeded. We studied propulsion in propeller planes and in jet aircraft and saw that these systems involve Bernoulli's equation and the forward force that comes from pushing air backward.

Explanation: A Vortex Cannon

The vortex cannon creates ring vortices—tiny tornadoes that are bent into loops so that they have no beginnings or ends. Air swirls forward in the middle of each ring and backward on its outer edge. This circular tornado structure is created when air flows through the hole of the vortex cannon. Air flows forward in the middle of the hole while the hole's edges create the backward flow around the outside of the ring. After it leaves the cannon, each ring vortex crawls forward through the surrounding air until its kinetic energy has been exhausted and it slows to a stop. It finishes its existence swirling in place until viscous forces bring its moving air to rest.

Chapter Summary

How Garden Watering Works: As water flows through a hose, viscous forces in the hose limit its speed, particularly near the walls of the hose. Each time the hose bends, the water pressure rises on the outside of the bend and drops on the inside of the bend. When the water reaches the nozzle, it accelerates through the narrow opening, increasing in speed and decreasing in pressure until it emerges at atmospheric pressure and arcs gracefully into the garden. Whenever the water flows rapidly around obstacles, both in the garden and at the faucet leading the hose, the flow becomes turbulent and noisy.

How Balls and Air Work: A ball traveling through the air experiences two major types of aerodynamic force—drag and lift. For a nonspinning ball traveling very slowly, the only drag force is viscous drag; it experiences no lift force. As the ball's speed through the air increases, a large turbulent wake appears behind the ball and the ball experiences pressure drag. At a still higher speed, the boundary layer of air near the ball's surface becomes turbulent and the size of the wake behind the ball shrinks. A ball with a turbulent boundary layer experiences less drag than one with a laminar boundary layer, which is why balls are designed to encourage turbulent boundary layers.

Rotating balls experience lift forces. These forces occur because bending and deflecting airstreams push asymmetrically on these balls. Lift forces can cause a ball to curve in flight or take surprisingly long to fall.

How Airplanes Work: An airplane is supported in flight by air passing across its wings. Air bending toward the top surface of a wing experiences a drop in pressure, while air bending away from the bottom surface of a wing experiences a rise in pressure. As a result of this pressure difference, the wing experiences an upward lift force. The airplane also experiences drag, which tends to slow it down. To keep the airplane moving forward, the plane employs propellers or jet engines. These devices push the air backward and the air reacts by pushing them forward. A propeller works directly in the oncoming air, increasing the air's energy and pushing it backward with rotating blades. A jet engine first slows the air, then increases its energy by burning fuel in it, and finally lets it accelerate backward to high speed.

Important Laws and Equations

1. Poiseuille's Law: The volume of fluid flowing through a pipe each second is equal to $(\pi/128)$ times the pressure difference across that pipe times the pipe's diameter to the fourth power, divided by the pipe's length times the fluid's viscosity, or

$$\text{volume} = \frac{\pi \cdot \text{pressure difference} \cdot \text{pipe diameter}^4}{128 \cdot \text{pipe length} \cdot \text{fluid viscosity}}. \quad (6.1.1)$$

2. Bends and Pressure Imbalances: When the path of a fluid in steady-state flow bends, the pressure on the outside of the bend is always higher than the pressure on the inside of the bend.

3. Reynolds Number: A measure of the relative importances of inertia and viscosity in the fluid flow around an obstacle is given by the Reynolds number:

$$\frac{\text{density} \cdot \text{obstacle length} \cdot \text{flow speed}}{\text{viscosity}}. \quad (6.1.2)$$

Check Your Understanding—Answers

Section 6.1 GARDEN WATERING

1. The air at the surfaces of the fibers is stationary, and the air's viscosity slows the motion of air in the vicinity of the fibers.
Why: Although air trying to pass through the gaps may not directly touch the fibers, viscous forces tend to keep all of the air moving together at the same velocity. As soon as some air is held up by a fiber, the air nearby is slowed by viscous forces. The fibers of a sweater are close enough together that viscous forces slow all the air trying to pass through the sweater. Imagine trying to pour honey through a sweater.

2. Air's viscosity slows its flow through ductwork. Moving large volumes of air rapidly without a large pressure difference between inlet and outlet requires a large diameter pipe.
Why: The airflow through long ductwork is dominated by viscous forces. The volume of air moved through ductwork is often enormous, and the pressure difference between the inlet and outlet is normally only a fraction of an atmosphere. To allow the air to move quickly through the ducts, their diameters must be large.

3. At the surface of the spoon.
Why: The pressure in the stream of water must be higher on the outside of the bend—at the spoon's surface—than on the inside of the bend.

4. Even at atmospheric pressure, air has pressure potential energy. As it accelerates toward even lower pressure in the attachment, air converts some of that pressure potential energy into kinetic energy.
Why: Vacuum cleaners employ nozzles to get air moving very quickly. Fast-moving air rushing into the vacuum cleaner draws dust with it and cleans your home.

5. The air flowing through the "canyons" created by the buildings becomes turbulent and forms vortices that swirl the leaves and papers.
Why: Whether an object moves through a fluid or a fluid moves past an object, a Reynolds number is associated with the situation. When the Reynolds number becomes high enough that viscosity is unable to keep the fluid flowing in an orderly fashion, turbulence appears. In wind blowing through a big city, turbulence and its whirling vortices are everywhere.

Section 6.2 BALLS AND AIR

1. The slow flow of water around the rock is laminar, so its pressure is highest at the front and back of the rock. The increased pressure behind the rock lifts the water level there.

Why: Laminar flow around an obstacle tends to create high-pressure regions at the front and back, and low-pressure regions on the sides. In this case, the pressure differences are visible as changes in the water level. At the front and back of the rock, the relatively high pressures push the water level upward while at the sides of the rock, the water level is depressed.

2. The slow-moving canoe experiences laminar flow in the water while the fast-moving canoe experiences turbulent flow.

Why: If the canoe's speed is less than about 1 cm/s, its Reynolds number will be less than 2000 and the water flow around it will be laminar. The water will pass smoothly around the canoe's sides and join back together behind it. But when the canoe is moving fast enough that its Reynolds number exceeds 2000, the water flow becomes turbulent and the canoe leaves a churning wake in the water behind it. This wake produces pressure drag on the canoe and extracts energy from it. Anyone who has paddled a canoe knows that overcoming this pressure drag can be exhausting.

3. As far back on the car as possible.

Why: As for all large, fast-moving objects in air, pressure drag is the main source of air resistance. You want to minimize this drag by keeping the air flowing smoothly over the car until it leaves the rear around a small turbulent wake. The smaller the air pocket behind a car, the better. Aerodynamically designed production cars leave a turbulent wake that is only about one-third as large in area as the thickest cross section of the car. While there is still room for improvement, these cars experience far less drag than the boxy cars of earlier times.

4. The lift force is directed downward, so the ball accelerates downward faster than it would by gravity alone.

Why: A ball with topspin falls faster than it would without a spin. In tennis, the topspin strokes appear to dive downward once they cross the net. Their downward curve means that they can travel very fast and still remain inside the court and are thus very hard to return.

Section 6.3 AIRPLANES

1. A fairing delays airstream separation in the airflow around a bicycle and thereby reduces pressure drag.

Why: Without a fairing, a bicycle racer must struggle against severe pressure drag. The sharply rising pressure gradients that follow wide parts of his body trigger flow separation and he develops a huge turbulent wake. A fairing makes the vehicle streamlined: the gently rising pressure gradient following its widest part doesn't trigger flow separation.

2. The air traveling around the outside of the sail has a lower pressure than that traveling across the inside of the sail. The sail experiences a lift force that pushes it and the boat across the water.

Why: Sails experience both lift and drag forces. The sail experiences an aerodynamic force that pushes it outward (lift) and slightly downwind (drag) just as an airplane wing experiences an aerodynamic force upward (lift) and slightly downwind (drag). The sailboat's keel, or centerboard, and its rudder provide additional forces so that the net force on the sailboat can be controlled and it can travel in a variety of directions.

3. The plane's wings are stalling.

Why: Tipping the plane's nose upward increases the wings' angle of attack. While this action increases lift up to a point, it can also cause the airflow to separate from the top surface of the wing. The sudden reduction in lift and increase in drag that accompany stalling can cause the plane to drop. A stall during takeoff or landing is extremely dangerous.

4. The inlet side of each fan blade experiences the lowest air pressure.

Why: Air blowing toward you from a fan is like the air blown back by a propeller. The lowest pressures experienced by a propeller are on its forward surfaces. Similarly, the lowest pressures experienced by a fan are on its inlet side surfaces. This pressure imbalance pushes the fan away from you while the fan pushes the air toward you.

5. As the air is slowed down, its pressure increases. The air's total energy remains constant.

Why: The remarkable result of Bernoulli's effect is that you can slow the air down without squandering its kinetic energy. That energy becomes pressure potential energy, and it passes through the jet engine in that form. As the air leaves the jet engine, its pressure potential energy becomes kinetic energy once again, so that no energy is wasted.

Check Your Figures—Answers

Section 6.1 GARDEN WATERING

1. About 0.20 L/s.

Why: Water flow in pipes obeys Poiseuille's law (Eq. 6.1.1) and is proportional to the pipe diameter to the fourth power. Reducing the diameter by 20%, a factor of 0.8, while leaving the pipe length, pressures, and viscosity unchanged, will reduce the volume flow rate by a factor of 0.8^4 or 0.41. The

pipes now deliver about 0.20 L/s (the product of 0.50 L/s times 0.41). It clearly doesn't take much mineral accumulation to dramatically reduce the flow through a pipe.

2. Turbulent.

Why: To calculate the Reynolds number for the airflow around the car, you need air's viscosity from Table 6.1.1 (0.0000183 Pa · s or 0.0000183 kg/m · s), air's density from Section 4.1

(1.25 kg/m^3), the car's size (roughly 3 m), and its speed through the air (roughly 55 mph or 25 m/s). You can then calculate its approximate Reynolds number, using Eq. 6.1.2:

$$\text{Reynolds number} = \frac{1.25 \text{ kg/m}^3 \cdot 3 \text{ m} \cdot 25 \text{ m/s}}{0.0000183 \text{ kg/m} \cdot \text{s}}.$$
$$= 5.1 \text{ million}$$

The convertible's Reynolds number is far above the threshold for turbulence (2300), so the air swirls chaotically around the vehicle. That explains why your hair flies around wildly.

Exercises

1. A favorite college prank involves simultaneously flushing several toilets while someone is in the shower. The cold water pressure to the shower drops and the shower becomes very hot. Why does the cold water pressure suddenly drop?

2. On hot days in the city people sometimes open up fire hydrants and play in the water. Why does this activity reduce the water pressure in nearby hydrants?

3. Why does a relatively modest narrowing of the coronary arteries, the blood vessels supplying blood to the heart, cause a dramatic drop in the amount of blood flowing through them?

4. Why does hot maple syrup pour more easily than cold maple syrup?

5. Why is "molasses in January" slower than "molasses in July," at least in the northern hemisphere?

6. Why is it so difficult to squeeze ketchup through a very small hole in its packet?

7. A baker is decorating a cake by squeezing frosting out of a sealed paper cone with the tip cut off. If the baker makes the hole at the tip of the cone too small, it's extremely difficult to get any frosting to flow out of it. Why?

8. Why is the wind stronger several meters above a flat field than it is just above the ground?

9. Pedestrians on the surface of a wind-swept bridge don't feel the full intensity of the wind because near the bridge's surface, the air is moving relatively slowly. Explain this effect in terms of a boundary layer.

10. An electric valve controls the water for the lawn sprinklers in your backyard. Why do the pipes in your home shake whenever this valve suddenly stops the water but not when this valve suddenly starts the water?

11. If you drop a full can of applesauce and it strikes a cement floor squarely with its flat bottom, what happens to the pressures at the top and bottom of the can?

12. When you mix milk or sugar into your coffee, you should move the spoon quickly enough to produce turbulent flow around the spoon. Why does this turbulence aid mixing?

13. If you start two identical paper boats from the same point, you can make them follow the same path down a quiet stream. Why can't you do the same on a brook that contains eddies and vortices?

14. When you swing a stick slowly through the air, it's silent. But when you swing it quickly, you hear a "whoosh" sound. What behavior of the air is creating that noise?

15. If you try to fill a bucket by holding it in a waterfall, you will find the bucket pushed downward with enormous force. How does the falling water exert such a huge downward force on the bucket?

16. You sometimes see paper pressed tightly against the front of a moving car. What holds the paper in place?

17. Fish often swim just upstream or downstream from the posts supporting a bridge. How does the water's speed in these regions compare with its speed in the open stream and at the sides of the posts?

18. Why do flyswatters have many holes in them?

19. You have two golf balls that differ only in their surfaces. One has dimples on it while the other is smooth. If you drop these two balls simultaneously from a tall tower, which one will hit the ground first?

20. If you ride your bicycle directly behind a large truck, you will find that you don't have to pedal very hard to keep moving forward. Why?

21. How does running directly behind another runner reduce the wind resistance you experience?

22. When a car is stopped, its flexible radio antenna points straight up. But when the car is moving rapidly down a highway, the antenna arcs toward the rear of the car. What force is bending the antenna?

23. To drive along a level road at constant velocity, your car's engine must be running and friction from the ground must be pushing your car forward. Since the net force on an object at constant velocity is zero, why do you need this forward force from the ground?

24. Racing bicycles often have smooth disk-shaped covers over the spokes of their wheels. Why would these thin wire spokes be a problem for a fast-moving bicycle?

25. A bullet slows very quickly in water but a spear doesn't. What force acts to slow these two objects, and why does the spear take longer to stop?

26. If you hang a tennis ball from a string, it will deflect downwind in a strong breeze. But if you wet the ball so that the fuzz on its surface lies flat, it will deflect even more than before. Why does smoothing the ball increase its deflection?

27. Explain why a parachute slows your descent when you leap out of an airplane.

28. Bicycle racers sometimes wear teardrop-shaped helmets that taper away behind their heads. Why does having this smooth taper behind them reduce the drag forces they experience relative to those they would experience with more ball-shaped helmets?

29. If you want the metal tubing in your bicycle to experience as little drag as possible while you're riding in a race, is cylindrical tubing the best shape? How should it be shaped?

30. In 1971, astronaut Alan Shepard hit a golf ball on the moon. How did the absence of air affect the ball's flight?

31. A water-skier skims along the surface of a lake. What types of forces is the water exerting on the skier, and what is the effect of these forces?

32. How would a Frisbee fly on the airless moon?

33. You can buy special golf tees that wrap around behind the ball to prevent you from giving it any spin when you hit it. These tees are guaranteed to prevent hooks and slices (i.e., curved flights). But how do these tees affect the distance the ball travels? Why?

34. A hurricane or gale force wind can lift the roof off a house, even when the roof has no exposed eaves. How can wind blowing across a roof produce an upward force on it?

35. A skillful volleyball player can serve the ball so that it barely spins at all. The ball dithers slightly from side to side as it flies over the net and is hard to return. What causes the ball to accelerate sideways?

36. Why does an airplane have a "flight ceiling," a maximum altitude above which it can't obtain enough lift to balance the downward force of gravity?

37. If you let a stream of water from a faucet flow rapidly over the curved bottom of a spoon, the spoon will be drawn into the stream. Explain this effect.

38. If you put your hand out the window of a moving car, so that your palm is pointing directly forward, the force on your hand is directly backward. Explain why the two halves of the airstream, passing over and under your hand, don't produce an overall up or down force on your hand.

39. If instead of holding your hand palm forward (see Exercise 38), you tip your palm slightly downward, the force on your hand will be both backward and upward. How is the airstream exerting an upward force on your hand?

40. When a hummingbird hovers in front of a flower, what forces are acting on it and what is the net force it experiences?

41. A poorly designed household fan stalls, making it inefficient at moving air. Describe the airflow through the fan when its blades stall.

42. When a plane enters a steep dive, the air rushes toward it from below. If the pilot pulls up suddenly from such a dive, the wings may abruptly stall, even though the plane is oriented horizontally. Explain why the wings stall.

Problems

1. About how fast can a small fish swim before experiencing turbulent flow around its body?

2. How much higher must your blood pressure get to compensate for a 5% narrowing in your blood vessels? (The pressure difference across your blood vessels is essentially equal to your blood pressure.)

3. If someone replaced the water in your home plumbing with olive oil, how much longer would it take you to fill a bathtub?

4. You are trying to paddle a canoe silently across a still lake and know that turbulence makes noise. How quickly can the canoe and the paddle travel through water without causing turbulence?

5. The pipes leading to the showers in your locker room are old and inadequate. While the city water pressure is 700,000 Pa, the pressure in the locker room when one shower is on is only 600,000 Pa. Use Eq. 6.1.1 to calculate the approximate pressure if three showers are on.

6. If the plumbing in your dorm carried honey instead of water, filling a cup to brush your teeth could take awhile. If the faucet takes 5 s to fill a cup with water, how long would it take to fill your cup with honey, assuming all the pressures and pipes remain unchanged?

7. How quickly would you have to move a 1-cm-diameter stick through olive oil to reach a Reynolds number of 2000, so that you would begin to see turbulence around the stick? (Olive oil has a density of 918 kg/m^3.)

8. The effective obstacle length of a blimp is its width—the distance to which the air is separated as it flows around the blimp. How slowly would a 15-m-wide blimp have to move in order to keep the airflow around it laminar? (Air has a density of 1.25 kg/m^3.)

CHAPTER 7

Heat and Phase Transitions

We can't see all of the motion that takes place around us. Some of it is hidden deep inside each object, where thermal energy keeps the individual atoms and molecules jiggling back and forth in an endless flurry of activity. We're usually aware of this thermal energy only because it determines an object's temperature; the more thermal energy an object contains, the higher its temperature and the hotter it feels.

However, thermal energy plays an important role in everyday life. In addition to moving from one place to another, thermal energy can transform a substance from solid to liquid to gas. What you're feeling when you touch a hot object is actually its thermal energy flowing into your colder hand and raising the temperature of your skin. When thermal energy is flowing in this manner, from a hotter object to a colder one, we call it heat. In this chapter, we'll examine temperature, heat, and the phases of matter in order to understand more about our hot and cold world.

EXPERIMENT: A Ruler Thermometer

One effect that a change in temperature has on a typical object is to change its size. While this size change is tiny and easily overlooked, you can use mechanical advantage to make it quite visible. Here's how to make a size-change thermometer using only a clear plastic ruler, a pin, a small weight, a piece of stiff paper, and some tape.

Lay the plastic ruler along the edge of a table and tape one end of it securely to the table. Cut a thin strip of stiff paper, about 3 mm wide (0.1 inches) and 15 cm long (6 inches), and push the pin carefully through one end of the strip. Use a dot of tape to stick the pin's head to the paper. When you're done, the paper strip should be securely attached to the pin so that as the pin turns, the strip turns. This strip is your thermometer's pointer.

Now slide the pin under the free end of the ruler and place the small weight above it. The weight is there to push the ruler and pin together so that they experience plenty of

static friction. That way, as the free end of the ruler moves left or right, the pin will rotate and turn the pointer.

Your thermometer is now complete. If you turn the pin and pointer carefully by hand so that the pointer is horizontal, you can "read" the thermometer by its angle relative to the tabletop. If you heat the plastic ruler by breathing on it, laying your hands on it, or warming it gently with a hair dryer, the ruler will become longer. Its free end will move away from the fixed end and will cause the pin to rotate. You will see a small change in the pointer's orientation as your thermometer reports its new temperature.

Predict what will happen if you don't heat the whole ruler uniformly and then *observe* what happens when you don't. Try to *measure* the effects of nonuniform heating. Did you *verify* your prediction?

You can also make the needle turn the other way by cooling the ruler. Placing a few ice cubes on the ruler will cause the ruler to contract and the needle will turn in the opposite direction.

Chapter Itinerary

In this chapter, we'll examine thermal energy, temperature, and heat in the context of three common types of objects: (1) *woodstoves,* (2) *water, steam, and ice,* and (3) *incandescent lightbulbs.* In *woodstoves,* we'll look at how to produce thermal energy and at the three principal means by which thermal energy is transferred as heat from hotter objects to colder ones: conduction, convection, and radiation. In *water, steam, and ice,* we'll look at the effects of heat and temperature on the three material phases of water and at how transformations occur between those phases. In *incandescent lightbulbs,* we'll study heat transfer by radiation to see that it can include the emission of visible light. For a more detailed preview, look ahead at the Chapter Summary on p. 232.

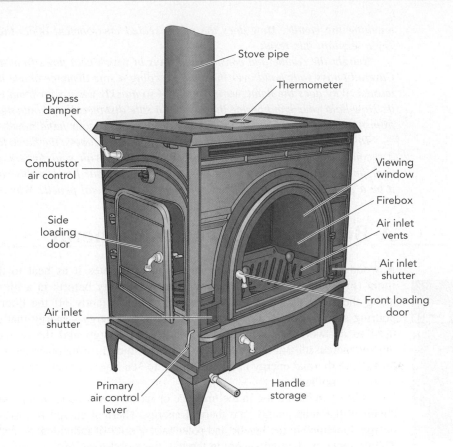

Labels: Stove pipe, Thermometer, Bypass damper, Combustor air control, Viewing window, Firebox, Air inlet vents, Air inlet shutter, Side loading door, Air inlet shutter, Front loading door, Primary air control lever, Handle storage

SECTION 7.1 Woodstoves

Winter would be pretty unpleasant for most of us were it not for heating. Heating keeps our rooms warm even when the weather outside is cold. One of the most fashionable types of heating is a woodstove, which burns logs in its firebox and sends thermal energy out into the room. In this section we'll look at how a woodstove produces thermal energy and how this thermal energy flows out of the stove to keep us warm.

Questions to Think About: *What happens to the chemical potential energy in a log when you burn it? Why do you get burned when you touch a hot object? Why does your hand feel hotter when you hold it above a hot surface than next to that hot surface? Why does your skin feel warm when you face a campfire, even if the air around you is cold? Why does it take time to warm up a cold object?*

Experiments to Do: *A burning candle produces thermal energy, providing both light and warmth to a small room. But where does this thermal energy come from and how does it flow into its surroundings?*

Light a short candle and look first at how it releases thermal energy. The flame slowly consumes the wax but it also needs air. Cover the candle with a tall glass, one that won't be touched or damaged by the flame. The glass should prevent room air from

reaching the candle. How does the newly sealed environment affect the candle flame? Try to explain this result.

Relight the candle and consider the ways in which heat flows from the flame to you. Carefully pass your hand over the flame, keeping a safe distance above it to avoid being burned. Why does the flame warm your skin so quickly when your hand is directly above it? Now hold your hand beside the flame at a safe distance. You should again feel warmth from the flame. How is heat flowing from the flame to your hand now?

Take a wooden pencil and hold it a few centimeters above the flame for no more than 2 seconds. Then carefully touch the pencil's surface with your fingers. It should warm your fingers. How is heat flowing from the pencil to your hand in this case? Why would it be a painful mistake to try this experiment with a metal pencil? Why only 2 seconds?

A Burning Log: Thermal Energy

Fig. 7.1.1 This woodstove transfers heat to the room by conduction through its metal walls, convection of air past its surfaces, and radiation from its black exterior.

A woodstove produces thermal energy and distributes it as heat to the surrounding room (Fig. 7.1.1). We've encountered thermal energy before: in a file cabinet sliding along the sidewalk, in an old ball bouncing inefficiently off the floor, and in honey pouring slowly from a jar. In each case, ordered energy—energy that could easily be used to do work—became disordered thermal energy and the temperatures of the objects increased. But now that we're going to study heating machines themselves, let's reexamine thermal energy and temperature to see how thermal energy moves from one object to another.

When you burn a log in the fireplace or woodstove, you're turning the log's ordered chemical potential energy into thermal energy. Thermal energy is a disordered form of energy contained in the kinetic and potential energies of individual atoms and molecules. The presence of thermal energy in the log, the woodstove, or the room air is what gives it a temperature; the more thermal energy it has, the higher its temperature.

The nature of thermal energy depends somewhat on what it's in. In the hot, burning log, thermal energy is mostly in the wood's atoms and molecules, which jitter back and forth rapidly relative to one another. When each of these particles moves, it has kinetic energy. When it pushes or pulls on its neighbors, it has potential energy.

In the air near the burning log, thermal energy is again mostly in the atoms and molecules. But since those particles are essentially free and independent, most of this thermal energy is kinetic energy. The air particles store potential energy only during the brief moments when they collide with one another.

And in the metal poker that you use to stir the fire, thermal energy is not only in the atoms and molecules, but also in the mobile electrons that move about the metal and allow it to conduct electricity.

While it's important to know what is thermal energy, it's also important to know what is not thermal energy. The log's thermal energy includes only its internal disordered energy and not the energy that's associated with the log as a whole. Moving the log with the poker to increase its kinetic energy, lifting it with tongs to increase its gravitational potential energy, and bending it with another log to increase its elastic potential energy all increase its energy as a whole and not its thermal energy.

> ► Check Your Understanding #1: A Warm-Up Pitch
>
> for answers, see page 233

If you drop a ball, will its thermal energy increase as it falls? (Neglect air resistance.)

Forces between Atoms: Chemical Bonds

To understand how a burning log produces thermal energy, let's take a look at bonds between atoms and the chemical potential energy that's stored in those bonds. Since both result from the forces between atoms, that's where we'll begin.

As you bring two atoms close together, they exert attractive forces on one another (Fig. 7.1.2a). These chemical forces are electromagnetic in origin and grow stronger as the atoms approach. But the attraction diminishes when the atoms start to touch and is eventually replaced by repulsion when the atoms are too close (Fig. 7.1.2b). The separation between atoms at which the attraction ends and the repulsion begins is their *equilibrium separation,* that is, the separation at which the atoms exert no forces on one another (Fig. 7.1.2c). Since atoms are tiny, this equilibrium separation is also tiny, typically only about a ten-billionth of a meter.

Imagine holding two atoms in tweezers and slowly bringing them together. They pull toward one another as they approach, doing work on you and increasing your energy. Since energy is conserved, their energy must be decreasing. They're giving up **chemical potential energy**—energy stored in the chemical forces between atoms.

Once the atoms reach their equilibrium separation, you can let go of them and they won't come apart. Like two balls attached by a spring, the atoms are in a stable equilibrium. Since they've given up some of their chemical potential energy, they can't separate unless that energy is returned to them. It takes work to pull them apart, so the atoms are held together by a chemical bond.

The bound atoms have become a molecule. The strength of their bond is equal to the amount of work the atoms did when they drew together or, equivalently, the work required to separate them. Bond strengths range from extremely strong in the case of two nitrogen atoms to extremely weak in the case of two neon atoms.

If they have a little extra energy, bound atoms can vibrate back and forth about their equilibrium separation (Fig. 7.1.2d). Whenever the atoms are moving quickly toward or away from one another, most of their energy is kinetic. Whenever the atoms are slowing to turn around, most of their energy is chemical potential. Overall, the molecule's total energy remains constant, and it vibrates back and forth until it transfers its extra energy elsewhere.

But many molecules have more than two atoms. In a large molecule, each pair of adjacent atoms has a chemical bond and an equilibrium separation. If you give this molecule excess energy, it will vibrate in a complicated manner as the energy moves among the various atoms and chemical bonds. The atoms in the molecule will continue to jiggle about until something removes the excess energy from the molecule.

Like all liquids and solids, our burning log is just a huge assembly of atoms and molecules, held together by chemical bonds of various strengths. These particles push and pull on one another as they vibrate about their equilibrium separations. Their motion is *thermal motion,* and the energy involved in this disorderly jiggling is thermal energy. Because thermal energy is fragmented among the atoms and exchanged between them unpredictably, it can't be used directly to do work.

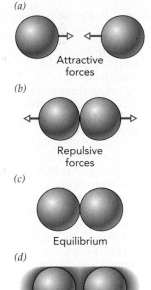

Fig. 7.1.2 (a) Two atoms attract one another at moderate distances but (b) repel when they're too close. (c) In between is their equilibrium separation, at which they neither attract nor repel and are thus in equilibrium. (d) Pairs of atoms with excess energy tend to vibrate about their equilibrium separations.

> Check Your Understanding #2: When Atoms Collide . . .

for answers, see page 233

As two independent nitrogen atoms collide with one another, what forces do they experience?

Heat and Temperature

Everything contains thermal energy, from the hot burning log to the cold metal poker you use to stir the fire. However, that doesn't mean that the thermal energy is equitably distributed. What happens to thermal energy when you push the log with the poker?

When they touch, the poker and the log begin to exchange thermal energy. In effect, the two objects become one larger object, and thermal energy that has been moving among the atoms in each individual object begins to flow across the junction between the two. Since each object starts with some amount of thermal energy, energy moves in both directions across this junction. Nonetheless, there may be some net flow from one object to the other. To allow us to predict the direction of this flow, we define a temperature for each of the objects.

Temperature is the quantity that indicates which way, if any, thermal energy will naturally flow between two objects. If no thermal energy flows when two objects touch, then those objects are in **thermal equilibrium** and their temperatures are equal. But if thermal energy flows from the first object to the second, then the first object is hotter than the second.

A temperature scale classifies objects according to which way thermal energy will flow between any pair. An object with a hotter temperature will always transfer thermal energy to an object with a colder temperature, and two objects with the same temperature will always be in thermal equilibrium. Thus the hot burning log will transfer thermal energy to the cold poker. We say that the burning log is hot because it tends to transfer thermal energy to most objects, while the poker is cold because most objects tend to transfer thermal energy to it.

Energy that flows from one object to another because of a difference in their temperatures is called **heat.** Heat is thermal energy on the move. Strictly speaking, the burning log doesn't *contain* heat; it contains thermal energy. However, when that log transfers energy to the cold poker because of their temperature difference, it's heat that *flows* from the log to the poker. (For a historical note about the understanding of heat, see ❏.)

Our present definition of temperature can order the objects around us from hottest to coldest, but it doesn't quantify temperature in any unique way. You could make your own temperature scale by comparing every pair of objects to see which way heat flows between them, but you probably wouldn't enjoy it. You do better to use a standard temperature scale such as Celsius, Fahrenheit, or Kelvin.

Standard temperature scales are based on an object's average thermal kinetic energy per atom. The more kinetic energy each atom has, on average, the more vigorous the object's thermal motion and the more thermal energy it transfers to the atoms in a second object by way of microscopic portions of work. Microscopic work is what actually passes heat between objects—a teeny shove here, a tiny yank there, all at the atomic scale. Since an object with more average thermal kinetic energy per atom will pass heat to an object with less, it makes sense to assign temperatures according to average thermal kinetic energies per atom.

The Celsius, Fahrenheit, and Kelvin scales all measure temperature in this manner. In each scale, a 1 degree or unit increase in temperature reflects a specific increase in average thermal kinetic energy per atom. The relationship between average thermal kinetic energy per atom and assigned temperature is based on several standard conditions: absolute zero, water's freezing temperature, and water's boiling temperature. (Recall from Section 5.1 that absolute zero is the temperature at which all thermal energy has been removed from an object.) Once specific temperatures have been assigned to two of these standard conditions, the whole temperature scale is fixed. For example, the Celsius scale

❏ Before the time of Benjamin Thompson, Count Rumford (American-born British physicist and statesman, 1753–1814), heat was believed to be a fluid called caloric that was contained within objects. Thompson disproved the caloric theory by showing that the boring of cannons produced an inexhaustible supply of heat. Among his scientific and technological contributions, Thompson improved cooking and heating methods. He reshaped fireplaces and developed the damper as ways to reduce smoking and improve heat transfer to the room. Thompson also had a life of sensational escapades and great rises and falls in fortune. He fled New Hampshire in 1775 because he was a British loyalist, he fled London in 1782 under suspicion of being a French spy, and he was, at the time of his studies of heat, among the most powerful people in Bavaria.

| Table 7.1.1 | Temperatures of Several Standard Conditions, as Measured in Three Temperature Scales: Celsius, Kelvin, and Fahrenheit |

STANDARD CONDITION	CELSIUS (°C)	KELVIN (K)	FAHRENHEIT (°F)
Absolute zero	−273.15	0	−459.67
Freezing water	0	273.15	32
Boiling water	100	373.15	212

is built around 0 °C being water's freezing temperature and 100 °C being water's boiling temperature. Temperatures for the three standard conditions appear in Table 7.1.1.

▶ Check Your Understanding #3: Frozen Fingers

for answers, see page 233

If you pick up an ice cube, your hand suddenly feels cold. Which way is heat flowing?

Open Fires and Woodstoves

Suppose you needed an easy way to heat your room. The oldest and simplest method would be to start a campfire in the middle of the floor. The burning wood would produce thermal energy, which would flow as heat into the colder room. But how does burning wood produce thermal energy?

This thermal energy is released by a **chemical reaction** between molecules in the wood and oxygen in the air. Recall that atoms do work as they join together in a chemical bond and that the amount of work done depends on which atoms are being joined. For example, while carbon and hydrogen atoms can bond with one another to form *hydrocarbon* molecules, these atoms form much stronger bonds with oxygen atoms. Thus while it may take work to disassemble a hydrocarbon molecule, the work done by its hydrogen and carbon atoms as they bind to oxygen atoms more than makes up for that investment. As a hydrocarbon molecule burns in oxygen, new, more tightly bound molecules are formed and chemical potential energy is released as thermal energy. The *reaction products* produced by burning hydrocarbons in air are primarily water and carbon dioxide.

Wood is composed mostly of cellulose, a long carbohydrate molecule. *Carbohydrates* contain carbon, hydrogen, and oxygen atoms. Despite the presence of a few oxygen atoms, carbohydrates still burn nicely to form water and carbon dioxide. When you light the wood with a match, you're supplying the energy needed to break the old chemical bonds so that the new bonds can form. This starting energy is called **activation energy**—the energy needed to initiate the chemical reaction. Heat from the match flame gives the wood enough thermal energy to break chemical bonds between various atoms and start the reaction.

Unfortunately, wood isn't pure cellulose. It also contains many complex resins that don't burn well and create smoke. If you plan to breathe the air in which you burn fuel, wood is an awful choice. You'd be better off with kerosene and natural gas, both of which are nearly pure hydrocarbons and burn cleanly. Actually, wood can be converted to a cleaner fuel by baking it in an airless oven to remove all of its volatile resins. This process converts the wood into charcoal, which burns to form nearly pure carbon dioxide, water vapor, and ash.

But even with clean burning fuels, the direct fire-in-the-room heating concept has its disadvantages: it consumes the room's oxygen and presents a safety hazard. Nonetheless, fires have heated dwellings for thousands of years. While fireplaces that burn wood or peat

Courtesy Bryant Heating and Cooling

Fig. 7.1.3 This modern furnace burns natural gas in an S-shaped firebox. A fan at the bottom of the furnace blows fresh air past the hot outer surfaces of the firebox and then circulates the heated air among the rooms.

have chimneys that carry away their noxious fumes, the rising smoke takes with it much of the fire's thermal energy and some of the room's air. That's why a room that's heated by a fireplace often feels drafty away from the fireplace itself—cold outside air is seeping in through cracks to replace air drawn up the chimney. Even when clean burning fuels are used without a chimney, there are no simple solutions to the oxygen or safety problems.

Like a fireplace, a woodstove sends fumes from its burning wood up a chimney. But before its thermal energy can follow the fumes outside, a well-designed woodstove transfers most of that energy into the room. A woodstove is an example of a **heat exchanger**—a device that transfers heat without transferring the hot molecules themselves. Its smoke never enters the room but heat from that smoke does. The gas furnace in Fig. 7.1.3 also employs a heat exchanger.

The burning coals and hot gases inside the woodstove contain a great deal of thermal energy and are much hotter than the room air. Because of this temperature difference, heat tends to flow from the fire to the room. What is not so clear yet is how that heat is transferred.

There are three principal mechanisms by which heat moves from the fire to the room: conduction, convection, and radiation. The woodstove makes wonderful use of all three so that most of the thermal energy released by the burning wood is transferred to the room. Let's examine these three mechanisms of heat transport, beginning with conduction.

> Check Your Understanding #4: Feeling the Heat

for answers, see page 233

You can make a heat pack by wrapping hot, wet towels in a plastic sheet. This pack will warm an injured muscle but will not get it wet. Is thermal energy moving in this case?

Heat Moving Through Metal: Conduction

Conduction occurs when heat flows through a stationary material. The heat moves from a hot region to a cold region but the atoms and molecules don't. For example, if you place the tip of a metal poker in the fire, the poker's handle will gradually become warm as the metal conducts heat.

Some of this heat is conducted by interactions between adjacent atoms. The vibrating atoms frequently push on one another, doing microscopic work in the process and exchanging miniscule amounts of thermal kinetic energy. In this fashion, thermal energy flows randomly from atom to neighboring atom.

But when the poker's tip is hotter than its handle, the flow is no longer completely random. The atoms at the hot tip have more thermal kinetic energy to exchange with their neighbors than atoms at the cold handle. The exchanges statistically favor the flow of thermal energy away from the hot tip and toward the cold handle. This flow of thermal energy from hot to cold through the poker is conduction (Fig. 7.1.4).

However, this atom-by-atom bucket brigade isn't the only way in which materials conduct heat. In a metal, the primary carriers of heat are actually mobile **electrons**—the tiny negatively charged particles that make up the outer portions of atoms. When atoms join together to form a metal, some of the electrons stop belonging to particular atoms and travel almost freely throughout the metal. These mobile electrons can carry electricity (as we'll discuss in Chapter 10) and are also good at transporting heat.

Mobile electrons participate in the bucket brigade of heat conduction because they, too, can push on vibrating atoms and exchange thermal kinetic energy with them. But while atoms can pass thermal energy only from one neighbor to the next, mobile electrons can travel great distances between exchange partners and can move thermal energy quickly from one place to another.

The ease with which electrons move heat about a metal explains why metals generally have higher thermal conductivities than nonmetals. **Thermal conductivity** is the measure of how rapidly heat flows through a material when it's exposed to a difference in temperatures. The best conductors of electricity—copper, silver, aluminum, and gold—are also the best conductors of heat. Poor conductors of electricity—stainless steel and insulators such as plastic, and glass—are also poor conductors of heat. There are a few exceptions to this rule. Diamonds, for example, are terrible conductors of electricity but wonderful conductors of heat. Of course, it would be silly to make a woodstove out of diamonds; after all, diamonds burn.

Conduction is what moves thermal energy from the woodstove's inside to its outside. No atoms move through the metal walls of the stove, just heat. So conduction serves as a filter, separating desirable thermal energy from the unwanted smoke and noxious gases that then go up the chimney.

Thus conduction makes the outside surface of the woodstove hot so that heat should flow from it to the colder room. But what carries heat into the room? If you touch the stove, conduction will immediately transfer a huge amount of heat to your skin and you'll be burned. But even without touching the stove, you're aware of its high temperature. It transfers heat into the room by convection and radiation.

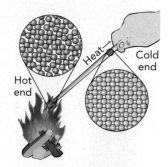

Fig. 7.1.4 When one end of a metal poker is hotter than the other, the atoms at the hot end vibrate more vigorously than those at the cold end. The poker then conducts heat from the hot end to the cold end. Some of this heat is conducted by interactions between adjacent atoms. In the metal poker, however, most of the heat is conducted by mobile electrons, which carry thermal energy long distances from one atom to another.

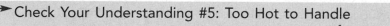

Check Your Understanding #5: Too Hot to Handle

for answers, see page **233**

Some pot handles remain cool during cooking while others become unpleasantly hot. What determines which handles remain cool and which become hot?

Heat Moving with Air: Convection

Convection occurs when a moving fluid transports heat from a hotter object to a colder object. The heat moves as thermal energy in the fluid so that the two travel together. The fluid usually follows a circular path between the two objects, picking up heat from the hotter object, giving it to the colder object, and then returning to the hotter object to begin again.

This circulation often develops naturally. As the fluid warms near the hotter object, its density decreases and it floats upward, lifted by the buoyant force. When the fluid cools near the colder object, its density increases and it sinks downward.

Thus air heated by contact with the woodstove rises toward the ceiling and is replaced by colder air from the floor (Fig. 7.1.5). Eventually, this heated air cools and descends. Once it reaches the floor, it's drawn back toward the hot woodstove to start the cycle over. This moving air is a **convection current,** and the looping path that it follows is a **convection cell.** Within the room, convection currents carry heat up and out from the woodstove to the ceiling and walls. When you put your hand over the stove, you feel this convection current rising upward as it transfers heat to your hand.

Natural convection is good at heating the air above the woodstove, but most of that hot air ends up near the ceiling. While some of it will eventually drift downward to where you're standing, convection sometimes needs help. Adding a ceiling fan will help move the hot air around the room and make the woodstove more effective. This forced convection still transfers heat from the hot stove to the colder occupants of the room, but it doesn't rely on the buoyant force to keep the air circulating. The faster the air moves, the more heat it can transport from hot objects to cold objects.

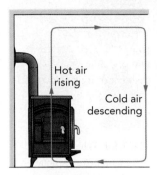

Fig. 7.1.5 When the woodstove is hot, convection carries heat from its surfaces to the ceiling and walls of the room. Warm air rises upward, supported by the buoyant force, and is replaced by cooler air from the floor. The warmed air eventually cools and descends. It then returns toward the stove to repeat the cycle.

→ Check Your Understanding #6: Heat and Wind

for answers, see page **233**

When sunlight warms the land beside a cool body of water, a breeze begins to blow from the water toward the land. Explain.

Heat Moving as Light: Radiation

There is one more important mechanism of heat transfer: radiation. As the particles inside a material jitter about with thermal energy, they emit and absorb electromagnetic radiation. This radiation consists of *electromagnetic waves,* which include radio waves, microwaves, and infrared, visible, and ultraviolet light.

We'll study electromagnetic radiation in Chapters 13 and 14. For now, what's most important is that this radiation can carry thermal energy. When heat flows from a hot object to a cold object as electromagnetic radiation, we say that heat is being transferred by thermal radiation or simply **radiation.** Unlike conduction and convection, which depend on atoms, molecules, or electrons to carry the heat, radiation occurs directly through space. Radiative heat transfer happens even when two objects have nothing at all between them.

The types of electromagnetic waves in an object's thermal radiation depend on its temperature. While a colder object emits only radio waves, microwaves, and infrared light, a hotter object can also emit visible or even ultraviolet light. The red glow of a hot coal in the woodstove is that coal's thermal radiation.

Since our eyes are only sensitive to visible light, we can't see all of the thermal radiation emitted by an object, even when it's hot. But whether we see it or not, electromagnetic radiation contains energy and transfers heat to whatever absorbs it. While everything emits thermal radiation, the amount of that emission depends on temperature: the hotter an object, the more thermal radiation it emits. When two objects face one another, thermal radiation will travel in both directions between them. However, the hotter object will dominate this radiant exchange of thermal energy, resulting in a net transfer of thermal energy to the colder object. Exchanges of thermal energy via radiation always transfers heat from a hotter object to a colder one.

Radiation transfers a great deal of heat from the woodstove's surface to the surrounding objects. The stove bathes the room in infrared light, which warms everything it reaches. To encourage such radiative heat transfer, the woodstove and its chimney are often painted black. Black not only absorbs light well, but it's also particularly good at emitting thermal light (Fig. 7.1.6). If you heat a black poker red hot, it will glow much more brightly than one that's white, silvery, or transparent. To see that a white, silvery, or transparent surface is a poor absorber of light, look at it when it's cold. To see that it is also a poor emitter of thermal light, look at it when it's hot.

Common Misconceptions:
Black Objects and Light

Misconception: A black object never emits light.

Resolution: While a black object absorbs all light that strikes it, it still emits thermal radiation and can glow brightly if it's hot enough.

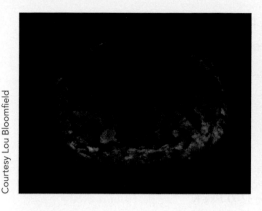

Fig. 7.1.6 A hot charcoal briquette glows brightly in the dark (*left*). However, a flash photograph reveals that its surface is actually gray and thus a partial absorber of light (*right*). If it weren't for the white ash, the briquette's black carbon would make it a nearly ideal emitter of blackbody radiation and absorber of light.

Courtesy Lou Bloomfield

Even if air in the room is cold, you can usually feel the invisible infrared light from a woodstove on your face. When you block this light with your hands, your face suddenly feels colder because less heat is reaching your skin. This thermal radiation effect is even more pronounced with a fireplace or campfire, where thermal radiation from the hot coals and flames is the primary mechanism for heat transfer to the surroundings.

Overall, a modern woodstove is an excellent heat exchanger. As convection draws hot smoke up the long black chimney pipe, the smoke heats the stove and the pipe. These metal components conduct heat to their outer surfaces, which then distribute it around the room by convection and radiation. Although the stove consumes some room air, it controls the airflow with dampers so that it draws in only enough air to completely burn the wood. Overall, the stove extracts heat efficiently, cleanly, and safely from the burning wood.

▶ Check Your Understanding #7: Keeping Warm

for answers, see page 233

When you stand under a heat lamp, you feel warm even though the lamp emits little visible light. How is heat reaching your skin?

Warming the Room

You light the woodstove and its heat begins flowing out into the cold room. Moving always from hotter to colder, heat enters each item in the room and gradually raises its temperature. For example, a once frigid brass bowl near the woodstove is soon pleasantly warm to the touch.

Let's take a look at the relationship between the heat added to that bowl and its temperature rise. Because the bowl's temperature increases steadily when heat is flowing into it steadily, its overall temperature rise must be proportional to the added heat. The constant of proportionality is called the bowl's **heat capacity** and is the amount of heat that must be added to the bowl to cause its temperature to rise by 1 unit. In effect, the bowl's heat capacity is the measure of its thermal sluggishness, its resistance to temperature changes.

But suppose that you have an assortment of bowls near the woodstove, each a different material. If you keep track of their temperatures, you'll find that some of them warm faster than others. Even when you take into account differences in their masses and in how much heat each one is receiving from the woodstove, you'll find that bowls made of different materials respond differently to added heat. Some materials are more thermally sluggishly than others.

Table 7.1.2 Specific Heats Measured Near Room Temperature (293 K) and Atmospheric Pressure

MATERIAL	SPECIFIC HEAT
Lead	128 J/kg·K
Brass	380 J/kg·K
Copper	386 J/kg·K
Air (at constant volume)	715 J/kg·K
Glass	840 J/kg·K
Aluminum	900 J/kg·K
Air (at constant pressure)	1001 J/kg·K
Wood	~1100 J/kg·K
Plexiglas or Lucite	1349 J/kg·K
Steam (at constant pressure)	2027 J/kg·K
Ice	2220 J/kg·K
Water	4190 J/kg·K

It makes sense to characterize each material by its heat capacity per unit mass, a quantity known as **specific heat.** The SI unit of specific heat is the **joule-per-kilogram-kelvin** (abbreviated J/kg · K). Each bowl's heat capacity is the product of its mass times the specific heat of the material from which it's made.

Table 7.1.2 contains specific heats for a number of common materials. The wide range of values indicates that different materials respond quite differently to added heat. Each material's specific heat depends principally on the number of microscopic ways it can store thermal energy per kilogram. Known as a *degree of freedom,* each independent way of handling thermal energy stores an average thermal energy equal to half the Boltzmann constant times the absolute temperature ($\frac{1}{2}kT$). The Boltzmann constant, which we first encountered in Section 5.1, has a measured value of 1.381×10^{-23} J/K and its units here are equivalent to those that appeared in Section 5.1.

Brass's relatively small specific heat explains why the brass bowl heats up so quickly when you place it directly on the woodstove; the bowl has relatively few degrees of freedom in which to store its thermal energy. But if you add even a modest amount of water to the bowl, the water's astonishingly large specific heat will dramatically slow the bowl's temperature rise. Water has a remarkable capacity for thermal energy.

Like brass or water, air also has a specific heat. However, air's specific heat depends on how you measure it. That's because gases tend to expand as they warm up. If you seal the air in a bottle, so that its volume doesn't change, it warms relatively easily; air's specific heat at constant volume is 715 J/kg · K. But if you allow the air to expand as its temperature rises, so that its pressure doesn't change, it's harder to warm. That's because it needs extra energy to push the surrounding air out of its way as it expands; air's specific heat at constant pressure is 1001 J/kg · K.

►Check Your Understanding #8: Hot Stuff

for answers, see page 233

When you pull a metal tray of moist cookies out of the oven, the tray initially feels much hotter than the cookies. A little while later, however, the cookies feel much hotter than the tray. Explain.

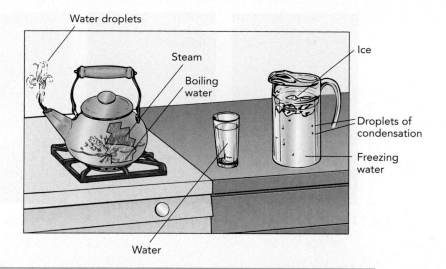

Water droplets

Steam

Boiling water

Ice

Droplets of condensation

Freezing water

Water

SECTION 7.2 | Water, Steam, and Ice

Water is probably the most important chemical in our daily lives. It is so crucial to biology, climate, commerce, industry, and entertainment that it merits a whole section of its own. Moreover, it exhibits the three classic phases of matter, solid, liquid, and gas, and illustrates the role that heat plays in transforming one phase into another. While most of what we can learn from water is applicable to any material, there are a few aspects of water that are almost unique in nature. Water really is a remarkable substance.

Questions to Think About: Why does adding ice make a drink cold? Why do icebergs float on water? How does perspiration cool you off? Why do humid days feel so muggy? What is the difference between evaporating and boiling? How does snow disappear from the ground even when it's very cold outside?

Experiments to Do: Experimenting with water is easy. Pick up an ice cube with wet fingers. What happens if the ice cube has just come out of the freezer? What if it has been melting on the table for a few minutes? Why is there a difference? Put both ice cubes in water. Why do they float?

   Now heat tap water in a pot. Shortly before the water starts to boil, you will see mist begin to form above it. What is this mist and why does it form? Carefully feel the mist, but don't burn yourself! The mist feels damp because it contains water. How can water be leaving the pot before the water boils? Notice the small gas bubbles on the walls of the pot. They're not steam; where did this gas come from?

   Once the water boils, gaseous water or steam will appear in the mist. Don't touch this steam because it can burn you quickly. Why does steam release so much thermal energy when it touches your skin?

Solid, Liquid, and Gas: The Phases of Matter

Like most substances, water exists in three distinct forms or **phases:** solid ice, liquid water, and gaseous steam (Fig. 7.2.1). These phases differ in how easily their shapes and volumes can change. Ice is **solid**—rigid and incompressible; you can't alter an ice cube's

Courtesy Lou Bloomfield

Fig. 7.2.1 The three phases of water: solid (ice), liquid (water), and gas (steam).

shape or its volume. Water is **liquid**—fluid but incompressible; you can reshape the water in a pitcher but you can't change its volume. Steam is **gaseous**—fluid and compressible; you can vary both the shape and the volume of the steam in a tea kettle.

These different characteristics reflect the different microscopic structures of steam, water, and ice. *Steam* or *water vapor* is a gas, a collection of independent molecules kept in motion by thermal energy. These water molecules bounce around their container, periodically colliding with one another or with the walls. The water molecules fill the container uniformly and can accommodate any changes in its shape or size. Enlarging the container simply decreases the steam's density and lowers its pressure.

When they are independent of one another as gaseous steam, water molecules have a substantial amount of chemical potential energy. They can release some of this energy by joining together to form ordinary water. *Water* is a liquid, a disorderly collection of molecules that cling to one another with chemical bonds. Because these bonds aren't very strong, the molecules in water can use thermal energy to break them temporarily and then change bonding partners. This rebonding process allows water to change shape so that it is a fluid. But despite their flexibility, these bonds manage to bunch the water molecules together so snuggly that even squeezing can't pack them much tighter. That's why water is incompressible.

The molecules in water can release still more chemical potential energy by linking together stiffly as ice. *Ice* is a solid, a rigid collection of chemically bound molecules. Like most solids, ice is **crystalline**—its water molecules are arranged in an orderly latticework that extends over long distances and gives rise to the beautiful crystal facets seen on snowflakes and frost. Ice's crystalline structure is so constraining that its water molecules can't use thermal energy to change bonding partners and consequently ice can't change shape.

Just as an orderly stack of oranges at the grocery store takes up less volume than a disorderly heap, a crystalline solid almost always occupies less volume than its corresponding disorderly liquid. The solid phase of a typical substance is thus denser than the liquid phase of that same substance, so the solid phase sinks in the liquid phase.

There is only one common substance that violates that rule: water. Ice's crystalline structure is unusually open and its density is surprisingly low. Almost unique in nature, solid ice is slightly less dense than liquid water and ice therefore floats on water. That's why icebergs float on the open ocean and why ice cubes float in your drink. In fact, water reaches its greatest density at about 4 °C (39 °F).

Check Your Understanding #1: Skating Anyone?

for answers, see page 233

As a pond freezes, where does the ice begin to form?

Melting Ice and Freezing Water

Ice in a freezer is extremely cold, typically $-18\ °C$ ($0\ °F$). When you place this ice on a warm countertop, heat flows into it and its temperature rises. The ice remains solid until its temperature reaches $0\ °C$ ($32\ °F$). At that point, the ice stops getting warmer and begins to melt. **Melting** is a **phase transition,** a transformation from the ordered solid phase to the disordered liquid phase. This transition occurs when heat breaks some of the chemical bonds between water molecules and permits the molecules to move past one another. The melting ice transforms into water, losing its rigid shape and crystalline structure.

Zero degrees Celsius is ice's **melting temperature,** the temperature at which heat goes into breaking bonds and converting ice into water, rather than into making the ice hotter. The ice–water mixture remains at $0\ °C$ until all of the ice has melted. When only water remains, heat once again causes its temperature to rise.

The heat used to transform a certain mass of solid into liquid, without changing its temperature, is called the **latent heat of melting** or, more formally, **latent heat of fusion.** The bonds between the water molecules in ice are strong enough to give ice an enormous latent heat of melting: it takes about 333,000 J of heat to convert 1 kg of ice at $0\ °C$ into 1 kg of water at $0\ °C$. Since water's specific heat is $4190\ J/kg\cdot K$, that same amount of heat would raise the temperature of 1 kg of liquid water by about $80\ °C$. Thus it takes almost as much heat to melt an ice cube as it does to warm the resulting water all the way to boiling.

The latent heat of melting reappears when you cool the water back to its melting temperature and it starts to freeze. **Freezing** is another phase transition, a transformation from the disordered liquid phase to the ordered solid phase. As you remove heat from water at $0\ °C$, the water freezes into ice rather than becoming colder. Because the water molecules release energy as they bind together to form ice crystals, the water releases heat as it freezes. The heat released when transforming a certain mass of liquid into solid, without changing its temperature, is again the latent heat of melting. You must add a certain amount of heat to ice to melt it and you must remove that same amount of heat from water to solidify it.

Check Your Understanding #2: Keeping Crops Warm

for answers, see page 233

Fruit growers often spray their crops with water to protect them from freezing in unusually cold weather. How does liquid water keep the fruit from freezing?

Phase Equilibrium: Leaving and Landing

We've seen that ice has a melting temperature, so now let's see *why* it has a melting temperature. To do that, we must look at the interface between solid ice and liquid water. Whenever both phases are in contact, they exchange water molecules across the interface between them. Water molecules regularly break free from the ice to enter the water

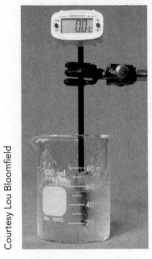

Courtesy Lou Bloomfield

Fig. 7.2.2 Ice and water can coexist only at 0 °C, ice's melting temperature.

and they often drop out of the water to stick to the ice. In other words, water molecules are leaving the ice and landing on it all the time, like airplanes at a busy airport.

While you can't see the individual leavings and landings, you can observe their net effect. If leavings outpace landings, the ice will gradually transform into water. If the landings outpace leavings, the water will gradually transform into ice. And if the two processes balance one another, the ice and water will coexist indefinitely—a situation known as **phase equilibrium.**

Temperature plays a crucial role in this balance because it affects the rate at which water molecules leave the ice. The warmer the ice, the more often water molecules at its surface can gather enough thermal energy to break free and leave. Below ice's melting temperature, water molecules leave the ice too infrequently to balance the landing process and the water transforms completely into ice. Above ice's melting temperature, water molecules leave the ice so often that they outstrip the landing process and the ice transforms completely into water. Only at the melting temperature do the leaving and landing rates balance so that ice and water can coexist in phase equilibrium.

Ice's huge latent heat of melting has a stabilizing effect on the phase equilibrium between ice and water. Whenever ice and water are mixed together, the mixture's temperature will shift rapidly toward 0 °C. That's because if the mixture's temperature is above 0 °C, the ice will melt—a phase transformation which absorbs the heat of melting and thereby lowers the mixture's temperature toward 0 °C. And if the mixture's temperature is below 0 °C, the water will freeze—a phase transformation which releases the heat of melting and thereby raises the mixture's temperature toward 0 °C.

As long as the mixture doesn't run out of ice or water, its temperature will soon reach 0 °C and remain there even as you add or remove heat from it. Any heat you add to the mixture goes into melting more ice, not into raising its temperature. Any heat you remove from the mixture comes from freezing more water, not from lowering its temperature. That's why your glass of ice water remains at 0 °C, even in the hottest or coldest weather (Fig. 7.2.2).

> ## Check Your Understanding #3: Ice Water
>
> *for answers, see page* **233**

At a restaurant, your glass of water contains 25% ice while your friend's glass of water contains 75% ice. Whose water is colder, or are their temperatures equal?

Evaporating Water and Condensing Steam

Water's open surface is another active interface between phases, but this time the liquid water is exchanging molecules with gaseous steam. Water molecules are feverishly leaving the water for the steam and landing on the water from the steam, once again like planes at a busy airport.

While this frantic exchange of molecules is interesting, what matters most is its net effect. If more molecules leave the water than return to it, then water gradually evaporates into steam. **Evaporation** is the phase transition from liquid to gas. On the other hand, if more molecules are landing on the water than leaving it, the steam gradually condenses into water. **Condensation** is the phase transition from gas to liquid. And if landing and leaving are balanced, water and steam coexist in phase equilibrium.

These two phase transitions have enormous thermal consequences. Since the molecules in water cling to one another with chemical bonds, it takes energy to separate them.

Although the bonds *between* water molecules are weaker than the bonds *within* water molecules, it still takes a great deal of energy to transform water into steam.

The heat needed to transform a certain mass of liquid into gas, without changing its temperature, is called the **latent heat of evaporation** or, more formally, **latent heat of vaporization.** Water's latent heat of evaporation is truly enormous because water molecules are surprisingly hard to separate. About 2,300,000 J of heat is needed to convert 1 kg of water at 100 °C into 1 kg of steam at 100 °C. That same amount of heat would raise the temperature of 1 kg of water by more than 500 °C!

You are most aware of this latent heat of evaporation on hot summer days, when perspiration that evaporates from your skin draws heat from you and lowers your temperature. As each water molecules leaves the perspiration to become steam, it gathers up more than its fair share of thermal energy from its environment and carries it away as chemical potential energy. The departing water molecules thus leave you bereft of thermal energy and you grow colder.

The latent heat of evaporation reappears when steam condenses and the gathering water molecules release their chemical potential energy as heat. The heat released when transforming a certain mass of gas into liquid, without changing its temperature, is again the latent heat of evaporation. You must add a certain amount of heat to water to evaporate it and you must remove that same amount of heat from steam to condense it.

The huge amount of heat released by condensing steam is often used to cook food or warm radiators in older buildings. When you steam vegetables, you are allowing steam to condense on the vegetables and transfer heat to them. A double-boiler uses condensing steam to transfer heat from a burner to a cooking container in a controlled manner.

➤ Check Your Understanding #4: Tea Time

for answers, see page 233

A kettle of water heats up rapidly on the stove but takes quite a while to boil away. Why does the water take so long to turn into steam?

Relative Humidity

Although we've examined the consequences of evaporating and condensing, we haven't yet seen what conditions determine when they'll occur. It all comes down to water molecules leaving and landing, so let's take a look at why one process wins out over the other.

The basic indicator of whether water will evaporate or steam will condense is relative humidity. **Relative humidity** measures the water molecule landing rate as a percentage of the leaving rate. When the relative humidity is !00%, the two rates are equal and water and steam are in phase equilibrium. But if the relative humidity is less than 100%, the landing rate is less than the leaving rate and the water evaporates. And if the relative humidity is more than 100%, the landing rate is more than the leaving rate and the steam condenses.

Relative humidity depends on the temperature and on the density of the steam. Temperature affects the leaving rate. The warmer the water, the more thermal energy it contains and the more frequently water molecules leave its surface to become gas. By itself, an increase in temperature will boost the leaving rate and thereby decrease the relative humidity. Rising temperatures thus favor evaporation.

The density of water molecules in the steam affects the landing rate. The denser the steam, the more often water molecules land on the water to become liquid. By itself, an

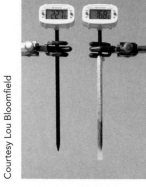

Fig. 7.2.3 You can determine the air's relative humidity using two thermometers—one of which has a wet cloth wrapped around its bulb (*right*). Evaporation cools the wet bulb thermometer by an amount related to the air's relative humidity. The dryer the air, the colder the wet bulb thermometer becomes.

increase in steam density will boost the landing rate and thereby increase the relative humidity. Rising steam densities thus favor condensation. Even when that steam is mixed with air, as it often is in everyday life, the air molecules act as passive bystanders. The density of water molecules alone determines the air's relative humidity.

Relative humidity plays an important role in countless experiences of everyday life. When the relative humidity is low, water evaporates quickly and the air feels dry. Perspiration cools you effectively. When the relative humidity is high (near 100%), water barely evaporates at all and the air feels damp. Perspiration clings to your skin and doesn't cool you much.

And when the relative humidity exceeds 100%, perhaps because of a sudden temperature drop, steam begins to condense everywhere. Water droplets grow on surfaces as dew or form directly in the air as fog, mist, or clouds. If the humidity remains high, these droplets grow larger and eventually fall as rain.

You can measure relative humidity by observing the cooling that accompanies evaporation. The most common scheme involves two thermometers (Fig. 7.2.3), one of which has a wet cloth wrapped around its bulb (right). Water evaporating from the wet cloth cools the thermometer until that water reaches phase equilibrium with steam in the air and evaporation ceases. The dryer the air, the colder the thermometer gets. The temperatures of the two thermometers can then be used to determine the relative humidity, usually with the help of tabulated values.

Check Your Understanding #5: Seeing Your Breath

for answers, see page **234**

When you breathe out on a cold day, you often see mist appearing from your mouth. Explain.

Subliming Ice and Depositing Steam

We've examined phase transitions between ice and water and between water and steam. That brings us to the phase transitions between ice and steam. Oddly enough, water molecules can leave ice to become steam and can land from steam to become ice. In fact, ice and steam regularly exchange water molecules even when there is no liquid water present at all.

As usual, this exchange of water molecules occurs at the surface of the ice, the interface between ice and steam. Since we're most interested in the net movement of molecules, it comes down to leaving and landing rates on the ice. If molecules leave the ice more often than they land, the ice sublimes. **Sublimation** is the phase transition from solid to gas. And if molecules land on the ice more often than they leave, the steam deposits. **Deposition** is the phase transition from gas to solid. Once again, relative humidity measures the landing rate as a percentage of the leaving rate. At 100% relative humidity, ice and steam are in phase equilibrium.

When the relative humidity is below 100%, ice sublimes. This effect gives rise to a number of familiar phenomena. When the weather is cold and dry, snow gradually disappears from the ground without ever melting. In the low relative humidity of a frostless freezer, the ice cubes slowly shrink to midget size. And when you leave food unprotected in that same frostless freezer, it eventually dries out. While this "freezer burn" is a nuisance at home, sublimation from frozen food is used commercially to prepare freeze-dried foods.

And when the relative humidity exceeds 100%, steam deposits. This process yields several more familiar effects. Frost forms on cold windows and lawns that are exposed to humid air. In the high relative humidity of a non-frostless freezer, frost and ice accumulate on the walls and require periodic defrosting. And snowflakes grow in clouds and then descend gracefully to the ground.

➤ Check Your Understanding #6: Disappearing Stink

for answers, see page **234**

Mothballs are smelly, crystalline pellets used to protect woolen clothes from attack by insects. If you sprinkle them into a closet and wait a few years, the mothballs will disappear. How do they vanish?

Boiling Water

With three phases and six phase transitions, we seem to be out of possibilities. So where does boiling fit into this picture? Boiling is simply an accelerated form of evaporation in which bubbles of pure steam grow by evaporation inside the water itself. To understand boiling, let's take a look at the interplay between water and steam.

Suppose we seal some water inside an airless container and keep it at a constant temperature. The water will evaporate as steam until the relative humidity inside the container reaches 100%. At that point, the water and steam will have reached phase equilibrium; the steam will have just the right density so that its water molecules will land on the water as often as they leave. Steam at its equilibrium density is said to be **saturated.**

Of course, the density of that saturated steam depends on the temperature of the container. If you warm up the container, water molecules will leave the water more frequently and the density of the steam will have to increase in order to match the landing rate to the leaving rate. The density of saturated steam is thus an increasing function of the temperature.

The saturated steam's density, together with its temperature, determine its pressure—the pressure inside our container. If we warm the container, the density of the saturated steam will increase and so will its pressure. Near room temperature, the pressure of saturated steam is a few percent of atmospheric pressure. As the temperature approaches 100 °C (212 °F), however, the pressure of saturated steam approaches atmospheric pressure.

With that background, suppose that we place a bubble of pure saturated steam in room temperature water. Because the pressure inside that bubble is much lower than atmospheric pressure, the surrounding water will rush inward and compress it. As the steam bubble's volume shrinks, the steam will exceed its saturated density and begin to condense. In almost no time, the bubble will be smashed out of existence.

Now suppose we begin warming the water on the stove. As the water temperature increases, steam's saturated density and pressure both increase. At first, saturated steam bubbles remain unstable; they're crushed quickly by atmospheric pressure. But when the water temperature is near 100 °C, something remarkable happens: saturated steam bubbles suddenly become stable and the water can begin to **boil** (Fig. 7.2.4). At that temperature, water's **boiling temperature,** the pressure of saturated steam reaches atmospheric pressure and bubbles of saturated steam can survive indefinitely within the water. Even more remarkably, these bubbles can grow by evaporation; each bubble's surface is

Fig. 7.2.4 Water boils at 100 °C, when atmospheric pressure can no longer smash the bubbles of water vapor.

an interface between water and steam, so when heat is added to the water, water can transform into steam and enlarge the bubble. Although the bubbles quickly float to the water's surface and pop, new bubbles can promptly take their place.

Boiling converts water to steam so rapidly that it consumes almost any amount of heat you add to the water. That's why it's so difficult to warm water above its boiling temperature. An open pot of water on the stove warms to water's boiling temperature and then remains at that temperature until all of the water has transformed into steam. Only then can the pot's temperature again begin to increase.

The constant, well-defined temperature of boiling water allows you to cook vegetables or an egg at a particular rate. When you place an egg in boiling water, it cooks in 3 minutes because it's in contact with water at its boiling temperature.

➤Check Your Understanding #7: Vanishing Bubbles

for answers, see page **234**

You are heating water in a pot on the stove. Shortly before the water boils properly, bubbles of steam begin rising from the bottom of the pot but vanish before they reach the water's surface. What happens to those bubbles?

Changing Water's Boiling Temperature

Water's boiling temperature depends on the ambient pressure. For an open pot or pan, that pressure is atmospheric pressure. However, atmospheric pressure decreases with altitude and depends somewhat on the weather. Water boils at 100 °C (212 °F) near sea level but at only 90 °C (194 °F) at an altitude of 3000 m (9800 ft). This reduction in water's boiling temperature with altitude explains why many recipes must be adapted for use at higher elevations. At 3000 m, an egg cooks slowly in boiling water because it's surrounded by 90 °C water, not 100 °C water. The same problem slows the cooking of rice, beans, and many other foods at high altitudes.

At sufficiently low pressures, water boils even at room temperature or below. At the other extreme, high pressures can prevent water from boiling until it's extremely hot. The boilers in steam engines and power plants often operate at such high pressures that water's boiling temperature inside them may exceed 300 °C (572 °F).

One way to decrease cooking times is to use a pressure cooker, a pot that seals in steam so that the pressure inside it can exceed atmospheric pressure. This increased pressure prevents boiling until the water temperature is well above 100 °C. If you subject water to twice sea-level atmospheric pressure, it won't boil until 121 °C (250 °F). An egg cooks very quickly at that temperature, as do vegetables and other foods.

However, just because water *can* boil, doesn't mean that it *will* boil. Boiling depends on tiny seed bubbles that subsequently grow by evaporation. Without seed bubbles, the water won't boil. Because **nucleation** or seed-bubble formation almost never occurs spontaneously in water below 300 °C, something else must create those seeds. Most nucleation occurs at defects or hotspots on the container or at contaminants in the liquid. Those nucleation sites usually trap air or other permanent gases and then serve as nurseries for seed bubbles of steam. Reliance on these specific nucleation sites explains why the bubbles of boiling water, like those in soda or champagne, often stream upward from specific spots on their containers.

When you heat water uniformly in a clean, glass container, it may not boil properly at its boiling temperature. Glass has a liquid-like smoothness even at the atomic scale and rarely aids seed bubble formation. Without any long-lived nucleation sites, the water may stop forming seed bubbles and cease boiling. Once boiling stops, the water's temperature can rise above the boiling temperature so that it becomes **superheated.**

Superheated water, which forms easily and often in a microwave oven, can be extremely dangerous. Touching it with a fork, adding sugar or salt, or even just tapping its container can initiate violent or even explosive boiling (Fig. 7.2.5). The more the water's temperature exceeds its boiling temperature, the more energy it can release suddenly if it abruptly boils. Be careful when you heat water in a microwave oven, particularly in a glass or glazed container. If it doesn't appear to be boiling properly despite being very hot, recognize that it may be superheated. Your safest bet is to stay away from it until it has cooled down.

There is one other interesting way to change water's boiling temperature: dissolve chemicals in it. A dissolved chemical keeps the water molecules busy so that they are less likely to leave the water to become steam, or ice for that matter. Since dissolved chemicals discourage water molecules from leaving water's liquid phase, they suppress any phase transitions that reduce the amount of liquid phase water. That's why dissolving sugar or salt in water slows its evaporation and raises its boiling temperature. It's also why saltwater freezes at a lower temperature than freshwater and why salt tends to melt ice. In contrast, sand doesn't melt ice because it doesn't dissolve in water.

Courtesy Lou Bloomfield

Fig. 7.2.5 The water in this glass measuring cup was superheated in a microwave oven and then disturbed with a fork. Explosive boiling blew all of the water out of the cup in a fraction of second.

Check Your Understanding #8: Lucky Accident

for answers, see page 234

You pull a ceramic mug of hot water out of the microwave oven and add a spoonful of instant coffee. It bubbles amusingly and splatters a little onto the table. What happened?

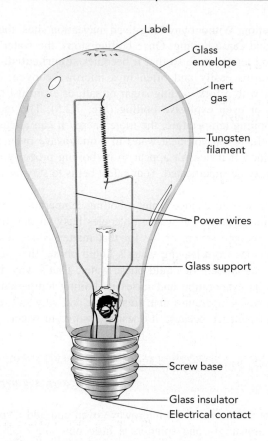

Label
Glass envelope
Inert gas
Tungsten filament
Power wires
Glass support
Screw base
Glass insulator
Electrical contact

SECTION 7.3 Incandescent Lightbulbs

For more than a century, incandescent lightbulbs have provided light at the flip of a switch. Their invention brought to a close the era of candles and gaslights and spurred the development of electric power. While the variety of incandescent bulbs has grown over the years to include everything from heat lamps to halogen headlights, all incandescent bulbs have at their hearts one simple object: an extremely hot filament.

Questions to Think About: What part of a lightbulb emits the light? How is a lightbulb similar to a fire or a candle? How is it different? What colors of light can a plain, unpainted lightbulb emit? Why does the top of a lightbulb darken with age? What happens when a lightbulb burns out?

Experiments to Do: Take a look at a few incandescent lightbulbs. Try turning one on and off. Are the transitions instantaneous? Stand in a darkened room with your eyes closed and open your eyes suddenly, just after you turn the bulb off. Can you see the bulb going dark? How do its brightness and color change with time?

Compare the color of the light from a conventional bulb with that from an extended life bulb. Which one produces a better simulation of sunlight? Which bulb should you use in your desk lamp? in an inaccessible ceiling fixture?

Now compare both bulbs with a halogen bulb. How do their colors differ? Which bulb do you expect to live the longest in normal use? Is it surprising that halogen bulbs live longer than conventional bulbs?

Light, Temperature, and Color

Courtesy Lou Bloomfield

Light from an incandescent lightbulb is part of the thermal radiation emitted by its hot wire filament. While most types of electromagnetic waves are invisible, our eyes are sensitive to a narrow range of waves that we call visible light. Any object that's hotter than about 400 °C (750 °F) emits enough visible light for us to see it in a dark room. At higher temperatures, that visible light brightens and shifts in color from red to orange to yellow to white. At 500 °C (930 °F), an object glows a dull red. At 1700 °C (3100 °F), it emits the orange light of a candle. And at 5800 °C (10,500 °F), the temperature of the sun's surface, it gives off the brilliant white light of the sun.

To reproduce pure white sunlight, the bulb's filament should be heated to 5800 °C. Unfortunately, nothing is solid at that high temperature. Even tungsten metal, the best filament material known, readily sublimes at temperatures above 2500 °C (4500 °F). Since incandescent lightbulbs must operate at lower temperatures, they can't really reproduce sunlight. Most give off the warm, yellow-white light that's characteristic of tungsten metal at about 2500 °C.

The filament's brightness and color both depend on its temperature (Fig. 7.3.1). Since light carries energy, we can measure its brightness as the number of watts of visible light it emits. But how do we characterize its color? Moreover, what distinguishes visible light from the invisible types of electromagnetic radiation? Although the full answers to those questions will have to wait until Chapters 13 and 14, we can make a few essential observations about them now.

Visible light is part of a continuous spectrum of electromagnetic radiation that extends from radio waves at one extreme to gamma rays at the other (Fig. 7.3.2). Different types of electromagnetic radiation are distinguished by their **wavelengths,** that is, the distance between their wave crests. Wavelength is easy to see in the waves on a lake or sea, where the crests are visible and you can directly measure the distance from one to the next. But while the wave crests of electromagnetic waves aren't so easy to observe, they exist and it's possible to measure the distance between them.

The electromagnetic radiation produced by a hot filament is mostly infrared, visible, and ultraviolet light. Although this light is just a tiny portion of the overall electromagnetic spectrum, it's particularly important to our everyday world. Figure 7.3.3 gives an expanded view of the visible portion of the electromagnetic spectrum. Various colors that we see correspond to specific wavelength ranges. For example, light with a wavelength of 530 nanometers (billionths of a meter, abbreviated nm) appears green to our eyes.

But the thermal radiation emitted by a filament isn't a single electromagnetic wave with one specific wavelength. Instead, it's many individual waves that cover a broad range of wavelengths. Some of these waves are red light, some green, some blue, and some are invisible.

Fig. 7.3.1 As you increase the power to an incandescent lightbulb, its filament becomes hotter and emits a brighter and whiter light. The cooler filament on the left is dim and red while the hotter one on the right is bright and yellow-white. Because these bulbs are frosted, you can't see their filaments directly.

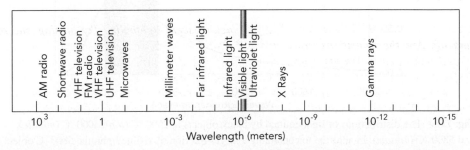

Fig. 7.3.2 The spectrum of electromagnetic radiation, arranged by wavelength. The scale here is logarithmic, meaning that the wavelength decreases by a factor of 10 with each tick mark to the right.

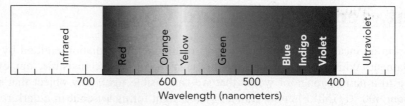

Fig. 7.3.3 A portion of the electromagnetic radiation spectrum around visible light. Wavelengths are measured in nanometers (nm, or billionths of a meter).

The distribution of wavelengths emitted by the filament depends on its temperature and surface properties, particularly its **emissivity**—the efficiency with which it emits and absorbs light. Emissivity is measured on a scale from 0 to 1, with 1 being ideal efficiency. A perfectly black object has an emissivity of 1; it absorbs all light that strikes it and emits thermal light as efficiently as possible. Although tungsten's emissivity is only 0.43, the filament wire is wound in such a way that it has many dark nooks and crannies. The filament is thus so nearly black that its emissivity is essentially 1.

The distribution of wavelengths emitted by a black object is determined by its temperature alone and is called a **blackbody spectrum.** As you can see from the examples in Fig. 7.3.4, the spectrum of a blackbody brightens and shifts toward shorter wavelengths as its temperature increases. An object that isn't black emits somewhat less thermal radiation, but that radiation still brightens and shifts toward shorter wavelengths as the object becomes hotter.

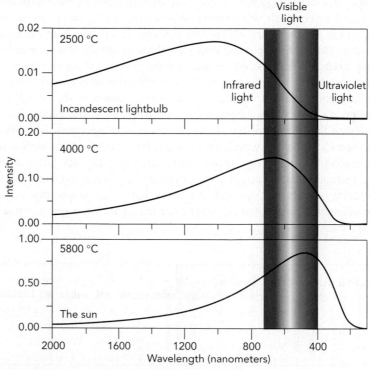

Fig. 7.3.4 The distributions of light emitted by black objects at 2500 °C (*top*), 4000 °C (*middle*), and 5800 °C (*bottom*). In addition to containing a larger fraction of visible light, the 5800 °C object is much brighter than the 2500 °C object (note the different intensity scales).

Table 7.3.1 Temperatures and Colors of Light Emitted by Hot Objects

OBJECT	TEMPERATURE	COLOR
Heat lamp	500 °C (930 °F)	Dull red
Candle flame	1700 °C (3100 °F)	Dim orange
Bulb filament	2500 °C (4500 °F)	Bright yellow-white
Sun's surface	5800 °C (10,500 °F)	Brilliant white
Blue star	6000 °C (10,800 °F)	Dazzling blue-white

Our eyes make an average assessment of the distribution of wavelengths emitted by a black object, and we observe reddish, orangish, yellowish, whitish, or bluish light, depending on the object's temperature (Table 7.3.1). The temperature associated with a particular distribution of wavelengths is the **color temperature** of that light.

We can already see two of the principal shortcomings of an incandescent lightbulb: its poor efficiency at converting electric energy into visible light and its low color temperature. At 2500 °C, only about 12% of its thermal radiation is visible light; the rest is invisible infrared light. Its filament would have to reach 5000 °C before the infrared fraction of its thermal radiation would drop below 50%. Moreover, its 2500 °C color temperature makes it look yellowish when compared to sunlight because it doesn't emit enough blue light. Most of the developments in lighting over the past half century have focused on improving energy efficiency and color temperature.

Check Your Understanding #1: Even Plants Can Look Cool

for answers, see page 234

Satellites measure the temperatures of agricultural regions from space, looking for signs of crop distress and disease. How do they make such measurements?

The Filament

The bulb's filament is heated by an electric current, which provides it with thermal power in the amount specified on the bulb. For example, the filament of a 60-watt bulb produces 60 watts of thermal power from the 60 watts of electric power it consumes. Since the filament can't accumulate thermal energy indefinitely, its temperature rises until thermal energy flows out of it as heat as quickly as it's produced from electricity. Thus a 60-watt bulb sends 60 watts of thermal power into its environment as heat. Much of that heat is thermal radiation.

The temperature at which this heat balance occurs is surprisingly specific. That's because the filament's thermal radiation increases so rapidly with temperature that even a small excursion above its normal operating temperature will cause the filament to radiate away more thermal energy than it produces from electricity and it will quickly cool back down to normal. Like any hot object, the power radiated by the filament is proportional to the fourth power of its absolute temperature. The precise relationship between temperature and emitted power can be written as a word equation:

$$\text{radiated power} = \text{emissivity} \cdot \text{Stefan–Boltzmann constant} \cdot \text{temperature}^4 \cdot \text{surface area},$$

$$(7.3.1)$$

❏ Lewis Howard Latimer (African-American scientist and inventor, 1848–1928) was only eight when the U.S. Supreme Court's *Dred Scott* decision made his escaped-slave father a fugitive and forced him to disappear. Left behind with his mother, Latimer did well in school and became a skilled draftsman and engineer. While working for Edison's rival, Hiram Maxim, Latimer became an expert in fabricating carbon filaments for incandescent lamps. When Latimer later joined Edison's team of inventors, the "Edison Pioneers," his sturdy carbon filaments quickly replaced Edison's own fragile bamboo ones and provided the crucial ingredient necessary for Edison's lamps to become a commercial success.

in symbols:

$$P = e \cdot \sigma \cdot T^4 \cdot A,$$

and in everyday language:

> *Warm skin radiates away lots of heat, so cover it up to avoid feeling cold.*

This relationship is called the **Stefan–Boltzmann law,** and the **Stefan–Boltzmann constant** (σ) that appears in it has a measured value of 5.67×10^{-8} J/(s·m^2·K^4). Remember that the temperature must be measured in kelvins.

Increasing the filament's temperature therefore increases both its color temperature and its brightness. Each of these characteristics is important to you, so it's fortunate that you can adjust them somewhat independently. Unless you're enjoying a romantic dinner and want the reddish glow of candlelight, you usually prefer higher color temperatures. For an incandescent bulb, that normally means a hotter filament. But once you've chosen the filament temperature, you can still make the filament brighter by increasing its surface area.

But while the concept of an incandescent bulb is simple, finding a material that can tolerate extremely high temperatures is not. Early filaments were made of carbon and platinum. Of these materials, carbon showed the most promise. In 1879, Thomas Edison developed an incandescent lamp with a carbon filament that operated for several hundred hours. His wasn't the first incandescent bulb ever made but rather the first practical one. (For more about the carbon filaments, see ❏.)

However, while carbon has the highest melting temperature of any element (3550 °C or 6422 °F), it sublimes relatively quickly even at much lower temperatures. Thus a carbon filament that is heated close to its melting temperature quickly disappears as a gas. When a gap appears in the filament, it stops carrying electricity and "burns out." Carbon is also flammable, so it must be enclosed in an airtight glass bulb that contains either inert gases or a vacuum.

A better choice for filaments, now used in virtually all incandescent bulbs, is tungsten metal. Tungsten melts at 3410 °C (6170 °F) and sublimes extremely slowly below that temperature. Tungsten filaments can thus operate at higher temperatures than carbon filaments, producing richer, whiter light. Like carbon, however, hot tungsten burns in air and must be protected in a glass bulb.

To ensure that most of the electric power passing through the filament is converted into thermal power, the filament must be long and thin. A typical 60-W lightbulb has about 0.5 m (20 inches) of 25-micron (0.001-inch) tungsten wire, coiled up into a filament only about 2 cm (0.8 inches) long. To minimize the filament's length, it's wound into a double spiral. First it's wound into a thin springlike coil about 0.25 mm wide. Then this coil is wound into another coil to form the actual filament (Fig. 7.3.5). Fabricating such a complicated tungsten filament is so difficult that it wasn't accomplished until 1937.

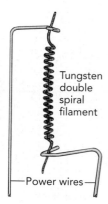

Tungsten double spiral filament

Power wires

Fig. 7.3.5 The tungsten filament of a modern incandescent lightbulb is a double spiral—a coil wound from a smaller coil of extremely fine tungsten wire. The double spiral allows the manufacturers to put a long length of wire in a small space.

Check Your Understanding #2: De-Lighted at the Blacksmith's Shop

for answers, see page 234

When a blacksmith pulls a piece of yellow-hot steel out of the furnace, its temperature plummets and it dims rapidly. However, it remains too hot to touch for several minutes. Why does it cool so much faster at first?

► Check Your Figures #1: The Cold of Deep Space

for answers, see page **234**

Suppose that an accident in the depths of space leaves you exposed to an environment near absolute zero. Since your surroundings radiate almost no heat toward you,
you are losing heat fast. If your surface area is 2 m², your skin temperature is 310 K,
and your emissivity is 0.5, how much power will you radiate?

The Glass Bulb

To keep the white-hot filament from burning, it's surrounded by a glass bulb that usually
contains oxygen-free inert gas. This gas, typically nitrogen and argon, slows sublimation
by bouncing some of the escaping tungsten atoms back onto the filament. Although the
gas extends the filament's life, it has at least two drawbacks. First, it allows conduction
and convection to carry some heat away from the filament. Second, tiny tungsten particles
that form in this gas rise with convection currents to produce a dark spot on the top of
the bulb.

The glass bulb presents another challenge: operating a filament inside it involves
passing wires right through the glass. This step isn't so easy because the glass and metal
must seal to one another perfectly. Complicating this sealing requirement is the fact that
materials expand as their temperatures increase. If the glass and metal don't expand
equally as the bulb warms up, the bulb may leak or even break.

A material's thermal expansion is caused by atomic vibrations. Because of thermal energy, adjacent atoms vibrate back and forth about their equilibrium separations
(Fig. 7.3.6). This vibrational motion isn't symmetric; the repulsive force the atoms
experience when they're too close together is stiffer than the attractive force they experience when they're too far apart. As a result of this asymmetry, they push apart more
quickly than they draw together and thus spend most of their time at more than their
equilibrium separation. On average, their actual separation is larger than their equilibrium separation, and the material containing them is bigger than it would be without
thermal energy.

As an object's temperature rises, its atoms move farther apart on average and the
object grows larger in all directions. The extent to which an object expands with
increasing temperature is normally described by its **coefficient of volume expansion:**
the fractional change in the object's volume per unit of temperature increase. Fractional
change in volume is the net change in volume divided by the total volume. Since most
materials expand only a small amount when becoming 1 °C (or 1 K) hotter, coefficients
of volume expansion are small, typically about $10^{-5}\,K^{-1}$ for metals, about $10^{-6}\,K^{-1}$
for special low-expansion glasses, and about $10^{-4}\,K^{-1}$ for liquids. In a lightbulb, the
metal wires and the glass are carefully selected to have similar coefficients of volume
expansion. As the bulb warms up, the wires and glass expand together and the seals
remain intact.

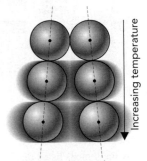

Average separation

Fig. 7.3.6 The thermal
kinetic energy of a solid
increases with temperature,
causing its atoms to bounce
against one another more
vigorously. As they vibrate,
the atoms repel more
strongly than they
attract so their average
separation increases
slightly.

► Check Your Understanding #3: Overcooked

for answers, see page **234**

If you place a pot or saucepan on the stove and fill it to the brim with cold water,
it will overflow as you heat it. From where does the extra water come?

Extended Life, Halogen, and Three-Way Bulbs

One way to prolong the life of the filament is to increase its surface area while providing it with the same electric power. With more surface area to radiate away thermal power, the filament doesn't get quite as hot and doesn't sublime as quickly. The result is an extended life bulb. Unfortunately, extended life bulbs are redder than conventional bulbs and also less energy efficient. Because an extended life bulb emits a smaller fraction of its input power as visible light, it must have a higher wattage to give equivalent lighting. As a result, extended life bulbs aren't always a bargain; the money you save on replacement bulbs may well be spent on increased energy costs. When you're choosing bulbs, it's worth looking at how many **lumens**—the measure of useful illumination—they produce per watt of electric power consumed (Fig. 7.3.7).

You're much better off buying a halogen bulb, which is both longer-lived and more energy efficient than a conventional bulb. A halogen bulb uses a chemical trick to rebuild its filament continuously during operation. This filament is enclosed in a small tube of quartz or aluminosilicate glass, which can tolerate high temperatures and reactive chemicals (Fig. 7.3.8). The tube contains molecules of the halogen elements bromine and/or iodine. During operation, the tube becomes extremely hot and the halogen reacts with any tungsten atoms on its inside surface. They form a gas of tungsten–halogen molecules that drift about the tube until they encounter the white-hot filament. The molecules then break apart and the tungsten atoms stick to the filament.

The halogens act as recycling agents, seeking out tungsten atoms that have sublimed from the filament and returning them to it. Unfortunately, this recycling process slowly changes the structure of the filament. The returning tungsten atoms deposit unevenly, so that the filament gradually develops thin spots and eventually burns out. Nonetheless, the filament lives more than 2000 hours, even when it runs several hundred degrees hotter than in a conventional bulb. Its higher filament temperature allows a halogen bulb to produce whiter light than a conventional bulb and increases its energy efficiency.

Bulbs with different power ratings have filaments with different amounts of surface area. The filament in a 100-W bulb has four times the surface area of the filament in a 25-W bulb and thus emits four times as much light. To support this fourfold increase in thermal power, the 100-W filament also consumes four times as much electric power. Both filaments operate at the same temperature and emit light with the same color temperature.

Courtesy Lou Bloomfield

Fig. 7.3.7 Different 100-W lightbulbs produce different amounts of useful light, as revealed by the lumens listed on their packaging.

Courtesy Lou Bloomfield

Fig. 7.3.8 These halogen bulbs operate at higher temperatures than normal incandescent bulbs and produce whiter light. The large glass envelope of the upper bulb protects a smaller lamp inside.

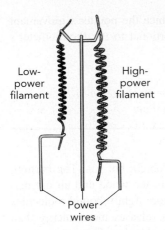

Low-power filament

High-power filament

Power wires

Fig. 7.3.9 A three-way bulb has two independent filaments. The filament on the left is shorter and thinner than the one on the right and emits about half as much light. The three different light levels correspond to having the left filament on, the right filament on, and both filaments on.

One way to make an incandescent bulb with a variable light output is to put several independent filaments in it. A "three-way bulb" has two filaments (Figs. 7.3.9 and 7.3.10) that can be turned on and off separately. In a 50-100-150 W bulb, one filament uses 50W of electric power and the other uses 100 W. If only the low-power filament is on, the bulb appears to be a 50-W bulb. If only the high-power filament is on, it appears to be a 100-W bulb. But when both filaments are on, the bulb appears to be a 150-W bulb. Since one of the filaments invariably burns out before the other, the bulb fails by going from having three light levels to only one.

Check Your Understanding #4: Double Wrapping
for answers, see page 234

While conventional lightbulbs have thin glass envelopes, halogen bulbs that replace them have surprisingly thick glass shells. If you cut one of these shells open, you'll find a second, much smaller glass bulb inside. Why does the manufacturer go to the trouble of putting two separate bulbs around the filament?

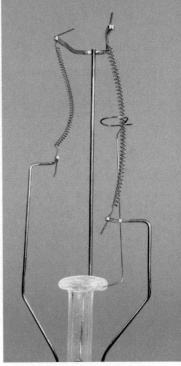

Courtesy Lou Bloomfield

Fig. 7.3.10 The glass envelope of this three-way lightbulb has been removed to expose its two filaments. The shorter, thinner filament (*left*) uses 50 W, while the longer, thicker one (*right*) uses 100 W.

Epilogue for Chapter 7

This chapter examined the roles of thermal energy and heat in a variety of common objects. In *woodstoves,* we saw how combustion converts ordered chemical potential energy into disordered thermal energy and we studied the ways in which this thermal energy flows into the woodstove's surroundings: conduction, convection, and radiation. In *water, steam, and ice,* we examined the three phases of matter and looked at how the transformations between those phases are influenced by temperature, heat, and other characteristics of their environment. In *incandescent lightbulbs,* we found that a sufficiently hot object radiates some of its heat as visible light and saw that the distribution of wavelengths in this light depends on the object's temperature.

Explanation: A Ruler Thermometer

Like most things, the clear plastic ruler expands when you heat it. Its length increases by an amount proportional to its increase in temperature. When you transfer heat to the ruler, by breathing on it, touching it, or exposing it to a hair dryer, its temperature rises, it expands, and its free end turns the needle and pointer. Since the pointer's movement is proportional to the ruler's length change, it's also proportional to the thermometer's change in temperature.

Chapter Summary

How Woodstoves Work: A woodstove burns wood in air to obtain hot gas. The burning process is actually a chemical reaction in which molecules in the wood and air are disassembled into fragments and then reassembled into new, more tightly bound molecules such as water and carbon dioxide. This reassembly process releases more energy than was required to disassemble the original wood and oxygen molecules. This extra energy appears as thermal energy within the reaction products, so that they're hot. Rather than distribute the hot burned gas directly to the room, a woodstove transfers heat from the burned gas to clean air or water. Heat is conducted through the walls of a woodstove and then flows into the room via convection and radiation.

How Water, Steam, and Ice Work: Water, steam, and ice are the liquid, gaseous, and solid phases of the chemical we call water. The water molecules in liquid water are bound together strongly enough to fix water's volume, but not strongly enough to give water a rigid shape. In steam, the water molecules are independent and the gas has neither a fixed shape nor fixed volume. The molecules in ice are bound together rigidly in solid crystals, so that ice is rigid and its volume is constant.

Ice can transform into water or vice versa at its melting temperature. Heat added to ice at this temperature causes it to melt without becoming warmer. Heat removed from water at this temperature causes it to freeze without becoming colder. Similarly, water and ice can transform into steam or vice versa through evaporation, condensation, sublimation, and deposition. Turning water into steam requires heat and this heat is released when the steam turns back into water. Boiling is a special case of evaporation, in which evaporation occurs within the body of the water itself. For boiling to occur, the water's vapor pressure must equal the ambient pressure on the water, which is normally atmospheric pressure.

How Incandescent Lightbulbs Work: An incandescent lightbulb produces light as thermal radiation from an extremely hot tungsten filament. The spectrum of light emitted by that filament depends on its temperature—the hotter the filament, the whiter its light. Atoms in the filament sublime during operation so the filament gradually disappears. Eventually it becomes so thin that it breaks. The tungsten filament of a normal bulb operates at 2500 °C because it would burn out too quickly at higher temperatures. Reducing the operating temperature, as is done in an extended life bulb, prolongs the filament's life at the expense of color temperature and energy efficiency. In contrast, adding halogen gas to recycle the tungsten atoms actually increases the filament's life and improves both its color temperature and energy efficiency.

Important Laws and Equations

1. **The Stefan–Boltzmann Law:** The power an object radiates is proportional to the product of its emissivity times the fourth power of its temperature times its surface area, or

$$\text{radiated power} = \text{emissivity} \cdot \text{Stefan–Boltzmann constant} \cdot \text{temperature}^4 \cdot \text{surface area}. \tag{7.3.1}$$

Check Your Understanding—Answers

Section 7.1 WOODSTOVES

1. No, its thermal energy will remain constant.
Why: As it falls, the ball's gravitational potential energy is transformed into kinetic energy. However both energies enable the ball to do work so they aren't included in its thermal energy.

2. At first, they experience attractive forces. But when they come too close, the forces become repulsive and they bounce. As they subsequently separate, the forces again become attractive.
Why: As they approach one another, the two nitrogen atoms experience attractive forces and a chemical bond begins to form between them. They accelerate toward one another, converting chemical potential energy into kinetic energy. However, when they are very close together, the forces become repulsive and they bounce off one another. They head apart and the forces again become attractive, but their kinetic energy breaks the bond and they separate forever.

3. From your hotter hand to the colder ice cube.
Why: Heat naturally flows from a hotter object to a colder object. Since the ice cube is colder than your hand, heat flows out of your hand and into the ice cube. Since your hand is losing thermal energy, its temperature drops and you sense cold. While it's tempting to think of cold as something that flows out of an ice cube, the only thing that really moves about is heat. Ice cubes are wonderful absorbers of heat and cool our drinks by reducing their thermal energies.

4. Yes, thermal energy is flowing from the hot towels, through the plastic, to the muscle.
Why: The plastic sheet is acting as a heat exchanger, allowing heat to flow from the hot towels to the cooler muscle but preventing any movement of the hot water itself.

5. The handle's thermal conductivity.
Why: Some handles are made of plastics or stainless steel, which are poor conductors of both electricity and heat. These handles normally remain cool, unless hot gases from the stove directly heat them. Other handles are made from aluminum, copper, or cast iron, which are good conductors of electricity and heat. These handles often become unbearably hot.

6. Convection occurs, with warmed air rising over the land and being replaced by cooler air from above the water. The air moving from over the water to over the land creates the breeze.
Why: Winds are giant convection currents caused by solar heating. Air rises over warm spots on the earth's surface, and surface winds blow toward those warm spots to replace the missing air.

7. Radiation transfers heat from the lamp's filament to your skin.
Why: A heat lamp emits large amounts of invisible infrared radiation. Although you can't see this radiation, you can feel it on your skin.

8. At first, both the tray and the cookies are hot. The tray feels hottest because its metal conducts heat well and can deliver more heat to your skin. But the tray has a relatively small heat capacity and its temperature plummets in the cool room air. The moist cookies have large heat capacities, so they cool relatively slowly.
Why: Metals often feel hotter or colder to the touch than insulators because they're so good at transferring heat in either direction. But materials containing water have much larger specific heats than metals and therefore maintain their temperatures better.

Section 7.2 WATER, STEAM, AND ICE

1. At the surface of the pond.
Why: Because ice is less dense than water, it floats at the surface of the pond. The pond therefore freezes from its top down. The insulating layer of ice at the top of the pond impedes the freezing of the remainder of the water so that the pond is unlikely to freeze solid. This behavior allows animals to live in the pond during winter. If ice sank, the pond would freeze solid and the animals would die. Thus ice's tendency to float has profound implications for life on our planet.

2. Liquid water releases a large amount of heat as it solidifies and this heat helps to protect the fruit from freezing.
Why: Fruit freezes at a temperature somewhat below 0 °C. As cold air removes heat from the water-coated fruit, the water begins to freeze. The liquid water gives up a great deal of heat as it solidifies and prevents the fruit's temperature from dropping below 0 °C until the water has completely turned to ice. At that point, ice's poor thermal conductivity insulates the fruit and slows its loss of heat to the colder air.

3. Their temperatures are equal.
Why: Both glasses contain mixtures of ice and water at 0 °C (32 °F). If either mixture were colder than that, its water would be freezing rapidly into ice and if either mixture were warmer than that, its ice would be melting rapidly into water.

4. The stove must operate for a long time to provide water's huge latent heat of evaporation.
Why: Converting a kettle of hot water into steam requires an enormous amount of heat. This heat separates the molecules without raising their temperature. Since the stove provides only a certain amount of heat each second, it may take many minutes to boil away the water.

5. As the warm, moist air from your mouth cools, its relative humidity increases above 100%. The water vapor condenses to form a mist of water droplets.

Why: Changes in temperature affect relative humidity. Although the warm air leaving your lungs can contain a large portion of water vapor, its relative humidity increases dramatically as it cools. At lower temperatures, condensation takes over and the invisible water vapor condenses into a mist of visible water droplets.

6. The mothballs sublime.

Why: Although mothballs do not melt at room temperature, they sublime readily. The air around them quickly becomes saturated with gaseous mothball molecules and that's why it smells so strongly.

7. The rising steam bubbles condense as they pass through water that is below its boiling temperature.

Why: Water at the bottom of the pot reaches its boiling temperature first, so bubbles of steam begin to rise from it. But water nearer the surface of the pot isn't quite hot enough to sustain those bubbles. Instead, the steam inside them condenses and they collapse completely.

8. The water in your mug was mildly superheated and adding the powered coffee nucleated boiling.

Why: Like glass itself, a glazed ceramic mug has no permanent nucleation sites to assist seed bubble formation and boiling. You're lucky. The water was only slightly superheated and when you triggered boiling by adding the powder, it boiled relatively gently and quickly cooled to its normal boiling temperature. Had the water been more severely superheated, you might be badly scalded by now.

Section 7.3 INCANDESCENT LIGHTBULBS

1. These satellites can measure the wavelength distributions of thermal radiation emitted by various patches of land and determine their temperatures.

Why: Even objects that are near room temperature emit thermal radiation, although this radiation is entirely in the infrared. While the land is not really black, the infrared light it emits is still an accurate indication of its temperature. A satellite can sense the exact form of the distribution and determine the temperature with great accuracy.

2. When it's yellow-hot, the metal loses heat quickly via radiation. However, its radiated power decreases drastically as it cools.

Why: Since the steel's radiated power is proportional to the fourth power of its absolute temperature, even a small decrease in that temperature can radically reduce how quickly it radiates away thermal energy and lowers its temperature.

3. As you heat the pot of water, the water expands more than the pot. Since the water no longer fits in the pot, it overflows.

Why: Both the pot and the water expand with increasing temperature. However, the water has a larger coefficient of volume expansion than the pot, so its volume increases more than that of the pot.

4. The inner bulb aids the tungsten recycling process, while the outer bulb diffuses the light and guards the inner bulb.

Why: The inner bulb must get hot enough for the halogen gas to react with and recycle the tungsten atoms on its surface. The outer bulb is cloudy to soften the glare and ensures that nothing touches the hot inner bulb.

Check Your Figures—Answers

Section 7.3 INCANDESCENT LIGHTBULBS

1. About 524 W.

Why: We can use Eq. 7.3.1 to obtain the radiated power:

$$\text{radiated power} = 0.5 \cdot 5.67 \times 10^{-8}\ \text{J/(s} \cdot \text{m}^2 \cdot \text{K}^4)$$
$$\cdot (310\ \text{K})^4 \cdot 2\ \text{m}^2$$
$$= 524\ \text{J/s} = 524\ \text{W}.$$

This power is much more than the thermal power your body produces while you're standing still. You'll get cold quickly.

Exercises

1. Your body is presently converting chemical potential energy from food into thermal energy at a rate of about 100 J/s, or 100W. If heat were flowing out of you at a rate of about 200 W, what would happen to your body temperature?

2. You can use a blender to crush ice cubes, but if you leave it churning too long, the ice will melt. What supplies the energy needed to melt the ice?

3. When you knead bread, it becomes warm. From where does this thermal energy come?

4. You throw a ball into a box and close the lid. You hear the ball bouncing around inside as the ball's energy changes from gravitational potential energy to kinetic energy, to elastic potential energy, and so on. If you wait a minute or two, what will have happened to the ball's energy?

5. Why do meats and vegetables cook much more quickly when there are metal skewers sticking through them?

6. Why do aluminum pans heat food much more evenly than stainless steel pans when you cook on a stove?

7. Use the concept of convection to explain why firewood burns better when it's raised above the bottom of a fireplace on a grate.

8. Explain how convection contributes to the shape of a candle flame.

9. When you hold a lighted match so that its tip is lower than its stick, the flame travels quickly up the stick. Why?

10. It's often a good idea to wrap food in aluminum foil before baking it near the red-hot heating element of an electric oven. Why does this wrapped food cook more evenly?

11. Why are black steam radiators better at heating a room than radiators that have been painted white or silver?

12. The space shuttle generates thermal energy during its operation in earth orbit. How is it able to get rid of that thermal energy as heat in an airless environment?

13. Some air fresheners are solid materials that have strong odors. If you leave these air fresheners out, they slowly disappear. What happens to the solid material?

14. Frozen vegetables will "freeze-dry" if they're left in cold, dry air. How can water molecules leave the frozen vegetables?

15. If you remove ice cubes from the freezer with wet hands, the cubes often freeze to your fingers. How can the ice freeze the water on your hands? Shouldn't they melt instead?

16. On a bitter cold day, the snow is light and powdery. This snow doesn't begin melting immediately when you bring it into a warm room. Why?

17. When you put a loaf of bread in a plastic bag, there is still air around the bread. Why doesn't the bread dry out as quickly in the bag as it does when there's no bag around it?

18. Even on a very humid day the hot air from your blow dryer can extract moisture from your hair. Why is heated air able to dry your hair when the air around you can't?

19. If you add a little hot tea to ice water at 0 °C, the mixture will end up at 0 °C so long as some ice remains. Where does the tea's extra thermal energy go?

20. If you put a warm bottle of wine in a container of ice water, the wine will cool but the ice water won't become warmer. Where is the wine's thermal energy going?

21. Why is it that you can put a metal pot filled with water on a red-hot electric stove burner without fear of damaging the pot?

22. Why does a pot of water heat up and begin boiling more quickly if you cover it?

23. Putting a clear plastic sheet over a swimming pool helps keep the water warm during dry weather. Explain.

24. If you try to cook vegetables in 100 °C air, it takes a long time. But if you cook those same vegetables in 100 °C steam, they cook quickly. Why does the steam transfer so much more heat to the vegetables?

25. Why does it take longer to cook pasta properly in boiling water in Denver (the "mile high city") than it does in New York City?

26. You're floating outside the space station when your fellow astronaut tosses you a cold bottle of mineral water. It's outside your space suit but you open it anyway. It immediately begins to boil. Why?

27. The molecules of antifreeze dissolve easily in water. Why does adding antifreeze to the water in a car radiator keep the water from freezing in the winter and boiling in the summer?

28. When the outside temperature is −2 °C (29 °F), you can melt the ice on your front walk by sprinkling salt on it. But if you are out of salt, will baking soda do?

29. A quality perfume is a mixture of essential oils and has a scent that changes with time and skin temperature. Explain the gradual change of the perfume's scent with time in terms of evaporation.

30. An icy sidewalk gradually loses its ice even when the weather stays cold. Where does the ice go?

31. How would you estimate the temperature of a glowing coal in a fireplace?

32. A lightbulb burns out when a small portion of its thinning filament overheats and vaporizes. Why is this event accompanied by a bright flash of blue-white light?

33. The strongest evidence for the Big Bang theory of the origin of the universe is the thermal radiation emitted by that explosion. This radiation has cooled over the years to only 3 K and is now mostly microwaves. Why should 3 K thermal radiation be microwaves?

34. You have a table lamp with a dimmer switch. The dimmer allows you to adjust the temperature and brightness of the lamp's incandescent bulb from very dim red up to brilliant yellow-white. How does the bulb's energy efficiency, the amount of visible light it produces per unit of power consumed, depend on the dimmer's setting?

35. When you operate a 50-100-150 W three-way bulb at its 50-W setting, it emits yellow-white light. When you use a dimmer to operate a regular 150-W bulb on only 50 W of electric power, it emits orangish light. Explain the difference.

36. Astronomers can tell the surface temperature of a distant star without visiting it. How is this done?

37. A doctor can study a patient's circulation by imaging the infrared light emitted by the patient's skin. Tissue with poor blood flow is relatively cool. What changes in the infrared emissions would indicate such a cool spot?

38. The filament of an incandescent bulb is quite small, yet it emits as much as 100 W of thermal radiation. That's almost as much as your whole body emits. What accounts for the strength of the filament's thermal radiation?

39. Why are concrete sidewalks divided into individual squares rather than being left as continuous concrete strips?

40. If you were laying steel track for a railroad, what influence would thermal expansion have on your work?

41. A difficult-to-open jar may open easily after being run under hot water for a moment. Explain.

42. Why do bridges have special gaps at their ends to allow them to change lengths?

Problems

1. If a burning log is a black object with a surface area of 0.25 m^2 and a temperature of 800 °C, how much power does it emit as thermal radiation?

2. When you blow air on the log in Problem 1, its temperature rises to 900 °C. How much thermal radiation does it emit now? Why did the 100 °C rise make so much difference?

3. If the sun has an emissivity of 1 and a surface temperature of 6000 K, how much power does each 1 m^2 of its surface emit as thermal radiation?

4. A dish of hot food has an emissivity of 0.4 and emits 20 W of thermal radiation. If you wrap it in aluminum foil, which has an emissivity of 0.08, how much power will it radiate?

5. A space vehicle uses a large black surface to radiate away waste heat. How does the amount of heat radiated away depend on the area of that radiating surface? If the surface area were doubled, how much more heat would it radiate, if any?

Thermodynamics

Heat normally flows from a hotter object to a colder object, which is why the hot sun warms your skin as you sit on the beach and why a cold winter breeze cools it as you ski down the mountain. But not everything in nature permits heat to flow passively. Our technological world includes many devices that actively transfer heat from colder objects to hotter objects or that use the flow of heat to do useful work. In this chapter, we will examine the rules governing the movement of heat, a field of physics known as thermodynamics.

EXPERIMENT: Making Fog in a Bottle

Fog occurs naturally when humid air experiences a sudden drop in temperature. To make fog in a bottle, you'll need to do the same thing: cool humid air quickly. In a room where everything is at the same temperature, cooling air sounds impossible. It's not.

To make fog in a bottle, obtain a clean 1- or 2-liter plastic soda bottle and put several spoonfuls of water in it. Cap the bottle tightly and shake it to help the water evaporate into the air. In a minute or two, the relative humidity in the bottle will reach 100% and you'll be ready to do the experiment.

Lay the bottle on its side on the floor and step on it with your shoe. You must press down hard enough to dent the bottle significantly, though you don't want to pop it or crack its sides. Wait like this for 20 or 30 seconds. Heat is flowing out of the bottle and that takes time. Now step quickly off the bottle and immediately illuminate its contents with a bright light. You should see a mist of tiny water droplets. Abruptly releasing the compression causes the air's temperature to plummet and fog forms in the bottle. *Predict*

Courtesy Lou Bloomfield

how changing the pressure or timing will affect the fog, then *observe* the result. Did you *verify* your prediction? What can you *measure* about the fog's formation and properties?

If no fog formed, despite a hard squeeze with your shoe and enough patience to let the heat flow out of the bottle, you can help the fog start by dropping a smoking match into the bottle before capping it. You want just enough tiny smoke particles to help the mist droplets form, but not enough smoke to obscure the fog itself. It doesn't take much smoke to do the trick. After capping the bottle, squeeze it hard and wait, then release the compression suddenly and look for the fog.

Chapter Itinerary

In this chapter, we'll examine the movement of heat in two everyday machines: (1) *air conditioners* and (2) *automobiles*. In *air conditioners,* we'll look at the rules governing the movement of thermal energy—the laws of thermodynamics—and see how those rules permit an air conditioner to use ordered electric energy to transfer heat from the cold room air to the hot outdoor air. In *automobiles,* we'll investigate the ways in which thermal energy can be used to do work and see how the engine is able to convert thermal energy into work as heat flows from the hot burning fuel to the cold outdoor air. For additional preview information, skip ahead to the Chapter Summary on p. 258.

The principles illustrated by these two machines are also found in many other places. Refrigerators and home heat pumps use ordered energy to transfer heat from cold objects to hot objects while steam engines and hot air balloons use the natural flow of heat to make things move. Once you understand the concepts behind these *heat pumps* and *heat engines,* you'll see that they're quite common in the world around you.

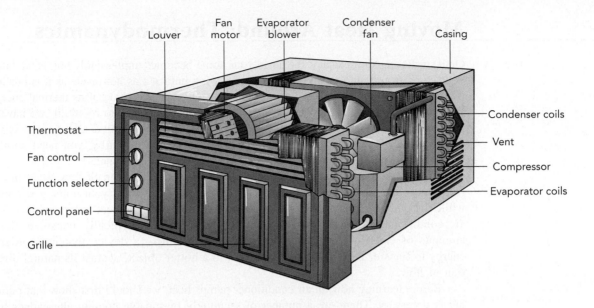

Louver   Fan motor   Evaporator blower   Condenser fan   Casing

Thermostat

Fan control

Function selector

Control panel

Grille

Condenser coils

Vent

Compressor

Evaporator coils

SECTION 8.1 **Air Conditioners**

On a summer day, your problem isn't staying warm; it's keeping cool. Instead of looking for something to burn in your woodstove, you turn on your air conditioner. An air conditioner is a device that cools room air by removing some of its thermal energy. But the air conditioner can't make thermal energy disappear. Instead, it transfers thermal energy from the cooler room air to the warmer air outside. Since the air conditioner transfers heat against its natural direction of flow, the air conditioner is a "heat pump." It's also a classic illustration of the laws of thermodynamics in action.

Questions to Think About: Why doesn't heat naturally flow from a colder object to a hotter object? An air conditioner removes thermal energy from room air, so why does the air conditioner require electric energy to operate? Where does this electric energy go? Why does an air conditioner always have an indoor component and an outdoor component? If you put a window air conditioner in the middle of a room and turn it on, what will happen to the temperature of the room?

Experiments to Do: Take a look at a window air conditioner. If you can't find one, examine a refrigerator instead because it's basically a powerful air conditioner cooling a food-storage closet. As the air conditioner (or refrigerator) operates, feel the air leaving the indoor air vent (or inside the refrigerator) and compare its temperature to that of air leaving the outdoor air vent (or near the metal coils on the back of the refrigerator). Which way does the cooling mechanism move heat? If that mechanism were absent, which way would heat flow? Turn off the air conditioner or refrigerator and observe the resulting heat flow. Did you confirm your prediction?

Moving Heat Around: Thermodynamics

On a sweltering summer day, the air in your home becomes unpleasantly hot. Heat enters your home from outdoors and doesn't stop flowing until it's as hot inside as it is outside. You can make your home more comfortable by getting rid of some of its thermal energy. But while we've already looked at ways to *add* thermal energy to room air, we haven't yet learned how to *remove* it. At present, the only cooling method we've discussed is contact with a colder object. Unless you have an icehouse nearby, you need another scheme for eliminating thermal energy. You need an air conditioner.

An air conditioner transfers heat against its natural direction of flow. Heat moves from the colder air in your home to the hotter air outside, so that your home gets colder while the outdoor air gets hotter. There's a cost to transferring heat in this manner. The air conditioner requires ordered energy to operate and typically consumes large amounts of electric energy. It's a type of **heat pump**—a device that uses ordered energy to transfer heat from a colder object to a hotter object, against its natural direction of flow.

Before learning how an air conditioner pumps heat, we should first show that pumping is necessary. There are a number of seemingly reasonable cooling alternatives that we should consider before turning to air-conditioning. Three such alternatives are

1. Letting heat flow from your home to your neighbor's home.
2. Destroying some of your home's thermal energy.
3. Converting some of your home's thermal energy into electric energy.

Unfortunately, these three alternatives can't be done. Still, it will be useful for us to examine them more closely because in doing so, we'll learn about the laws governing the movement of thermal energy, the **laws of thermodynamics.**

The first alternative raises an interesting issue. Your home is in thermal equilibrium with the outdoor air, meaning that no heat flows from one to the other and they're at the same temperature. Your neighbor's home is also in thermal equilibrium with the outdoor air. What will happen if you permit heat to flow between your home and your neighbor's home? Nothing. Since both homes are simultaneously in thermal equilibrium with the outdoor air, they're also in thermal equilibrium with one another. All three are at the same temperature.

This observation is an example of the **zeroth law of thermodynamics,** which says that two objects that are each in thermal equilibrium with a third object are also in thermal equilibrium with one another. This seemingly obvious law is the basis for a meaningful system of temperatures. If you had a roomful of objects at 35 °C (95 °F) and some were in thermal equilibrium with one another while others were not, then "being at 35 °C" wouldn't mean much. However, every object that has a temperature of 35 °C is in thermal equilibrium with every other object at 35 °C. The zeroth law is observed to be true in nature so that temperature does have meaning. And since your neighbor's home is just as hot as yours, they can relax because you're not going to be sending them any extra heat.

> **The Zeroth Law of Thermodynamics**
> Two objects that are each in thermal equilibrium with a third object are also in thermal equilibrium with one another.

The second alternative sounds unlikely from the outset. We've known since the first chapter that energy is special, that it's a conserved quantity. You can't cool your home by destroying thermal energy because energy can't be destroyed. To eliminate thermal energy, you must convert it to another form or transfer it elsewhere.

This concept of energy conservation is the basis for the **first law of thermodynamics,** which states that the change in a stationary object's internal energy is equal to the heat transferred into that object minus the work that object does on its surroundings. An object's **internal energy** is the sum of its thermal energy and any additional potential energy stored entirely within the object. This law says that heat added to the object increases its internal energy while work done by the object decreases its internal energy. In other words, since energy is conserved, the only way the object's internal energy can change is by transferring energy as heat or work. The first law of thermodynamics can be written as a word equation:

change in object's internal energy = heat added to object − work done by object, **(8.1.1)**

in symbols,

$$\Delta U = Q - W,$$

and in everyday language:

You can add energy to a ball as heat by cooking it or as work by squeezing it.

The First Law of Thermodynamics
The change in a stationary object's internal energy is equal to the heat transferred into that object minus the work that object does on its surroundings.

▶ Check Your Understanding #1: Stirring Up Trouble

for answers, see page 259

If you put cold water into a blender and mix it rapidly for several minutes, the water will become warm. From where does the additional thermal energy come?

▶ Check Your Figures #1: Powerful You

for answers, see page 261

You have been pedaling the exercise bicycle steadily for 20 minutes at a power of 200 watts. How much thermal power is the bicycle emitting into the room?

Disorder, Entropy, and the Second Law

The third alternative looks much more promising than the first two. It seems as though you should be able to convert thermal energy into electricity (or some other ordered form of energy). You could then sell it back to the electric company and get credit on your bill. Wouldn't that be great?

But there's a problem. Ordered energy and thermal energy aren't equivalent. You can easily convert ordered energy into thermal energy but the reverse is much harder. For example, you can burn a log to convert its chemical potential energy into thermal energy, but you'll have trouble converting that thermal energy back into chemical potential energy to recreate the log.

The basic laws of motion are silent on this issue. It isn't that the smoke doesn't have the energy to recreate the log. It's that the individual smoke particles must pool their thermal energies together to carry out the reassembly, a remarkably unlikely event. The particles would all have to move in just the right ways to turn the burned gases back into wood and oxygen, an incredible coincidence that simply never happens. Similarly, all of the air particles in your home would have to act together to convert their thermal energy into electricity. Since that coordinated behavior is ridiculously improbable, you're not going to be selling power to the electric company any time soon.

Once energy has been scattered randomly among the individual air particles, you can't collect that energy back together again. Creating disorder out of order is easy, but recovering order from disorder is nearly impossible. As a result, systems that begin with some amount of order gradually become more and more disordered, never the other way around. The best they can do is to stay the same for a while so that their disorder doesn't change. From these observations, we can state that the disorder of an isolated system *never decreases.*

This notion of never decreasing disorder is one of the central concepts of thermal physics. There is even a formal measure of the total disorder in an object: **entropy.** All disorder contributes to an object's entropy, including its thermal energy and its structural defects. Breaking a window or heating it both increase its entropy. Although its name sounds similar to energy, don't confuse energy and entropy. Energy is a conserved quantity, while entropy is a quantity that can and generally does increase. It's easy to make more entropy.

Because disorder never decreases, the third cooling alternative is impossible. Turning your home's thermal energy into electric energy would reduce its disorder and decrease its entropy. But our observations about entropy aren't yet complete. There's one way to decrease your home's entropy: you can export that entropy somewhere else. In fact, you export entropy every time you take out the garbage, though that action also changes the contents of your home. You can also export entropy without modifying your home's contents by transferring heat somewhere else. Heat carries disorder and entropy with it, so getting rid of heat also gets rid of entropy.

Our rule about entropy never decreasing is weakened by the possibility of exchanging heat and entropy between objects. Before asserting that an object or system of objects can't decrease its entropy, we must ensure that it's thermally isolated from its surroundings so that it can't export its entropy. With that in mind, the strongest statement that we can make concerning entropy is that the entropy of a thermally isolated system of objects never decreases. This observation is the **second law of thermodynamics.**

The Second Law of Thermodynamics
The entropy of a thermally isolated system of objects never decreases.

Because of the second law, the only way to cool your home is to export its thermal energy and entropy elsewhere. Such a transfer would be easy if you had a cold object nearby to receive the heat. But lacking a cold object, you must use an air conditioner. Like all heat pumps, an air conditioner transfers heat and entropy in such a way that the second law of thermodynamics is never violated and the entropy of each thermally isolated system of objects never decreases. As we'll see, the air conditioner lowers the entropy of your home but raises the entropy of the outdoor air even more so that, overall, the entropy of the world actually increases.

There's a limit to how much entropy an air conditioner can remove from your home. As it exports thermal energy and entropy, the air conditioner lowers your home's temperature. In principle, your home will eventually approach absolute zero and, as it does, its entropy will approach zero. This relationship between the zero of temperature and the zero of entropy is the **third law of thermodynamics,** which states that as an object's temperature approaches absolute zero, its entropy approaches zero. The third law establishes absolute zero as a destination with no disorder left, but the second law ultimately makes it impossible to extract all the disorder from an object. Because absolute zero is unattainable, the third law refers to *approaching* it rather than to *arriving* at it.

> **The Third Law of Thermodynamics**
> As an object's temperature approaches absolute zero, its entropy approaches zero.

Check Your Understanding #2: Something for Nothing

for answers, see page 260

People have tried for centuries to build machines that provide endless outputs of useful, ordered energy without any inputs of ordered energy. Unfortunately, such perpetual motion machines violate the laws of thermodynamics. If such a machine is thermally isolated, which law does it violate? What about if it's not thermally isolated?

Pumping Heat against Its Natural Flow

While the second law of thermodynamics doesn't allow the entropy of a thermally isolated system to decrease, it does permit the objects in that system to redistribute their individual entropies. One object's entropy can decrease as long as the entropy of the rest of the system increases by at least as much. Such entropy redistribution allows part of the system to become colder if the rest of the system becomes hotter.

For example, suppose that there's a pond of cold water behind your home. You pump that water through your bathtub and let it draw heat out of the room air. Your home becomes colder while the pond becomes warmer. This transfer of heat from the hot air in your home to the cold water in the pond satisfies the second law of thermodynamics. The entropy of the combined system—your home and the pool of water—doesn't decrease. In fact, it actually increases!

This entropy increase occurs because heat is more disordering to cold objects than it is to hot objects. Each joule of heat that flows from your home to the pool creates more disorder in the pool than it creates order in your home.

A useful analog for this effect involves two parties taking place simultaneously: the garden society's annual tea party and a 4 year old's birthday party. The orderly tea party represents the cold pool while the disorderly birthday party represents your hot home. The analogy to letting heat flow from your hot home to the cold pool is to trade one lively 4 year old from the disorderly birthday party for one quiet octogenarian from the orderly tea party. This exchange will reduce the birthday party's disorder only slightly, but it will dramatically increase the disorder of the tea party. The attendance at each party will be unchanged but their total disorder will increase.

When heat flows from your home to the pool, the overall entropy increases and the second law is more than satisfied. A similar increase in entropy occurs whenever heat flows from a hot object to a cold object, which is why heat normally flows in that direction.

But an air conditioner does the seemingly impossible: it transfers heat from a cold object, your home, to a hot object, the outdoor air. This heat flows in the wrong direction and the disorder it creates by entering the hot outdoor air is less than the disorder it removes by leaving the cold indoor air. It's like returning the tea party's lone 4 year old to the birthday party in exchange for the elderly garden fancier—the birthday party becomes only a tiny bit more disorderly while the garden party becomes much more orderly, so the net disorder of the two gatherings decreases substantially. Similarly, if nothing else happened when the air conditioner moved heat from the cold indoor air to the hot outdoor air, the entropy of the combined system would decrease and the second law of thermodynamics would be violated!

However, we've omitted an important feature of the air conditioner's operation: the electric energy it consumes. The air conditioner converts this ordered energy into thermal energy and delivers it as additional heat to the outdoor air (Fig. 8.1.1). In doing so, the air conditioner creates enough extra entropy to ensure that the overall entropy of the combined system increases. The second law is satisfied after all.

The amount of ordered energy the air conditioner consumes depends on the indoor and outdoor temperatures. If the two are close in temperature, the transfer of heat reduces the entropy only slightly so the air conditioner doesn't need to convert much ordered energy into thermal energy. But if they are far apart in temperature, the air conditioner must create lots of extra entropy to make up for the entropy lost in the transfer.

This requirement that entropy not decrease explains why an air conditioner works best when it's cooling your home the least. The greater the temperature difference between the indoor air and the outdoor air, the more electric energy or other form of work the air conditioner must consume to transfer each joule of heat. For an ideally efficient air conditioner or other heat pump, the relationships between the work consumed, the heat removed from the cold object, and the heat added to the hot object can be written as word equations:

$$\text{heat removed from cold object} = \text{work consumed} \cdot \frac{\text{temperature}_{cold}}{\text{temperature}_{hot} - \text{temperature}_{cold}}$$

$$\text{(8.1.2)}$$

$$\text{heat added to hot object} = \text{heat removed from cold object} + \text{work consumed,}$$

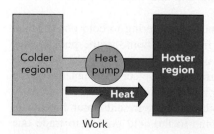

Fig. 8.1.1 A heat pump transfers heat from a colder region to a hotter region. In doing so, it converts some work (ordered energy) into heat (thermal energy in the hotter region). The larger the temperature difference between the two regions, the more work is required to transfer each joule of heat.

in symbols:

$$-Q_c = W \frac{T_c}{T_h - T_c},$$

$$Q_h = -Q_c + W,$$

and in everyday language:

The closer the two temperatures are, the easier it is to move heat from cold to hot.

The hot object receives not only the heat removed from the cold object, but also an amount of heat equal to the work consumed by the transfer. Note also that the work needed to remove heat from your home approaches infinity as its temperature approaches absolute zero; that's why absolute zero is unattainable.

Sadly, a practical air conditioner never reaches ideal efficiency, so it moves less heat than promised by Eqs. 8.1.1. Moreover, heat leaks back into your home through its walls at a rate that's roughly proportional to the temperature difference. No wonder your electric bill soars when you turn down the thermostat too far!

▶ Check Your Understanding #3: Heat Pumps in Cold Weather

for answers, see page 260

Homes located in mild climates are often heated by heat pumps during the winter. Home heat pumps are essentially air conditioners run backwards. They extract heat from the cold outdoor air and release it to the warm indoor air. Why are heat pumps most effective in mild weather, when the outdoor air isn't too cold?

▶ Check Your Figures #2: A Hot House for Cheap

for answers, see page 261

Your house is heated by an ideally efficient heat pump that uses ordered energy to pump heat from the colder outdoor air to the warmer indoor air. On a winter day, the outdoor temperature is 270 K ($-3\,°C$ or 26 °F) and the indoor temperature is 300 K (27 °C or 80 °F). To deliver 1000 J of heat to the indoor air, how much electric energy must the heat pump consume?

How an Air Conditioner Cools the Indoor Air

Having determined the air conditioner's goals, it's time for us to look at how a real air conditioner meets them. In most cases, the air conditioner uses a fluid to transfer heat from the colder indoor air to the hotter outdoor air. Known as the *working fluid,* this substance absorbs heat from the indoor air and releases that heat to the outdoor air.

The working fluid flows in a looping path through the air conditioner's three main components: an evaporator, a condenser, and a compressor (Fig. 8.1.2). The evaporator

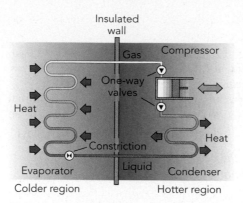

Fig. 8.1.2 A typical air conditioner transfers heat from colder air to hotter air by condensing a gas to a liquid in the hotter region and evaporating the liquid to a gas in the colder region. A compressor provides the necessary input of ordered energy.

is located indoors, where it transfers heat from the indoor air to the working fluid (Fig. 8.1.3). The condenser is located outdoors, where it transfers heat from the working fluid to the outdoor air. And the compressor is also located outdoors, where it squeezes the working fluid and does the work needed to move heat against its natural flow. To see how these three components pump heat out of your home, let's look at them individually.

We'll begin with the evaporator, a long metal pipe that's decorated with thin metal fins. The evaporator is a heat exchanger that allows heat to flow from the warm air around it to the cool working fluid inside it. Its fins provide additional surface area to facilitate that heat flow and a fan blows the indoor air rapidly past the fins so that heat moves quickly into the working fluid.

True to its name, the evaporator allows the working fluid inside it to evaporate. That's why the working fluid becomes so cool and absorbs so much heat. Like any evaporating substance, the liquid working fluid needs its latent heat of evaporation to separate its molecules from one another and become a gas. The working fluid obtains that latent heat of evaporation from its own thermal energy, so its temperature drops and heat then flows into it through the walls of the evaporator. By the time the gaseous working fluid leaves the evaporator, it has absorbed a great deal of the indoor air's thermal energy and carries that energy with it as chemical potential energy.

To make the liquid working fluid evaporate in the evaporator, the air conditioner abruptly reduces the fluid's pressure. Recall from Chapter 7 that phase transitions such

Courtesy Bryant Heating and Cooling

Fig. 8.1.3 The evaporator of this central air-conditioning unit extracts heat from the indoor air. This heat is transferred outside, where the compressor and condenser release it into the outdoor air.

as evaporation depend on molecular landing and leaving rates. When both the liquid and the gaseous phases are present, molecules are continually leaving the liquid for the gas and landing on the liquid from the gas. If leaving dominates, the liquid evaporates and the gas is stable. If landing dominates, gas condenses and the liquid is stable. And if the two balance one another, the liquid and gas can coexist in phase equilibrium.

As it flows through a pipe toward the evaporator, the working fluid's pressure is high and it's stable as a liquid. At this high pressure, any gaseous working fluid would be so dense that landing would dominate leaving and the gas would condense. Thus the pipe carries only high-pressure liquid working fluid to the evaporator.

But just before reaching the evaporator, the liquid working fluid passes through a narrow constriction in the pipe and its pressure drops dramatically. The low-pressure working fluid that results isn't stable as a liquid. At low pressure, any gaseous working fluid is so dilute that leaving dominates landing and the liquid evaporates. Thus as the low-pressure liquid working fluid emerges from the constriction and pours into the evaporator, it's evaporating rapidly. It continues to evaporate even though its temperature decreases as it absorbs its latent heat of evaporation.

By the time the working fluid emerges from the evaporator, it has evaporated completely and has absorbed considerable thermal energy from the indoor air. It leaves the evaporator as a cool low-pressure gas and travels through a pipe toward the compressor.

Half the air conditioner's job is done: it has removed heat from the indoor air. But the remaining half of its job is more complicated: it must add heat to the outdoor air while ensuring that the total entropy of the combined system doesn't decrease. After all, there's no getting around the second law of thermodynamics.

► Check Your Understanding #4: Cooling at the Gas Grill

for answers, see page 260

When liquid propane evaporates into a gas in the tank of a propane grill, the tank that contains that liquid propane becomes colder. Why?

How an Air Conditioner Warms the Outdoor Air

Satisfying the second law is the task of the compressor. The compressor receives low-pressure gaseous working fluid from the evaporator, compresses it to much higher density, and delivers it as a high-pressure gas to the condenser. The compressor may use a piston and one-way valves, like the water pump in Fig. 5.2.3, or it may use a rotary pumping mechanism. But regardless of how it functions, the result is the same: the gaseous working fluid undergoes a dramatic increase in density and pressure as it passes through the compressor.

Compressing a gas requires work because the compressor must push the gas inward while moving it inward—force times distance. Since work transfers energy, the compressor increases the energy of the gas. The air conditioner usually obtains this energy from the electric company and converts it into mechanical work with an electric motor.

In accordance with the first law of thermodynamics, this work increases the internal energy of the working fluid. The only way that the gaseous working fluid can store this additional energy is as thermal energy in its individual particles. These particles begin to move about more and more rapidly, so that the gaseous working fluid leaves the compressor much hotter than when it arrived. There is no getting around that temperature rise; compressing the working fluid unavoidably raises its temperature.

This hot, high-pressure working fluid then flows into the condenser. Like the evaporator, the condenser is a long metal pipe with fins attached to it. It acts as a heat exchanger and its metal fins provide extra surface area to speed the flow of heat from the hotter working fluid inside it to the less hot outdoor air. There may also be a fan to move outdoor air quickly past the condenser and speed up the heat transfer.

As its name suggests, the condenser allows the gaseous working fluid inside it to condense into a liquid. Compression prompts that condensation. While it flows through a pipe toward the compressor, the low-pressure working fluid is stable as a gas. But in the high-pressure, high-density working fluid that emerges from the compressor, the landing rate far outpaces the leaving rate and the gas condenses.

Like any condensing substance, the gaseous working fluid releases its latent heat of evaporation as its molecules bind together and it transforms into a liquid. This latent heat of evaporation becomes thermal energy in the working fluid, so its temperature rises still further and heat flows out of it through the walls of the condenser.

By the time the liquid working fluid leaves the condenser, it has transformed a great deal of chemical potential energy into thermal energy and released that energy into the outdoor air. The outdoor air receives as heat not only the thermal energy extracted from the indoor air, but also the electric energy consumed by the compressor. The working fluid leaves the condenser as a warm, high-pressure liquid and travels through a pipe toward the evaporator.

The second half of the air conditioner's job is now complete: it has released heat into the outdoor air and, in the process, converted ordered energy into thermal energy. From here, the working fluid returns to the evaporator to begin the cycle all over again. The working fluid passes endlessly around the loop, extracting heat from the indoor air in the evaporator and releasing it to the outdoor air in the condenser. The compressor drives the whole process and thereby satisfies the second law of thermodynamics. The same technique is used to extract heat from the air inside a refrigerator and to release that heat to the room air (Fig. 8.1.4).

Before leaving air conditioners, we should take a moment to look at the working fluid itself. This fluid must become a gas at low pressure and a liquid at high pressure, over most of the temperature range encountered by the air conditioner. For decades, the standard working fluids were *chlorofluorocarbons* such as the various Freons. These compounds replaced ammonia, a toxic and corrosive gas used in early refrigeration.

Chlorofluorocarbons are ideally suited to air conditioners because they transform easily from gas to liquid and back again over a broad range of temperatures. They're also chemically inert and inexpensive. Unfortunately, chlorofluorocarbon molecules contain chlorine atoms and, when released into the air, can carry those chlorine atoms to the upper atmosphere. There they promote the destruction of ozone molecules, essential atmospheric constituents that absorb portions of the sun's ultraviolet radiation. Recently, chlorine-free *hydrofluorocarbons* have replaced chlorofluorocarbons as the working fluids in most air conditioners. Though not as energy efficient and chemically inert as the materials they replace, hydrofluorocarbons appear to be safe for the environment.

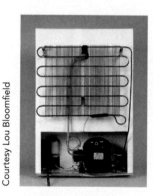

Courtesy Lou Bloomfield

Fig. 8.1.4 The compressor (bottom) and condenser coils (top) are visible on the back of this refrigerator. The compressor squeezes the working fluid into a hot, dense gas and delivers it to the condenser. There it gives up heat to the room air and condenses into a liquid. The working fluid evaporates inside the refrigerator, extracting heat from the food.

▶ Check Your Understanding #5: Air-Conditioning or Space Heating?
for answers, see page 260

What would happen to the average air temperature in your room if you placed a window air-conditioning unit in the middle of the room and turned it on?

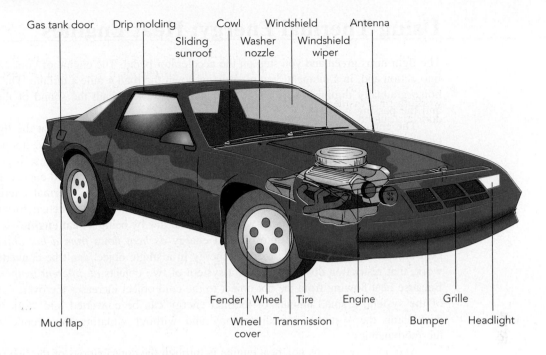

Gas tank door   Drip molding   Cowl   Windshield   Antenna
                        Sliding   Washer   Windshield
                        sunroof   nozzle   wiper

Fender | Wheel | Tire   Engine   Grille
       Wheel   Transmission   Bumper   Headlight
       cover

Mud flap

SECTION 8.2 **Automobiles**

Nothing is more symbolic of freedom and personal independence than an automobile. With its keys in your hand, you can go almost anywhere at a moment's notice. The mechanism that makes this instant transportation possible is the internal combustion engine. Though it has been refined over the years, this engine's basic design has changed little since it was invented more than a century ago. It uses thermal energy released by burning fuel to do the work needed to propel the car forward. That thermal energy can do work at all is one of the marvels of thermal physics and the primary focus of this section.

Questions to Think About: What obstacles stand in the way of using burning fuel's thermal energy to propel a car? Why are two objects, one hot and one cold, required in order to convert any thermal energy into useful work? What hot and cold objects does a car have? Why does a car have a cooling system to get rid of waste heat, rather than just converting it all into useful work?

Experiments to Do: The recent advances in automobile technology and the increasing demands for pollution control equipment have made automobiles exceedingly complicated. Nonetheless, take a moment to look under the hood of your or a friend's car. You should be able to identify the engine and its electric support system. You should be able to count four or more spark plug wires heading toward the engine's cylinders. These cylinders convert thermal energy from burning fuel into work to propel the car. Why does the engine need so many cylinders, rather than relying on one larger cylinder?

At the front of the engine compartment, you'll find the radiator. How does this device extract waste heat from the engine? Does heat flow naturally into the radiator and then into the outdoor air, or is there a heat pump involved?

Using Thermal Energy: Heat Engines

The light turns green and you step on the accelerator pedal. The engine of your car roars into action and, in a moment, you're cruising down the road a mile a minute. The engine noise gradually diminishes to a soft purr and vanishes beneath the sound of the radio and the passing wind.

The engine is the heart of the automobile, pushing the car forward at the light and keeping it moving against the forces of gravity, friction, and air resistance. It's not simply a miracle of engineering. It's also a wonder of thermal physics because it performs the seemingly impossible task of converting thermal energy into ordered energy. But the second law of thermodynamics forbids the direct conversion of thermal energy into ordered energy, so how can a car engine use burning fuel to propel the car forward?

The car engine avoids conflict with the second law by being a **heat engine**—a device that converts thermal energy into ordered energy *as heat flows from a hot object to a cold object* (Fig. 8.2.1). While thermal energy in a single object can't be converted into work, that restriction doesn't apply to a system of two objects *at different temperatures*. Because heat flowing from the hot object to the cold object increases the overall entropy of the system, a small amount of thermal energy can be converted into work without decreasing the system's overall entropy and without violating the second law of thermodynamics.

Another way to look at a heat engine is through the contributions of the two objects. The hot object provides the thermal energy that's converted into work. The cold object provides the order needed to carry out that conversion. As the heat engine operates, the hot object loses some of its thermal energy and the cold object loses some of its order. The heat engine has used them to produce ordered energy. Since the heat engine needs both thermal energy and order, it can't operate if either the hot or the cold object is missing.

In a car engine, the hot object is burning fuel and the cold object is outdoor air. Some of the heat passing from the burning fuel to the outdoor air is diverted and becomes the ordered energy that propels the car. But what limits the amount of thermal energy the engine can convert into ordered energy?

To answer that question, let's examine a simplified car engine. We'll treat the burning fuel and outdoor air as a single, thermally isolated system and look at what happens to their total entropy as the engine operates. In accordance with the second law of thermodynamics, this total entropy can't decrease while the engine is transforming some thermal energy into ordered energy.

When the car is idling at a stoplight, its engine is doing no work and heat is simply flowing from the hot burning fuel to the cold outdoor air. The system's total entropy increases because this heat is more disordering to the cold air it enters than to the hot burning fuel it leaves. In fact, the system's entropy increases dramatically because the burning fuel is extremely hot compared to the cold outdoor air.

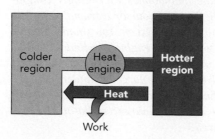

Fig. 8.2.1 A heat engine converts heat (thermal energy from the hotter region) into work (ordered energy) as heat flows from a hotter region to a colder region. The larger the temperature difference between the two regions, the larger the fraction of heat that can be converted into work.

This increase in the system's entropy is unnecessary and wasteful. The second law of thermodynamics only requires that the engine add as much entropy to the cold outdoor air as it removes from the hot burning fuel. Since a little heat is quite disordering to cold air, the car engine can deliver much less heat to the outdoor air than it removes from the burning fuel and still not cause the system's total entropy to decrease. As long as the engine delivers enough heat to the outdoor air to keep the total entropy from decreasing, there's nothing to prevent it from converting the remaining heat into ordered energy!

This conversion starts as soon as you remove your foot from the brake and begin to accelerate forward. Instead of transferring all of the thermal energy in the burning fuel to the outdoor air, your car then extracts some of it as ordered energy and uses it to power the wheels. The car engine can convert thermal energy into ordered energy, as long as it passes along enough heat from the hot object to the cold object to satisfy the second law of thermodynamics.

Obeying the second law becomes easier as the temperature difference between the two objects increases. When the temperature difference is extremely large, as it is in an automobile engine, a large fraction of the thermal energy leaving the hot object can be converted into ordered energy—at least in theory. For an ideally efficient automobile engine or other heat engine, the relationships between the heat removed from the hot object, the heat added to the cold object, and the work provided can be written as word equations:

$$\text{work provided} = \text{heat removed from hot object} \cdot \frac{\text{temperature}_{hot} - \text{temperature}_{cold}}{\text{temperature}_{hot}}$$

$$(8.2.1)$$

$$\text{heat added to cold object} = \text{heat removed from hot object} - \text{work provided},$$

in symbols:

$$W = -Q_h \frac{T_h - T_c}{T_h}$$

$$Q_c = -Q_h - W,$$

and in everyday language:

The greater the temperature difference between hot and cold, the larger the fraction of heat you can divert and transform into work.

Unfortunately, theoretical limits are often hard to realize in actual machines, and the best automobile engines extract only about half the ordered energy specified by Eq. 8.2.1. Still, obtaining even that amount is a remarkable feat and a tribute to scientists and engineers who, in recent years, have labored to make automobile engines as energy efficient as possible.

▶ Check Your Understanding #1: Heat Pumps and Heat Engines
for answers, see page 260

An air conditioner uses electric energy to make the air in your home colder than the outdoor air. Could you use this difference in temperatures to operate a heat engine and generate electric energy?

Check Your Figures #1: Back to the Steam Age

for answers, see page **261**

A train locomotive is powered by an ideal steam engine. If the steam boiler operates at 450 K (177 °C or 350 °F) and the outdoor temperature is 300 K (26 °C or 80 °F), how much work can the steam engine obtain by removing 1200 J of heat from the boiler?

The Internal Combustion Engine

Invented by the German engineer Nikolaus August Otto in 1867, an internal combustion engine burns fuel directly in the engine itself. Gasoline and air are mixed and ignited in an enclosed chamber. The resulting temperature rise increases the pressure of the gas and allows it to perform work on a movable surface.

To extract work from the fuel, the internal combustion engine must perform four tasks in sequence:

1. It must introduce a fuel–air mixture into an enclosed volume.
2. It must ignite that mixture.
3. It must allow the hot burned gas to do work on the car.
4. It must get rid of the exhaust gas.

In the standard, four-stroke fuel-injected engine found in modern gasoline automobiles, this sequence of events takes place inside a hollow cylinder (Fig. 8.2.2). It's called a "four-stroke" engine because it operates in four distinct steps or strokes: induction, compression, power, and exhaust. "Fuel-injected" refers to the technique used to mix the fuel and air as they're introduced into the cylinder.

Automobile engines usually have four or more of these cylinders. Each cylinder is a separate energy source, closed at one end and equipped with a movable piston, several valves, a fuel injector, and a spark plug. The piston slides up and down in the cylinder, shrinking or enlarging the cavity inside. The valves, located at the closed end of the cylinder, open to introduce fuel and air into the cavity or to permit burned exhaust gas

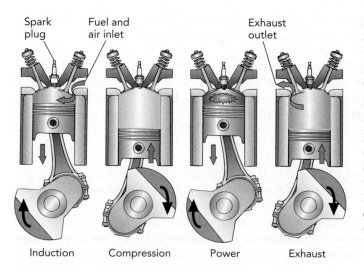

Spark plug   Fuel and air inlet   Exhaust outlet

Induction   Compression   Power   Exhaust

Fig. 8.2.2 A four-stroke engine cylinder. During the induction stroke, fuel and air enter the cylinder. The compression stroke squeezes that mixture into a small volume. The spark plug ignites the mixture and the power stroke allows the hot gas to do work on the automobile. Finally, the exhaust stroke ejects the exhaust gas from the cylinder.

to escape from the cavity. The fuel injector adds fuel to the air as it enters the cylinder. And the spark plug, also located at the closed end of the cylinder, ignites the fuel–air mixture to release its chemical potential energy as thermal energy.

The fuel–air mixture is introduced into each cylinder during its induction stroke. In this stroke, the engine pulls the piston away from the cylinder's closed end so that its cavity expands to create a partial vacuum. At the same time, the cylinder's inlet valves open so that atmospheric pressure can push fresh air into the cylinder. The cylinder's fuel injector adds a mist of fuel droplets to this air so that the cylinder fills with a flammable fuel–air mixture. Because it takes work to move air out of the way and create a partial vacuum, the engine does work on the cylinder during the induction stroke.

At the end of the induction stroke, the inlet valves close to prevent the fuel–air mixture from flowing back out of the cylinder. Now the compression stroke begins. The engine pushes the piston toward the cylinder's closed end so that its cavity shrinks and the fuel–air mixture becomes denser. Because it takes work to compress a gas, the engine does work on the mixture during the compression stroke. In accordance with the first law of thermodynamics, this work increases the internal energy of the fuel–air mixture. That gaseous mixture can't store the added energy in potential form, so its thermal energy rises and it becomes hotter. Since increases in a gas's density and temperature both increase its pressure, the pressure in the cylinder rises rapidly as the piston approaches the spark plug.

At the end of the compression stroke, the engine applies a high-voltage pulse to the spark plug and ignites the fuel–air mixture. The mixture burns quickly to produce hot, high-pressure burned gas, which then does work on the car during the cylinder's power stroke. In that stroke, the gas pushes the piston away from the cylinder's closed end so that its cavity expands and the burned gas becomes less dense. Since the hot gas exerts a huge pressure force on the piston as it moves outward, it does work on the piston and ultimately propels the car. As it does work, the burned gas gives up thermal energy and cools in accordance with the first law of thermodynamics. Its density and pressure also decrease. At the end of the power stroke, the exhaust gas has cooled significantly and its pressure is only a few times atmospheric pressure. The cylinder has extracted much of the fuel's chemical energy as work.

The cylinder gets rid of the exhaust gas during its exhaust stroke. In this stroke, the engine pushes the piston toward the closed end of the cylinder while the cylinder's outlet valves are open. Because the burned gas trapped inside the cylinder at the end of the power stroke is well above atmospheric pressure, it accelerates out of the cylinder the moment the outlet valves open. These sudden bursts of gas leaving the cylinders create the "poof-poof-poof" sound of a running engine. Without a muffler on its exhaust pipes, the engine would be loud and unpleasant.

Just opening the outlet valves releases most of the exhaust gas, but the rest is squeezed out as the piston moves toward the cylinder's closed end. The engine again does work on the cylinder as it squeezes out the exhaust gas. At the end of the exhaust stroke, the cylinder is empty and the outlet valves close. The cylinder is ready to begin a new induction stroke.

Check Your Understanding #2: Getting Out More than You Put In
for answers, see page 260

Why does the burned gas do more work on the piston during the power stroke than the piston does on the unburned fuel–air mixture during the compression stroke?

Engine Efficiency

The goal of an internal combustion engine is to extract as much work as possible from a given amount of fuel. In principle, all of the fuel's chemical potential energy can be converted into work because both are ordered energies. But it's difficult to convert chemical potential energy directly into work, so the engine burns the fuel instead. This step is unfortunate, for in burning the fuel, the engine converts the fuel's chemical potential energy directly into thermal energy and produces lots of unnecessary entropy.

But all is not lost. Since the burned fuel is extremely hot, a good fraction of its thermal energy can be converted into ordered energy by diverting some of the heat that flows from the burned fuel to the outdoor air. As we noted earlier, the hotter the burned fuel and the colder the outdoor air, the more ordered energy the engine can extract. To maximize its fuel efficiency, an internal combustion engine obtains the hottest possible burned gas, lets that gas do as much work as it can on the piston, and releases the gas at the coldest possible temperature.

It would be wonderful if, during the power stroke, the burned gas expanded and cooled until it reached the temperature of the outdoor air. The exhaust gas would then leave the engine with the same amount of thermal energy it had when it arrived, and the engine would have extracted all of the fuel's chemical potential energy as work. Unfortunately, that would violate the second law of thermodynamics by converting thermal energy completely into ordered energy. As Eqs. 8.2.1 indicate, an operating heat engine always adds some heat to its cold object. In this case, the engine releases the burned gas before it cools to the temperature of the outdoor air. It has no choice; the engine's exhaust must be hot!

But a real internal combustion engine wastes energy and extracts less work than the second law allows. For example, some heat leaks from the burned gas to the cylinder walls and is removed by the car's cooling system. This wasted heat isn't available to produce work. Similarly, sliding friction in the engine wastes mechanical energy and necessitates an oil-filled lubricating system. Overall, a real internal combustion engine converts only about 20% to 30% of the fuel's chemical potential energy into work.

Check Your Understanding #3: Avoiding the Burn

for answers, see page 260

Fuel cells are essentially batteries that convert a fuel's chemical potential energy directly into electric energy, without burning the fuel first. Though harder to build and operate, fuel cells are potentially more energy efficient than combustion engines. Explain.

Improving Engine Efficiency

To obtain the hottest possible burned gas, the compression stroke should squeeze the fuel–air mixture into as small a volume as possible. The more tightly the piston compresses the mixture, the higher its density, pressure, and temperature will be before ignition and the hotter the burned gases will be after ignition. Since the efficiency of any heat engine increases as the temperature of its hot object increases and since the hot burned gas is the automobile engine's "hot object," its high temperature after ignition is good, the hotter the better.

The extent to which the cylinder's volume decreases during the compression stroke is measured by its compression ratio—its volume at the start of the compression stroke divided by its volume at the end of the compression stroke. The larger this compression ratio, the hotter the burned gas and the more energy efficient the engine. While normal compression ratios are between 8 : 1 and 12 : 1, those in high-compression engines may be as much as 15 : 1.

Unfortunately, the compression ratio can't be made arbitrarily large. If the engine compresses the fuel–air mixture too much, the flammable mixture will become so hot that it will ignite all by itself. This spontaneous ignition due to overcompression is called preignition or knocking. When an automobile knocks, the gasoline burns before the engine is ready to extract work from it and much of the energy is wasted.

There are two ways to reduce knocking. First, you can mix the fuel and air more uniformly. In a non-uniform mixture, there may be small regions of gas that get hotter or are more susceptible to ignition than others. The fuel-injection technique used in all modern cars provides excellent mixing and also allows a car's computer to adjust the fuel–air mixture for complete combustion and minimal pollution. So unless a car is seriously out of tune, there isn't much room for improvement as far as mixture uniformity is concerned.

Second, you can use the most appropriate fuel. Not all fuels ignite at the same temperature, so you should select a fuel that is able to tolerate your car's compression process without igniting spontaneously. That's exactly what you do when you purchase the proper grade of gasoline. Regular gasoline ignites at a relatively low temperature and is most susceptible to knocking. Premium gasoline ignites at a relatively high temperature and is most resistant to knocking.

Fuels that are more difficult to ignite and more resistant to knocking are assigned higher "octane numbers." Regular gasoline has an octane number of about 87 while premium has an octane number of about 93 (Table 8.2.1). Choosing the proper fuel is simply a matter of finding the lowest octane gasoline that your car can use without excessive knocking. A little knocking in the most demanding circumstances is quite acceptable. Most modern well-tuned automobiles work beautifully on regular gasoline. Since only high-performance cars with high-compression engines need premium gasoline, putting anything other than regular gasoline in a normal car is usually a waste of money.

Check Your Understanding #4: The Only Premium is the Price

for answers, see page 260

As part of its aggressive advertising campaign, one oil company has begun calling its 93-octane gasoline "the auto elixir." People are flocking to the pumps to fill up even the most ordinary cars with it. Should you join them or merely chuckle?

Table 8.2.1 Approximate Ignition Temperatures for the Three Standard Grades of Gasoline during Compression

OCTANE NUMBER	APPROXIMATE IGNITION TEMPERATURE
87 (regular)	750 °C (1382 °F)
90 (plus)	800 °C (1472 °F)
93 (premium)	850 °C (1562 °F)

Diesel Engines and Turbochargers

Since knocking sets the limit for compression ratio, it also sets the limit for efficiency in a gasoline engine. However, diesel engines avoid the knocking problem by separating the fuel and air during the compression stroke (Fig. 8.2.3). Invented by German engineer Rudolph Christian Karl Diesel (1858–1913) in 1896, the diesel engine has no spark plug to ignite the fuel. Instead, it compresses pure air with an extremely high compression ratio of perhaps 20 : 1 and then injects diesel fuel directly into the cylinder just as the power stroke begins. The fuel ignites spontaneously as it enters the hot, compressed air.

Because of its higher compression ratio, a diesel engine burns its fuel at a higher temperature than a standard gasoline engine and can therefore be more energy efficient. It effectively has a hotter "hot object" and can convert a larger fraction of heat into work. Unfortunately, a diesel engine is also harder to start than a gasoline engine and requires carefully timed fuel injection.

Some gasoline or diesel engines combine fuel injection with a turbocharger. A *turbocharger* is essentially a fan that pumps outdoor air into the cylinder during the induction stroke. By squeezing more fuel–air mixture into the cylinder, a turbocharger increases the engine's power output. The engine burns more fuel each power stroke and behaves like a larger engine. The fan of a turbocharger is powered by pressure in the engine's exhaust system. A nearly identical device called a *supercharger* is driven directly by the engine's output power.

The downside of a turbocharger, other than being expensive and wearing out rather quickly, is that it encourages knocking. As it squeezes air into the cylinder, it does work on that air and the air becomes hot. Since the fuel–air mixture enters the engine hot, it may ignite spontaneously during the compression stroke. To avoid knocking in a car equipped with a turbocharger, you may need to use premium gasoline. Some turbocharged cars are equipped with an *intercooler,* a device that removes heat from the air passing through the turbocharger. By providing cool, high-density air to the cylinders, the intercooler reduces the peak temperature of the compression stroke and avoids knocking.

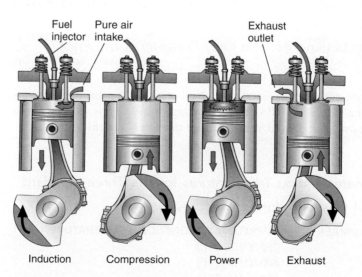

Fig. 8.2.3 A diesel engine cylinder contains pure air during the compression stroke. As the piston does work on it, this air becomes extremely hot. At the start of the power stroke, diesel fuel is injected into the cylinder. The fuel ignites spontaneously, and the hot burned gas does work on the piston and engine during the power stroke.

Fuel injector   Pure air intake   Exhaust outlet

Induction   Compression   Power   Exhaust

► Check Your Understanding #5: Steam Heat

for answers, see page 260

How can a steam engine be more energy efficient when it operates on 325 °C steam than when it uses 300 °C steam?

Multicylinder Engines

Since the purpose of the engine is to extract work from the fuel–air mixture, it's important that each cylinder do more work than it consumes. Three of the strokes require the engine to do work on various gases, and only one of the strokes extracts work from the burned gas. During the induction stroke, the engine does work drawing the fuel–air mixture into the cylinder. During the compression stroke, the engine does work compressing the fuel–air mixture. During the exhaust stroke, the engine does work squeezing the exhaust gas out of the cylinder. Fortunately, the work done on the engine by the hot burned gas during the power stroke more than makes up for the work the engine does during the other three strokes.

Still, the engine has to invest a great deal of energy into the cylinder before each power stroke. To provide this initial energy, most four-stroke engines have four or more cylinders (Fig. 8.2.4), timed so that there is always one cylinder going through the power stroke. The cylinder that is in the power stroke provides the work needed to carry the other cylinders through the three nonpower strokes, and there is plenty of work left over to propel the car itself.

While the pistons move back and forth, the engine needs a rotary motion to turn the car's wheels. The engine converts each piston's reciprocating motion into rotary motion by coupling that piston to a crankshaft with a connecting rod. The crankshaft is a thick steel bar, suspended in bearings, that has a series of pedal-like extensions, one for each cylinder. As the piston moves out of the cylinder during the power stroke, it pushes on the connecting rod and the connecting rod pushes on its crankshaft pedal. The connecting rod thus produces a torque on the crankshaft. The crankshaft rotates in its bearings and transmits this torque out of the engine so that it can be used to propel the car. So, while each cylinder initially exerts a force, the crankshaft uses that force to produce a torque.

The spinning crankshaft conveys its rotary power to the car's transmission and from there the power moves on to the wheels. Overall, a significant portion of the heat flowing

Courtesy of BMW Corporation

Fig. 8.2.4 This cutaway view of a BMW automobile engine shows the six pistons arranged in a single row. The cylinders have been omitted from the drawing.

out of the burning fuel–air mixture is being converted into work and used to spin the car's wheels. Assisted by friction with the pavement, the wheels push the car forward and you cruise down the highway toward your destination.

Check Your Understanding #6: Hard Starting

for answers, see page 260

Modern cars use an electric motor to start the engine turning but early cars were started with a hand crank. Why was it so hard to turn the crank?

Epilogue for Chapter 8

This chapter examined two devices that control the flow of heat to accomplish challenging tasks. In *air conditioners,* we learned how heat pumps use ordered energy to pump heat against its natural flow and observed that the only way to get rid of thermal energy is to transfer it to something else. And in *automobiles,* we saw that heat engines are able to divert some of the heat flowing from a hot object to a cold object and convert it into useful work. We also examined the roles of the two objects in a heat engine, hot and cold, finding that the hot object provides the energy needed to do the work while the cold object provides the order that makes the conversion of thermal energy into ordered energy possible.

Explanation: Making Fog in a Bottle

Even when everything in the room is at a single temperature, you can still use thermodynamics to cool the air in the bottle below room temperature. When you step on the bottle, you compress the air inside it and do work on that air. Since air can accommodate added energy only as thermal energy, its temperature rises. The air in the bottle becomes hotter and heat flows out of the warmer bottle into the cooler room. After half a minute or so, the temperature of air in the bottle returns to the temperature of the room.

When you step off the bottle, the air inside it expands and does work on you. Since air can obtain that work only from its thermal energy, its temperature drops. The air in the bottle becomes colder and its relative humidity suddenly exceeds 100%. As a result, moisture inside the bottle condenses into droplets and the air becomes foggy.

Smoke particles assist the droplet formation by acting as seeds for the droplets. We saw in Chapter 7, water can't boil without seed bubbles. Similarly, steam can't condense in air without seed droplets. When the seed droplets fail to form spontaneously, the smoke gives them a little help.

Chapter Summary

How Air Conditioners Work: An air conditioner moves heat against its natural direction of flow, using a working fluid that passes endlessly through an evaporator, compressor, and condenser. The working fluid flows toward the evaporator as a stable high-pressure liquid. Just before pouring into the evaporator, this liquid passes through a constriction in the pipe and experiences a large drop in pressure. Since the working fluid is not stable as a low-pressure liquid, it evaporates rapidly in the evaporator and thereby absorbs a great deal of heat from the indoor air.

The working fluid then flows to the compressor as a stable low-pressure gas. The compressor squeezes it into a hot, high-density, high-pressure gas and blows it into the condenser. Since the working fluid is not stable as a high-pressure gas, it condenses rapidly in the condenser and thereby releases a large amount of heat to the outdoor air. The resulting liquid working fluid then returns toward the evaporator to begin the cycle again.

To propel this process, the compressor consumes ordered energy and delivers it as heat to the outdoor air. Without this input of ordered energy, the air conditioner could not move heat against its natural direction of flow.

How Automobiles Work: An automobile engine extracts work from its chemical fuel by burning that fuel inside its cylinders and making the resulting burned gas do work on the engine. Most engines have at least four cylinders, each of which requires four strokes to extract work from the fuel. During the induction stroke, a piston moves out of the cylinder and draws a mixture of fuel and air into the resulting cavity. During the compression stroke, the piston moves into the cylinder, compressing this fuel–air mixture to high density, pressure, and temperature. An electric spark then ignites the mixture and it burns to form extremely hot burned gas. During the power stroke, the piston again moves out of the cylinder while the hot gas does work on it. This work is what powers the car. Finally, during the exhaust stroke, the piston moves into the cylinder and squeezes out the burned gas. The cylinder then begins again with fresh fuel and air.

Important Laws and Equations

1. **The Zeroth Law of Thermodynamics:** Two objects that are each in thermal equilibrium with a third object are also in thermal equilibrium with one another.

2. **The First Law of Thermodynamics:** The change in a stationary object's internal energy is equal to the heat transferred into that object minus the work that object does on its surroundings or

internal energy change = heat added − work done.

3. **The Second Law of Thermodynamics:** The entropy of a thermally isolated system of objects never decreases.

4. **The Third Law of Thermodynamics:** As an object's temperature approaches absolute zero, its entropy approaches zero.

5. **Efficiency of a Heat Pump:** The ideal efficiency of a heat pump depends on the temperatures of its hot and cold objects:

$$\text{heat from cold object} = \text{work} \cdot \frac{\text{temp}_{cold}}{\text{temp}_{hot} - \text{temp}_{cold}} \quad (8.1.2)$$

heat to hot object = heat from cold object + work.

6. **Efficiency of a Heat Engine:** The ideal efficiency of a heat engine depends on the temperatures of its hot and cold objects:

$$\text{work} = \text{heat from hot object} \cdot \frac{\text{temp}_{hot} - \text{temp}_{cold}}{\text{temp}_{hot}} \quad (8.2.1)$$

heat to cold object = heat from hot object − work.

Check Your Understanding—Answers

Section 8.1 AIR CONDITIONERS

1. The blender's blade does work on the water, and this work becomes thermal energy.

Why: The first law of thermodynamics states that the change in the water's internal energy is equal to the heat flowing into it minus the work it does on its surroundings. In this case, the water's surroundings are doing work on it by stirring it so its internal energy increases. Since the water can't store this new internal energy as potential energy, the energy becomes thermal energy and the water gets hotter.

2. A thermally isolated perpetual motion machine violates the first law, while one that is not thermally isolated violates the second law.

Why: A thermally isolated perpetual motion machine clearly violates the conservation of energy aspect of the first law of thermodynamics. This isolated machine simply can't export energy forever because it will eventually run out. A perpetual motion machine that is not thermally isolated may not violate conservation of energy because it can absorb heat energy from its surroundings. Instead, it violates the second law of thermodynamics. This machine can't endlessly absorb heat energy and then export it as ordered energy. In doing so, the machine will eventually begin to reduce the entropy of the universe and violate the second law. Sad though it may be, perpetual motion machines can't exist.

3. A heat pump requires more ordered energy to pump heat from a cold object to a hot object when the temperature difference between them is large.

Why: A heat pump becomes less efficient at pumping heat when the temperature of the heat's source becomes much colder than the heat's destination. The colder it is outside, the more ordered energy it takes to move each joule of heat. On bitter cold days, heat pumps aren't able to move enough heat to keep their homes warm, which is why most home heat pumps have built-in electric or gas furnaces to assist them during unusually cold weather.

4. The tank's liquid propane needs heat to evaporate into a gas, and it extracts that heat from its surroundings.

Why: Just as in the evaporator of an air conditioner, the evaporating liquid propane absorbs heat.

5. The room air would become warmer, on average.

Why: The air conditioner would begin to pump heat from its front to its back. The air right in front of the unit would become colder, while the air behind the unit would become hotter. Since the unit would deliver more heat to the hotter air than it would absorb from the colder air, it would increase the total amount of thermal energy in the room. On average, the room would become warmer.

Section 8.2 AUTOMOBILES

1. Yes.

Why: A heat engine is essentially a heat pump operating backwards. The air conditioner (a heat pump) uses ordered electric energy to pump heat from the cold air in your home to the hot outdoor air. The heat engine we are considering would use the flow of heat from the hot outdoor air to the cold air in your home to produce ordered electric energy.

2. The pressure is much higher in the burned gas than in the unburned fuel–air mixture.

Why: The amount of work done on the piston by the gas or done on the gas by the piston depends on the pressure inside the cylinder. The higher that pressure, the more outward force the piston experiences and the more work is done on it as it moves. The sudden rise in pressure that occurs when the fuel–air mixture burns explains why the burned gas does so much work on the piston as it moves outward. That pressure rise is due partly to the rise in temperature and partly to the fragmentation of the fuel and oxygen molecules into more, smaller molecules.

3. Because fuel cells don't turn the fuel's ordered energy into thermal energy, they don't have to operate as heat engines. In principle, they can convert all of the fuel's chemical potential energy into electric energy, unlike combustion engines.

Why: Fuel cells remain a promising alternative to combustion engines because they avoid the wasted energy that comes with burning fuel. However, making fuel cells that are efficient and robust is difficult and expensive. Fuel cell-powered vehicles are only just beginning to appear on the market.

4. Chuckle away.

Why: Like any high-octane gasoline, the elixir is an expensive fuel that has been carefully formulated to be hard to ignite. While it works wonders for a high-compression engine that would otherwise overheat the fuel–air mixture and cause it to knock, its resistance to ignition is wasted on most ordinary engines.

5. Like all heat engines, the steam engine can convert more thermal energy into work when the temperature of its hot object (the steam) increases.

Why: The steam engine converts thermal energy into work as heat flows from the hot steam to the outdoor air. The greater the temperature difference between those two objects, the more efficient the steam engine can be at turning thermal energy into work. That is why most steam engines use extremely hot steam.

6. The person turning the crank had to do all the work needed to move the engine's pistons through the three nonpower strokes.

Why: Before the engine started running on its own, it couldn't provide any of the energy the cylinders needed during the induction, compression, and exhaust strokes. The person turning the crank had to provide this energy. Once the fuel started burning, the power strokes could take over, but up to that point, turning the crank was hard work.

Check Your Figures—Answers

Section 8.1 AIR CONDITIONERS

1. 200 watts.
Why: Since the bicycle's internal energy can't increase or decrease forever, it must be emitting as much thermal power into the room as it is receiving mechanical power from you. As required by Eq. 8.1.1, the 200 watts of mechanical power you supply to the bicycle is flowing into the room as 200 watts of thermal power.

2. 100 J.
Why: The difference in temperature between indoor air and outdoor air is 30 K, so according to Eqs. 8.1.2, the heat pump can remove 9 times as much heat from the outdoor air as it consumes in work and deliver 10 times as much heat to the indoor air. Thus it takes only 100 J of electric energy to remove 900 J of heat from the outdoor air and deliver all 1000 J as heat to the indoor air. What a bargain!

Section 8.2 AUTOMOBILES

1. 400 J.
Why: Since the hot object is at 450 K and the temperature difference is 150 K, Eqs. 8.2.1 allow 1/3 of the heat removed from the boiler to be converted into work. The remaining 800 J of heat must flow into the outdoor air.

Exercises

1. Drinking fountains that actively chill the water they serve can't work without ventilation. They usually have louvers on their sides so that air can flow through them. Why do they need this airflow?

2. If you open the door of your refrigerator with the hope of cooling your room, you will find that the room's temperature actually increases somewhat. Why doesn't the refrigerator remove heat from the room?

3. The outdoor portion of a central air-conditioning unit has a fan that blows air across the condenser coils. If this fan breaks, why won't the air conditioner cool the house properly?

4. If you block the outlet of a hand bicycle pump and push the handle inward to compress the air inside the pump, the pump will become warmer. Why?

5. When the gas that now makes up the sun was compressed together by gravity, what happened to the temperature of that gas? Why?

6. Why is a car more likely to knock on a hot day than on a cold day?

7. A soda siphon carbonates water by injecting carbon dioxide gas into it. The gas comes compressed in a small steel container. As the gas leaves the container and pushes its way into the water, why does the container become cold?

8. A high-flying airplane must compress the cold, rarefied outside air before delivering it to the cabin. Why must this air be air conditioned after the compression?

9. If you drop a glass vase on the floor, it will become fragments. If you drop those fragments on the floor, however, they will not become a glass vase. Why not?

10. When you throw a hot rock into a cold puddle, what happens to the overall entropy of the system?

11. What prevents the bottom half of a glass of water from spontaneously freezing while the top half becomes boiling hot?

12. Suppose someone claimed to have a device that could convert heat from the room into electric power continuously. You would know that this device was a fraud because it would violate the second law of thermodynamics. Explain.

13. Why does snow blanket the ground almost uniformly rather than creating tall piles in certain areas and bare spots in others?

14. If you transfer a glass baking dish from a hot oven to a cold basin of water, that dish will probably shatter. What produces the ordered mechanical energy needed to tear the glass apart?

15. Freezing and thawing cycles tend to damage road pavement during the winter, creating potholes. What provides the mechanical work that breaks up the pavement?

16. The air near a woodstove circulates throughout the room. What provides the energy needed to keep the air moving?

17. Winds are driven by differences in temperature at the earth's surface. Air rises over hot spots and descends over cold spots, forming giant convection cells of circulating air. Near the

ground, winds blow from the cold spots toward the hot spots. Explain how the atmosphere is acting as a heat engine.

18. Hurricanes are giant heat engines powered by the thermal energy in warm ocean regions and the order in colder surrounding areas. Why are hurricanes most violent when they form over regions of unusually hot water at the end of summer?

19. On a clear sunny day, the ground is heated uniformly and there is very little wind. Use the second law of thermodynamics to explain this absence of wind.

20. A plant is a heat engine that operates on sunlight flowing from the hot sun to the cold earth. The plant is a highly ordered system with relatively low entropy. Why doesn't the plant's growth violate the second law of thermodynamics?

21. A diesel engine burns its fuel at a higher temperature than a gasoline engine. Why does this difference allow the diesel engine to be more efficient at converting the fuel's energy into work?

22. A chemical rocket is a heat engine, propelled forward by its hot exhaust plume. The hotter the fire inside the chemical rocket, the more efficient the rocket can be. Explain this fact in terms of the second law of thermodynamics.

23. An acquaintance claims to have built a gasoline-burning car that doesn't release any heat to its surroundings. Use the second law of thermodynamics to show that this claim is impossible.

Problems

1. You stir 1 kg of water until its temperature rises by 1 °C. How much work did you do on the water?

2. While polishing a 1-kg brass statue, you do 760 J of work against sliding friction. Assuming all of the resulting heat flows into the statue, how much does its temperature rise?

3. You drop a lead ball on a cement floor from a height of 10 m. When the ball stops bouncing, how much will its temperature have risen?

4. Roughly how high could a 300 K copper ball lift itself if it could transform all of its thermal energy into work?

5. Drilling a hole in a piece of wood takes 1000 J of work. How much does the total internal energy of the wood and drill increase as a result of this process?

6. An ideally efficient freezer cools food to 260 K. If room temperature is 300 K, how much work does this freezer consume when removing 100 J of heat from the food?

7. An ideally efficient refrigerator removes 900 J of heat from food at 270 K. How much heat does it then deliver to the 300 K room air?

8. An ideally efficient heat pump delivers 1000 J of heat to room air at 300 K. If it extracted heat from 260 K outdoor air, how much of that delivered heat was originally work consumed in the transfer?

9. An ideally efficient air conditioner keeps the room air at 300 K when the outdoor air is at 310 K. How much work does it consume when delivering 1240 J of heat outside?

10. An ideally efficient airplane engine provides work as heat flows from 1500 K burned gases to 300 K air. What fraction of the heat leaving the burned gases is converted into work?

11. An ideally efficient steamboat engine operates on 500 K steam in 300 K weather. How much work can it obtain when 1000 J of heat leaves the steam?

12. An offshore breeze at the beach is powered by heat flowing from hot land (310 K) to cool water (290 K). Assuming ideal efficiency, how much work can this breeze provide for each 1000 J of heat it carries away from the land?

13. An ideally efficient solar energy system produces work as heat flows from the 6100 K surface of the sun to the 300 K room air. What fraction of the solar heat can it transform into work?

Resonance and Mechanical Waves

Many fascinating motions in the world around us are repetitive ones. Our lives are filled with cycles, from the sun's daily passage overhead to a pond's undulating ripples on a rainy day. These cyclic motions are governed by the physical laws and steadily mark our journey through time and space. Some of these cycles structure our lives out of necessity or tradition, while others are simply there to be observed. Still other cycles have become part of our everyday world because they're useful or enjoyable. This chapter is about cyclic motions in three contexts: in clocks, in musical instruments, and at the seashore.

EXPERIMENT: A Singing Wineglass

One simple experiment with cyclic motion involves a crystal wineglass, a little water, and a delicate touch. A crystal wineglass is hard and thin and easily supports a repetitive

Courtesy Lou Bloomfield

mechanical motion called a vibration. The glass's hardness allows it to retain energy in this vibrating motion for a long time so that it rings clearly when you tap it gently with a spoon and can acquire energy slowly in the manner described below.

The experiment itself is an activity discovered by many a bored youth, sitting too long in a fancy restaurant with only the place settings as entertainment. If you wet your finger slightly and run it gently around the lip of the wineglass at a slow, steady pace, you should be able to get the wineglass to sing loudly. The walls of the glass will vibrate back and forth and emit a clear tone.

If you can't find a crystal wineglass, you may be out of luck. An ordinary glass doesn't work as well because it converts the energy from your finger into thermal energy rapidly and doesn't emit much sound. In any case, be sure that the glass has no sharp edges on its lip so that you don't cut your finger.

What will happen to the sound if you add some water to the glass? Try to *predict* how its pitch will change; how its volume will change. Now *observe* what happens. Can you *measure* the changes? Did you *verify* your predictions? Is anything visible on the surface of the water as you make the glass vibrate?

Chapter Itinerary

The wineglass's vibration is an example of a natural resonance, a cyclic motion of the glass itself. Whether you tap the glass gently with a spoon or rub it with your finger, you're causing it to undergo this characteristic motion. In this chapter we'll examine several other objects that exhibit natural resonances: (1) *clocks,* (2) *musical instruments,* and (3) *the sea.*

In *clocks,* we'll study pendulums, balance rings, and quartz crystals to see how their natural resonances are used to measure the passage of time. In *musical instruments,* we'll look at the vibrations of strings, air columns, and drumheads to find out how these instruments produce their musical sounds and learn how those sounds travels to our ears. And in *the sea,* we'll look at the nature of waves on the surface of water and explore such issues as tides, tsunamis, and surf. For a more detailed preview, turn to the Chapter Summary on p. 297.

SECTION 9.1 Clocks

People measure their lives according to the sky, dividing existence into days, months, and years according to the celestial motions of the sun, moon, and stars. But on the less romantic scale of daily life, the sky offers little help. Since it provides no easy way to measure short periods of time, people invented clocks.

Early clocks were based on the time it took to complete simple processes—the flow of sand or water, or the burning of candles. However, these clocks had limited accuracy and required constant attention. Better clocks measure time with repetitive motions such as swinging or rocking. In this section, we'll examine the workings of modern clocks based on repetitive motions. As we do, we'll see that repetitive motions are interesting in their own right and appear throughout nature in countless objects besides clocks.

Questions to Think About: What exactly is time? Why do some objects swing or rock back and forth repetitively? How would you use a repetitive motion to measure time? How can you change the rate at which an object swings or rocks? Since repetitive motions don't normally continue forever, how can you keep them going without upsetting their timekeeping ability?

Experiments to Do: You can build the timekeeping portion of a pendulum clock by attaching a small, dense object to the end of a string and hanging that string from a table or doorway. Push the object gently so that it swings back and forth. You'll find that it completes this repetitive motion with great regularity. What limits its regularity?

If the string is about 25 cm (10 inches) long, a complete swing (back and forth) will take almost exactly 1 s. Change the length of the string and observe its effect on the swing. Do you think that the object's weight affects the swing? Try a different object and see if you're right.

Now vary the extent of the swing to see if it affects the time each swing takes. How can it be that a small swing takes the same time as a large swing? Notice that you can get the object swinging by pushing on it rhythmically. Randomly timed pushes won't do; you must push the object in synchrony with its motion. At what times should you push it to make it swing farther? To make it swing less far? Rhythmic pushes of this sort are what keep the timekeeper in a clock moving, hour after hour.

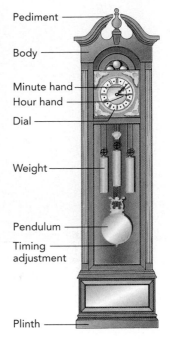

Pediment

Body

Minute hand
Hour hand
Dial

Weight

Pendulum

Timing
adjustment

Plinth

Time

Before examining clocks, we should take a brief look at time itself. Scientists treat time as a dimension, similar but not identical to the three spatial dimensions that we perceive in the world around us. In total, our universe has four dimensions: three spatial dimensions and one temporal dimension. Thus it takes four numbers to completely specify when and where an event occurs; three numbers identify the event's location and one identifies its moment in time.

An obvious difference between space and time is that, while we can see space stretched out around us, we can only observe the *passage* of time. Though we occupy only one location in space at a given moment, we are somehow more aware of the expanse of space around us. It's much harder to sense the whole framework of time stretching off into the past and future; you must use your imagination.

Our perception of space is ultimately based on the need for forces, accelerations, and velocities to travel from one place to another. A city seems far away because we know that traveling there with reasonable forces, accelerations, and velocities would take a long time. Our perception of time is based on the same mechanical principles. If two moments are separated by a long time, then reasonable forces, accelerations, and velocities will permit us to travel large distances between the two moments. In short, our perceptions of space and time are interrelated, and measurements of time and space are connected as well.

We measure space with rulers and time with clocks. But how would you make a ruler? You could construct a rather large ruler by driving a car at constant velocity and marking the pavement with paint once each second. Your ruler wouldn't be very practical, but it would fit the definition of a ruler as having spatial markings at uniform distances. You would be using your movement through time to measure space.

How would you make a clock? You could make a rather strange clock by driving a car at constant velocity down your giant ruler and counting each time you saw one of your marks go by. You would then be using your movement through space to measure time. Most clocks really do use motion to measure time. As we are about to see, however, they use motions that are a bit more compact than a car ride.

Check Your Understanding #1: The Ultimate Moon-Bounce

for answers, see page 298

To measure the distance from the earth to the moon, scientists bounce light from reflectors placed on the moon by the Apollo astronauts. Light travels at a constant speed. How can a measurement of light's travel time to and from the moon be used to determine the distance from the earth to the moon?

Natural Resonances

☐ The daughter of a planetarium architect, Jocelyn Bell (British astronomer, 1943–) acquired an early interest in radio astronomy. Advised to study physics first, she became the only woman in a class of 50 at Glasgow University. While working on her Ph.D. at Cambridge, Bell discovered an extraterrestrial source of radio bursts, occurring precisely 1.33730113 seconds apart. She had discovered the first pulsar, a collapsed star whose angular momentum keeps it turning at an extraordinarily uniform rate. Each burst coincided with one rotation of the star remnant.

An ideal timekeeping motion should offer both accuracy and convenience. That rules out some of the obvious choices. The sun, moon, and stars keep excellent time but fail the convenience test. Sure, conservation of energy, momentum, and angular momentum so dominate their motions that these celestial bodies move steadily and predictably through the heavens, century after century, but what do you do on a cloudy day? And while simple interval timers like sandglasses and burning candles are easy to make and use, they're not very accurate. Besides, who's going to stay up all night lighting fresh candles just to keep the "clock" running? (For an interesting astronomical clock, see ☐.)

Instead, practical clocks are based on a particular type of repetitive motion called a **natural resonance.** In a natural resonance, the energy in an isolated object or system of objects causes it to perform a certain motion over and over again. Many objects in our world exhibit natural resonances, from tipping rocking chairs, to sloshing basins of water, to waving flagpoles, and those natural resonances usually involve motion about a stable equilibrium. Like the bouncing spring scale in Section 3.1, an object that has been displaced from its stable equilibrium accelerates toward that equilibrium but then overshoots; it coasts right through equilibrium and must turn around to try again. As long as it has excess energy, this object continues to glide back and forth through its equilibrium and thus exhibits a natural resonance.

Some resonances, such as those of bouncing balls and teetering bottles, don't maintain a steady beat and aren't suitable for clocks. But we are about to encounter a group of resonances that are extremely regular and that can be used to measure the passage of time with remarkable accuracy. Those resonances belong to an important class of mechanical systems known as *harmonic oscillators*.

► Check Your Understanding #2: An Egg-Timer Clock

for answers, see page **298**

Although a sandglass can be made repetitive by turning it over every time the sand runs out, this manual restarting process introduces timing errors. If a 3-minute sandglass is always turned over within 10 s after the sand runs out, how accurately will it measure time over the course of a day?

Pendulums and Harmonic Oscillators

One of the first natural resonances to find its way into clocks is the swing of a pendulum—a weight hanging from a pivot (Fig. 9.1.1). When the pendulum's center of gravity is directly below its pivot, it's in a stable equilibrium. Its center of gravity is then as low as possible, so displacing it raises its gravitational potential energy and a restoring force begins pushing it back toward that equilibrium position (Fig. 9.1.2). For geometrical reasons, this restoring force is almost exactly proportional to how far the pendulum is from equilibrium. As you displace the pendulum steadily from equilibrium, the restoring force on it also increases steadily.

When you release the displaced pendulum, its restoring force accelerates it back toward equilibrium. But instead of stopping, the pendulum swings back and forth about its equilibrium position in a repetitive motion called an **oscillation.** As it swings, its energy alternates between potential and kinetic forms. When it swings rapidly through its equilibrium position in the middle of a swing, its energy is all kinetic. When it stops momentarily at the end of a swing, its energy is all gravitational potential. This repetitive transformation of excess energy from one form to another is part of any oscillation and keeps the oscillator—the system experiencing the oscillation—moving back and forth until that excess energy is either converted into thermal energy or transferred elsewhere.

But the pendulum isn't just any oscillator. Because its restoring force is proportional to its displacement from equilibrium, the pendulum is a **harmonic oscillator**—the

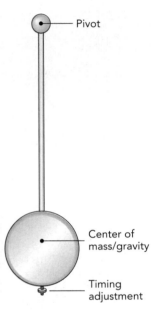

Fig. 9.1.1 A pendulum consists of a weight hanging from a pivot. The pendulum is in a stable equilibrium when its center of gravity is directly below the pivot.

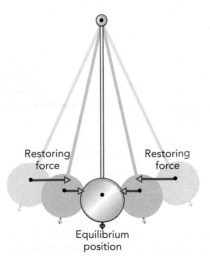

Fig. 9.1.2 If you tilt a pendulum's center of gravity away from its equilibrium position, it experiences a restoring force proportional to its distance from that equilibrium position.

simplest and best understood mechanical system in nature. As a harmonic oscillator, the pendulum undergoes **simple harmonic motion,** a regular and predictable oscillation that makes it a superb timekeeper.

The **period** of any harmonic oscillator—the time it takes to complete one full cycle of its motion—depends only on how stiffly its restoring force pushes it back and forth and on how stubbornly its mass resists that back-and-forth motion. **Stiffness** is the measure of how sharply the restoring force strengthens as the oscillator is displaced from equilibrium; stiff restoring forces are associated with firm or hard objects, while less stiff restoring forces are associated with soft objects. The stiffer the restoring force, the more forcefully it pushes the oscillator back and forth and the shorter the oscillator's period. On the other hand, the larger the oscillator's mass, the less it accelerates and the longer its period.

However, the most remarkable and important characteristic of a harmonic oscillator is not that its period depends on stiffness and mass, but that its period *doesn't* depend on **amplitude**—its furthest displacement from equilibrium. Whether that amplitude is large or small, the harmonic oscillator's period remains exactly the same. This insensitivity to amplitude is a consequence of its special restoring force, a restoring that is proportional to displacement from equilibrium. At larger amplitudes, the oscillator travels farther each cycle, but the forces accelerating it through that cycle are stronger as well. Overall, the harmonic oscillator completes a large cycle of motion just as quickly as it completes a small cycle of motion.

Harmonic oscillators with stiff restoring forces and small masses have short periods, while those with soft restoring forces and large masses have long periods. Their amplitudes of oscillation simply don't affect their periods, which is why harmonic oscillators are so ideal for timekeeping. Because practical clocks can't control the amplitudes of their timekeeping oscillators perfectly, virtually all of them are based on harmonic oscillators.

> **Harmonic Oscillators**
> A harmonic oscillator is one with a restoring force proportional to displacement. Its period of oscillation depends only on the stiffness of that restoring force and on its mass, not on its amplitude of oscillation.

Actually, a pendulum is an unusual harmonic oscillator because its period is independent of its mass. That's because increasing the pendulum's mass also increases its weight and therefore stiffens its restoring force. These two changes compensate for one another perfectly so that the pendulum's period is unchanged.

A pendulum's period does, however, depend on its length and on gravity. When you reduce the pendulum's length—the distance from its pivot to its center of mass—you stiffen its restoring force and shorten its period. Similarly, when you strengthen gravity (perhaps by traveling to Jupiter), you increase the pendulum's weight, stiffen its restoring force, and reduce its period. Though we won't try to prove it, the period of a pendulum is

$$\text{period of pendulum} = 2\pi\sqrt{\frac{\text{length of pendulum}}{\text{acceleration due to gravity}}}.$$

Thus a short pendulum swings more often than a long one and any pendulum swings more often on the earth than it would on the moon.

On the earth's surface, a 0.248-m (10-inch) pendulum has a period of 1 s (Fig. 9.1.3), making it suitable for a wall clock that advances its second hand by 1 second each time

Robert Mathena/Fundamental Photographs

Fig. 9.1.3 The swinging pendulum controls the movement of the clock's hands. The pendulum is 0.248 m long, from pivot to center of mass/gravity, so each cycle takes 1 s to complete and advances the hands by 1 second.

the pendulum completes a cycle. Since a pendulum's period increases as the square root of its length, a 0.992-m (40-inch) pendulum (four times as tall as a 0.248-m pendulum) takes 2 s to complete its cycle and is appropriate for a floor clock that advances its second hand by 2 seconds per cycle.

Because a pendulum's period depends on its length and gravity, a change in either one causes trouble. As we learned in Chapter 7, materials expand with increasing temperature, so a simple pendulum slows down as it heats up. A more accurate pendulum is thermally compensated by using several different materials with different coefficients of volume expansion to ensure that its center of mass remains at a fixed distance from its pivot.

While gravity doesn't change with time, it does vary slightly from place to place. To correct for differences in gravity between the factory and a clock's final destination, its pendulum has a threaded adjustment knob. This knob allows you to change the pendulum's length to fine-tune its period.

Check Your Understanding #3: Swing Time

for answers, see page **298**

A child swinging on a swing set travels back and forth at a steady pace. What determines the period of his motion?

Pendulum Clocks

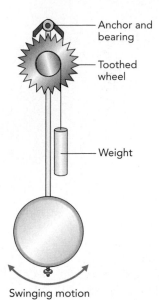

Anchor and bearing

Toothed wheel

Weight

Swinging motion

Fig. 9.1.4 A pendulum clock uses a swinging pendulum to determine how quickly a toothed wheel turns and advances a series of gears that control the hands of the clock. The anchor permits the toothed wheel to advance by one tooth each time the pendulum completes a full cycle.

While a pendulum maintains a steady beat, it's not a complete clock. Something must keep the pendulum swinging and use that swing to determine the time. A pendulum clock does both. It sustains the pendulum's motion with gentle pushes, and it uses that motion to advance its hands at a steady rate.

The top of the pendulum has a two-pointed *anchor* that controls the rotation of a *toothed wheel* (Fig. 9.1.4). This mechanism is called an *escapement*. A weighted cord wrapped around the toothed wheel's shaft exerts a torque on that wheel, so that the wheel would spin if the anchor weren't holding it in place. Each time the pendulum reaches the end of a swing, one point of the anchor releases the toothed wheel while the other point catches it. The wheel turns slowly as the pendulum rocks back and forth, advancing by one tooth for each full cycle of the pendulum. This wheel turns a series of gears, which slowly advance the clock's hands. Although these hands are actually counting the number of pendulum swings since midnight, their movement is calibrated so that their positions indicate the current time.

The toothed wheel also keeps the pendulum moving by giving the anchor a tiny forward push each time the pendulum completes a swing. Since the anchor moves in the direction of the push, the wheel does work on the anchor and pendulum, and replaces energy lost to friction and air resistance. This energy comes from the weighted cord, which releases gravitational potential energy as its weight descends. When you wind the clock, you rewind this cord around the shaft, lifting the weight and replenishing its potential energy.

While these pushes from the toothed wheel can keep even the clumsiest pendulum swinging, a clock works best when its pendulum swings with almost perfect freedom. That's because any outside force—even the push from the toothed wheel—will influence the pendulum's period. The most accurate timekeepers are those that can oscillate without any assistance or energy replacement for thousands or millions of cycles. These precision timekeepers need only the slightest pushes to keep them moving and thus have extremely precise periods. That's why a good pendulum clock uses an aerodynamic pendulum and low-friction bearings.

Finally, the clock must keep the oscillation amplitude of its pendulum relatively constant. From a practical perspective, drastic changes in that amplitude will make the toothed wheel turn erratically. But there is a more fundamental reason to keep the pendulum's amplitude steady: it's not really a perfect harmonic oscillator. If you displace the pendulum too far, it becomes an **anharmonic oscillator**—its restoring force ceases to be proportional to its displacement from equilibrium, and its period begins to depend on its amplitude. Since a change in period would spoil the clock's accuracy, the pendulum's amplitude must be kept small and steady. That way, the amplitude has almost no effect on the pendulum's period.

Check Your Understanding #4: Swing High, Swing Low

for answers, see page 298

When pushing a child on a playground swing, you normally push her forward as she moves away from you. What happens if you push her forward each time she moves toward you?

Balance Clocks

Because it relies on gravity for its restoring force, a swinging pendulum mustn't be tilted or moved. That's why there are so few pendulum-based wristwatches. To make use of the excellent timekeeping characteristics of a harmonic oscillator, a portable clock needs some other restoring force that's proportional to displacement but independent of gravity. It needs a spring!

As we saw in Section 3.1, the force a spring exerts is proportional to its distortion. The more you stretch, compress, or bend a spring, the harder it pushes back toward its equilibrium shape. Attach a block of wood to the free end of a spring, stretch it gently, and let go and you'll find you have a harmonic oscillator with a period determined only by the stiffness of the spring and the block's mass (Fig. 9.1.5). Since the period of a harmonic oscillator doesn't depend on the amplitude of its motion, the block oscillates steadily about its equilibrium position and makes an excellent timekeeper.

Unfortunately, gravity complicates this simple system. Although gravity doesn't alter the block's period, it does shift the block's equilibrium position downward. That shift is a problem for a clock that might be tilted sometimes. However, there's another spring-based timekeeper that marks time accurately in any orientation or location. This ingenious device, used in most mechanical clocks and watches, is called a balance ring or simply a balance.

A balance ring resembles a tiny metal bicycle wheel, supported at its center of mass/gravity by an axle and a pair of bearings (Fig. 9.1.6). Any friction in the bearings is exerted so close to the ring's axis of rotation that it produces little torque and the ring turns extremely easily. Moreover, the ring pivots about its own center of gravity so that its weight produces no torque on it.

The only thing exerting a torque on the balance ring is a tiny coil spring. One end of this spring is attached to the ring while the other is fixed to the body of the clock. When the spring is undistorted, it exerts no torque on the ring and the ring is in equilibrium. But if you rotate the ring either way, torque from the distorted spring will act to restore it to its equilibrium orientation. Since this restoring torque is proportional to the ring's rotation away from a stable equilibrium, the balance ring and coil spring form a harmonic oscillator!

Because of the rotational character of this harmonic oscillator, its period depends on the *torsional* stiffness of the coil spring—how rapidly the spring's torque increases as you twist it—and on the balance ring's *rotational* mass. Since the balance ring's period doesn't depend on the amplitude of its motion, it keeps excellent time. And because gravity exerts no torque on the balance ring, this timekeeper works anywhere and in any orientation.

The rest of a balance clock is similar to a pendulum clock (Fig. 9.1.7). As the balance ring rocks back and forth, it tips a lever that controls the rotation of a toothed wheel. An anchor attached to the lever allows the toothed wheel to advance one tooth for each complete cycle of the balance ring's motion. Gears connect the toothed wheel to the clock's hands, which slowly advance as the wheel turns.

Because the balance clock is portable, it can't draw energy from a weighted cord. Instead, it has a main spring that exerts a torque on the toothed wheel. This main spring is a coil of elastic metal that stores energy when you wind the clock. Its energy keeps the balance ring rocking steadily back and forth and also turns the clock's hands. Since the main spring unwinds as the toothed wheel turns, the clock occasionally needs winding. (For an interesting example of a balance clock, see ❏.)

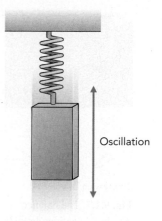

Fig. 9.1.5 A block attached to a spring is a harmonic oscillator. The oscillator's period is determined only by the stiffness of the spring and the mass of the block.

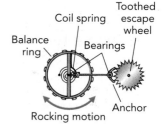

Fig. 9.1.6 A balance clock uses a rocking balance ring to control the rotation of a toothed wheel and the gears that advance the clock's hands. The anchor permits the toothed wheel to advance by one tooth each time the ring completes a full cycle. The clock's energy comes from a main spring (not shown) that exerts a steady torque on the toothed wheel.

➤ Check Your Understanding #5: A Little Light Entertainment

for answers, see page **298**

When you accidentally strike a chandelier with a broom, this hanging lamp begins to twist back and forth with a regular period. What determines its period of oscillation?

Electronic Clocks

The potential accuracy of pendulum and balance clocks is limited by friction, air resistance, and thermal expansion to about ten seconds per year. To do better, a clock's timekeeper must avoid these mechanical shortcomings. That's why so many modern clocks use quartz oscillators as their timekeepers.

A quartz oscillator is made from a single crystal of quartz, the same mineral found in most white sand. Like many hard and brittle objects, a quartz crystal oscillates strongly after being struck. In fact, it's a harmonic oscillator because it acts like a spring with a block at each end (Fig. 9.1.8*a,b*). The two blocks oscillate in and out symmetrically about their combined center of mass, with a period determined only by the blocks' masses and the spring's stiffness. In a quartz crystal, the spring is the crystal itself and the blocks are its two halves (Fig. 9.1.8*c,d*). Since the forces on the blocks are proportional to their displacements from equilibrium, they're harmonic oscillators.

Because of its exceptional hardness, a quartz crystal's restoring force is extremely stiff. Even a tiny distortion leads to a huge restoring force. Since the period of a harmonic oscillator decreases as its spring becomes stiffer, a typical quartz oscillator has an extremely short period. Its motion is usually called a **vibration** rather than an oscillation because vibration implies a fast oscillation in a mechanical system. Oscillation itself is a more general term for any repetitive process and can even apply to such nonmechanical processes as electric or thermal oscillations.

Because of its rapid vibration, a quartz oscillator's period is a small fraction of a second. We normally characterize such a fast oscillator by its **frequency**—the number of cycles it completes in a certain amount of time. The SI unit of frequency is the **cycle-per-second,** also called the **hertz** (abbreviated Hz) after German physicist Heinrich Rudolph Hertz. Period and frequency are reciprocals of one another (period equals 1/frequency and vice versa) so that an oscillator with a period of 0.001 s has a frequency of 1000 Hz.

Courtesy Lou Bloomfield

Fig. 9.1.7 The balance ring in this antique French carriage clock twists back and forth rhythmically under the influence of the spiral spring near its center. The tiny ruby bearings that support the ring minimize friction and permit this clock to keep very accurate time.

❑ The son and grandson of freed slaves, Benjamin Banneker (African-American mathematician, astronomer, and writer, 1731–1806) grew up on a Maryland tobacco farm. He was fascinated by mathematics and science, and supplemented his limited schooling with borrowed books. Though he is best remembered for his work in astronomy and for compiling six almanacs, he also produced one of the first clocks made entirely in America. With only a borrowed pocket watch as a guide, Banneker built his wooden balance clock by hand, using a knife to shape the parts. The clock kept accurate time for half a century and even struck the hours.

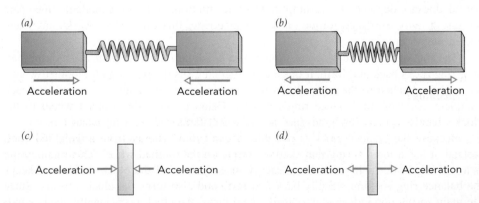

Fig. 9.1.8 A quartz crystal acts like a spring with masses at each end. Just as the two masses alternately accelerate (*a*) toward and (*b*) away from one another, so the two halves of the vibrating crystal alternately accelerate (*c*) together and (*d*) apart.

Because the vibrating crystal isn't sliding across anything or moving quickly through the air, it loses energy slowly and vibrates for a long, long time. And because quartz's coefficient of thermal expansion is extremely small, the crystal's period is nearly independent of its temperature. With its exceptionally steady period, a quartz oscillator can serve as the timekeeper for a highly accurate clock, one that loses or gains less than a tenth of a second per year.

Of course, a quartz crystal isn't a complete clock. Like the pendulum and balance, it needs something to keep it vibrating and use that vibration to determine the time. While these tasks could conceivably be done mechanically, quartz clocks are normally electronic. There are two reasons for this choice. First, the crystal's vibrations are too fast and too small for most mechanical devices to follow. Second, a quartz crystal is intrinsically electronic itself; it responds mechanically to electrical stress and electrically to mechanical stress. Because of this coupling between its mechanical and electrical behaviors, crystalline quartz is known as a *piezoelectric* material and is ideal for electronic clocks.

The clock's circuitry uses electrical stresses to keep the quartz crystal vibrating (Fig. 9.1.9). Just as carefully timed pushes keep a child swinging endlessly on a playground swing, carefully timed electrical stresses keep the quartz crystal vibrating endlessly in its holder. Because the crystal loses so little energy with each vibration, only a tiny amount of work is required each cycle to maintain its vibration.

The clock also detects the crystal's vibrations electrically. Each time its halves move in or out, the crystal experiences mechanical stress and emits a pulse of electricity. These pulses may control an electric motor that advances clock hands or may serve as input to an electronic chip that measures time by counting the pulses.

The quartz crystals used in clocks and watches are carefully cut and polished to vibrate at specific frequencies. The thinner the crystal, the faster it vibrates—less mass and a stiffer restoring force. In effect, these crystals are tuned like musical instruments to match the requirements of their clocks.

While most tiny quartz crystals vibrate millions of times each second, common watch crystals vibrate at 2^{15} Hz or 32,768 Hz. This low frequency prolongs a watch's battery life because counting each pulse consumes some of the battery's energy. To make a small crystal vibrate this slowly, the manufacturer cuts away most of the center of the crystal to weaken its restoring force and slow its oscillations (Fig 9.1.10). The resulting quartz "tuning fork" oscillator is carefully metalized to permit the watch to interact with it electrically, and it's then tuned to exactly 32,768 Hz by burning away some of the metal mass with a laser beam.

Courtesy Lou Bloomfield

Fig. 9.1.9 The quartz crystal in this wristwatch is located inside the silver cylinder at the bottom. Carefully polished to vibrate at a precise frequency, the crystal keeps the watch accurate to a few seconds a month.

Check Your Understanding #6: Heavy Metal Music

for answers, see page **298**

If you drop a metal rod on the floor, end first, you hear a high-pitched tone. What's happening?

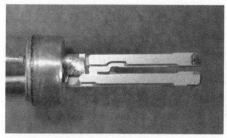

Courtesy Lou Bloomfield

Fig. 9.1.10 Shaped like a tiny tuning fork, this watch crystal vibrates almost exactly 32,768 times per second. Burn marks at its tips were created when its vibrational frequency was tuned by a laser beam.

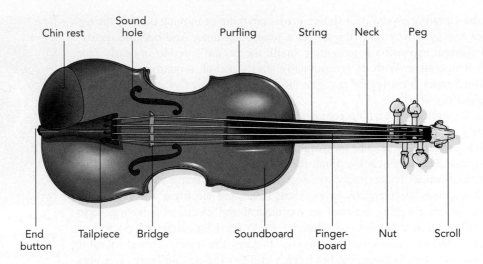

Chin rest   Sound hole   Purfling   String   Neck   Peg

End button   Tailpiece   Bridge   Soundboard   Finger-board   Nut   Scroll

SECTION 9.2   **Musical Instruments**

Music is an important part of human expression. While what qualifies as music is a matter of taste, it always involves sound and often involves instruments. In this section, we'll examine sound, music, and several instruments: violins, pipe organs, and drums. As examples of the three most common types of instruments, strings, winds, and surfaces, this trio will help us to understand many other instruments as well.

Questions to Think About: *Why are the low-pitch strings on a violin thicker than the high-pitch strings? How does pressing a violin's string against the fingerboard change its pitch? Why does a violin sound different when it's plucked rather than bowed? What purpose does the violin's body serve? What is vibrating inside a pipe organ? Why are some organ pipes longer than others? Why do most drums sound toneless, providing more rhythm than pitch?*

Experiments to Do: *Find a violin or guitar, or stretch a strong string between two rigid supports. Even a rubber band will do in a pinch. Pluck the string with your finger and listen to the tone it makes. The string vibrates back and forth at a particular frequency or pitch even as its amplitude of motion gradually decreases. What kind of oscillator has that behavior?*

Change the string's frequency of vibration by changing its tension or length. What happens when you shorten the string or prevent part of it from moving? What happens if you increase its tension by pulling it tighter? You can also increase the string's mass by wrapping it with tape. How does that affect its pitch?

You can imitate a pipe organ by blowing gently across the mouth of a bottle or soda straw. If done properly, you'll get the air inside the bottle vibrating up and down rhythmically and you'll hear a tone. What happens to its pitch when you add water to the bottle or pinch off the straw at various points? Why does this tone sound different from that of the string, even when the two have the same pitch?

Finally, a table or plate will act like a drum when you tap it. Compare the sound it makes with those of the previous two "instruments." Does it have a pitch? What distinguishes the sounds of different tables or plates?

Sound and Music

To understand how instruments work, we'll need to know a bit more about sound and music. In air, **sound** consists of density waves—patterns of compressions and rarefactions that travel outward rapidly from their source. When a sound passes by, the air pressure in your ear fluctuates up and down about normal atmospheric pressure. Even when these fluctuations have amplitudes less than a millionth of atmospheric pressure, you hear them as sound.

When the fluctuations are repetitive, you hear a *tone* with a pitch equal to the fluctuation's frequency. **Pitch** is the frequency of a sound. A bass singer's pitch range extends from 80 Hz to 300 Hz, while that of a soprano singer extends from 300 Hz to 1100 Hz. Musical instruments can produce tones over a much wider range of pitches, but we can only hear those between about 30 Hz and 20,000 Hz, and that range narrows as we get older.

Most music is constructed around *intervals*—the frequency ratio between two different tones. This ratio is found by dividing one tone's frequency by that of the other. Our hearing is particularly sensitive to intervals, with pairs of tones at equal intervals sounding quite similar to one another. For example, a pair of tones at 440 Hz and 660 Hz sounds similar to a pair at 330 Hz and 495 Hz because they both have the interval 3/2.

The interval 3/2 is pleasing to most ears and is common in Western music, where it's called a *fifth*. A fifth is the interval between the two "twinkles" at the beginning of "Twinkle, Twinkle, Little Star." If your ear is good, you can start with any tone for the first twinkle and will easily find the second tone, located at 3/2 the frequency of the first. Your ear hears that factor of 3/2 between the two frequencies.

The most important interval in virtually all music is 2/1 or an *octave*. Tones that differ by a factor of 2 in frequency sound so similar to our ears that we often think of them as being the same. When men and women sing together "in unison," they often sing an octave or two apart and the differences in the tones, always factors of 2 or 4 in frequency, are only barely noticeable.

The octave is so important that it structures the entire range of audible pitches. Most of the subtle interplay of tones in music occurs in intervals of less than an octave, less than a factor of 2 in frequency. Thus most traditions build their music around the intervals that lie within a single octave, such as 5/4 and 3/2. They pick a particular standard pitch and then assign *notes* at specific intervals from this standard pitch. This arrangement repeats at octaves above and below the standard pitch to create a complete *scale* of notes. (For a history of scales, see ❏.)

The scale used in Western music is constructed around a note called A_4 or Concert A which has a standard pitch of 440 Hz. At intervals of 9/8, 5/4, 4/3, 3/2, 5/3, and 15/8 above A_4 lie the six notes B_4, $C^{\#}_5$, D_5, E_5, $F^{\#}_5$, and $G^{\#}_5$. Similar collections of six notes are built above A_5 (880 Hz), which has a frequency twice that of A_4, and above A_3 (220 Hz), which has a frequency half that of A_4. In fact, this pattern repeats above A_1 (55 Hz) through A_8 (7040 Hz).

Actually, Western music is built around 12 notes and 11 intervals that lie within a single octave. Five more intervals account for five additional notes, B^{b}_4, C_5, $D^{\#}_5$, F_5, and G_5. It's also not quite true that every note is based exclusively on its interval from A_4. While A_4 remains at 440 Hz, the pitches of the other 11 notes have been modified slightly so that they're at interesting and pleasing intervals from one another as well as from A_4. This adjustment of the pitches led to the *well-tempered scale* that has been the basis for Western music for the last several centuries.

❏ In addition to his contributions to mathematics, geometry, and astronomy, the Greek mathematician Pythagoras (ca 580–500 B.C.) was perhaps the first person to use mathematics to relate intervals, pitches, and the lengths of vibrating strings. He and his followers laid the groundwork for the scale used in most Western music.

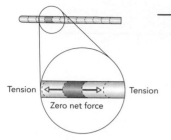

Tension ← Zero net force → Tension

Fig. 9.2.1 A taut violin string can be viewed as composed of many individual pieces. When the string is straight, the two forces exerted on a given piece by its neighbors cancel perfectly.

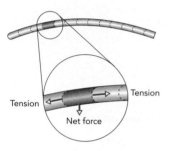

Tension ← Net force → Tension

Fig. 9.2.2 When a violin string is curved, the two forces exerted on a given piece by its neighbors don't point in exactly opposite directions and don't balance one another. The piece experiences a net force.

Fundamental

Antinode

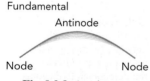

Node   Node

Fig. 9.2.3 A string vibrating between two fixed points in its fundamental vibrational mode. The whole string moves together, traveling up and down as a single harmonic oscillator.

→ **Check Your Understanding #1: A Night at the Opera**

for answers, see page **298**

A typical singing voice can cover a range of about two octaves, for example, from C_4 to C_6. How broad is this range of frequencies?

A Violin's Vibrating String

The tones produced by a violin begin as vibrations in its strings. But these strings are limp and shapeless on their own and rely on the violin's rigid body and neck for structure. The violin subjects its strings to **tension**—outward forces that act to stretch it—and this tension gives each string an equilibrium shape: a straight line.

To see that a straight violin string is in equilibrium, think of it as composed of many individual pieces that are connected together in a chain (Fig. 9.2.1). Tension exerts a pair of outward forces on each piece of the string; its neighboring pieces are pulling that piece toward them. Since the string's tension is uniform, these two outward forces sum to zero; they have equal magnitudes but point in opposite directions. With zero net force on each of its pieces, the straight string is in equilibrium.

When the string is curved, however, the pairs of outward forces no longer sum to zero (Fig. 9.2.2). Although those outward forces still have equal magnitudes, they now point in slightly different directions. As a result, each piece experiences a small net force.

The net forces on its pieces are restoring forces because they act to straighten the string. If you distort the string and release it, these restoring forces will cause the string to vibrate about its straight equilibrium shape in a natural resonance. But the string's restoring forces are special: the more you curve the string, the stronger the restoring forces on its pieces become. In fact, the restoring forces are **springlike forces**—they increase in proportion to the string's distortion—so the string is a form of harmonic oscillator!

Actually, the string is much more complicated than a pendulum or a balance ring. It can bend and vibrate in many distinct **modes** or basic patterns of distortion, each with its own period of vibration. Nonetheless, the string retains the most important feature of a harmonic oscillator: the period of each vibrational mode is independent of its amplitude. Thus a violin string's pitch doesn't depend on how hard it's vibrating. Think how tricky it would be to play a violin if its pitch depended on its volume!

A violin string has a simplest vibration: its **fundamental vibrational mode.** In this mode, the entire string arcs alternately one way then the other (Fig. 9.2.3). Its kinetic energy peaks as it rushes through its straight equilibrium shape and its potential energy (elastic potential energy in the string) peaks as it stops to turn around. The string's midpoint travels the farthest (the **vibrational antinode**) while its ends remain fixed (the **vibrational nodes**). At each moment its shape is the gradual curve of the trigonometric sine function.

In this fundamental mode, the violin string behaves as a single harmonic oscillator. As with any harmonic oscillator, its vibrational period depends only on the stiffness of its restoring forces and on its inertia. Stiffening the violin string or reducing its mass both quicken its fundamental vibration and increase its fundamental pitch.

A violin has four strings, each with its own stiffness and mass and therefore its own fundamental pitch. In a tuned violin, the notes produced by these strings are G_3 (196 Hz), D_4 (294 Hz), A_4 (440 Hz), and E_5 (660 Hz). The G_3 string, which vibrates rather slowly, is the most massive. It's usually made of gut, wrapped in a coil of heavy metal wire.

The E$_5$ string, on the other hand, must vibrate quite rapidly and needs to have a low mass. It's usually a thin steel wire.

You tune a violin by adjusting the tension in its strings, using pegs in its neck and tension adjusters on the tailpiece. Tightening the string stiffens it by increasing both the outward forces on its pieces and the net forces they experience during a distortion. Since temperature and time can alter a string's tension, you should always tune your violin just before a concert.

A string's fundamental pitch also depends on its length. Shortening the string both stiffens it and reduces its mass, so its pitch increases. That stiffening occurs because a shorter string curves more sharply when it's displaced from equilibrium and therefore subjects its pieces to larger net forces. This dependence on length allows you to raise a string's pitch by pressing it against the fingerboard in the violin's neck and effectively shortening it. Part of a violinist's skill involves knowing exactly where on the string to press it against the fingerboard in order to produce a particular note.

If the arc of a string vibrating in its fundamental mode reminds you of a wave, that's because it is one. It's a **mechanical wave**—the natural motions of an extended object about its stable equilibrium shape or situation. An *extended* object is one like a string, stick, or lake surface that has many parts that move with limited independence. Since its parts influence one another, an extended object with a stable equilibrium exhibits fascinating natural motions that involve many parts moving at once; it exhibits mechanical waves.

With its innumerable linked pieces and its stable equilibrium shape, the violin string exhibits such waves. And the string's fundamental mode is a particularly simple wave, a **standing wave**—a wave in which all the nodes and antinodes remain in place. A standing wave's basic shape doesn't change with time, it merely scales up and down rhythmically at a particular frequency and amplitude—its peak extent of motion. Most importantly, the standing wave doesn't travel along the string.

Although this wave extends along the string, its associated oscillation is *perpendicular* to the string and therefore *perpendicular* to the wave itself. A wave in which the underlying oscillation is perpendicular to the wave itself is called a **transverse wave.** Waves on strings, drums, and the surface of water are all transverse waves.

Second harmonic

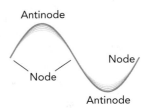

Third harmonic

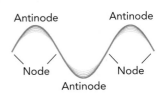

Fig. 9.2.4 A string vibrating between two fixed points in its second and third harmonic modes. The string vibrates as two or three segments, completing cycles at two or three times the fundamental frequency, respectively.

→ Check Your Understanding #2: Feeling Tense?

for answers, see page **298**

A common way to determine the tension in a cord is to pluck it and listen for how fast it vibrates. Why does this technique measure tension?

The Violin String's Harmonics

The fundamental vibrational mode isn't the only way in which a violin string can vibrate. The string also has **higher-order vibrational modes** in which the string vibrates as a chain of shorter strings arcing in alternate directions (Fig. 9.2.4). Each of these higher-order vibrational modes is another standing wave, with a fixed shape that scales up and down rhythmically at its own frequency and amplitude.

For example, the string can vibrate as two half-strings arcing in opposite directions and separated by a motionless vibrational node. In this mode, the violin string not only vibrates as half-strings, it has the pitch of half-strings as well. Remarkably, that half-string pitch is exactly twice the whole-string (i.e., fundamental) pitch! In general, a string's vibrational frequency is inversely proportional to its length, so halving its length doubles its frequency. Frequencies that are integer multiples of the fundamental pitch are called **harmonics,** so this half-string vibration occurs at the *second harmonic pitch* and is called the *second harmonic mode.*

A violin string can also vibrate as three third-strings, with a frequency that's three times the fundamental. The interval between this *third harmonic pitch* and the fundamental pitch is an octave and a fifth (2/1 times 3/2). Overall, the fundamental and its second and third harmonics sound very pleasant together.

While the violin string can vibrate in even higher harmonics, what's more important is that the string often vibrates in more than one mode at the same time. For example, a violin string vibrating in its fundamental mode can also vibrate in its second harmonic and emit two tones at once.

Harmonics are important because bowing a violin excites many of its vibrational modes. The violin's sound is thus a rich mixture of the fundamental tone and the harmonics. Known as **timbre,** this mixture of tones is characteristic of a violin, which is why an instrument producing a different mixture doesn't sound like a violin.

When a violin string is vibrating in several modes at once, its shape and motion are complicated. The individual standing waves add on top of one another, a process known as **superposition.** Each vibrational mode has its own amplitude and therefore its own volume contribution to the string's timbre.

While these individual waves coexist beautifully on the string, with virtually no effect on one another, the string's overall distorted shape is now the superposition of the individual waveshapes. Not only is that overall shape quite complicated, it actually changes substantially with time. That's because the different harmonic waves vibrate at different frequencies and their superposition changes as they change. The string's overall wave is not a standing wave and its features can even move along the string!

→ Check Your Understanding #3: Swinging High and Low Together

for answers, see page **298**

When two people swing a long jump rope, they can make it swing as a single arc or as two half-ropes arcing in opposite directions. To make the rope swing as two half-ropes, it must be turned faster or with less tension. Why?

Bowing and Plucking the Violin String

You play a violin by drawing a bow across its strings. The bow consists of horsehair, pulled taut by a wooden stick. Horsehair is rough and exerts frictional forces on the strings as it moves across them. But most importantly, horsehair exerts much larger static frictional forces than sliding ones.

As the bow hairs rub across a string, they grab the string and push it forward with static friction. Eventually the string's restoring force overpowers static friction, and the string suddenly starts sliding backward across the hairs. Because the hairs exert little sliding friction, the string completes half a vibrational cycle with ease. But as it stops to reverse direction, the hairs grab the string again and begin pushing it forward. This process repeats over and over.

Each time the bow pushes the string forward, it does work on the string and adds energy to the string's vibrational modes. This process is an example of **resonant energy transfer,** in which a modest force doing work in synchrony with a natural resonance can transfer a large amount of energy to that resonance. Just as gentle, carefully timed pushes can get a child swinging high on a playground swing, so gentle, carefully timed pushes from a bow can get a string vibrating vigorously on a violin. Similar pushes can cause other objects to vibrate strongly, notably a crystal wineglass (Fig. 9.2.5) and the Tacoma Narrows Bridge near Seattle, Washington (Fig. 9.2.6). The wineglass's response to a certain tone is also an example of **sympathetic vibration**—the transfer of *vibrational energy* between two systems that share a common vibrational frequency.

The amount of energy the bow adds to each vibrational mode depends on where it crosses the violin string. When you bow the string at the usual position, you produce a strong fundamental vibration and a moderate amount of each harmonic. Bowing the string nearer its middle reduces the string's curvature, weakening its harmonic vibrations and giving it a more mellow sound. Bowing the string nearer its end increases the string's curvature, strengthening its harmonic vibrations and giving it a brighter sound.

The sound of a plucked violin string also depends on harmonic content and thus on where that string is plucked. But this sound is quite different from that of a bowed string. The difference lies in the sound's *envelope*—the way the sound evolves with time. This envelope can be viewed as having three time periods: an initial *attack,* an intermediate *sustain,* and a final *decay.* The envelope of a plucked string is an abrupt attack followed immediately by a gradual decay. In contrast, the envelope of a bowed string is a gradual attack, a steady sustain, and then a gradual decay. We learn to recognize individual instruments not only by their harmonic content but also by their sound envelopes.

Steve Bronstein/The Image Bank/Getty Images

Fig. 9.2.5 Resonant energy transfer makes it possible for sound to shatter a crystal wineglass. When the sound pushes on the glass rhythmically, the sound slowly transfers energy to the glass, until it finally shatters. Because the sound must be extremely loud and at exactly the resonant frequency of the glass, only the most extraordinary opera singers can break a crystal wineglass.

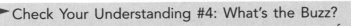

▶ Check Your Understanding #4: What's the Buzz?

for answers, see page **299**

Sometimes a tone from an instrument or sound system will cause some object in the room to begin vibrating loudly. Why does this happen?

AP/Wide World Photos

Fig. 9.2.6 The Tacoma Narrows Bridge collapsed in November 1940, as the result of resonant energy transfer between the wind and the bridge surface. Shortly after construction, the automobile bridge began to exhibit an unusual natural resonance in which its surface twisted slowly back and forth so that one lane rose as the other fell. During a storm, the wind slowly added energy to this resonance until the bridge ripped itself apart.

An Organ Pipe's Vibrating Air

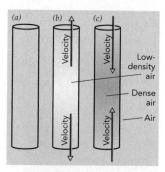

Fig. 9.2.7 In a pipe that's open at both ends (*a*), the air vibrates in and out about the middle of the pipe. (*b*) For half a cycle the air moves outward and creates a low-pressure region in the middle and (*c*) for half a cycle the air moves inward and creates a high-pressure region there.

Like a violin, a pipe organ uses vibrations to create sound. However, its vibrations take place in the air itself. An organ pipe is essentially a hollow cylinder, open at each end and filled with air. Because that air is isolated, its pressure can fluctuate up and down relative to atmospheric pressure and it can exhibit natural resonances.

In its fundamental vibrational mode, air moves alternately toward and away from the pipe's center (Fig. 9.2.7), like two blocks on a spring. As air moves toward the pipe's center, the density there rises and a pressure imbalance develops. Since the pressure at the pipe's center is higher than at its ends, air accelerates *away* from the center. The air eventually stops moving inward and begins to move outward. As air moves away from the pipe's center, the density there drops and a reversed pressure imbalance occurs. Since the pressure at the pipe's center is lower than at its ends, air now accelerates *toward* the center. It eventually stops moving outward and begins to move inward, and the cycle repeats. The air's kinetic energy peaks each time it rushes through that equilibrium and its potential energy (pressure potential energy in the air) peaks each time it stops to turn around.

This air is vibrating about a stable equilibrium of uniform atmospheric density and pressure, and is clearly experiencing restoring forces. It should come as no surprise that those restoring forces are springlike and that the air column is yet another harmonic oscillator. As such, its vibrational frequency depends only on the stiffness of its restoring forces and on its inertia. Stiffening the air column or reducing its mass both quicken its vibration and increase its pitch.

These characteristics depend on the length of the organ pipe. A shorter pipe not only holds less air mass than a longer pipe, it also offers stiffer opposition to any movements of air in and out of that pipe. With less room in the shorter pipe, the pressure inside it rises and falls more abruptly, leading to stiffer restoring forces on the moving air. Together, these effects make the air in a shorter pipe vibrate faster than that in a longer pipe. In general, an organ pipe's vibrational frequency is inversely proportional to its length.

Unfortunately, the mass of vibrating air in a pipe also increases with the air's average density, so that even a modest change in temperature or weather will alter the pipe's pitch. Fortunately, all of the pipes shift together so that an organ continues to sound in tune. Nonetheless, this shift may be noticeable when the organ is part of an orchestra.

As you may suspect, the fundamental vibrational mode of air in the organ pipe is another standing wave. Air in the pipe is an extended object with a stable equilibrium and the disturbance associated with its fundamental vibrational mode has a basic shape that doesn't change with time; it merely scales up and down rhythmically.

However, the shape of the wave in the pipe's air now has to do with back-and-forth compressions and rarefaction, not with side-to-side displacements as it did in the violin string. In fact, all of the wave's associated oscillation is *along* the pipe and therefore *along* the wave itself. A wave in which the underlying oscillation is parallel to the wave itself is called a **longitudinal wave**. Waves in the air, including those inside organ pipes and other wind instruments and sound waves in the open air, are all longitudinal waves.

▶ Check Your Understanding #5: A Pop Organ

for answers, see page **299**

If you blow across a soda bottle, it emits a tone. Why does adding water to the bottle raise the pitch of that tone?

Playing an Organ Pipe

The organ uses resonant energy transfer to make the air in a pipe vibrate. It starts this transfer by blowing air across the pipe's lower opening (Fig. 9.2.8a), although for practical reasons that lower opening is usually found on the pipe's side (Fig. 9.2.8b). As the air flows across the opening, it's easily deflected and tends to follow any air that's already moving into or out of the pipe. If the air inside the pipe is vibrating, the new air will follow it in perfect synchrony and strengthen the vibration.

This following process is so effective at enhancing vibrations that it can even initiate a vibration from the random noise that's always present in a pipe. That's how the sound starts when the organ's pump first blows air across the pipe. Once the vibration has started, it grows quickly in amplitude until energy leaves the pipe as sound and heat as quickly as it arrives via compressed air. The more air the organ blows across the pipe each second, the more power it delivers to the pipe and the louder the vibration.

Like a violin string, an organ pipe can support more than one mode of vibration. In its fundamental vibrational mode, the pipe's entire column of air vibrates together. In the higher-order vibrational modes, this air column vibrates as a chain of shorter air columns moving in alternate directions. If the pipe has a constant width, these vibrations occur at harmonics of the fundamental. When the air column vibrates as two half-columns, its pitch is exactly twice that of the fundamental mode. When it vibrates as three third-columns, its pitch is exactly three times that of the fundamental. And so on.

But the air column inside a pipe can vibrate in more than one mode simultaneously. As with a violin string, the standing waves superpose and the fundamental and harmonic tones are produced together. The shape of the organ pipe and the place where air is blown across it determine the pipe's harmonic content and thus its timbre. Different pipes can imitate different instruments. To sound like a flute, the pipe should emit mostly the fundamental tone and keep the harmonics fairly quiet. To sound like a clarinet, its harmonics should be much louder. An organ pipe's volume always builds slowly during the attack, so it can't pretend to be a plucked string. However, a clever designer can make the organ imitate a surprising range of instruments.

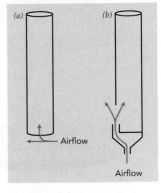

Fig. 9.2.8 (*a*) Air blown across the bottom of an open pipe will follow any other air that's moving into the pipe. If the air in the pipe is vibrating, this effect will add energy to that vibration. (*b*) The lower opening in an organ pipe is cut in its side for practical reasons.

> **Check Your Understanding #6: An Across Bow**
> *for answers, see page* **299**

To make the air in a soda bottle vibrate, you must blow *across* the bottle's mouth. Why doesn't blowing *into* that mouth work?

A Drum's Vibrating Surface

After examining violin strings and organ pipes, it might seem that drums offer little new. But while a drumhead is yet another extended object with a stable equilibrium and springlike restoring forces, its overtone vibrations have an important difference: they *aren't* harmonics.

Violin strings and organ pipes are effectively one-dimensional objects, dividing easily into half-objects or third-objects that then vibrate at second or third harmonic pitches. Together with the many other one-dimensional instruments in an orchestra or band, they blend seamlessly when they're playing the same fundamental pitch because they share the same harmonics.

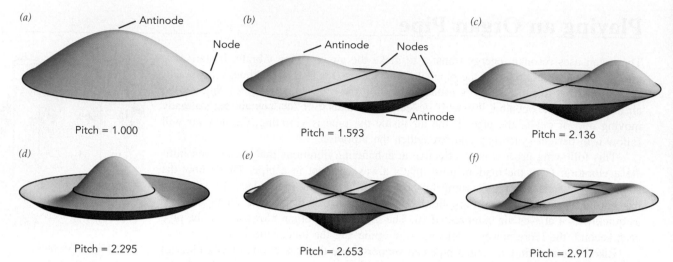

Fig. 9.2.9 The six lowest-pitch vibrational modes of a drumhead, including (*a*) the fundamental vibrational mode and (*b–f*) overtone modes. Pitches are shown relative to the fundamental pitch.

❑ In 1809, the French Academy of Sciences announced a competition to explain of the intricate patterns observed on vibrating surface plates. The only respondent was French mathematician Sophie Germain (1776–1831). As a woman, Germain was barred from formal education in mathematics and struggled to learn the subject from books and via correspondences with leading mathematicians, which she conducted under the pseudonym Antoine-August Le Blanc. It took her three tries, but in 1816 she was awarded the prize. Because she was a woman, however, she did not attend the ceremony. Her analysis of surface vibrations, though imperfect, was a visionary effort, made all the more extraordinary by her circumstances. Although her mentor, Carl Fredrick Gauss, managed to convince the University of Göttingen to award her an honorary degree, she died of breast cancer before she could receive it.

But because a drumhead is effectively two-dimensional, it doesn't divide easily into pieces that resemble the entire drumhead. As a result, the pitches of its overtone vibrations have no simple relationship to its fundamental pitch. A kettledrum or timpani stands out relative to other instruments in part because of the unique overtone pitches.

Figure 9.2.9 illustrates the fundamental (*a*) and five lowest-pitched overtone vibrational modes (*b–f*) for a drumhead. Each vibrational mode is a standing wave, but with vibrational nodes that are curves or lines rather than points. The fundamental mode (*a*) has only one node on its outer edge, while the overtone modes have additional nodes within the surface. In each vibrational mode, these nodes remain motionless as the rest of the surface vibrates up and down, its peaks and valleys interchanging alternately. The pitches of the overtone vibrations are indicated relative to that of the fundamental vibration. (For a historical note on the understanding of surface modes, see ❑).

Because striking a drumhead causes it to vibrate in several modes at once, the drum emits several pitches simultaneously. The amplitude of each mode, and consequently its volume, depends not only on *how hard* you hit the drumhead, but also on *where* you hit it. If you hit it at its center, it vibrates primarily in circular modes like (*a*) and (*d*). If you hit it nearer its edge, it also vibrates in noncircular modes like (*b*), (*c*), (*e*), and (*f*).

A timpani sounds most musical when it's struck off-center in such a way that the amplitude of its fundamental vibrational mode is nearly zero and its overtones, particularly (*b*), dominate its sound. That's because the fundamental vibrational mode emits sound so well that its vibrational energy dissipates before it can produce a discernible tone. Unless all you want is a loud "thump," you must hit the timpani off-center so that its long-lived overtone vibrations receive most of the energy and emit most of the sound. The dominant pitch of a properly played timpani is that of its first overtone vibration and it is tuned with that pitch in mind.

In truth, the pitches shown in Fig. 9.2.9 neglect the effects of air's inertia on the drumhead's vibrations. Since air adds inertia to the drumhead, it lowers the pitches of all the vibrational modes, some more than others. Because of air's influence on pitch, a drum must be tuned to accommodate changes in temperature and weather.

Check Your Understanding #7: Space Kid One

for answers, see page 299

A trampoline is hazardous with several children on it because a child landing on one side of its surface can launch skyward a second child standing on the other side of the surface. How does a downward impact on one side of the trampoline produce a sudden rise of the other side?

Sound in Air

All these vibrations would serve little purpose if we couldn't hear them, so it's time to look at how instruments produce sound. We'll start by looking at sound itself.

We noted at the beginning of this section that sound in air consists of density waves—patterns of compressions and rarefactions that travel outward rapidly from their source. While that observation was mysterious at the time, we can now understand those waves as vibrations in an extended object with a stable equilibrium. That extended object is air.

Neglecting gravity, air is in a stable equilibrium when its density is uniform. If we disturb it from equilibrium, the resulting pressure imbalances will provide springlike restoring forces. These forces, together with air's inertia, lead to rhythmic vibrations—the vibrations of harmonic oscillators. In open air, the most basic vibrations are waves that move steadily in a particular direction and are therefore called **traveling waves.** Like the standing waves inside an organ pipe, these traveling waves in open air are longitudinal—air vibrates along the same direction as the sound wave travels.

As it moves through the open air, a basic traveling sound wave consists of an alternating pattern of high density regions we'll call **crests** and low density regions we'll call **troughs** (Fig. 9.2.10). While those names will seem more appropriate when we examine water surface waves in the next section, it's customary to refer to the alternating highs and lows of any wave as crests and troughs, respectively. Whether a wave is standing or traveling, the shortest distance between two adjacent crests is known as the **wavelength.**

While a standing wave's crests and troughs merely flip back and forth in place, crests becoming troughs and troughs becoming crests, a traveling wave's crests and troughs move steadily in a particular direction at a particular speed. That speed and direction of travel together constitute the traveling wave's **wave velocity.**

Figure 9.2.11 shows five snapshots of a simple sound wave that's heading toward the right. If we watch air's density at the same point in space (green line), it begins as a crest (*a*), decreases (*b*) to a trough (*c*), then rises (*d*) back to a crest (*e*) during one complete vibration cycle. However, if we follow the same crest (red line) over time, it travels one wavelength to the right during one complete vibration cycle (*a–e*). Since a crest moves one wavelength per vibration cycle and frequency is the number of vibration cycles per second, the speed at which the crest moves is the product of the wavelength times the frequency. That relationship can be written as a word equation:

$$\text{wave speed} = \text{wavelength} \cdot \text{frequency} \qquad (9.2.1)$$

in symbols:

$$v = \lambda \nu,$$

and in everyday language:

Broad waves that vibrate quickly travel fast.

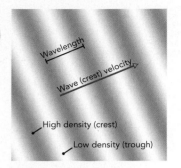

Fig. 9.2.10 A traveling sound wave in air consists of a washboard pattern of high density (blue) and low density (white) regions. The distance separating adjacent crests is the wavelength and the speed and direction of crest motion are the wave velocity.

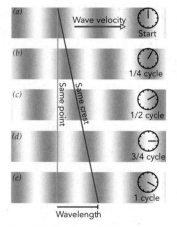

Fig. 9.2.11 A sound wave at five evenly spaced times (*a–e*) showing one complete cycle of oscillation. During that cycle, the pressure at a specific point in space goes from high to low to high (green line) and a specific crest moves one wavelength to the right (red line).

Remarkably enough, all sound waves travel at the same speed through air, regardless of wavelength or frequency. That's because a sound's wavelength is always inversely proportional to its frequency. We saw this same inverse relationship for the standing waves in organ pipes: if you double the length of a pipe, and therefore double the wavelength of its fundamental vibrational mode, you halve the frequency of that vibration. Even when it's not confined by a pipe, vibrating air has a wavelength that's inversely proportional to the frequency of that vibration.

Because of this extraordinary inverse relationship between a traveling sound wave's wavelength and its frequency, Eq. 9.2.1 yields the same wave speed for any sound wave. Known as the **speed of sound** in air, it's about 331 m/s (1086 ft/s) in standard conditions at sea level (0 °C, 101,325 Pa pressure). While that's fast, there is still a noticeable delay between when a percussionist strikes the cymbals and when you hear them from across the concert hall. Fortunately, because the speed of sound doesn't depend on frequency, when the entire orchestra plays in unison you hear all its different pitches simultaneously.

This discussion of sound assumes that the instruments and listener maintain a constant separation, as they usually do at an orchestra concert. But when a marching band dashes toward or away from the listener, something odd happens: the listener hears its music shifted up or down in pitch. Known as the **Doppler effect,** this frequency shift occurs because the listener encounters sound wave crests at a rate that's different from that at which those crests were created. If an instrument and the listener are approaching one another, the listener encounters the crests at an increased rate and the pitch increases. If the two are separating from one another, the listener encounters the crests at a decreased rate and the pitch decreases. Fortunately, the Doppler effect is subtle at speeds that are small compared to the speed of sound, so you can listen to parades without them sounding flat or sharp.

Check Your Understanding #8: Sounds Faster

for answers, see page 299

Helium in a toy balloon has the same stiffness as ordinary air, but its density and inertia are smaller. How does this difference affect the speed of sound in helium?

Check Your Figures #1: Underwater Sound

for answers, see page 299

Although water is about 800 times as dense as ordinary air, water is also about 15,000 times as stiff and its sound vibrations therefore have increased frequencies relative to those in air. When two sound waves have equal wavelengths, the wave in water has a frequency about 4.3 times greater than the wave in air. What is the speed of sound in water?

Turning Vibrations into Sound

Anything that disturbs air's otherwise uniform density can produce traveling sound waves. Instruments emit sound by compressing and rarefying the nearby air in synch with their own vibrations. How they accomplish this task differs from instrument to instrument, so we'll have to look at them individually. As we'll see, some instruments find it easier to produce sound than others.

A drum produces sound when its vibrating drumhead alternately compresses and rarefies the nearby air. As portions of that drumhead rise and fall, they upset the air's uniform density and thereby produce sound waves. But whenever it can, air simply flows silently out of the drumhead's way, leading to smaller density fluctuations and less intense sound. For example, when the drumhead is experiencing one of the five overtone vibrations shown in Fig. 9.2.9, air flows away from each rising peak in the undulating surface and toward each falling valley. The overtone vibrations still manage to produce sound, but it's less intense and the vibrational energy in the drumhead transforms relatively slowly into sound energy.

Air's partial success in dodging the drumhead's overtone vibrations allows those overtones to complete many vibrational cycles before running out of vibrational energy. Their vibrations are therefore long-lived and have distinct pitches. In contrast, air has difficulty dodging the drumhead's fundamental vibrational mode, which alternately compresses and rarefies the air so effectively that it transfers all of its vibrational energy to the air in just a few cycles. That's why the fundamental vibrational mode produces an intense and nearly pitchless "thump" sound.

If air can dodge a vibrating surface, it can certainly dodge a vibrating string. Little of a violin's sound comes directly from its vibrating strings; the air simply skirts around them. Instead, the violin creates sound with its top plate or *belly* (Fig. 9.2.12). The strings transfer their vibrational motions to the belly and the belly pushes on the air to create sound.

Most of this vibrational energy flows into the belly through the violin's *bridge,* which holds the strings away from the violin's body (Fig. 9.2.13). Beneath the G_3 string side of the bridge is the *bass bar,* a long wooden strip that stiffens the belly. Beneath the E_5 side of the bridge is the *sound post,* a shaft that extends from the violin's belly to its back.

As a bowed string vibrates across the violin's belly, it exerts a torque on the bridge about the sound post. The bridge rotates back and forth, causing the bass bar and belly to move in and out. The belly's motion produces most of the violin's sound. Some of this sound comes directly from the belly's outer surface, and the rest comes from its inner surface and must emerge through its f-shaped holes.

An organ pipe doesn't have to produce sound because that sound already exists. In effect, the pipe's vibrating column of air is a standing sound wave that gradually leaks out of the pipe as a traveling one. Trapped sound is escaping from its container.

This conversion of a standing wave into a traveling wave isn't so remarkable because the two types of waves are closely related. The pipe's standing wave can be thought of as a reflected traveling wave, a traveling wave that's bouncing back and forth between the two ends of the pipe. Because of the reflections, the traveling wave is superposed with itself heading in the opposite direction and the sum of two equal but oppositely directed traveling waves *is* a standing wave!

The fact that sound reflects from the open end of an organ pipe is rather surprising. If that end were closed, you'd probably expect a reflection. After all, sound echos from cliffs and other rigid surfaces. But sound partially reflects from a surprising range of other transitions, including the transition from inside a pipe to outside it. If you don't believe that, clap your hands inside a long pipe and listen for the decaying echos.

The reflections at the organ pipe's open ends aren't perfect, so the trapped sound wave gradually leaks out and becomes the sound you hear. This process of letting a standing sound wave emerge slowly as a traveling wave is typical of woodwind and brass instruments. The reflection at an open pipe end depends on the shape of that end. Flaring it into the horn shape common in brass instruments reduces the reflection and eases the transition from standing wave to traveling wave. That's why horns project sound so well.

Fig. 9.2.12 A violin's bridge transfers energy from its vibrating strings to its belly. The belly moves in and out, emitting sound. Some of this sound leaves the violin through the f-holes in its body.

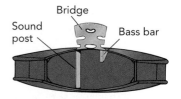

Fig. 9.2.13 The bridge is supported by the bass bar on one side and the sound post on the other. As the strings vibrate back and forth, the bridge experiences a torque that causes the belly of the violin to move in and out and emit sound.

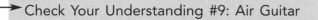

Check Your Understanding #9: Air Guitar

for answers, see page **299**

Why does an acoustic guitar have a sound box?

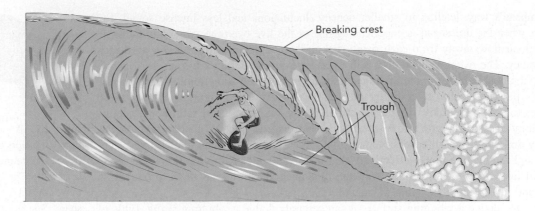

Breaking crest

Trough

SECTION 9.3 **The Sea**

The sea is never still. If you've visited the seashore, you've probably noticed two of the sea's most important motions: tides and surface waves. In this section, we'll examine the cycle of tides and look at how surface waves travel across water. These water waves can help us understand other wave phenomena, including the electromagnetic waves that are responsible for light and the density waves that are the basis for sound.

Questions to Think About: *Why do the tides vary in height from place to place? Why are there no significant tides in a lake or swimming pool? Why does high tide occur about every 12 hours? What moves in a water wave? Do all waves travel at the same speed? How deep is a wave? Why do waves break near shore? Why do waves always seem to head almost directly toward shore? Why is there often a rhythm to the surf?*

Experiments to Do: *While you can make water waves in a basin or tub, they move too quickly to see clearly. You do better to watch waves at the beach. On a calm day, the sea is in its flat equilibrium shape. But when the wind picks up, look out. How does wind affect the sea's surface? If water waves contain energy, where does that energy come from and what forms does it take in the water?*

Watch a floating object move as a wave passes it. Does the object travel with the wave? Does the water itself travel with the wave? Can you think of other cases in which a disturbance moves steadily forward but the material carrying the disturbance remains behind?

Now watch as waves break near the shore. If you look around at different beaches, you'll find waves breaking in two different ways. In some cases a wave will simply collapse into churning froth, while in others its top will dive forward over the water ahead of it. Can you see anything about the beaches or water that might account for this difference?

The Tides

If you watch the sea for a few days, you'll notice the *tides.* In a cycle as old as the oceans themselves, the water level rises for about 6¼ hours to reach *high tide,* drops for about 6¼ hours to arrive at *low tide,* and then begins rising again. Once a wonderful mystery, we now know that the tides are caused by the earth's rotation, the moon's gravity, and, to a lesser extent, the gravity of the sun.

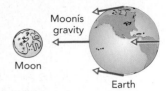

Fig. 9.3.1 The moon's gravity varies over the surface of the earth. The closer a point is to the moon, the stronger the gravity it experiences. This variation in the moon's gravity produces the tides.

On earth, the moon's gravity is so weak that we normally don't notice it. The moon is far away and, as we learned in Section 4.2, gravity diminishes with distance. But this dependence on distance also means that the moon's gravity is stronger on one side of the earth than the other; you experience a stronger pull when you're on the side of the earth nearest the moon than you do when you're on the side opposite it (Fig. 9.3.1). While you can't feel these variations in moon gravity yourself, the earth's oceans respond to them. The oceans are deformed by the moon's gravity (Fig. 9.3.2) and this deformation produces the tides.

The differences between the moon's gravity at particular locations on earth and its average strength for the entire earth give rise to **tidal forces**—residual gravitational forces that act to displace those locations relative to the earth as a whole. The near side of the earth is pulled toward the moon more strongly than average so it experiences a tidal force toward the moon. The far side of the earth is pulled toward the moon less strongly than average so it experiences a tidal force away from the moon.

If the earth were less rigid, these tidal forces would stretch it into an egg shape. The near side of the earth would bulge outward toward the moon, while the far side of the earth would bulge outward away from the moon. But while the earth itself is too stiff to deform much, the oceans are not and they bulge outward in response to the tidal forces. Two *tidal bulges* appear: one closest to the moon and one farthest from the moon (Fig. 9.3.2). A beach located in one of these tidal bulges experiences high tide, while one in the ring of ocean between the bulges experiences low tide.

As the earth rotates, the locations of the two tidal bulges move westward around the equator. Since a particular beach experiences high tide whenever it's closest to or farthest from the moon, the full cycle from high to low to high tide occurs about once every 12 hours and 25 minutes. The extra 25 minutes reflects the fact that the moon isn't stationary; it orbits the earth every 27.3 days and thus passes overhead once every 24 hours and 50 minutes, rather than every 24 hours.

But the moon isn't the only source of tidal forces on the earth's oceans. Although the sun is much farther away than the moon, it's so massive that the tidal forces it exerts are almost half as large as those exerted by the moon. The sun's principal effect is to increase or decrease the strength of the tides caused by the moon (Fig. 9.3.3). When the moon and the sun are aligned with one another, their tidal forces add together and produce extra large tidal bulges. When the moon and the sun are at right angles to one another, their tidal forces partially cancel and produce tidal bulges that are unusually small.

Twice each lunar month, the time it takes for the moon to orbit the earth, the tides are particularly strong. These *spring tides* occur whenever the moon and sun are aligned with one another (full moon and new moon). Twice each lunar month the tides are particularly weak. These *neap tides* occur whenever the moon and sun are at right angles to one another (half moon).

Fig. 9.3.3 The tides vary over a lunar month. They're strongest when the sun and the moon are aligned (spring tides) and weakest when they are at 90° from one another (neap tides).

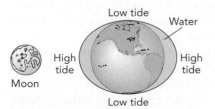

Fig. 9.3.2 Variations in the moon's gravity produce tidal forces on the earth's surface and cause the earth's oceans to bulge outward in two places. These bulges, which are located nearest and farthest from the moon, move over the earth's surface as it rotates.

Because of this interplay between lunar and solar tidal effects, the cycle of tides varies slightly from day to day. While the average cycle is 12 hours and 25 minutes, it fluctuates over a lunar month. Moreover, the exact moment of high or low tide at a particular location is influenced by water's inertia, the earth's rotation, and the environment through which water must flow to form the tidal bulge. That's why shore areas often publish tables of the local tides.

> ### Check Your Understanding #1: Seaside Vacations of the Tidal Rich
> for answers, see page 299

The Atlantic and Pacific Oceans come very close to one another in Central America. If the Atlantic side of this isthmus is experiencing high tide, what tide is the Pacific side experiencing?

Tidal Resonances

The sizes of the tides depend on where you are, but high tide is typically a meter or two above low tide. Because the tidal bulges are located near the equator, tides far to the north or south are smaller than that; tides in isolated lakes or seas are smaller still because water can't flow in to create the bulges. However, there are a few special places that have enormous tides. For example, tides in the Bay of Fundy, an estuary between New Brunswick and Nova Scotia, can change the water level by as much as 15 m. How can tides ever get this large?

Giant tides result from natural resonances in channels and estuaries. Just as air in an organ pipe can be made to vibrate strongly by a series of carefully timed pushes from a pump, water in a channel or estuary can be made to oscillate strongly by a series of carefully timed pushes from the tides. The water in that channel is another extended object with a stable equilibrium, so it will oscillate about its equilibrium after being disturbed. Giant tides occur through resonant energy transfer when the cycle of the tides gradually builds up the amplitude of a suitable standing wave in a channel.

However, while standing waves in an organ pipe involve the column of air as a whole, standing waves in a channel involve primarily the water's open surface. That water is at equilibrium when its surface is smooth and horizontal, and it experiences spring-like restoring forces whenever its surface is disturbed. For the broad waves we're considering in this section, those restoring forces are due to gravity and they are known as *gravity waves*. For the miniature waves in a drinking glass, however, water's springy, elastic surface contributes significantly to the restoring forces. Waves that involve this *surface tension* are known as *capillary waves*.

You can observe standing gravity waves on the surface of water in a large basin. If you begin pushing the water back and forth rhythmically with your hand, you'll produce waves on its surface. And if you time your pushes carefully so that they're synchronized with the natural rhythm of a particular standing wave, you'll build up the amplitude of that wave by resonant energy transfer. Countless children have discovered this phenomenon as a way to amuse themselves at bath time—to the dismay of their parents and the delight of people who repair water-damaged floors and ceilings.

Like standing waves on violin strings and drumheads, standing **surface waves** on water are transverse—the water's vertical surface vibrations are perpendicular to the horizontal waves themselves. In a basin, the fundamental vibrational mode (Fig. 9.3.4) has a node along the middle of the basin and antinodes at either end. At one antinode, water arcs up to a crest—a maximum upward displacement from equilibrium. At the other

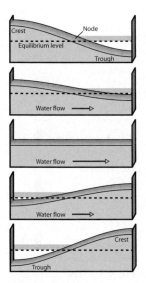

Fig. 9.3.4 Water sloshing in a basin in its fundamental mode. Its surface oscillates up and down, its crest becoming a trough and vice versa, over and over again.

Fig. 9.3.5 The giant tides in the Bay of Fundy can cause its water level to change by as much as 15 m between high and low tide.

antinodes, water arcs down to a trough—a maximum downward displacement from equilibrium. Over time, the crest drops to become a trough while the trough rises to become a crest. That process reverses and then repeats, over and over again until the sloshing water turns all of its vibrational energy into thermal energy or transfers it elsewhere.

The giant tides at the end of an estuary are simply the antinode of a standing wave fluctuating up and down between a crest and a trough. While the standing waves in an ordinary washbasin have periods measured in seconds or less, large bodies of water can sustain standing waves known as *seiches* that have periods of minutes or even hours.

Water in the Bay of Fundy has a fundamental seiche mode with a period of roughly 13.3 hours. Since this period nearly matches the 12.5 hour cycle of the tides, there's a resonant transfer of energy from the moon to the water oscillating in the estuary. The tides drive water back and forth in this estuary until, after many cycles, that water is moving so strongly that its height varies dramatically with time (Fig. 9.3.5).

▶ Check Your Understanding #2: Potential Profits

for answers, see page 299

People occasionally propose using giant tides to generate electric power. But there's a problem with this idea. If you extract all of the gravitational potential energy from the water at high tide in the Bay of Fundy, how long will it take for a giant high tide to appear again?

Traveling Waves on the Surface of Water

As you sit at the seashore on a cloudless day, enjoying a warm, steady breeze, you can't help but notice that the sea in front of you is covered with ridges. These ridges move steadily toward land and finally crash on the beach. Though it's customary to think of each breaking swell as a separate wave, we'll find it useful to view the entire moving pattern of evenly spaced ridges as a single wave—a traveling surface wave on water.

Traveling surface waves are the basic modes of oscillation on the open ocean, the simplest waves on that effectively *limitless* surface. Despite their steady progress across the water, these traveling waves actually involve oscillation. You can see this oscillation by watching a fixed point on the water's surface. That point fluctuates up and down as the crests and troughs of a traveling wave pass through it. The period of this oscillation is the time required for one full cycle of rise and fall, and the frequency is the number of crests passing through that fixed point each second.

The ocean's surface can host an incredible variety of traveling waves, each moving in its own direction with its own period and frequency. Moreover, these basic waves can coexist, adding together on the ocean's surface to create ever more complicated patterns. Like primary colors, which when blended in proper proportions can produce any possible color, traveling waves can be superposed in proper proportions to produce any possible surface pattern or wave. When the ocean is rough and its surface features ripples layered upon swells layered upon broad undulations, you're seeing this superposition of "primary" traveling waves in all its glory.

In contrast, the basic modes of oscillation on a channel or lake are standing surface waves—the simplest waves on that *limited* surface. In a standing wave, the water's surface oscillates up and down vertically, with its crests and troughs interchanging periodically: crests become troughs and troughs become crests. The standing wave's pattern of crests and troughs doesn't move anywhere; it simply flips up and down in place at a certain frequency.

On their limited surface, these standing waves can be superposed to produce any possible surface pattern or wave, so they, too, are like primary colors. Overall, traveling waves constitute the primary palette of waves for a limitless ocean and standing waves constitute the primary palette of waves for a limited channel or lake.

> **Standing and Traveling Waves**
> The most basic waves on an extended object of limited dimensions are standing waves. With their different periods and/or patterns, these standing waves can be superposed to form any possible wave on that limited object.
>
> The most basic waves on an extended object of limitless dimensions are traveling waves. With their different periods and/or directions of travel, these traveling waves can be superposed to form any possible wave on that limitless object.

Actually, we've seen these ideas before in the context of musical instruments and sound. Since instruments are limited objects, their basic vibrations are standing waves—the fundamental and overtone vibrations. And because air is effectively limitless, its basic vibrations are traveling waves—the sound waves it carries. The timbre of an instrument reveals the superposition of its many standing waves, while the full sound of an orchestra or band displays the superposition of its many traveling waves.

Both standing and traveling water surface waves carry energy, energy which they typically obtain from the wind, the tide, or occasionally seismic activity. Each wave's energy consists of kinetic energy in moving water and gravitational potential energy in water that has been displaced from level.

In a standing wave, the energy of the entire wave fluctuates back and forth between kinetic and gravitational potential; the wave's kinetic energy peaks as the surface rushes through its level equilibrium, and its potential energy peaks as the surface stops to turn around at its maximum displacement from equilibrium.

A traveling wave's crests and troughs move steadily forward, so the water is never level or motionless. The energy in a traveling wave is therefore always an even mixture of kinetic and potential energies. And because of its directed motion, a traveling wave carries momentum pointing in the same direction as the wave velocity.

Check Your Understanding #3: The Frequency of Seasickness
for answers, see page 299

You're sitting in a small boat on the open ocean, rising and falling with each passing wave crest. How could you measure the frequency of the wave passing under your boat?

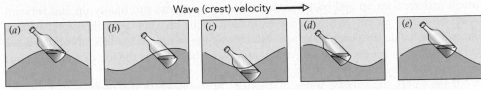

Wave (crest) velocity ⟶

Fig. 9.3.6 You can see that water doesn't move with a wave by watching a bottle floating on the water. The bottle moves in a circle as each crest passes by. Starting at (*a*) a crest, the bottle moves (*b*) down and right, (*c*) down and left, (*d*) up and left, and then (*e*) up and right. It returns to its original position just as the next crest arrives.

The Structure of a Water Wave

We've seen that a traveling surface wave moves across the open ocean with a certain velocity, wavelength, and frequency. But what is the water itself doing as the wave passes?

You can begin to answer that question by watching a bottle floating on the water's surface (Fig. 9.3.6). As a wave passes it, that bottle rises and falls with the crests and troughs, but it makes no overall progress in any direction. Instead, the bottle travels in a circle. Like the bottle, the water itself doesn't actually move along with the passing wave. Although this water bunches up to create each crest and spreads out to create each trough, it returns to its starting point once the wave has left.

Like the bottle in Fig. 9.3.6, a patch of water on the ocean's surface moves in a circular pattern (Fig. 9.3.7) as a traveling wave passes. Water that starts out on top of a crest moves down and forward as the crest departs. It moves down and backward as the

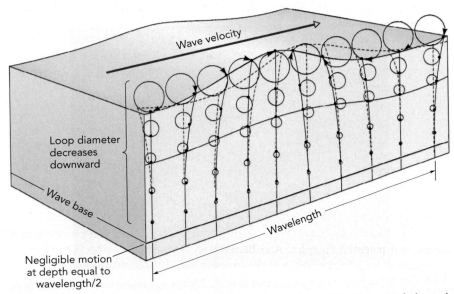

Fig. 9.3.7 Surface water moves in a circular motion as a wave passes. The water currently located at each dark dot will follow the circular path outlined around it as time passes. The circles are largest at the surface and become relatively insignificant once you look more than half a wavelength below the surface. The sense of the circular motion (clockwise or counterclockwise) determines the direction in which the wave travels. This wave travels toward the right.

trough arrives, then up and backward as the trough departs, and finally up and forward as the next crest arrives. When it reaches the top of the arriving crest, this water is back where it started on the ocean's surface. Which way the water circles depends on the wave velocity's direction—the wave's direction of travel. The water at the top of a crest always moves in the same direction as the wave itself.

It isn't only the surface water that moves; water beneath the surface also circles. However, the circles diminish in radius gradually with depth and become negligible at a depth roughly equal to half the wave's wavelength. Thus while it's called a *surface* wave, it has a *depth* to it and therefore a sensitivity to shallow water, as we'll soon see.

These surface waves have another interesting characteristic: their wave speed increases with wavelength. As you may have noticed, long-wavelength swells travel faster than short-wavelength ripples. That's quite different from sound waves, which all have the same wave speed regardless of wavelength.

Such dependence of wave speed on wavelength is known as **dispersion.** It occurs in this case because water's surface is surprisingly stiff when carrying long-wavelength waves. Unlike a tense string, which opposes short-wavelength distortions much more stiffly than long-wavelength ones, water's surface uses its weight to oppose long-wavelength disturbances almost as vigorously as it opposes short-wavelength ones. That heightened stiffness for long-wavelength waves boosts their frequencies and therefore increases their wave speeds (see Eq. 9.2.1).

Little ripples have short wavelengths and travel slowly, while large ocean swells have long wavelengths and travel much more rapidly. Giant waves produced by earthquakes and volcanic eruptions, known as *tsunamis,* have extremely long wavelengths and can travel at hundreds of kilometers-per-hour. Because these giant waves travel so fast and move water so deep in the ocean, they carry enormous amounts of energy and momentum and are potentially disastrous to shore areas (Fig. 9.3.8).

© Reuters/Corbis Images

Fig. 9.3.8 The Indian Ocean tsunami of December 26, 2004, was initiated by a sudden rise in the ocean floor off the northern coast of Sumatra. This long-wavelength traveling wave moved so quickly through the surrounding Indian Ocean that most shore inhabitants received no warning and were caught unprepared. Moreover, the wave's phenomenal troughs exposed vast stretches of seabed, attracting inquisitive people offshore where they were then unprotected from the destructive crests that followed. Roughly a quarter-million people perished.

► Check Your Understanding #4: Beneath the Waves
for answers, see page **299**

If you're swimming and want to dive beneath a wave, how deep must you go to avoid any significant motion in the water?

Waves at the Shore

As a wave approaches the shore, it begins to travel through shallow water. Since the water's circular motion extends below the surface, there comes a point at which the wave begins to encounter the seabed. Once the water is shallower than half the wavelength of the wave, the seabed distorts the water's circular motion so that it becomes elliptical.

That change has a number of interesting effects on the wave. First, its wave speed gradually decreases so that the crests begin to bunch together. Second, its amplitude—the height of its crests and depth of its troughs—increases as the wave acts to keep its overall forward momentum constant despite its decrease in speed. These two effects explain why waves that look broad and gradual on the open ocean look quite steep and dangerous nearer the beach. Their crests have bunched together and grown taller, so the slopes between crests and troughs really have become steeper.

A third effect of the shallowing water is a gradual change in the wave's direction of travel. Known as **refraction,** this bending occurs whenever a wave's speed changes as it passes from one environment to another. Since a water surface wave slows as it approaches the shore, that wave refracts—it bends—so as to head more directly toward the shore (Fig. 9.3.9). Because of refraction, waves approach the beach almost perpendicular to it, even if they were traveling at relatively oblique angles far from shore.

A fourth and final effect is the destruction of the wave—it eventually runs out of water and crashes onto the beach. The wave builds each of its crests using local water. When the crest enters very shallow water, there isn't enough water in front of it to construct its forward side. The crest becomes incomplete and begins to "break."

The form of its demise depends on the slope of the seabed. If that slope is gradual, the wave breaks slowly to form a smooth, rolling, "boiling" surf (Fig. 9.3.10). But if the slope of the seabed is steep, the wave breaks quickly by having the top of its crest plunge

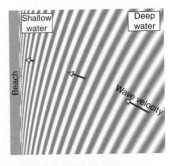

Fig. 9.3.9 When a traveling water surface wave encounters shallow water, it slows down and its direction of travel changes. This refraction process bends the wave velocity so that it points more directly toward the beach.

Courtesy Lou Bloomfield

Fig. 9.3.10 When the slope of the sea bottom is very gradual, the wave crests crumble gently into rolling surf.

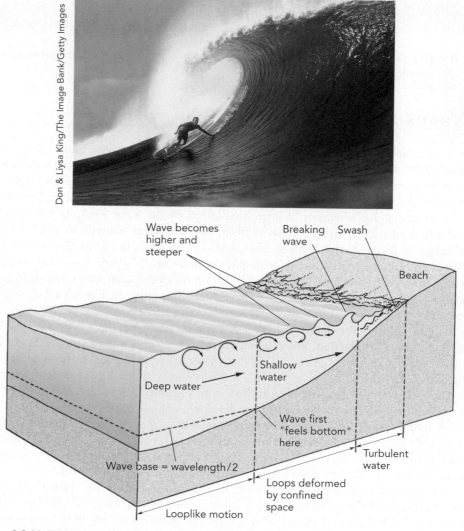

Don & Liysa King/The Image Bank/Getty Images

Fig. 9.3.11 When the water becomes too shallow to form a complete crest for the wave, the wave "breaks." If the slope of the shore is steep enough, the crest will be quite incomplete on its shore side and will plunge forward over the trough in front of it.

forward over the trough in front of it (Fig. 9.3.11). The steep slope essentially prevents the forward half of the crest from forming. The rearward half continues through its normal circular motion and dives over the missing half-crest in front of it.

However, the wave can avoid this violent end by colliding with a seawall or cliff instead of a beach. Rather than breaking, the wave will then bounce off the wall and continue on in a new direction. Known as **reflection,** this bouncing effect occurs whenever certain dynamical properties of a wave, particularly its speed, change abruptly as it passes from one environment to another. In this case, the water surface wave would have to change so radically to enter the seawall or cliff that it instead reflects almost perfectly. But even less severe changes in environment can make the wave reflect, if only partially. Thus when a water surface wave passes over a sandbar or coral reef and its wave speed changes, it may experience both reflection and refraction. These effects contribute to the complicated dynamics of waves near the shore.

► Check Your Understanding #5: Out Past the Surf

for answers, see page **299**

As you swim away from shore at the ocean, you go through a region where the wave crests are breaking and then reach a more distant region where they don't break. What distinguishes the two regions?

The Rhythm of the Surf: Wave Interference

If the ocean were carrying only one pure traveling wave toward shore, every breaking swell would look and sound the same. However, there is often a complicated rhythm to the crashing surf; its volume fluctuates with an overall pattern known as *surf beat*. Surf beat is a sign that the ocean's surface is a busy place: it's actually carrying more than one traveling wave at a time and these various waves all contribute to the surf.

To understand how multiple traveling waves produce surf beat, let's consider a simple case. Suppose that two traveling waves are heading toward shore and that they have equal amplitudes but different wavelengths (Fig. 9.3.12). Such a situation can easily arise when winds over two portions of the ocean produce two different traveling waves that later overlap. Since they are sharing the ocean's surface, they are superposed on one another.

Because these traveling waves are different, their patterns of crests and troughs can't coincide everywhere. Instead, these waves experience **interference**—their overlying patterns enhance one another at some locations and cancel one another at other locations. Wherever their crests or troughs coincide, they experience **constructive interference**—the waves act together to produce enormous crests or troughs. But wherever the crest of one wave coincides with the trough of the other wave, they experience **destructive interference**—the waves oppose one another to produce muted or absent crests or troughs.

The result is an **interference pattern**—an intricate structure that spreads across space and time when waves are superposed. This interference pattern on the ocean's surface moves and evolves as the traveling waves head toward shore and it leaves its impression on the crests that eventually break on the beach. Since these crests are no longer equal in height, they exhibit the complicated rhythm of surf beat. When you listen to that beat, you're hearing the consequence of superposition and the interference of waves.

Of course, the real ocean carries many traveling waves, each with its own amplitude, wavelength, and direction of travel. But no matter how complicated the ocean's surface or how intricate the surface beat, you are still just observing the interference of waves.

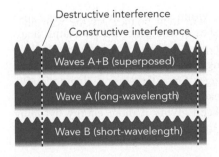

Fig. 9.3.12 When two traveling waves A and B are superposed on the surface of the ocean, they produce interference pattern A + B. As this moving pattern encounters the shore, its varying crest heights give rise to surf beat.

Summary of Important Wave Phenomena

Reflection: The complete or partial mirror redirection of a wave that occurs when certain dynamical properties of that wave, particularly its speed, change abruptly as it passes from one environment to another.

Refraction: The bending of a wave that occurs when that wave's speed changes as it passes from one environment to another.

Dispersion: The dependence of a wave's speed on its wavelength.

Interference: When two or more waves are superposed, their crests and troughs can enhance one another (constructive interference) or cancel one another (destructive interference) and produce an interference pattern.

Check Your Understanding #6: Sound Beats

for answers, see page 299

Many notes on a piano are played by two or even three separate strings. If those strings aren't tuned to exactly the same pitch, the resulting sound has a pulsing character—it grows louder and softer rhythmically. What causes this strange pulsation or beat?

Epilogue for Chapter 9

In this chapter, we looked at natural resonances in a variety of objects. In *clocks,* we examined resonances in pendulums, balance rings, and quartz crystals and found that these objects are harmonic oscillators—their restoring forces are proportional to displacement. As such, these timekeepers have periods that don't depend on the amplitudes of their motions.

In *musical instruments,* we explored the vibrations of strings and air columns to see that they, too, behave as harmonic oscillators, but with many modes of vibration and thus more complicated behaviors. We found that we could view their motions as waves.

In *the sea,* we further explored two different types of waves: standing waves and traveling waves. We saw that both types of waves are found in water, with standing waves appearing in tidal resonances and traveling waves heading out over the open ocean.

Explanation: A Singing Wineglass

Rubbing the lip of the glass is similar to bowing a violin string; both procedures cause resonant energy transfer. Your finger alternately pushes on the glass and then slips, helping the glass itself to vibrate back and forth beneath your finger in its fundamental vibrational mode. As the glass begins to vibrate, your finger does a little work on it each time the glass is moving in the same direction as your finger. Once the glass is singing loudly, you can keep it singing by circling your finger steadily around the glass. Although the glass continues to emit its energy as sound, you keep giving it more energy by doing work on it with your finger. Adding more water to the glass increases its inertia and slows its vibrations.

Chapter Summary

How Clocks Work: Clocks are usually based on harmonic oscillators because of their extremely steady periods. Most importantly, the period of a harmonic oscillator doesn't depend on the amplitude of its motion. Common harmonic oscillator timekeepers include pendulums, balance rings, and quartz crystals.

In a pendulum clock, a swinging pendulum controls the rotation of a toothed wheel, which in turn controls the rotation of the clock's hands. Energy needed to keep the pendulum swinging and to advance the hands comes from the descent of a weighted cord. In a balance clock, a rocking balance ring controls the toothed wheel. Energy for this clock comes from a wound coil spring. A quartz crystal clock detects the vibration of its crystal electronically and uses that vibration to control a motor that advances its clock hands or an electronic circuit that measures time by counting the crystal's vibrations. Small, carefully timed electric pulses from the clock provide the energy that keeps the crystal vibrating.

How Musical Instruments Work: A violin string exhibits natural resonances by vibrating back and forth about its equilibrium shape, a straight line. Bowing the string causes it to vibrate by pushing it away from its equilibrium shape and then allowing it to slip back. The bow transfers energy to the vibrating string by pushing it each time it moves in the bow's direction. The string's fundamental pitch is determined by its mass, tension, and length, and corresponds to a standing wave on the string. After selecting a string of the correct mass, you tune that string by adjusting its tension. During play, you change the string's length and pitch by pressing it against the fingerboard. The belly of the violin serves to convert the string's vibration into sound by moving in and out as the string vibrates.

The air inside an organ pipe also exhibits natural resonances. As the organ blows air across the opening at the bottom of a pipe, the air inside that pipe begins to vibrate as a standing wave. The frequency of this vibration is determined principally by the pipe's length, so that a short pipe has a higher pitch than a long one. Pipes with different shapes produce different amounts of harmonics and yield different sounds.

The surface of a drum has natural resonances, but with a difference: they're not harmonics of the fundamental pitch. The complicated standing waves on the drumhead contribute to its unique sound.

How the Sea Works: The sea exhibits two interesting motions: tides and waves. Tides are caused by bulges in the earth's oceans, created by gravitational tidal forces from the moon and sun. The moon dominates the tides so that bulges appear on regions of the earth closest to and farthest from the moon. Since the earth is rotating, these bulges move with respect to land and give the tides their 12½-hour cycle.

Waves occur because the water's surface has a stable equilibrium. When it's disturbed from that flat, level equilibrium, this surface will oscillate. Its most basic oscillations are standing waves on confined bodies of water and traveling waves on open water.

The familiar ripples on the ocean's surface are traveling waves that head toward shore at speeds that increase with wavelength. As they approach the shore, these waves form incomplete crests, which eventually break. The shallowing water also slows the waves, so that they bend more directly toward shore. And interference effects between multiple traveling waves give rise to intricate patterns on the water's surface and to complicated rhythms in the crashing surf.

Important Laws and Equations

1. Harmonic Oscillator: An oscillator with a restoring force proportional to displacement. Its period of oscillation depends only on the stiffness of its restoring force and on its mass, not on its amplitude of oscillation.

2. Relationship between Wave Speed, Wavelength, and Frequency: Wave speed equals the product of wavelength times frequency, or

$$\text{wave speed} = \text{wavelength} \cdot \text{frequency.} \qquad (9.2.1)$$

Check Your Understanding—Answers

Section 9.1 CLOCKS

1. Since light travels at a constant speed, the distance it travels is equal to the product of its speed times its travel time. If you know the travel time and the speed, you can determine that distance.

Why: Many distance measurements are made by measuring time. Surveyors routinely use light's travel time to measure distances. Decorators and architects often use sound's travel time to measure the distances between walls. In general, the motion of an object at constant velocity can be used either to measure the distance traveled if you know the elapsed time or to measure the elapsed time if you know the distance traveled.

2. It may have lost as much as 80 minutes after 24 hours.

Why: If the operator of a sandglass waits 10 s before turning it over each time the sand runs out, the clock will lose 10 s every 3 minutes, 200 s every hour, and 4800 s every day. If you could be sure that the operator would always wait 10 s, you could include it in the clock's design. However, the operator might be faster one time than the next, and this uncertainty makes the clock unreliable and inaccurate. For a repetitive clock to keep accurate time, it must repeat its motion with almost perfect regularity.

3. The strength of the earth's gravity and the length of the swing's chains.

Why: A child swinging on a swing set is a form of pendulum. As with any pendulum, the child's period of motion is determined only by the strength of gravity and the length of the pendulum. In this case, the length of the pendulum is approximately the length of the swing's supporting chains. Thus a tall swing has a longer period than a short swing.

4. The amplitude of her motion will gradually decrease so that she comes to a stop.

Why: To keep her swinging, you must make up for the energy she loses to friction and air resistance. By pushing her forward each time she moves away from you, you do work on her and increase her energy. But when you push her as she moves toward you, she does work on you and you extract some of her energy. You are then slowing her down rather than sustaining her motion.

5. The torsional stiffness of the chandelier's supporting cord and the chandelier's rotational mass.

Why: The hanging chandelier is a harmonic oscillator. Its supporting cord opposes any twists by exerting a restoring force on the chandelier. Once you twist it away from its equilibrium orientation, the chandelier oscillates back and forth with a period determined only by the cord's torsional stiffness (its stiffness with respect to twists) and the chandelier's rotational mass. As with any harmonic oscillator, the amplitude of the chandelier's motion doesn't affect its period.

6. The rod is vibrating as a harmonic oscillator, with its two halves first approaching one another and then moving apart.

Why: The metal rod vibrates in the same manner as a quartz crystal. The body of the rod exerts restoring forces on its two halves. After hitting the floor, these halves move toward and away from one another rapidly, emitting the tone that you hear. Because it's a harmonic oscillator, the frequency (and pitch) of the tone doesn't change as the amplitude of motion decreases.

Section 9.2 MUSICAL INSTRUMENTS

1. There is a factor of 4 in frequency between the lowest and the highest notes that voice can sing.

Why: Since notes separated by an octave are separated by a factor of 2 in frequency, notes separated by two octaves are separated by a factor of 4.

2. The frequency of a string's fundamental vibrational mode increases with its tension.

Why: Any cord that is drawn taut from its ends will exhibit natural resonances like those in a violin string. The tauter the string, the higher the frequencies of those resonances.

3. A jump rope is essentially a vibrating string. The two half-rope pattern is the second harmonic mode, with a vibrational frequency twice that of the normal fundamental arc.

Why: Although it swings around in a circle, the jump rope is actually vibrating up and down at the same time it's vibrating forward and back. Together, these two vibrations create the circular motion. To make the rope vibrate in its second harmonic mode (as two half-ropes) without changing its tension, it must be swung twice as fast as normal.

4. The object has a natural resonance at the tone's frequency, and sympathetic vibration is transferring energy to the object.
Why: Energy moves easily between two objects that vibrate at the same frequency. A note played on one instrument will cause the same note on another instrument to begin playing. Even everyday objects will exhibit sympathetic vibration when the right tone is present in the air.

5. The water shortens the column of moving air inside the bottle and increases the frequency of its fundamental vibrational mode.
Why: A water bottle is essentially a pipe that is open at only one end. It has a fundamental vibrational mode with a frequency that is half that of an open pipe of equal length. As you add water to the bottle, you shorten the effective length of the pipe and raise its pitch.

6. By blowing across the mouth, you let air that is already vibrating in the bottle redirect your breath so that it enhances the vibration. Blowing into the bottle's mouth merely compresses the air inside the bottle.
Why: Like the bow of a violin moving across its strings, your breath moving across the bottle's mouth enhances the air's vibration via resonant energy transfer. The spontaneous redirection of your breath when you blow across the bottle's mouth leads to rhythmic pushes that are perfectly synchronized with the air's vibration.

7. The off-center impact causes the surface to vibrate in its noncircular overtone modes. The simplest such mode, Fig. 9.2.9b, has its two sides moving in alternate directions.
Why: The trampoline is essentially a drumhead and the children are riding its vibrational modes. Off-center impacts can cause the surface to vibrate in its overtone modes and these can toss the children in unexpected directions.

8. Sound travels faster in helium than it does in ordinary air.
Why: With its reduced density and inertia, helium vibrates faster than air when the two gases carry sound waves of equal wavelengths. Since the speed at which sound of a specific wavelength travels is proportional to the frequency of that sound, the wave speed in helium is greater than that in air.

9. To transfer the vibrational energy of the strings to the air.
Why: Guitar strings are too narrow to push effectively on the air and emit sound. They do better to transfer their energy to the body of the acoustic guitar so that its flat surfaces can push on the air. An electric guitar avoids the need for a sound box by converting the string's vibrations directly into electric currents and from there into movements of an audio speaker.

Section 9.3 THE SEA

1. High tide, too.
Why: When the Atlantic side of the Central American isthmus is experiencing high tide, there is a tidal bulge over all of Central America. Both sides of the isthmus experience the same (high) tide.

2. Many days.
Why: The sloshing water in the Bay of Fundy acquires its energy via resonant energy transfer. The tides do work on the water, slowly increasing its energy over many cycles of its periodic motion. Since each cycle takes half a day, many days are required to build up the energy needed for a giant tide. If you were to extract all this stored energy at once, the bay would have to start sloshing all over again.

3. Count the number of up and down cycles you complete in a certain amount of time.
Why: Your boat's up and down motion is caused by the oscillation of the ocean's surface. You are rising and falling at the frequency of the passing wave.

4. You must dive about half a wavelength beneath the water's surface.
Why: A wave causes motion in the water up to a depth of roughly half its wavelength. A typical wave has a wavelength of about 5 m, so you must dive 2.5 m under water before the water will remain essentially still.

5. The breaking region is too shallow for complete crests to form. The nonbreaking region is deep enough to form complete crests.
Why: Wave crests break in shallow regions, where complete crests can't be formed. As long as the water is deep enough, the waves travel intact. But if there is a shallow region such as a sand bar, even quite far from shore, the crests may break as the wave passes through it.

6. The slightly different sound waves produced by the separate strings are experiencing interference.
Why: Like slightly different water waves, slightly different sound waves exhibit interference effects. The resulting pulsation reflects alternating constructive and destructive interference between the sound waves with slightly different pitches.

Check Your Figures—Answers

Section 9.2 MUSICAL INSTRUMENTS

1. About 1420 m/s (4700 ft/s).
Why: From Eq. 9.2.1, the wave speed in water must be 4.3 times the wave speed in air. Since the sound waves have a wavelength speed of about 331 m/s in air, they must have a speed of about 331 m/s times 4.3 or 1420 m/s in water.

Exercises

1. The acceleration due to gravity at the moon's surface is only about one-sixth that at the earth's surface. If you took a pendulum clock to the moon, would it run fast, slow, or on time?

2. A clothing rack hangs from the ceiling of a store and swings back and forth. Why doesn't the period of this motion depend on how many dresses the rack is holding? (Neglect the rack's own mass.)

3. If a child stands up on the seat of a playground swing, how will the swing's period be affected?

4. If you pull a small tree to one side and suddenly let go, it will swing back and forth several times. The period of this motion won't depend on how far you bend the tree. How does the restoring force that returns the tree to its upright position depend on how far you bend the tree?

5. A flagpole is a harmonic oscillator, flexing back and forth with a steady period. If you want to increase the amplitude of the pole's motion by pushing it near its base, when should you push—as it flexes toward you or away from you?

6. Depending on how the base of a rocking chair was cut, the period of its motion may or may not depend on how hard you're rocking it. What can you say about the restoring forces acting in these two cases?

7. An electronic ruler measures the distance to a wall by bouncing sound off it. How can a ruler use a sound emitter, a sound receiver, and a timer to measure how far away the wall is?

8. Which of the following clocks would keep accurate time if you took them to the moon: a pendulum clock, a balance clock, and a quartz watch? Why?

9. To modify the pitch of a guitar string you could change its mass, tension, or length. To raise its pitch, how should you change each of these three characteristics?

10. Why do changes in temperature affect a violin string's pitch?

11. The strings that play the lowest notes on a piano are made of thick steel wire wrapped with a spiral of heavy copper wire. The copper wire doesn't contribute to the tension in the string, so what is its purpose?

12. Why are the highest pitched strings on most instruments, including guitars, violins, and pianos, the most likely strings to break?

13. Some wind chimes consist of sets of metal rods that emit tones when they're struck by wind-driven clappers. These rods vibrate like the baseball bat in Fig. 3.2.7. Why do the longer rods emit lower pitched tones than the shorter rods?

14. Why would replacing the air in an organ pipe with helium raise its pitch?

15. A flute and a piccolo are both effectively pipes that are open at both ends, with holes in their sides to allow them to produce more tones. The piccolo is very nearly a half-size version of the flute. How does this fact explain why the piccolo's tones are one octave above those of a flute?

16. The most important difference between a trumpet and a tuba is in the lengths of their pipes. The tuba's pipe is much longer than that of the trumpet. How does this difference affect the relative pitches of the two instruments?

17. In the Mediterranean Sea, high tide is only 30 cm above low tide. Why?

18. Why are the tides relatively weak near the north and south poles?

19. If you're carrying a full cup of coffee and take your steps at just the wrong frequency, the coffee will begin to slosh wildly in the cup. What is causing this energetic motion?

20. When you throw a stone into a pool of still water, small ring-shaped ripples begin to spread outward at a modest pace. Why do these ripples travel so much more slowly than waves on the ocean?

21. When you throw a rock into calm water, ripples head outward as several concentric circles. As the circumferences of these circles grow larger, their heights grow smaller. Explain this effect in terms of conservation of energy.

22. If you pull downward on the middle of a trampoline and let go, the surface will fluctuate up and down several times. Why is this motion an example of a standing wave?

23. Why is a violin string vibrating up and down in its fundamental vibrational mode an example of a standing wave rather than a traveling wave?

24. When you pluck the end of a kite string, a ripple will head up the string toward the kite. Why is this motion an example of a traveling wave rather than a standing wave?

25. As waves pass over a shallow sand bar, the largest ones break. What causes these waves to break and why only the largest ones?

26. Even when waves don't break as they pass over a sand bar (Exercise 25), the sand bar is noticeable because you can see the wave crests move closer together. What is happening to cause this bunching?

27. Sound can travel from one paper cup to another through a long, taut string that connects their bottoms. Is the wave passing through the string longitudinal or transverse?

28. If you stamp on a wooden floor, you can make objects on a nearby table jump slightly. Are the waves traveling through the floor longitudinal or transverse?

29. The crash of brass cymbals is rich with overtones that are not harmonics of the fundamental. Why aren't the overtones harmonics?

30. A Chinese gong produces a loud ringing sound which has nonharmonic overtones. Why aren't the overtones harmonic?

31. A harp is not a loud instrument, but it would be even softer if its strings were not attached to a wide wooden base. What purpose does the wooden surface serve?

32. A string bass is an enormous instrument to carry around. Why can't you just support the four strings with a sturdy metal bar and leave out the whole wooden structure?

33. You clap your hands and send out a sound wave. Is there anything you can do one second later to send out a sound wave that will overtake the first one you sent out?

34. You drop a coin in a calm pool and send out a water surface wave. Is there anything you can do one second later to send out a water surface wave that will overtake the first one you sent out?

35. If you stand in front of a stone building and clap your hands, you hear an echo. What is happening to the sound wave to cause this echo?

36. If you stand behind a massive stone wall, you have trouble hearing a person on the other side clapping her hands. What is happening to the sound wave to make it difficult for you to hear?

37. A harbor breakwater is a stone wall in the water that prevents waves from entering the harbor. Where does a wave's energy go after it hits the breakwater?

38. Waves tend to bend toward points of land projecting into the ocean and erode those points. What causes the wave to bend toward the points?

39. Surfers are well aware that waves bend as they pass over coral reefs. What causes this bending?

40. Musical tones can linger for many seconds in a stone cathedral. Why?

41. If two violinists play slightly different notes at the same time, the combined sound has a pulsing character. What causes that pulsation?

42. When an airplane's two propellers are turning at almost the same rate, their combined sound can have a pulsing character to it. Explain that pulsing.

Problems

1. What is the wavelength of a tuba's A_2 (110 Hz) tone in air at standard conditions?

2. A piccolo is playing A_6 (1760 Hz). What is the wavelength of that tone in air at standard conditions?

3. When a piano plays C_4 (264 Hz) in a room containing a somewhat unusual mixture of gases, the wavelength of the sound in that gas is 1.00 m. What is the speed of sound in that gas?

4. At an altitude of 3000 m and standard temperature (0 °C), a violin's A_4 (440 Hz) has a wavelength of 0.725 m. What is the local speed of sound?

5. A water surface wave with a frequency of 0.3 Hz has a wavelength of 17.3 m/s. What is its wave speed?

6. A water surface wave has a wave speed of 15.6 m/s and a frequency of 0.1 Hz. What is its wavelength?

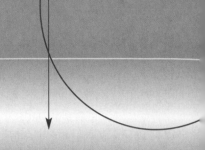

CHAPTER 10

Electricity

Electricity is all around us, in the sparks and shocks of a cold winter day and the illumination of a flashlight when you turn on its switch. While it's hard to see the electric charges that are responsible for electricity, it's easy to see their effects. These charges serve so many purposes in modern life that we take them for granted. If there were no charges or electricity, we'd be sitting around campfires at night, trying to think of things to do without television, cell phones, or computer games. Of course, we wouldn't exist either. Whether it's motionless as static charge or moving as electric current, electricity really does make the world go 'round.

EXPERIMENT: Moving Water without Touching It

Unlike gravity, which always pulls objects toward one another, electric forces can be either attractive or repulsive. You can experiment with electric forces using a thin

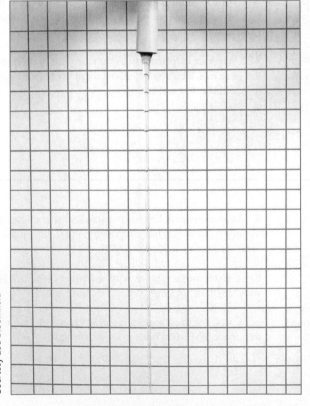

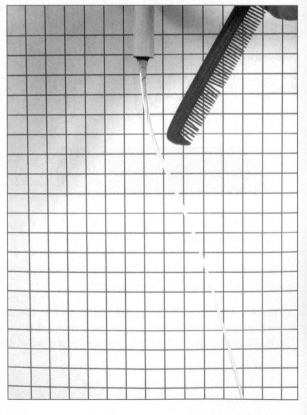

Courtesy Lou Bloomfield

stream of water and an electrically charged comb. First, open a water faucet slightly so that the flow of water forms a thin but continuous strand below the mouth of the faucet. Next, give your rubber or plastic comb an electric charge by passing it rapidly through your hair or rubbing it vigorously against a wool sweater. Finally, hold the comb near the stream of water, just below the faucet, and watch what happens to the stream. Is the electric force that you're observing attractive or repulsive? Why does this force change the path of the falling water?

Rubbing the comb through your hair makes it electrically charged. Find other objects that can acquire and hold a charge when you rub them across hair or fabric. Try to *predict* which of these objects will work best. Now rub these objects to charge them and *observe* their behaviors. You can use the water stream to *measure* their charges. Did you *verify* your predictions? Which works better: a metal object or one that's an insulator? Why?

Chapter Itinerary

While we often experience electric forces and currents as novelties or nuisances, there are also many devices that depend on them. In this chapter, we'll examine the mysteries of (1) *static electricity* and study two modern devices that are based on electricity: (2) *xerographic copiers* and (3) *flashlights*. In *static electricity*, we'll look at how clothes and other objects acquire charges and how they exert forces on one another as result. In *xerographic copiers*, we'll see how these same electric forces work together with light to control the placement of black powder in order to reproduce images on sheets of paper. In *flashlights*, we'll look at how a current of electric charges conveys power from batteries to a lightbulb. For a more complete preview of the chapter, turn ahead to the Chapter Summary on p. 335.

Although this chapter concentrates on electricity and its charges, we will see in Chapter 11 that magnetism and its poles are closely related. While we'll leave the relationships between electricity and magnetism for that chapter, you may already begin seeing similarities between those two seemingly separate phenomena as you read Chapter 10.

| SECTION 10.1 | **Static Electricity** |

Electricity may be difficult to see, but you can easily observe its effects. How often have you found socks clinging to a shirt as you remove them from a hot dryer or struggled to throw away a piece of plastic packaging that just won't leave your hand or stay in the trash can? The forces behind these familiar effects are electric in nature and stem from what we commonly call "static electricity." But static electricity does more than just push things around, as you've probably noticed while reaching for a doorknob or a friend's hand on a cold, dry day. In this section, we'll examine static electricity and the physics behind its intriguing forces and often painful shocks.

Questions to Think About: How does a dryer produce static electricity and why do some clothes cling while others repel? Why does walking across a carpet on a cold, dry day put you at great risk of a shock as you reach for a doorknob? Why do you get only a single, brief shock from that knob and not a long, sustained one? When touching a friend results in a shock, did one of you cause that shock or are you both responsible? If rubbing is required to develop static electricity, why does the plastic wrap produce so much of it when you simply open a new CD? Why do moist air and antistatic chemicals reduce static electricity?

Experiments to Do: You can study static electricity by rubbing a toy balloon vigorously through your hair or against a wool sweater. Though its appearance won't change, the balloon will begin to attract other things, particularly your hair. What has happened to the balloon? to your hair? Why does the balloon also attract things that weren't rubbed?

Try to get rid of the balloon's attractiveness by letting a thick stream of water flow over its surface. Why does this process return the balloon to normal? What did you "wash" off the balloon? Now rub two identical balloons through your hair and see whether they attract or repel one another. Does the result make sense?

Finally, draw two long strips of transparent tape from a dispenser without rubbing them on anything and see if they attract or repel. Is rubbing essential to the development of static electricity?

Electric Charge and Freshly Laundered Clothes

Unless you have always lived in a damp climate and avoided synthetic materials, you have experienced the effects of static electricity. Seemingly ordinary objects have pushed or pulled on one another mysteriously and you've received shocks while reaching for light switches, car doors, or friends' hands. But static electricity is more than an interesting nuisance; it's a simple window into the inner workings of our universe and worthy of a serious look. It will take some time to lay the groundwork, but soon you'll be able to explain most of the effects of static electricity and even to control it to some extent.

The existence of static electricity has been known for several thousand years. About 600 B.C., the Greek philosopher Thales of Miletus (ca 624–546 B.C.) observed that when amber is rubbed vigorously with fur, it attracts light objects such as straw and feathers. Known in Greek as *elektron,* amber is a fossil tree resin with properties similar to those of modern plastics. The term "static electricity," like many others in this chapter, derives from that Greek root.

Static electricity begins with **electric charge,** an intrinsic property of matter. Electric charge is present in many of the **subatomic particles** from which matter is constructed, and these particles incorporate their charges into nearly everything. No one knows why charge exists; it's simply one of the basic features of our universe and something that people discovered through observation and experiment. Because electric charge has so much influence on objects that contain it, we'll sometimes refer to those objects as **electric charges** or simply as **charges.**

Charges exert forces on one another and it is these forces that you observe with static electricity. Next time you're doing laundry, experiment with your clothes as they come out of the dryer. You'll find that some electrically charged garments attract one another while others repel. Evidently, there are two different types of charge. But while this dichotomy has been known since 1733, when it was discovered by French chemist Charles-François de Cisternay du Fay (1698–1739), it was Benjamin Franklin ❑ who finally gave the two charges their present names. Franklin called what appears on glass when it's rubbed with silk "positive charge" and what appears on hard rubber when it's rubbed with animal fur "negative charge."

Two like charges (both positive or both negative) push apart, each experiencing a repulsive force that pushes it directly away from the other (Fig. 10.1.1*a,b*). Two opposite charges (one positive and one negative) pull together, each experiencing an attractive force that pulls it directly toward the other (Fig. 10.1.1*c*). These forces between stationary electric charges are called **electrostatic forces.**

When you find that two freshly laundered socks push apart, it's because they both have the same type of charge. Whether that charge is positive or negative depends on

❑ Although best remembered for his political activities, American statesman and philosopher Benjamin Franklin (1706–1790) was also the preeminent scientist in the American colonies during the mid-1700s. His experiments, both at home and in Europe, contributed significantly to the understanding of electricity and electric charge. In addition to demonstrating that lightning is a form of electric discharge, Franklin invented a number of useful devices, including the Franklin stove, lightning rods, and bifocals.

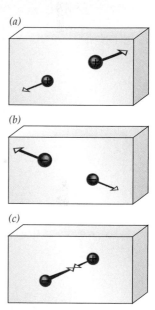

Fig. 10.1.1 (*a*) Two positive charges experience equal but oppositely directed forces exactly away from one another. (*b*) The same effect occurs for two negative charges. (*c*) Two opposite charges experience equal but oppositely directed forces exactly toward one another.

the fabrics involved (more on that later), so let's just suppose that the dryer has given each sock a negative charge. Since like charges repel, the socks push apart. But what does it mean for the dryer to give each sock a negative charge?

The answer to that question has several parts. First, the dryer didn't create the negative charge that it gave to a sock. Like momentum, angular momentum, and energy, electric charge is a conserved physical quantity—it cannot be created or destroyed, only transferred. The negative charge that the dryer gave to the sock must have come from something else, perhaps a shirt.

Second, positive charge and negative charge aren't actually separate entities—they're just positive and negative amounts of the same physical quantity: electric charge. Positive charges have positive amounts of electric charge, while negative charges have negative amounts. Like most physical quantities, we measure charge in standard units. The SI unit of electric charge is the **coulomb** (abbreviated C). Small objects rarely have a whole coulomb of charge and your sock's charge is only about -0.0000001 C.

Third, the sock's negative charge refers to the sock as a whole, not to its internal pieces. As with all ordinary matter, the sock contains an enormous number of positively and negatively charged particles. Each of the sock's atoms consists of a dense central core or **nucleus,** containing positively charged **protons** and uncharged **neutrons,** surrounded by a diffuse cloud of negatively charged **electrons.** The electrostatic forces between those tiny charged particles hold together not only the atoms, but also the entire sock. However, in giving the sock a negative charge, the dryer saw to it that the sock's **net electric charge**—the sum of all its positive and negative amounts of charge—is negative. With its negative net charge, the sock behaves much like a simple, negatively charged object.

Finally, the sock became negatively charged when it contained more electrons than protons. Underlying that seemingly simple statement is a great deal of painstaking scientific study. To begin with, experiments have shown that electric charge is **quantized:** charge always appears in integer multiples of the **elementary unit of electric charge.** This elementary unit of charge is extremely small, only about 1.6×10^{-19} C, and is the magnitude of the charge found on most subatomic particles. An electron has -1 elementary unit of charge, while a proton has $+1$ elementary unit of charge. Since the only charged subatomic particles in normal matter are electrons and protons, the sock becomes negatively charged simply by having more electrons than protons.

Returning to the original question, we now know what the dryer did in order to give a sock a negative charge. Assuming the sock started electrically **neutral**—it had zero net charge—the dryer must have added electrons to the sock or removed protons from the sock or both. These transfers of charge upset the sock's charge balance and gave it a negative net charge.

In keeping with our convention regarding conserved quantities, all unsigned references to charge in this book imply a positive amount. For example, if the dryer gave charge to a jacket, we mean it gave a positive amount of charge to that jacket. We follow this same convention with money: when you say that you gave money to a charity, we assume that you gave a positive amount.

Finally, Franklin's charge-naming scheme was brilliant in concept but unlucky in execution. While it reduced the calculation of net charge to a simple addition problem, it required Franklin to choose which type of charge to call "positive" and which to call "negative." Unfortunately, his seemingly arbitrary choice made electrons, the primary constituents of electric current in wires, negatively charged. Scientists and engineers have had to deal with negative amounts of charge flowing through wires ever since. Imagine the awkwardness of having to carry out business using currency printed only in negative denominations!

Check Your Understanding #1: In Charge of Opening Gifts

for answers, see page 336

The gift you are about to unwrap is electrically neutral. You tear off the clingy wrapper and find that it has a large negative charge. What charge does the gift itself have, if any?

Coulomb's Law and Static Cling

Although your sock and shirt pull together strongly when they're only inches apart, you can put on your shirt and go to the movies without fear of being attacked by your sock from across town. Evidently, the forces between charges weaken with distance.

Over two centuries ago, French physicist Charles-Augustin de Coulomb ❏ studied electrostatic forces experimentally and determined that the forces between two electric charges are inversely proportional to the square of their separation (Fig. 10.1.2). For example, doubling the separation between your shirt and sock reduces their attraction by a factor of four, which explains your uneventful night out on the town.

Coulomb's experiments also showed that the forces between electric charges are proportional to the amount of each charge. That means that doubling the charge on either your shirt or your sock doubles the force each garment exerts on the other. Finally, changing the sign of either charge turns attractive forces into repulsive ones or vice versa. If both garments were either positively charged or negatively charged, they'd repel instead of attracting.

These ideas can be combined to describe the forces acting on two charges and written as a word equation:

$$\text{force} = \frac{\text{Coulomb constant} \cdot \text{charge}_1 \cdot \text{charge}_2}{(\text{distance between charges})^2}, \qquad (10.1.1)$$

in symbols:

$$F = \frac{k \cdot q_1 \cdot q_2}{r^2},$$

and in everyday language:

When two clouds with large opposite charges approach, expect lightning.

The force on charge$_1$ is directed toward or away from charge$_2$, and the force on charge$_2$ is directed toward or away from charge$_1$.

This relationship is called **Coulomb's law,** after its discoverer. The **Coulomb constant** is about $8.988 \times 10^9 \, \text{N} \cdot \text{m}^2/\text{C}^2$ and is one of the physical constants found in nature. Consistent with Newton's third law, the force exerted on charge$_1$ by charge$_2$ is equal in amount but oppositely directed from the force exerted on charge$_2$ by charge$_1$.

❏ In 1781, after a career as a military engineer in the West Indies, French physicist Charles-Augustin de Coulomb (1736–1806) returned to his native Paris in poor health. There he conducted scientific investigations into the nature of forces between electric charges and published a series of memoirs on the subject between 1785 and 1789. His research came to a close in 1789 when he was forced to leave Paris because of the French Revolution.

(a)

(b)

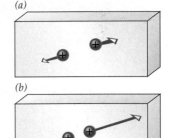

Fig. 10.1.2 The electrostatic forces between two charges increase dramatically as they become closer. As the distance separating two positive charges decreases by a factor of 2 between (a) and (b), the forces those two charges experience increase by a factor of 4.

Coulomb's Law
The magnitudes of the electrostatic forces between two objects are equal to the Coulomb constant times the product of their two electric charges divided by the square of the distance separating them. If the charges are like, then the forces are repulsive. If the charges are opposite, then the forces are attractive.

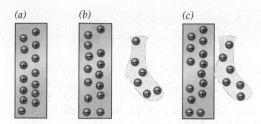

Fig. 10.1.3 (*a*) A neutral wall contains countless positive and negative charges. (*b*) As a negatively charged sock approaches the wall, the positive charges move toward it and the negative charges move away. (*c*) The polarized wall continues to attract the sock and holds it in place.

In addition to protecting you from distant charged socks, this relationship between electrostatic forces and distance gives rise to another intriguing feature of laundry static: charged clothes can cling to objects that are electrically neutral! For example, a negatively charged sock can stick to a neutral wall.

The origin of this attraction is a subtle rearrangement of charges within the wall. Even though the wall has zero net charge, it still contains both positively and negatively charged particles. When the negatively charged sock is near the wall, it pulls the wall's positive charges a little closer and pushes the wall's negative charges a little farther away (Fig. 10.1.3). Although each individual charge shifts just a tiny distance, the wall contains so many charges that together they produce a dramatic result. The wall develops an **electric polarization**—it remains neutral overall, but with a positively charged region nearest the sock and a negatively charged region farthest from the sock.

The wall's positive region attracts the sock while its negative region repels the sock. Though you might expect those two opposing forces to balance, Coulomb's law says otherwise. Since electrostatic forces grow weaker with distance, the sock is attracted more strongly to the nearer positive region than it is repelled by the more distant negative region. Overall, there is a net electrostatic attraction between the charged sock and the polarized wall, so the sock clings to the wall!

> **Check Your Understanding #2: Wrapper Recycling**
>
> *for answers, see page* 336

After opening your gift, you try to throw away its negatively charged wrapper. However, the wrapper keeps returning to your hand. What attracts it to your electrically neutral hand?

> **Check Your Figures #1: Moving Out**
>
> *for answers, see page* 337

You have two positively charged balls, each of which is experiencing a force of 1 N away from the other. If you halve the distance separating the balls, what force will each exert on the other?

Transferring Charge: Sliding Friction or Contact?

While it's clear that the dryer transfers charge between the clothes, why does that charge move and what determines which garments gain charge and which lose it?

You might suppose that sliding friction is responsible for the transfer—that the dryer rubs the clothes together and somehow wipes charge from one garment to the other. After all, friction seems to help you charge a balloon as you rub it through your hair or against a wool sweater. But be careful—there are other cases of charge transfer that don't involve rubbing at all. For example, the plastic wrap you remove from a store package can acquire a charge no matter how careful you are not to rub it against its contents. And an antique car can build up enough charge to give you a nasty shock even when its pale rubber tires never skid across the pavement.

Charge transfer is less the result of rubbing than it is of contact between dissimilar surfaces. When two different materials touch one another, a few electrons normally shift from one surface to the other. That transfer results from the chemical differences between the two touching surfaces and the associated change in an electron's potential energy when it shifts. In effect, some surfaces are "hungrier" for electrons than others and whenever two dissimilar surfaces touch, the hungrier surface steals a few electrons from its less hungry partner.

The physics behind this theft has to do with **chemical potential energy**—energy stored in the chemical forces that bind together a material's constituent atoms and electrons. To hold onto its electrons, a surface reduces their chemical energies to less than zero, meaning that it would take additional energy to free those electrons from the surface. However, some surfaces reduce the electron chemical potential energies more than others and thus bind their electrons more tightly. If an electron on one surface can reduce its chemical potential energy by shifting to the other surface, it will accelerate toward that "hungrier" surface and eventually stick there. You can picture the electron as "rolling downhill" from a chemical "valley" on one surface to an even deeper valley on the other surface.

This transfer of electrons is self-limiting. As electrons accumulate on the lower energy surface, they begin to repel any electrons that try to follow and the transfer process soon grinds to a halt. It stops altogether when the electrons reach equilibrium—when the forward chemical force an electron experiences is exactly balanced by the backward electrostatic force. The transfer won't resume until you bring fresh, uncharged surface regions into contact.

That's where rubbing enters the picture. Rubbing involves lots of surface contact and almost endless opportunities for charge transfer between those surfaces. As clothes tumble about in the dryer, touching one another and often rubbing, some fabrics steal electrons and become negatively charged while other fabrics lose electrons and become positively charged.

That said, you should be aware that the details of contact charging are messy. For starters, the surfaces that actually touch one another are neither chemically pure nor free of microscopic defects. While it's generally true that whichever fabric binds electrons most tightly is the one most likely to develop a negative net charge, surface contamination and defects can change the outcome radically. Even your choice of laundry detergent may affect the fabric's surface chemistry and thus how it charges. Furthermore, water molecules cling to most surfaces and influence the contact charging process. Finally, while we've concentrated on the exchange of electrons, it's also possible for certain surfaces to exchange **ions**—that is, electrically charged atoms, molecules, or small particles—along with electrons and acquire net charges as a result.

Check Your Understanding #3: Sticky Tape

for answers, see page 336

When you peel a piece of adhesive tape off a glass window, you find that the tape is attracted toward the spot it left behind. How did the tape and glass acquire electric charges?

Separating Your Clothes: Producing High Voltages

The dryer stops and you take out your favorite shirt. It has several socks clinging to it, so you begin to remove them. As you separate the garments, they crackle and spark. Their attraction is obviously due to opposite charges, but why does separating them make them spark?

To answer that question, let's think about energy as you pull the negatively charged sock steadily away from the positively charged shirt. Since the sock would accelerate toward the shirt if you let go, you are clearly exerting a force on the sock. And because that force and the sock's movement are in the same direction, you are also doing work on the sock. You are transferring energy to it.

That energy is stored in the electrostatic forces—the shirt and sock accumulate **electrostatic potential energy.** Electrostatic potential energy is present whenever opposite charges have been pulled apart or like charges have been pushed together. With the negatively charged sock now far from the positively charged shirt, both attraction and repulsion contribute to the electrostatic potential energy: opposite charges are separated on the two garments and like charges are assembled together on each garment.

The total electrostatic potential energy in the shirt and sock is the work you did to separate them. But that potential energy isn't divided equally among the individual charges on these garments. Depending on their locations, some charges have more electrostatic potential energy than others and are therefore more important when it comes to sparks. In recognition of those differences, we need a proper way to characterize the electrostatic potential energy available to a charge at a particular location. The measure we're seeking is **voltage**—the electrostatic potential energy available per unit of electric charge at a given location.

Voltage is a difficult quantity to conceptualize because you can't see charge or sense its stored energy. To help you understand voltage, we'll use a simple analogy. In this analogy, the role of charge will be played by water and the role of voltage will be played by altitude. Where voltage is high, visualize water far above you. Where voltage is low, picture water at a lesser height. And just as water tends to flow from higher altitude to lower altitude, so charge tends to flow from higher voltage to lower voltage.

This analogy works well because both voltage and altitude measure the energy in a unit of something. Voltage is the electrostatic potential energy per unit of charge and altitude can be construed as the gravitational potential energy per unit of weight. Though thinking of altitude this way may seem strange, both water at high altitude and charge at high voltage are loaded with energy per unit and likely to cause trouble!

Since the SI unit of energy is the joule and the SI unit of electric charge is the coulomb, the SI unit of voltage is the joule-per-coulomb, more commonly called the **volt** (abbreviated V). Where the voltage is positive, (positive) charge can release electrostatic potential energy by escaping to a distant neutral place. Charge at positive voltage is analogous to water on a hill, which can release gravitational potential energy by flowing down to a distant level place. Where the voltage is negative, charge needs energy to escape to a distant neutral place. Charge at negative voltage is analogous to water in a valley, which needs energy in order to flow up to a distant level place.

As you can see, this voltage–altitude analogy is quite a boon. But while you should find it helpful now and throughout this book, please remember that the ups and down in altitude that you're using to visualize voltage occur only in your mind's eye and not in the real world. Your clothes don't necessary move up or down as their voltages change!

Returning to those clothes, you'll find that each point on the shirt or sock has its own voltage. You can determine that voltage by taking a tiny amount of (positive) charge at that point—a "test charge"—and moving it to a distant neutral place. The point's voltage is simply the electrostatic potential energy the test charge releases during that trip divided by the amount of its charge. If the point you examine is on the positively charged shirt, you'll obtain a large positive voltage—probably several thousand volts. If it's on the negatively charged sock, you'll obtain a negative voltage of similar magnitude. Whether positive or negative, these high voltages tend to cause sparks.

We'll look at the physics of sparks and discharges soon, but you can already see why oppositely charged clothes spark as you separate them: that's when the high voltages develop. As long as your sock is clinging tightly to your shirt, there isn't much electrostatic potential energy available. But as soon as you begin to separate them, watch out!

Check Your Understanding #4: High Altitude Voltage

for answers, see page 336

While any cloud may contain opposite charges, only the violent updrafts inside thunderheads are able to separate those charges and produce lightning. Why does such separation lead to lightning?

Accumulating Huge Static Charges

We've seen that touching two different materials together causes a small transfer of charge from one surface to the other and that separating those oppositely charged surfaces produces elevated voltages and perhaps sparks. However, the quiet crackling and snapping of items in your laundry basket is nothing compared to the miniature lightning bolts you can unleash after walking across a carpet on a dry winter day, stepping out of an antique car, or playing with a static generator. To get a really big spark, you need to separate lots of charge and that usually requires repeated effort.

Walking across a carpet is just such a repetitive process. Each time your rubber-soled shoe lands on an acrylic carpet, some (positive) charge shifts from the carpet to your shoe. Although the transfer is brief and self-limiting, you now have a little extra charge on your shoe. When you lift that shoe off the carpet, you do work on its newfound charge and your shoe's voltage surges to a high positive value. High voltage charge tends to leak from one place to another and the shoe's charge quickly spreads to the rest of your body. By the time your foot lands again on a fresh patch of carpet, the shoe has given away most of its charge and is ready to begin the process all over again.

Each time your foot lands on the carpet, it picks up some charge. And each time it lifts off the carpet, that charge spreads out on your body. By the time you finally reach for the doorknob, you are covered with charge and have an enormous positive voltage. As your hand draws close to the doorknob, it begins to influence the doorknob's charges—pulling the doorknob's negative charges closer and pushing its positive charges away. You are polarizing the doorknob.

As we saw while separating your freshly laundered sock from your shirt, oppositely charged objects that are close but not touching can have both large electrostatic potential energies and strong electrostatic forces. That's the situation here. The closer your hand gets to the doorknob, the stronger the electrostatic forces become until finally the air itself cannot tolerate the forces and a spark forms. In an instant, most of your

Courtesy Lou Bloomfield

Fig. 10.1.4 Static electricity can be produced by mechanical processes. In this Van de Graaff generator, a moving rubber belt transfers negative charges from the base to the shiny metal sphere. This negative charge creates dramatic sparks as it returns through the air toward the positive charge it left behind.

accumulated electrostatic potential energy is released as light, heat, and sound. And that doesn't include any screams.

But as good as walking is at building up charge, an antique car is even better. Its pale rubber tires gather negative charge when they touch the pavement and develop large negative voltages as they roll away from it. This charge migrates onto the car body so that after a few seconds of driving, the car accumulates enough charge to give anyone who touches it a painful shock. Collecting tolls used to be hazardous work! Fortunately, modern tires are formulated to allow this negative charge to return safely to the pavement, so that cars rarely accumulate much charge. Instead, most shocks associated with cars now come from sliding across the seat as you step in or out.

While cars try to avoid static charging, there are machines that deliberately accumulate separated charge to produce extraordinarily high voltages. The most famous of these static machines is the Van de Graaff generator (Fig. 10.1.4). It uses a rubber belt to lift positive or negative charges onto a metal sphere until the magnitude of that sphere's voltage reaches hundreds of thousands or even millions of volts.

A typical classroom Van de Graaff uses a motor-driven rubber belt to carry negative charges from its base to its spherical metal top. Once inside the sphere, the belt's negative charges flow outward onto the sphere's surface, where they can be as far apart as possible. There they remain until something releases them.

Suspended at the top of a tall, insulating column, the Van de Graaff generator's sphere can accumulate an enormous negative charge. You may hear the motor struggling as it pushes the belt's negative charges up to the sphere, a reflection of how much negative voltage the sphere is developing. Eventually it releases its negative charge via an immense spark.

But even without sparks, the Van de Graaff is an interesting novelty. If you isolate yourself from the ground and touch the metal sphere while it's accumulating negative charges, some of those negative charges will spread onto you as well. If your hair is long and flexible, and permits the negative charges to distribute themselves along its length, it may stand up, lifted by the fierce repulsions between those like charges.

Check Your Understanding #5: Stop the Presses!

for answers, see page **336**

The paper in some printing presses moves through the rollers at half a kilometer per minute. If no care is taken, dangerous amounts of static charge can accumulate on parts of the press. How does the moving paper contribute to that charging process?

Controlling Static Electricity:
Fabric Softeners and Conditioners

Now that we've seen what static electricity is and how to produce it, we're ready to see how to tame it. Static cling, flyaway hair, and electrifying handshakes aren't everyone's cup of tea. The basic solution to static charge is mobility: if charges can move freely, they'll eliminate static electricity all by themselves. Opposite charges attract, so any separated positive and negative charges will join up as soon as they're allowed to move.

Materials such as metals that permit free charge movement are called **electrical conductors.** Those such as plastic, hair, and rubber that prevent free charge movement are called **electrical insulators.** Since charge movement eliminates static electricity, our troubles with static electricity stem mostly from insulators. If you wore metal clothing, you wouldn't have static problems with your laundry.

The simplest way to reduce static electricity is to turn the insulators into conductors. Even slight conductors, ones that just barely let charges move, will gradually get rid of any accumulations of separated charge. That's one of the main goals of fabric softeners, dryer sheets, and hair conditioners. They all turn insulating materials—fabrics and hair—into slight electrical conductors. The result is the near disappearance of static electricity and all its fashion inconveniences.

How these three items work is an interesting tale. They all employ roughly the same chemical: a positively charged detergent molecule. A detergent molecule is a long molecule that is electrically charged at one end and electrically neutral at the other end. Its charged end clings electrostatically to opposite charges and is chemically "at home" in water. Its neutral end is oil-like, slippery, and "at home" in oils and greases. This dual citizenship is what makes detergents so good for cleaning.

But while it might seem that positively and negatively charged detergent molecules would clean equally well, that's not the case. Since cleaning agents shouldn't cling to the materials they're cleaning, it's important that the two not have opposite charges. Fabrics and hair generally become negatively charged when wet—another example of a charge shift when two different materials touch—so negatively charged detergent molecules clean much better than positively charged ones.

Positively charged detergents are still useful, however, although you mustn't apply them until after you've cleaned your clothes or hair. Because they cling so well to wet fibers, these slippery detergent molecules will remain in place long after washing and give fabrics and hair a soft, silky feel. And they'll allow those materials to conduct electricity, albeit poorly, so as to virtually eliminate static electricity!

This conductivity is due principally to their tendency to attract moisture. Water is a slight electrical conductor and damp surfaces allow charges to move around. That's why moist air decreases static electricity. By making fabrics and hair almost imperceptibly damp, the positively charged detergents allow separated charges to get back together and do away with static hair problems and laundry cling. That's why they're the main ingredients in fabric softeners, dryer sheets, hair conditioners, and even many antistatic sprays.

Check Your Understanding #6: No Lightning at Work

for answers, see page 336

The conveyor belts used to move flammable materials often have metal threads woven into their fabric. Why are such conducting belts important for fire safety?

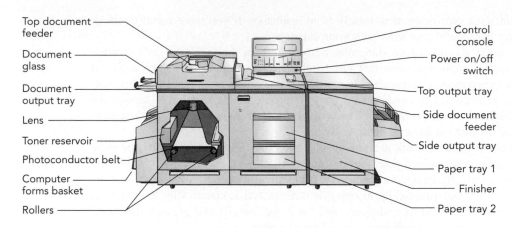

SECTION 10.2 **Xerographic Copiers**

The days of carbon paper and mimeograph machines are long gone. What modern office could operate without a xerographic copier? Advertisements for copiers are everywhere, and while each manufacturer claims to make the best copiers, that's mostly just salesmanship. In reality, all xerographic copiers are based on the same principles, discovered in 1938 by Chester Carlson. In this section, we'll examine xerographic copiers and the ideas that make them possible.

Questions to Think About: How could you use static electricity to position black powder on a sheet of paper? How would you put that static electricity on the paper? In order for characters to appear on the sheet, how should its static electricity be distributed? In a copier, what should light do to the static electricity to produce a copy of the original?

Experiments to Do: To get a feel for how a copier works, cut a small sheet of paper into tiny squares, about 1 mm on a side. Put the squares on a table and suspend a thin plate of clear plastic above them, a few millimeters away. The top of a clear plastic box will do. Now run a plastic comb through your hair or against a sweater several times and touch it to the top of the plastic plate. Squares of paper will leap off the table and stick to the plastic plate. What's holding the squares against the plastic? If the paper were black, how could you form letters on the surface of the plastic?

Xerography: Using Light to Print Copies

The image that a xerographic copier prints on a sheet of paper begins as a pattern of tiny black particles or *toner* on a smooth, light-sensitive surface. The copier uses static electricity and light reflected from the original document to arrange this toner on the surface and then carefully transfers the toner to the paper (Fig. 10.2.1). Invented in 1938 by Chester Carlson ❏, this process is basically our old friend static electricity doing something useful.

❏ Impoverished as a youth, American inventor Chester F. Carlson (1906–1968) supported his family by washing windows and cleaning offices after school. His work in a print shop as a teenager started him thinking about copying and he began to experiment with electrophotography. After attending Caltech, he worked for Bell Laboratories but was laid off in the Depression. While attending law school, he continued his experiments and invented the xerographic copying process in 1937–1938. Development of commercial copiers was slow and it wasn't until 1960 that Haloid Xerox Corporation produced its first successful copier, Model 914. Carlson became extremely wealthy but gave most of his money away anonymously.

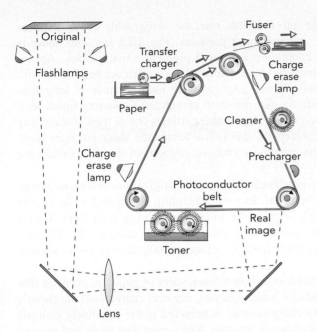

Fig. 10.2.1 **Fig. 10.2.1** This xerographic copying machine uses a photoconductor belt to form black-and-white images of an original document. The copying process begins with the precharger, which coats the photoconductor with charge. The optical system then forms a real image on a flat region of the photoconductor belt, producing a charge image. After the charge image picks up toner particles, the first charge erase lamp eliminates the charge image and weakens the toner's attachment to the belt. The toner is then transferred and fused to the paper.

At the heart of the xerographic copier is a thin, light-sensitive surface made from a **photoconductor**—a normally insulating material that becomes a conductor while exposed to light. Although the darkened photoconductor can keep positive and negative charges apart, these charges quickly draw together when light hits the photoconductor (Fig. 10.2.2). That flexibility allows light from the original document to determine the pattern of static electricity on the photoconducting surface and consequently the placement of toner on the piece of paper.

Each copying cycle begins in the dark with the copier spraying negative charges onto its photoconductor. On the other side of the photoconductor is a grounded metal surface—grounded in the sense that it's electrically connected to the earth so that charges are free to flow between the two. As negative charges land on the open surface of the photoconductor, they attract positive charges onto the metal surface beneath it. When the charge-spraying process is complete, the open surface of the photoconductor is uniformly coated with negative charges while the underlying metal surface is uniformly coated with positive charges (Fig. 10.2.3a).

After this precharging, the copier uses a lens to cast a sharp image of the original document onto the photoconducting surface. We'll examine lenses and the formation of images when we study cameras in Chapter 15. For now, what matters is that light hits the photoconductor only in certain places, corresponding to white parts of the original document.

There are two standard techniques for exposing the photoconductor to light. Some copiers illuminate the whole original document with the brilliant light of a flash lamp and cast a complete image onto a flattened portion of a photoconductor belt. In other copiers, a moving lamp or mirror illuminates the original a little at a time and the image is cast as a moving stripe on a rotating photoconductor drum. Either way, charges move through any regions of the photoconductor that are exposed to light, leaving these regions electrically neutral (Fig. 10.2.3b). The result is a *charge image*—a pattern of electric charge on the photoconductor's surface that exactly matches the pattern of ink on the original document (Fig. 10.2.3c).

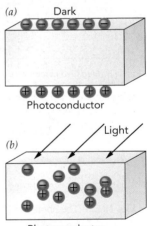

Fig. 10.2.2 (*a*) In the dark, a photoconductor is an electrical insulator so that separated electric charges on its surfaces remain there indefinitely. (*b*) When the photoconductor is exposed to light, it becomes an electrical conductor and the opposite electric charges soon join one another.

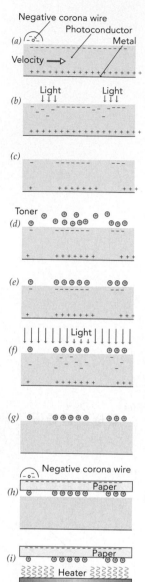

Fig. 10.2.3 The photoconductor is first coated (*a*) with a uniform layer of negative charge. Exposure to light (*b*) erases some charge to form a charge image (*c*). The charge image attracts (*d*) positively charged toner particles (*e*). The charge image is erased (*f*) to release the toner particles (*g*). The toner is transferred to the negatively charged paper (*h*) and fused to the paper with heat (*i*).

To develop this charge image into a visible one, the xerographic copier exposes the photoconductor to positively charged toner particles (Fig. 10.2.3*d*). This toner is a fine, insulating plastic powder containing a colored pigment, usually black. Applying toner to the photoconductor must be done gently and it's often accomplished with the help of Teflon-coated iron balls. These tiny balls are held together in long filaments by a rotating magnetic shaft, so that the shaft resembles a spinning brush with extraordinarily soft bristles. These bristles wipe toner particles out of their storage tray and onto the photoconductor. Contact with the Teflon leaves the toner particles positively charged, so they stick to the negatively charged portions of the photoconductor (Fig. 10.2.3*e*).

The photoconductor now carries a black image of the original document, an image that the copier must transfer to the paper. Before attempting that transfer, the copier first weakens the photoconductor's grip on the toner by exposing it to light from a charge erase lamp. This light eliminates the photoconductor's charge (Fig. 10.2.3*f*) and leaves the positively charged toner particles clinging only loosely to its surface (Fig. 10.2.3*g*).

The copier then transfers the toner image to a blank sheet of paper by pressing that paper lightly against the photoconductor while spraying negative charge onto the paper's back (Fig. 10.2.3*h*). The positively charged toner is attracted to the negatively charged paper and the two leave the photoconductor together. The copier then heats and presses the copy, permanently fusing the toner onto the paper (Fig. 10.2.3*i*). Once the image has been transferred to the paper, the copier cleans its photoconducting surface in preparation for the next copy: a second charge erase lamp eliminates any remaining charge and a brush or squeegee mops up any residual toner.

With that introduction to xerography, you can already explain many things about copiers. For example, while fixing a copier jam, you may find that you have removed unfinished copies—ones bearing toner images that haven't yet been fused onto the paper. The toner of an unfused copy comes off on your hand because it's held in place only by electrostatic forces. And when you replace the "toner cartridge" in a personal copier, in addition to adding new toner, you're also installing a new precharge system, photoconductor drum, and toner applicator (Fig. 10.2.4).

However, we've glossed over three important physics issues. Two we'll leave for later chapters: why a photoconductor becomes conducting when exposed to light (Chapter 12 on Electronics) and how a lens projects an image of the document onto the photoconductor (Chapter 15 on Optics). But the third issue is relevant now and so we'll examine it carefully: how the copier sprays charges onto surfaces.

Fig. 10.2.4 This xerographic copier places the photoconductor drum, toner supply, and a corona wire inside a disposable cartridge. After the paper passes through the cartridge, toner is fused onto its surface and it leaves the copier.

→ Check Your Understanding #1: Sticky Copies

for answers, see page 336

When the copies emerge from a xerographic copier, they tend to stick to things and attract lint. What causes this effect?

Discharges and Electric Fields

At the start of the copy cycle, the xerographic copier coats its photoconducting surface uniformly with electric charges. Because this precharging process is done in the dark, while the surface is an electrical insulator, the charges must be sprayed onto it like paint. The copier's charge sprayer is a **corona discharge**—a gentle, sustained spark that forms in the air near a needle or fine wire that's maintained at high voltage.

It's a type of **discharge**—a flow of electric charge through a gas. Air is normally an insulator because its atoms and molecules are neutral and can't convey charge from one place to another. However, by seeding air liberally with individual charged particles, the copier manages to turn that air into a conductor and then to produce a discharge in it. But how does the copier seed the air with charges and produce its discharge? And how does it use that discharge to coat its photoconducting surface? To answer those questions, we'll need to know more about electrostatic forces and voltages, and about a related concept: electric fields.

Because free charges are hard to come by in the air, the copier begins with just a few charged particles and uses them to generate more. The idea is simple: the copier uses electrostatic forces to accelerate those initial charges to enormous speeds and lets them smash into air's neutral particles. When hit hard enough, a neutral air particle breaks into oppositely charged fragments and thus adds two more free charges to the air. These new charges join the mix, accelerating, colliding, and breaking up still more air particles. A cascade of collisions ensues and the air "breaks down"—transforming from an insulator to a conductor. The copier then uses this conducting air to spray the photoconductor with charges.

So where do those initial charges come from? Surprisingly, they're already there, the products of cosmic rays and natural radioactivity! Every cubic centimeter of ordinary air contains almost two thousand charged particles, roughly half positive and half negative. Considering that this same volume of air contains almost 3×10^{19} neutral particles, that's not many charges. But it's enough to get the discharge started.

To parlay those initial charges into the vast numbers it needs, the copier must accelerate them aggressively. The neutral air particles are so densely packed that it's difficult for the charged ones to pick up much speed before they hit something and slow down. To give each initial charge a good shot at breaking the first neutral particle it hits, the copier must accelerate that charge very quickly.

The copier accelerates its free charges using electrostatic forces. Up until now, we've associated electrostatic forces with pairs of charges, each charge pushing or pulling on the other. Since the individual forces on an object sum to give the net force on that object, it's easy to figure out how three charges affect one another, or four, or five . . . . But in the copier's wires and discharge, there are so many individual charges that adding up their forces is virtually impossible. We need some other way to characterize the overall electrostatic force on a particular charge.

Instead of thinking about the many interactions between that charge and all the other charges around it (Fig. 10.2.5a), we can view the electrostatic force on our charge as the

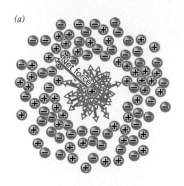

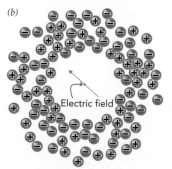

Fig. 10.2.5 (*a*) When a charge interacts with many other charges, adding up the individual electrostatic forces on that charge to obtain the net force on it becomes a daunting task. (*b*) It is often simpler to introduce an electric field as an intermediary. The other charges produce this electric field, which in turn exerts an electrostatic force on the charge (omitted for clarity). The electric field arrow passing through the red dot indicates the magnitude and direction of the force a positive test charge would experience at the dot's location.

result of its one interaction with something local: an **electric field**—an attribute of space that exerts an electrostatic force on a charge (Fig. 10.2.5*b*). From this new perspective, our charge accelerates forward because it interacts with the local electric field, a field that's created by all of the surrounding electric charges.

This electric field appears to be nothing more than an intermediary: the surrounding charges produce the electric field and this electric field pushes on our charge. But in later sections, we'll see that an electric field is more than that, more than a seemingly unnecessary fiction. That's because an electric field truly exists in space, independent of the charges that produce it. In fact, electric fields are often created by things other than charges and can influence things other than charges as well.

The copier's electric field varies with location, meaning that the electrostatic force on our charge depends on where it is. This force is equal to the product of the charge times the electric field and points in the direction of that electric field. We can write this relationship as a word equation:

$$\text{force} = \text{charge} \cdot \text{electric field}, \tag{10.2.1}$$

in symbols:

$$\mathbf{F} = q\mathbf{E},$$

and in everyday language:

Charged lint accelerates quickly in a region full of static electricity,

where the force is in the direction of the electric field. Note, however, that a particle carrying a negative amount of charge (e.g., an electron) experiences a force opposite the electric field. The SI unit of electric field is the **newton-per-coulomb** (abbreviated N/C).

The copier employs a very strong electric field to "break down" the air so that it can operate its discharge. That field accelerates charges so rapidly that collision cascades occur and fill the air with free charges. Unfortunately, you can't sense electric fields directly, so it's hard to visualize a strong one. We'll work on that problem, but for now just remember that strong electric fields can initiate discharges in air. That's how thunderstorms produce lightning!

Check Your Understanding #2: Medical Electrons

for answers, see page 336

A medical linear accelerator uses a strong electric field to accelerate electrons forward and give them enormous kinetic energies. These high-energy electrons enter the patient and kill cancer cells. In which direction does the accelerator's electric field point?

Check Your Figures #1: Lint Floats

for answers, see page 337

A piece of charged lint refuses to fall because an electric field is exactly supporting its weight. If the lint weighs 10^{-8} N and has a positive charge of 10^{-11} C, what electric field is supporting it?

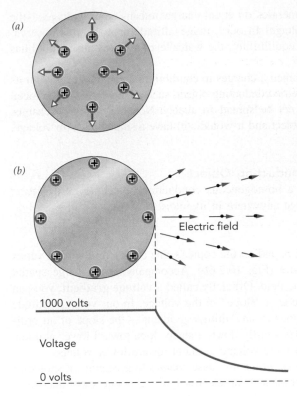

(a)

(b)

Electric field

1000 volts

Voltage

0 volts

Fig. 10.2.6 (*a*) When like charges are placed inside a conducting sphere, they repel one another and accelerate toward the sphere's surface. (*b*) When those charges have reached equilibrium on the sphere's surface, the sphere has a single, uniform voltage and zero electric field inside it. Outside the sphere, the voltage decreases toward zero and there is an electric field.

Conductors and Voltage Gradients

A copier's precharging system uses the gentle corona discharge that develops in the strong electric field just outside a fine, high-voltage wire. This discharge ferries charges to the photoconductor surface and coats it uniformly. But to understand why a strong electric field exists outside a fine, high-voltage wire and why the discharge it produces is "gentle," we need some background. Let's start by looking at electric fields inside and outside electrical conductors.

Consider the simplest conducting object: a solid metal ball. If you release some positive charges inside that ball (Fig. 10.2.6*a*), what happens to them? Because they repel one another, those charges accelerate outward and move apart. In fact, they'd leave the ball altogether if they weren't chemically bound to its metal. After spending a moment or two ridding themselves of extra electrostatic potential energy, principally as heat, the charges settle down in stable equilibria on the ball's surface (Fig. 10.2.6*b*). At its equilibrium point, the outward electrostatic force each charge experiences from its fellow charges is perfectly balanced by the inward chemical force it experiences from the metal. The net force on it is zero.

At equilibrium, each charge has also minimized its total potential energy. After all, it can't stop accelerating until there is no direction in which it can move to lower its potential energy further. But what's amazing about how the charges arrange themselves on the ball's surface is that each one ends up with the *same* total potential energy. That's because if one of them had less total potential energy than the rest, the other charges would accelerate toward it to lower their total potential energies as well!

Since the only potential energy that significantly affects charges in our small, homogeneous ball is electrostatic potential energy, every charge in our ball has essentially the same electrostatic potential energy. Because voltage is the electrostatic potential energy per unit of charge, equal potential energies on equal charges means equal voltages—the entire ball has a single, uniform voltage! In our voltage–altitude analogy, this observation is analogous to the fact that, at equilibrium, the water level in a swimming pool has a single, uniform altitude.

Because of the ball's perfect symmetry, charges in equilibrium are spread evenly on its surface. Had we chosen a less symmetric conducting object, such as the copier's fine metal wire, charges in equilibrium would not be spread so evenly. Nonetheless, those charges would still be on the outside of the object and it would still have a single, uniform voltage.

Voltage and Charge on a Conducting Object
With its charges at equilibrium, a homogeneous conducting object has a single, uniform voltage and the net charge anywhere in its interior is zero.

But while the voltage is uniform *in* and *on* the copier's fine conducting wire, it varies rapidly with position *outside* that wire (Fig. 10.2.6*b*). Accompanying this large spatial variation in voltage is a strong electric field. Officially called a **voltage gradient,** you can think of a spatial variation in voltage as a "slope" in the voltage. In our voltage–altitude analogy, a voltage gradient is analogous to an "altitude gradient"—the slope of an ordinary hill. And just as water accelerates swiftly down a steep slope toward lower altitude, so charge accelerates swiftly down a large voltage gradient toward lower voltage.

Since both electric fields and voltage gradients cause charges to accelerate, it shouldn't surprise you to learn that a voltage gradient *is* an electric field. Although we'll uncover a second source of electric fields in the next chapter, we'll treat a voltage gradient and an electric field as equivalent for now. Their relationship can be written as a word equation:

$$\text{electric field} = \text{voltage gradient} = \frac{\text{voltage drop}}{\text{distance}}, \tag{10.2.2}$$

in symbols:

$$\mathbf{E} = \mathbf{Gradient}(V),$$

and in everyday language:

Charges rush down steep drops in voltage, much as bicycles rush down steep drops in height,

where the electric field points in the direction of most rapid voltage decrease.

This relationship gives us a second way to look at an electric field. In addition to being the electrostatic force exerted per unit of charge, electric field is also the voltage drop per unit of distance. The SI unit of electric field therefore has a second form: the **volt-per-meter** (abbreviated V/m). The V/m is exactly the same unit as the N/C.

➤ Check Your Understanding #3: Don't Get Out of a Hot Car!
for answers, see page 336

During a thunderstorm, a lightning strike places a huge static charge on your car. Why don't you notice this charge as long as you remain inside the car?

► Check Your Figures #2: Riding the Field

for answers, see page **337**

An electric field pushes charged particles through the tube of a fluorescent lamp, allowing it to produce light. To operate properly, a typical fluorescent tube needs an electric field of about 100 V/m. If the average voltage difference between the two ends of a tube is 120 volts, how long can that fluorescent tube be?

Fine Wires and High Voltages: Corona Discharges

Ordinary air breaks down in an electric field of about 3×10^6 volts-per-meter or, in customary units, about 30,000 volts-per-centimeter. At that field, free charges accelerate so rapidly that a cascade of charge-freeing collisions suddenly transforms air from a nearly perfect insulator into a reasonably good conductor.

Courtesy Lou Bloomfield

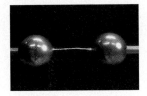

You can produce such a strong field all by yourself. On a dry winter day, you can coat yourself with positive charges and raise your voltage to about 30,000 volts simply by scuffing your rubber-soled shoes across an acrylic carpet. As you then approach a grounded doorknob at 0 volts, the voltage difference between the doorknob and your hand will be 30,000 volts. When your hand is about 1 cm from the doorknob, the electric field will reach 30,000 volts-per-centimeter and the air will break down with a brilliant spark (Fig. 10.2.7).

Fig. 10.2.7 These two metal spheres are 1 cm apart. When their difference in voltage is about 30,000 V, the air between them breaks down and forms a spark.

Because your hand and the doorknob are similar in size and shape, the voltage changes smoothly between them (Fig. 10.2.8*a*). It varies steadily from 0 volts on the doorknob to 30,000 volts on your hand, so the voltage gradient or electric field is nearly uniform. But when two objects differ significantly in size, the larger object dominates voltages in the space between them. For example, if you hold a long pin in your hand as you approach the doorknob, the doorknob will control the voltage most of the way to the pin and nearly all the increase in voltage will occur just outside the pin's point (Fig. 10.2.8*b*). Rather than being uniform, the voltage gradient or electric field will be strongest near that point.

The copier makes good use of this nonuniform field. Its fine, high-voltage wire is nearly surrounded by a much larger metal shroud. The wire is so thin that its influence fades just a hair's breadth from its surface and the grounded shroud dominates voltage almost all the way to the wire. Although the wire's negative voltage is only −3000 volts and it's about 1 cm from the shroud, the voltage changes so rapidly in the air just outside this wire that the electric field there easily exceeds 30,000 volts-per-centimeter and breaks down the air.

The discharge that forms near the fine wire is a special, self-regulating one—a corona discharge (Fig. 10.2.9). While most discharges can't control how many free charges they produce, a corona discharge automatically maintains a steady production. Because free charges form only in the strong electric field near its thin conductor, their production rate is very sensitive to changes in that conductor's effective thickness. If there are too many free charges in the air near the conductor, their ability to conduct electricity effectively thickens the conductor, weakens the electric field, and slows the production of free charges. The discharge is correcting its own mistake.

Because of this stabilizing effect, the air in a corona discharge maintains a steady electrical conductivity that's ideal for charging a copier's photoconductor. However,

Fig. 10.2.9 The electric field near this sharp, high-voltage pin is so strong that it breaks down the air and forms a corona discharge. The resulting glow is produced by air particles that receive energy from the discharge.

(a)

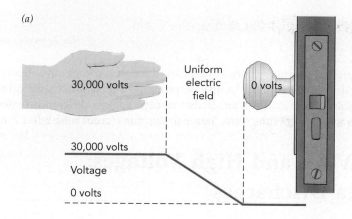

(b)

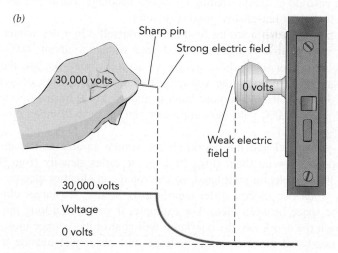

Fig. 10.2.8 Your voltage is 30,000 V when you reach for the 0 V doorknob. (*a*) Since your hand and the doorknob are similar in size, the voltage decreases steadily between them and the electric field is approximately uniform. (*b*) But when you hold out a pin, the voltage plummets near its sharp point and the electric field there is extremely strong.

❏ Contrary to popular belief, lightning rods don't simply attract lightning strikes so as to protect the surrounding roof. Instead, they produce corona discharges that diminish any local buildups of electric charge. By neutralizing the local electric charge, the lightning rod reduces the chances that lightning will strike the house. Similar devices, called static dissipaters, are found near the tips of airplane wings and protect planes from lightning strikes.

corona discharges were common long before copiers. They often occur spontaneously near sharp points or fine wires at high voltages, leading to charge leakage from power transmission lines and occasionally producing a glow called St. Elmo's fire on the masts and rigging of sailing ships (see ❏).

Check Your Understanding #4: A Safety Pin

for answers, see page 337

You could avoid the shock of static electricity by holding out a sharp needle as you reach for a metal doorknob or wall. How does that needle protect you from static electricity?

Getting Ready to Copy:
Charging by Induction

A corona discharge does more than just turn air into a conductor; it also produces an outward spray of electric charges. Those charges are pushed outward by the electric field surrounding the corona wire. Since the copier's corona wire has a negative voltage, the surrounding electric field points toward that wire. And because negative charges accelerate opposite an electric field, the copier's corona produces a shower of outgoing negative charges. They spray onto the photoconducting surface as it moves steadily past the corona and the photoconductor thus acquires a uniform coating of negative charges.

As each negative charge lands, it draws a positive charge onto the grounded metal surface beneath the photoconductor and the attraction between those two opposite charges holds them firmly in place. While the photoconductor's open surface is acquiring its uniform negative charge, the metal layer underneath is acquiring an equivalent positive charge (Fig. 10.2.3a). This process, whereby a grounded conductor acquires a charge through the attraction of nearby opposite charge, is called "charging by induction."

The induced positive charge on the metal side of the photoconductor is important to the xerographic process for several reasons. First, it lowers the electrostatic potential energy of the negative charge so that the surface's negative voltage isn't so enormous. Second, without that positive layer nearby, repulsion between like charges would tend to push negative charges on the open surface toward the edges of the photoconductor and distort the resulting images.

But most significantly, the positive charge layer gives the negative charge layer somewhere to go when the photoconductor is exposed to light! Wherever light from the original document turns a patch of photoconductor into a conductor, the negative and positive charge layers rush together and cancel. The resulting uncharged portion of photoconductor subsequently attracts no toner and produces a white patch on the finished copy.

Having come full circle, we can now see how the copier achieves its goal. It uses a corona discharge to coat a photoconducting surface with negative charge and then selectively erases portions of that charge layer with light from the original document. The remaining charged patches on the photoconductor attract positively charged black toner, which is then transferred permanently to the paper.

The only additional detail worth mentioning is that, for technical reasons, some copiers precoat their photoconductors with positive rather than negative charges and then use that charge to attract negatively charged toner. These copiers put high positive voltages on their fine wires so that their coronas spray positive charges.

Check Your Understanding #5: Hot Rod

for answers, see page 337

A thick wire connects the lightning rod on the courthouse steeple to the ground and normally ensures that the rod is electrically neutral. However, when a negatively charged cloud floats overhead, what charge does that rod acquire?

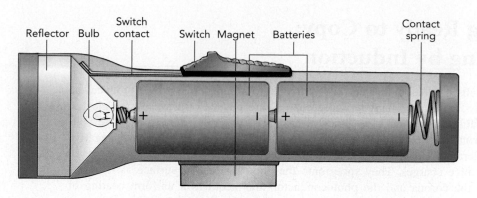

SECTION 10.3 Flashlights

There isn't much to a typical flashlight; you can see what few parts it has when you open it to replace its batteries. But a flashlight isn't a mechanical device, it's an electrical one: it contains an electric circuit and most of its components are involved in the flow of electricity. Understanding how a flashlight works means understanding how an electric circuit works and how electricity carries power from batteries to a bulb. As we'll see, flashlights aren't as simple as they appear.

Questions to Think About: Why are some flashlights brighter than others? Why is it important that all of the batteries point in the same direction? What is the difference between old batteries and new? What makes a flashlight suddenly become dim or bright when you shake it?

Experiments to Do: Find a flashlight that uses two or more removable batteries. Turn it on. What did the switch do to make the flashlight produce light? With the flashlight turned on, slowly open the battery compartment. The lamp will probably become dark. You should be able to turn the flashlight on and off by opening and closing the battery compartment. Why does this method work?

Replace the flashlight's batteries with older or newer ones and compare its brightness. Turn one or more of the batteries around backward and see how that change affects the flashlight. What happens when you put a piece of paper or tape between two of the batteries? What happens when you carefully clean the metal surfaces of each battery with a pencil eraser before putting them in the flashlight?

Electricity and the Flashlight's Electric Circuit

A basic flashlight has just three components—a battery, a bulb, and a switch—connected together by metal strips. When the switch is on, the strips transfer energy from the batteries to the bulb. But how does energy move through the strips and why does the switch start or stop that energy transfer? To answer these questions, we must first understand electricity and electric circuits, so that's where we'll begin.

When you turn on a flashlight, electricity conveys energy from the batteries to the bulb. An **electric current**—a current of electric charges—flows through these components, carrying the energy with it. Though we'll examine the exact nature of this current soon, you can picture it as a steady stream of tiny positive charges following a circular route that takes them through the batteries, through the bulb, then back to the batteries for another trip (Fig. 10.3.1). As long as the flashlight is on, charges flow around this loop, receiving energy from the batteries and delivering it to the bulb, over and over again. En route, the charges carry this energy mostly as electrostatic potential energy.

The looping path taken by charges in a flashlight is called an **electric circuit.** Because a circuit has no beginning or end, charges can't accumulate in one place, where their mutual repulsion would eventually stop them from flowing. Circuits are present in virtually all electric devices and explain the need for at least two wires in the power cord of any home appliance: one wire carries charges to the appliance to deliver energy and the other wire carries those charges back to the power company to receive some more.

But what role does the switch play in all of this? As part of one conducting path between the batteries and bulb, the switch can make or break the flashlight's circuit (Fig. 10.3.2). When the flashlight is on, the switch completes the loop so that charges can flow continuously around the **closed circuit** (Fig. 10.3.2a). A closed circuit appears on the left side of Fig. 10.3.3.

However, when you turn off the flashlight, the switch breaks the loop to form an **open circuit** (Fig. 10.3.2b). Although one conducting path still connects the batteries and bulb, the loop now has a gap in it and can no longer carry a continuous current. Instead, charges accumulate at the gap and current stops flowing through the flashlight. Since energy can no longer reach the bulb, it goes dark. An open circuit appears on the right side of Fig. 10.3.3.

There's one other type of circuit worth mentioning. A **short circuit** forms when the two separate paths connecting the batteries to the bulb accidentally touch one another (Fig. 10.3.4). This unintended contact creates a new, shorter loop around which the charges can flow. Because the bulb is supposed to extract energy from the charges, it's designed to impede their flow and to convert their electrostatic potential energy into thermal energy and light. This opposition to the flow of electricity is called **electrical resistance.** Since the shortened loop offers little resistance, most of the charges flow through it and bypass the bulb. The bulb dims or goes out altogether.

Since the bulb is the only part of the flashlight that's designed to get hot, a short circuit leaves the charges without a safe place to get rid of their electrostatic potential energies.

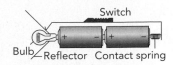

Fig. 10.3.1 A flashlight contains one or more batteries, a lightbulb, a switch, and several metal strips to connect them all together. When the switch is turned on (as shown), the components in the flashlight form a continuous loop of conducting materials. Electrons flow around this loop counterclockwise.

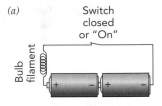

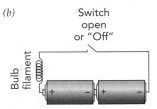

Fig. 10.3.2 (*a*) When the flashlight's switch is on, it closes the circuit so that current can flow continuously from the batteries, through the bulb's filament, and back through the batteries. It follows this circuit over and over again. (*b*) When the flashlight's switch is off, it opens the circuit so that current stops flowing.

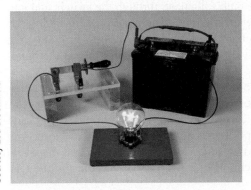
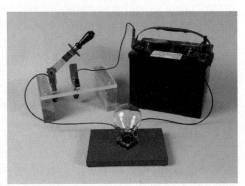

Fig. 10.3.3 When the switch is on (*left*), current flows around the closed circuit and carries power from the battery to the lightbulb. But when the switch is off (*right*), no current flows through the open circuit.

They deposit it dangerously in the batteries and the metal paths, making them hot. Since short circuits can start fires, flashlights and other electric equipment try to avoid them.

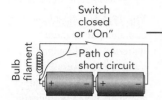

Switch
closed
or "On"

Bulb filament

Path of
short circuit

Fig. 10.3.4 When an unwanted conducting path allows current to bypass the flashlight's filament, it forms a short circuit. Because it has no proper place for electrons to deposit their energy, the short circuit becomes hot.

Check Your Understanding #1: Cutting the Power

for answers, see page 337

If you remove just one of the wires from an automobile's battery, the vehicle will not start at all. Why doesn't the other wire supply any energy?

The Electric Current in the Flashlight

Each of the tiny charged particles flowing through the flashlight's circuit carries with it just a single elementary unit of electric charge and a miniscule amount of electrostatic potential energy. However, because those charges flow in astonishing numbers, they convey a considerable amount of energy per second—the quantity we know as power (see Section 2.2) and measure in watts (abbreviated W). The bulb needs a certain amount of power to keep its filament glowing brightly and you can determine how much power is reaching the bulb by multiplying the number of elementary charges passing through the bulb each second by the amount of energy each one delivers.

But there are too many elementary charges to count. You'll do much better to measure the circuit's **current,** that is, the amount of charge passing a particular point in the circuit per unit of time. The SI unit of current is the **ampere** (abbreviated A) and corresponds to 1 C (1 coulomb) of charge passing by the designated point each second. One coulomb is roughly 6.25×10^{18} or 6,250,000,000,000,000,000 elementary charges, so even a 1-A current involves a tremendous flow of elementary charges.

Using electric current instead of counting charges, you can determine how much power is reaching the bulb by multiplying that current by its electrostatic energy per coulomb—the quantity we already know as voltage. For example, a current of 2 amperes (2 coulombs-per-second) at a voltage of 3 volts (3 joules-per-coulomb) would bring 6 watts of power (6 joules-per-second) to the bulb. Brighter flashlights involve larger currents, greater voltages, or both.

Current has a direction, pointing along the route of positive charge flow. When you turn on the flashlight in Fig. 10.3.1, charge flows around the circuit clockwise—from the battery chain's positive terminal, through the bulb's filament, through the switch, and into the battery chain's negative terminal. However, it's time to address an awkward issue: the positive charges that flow clockwise around this circuit are fictitious. In reality, the electric current is actually carried by negatively charged electrons heading in the opposite direction!

This issue dates back to Franklin's unfortunate choice of which charge to call positive and which to call negative. By the time scientists discovered the electron and realized that these negatively charged particles carry currents in wires, current had already been defined as pointing in the direction of positive charge flow. Since it was far too late to make current and electron-flow point in the same direction, scientists and engineers simply pretend that current is carried by fictitious positive charges heading in the current's direction.

This fiction works extremely well, as illustrated by a simple example. When negatively charged electrons flow to the right through a neutral piece of wire, the wire's right end becomes negatively charged and its left end becomes positively charged

(Fig. 10.3.5a). But exactly the same thing would happen if a current of fictitious positively charged particles were to flow to the left through that same piece of wire (Fig. 10.3.5b). Without sophisticated equipment, you can't tell whether negative charges are flowing to the right or positive charges are flowing to the left because the end results are essentially indistinguishable.

We, too, adopt this fiction and pretend that current is the flow of positively charged particles. In this and subsequent chapters, we'll stop thinking about electrons and imagine that electricity is carried by positive charges moving in the direction of the current. There are only a few special cases in which the electrons themselves are important, and we'll consider those situations separately when they arise.

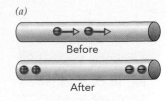

(a)

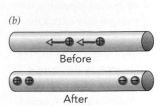

(b)

Check Your Understanding #2: Don't Touch That Pipe

for answers, see page 337

A walk across wool carpeting in rubber-soled shoes has left you covered with negative charges. When you bring your hand near a large piece of metal, the negative charges will leap through the air as a spark in order to reach the metal. Which way is current flowing in this spark?

Fig. 10.3.5 A current of negatively charged particles flowing to the right through a piece of wire (a) can't easily be distinguished from a current of positively charged particles flowing to the left (b). The end result of both processes is an accumulation of positive charge on the left end of the wire and negative charge on its right end.

Batteries, . . .

While a battery is basically a portable source of electric power, here are two other interesting ways to think of it. The first is rather abstract: a battery is a type of pump. It "pumps" charge from low voltage to high voltage, much as a water pump pumps water from low altitude to high altitude. Once again, our voltage–altitude analogy is helpful. Each of these pumps moves something against its natural direction of flow, pushing it forward and doing work on it in the process. The battery increases a charge's electrostatic potential energy by pushing it up a voltage gradient, while the water pump increases water's gravitational potential energy by pushing it up an altitude gradient.

The second perspective on batteries is more mechanical: a battery is a chemically powered machine. It uses chemical forces to transfer charges from its negative terminal to its positive terminal. As positive charges accumulate on the battery's positive terminal, the voltage there rises and as negative charges accumulate on the battery's negative terminal, the voltage there drops. Since the battery does work transferring charges from low voltage to high voltage, it is converting its chemical potential energy into electrostatic potential energy in these separated charges.

A battery's rated voltage reflects its chemistry, specifically the amount of chemical potential energy it has available for each charge transfer. As the voltage difference between its terminals increases, so does the energy required for each charge transfer. Eventually, the chemicals can't do enough work on a charge to pull it away from the negative terminal and push it onto the positive terminal, so the transfers stop. The battery is then in equilibrium—the electrostatic forces opposing the next charge transfer exactly balance the chemical forces promoting it. A typical alkaline battery reaches this equilibrium when the voltage of its positive terminal is 1.5 V above the voltage of its negative terminal. Lithium batteries, with their more energetic chemistries, can achieve voltage differences of 3 V or more.

When you turn on the flashlight, you upset its equilibrium by allowing charges to leave the battery's positive terminal for its negative terminal. With fewer separated charges now on its terminals, the battery's voltage difference decreases slightly and it

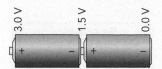

Fig. 10.3.6 When two 1.5-V batteries are connected in a chain, their voltages add so that the chain's positive terminal has a voltage that is 3.0 V higher than the chain's negative terminal. If the chain's negative terminal is chosen to be at 0 V, then the chain's positive terminal is at 3.0 V.

Courtesy Lou Bloomfield

Fig. 10.3.7 A 9-V battery actually contains six small 1.5-V cells, connected in a chain. Positive charges that enter the chain at the battery's negative terminal pass through all six cells before arriving at the battery's positive terminal.

Fig. 10.3.8 When one battery in a chain of three is reversed, the reversed battery's voltage is subtracted from the sum of the others. The chain's positive terminal has a voltage only 1.5 V higher than its negative terminal. The reversed battery recharges.

begins pumping charges again. That renewed charge transport replenishes the terminals' separated charges and opposes any further decrease in the battery's voltage. In this manner, a 1.5-V alkaline battery maintains a nearly steady voltage difference of 1.5 V between its terminals, whether its flashlight is on or off.

That alkaline battery is powered by an electrochemical reaction in which powdered zinc at its negative terminal reacts with manganese dioxide paste at its positive terminal. This reaction resembles controlled combustion. In effect, the battery "burns" zinc to obtain the energy it needs to pump charges from its negative terminal to its positive terminal. However, as the battery consumes its chemical potential energy, its ability to pump charges diminishes. When its chemicals are nearly exhausted, the battery's increasing disorder reduces its voltage. An aging battery can pump less current than a fresh one and provides that current with less voltage. Ultimately, less power reaches the flashlight's bulb and it goes dim.

Most flashlights use more than one battery. When two alkaline batteries are connected together in a chain, so that the positive terminal of one battery touches the negative terminal of the other, the two batteries work together to pump charges from the chain's negative terminal to its positive terminal (Fig. 10.3.6). Each battery pumps charges until its positive terminal is 1.5 V above its negative terminal, so the chain's positive terminal is 3.0 V above its negative terminal. Because charges never leave the flashlight's circuit, only relative voltages matter in that circuit. We'll find it convenient to ignore the flashlight's absolute voltages and define the voltage of the battery chain's negative terminal to be 0 V (Fig. 10.3.6). With that choice, the voltage of its positive terminal becomes 3.0 V.

The more batteries in the flashlight's chain, the more energy a charge receives overall and the more the voltage will increase from the chain's negative terminal to its positive terminal. A flashlight that uses six alkaline batteries in its chain has a positive terminal that is 9 V above its negative terminal. A typical 9-V battery actually contains a chain of six miniature 1.5-V batteries, arranged so that their voltages add up to 9 V (Fig. 10.3.7).

If you reverse one of the batteries in a chain, the reversed battery will extract energy from any charge passing through it (Fig. 10.3.8). While the chain may still pump charge from its negative terminal to its positive terminal, its overall voltage will be reduced because instead of adding 1.5 V to the chain's overall voltage, the reversed battery will subtract that amount. If the chain has three batteries, two will add energy to the charge while one will subtract it and the chain's overall voltage will be only 1.5 V.

As the reversed battery extracts energy from the charges passing through it, at least some of that extracted energy is converted into chemical potential energy. The reversed battery is recharging! Battery chargers follow that concept, pushing current backward through a battery—from its positive terminal to its negative terminal—to restore the chemical potential energy in a rechargeable battery. However, normal alkaline batteries are "nonrechargeable," meaning that they turn most of the recharging current's energy into thermal energy instead of chemical potential energy. Nonrechargeable batteries may overheat and explode during recharging.

Check Your Understanding #3: Car Batteries

for answers, see page 337

A lead–acid automobile battery provides 12 V between its negative terminal and its positive terminal. It actually contains six individual batteries, connected in a chain. How much voltage does each individual battery provide?

. . . Bulbs, and Metal Strips

While a battery gives charges electrostatic potential energy by pushing them *up* a voltage gradient, a bulb releases that electrostatic potential energy by letting charges slide *down* another voltage gradient. Those two devices make a perfect pair: the battery provides electric power and the bulb consumes it. We saw back in Chapter 7 that the bulb uses this power to heat its tungsten filament to incandescence, thereby producing light. Now it's time to look at how electricity heats the filament.

We'll consider a flashlight with two alkaline batteries (Fig. 10.3.9a). The bulb's filament is a fine wire and its two ends are electrically connected to battery chain's terminals. With one end at 3.0 V and the other at 0.0 V, the filament has a voltage gradient across it and therefore an electric field. But how is that possible? While discussing copiers, we observed that a conductor has a uniform voltage throughout. Isn't the filament violating that rule?

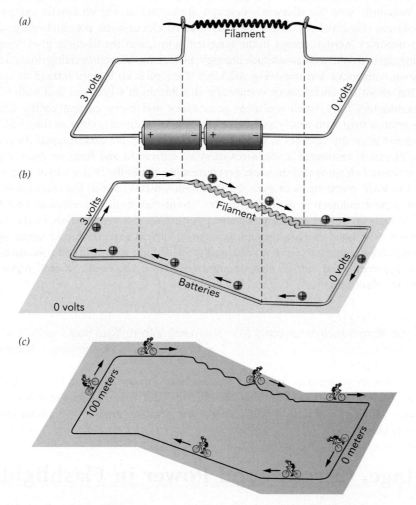

Fig. 10.3.9 (*a*) Current in a flashlight's circuit conveys power from the batteries and to the filament. (*b*) Its voltage rises in the batteries and decreases in the filament. Despite an electric field inside the filament, pushing the charges forward, they travel at constant velocity because of collisions. (*c*) This behavior is analogous to bicyclists pedaling up a smooth hill and then rolling down a rough one at constant velocity.

No. While a conductor has a uniform voltage *when its charges are in equilibrium,* the charges in the bulb are only in equilibrium when the flashlight is off. When you switch the flashlight on, you impose a 3.0-V difference between the two ends of the filament and the filament's charges immediately began accelerating down the voltage gradient toward the 0.0-V end.

In our voltage–altitude analogy, it's as though you suddenly tilted a level field to create a hill and water that was lying motionless on that field now accelerates downhill. However, a better analog to the individual charges that we're considering at the moment would be bicyclists: picture hundreds of bicyclists on a level field that suddenly tilts to form a hill. All the bicyclists that were at equilibrium on the level field now accelerate downhill.

If the filament were a perfect conductor of electricity, each charge would accelerate steadily down the voltage gradient and convert its electrostatic potential energy into kinetic energy. But the filament has a large electrical resistance and significantly impedes the flow of electric current. Each charge bounces its way down the voltage gradient, colliding frequently with the filament's tungsten atoms and giving up kinetic energy with each collision (Fig. 10.3.9*b*). What began as ordered electrostatic potential energy in the charges becomes thermal energy in the tungsten atoms, and the filament glows brightly. Referring again to our voltage–altitude analogy, picture the bicyclists riding down a rough hill strewn with rocks and trees (Fig. 10.3.9*c*). They pick up bruises instead of speed.

What about the metal strips connecting the flashlight's batteries and bulb? These thick conductors have small electrical resistances and carry current easily. Charges emerge from a strip with nearly as much electrostatic potential energy as they had when they entered it, so the voltages at the two ends of that strip are almost equal. In general, the less electrical resistance in the wires carrying current to and from the bulb, the less power is wasted en route and the more power reaches the bulb. That's why it's so important to use thick metal strips or even the flashlight's metal case in the connections.

But a poor connection anywhere in the circuit can spoil this efficient transfer of power. If there is dirt or grease on a battery terminal or worn materials in the switch, the current will have to pass through a large electrical resistance and waste power. Improving that connection, either by shaking the flashlight or by cleaning the metal surfaces, will increase the current flow through the circuit, reduce the wasted power, and brighten the flashlight.

Check Your Understanding #4: You Get What You Pay For

for answers, see page *337*

The battery in your car is dead, so you use cheap jumper cables to connect the electric system in your car to the electric system in your friend's car. But when you try to start your car, too little power reaches it to start its engine. What's wrong with those cables?

Voltage, Current, and Power in Flashlights

When you turn on the flashlight, an electric current carries power from its two alkaline batteries to its bulb. Let's suppose that a current of 1 A is flowing through the flashlight's circuit and take a look at how much power is being transferred.

A bulb consumes electric power because the current passing through it slides down the voltage gradient and experiences an overall drop in voltage. This **voltage drop** measures the electrostatic potential energy each unit of charge loses while struggling through the filament. Multiplying the voltage drop by the current passing through the bulb gives you the power consumed by the bulb. This observation can be written as a word equation:

$$\text{power consumed} = \text{voltage drop} \cdot \text{current}, \qquad (10.3.1)$$

in symbols:

$$P = V \cdot I,$$

and in everyday language:

When a big current loses a lot of voltage, something will probably get hot.

Since the voltage drop across the bulb is 3.0 V and the current passing through it is 1.0 A, it's consuming 3.0 W of power.

A battery chain produces electric power because the current passing through it is pushed up a voltage gradient and experiences an overall rise in voltage. This **voltage rise** measures the electrostatic potential energy each unit of charge gains while being pumped through the batteries. Multiplying the voltage gain by the current passing through the batteries gives you the power provided by the batteries. This observation can be written as a word equation:

$$\text{power provided} = \text{voltage rise} \cdot \text{current}, \qquad (10.3.2)$$

in symbols:

$$P = V \cdot I,$$

and in everyday language:

It takes a great deal of power to raise the voltage of a large electric current.

Since the voltage rise across the chain is 3.0 V and the current passing through it is 1.0 A, it's providing 3.0 W of power.

Check Your Understanding #5: Current Trends in Music

for answers, see page 337

A large battery powers your portable radio. Current enters the radio through one wire and leaves through another. Which wire has a higher voltage?

Check Your Figures #1: When You Turn the Ignition Key

for answers, see page 337

When you first start your car, its cold engine is difficult to turn and a 200-A current flows through its starter motor. If that current experiences a voltage drop of 12 V, how much power is being consumed?

➤ Check Your Figures #2: When You Turn the Ignition Key Again
for answers, see page **338**

When you restart the car, its warm engine is easier to turn and a smaller current of 150 A flows through the car's starter motor. The battery is supplying power to this current, providing it with a voltage rise of 12 V. How much power is the battery providing?

Choosing the Bulb: Ohm's Law

Our flashlight's bulb is designed to operate properly with a voltage drop of 3.0 V. Subjected to that voltage drop, it will carry a current of 1 A and thus consume 3 W of electric power: just enough to make it glow properly. If you were to use the wrong bulb in this flashlight, one designed for a different voltage drop, its filament would carry the wrong amount of current and receive the wrong amount of power. Too much power would quickly burn out its filament while too little power would make the filament glow dimly.

The bulb's filament must clearly match the flashlight, particularly the voltage of flashlight's battery chain. For example, flashlights that use many batteries require bulb filaments that are designed to operate with large voltage drops. But why is the current carried by a particular bulb filament related to the voltage drop across it and why do different bulbs respond differently to a particular voltage drop?

The relationship between current and voltage drop is the result of collisions. Charges effectively stop each time they crash into tungsten atoms, so they need the push of an electric field to keep them moving forward (Fig. 10.3.10). Doubling that electric field doubles each charge's average speed and, because the number of mobile charges in the filament is fixed, also doubles the overall current flowing through the filament. Since the electric field that propels this current is the filament's voltage gradient, doubling the voltage drop through the filament doubles the current as well.

In our voltage–altitude analogy, picture bicyclists riding on extremely rocky terrain without pedaling. These lazy bicyclists effectively stop each time they crash into rocks, so they need the push of a slope to keep them moving forward. Doubling the slope's altitude gradient—the altitude drop per meter of downhill travel—doubles each bicyclist's average speed and, because the number of bicyclists who can fit on the hill at once is fixed, also doubles the overall current of bicyclists rolling down the hill. Since the slope that propels this bicyclist-current is the hill's altitude gradient, doubling the hill height doubles the current of bicyclists as well.

The influence of filament choice on current flow reflects the different electrical resistances of those filaments. Anything that increases the number of mobile electric charges across the filament's width or allows those charges to maintain a higher average speed for a given voltage drop will decrease the filament's electrical resistance and increase the current flowing through it. In fact, electrical resistance is defined as the voltage drop through the filament divided by the current that arises as a result. Making the filament thicker or shorter will lower its resistance, as will changing its composition to make collisions less frequent.

Our voltage–altitude analogy with bicyclists on a hill is again helpful. Anything that increases the number of bicyclists across the hill's width or allows those bicyclists to maintain a higher average speed for a given hill height will decrease the hill's "bicycle resistance" and increase the current of bicyclists rolling down it. In fact, "bicycle resistance" is defined as the hill height divided by the current of bicyclists it produces. Making

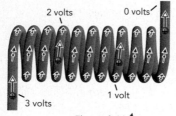

2 volts   0 volts

3 volts   1 volt

Electric field: ↑

Fig. 10.3.10 Charges moving through this filament experience a voltage drop of 3 V and are pushed forward by the resulting electric field. They maintain a constant speed despite frequent collisions with tungsten atoms.

the hill wider or shorter will lower its bicycle resistance, as will changing its rockiness to make collisions less frequent.

Combining these observations, we see that the current flowing through the filament is proportional to the voltage drop through it and inversely proportional to the filament's electrical resistance, which can be written:

$$\text{current} = \frac{\text{voltage drop}}{\text{electrical resistance}}. \qquad \textbf{(10.3.3)}$$

This relationship is called **Ohm's law,** after its discoverer Georg Simon Ohm ❏. Structuring it this way separates the causes (voltage drop and electrical resistance) from their effect (current flow). However, this equation is often rearranged to eliminate the division. The relationship then takes its customary form, which can be written as a word equation:

$$\text{voltage drop} = \text{current} \cdot \text{electrical resistance}, \qquad \textbf{(10.3.4)}$$

in symbols:

$$V = I \cdot R,$$

and in everyday language:

Long, skinny jumper cables lose so much voltage carrying current to your car that it probably won't start.

The SI unit of electrical resistance, the volt-per-ampere, is called the **ohm** (abbreviated Ω, the Greek letter omega). Despite its simplicity, Ohm's law is extremely useful in physics and electrical engineering. It applies to so many systems that nearly everything can be characterized by an electrical resistance. Once an object's electrical resistance is known, the current flowing through it can be calculated from its voltage drop or its voltage drop can be calculated from the current flowing through it. An object that obeys Ohm's law is often described as **ohmic.**

> **Ohm's Law**
> The voltage drop through a wire is equal to the current flowing through that wire times the wire's electrical resistance.

Finally, an object's electrical resistance is typically temperature-dependent. Rising temperature increases the number of mobile charges in an object but also makes them collide more frequently with the jiggling atoms. If the increasing collision frequency dominates, as it does in metals, an object's resistance increases with temperature. For example, a filament carries less and less current as it approaches operating temperature, a behavior that helps it avoid overheating. However, if the increase in mobile charges dominates, as it does in semiconductors, an object's resistance decreases with temperature. This explains why semiconductor-based computer chips carry more and more current as they get hotter and will self-destruct at excessive temperatures.

❏ German physicist Georg Simon Ohm (1787–1854) served as a professor of mathematics, first at the Jesuits' college of Cologne and then at the polytechnic school of Nuremberg. His numerous publications were undistinguished, with the exception of one pamphlet on the relationship between current and voltage. This extraordinary document, written in 1827, was initially dismissed by other physicists, even though it was based on good experimental evidence and explained many previous observations by others. In despair, Ohm resigned his position at Cologne, and it was not until the 1840s that his work was accepted. He was finally appointed professor of physics at Munich only two years before his death.

►Check Your Understanding #6: Skin Protection

for answers, see page 337

Your skin has a much larger electrical resistance than your tissues. If you touch the two terminals of a battery with your fingers, where is the larger voltage drop—in your skin or in your tissues?

►Check Your Figures #3: Light Resistance

for answers, see page 338

Two 3-W flashlight bulbs have different resistances: 3 Ω and 12 Ω. Which bulb is meant to operate from two 1.5-V alkaline batteries?

Epilogue for Chapter 10

This chapter dealt with three venues in which charge and electricity play important roles. In *static electricity,* we introduced the concept of electric charge and discussed the attractive and repulsive forces that charged particles exert on one another. We studied the electrostatic potential energies stored in those charge forces and the relationship between this energy and the voltages of various locations. We learned how different objects acquire net charges through contact and how high voltages can be produced by separating large quantities of opposite charge.

In *xerographic copiers,* we studied electric fields and saw how those fields can be used to move charges from one place to another. We examined the corona discharges that form in the strong electric fields around sharp points or thin wires at high voltages. And we saw how placing a charge on one object can induce an opposite charge on a grounded object nearby.

In *flashlights,* we examined circuits to see how a current of electric charges can transfer power from batteries to a lightbulb. We found that both voltage and current contribute to this power transfer and we learned how the bulb's electrical resistance governs the power it consumes.

Explanation: Moving Water without Touching It

This experiment uses contact between dissimilar materials to give the comb a net electric charge. Electrons shift from your hair to the comb, leaving your hair positively charged and the comb negatively charged. Because the comb is an electrical insulator, its negative charge remains trapped on its surface for a long time.

As you hold the negatively charged comb near the water stream, the comb attracts positive charges present in the water and repels negative charges. Because the water conducts electricity somewhat, positive charges can travel down the water stream from the faucet while negative charges travel the other way. The water thus acquires a net positive charge in the presence of the negatively charged comb, an example of charging by induction. Since the oppositely charged water and comb attract one another, the water accelerates toward the comb and arcs sideways as it falls.

Chapter Summary

How Static Electricity Works: When clothes tumble about in the dryer, contact between dissimilar materials transfers negatively charged electrons from one item to the other. As a result of this contact charging effect, the various garments acquire net charges, some positive and others negative. When the clothes are subsequently separated, the work done in pulling them apart becomes electrostatic potential energy and the clothes develop high voltages. High voltages can also develop when you walk across a carpet or drive you car along the road. Since high voltage tends to push charges into the air as leaks and sparks, static charging can be nuisance. You can control it with the help of conducting materials: allowing charge to move spontaneously from high voltage to low voltage prevents large quantities of separated charge from accumulating so that high voltages can't develop.

How Xerographic Copiers Work: The photoconductor in a xerographic copier allows light to control the distribution of electric charge on its surface. This photoconductor is uniformly precoated with charge by passing it near a corona discharge. An optical image of the original document is then projected onto this charged photoconductor. Wherever light hits the photoconductor, the surface charge escapes. The result is a charge image on the photoconductor. Tiny black toner particles, oppositely charged from the unilluminated portions of the photoconductor, are brought near the charge image. These toner particles stick to the charged portions of the photoconductor, forming a visible image of the document. These toner particles are then transferred to and fused onto a sheet of paper to create a finished copy.

How Flashlights Work: A flashlight produces light when a current flows through its electric circuit. This circuit consists of a chain of batteries, a lightbulb, a switch, and several metal strips, all connected in a continuous loop. The current obtains power as it passes through the battery, and it releases this power as it passes through the filament of the lightbulb, heating that filament white-hot.

The switch provides a way to control the flow of current in the circuit. When the switch is off, it breaks the circuit and prevents current from flowing completely around the circuit. Without a steady current to carry power from the batteries to the bulb, the bulb is dark. When the switch is on, it completes the circuit so that power can reach the bulb and the bulb lights up.

Important Laws and Equations

1. Coulomb's Law: The magnitudes of the electrostatic forces between two objects are equal to the Coulomb constant times the product of their two electric charges divided by the square of the distance separating them, or

$$\text{force} = \frac{\text{Coulomb constant} \cdot \text{charge}_1 \cdot \text{charge}_2}{(\text{distance between charges})^2}. \qquad (10.1.1)$$

If the charges are like, then the forces are repulsive. If the charges are opposite, then the forces are attractive.

2. The Force Exerted on a Charge by an Electric Field: A charge experiences a force equal to the product of its charge times the electric field, or

$$\text{force} = \text{charge} \cdot \text{electric field}, \qquad (10.2.1)$$

where the force points in the direction of the electric field.

3. The Electric Field Due to a Voltage Drop: A voltage drop produces an electric field equal to that voltage drop divided by the distance over which the drop occurs, or

$$\text{electric field} = \text{voltage gradient} = \frac{\text{voltage drop}}{\text{distance}}, \quad (10.2.2)$$

where the field points in the direction of most rapid voltage decrease.

4. Power Consumption: The electric power consumed by a device is the voltage drop across it times the current flowing through it, or

$$\text{power consumed} = \text{voltage drop} \cdot \text{current}. \quad (10.3.1)$$

5. Power Production: The electric power provided by a device is the voltage rise across it times the current flowing through it, or

$$\text{power provided} = \text{voltage rise} \cdot \text{current}. \quad (10.3.2)$$

6. Ohm's Law: The voltage drop through an ohmic object is equal to the current flowing through it times its electrical resistance, or

$$\text{voltage drop} = \text{current} \cdot \text{electrical resistance}. \quad (10.3.4)$$

This equation does not apply to non-ohmic devices.

Check Your Understanding—Answers

Section 10.1 STATIC ELECTRICITY

1. It has a large positive charge equal in amount to the wrapper's negative charge.
Why: Since charge is a conserved physical quantity, the wrapper and gift must remain neutral overall even after you separate them. The wrapper's negative charge must be balanced by the gift's positive charge.

2. Its negative charge polarizes your hand and is then attracted to your hand's nearby positive charge.
Why: Although your hand is neutral, its charges rearrange in response to the nearby wrapper's negative charge. Positive charge in your hand shifts toward the wrapper and attracts it.

3. While the tape and glass were in contact, charge was unevenly distributed between their surfaces. Removing the tape merely made that imbalance more obvious.
Why: The tape and glass have different chemical affinities for electrons and become oppositely charged whenever they touch. In fact, the tape's stickiness itself comes from electrostatic attraction.

4. That separation takes work, which appears as electrostatic potential energy in the separated charges. The positively charged regions of the thunderhead acquire huge positive voltages and the negatively charged regions acquire huge negative voltages.
Why: When opposite charges are nearby, they don't necessarily have much electrostatic potential energy per charge and the voltages may be small. Separating those charges to great distances dramatically increases their stored energy and produces high voltages.

5. Contact between dissimilar materials puts charge on the paper, which then carries that charge with it to isolated parts of the press. Enough charge can accumulate on those parts to be dangerous.
Why: Nonconductive paper is an excellent transporter of electric charge. Once the paper picks up a static charge by touching a dissimilar material, it can carry that charge with it as it moves through the press. Not surprisingly, printing presses use various tools to suppress this static charging.

6. An insulating conveyor belt can separate enormous amounts of charge, leading to high voltages, sparks, and possibly fire. A conductive belt can't carry charge with it as it moves, so no charge accumulates.
Why: When an insulating belt has charge on its surface, that charge must move with the belt. But charges are mobile in a conductive belt and don't normally move with it.

Section 10.2 XEROGRAPHIC COPIERS

1. The charge that was placed on the paper to attract the toner isn't always removed completely. Moreover, the toner itself is charged.
Why: The final transfer process, lifting the toner particles from the photoconductor to the paper, is done by charging the paper, and some of this charge remains on the paper when it leaves the copier. Copier transparencies are particularly clingy because plastic retains charge so well.

2. It points backward, away from the patient.
Why: Since electrons are negatively charged, they accelerate in the direction opposite to the field. Since the accelerator must push the electrons forward, toward the patient, its field must point away from the patient.

3. The accumulated charge is all on the outside of the conducting car, so it will only affect you if you step outside and offer it a conducting path to the ground.

Why: Since the car body is an electrical conductor and its charges are in equilibrium, the car has a uniform voltage and there is no electric field inside the car. But outside the car, there is a substantial electric field. If your body provides a conducting path to the ground, that electric field will push charges through you and you will experience a shock. Similarly, if electric power lines ever fall on your car during a storm, stay inside the car to avoid a potentially lethal shock.

4. The needle emits charge into the air via a corona discharge.
Why: The needle acts as your personal lightning rod. When you are carrying a net electric charge, some of that charge settles onto the needle. The strong electric field near the needle's tip initiates a corona discharge and much of your accumulated charge leaves through it. This discharge limits your net electric charge and thus the size of any shock you experience.

5. The rod becomes positively charged.
Why: The lightning rod charges by induction—the cloud's negative charge attracts positive charge up the wire and onto the rod. The rod's sharp point initiates a corona discharge, spraying positive charge toward the cloud and gradually decreasing the cloud's charge. In that fashion, the lightning rod acts to suppress lightning strikes.

Section 10.3 FLASHLIGHTS

1. Removing a single wire from the battery breaks the circuit and prevents a steady flow of electric current through the car's electric system.
Why: Neither the battery nor the rest of the car can accumulate charges indefinitely. Without one wire to carry charges from the battery to the car and a second wire to return those charges from the car to the battery, accumulation will occur and charge movement will stop.

2. The current flows from the metal toward your hand.
Why: Because current is defined as the flow of positive charges, it points in the direction opposite the flow of negative charges. Thus the current flows from the metal toward your hand. These charges move only briefly because there is no circuit. For the charges to move continuously they would have to be recycled and a circuit would be essential.

3. Each individual battery provides 2 V.
Why: The voltages of the individual batteries add so that it takes six of the 2-V batteries to yield a 12-V chain. If one of the batteries becomes weak, through damage, loss of fluid, or consumption of its chemical potential energy, the voltage of the chain will drop below 12 V. Pushing current backward through a lead–acid battery recharges it.

4. The cables have relatively large electrical resistances.
Why: Starting your car requires a huge current and the wires supplying that current must not limit it or waste its energy. Cheap jumper cables have too much electrical resistance to fulfill those requirements. There is no substitute for good, thick jumper cables—they're worth the extra money.

5. The wire through which current enters the radio.
Why: The radio is consuming power, so the current passing through it is experiencing a voltage drop. The current has a higher voltage when it enters the radio than when it leaves.

6. In your skin.
Why: The fluids in your body resemble salt water when exposed to voltages; they conduct current relatively well. If it weren't for your skin's large electrical resistance, even battery voltages would be capable of pushing large currents through you and might disrupt your heart and other functions. But your skin's high resistance protects you from battery voltages. As long as your skin is dry and intact, it usually takes higher voltages to push enough current through you to cause injury.

Check Your Figures—Answers

Section 10.1 STATIC ELECTRICITY

1. 4 N.
Why: According to Coulomb's law, the force on each charge varies inversely with the square of their separation. By halving that separation, you increase the electrostatic force by a factor of 4.

Section 10.2 XEROGRAPHIC COPIERS

1. An electric field of 1000 N/C in the upward direction.
Why: The electric field is equal to the force it exerts divided by the charge, or 10^{-8} N divided by 10^{-11} C. It must point

upward to support the positively charged lint against the downward pull of gravity.

2. About 1.2 m (4 feet).
Why: With a voltage difference of 120 volts between its ends and a length of 1.2 meters, the tube's electric field will be 120 V divided by 1.2 m, or about 100 V/m. This result explains why so many fluorescent lamps in the United States are about 1.2 m (4 feet) long.

Section 10.3 FLASHLIGHTS

1. 2400 W.
Why: The voltage drop across the starter motor is 12 V (each coulomb of charge loses 12 J of energy in passing through it)

and the current through it is 200 A (200 C of charge pass through it each second). We can use Eq. 10.3.1 to determine the power the motor is consuming:

$$\text{power consumed} = \text{voltage drop} \cdot \text{current}$$
$$= 12\,\text{V} \cdot 200\,\text{A} = 2400\,\text{W}.$$

2. 1800 W.

Why: The voltage rise across the battery is 12 V (each coulomb of charge gains 12 J of energy in passing through it) and the current through it is 150 A (150 C of charge pass through it each second). We can use Eq. 10.3.2 to determine the power the battery is providing:

$$\text{power provided} = \text{voltage rise} \cdot \text{current}$$
$$= 12\,\text{V} \cdot 150\,\text{A} = 1800\,\text{W}.$$

3. The 3-Ω, bulb.

Why: Equation 10.3.3 indicates that, with the voltage drop of 3 V supplied by the two alkaline batteries, the 3-Ω bulb will carry:

$$\text{current} = \frac{\text{voltage drop}}{\text{electrical resistance}}$$
$$= \frac{3\,\text{V}}{3\,\Omega} = 1\,\text{A},$$

and Eq. 10.3.1 then shows that this bulb consumes the specified 3 W:

$$\text{power consumed} = \text{voltage drop} \cdot \text{current}$$
$$= 3\,\text{V} \cdot 1\,\text{A} = 3\,\text{W}.$$

With a voltage drop of 3 V, the 12-Ω, bulb will carry a current of 0.25 A and consume just 0.75 W of power. It needs a voltage drop of 6 V to consume 3 W.

Exercises

1. If two objects repel one another, you know they have like charges on them. But how would you determine whether they were both positive or negative?

2. You have an electrically neutral toy which you divide into two pieces. You notice that at least one of those pieces has an electric charge. Do the two pieces attract or repel one another, or neither?

3. Suppose that you had an electrically charged stick. If you divided the stick in half, each half would have half the original charge. If you split each of these halves, each piece would have a quarter of the original charge. Can you keep on dividing the charge in this manner forever? If not, why not?

4. A bowling ball contains an enormous number of electrically charged particles. Why don't two bowling balls normally exert electrostatic forces on each other?

5. In industrial settings, neutral metal objects are often coated by spraying them with electrically charged paint or powder particles. How does placing charge on the particles help them to stick to an object's surface?

6. The paint or powder particles discussed in Exercise 5 are all given the same electric charge. Why does this type of charging ensure that the coating will be highly uniform?

7. If the forces between electric charges didn't diminish with distance, an electrically charged balloon wouldn't cling to an electrically neutral wall. Why not?

8. An ion generator clears smoke from room air by electrically charging the smoke particles. Why will those charged smoke particles stick to the walls and furniture?

9. After you extract two pieces of adhesive tape from a tape dispenser, those pieces will repel one another. Explain their repulsion.

10. After you peel a sticker from its paper backing, the two attract one another. Explain their attraction.

11. You're holding two oppositely charged balloons in your hands. Compare their voltages.

12. You begin to separate the two balloons in Exercise 11. You do work on the balloons, so your energy decreases. Where does your energy go?

13. When you separate the balloons in Exercise 12, do their voltages change? If so, how do those voltages change?

14. A car battery is labeled as providing 12 V. Compare the electrostatic potential energy of positive charge on the battery's negative terminal with that on its positive terminal.

15. When technicians work with static-sensitive electronics, they try to make as much of their environment electrically conducting as possible. Why does this conductivity diminish the threat of static electricity?

16. Antistatic fabric treatments render the fabrics slightly conducting. How does this treatment help diminish static electricity?

17. Holding your hand on a static generator (e.g., a van de Graff generator) can make your hair stand up, but only if you are standing on a good electrical insulator. Why is that insulator important?

18. In Exercise 17, having used a hair conditioner recently actually helps your hair stand up. Why?

19. The electric field around an electrically charged hairbrush diminishes with distance from that hairbrush. Use Coulomb's law to explain this decrease in the magnitude of the field.

20. Which way does the electric field point around the positive terminal of an alkaline battery?

21. Which has the stronger electric field between its two terminals: a 1.5-V AA battery or a standard 9-V battery? Explain.

22. You have 100 AA batteries. How should you arrange those batteries to make the strongest electric field?

23. It may seem dangerous to be in a car during a thunderstorm, but it's actually relatively safe. Since the car is essentially a metal box, the inside of the car is electrically neutral. Why does any charge on the car move to its outside surface?

24. Delicate electromagnetic experiments are sometime performed inside metal walled or screen rooms. Why does that enclosure minimize stray electric fields?

25. When a positively charged cloud passes overhead during a thunderstorm, which way does the electric field point?

26. The cloud in Exercise 25 attracts a large negative charge to the top of a tree in an open meadow. Why is the magnitude of the electric field larger on top of this tree than elsewhere in the meadow?

27. Corona discharges can occur wherever there is a very strong electric field. Why is there a strong electric field around a sharp point on an electrically charged metal object?

28. To minimize corona discharges, electric power pylons sometimes shroud connectors and other sharp features with smoothly curving metal rings or shells. How do those broad, smooth structures prevent corona discharges?

29. If you're ever standing on a mountaintop when a dark cloud passes overhead and your hair stands up, get off the mountain fast. How would your hair have acquired the charge to make it stand up?

30. The power source for an electric fence pumps charge from the earth to the fence wire, which is insulated from the earth. The earth can conduct electricity. When an animal walks into the wire, it receives a shock. Identify the circuit through which the current flows.

31. A bird can perch on a high-voltage power line without getting a shock. Why doesn't current flow through the bird?

32. The two prongs of a power cord are meant to carry current to and from a lamp. If you were to plug only one of the prongs into an outlet, the lamp wouldn't light at all. Why wouldn't it at least glow at half its normal brightness?

33. The rear defroster of your car is a pattern of thin metal strips across the window. When you turn the defroster on, current flows through those metal strips. Why are there wires attached to both ends of the metal strips?

34. If you touch only the tip of a headphone plug to the headphone jack of a portable audio player, what volume will the headphones produce?

35. If you transfer some (positive) charge to a battery's negative terminal, some of that charge will quickly move to the battery's positive terminal. Will the battery's store of chemical potential energy have changed and, if so, will it have increased or decreased?

36. If you transfer some positive charge to a battery's positive terminal, some of that charge will quickly move to the battery's negative terminal. Will the battery's store of chemical potential energy have changed and, if so, will it have increased or decreased?

37. When you plug a portable appliance into the power socket inside an automobile, current flows to the appliance through the central pin of that socket and returns to the car through the socket's outer ring. Which of the socket's contacts has the higher voltage?

38. When current is flowing through a car's rear defroster (Exercise 33), the voltage at each end of the metal strips is different. Which end of each strip has the higher voltage, the one through which current enters the strip or the one through which current leaves, and what causes the voltage drop?

39. You're given a sealed box with two terminals on it. You use some wires and batteries to send an electric current into the box's left terminal and find that this current emerges from the right terminal. If the voltage of the right terminal is 6 V higher

than that of the left terminal, is the box consuming power or providing it? Is it more likely to contain batteries or lightbulbs? How can you tell?

40. One time-honored but slightly unpleasant way in which a hobbyist determines how much energy is left in a 9-V battery is to touch both of its terminals to his tongue briefly. He experiences a pinching feeling that's mild when the battery is almost dead but one that's startlingly sharp when the battery is fresh. (Don't try this technique yourself.) What is the circuit involved in this taste test? How is energy being transferred?

41. Spot welding is used to fuse two sheets of metal together at one small spot. Two copper electrodes pinch the sheets together at a point and then run a huge electric current through that point. The two sheets melt and flow together to form a spot weld. Why does this technique work only with relatively poor conductors of electricity such as stainless steel and not with excellent conductors such as copper?

42. Why is it important that the filament of a lightbulb have a much larger electrical resistance than the supporting wires that carry current to and from that filament?

Problems

1. You remove two socks from a hot dryer and find that they repel with forces of 0.001 N when they're 1 cm apart. If they have equal charges, how much charge does each sock have?

2. If you separate the socks in Problem 1 until they're 5 cm apart, what force will each sock exert on the other?

3. If you were to separate all of the electrons and protons in 1 g (0.001 kg) of matter, you'd have about 96,000 C of positive charge and the same amount of negative charge. If you placed these charges 1 m apart, how strong would the attractive forces between them be?

4. If you place 1 C of positive charge on the earth and 1 C of negative charge 384,500 km away on the moon, how much force would the positive charge on the earth experience?

5. How close would you have to bring 1 C of positive charge and 1 C of negative charge for them to exert forces of 1 N on one another?

6. The upward net force on the space shuttle at launch is 10,000,000 N. What is the least amount of charge you could move from its nose to the launch pad, 60 m below, and thereby prevent it from lifting off?

7. What force will a 0.01-C charge experience in a 5-N/C electric field pointing upward?

8. A sock with a charge of −0.0005 C is in a 1000-N/C electric field pointing toward the right. What force does the sock experience?

9. A piece of plastic wrap with a charge of 0.00005 C experiences a forward force of 0.0010 N. What is the local electric field?

10. A Styrofoam ball with a charge of -1.0×10^{-6} C experiences an upward force of 0.01 N in an electric field. What is that electric field?

11. If you place a 0.0001-C charge halfway between the terminals of a common 9-V battery, 5 mm apart, what force will that charge experience?

12. If you place a 0.0001-C charge halfway between the terminals of a 1.5-V AA battery, 5 cm apart, what force will that charge experience?

13. An automobile has a 12-W reading lamp in the ceiling. This lamp operates with a voltage drop of 12 V across it. How much current flows through the lamp?

14. The rear defroster of your car operates on a current of 5 A. If the voltage drop across it is 10 V, how much electric power is it consuming as it melts the frost?

15. Your portable FM radio uses two 1.5-V batteries in a chain. If the batteries send a current of 0.05 A through the radio, how much power are they providing to the radio?

16. You have two flashlights that operate on 1.5-V D batteries. The first flashlight uses two batteries in a chain while the second uses five batteries in a chain. Each flashlight has a current of 1.5 A flowing through its circuit. What power is being transferred to the bulb in each flashlight?

17. You have two flashlights that have 2-A currents flowing through them. One flashlight has a single 1.5-V battery in its circuit, while the second flashlight has three 1.5-V batteries connected in a chain that provides 4.5 V. How much power is the battery in the first flashlight providing? How much power is each battery in the second flashlight providing?

18. How much power is the bulb of the first flashlight in Problem 17 consuming? How much power is the bulb of the second flashlight consuming?

19. A 1.5-V alkaline D battery can provide about 40,000 J of electric energy. If a current of 2 A flows through two D batteries while they're in the circuit of a flashlight, how long will the batteries be able to provide power to the flashlight?

20. A radio-controlled car uses four AA batteries to provide 6 V to its motor. When the car is heading forward at full speed, a current of 2 A flows through the motor. How much power is the motor consuming at that time?

21. Your car battery is dead, and your friends are helping you start your car with cheap jumper cables. One cable carries current from their car to your car, and a second cable returns that current to their car. As you try to start your car, a current of 80 A flows through the cables to your car and back, and a voltage drop of 4 V appears across each cable. What is the electric resistance of each jumper cable?

22. If you replace the cheap cables in Problem 21 with cables having half their electric resistance, what voltage drop will appear across each new cable if the current doesn't change?

23. Each of the two wires in a particular 16-gauge extension cord has an electric resistance of 0.04 Ω. You're using this extension cord to operate a toaster oven, so a current of 15 A is flowing through it. What is the voltage drop across each wire in this extension cord?

24. How much power is wasted in each wire of the extension cord in Problem 23?

25. The two wires of a high-voltage transmission line are carrying 600 A to and from a city. The voltage between those two wires is 400,000 V. How much power is the transmission line delivering to the city?

26. In bringing electricity to individual homes, the power in Problem 25 is transferred to low-voltage circuits so that the current passing through homes experiences a voltage drop of only 120 V. How much total current is passing through the homes of the city?

27. How much current flows through a 100-Ω heating filament when the voltage drop across it is 5 V?

28. If you subject a 2500-Ω heating filament to a voltage drop of 100 V, how much current will flow through it?

29. If a 10-A current experiences a 1-V voltage drop while flowing through a long wire, what is the electrical resistance of that wire?

30. The 2-A current flowing through a wire to a distant buzzer experiences a 2-V voltage drop. What is the electrical resistance of that wire?

31. If you send a 5-A current through a 1-Ω wire to the doorbell, what voltage drop will that current experience?

32. If a 1000-Ω heating filament is carrying a current of 0.120 A, what voltage drop is that current experiencing?

CHAPTER 11

Magnetism and Electrodynamics

Like electricity, magnetism is an important part of daily life. We use it to post notes on the refrigerator and to figure out which way is north. But the story of magnetism wouldn't be complete without including electricity. As we'll see, these two topics are related to one another through change and motion. For example, moving electric charges give rise to magnetism, and changing magnetism gives rise to electricity.

In this chapter, we'll examine magnetism itself, as well as several objects that use the relationships between electricity and magnetism to perform useful tasks. Since the word "dynamics" covers change and motion, these relationships are part of a field known as "electrodynamics." Brevity isn't the only reason for omitting reference to magnetism in that title; the other reason is that most magnetism is actually produced by electricity.

EXPERIMENT: A Nail and Wire Electromagnet

To explore the relationships between electricity and magnetism, try building a simple electromagnet. For this project, you'll need a large steel nail or bolt, a meter or so of insulated wire, a fresh 1.5-volt AA battery, and some small steel objects such as paper clips. The wire's metal conductor should be at least 0.65 mm in diameter (22 gauge or larger) to carry the current you'll send through it without becoming too hot.

Wind the wire tightly around the nail or bolt to form a coil. You should complete at least 50 turns of wire, all in the same direction. The exact number of turns isn't important, and you can make several layers. Be sure that the two ends of the wire are

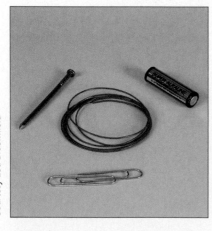
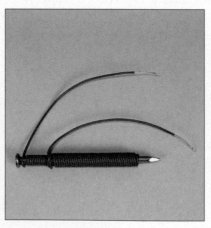

Courtesy Lou Bloomfield

still accessible and remove the insulation from each end so that you can connect them to the battery.

> **Warning**
> The electromagnet that you'll construct in this experiment **will become hot during use.** Be prepared to drop the electromagnet if it becomes uncomfortably hot. **Don't work near flammable materials.**

Now test your electromagnet. Connect one uninsulated end of the coil to each terminal of the battery. You can either hold the wires on the terminals with your fingers or use tape. A 1.5-volt battery can't give you a shock unless your skin is broken, but you should be prepared for the wire to get hot as current flows through it. If it gets too hot to hold, let go and make sure that the wire detaches from the battery so that it doesn't start a fire. Don't use a battery larger than AA or the wire may get dangerously hot.

While current is flowing through the wire, the nail will act as a strong magnet, an electromagnet. Try picking up steel objects with this electromagnet. As you touch each object with the nail, it should stick to the nail's surface. Your electromagnet will temporarily magnetize the steel object and attract it. Try to *predict* what will happen when you touch this magnetized steel object to a second object. *Observe* what happens and see if you can *verify* your prediction. Can you think of a way to *measure* how strong the magnet is? What will happen when you stop the flow of electric current through the coil of the electromagnet? Why does the coil get hot while current is flowing through it?

Chapter Itinerary

This chapter will examine (1) *household magnets,* (2) *electric power distribution,* and (3) *electric generators and motors.* In *household magnets,* we'll look at the forces that bind magnets to refrigerators and see why compasses point northward. We'll also examine electromagnets in order to explain how electric doorbells work. In *electric power distribution,* we'll see how electricity and magnetism are used to transport electric power from a distant power plant to your home and how that electric power differs from the power supplied by batteries. In *electric generators and motors,* we'll look at the ways in which mechanical power is used to produce electric power and electric power is used to produce mechanical power. While there are many other magnetic and electromagnetic objects that we encounter daily, these three topics are representative of most of the basic electromagnetic phenomena. For a more complete preview, turn ahead to the Chapter Summary on p. 382.

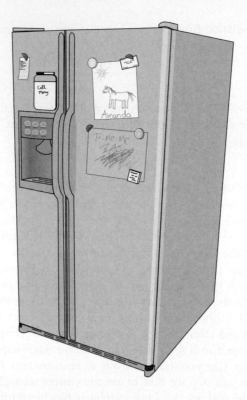

SECTION 11.1   Household Magnets

How would a family stay organized without refrigerator magnets? How would the door-bell ring if it couldn't use a magnet to strike its bells? How would a scout navigate in the woods without a compass? And how would you get cash or charge purchases without magnetic strips on plastic cards?

We're so used to having magnets around that we take them for granted. But along with being useful, household magnets let us experiment with another of the basic forces in nature. Though we'll see that magnetism is so intimately related to electricity that the two are ultimately a single, unified whole, we'll find it helpful to begin our study of magnetism as a separate phenomenon and bring electricity into the picture gradually.

Questions to Think About: Why do two magnets attract or repel, depending on how they're oriented? If a magnet is attracted to two different refrigerators, why don't those two refrigerators attract or repel one another? How can two strong magnets grip one another from opposite sides of your hand? Why isn't your hand involved in that magnetic attraction? How can some magnets be turned on or off using electricity?

Experiments to Do: Find two button-type refrigerator magnets—simple cylinders that resemble small hockey pucks. If you try to stack these magnets, you'll find that they either attract or repel. How do those forces depend on the orientations of the magnets? on their separation? See if you can float one magnet on top of the other using the repulsive force. What happens when you let go of the top magnet?

Now hold one of the button magnets near a refrigerator or another steel object and study the forces that arise. Can you find a way to make the two objects repel one another? Will a magnet stick to things that aren't made of steel? What about stainless steel?

Now find two identical sheet-type refrigerator magnets—flexible strips that may have advertising printed on their surfaces. Experiment with their interactions. You'll find that simply flipping one over doesn't change the forces from attraction to repulsion. Instead, you'll have to slide the strips across one another. As they slide, they'll alternately attract and repel. How is that possible?

Finally, obtain iron powder or make it by filing down a piece of steel (steel is mostly iron and they are magnetically quite similar). Sprinkle some of this powder on your collection of magnets. It forms strands that seem to bridge the gaps between various points on the magnets. What is the powder bridging? If you sprinkle your powder on a credit card or magnetic ID card, you'll find that it also forms bridges. However, those bridges are tiny and spaced at irregular intervals. Could there be information encoding in these erratically spaced magnetic features?

Button-Shaped Refrigerator Magnets

Refrigerator magnets come in all shapes and sizes, and some are more magnetically complicated than others. It's always best to start simple, so we'll begin with button-shaped magnets.

As you bring two button magnets together, they'll begin exerting forces on one another. You'll find that those forces can be either attractive or repulsive, depending on how the magnets are oriented, but always grow weaker with distance. Such magnetic forces resemble the electric ones you encounter while removing clothes from a hot dryer, but there are at least two important differences. First, reorienting two electrically charged garments won't turn their attraction into repulsion or vice versa. Second, no matter how you arrange two button magnets, you can't get a magnetic spark to jump from one to the other. Electricity and magnetism are evidently similar yet different. What's the story?

Magnetism is a phenomenon that closely resembles electricity. Just as there are two types of electric charges which exert electrostatic forces on one another, so there are two types of **magnetic poles** which exert **magnetostatic forces** on one another. The word "pole" serves to distinguish magnetism from electricity; **poles** are magnetic while charges are electric.

The two types of poles are called north and south, respectively, and in keeping with this geographical naming, they're exact opposites of one another. Both types of poles carry just one physical quantity: **magnetic pole.** North poles carry positive amounts of magnetic pole while south poles carry negative amounts. It should come as no surprise that like poles repel while opposite poles attract. Furthermore, the magnetostatic forces between two poles grow weaker as they move apart and are inversely proportional to the square of the distance between them. So far, the similarities between electricity and magnetism are striking.

However, we now come to a crucial difference between electricity and magnetism: while subatomic particles that carry pure positive or negative electric charges are common, particles that carry pure north or south magnetic poles have never been found. Called **magnetic monopoles,** such pure magnetic particles may not even exist in our universe. That cosmic omission explains why there are no magnetic sparks: without monopoles, there is no magnetic equivalent of an electric charge that can leap from one place to another as a magnetic current, let alone a magnetic spark.

But while isolated magnetic poles aren't available in nature, *pairs* of magnetic poles are. These pairs consist of equal north and south poles, spatially separated from one

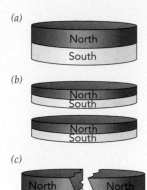

Fig. 11.1.1 (*a*) A typical button magnet has a north pole on one face and a south pole on the other. Its net pole is zero. (*b*) Slicing it between its poles or (*c*) breaking it through its poles always yields a pair of magnets, each with zero net pole.

another in an arrangement called a **magnetic dipole.** Since the two opposite poles have equal magnitudes, they sum to zero and the magnetic dipole has zero **net magnetic pole.**

A simple button magnet has both a north pole *and* a south pole, usually on opposite faces of the button (Fig. 11.1.1*a*). There are no purely north buttons or purely south buttons. Amazingly enough, even slicing that button magnet in half won't yield separated north and south poles (Fig. 11.1.1*b*). Instead, new poles will appear at the cut edges and each piece of the original magnet will end up with zero net pole! Breaking the button magnet in half (Fig. 11.1.1*c*) will also produce pieces with zero net pole.

We can now explain why two of these magnets sometimes attract and sometimes repel. With two poles on each magnet, we have to consider four interactions: two repulsive interactions between like poles (north–north and south–south) and two attractive interactions between opposite poles (north–south and south–north). While it might seem that all these forces should cancel, the distances separating the various poles and therefore the forces between them depend on the magnets' orientations. Since the closest poles experience the strongest forces, they dominate. If you turn two like poles toward one another, the two magnets will push apart (Fig. 11.1.2*a*). If you turn their opposite poles toward one another, they'll pull together (Fig. 11.1.2*b*). And if you tip them at an angle, they'll experience torques which tend to twist opposite poles together and like poles apart.

Without monopoles, we're going to need some imagination to understand magnetism well. Let's start with units. Even though we can't collect a unit of pure north pole, we can still define such a unit and understand its behavior. The SI unit of magnetic pole is the **ampere-meter** (abbreviated A·m). That astonishing choice, an *electric* unit appearing in a *magnetic* unit, foreshadows the profound connections between electricity and magnetism that we'll soon encounter.

Just as there is a Coulomb's law for electric charges, there is a Coulomb's law for magnetic poles. Coulomb's magnetic experiments, which were complicated by the fact that he had to work with magnetic dipoles rather than individual poles, showed that the forces between magnetic poles are proportional to the amount of each pole and inversely proportional to the square of their separation. The exact relationship can be written as a word equation:

$$\text{force} = \frac{\text{permeability of free space} \cdot \text{pole}_1 \cdot \text{pole}_2}{4\pi \cdot (\text{distance between poles})^2}, \tag{11.1.1}$$

in symbols:

$$F = \frac{\mu_0 \cdot p_1 \cdot p_2}{4\pi r^2},$$

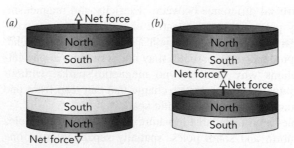

Fig. 11.1.2 (*a*) When like poles of two button magnets are turned toward one another, the magnets repel. (*b*) When opposite poles are turned toward one another, they attract.

and in everyday language:

Don't hold two strong magnetic poles nearby unless you're prepared to be
pushed around.

The force on pole$_1$ is directed toward or away from pole$_2$, and the force on pole$_2$ is directed toward or away from pole$_1$. The **permeability of free space** is $4\pi \times 10^{-7}$ N/A^2. Consistent with Newton's third law, the force exerted on pole$_1$ by pole$_2$ is equal in amount but oppositely directed from the force exerted on pole$_2$ by pole$_1$.

Coulomb's Law for Magnetism

The magnitudes of the magnetostatic forces between two magnetic poles are equal to the permeability of free space times the product of the two poles divided by 4π times the square of the distance separating them. If the poles are like, then the forces are repulsive. If the poles are opposite, then the forces are attractive.

Check Your Understanding #1: Two Halves Make a Whole

for answers, see page 383

You have a disk-shaped permanent magnet. The top surface is its north pole and the bottom surface is its south pole. If you crack the magnet into two half-circles, the two halves will push apart. Why?

The Refrigerator: Iron and Steel

While two button magnets can push or pull on one another, what if you have only one? The easiest way to observe magnetic forces is then to hold that single magnet near your refrigerator or another piece of iron or steel. The magnet is attracted to the refrigerator. But if you flip the button magnet over, thinking that it will now be repelled by the refrigerator, you'll be disappointed. Although the refrigerator is clearly magnetic, its magnetism somehow responds to the button magnet so that the two always attract.

Actually, the refrigerator's behavior isn't all that mysterious. Its steel is composed of countless microscopic magnets, each with a matched north pole and south pole (Fig. 11.1.3). Normally those individual magnetic dipoles are oriented semi-randomly (Fig. 11.1.3*a*), so the refrigerator exhibits no overall magnetism. However, as you bring one pole of a button magnet near the refrigerator, its tiny magnets evolve in size, shape, and orientation (Fig. 11.1.3*b*). Overall, opposite poles shift closer to the button magnet's pole and like poles shift farther from the button magnet's pole. The steel develops a **magnetic polarization** and consequently attracts the pole of the button magnet.

This polarization remains strong only as long as the button magnet's pole is nearby. When you remove the button magnet, most of the tiny magnets in the steel resume their semi-random orientations and the steel's magnetic polarization shrinks or disappears. When you then bring the button magnet's other pole close to the refrigerator, its steel develops the opposite magnetic polarization and again attracts the button magnet's pole. No matter which pole or assortment of poles you bring near the refrigerator, its steel will polarize in just the right way to attract those poles.

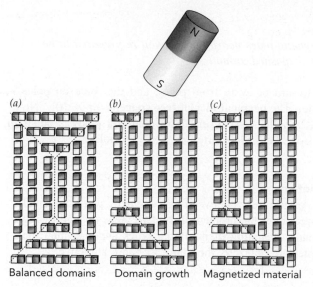

(a)   *(b)*   *(c)*

Balanced domains   Domain growth   Magnetized material

Fig. 11.1.3 (*a*) The countless microscopic magnets in iron or steel are normally oriented somewhat randomly. (*b*) But when a strong magnetic pole is nearby, those tiny magnets reorient to attract it. In soft magnetic materials, this reorientation is only temporary. (*c*) Hard magnetic materials, however, remain magnetized long after the external pole has departed.

If you try this trick with a plastic or aluminum surface, the button magnet won't stick. What's special about steel that allows it to develop such a strong magnetic polarization? The answer is that ordinary steel, like its constituent iron, is a **ferromagnetic** material—it is actively and unavoidably magnetic on the size scale of atoms.

To understand ferromagnetism, we must start by looking at atoms and the subatomic particles from which they're constructed: electrons, protons, and neutrons. For complicated reasons, all of those subatomic particles have magnetic dipoles, particularly the electrons, and the atoms they form often display this magnetism. Despite a tendency for the subatomic particles to pair up with opposite orientations so that their magnetic dipoles cancel one another, most isolated atoms have significant magnetic dipoles.

But while most atoms are intrinsically magnetic, most materials are not. That's because another round of pairing and canceling occurs when atoms assemble into materials. This second round of cancellation is usually so effective that it completely eliminates magnetism at the atomic scale. Materials such as glass, plastic, skin, copper, or aluminum retain no atomic-scale magnetism at all and your button magnet won't stick to them. Even most stainless steels are nonmagnetic.

However, there are a few materials that avoid this total cancellation and thus manage to remain magnetic at the atomic scale. The most important of these are the ferromagnets, a class of magnetic materials which includes ordinary steel and iron. If you examine a small region of ferromagnetic steel, you'll find that it is composed of many microscopic regions or **magnetic domains** that are naturally magnetic and cannot be demagnetized (Fig. 11.1.3*a*). Within a single domain, all of the atomic scale magnetic dipoles are aligned and together they give the overall domain a substantial net magnetic dipole.

While common steel always has these magnetic domains, magnetic interactions orient nearby domains so that their magnetic dipoles oppose one another and cancel. The microscopic magnets balance one another so well that the steel appears nonmagnetic. That's too bad; the appliance showroom would be a much more exciting place to visit if the cancellation weren't so good.

But when you bring a strong magnetic pole near steel (Fig. 11.1.3*b*), the individual domains grow or shrink, depending on which way they're oriented magnetically, and the steel becomes **magnetized** (Fig. 11.1.3*c*). The atoms themselves don't move during this process; the change is purely a reorientation of the atomic-scale magnetic dipoles.

Domains that attract your button magnet's pole grow while those that repel it shrink and the button magnet sticks to the refrigerator.

► Check Your Understanding #2: Chain Links

for answers, see page **383**

If you touch the north pole of a permanent magnet to one end of a steel paper clip, the clip's other end will become magnetic. What pole will that other end have?

Plastic Sheet Magnets and Credit Cards

When you remove your button magnet from the refrigerator, the steel returns to its original nonmagnetic state—it becomes **demagnetized.** Well, *almost* demagnetized. The demagnetization process isn't perfect because some of the domains get stuck. While magnetic forces within the steel favor complete return to apparent nonmagnetism, chemical forces can make it hard for the domains to grow or shrink. Adjacent domains are separated by **domain walls**—boundary surfaces between one direction of magnetic orientation and another. These domain walls must move if the domains are to change size. However, flaws and impurities in the steel can interact with a domain wall and keep it from moving. When that happens, the steel fails to demagnetize itself completely (Fig. 11.1.4). To remove the last bit of residual magnetism from steel, you must help the domain walls move, typically with heat or mechanical shock.

Courtesy Lou Bloomfield

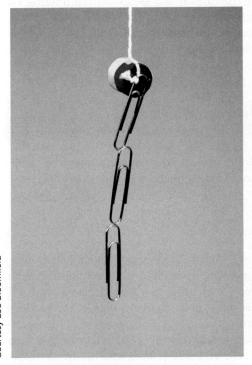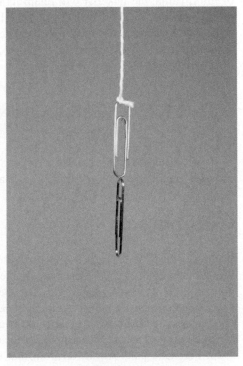

Fig. 11.1.4 (*a*) Although these paper clips were initially unmagnetized, the pole of a strong permanent magnet magnetizes them as a chain. (*b*) After the magnetizing magnet is removed, the clips retain some of their magnetization.

Fig. 11.1.5 Iron powder forms bridges between the magnetic poles of this plastic sheet magnet.

Fig. 11.1.6 Iron powder discloses the locations of magnetic poles on this credit card's magnetic strip. Information is stored as the locations of those poles.

A **soft magnetic material** is one that demagnetizes itself easily when all nearby poles are removed. Chemically pure iron, which has few flaws or impurities, is a soft magnetic material—easy to magnetize and easy to demagnetize. A **hard magnetic material** is one that does not demagnetize itself easily and that tends to retain whatever domain structure is imposed on it by its most recent exposure to strong nearby poles (Fig. 11.1.3c). Your button magnet is made from a hard magnetic material!

Like steel, the material in your button magnet is ferromagnetic (or closely related to ferromagnetic). But unlike steel, your button magnet's domains do not shrink or grow easily. During its manufacture, the button magnet was magnetized by exposing it to such strong magnetic influences that its domains rearranged to give it permanent magnetic poles. It now has a north pole on one face and a south pole on the other. Unless you expose the button to extremely strong magnetic influences or heat it or pound it, it will retain its present magnetization almost indefinitely. In that respect, the button is a **permanent magnet.**

Not all permanent magnets are as simple as button magnets. Depending on how they were magnetized, they can have their north and south poles located in unexpected places and or even have more than one pair of poles. Plastic sheet magnets are a good example of multiple-pole magnets: each has a repeating pattern of alternate poles along its length. The exact patterns vary, but most have poles that form alternating parallel stripes. You can find these stripes by letting them polarize and attract iron powder (Fig. 11.1.5), or by sliding two identical sheet magnets across one another. The sheets will attract and bind together most strongly when opposite poles are aligned across from each other. And they'll repel when you shift one of the magnets so that like poles are aligned.

A hard magnetic material's ability to "remember" its magnetization can be useful for saving information. Once magnetized in a particular manner so as to represent a piece of information, the material will retain its magnetization and the associated information until something magnetizes it differently. Information retention in hard magnetic materials forms the basis for most magnetic recording and storage, including the magnetic stripes on credit cards, magnetic tapes, computer disks, and magnetic random access memory (MRAM) (Fig. 11.1.6).

> ► Check Your Understanding #3: Now for the Flip Side
>
> *for answers, see page* **383**

If you bring the north pole of a large, strong magnet near the north pole of a small, weak magnet that you are holding in place, what will happen to that small magnet?

Compasses

If you've spent time hiking, you may well own a magnetic compass. Like a button magnet, the needle of that compass is a simple permanent magnet with one north magnetic pole and one south magnetic pole. This needle aids navigation because the earth itself has a magnetic dipole and that dipole affects the orientation of the needle: the needle's north magnetic pole tends to point northward.

Already, we can see what must be located near the earth's *north* geographic pole: a *south* magnetic pole. Attraction from that south magnetic pole is what draws the compass's north magnetic pole toward the north. However, the full story is more complicated. To begin with, the earth's magnetic poles are actually located far beneath its surface and aren't perfectly aligned with the geographic poles. To make matters worse, magnetically active materials in everything from distant mountains to nearby buildings assert their own magnetic influences on the compass needle. Overall, the compass needle is responding to the influences of countless magnetic poles, both near and far. Given how difficult it would be to sum up all those separate influences, we do better to view the compass needle as interacting with something local: a **magnetic field**—an attribute of space that exerts a magnetostatic force on a pole. According to this new perspective, the compass needle responds to the local magnetic field, a field that's created by all of the surrounding magnetic poles.

As with an electric field, the magnetic field here appears to be acting merely as an intermediary: various poles produce the magnetic field and this magnetic field affects our compass needle. But as we'll see, a magnetic field is more than just an intermediary or fiction. It is quite real and can exist in space, independent of the poles that produce it. And just as electric fields can be created by things other than charge, so magnetic fields can be created by things other than pole.

The magnetic field at a given location measures the magnetostatic force that a unit of pure north pole would experience if it were placed at that point. More specifically, the magnetostatic force is equal to the product of the pole times the magnetic field and points in the direction of that magnetic field. We can write this relationship as a word equation:

$$\text{force} = \text{pole} \cdot \text{magnetic field}, \qquad (11.1.2)$$

in symbols:

$$\mathbf{F} = p\mathbf{B},$$

and in everyday language:

If you place a strong magnet in a big magnetic field, expect to be pushed around,

where the force is in the direction of the magnetic field. Note, however, that a negative amount of pole (a south pole) experiences a force opposite the magnetic field. The SI unit of magnetic field is the **newton-per-ampere-meter,** also called the **tesla** (abbreviated T).

The earth's magnetic field is relatively weak, about 0.00005 T in a roughly northward direction. (For comparison, the field near your button magnet may be 0.1 T or more.) The earth's field pushes the compass needle's north pole northward and south pole southward (Fig. 11.1.7). Unless the compass needle is perfectly aligned with that field, it experiences a torque and undergoes angular acceleration. Since its mount only allows the needle to rotate horizontally and it experiences mild friction as it does, the needle soon settles down with its north pole pointing roughly northward. If its mount allowed it to rotate vertically as well as horizontally, the needle's north pole would dip downward in the northern hemisphere and upward in the southern hemisphere. In general, the needle minimizes its magnetostatic potential energy by pointing along the direction of the local magnetic field and is thus in a stable equilibrium when orientated that way. After a few swings back and forth, your compass needle points along the local magnetic field, which hopefully points northward.

Because the earth's magnetic field is so uniform in the vicinity of your compass, its northward push on the needle's north pole exactly balances its southward push on the needle's south pole and the needle experiences zero net force. But if you bring your compass near a button magnet, the local magnetic field will not be uniform and the needle

Earth's magnetic field

Fig. 11.1.7 A compass needle aligns with the local magnetic field. Its north pole experiences a magnetostatic force in the direction of the field and its south pole experiences a force opposite the field.

may experience a net force. The magnetic field gets stronger near one of the button's poles and the compass needle will experience a net force toward or away from that pole, depending on which way it's orientated.

When the needle is aligned with a nonuniform field—its north pole pointing in the same direction as the local field—the forces on its two opposite poles won't balance and it will experience a net force in the direction of increasing field. If it is aligned against the field, it will experience a net force in the direction of decreasing field. In practice, as you bring the compass near your button magnet, its needle will first pivot into alignment with the local field and then find itself pulled toward increasing field, toward the nearest pole of the button magnet. The same thing happens when you bring two button magnets together: each pivots into alignment with the other's magnetic field and the two then leap at each other. Watch out for your fingers!

A piece of steel exhibits similar behavior when you hold it near a button magnet: it becomes magnetized along the direction of the local magnetic field and then finds itself pulled toward increasing field, toward the button magnet's nearest pole. That's how the button magnet holds your notes to the refrigerator!

Check Your Understanding #4: Crazy Compass

for answers, see page 383

You lock the needle of your compass and move its north pole near the north pole of a powerful button magnet. Will the needle experience a magnetostatic force toward strong field or weak field?

Check Your Figures #1: Careful with That Wrench!

for answers, see page 385

You mistakenly place a long steel wrench in the 1-T field near a strong magnet. The field magnetizes the wrench and it develops a north pole of 1000 A·m at its near end and an equal south pole at its distant end. Only the near end of the wrench is in the 1-T field and experiences a magnetic force. What force does the field exert on the wrench and its north pole?

Iron Filings and Magnetic Flux

Magnetic fields seem so abstract; it would be helpful if you could see them. Remarkably enough, you can: just sprinkle iron filings into the field! Though you'll need to support their weight with paper or a liquid, an interesting pattern will form. Like tiny compass needles, iron particles magnetize along the local magnetic field and then stick together, north pole to south pole, in long strands that delineate the magnetic field (Fig. 11.1.8)!

These strands map the magnetic field in an interesting way. First, at each point on a strand, the strand points along the local magnetic field. Second, the strands are most tightly packed where the local magnetic field is strongest. In other words, the strands follow along the local magnetic field direction and have a density proportional to that local field. The lines highlighted by these strands are so useful that they have their own names: **magnetic flux lines.**

Flux lines are often helpful when exploring a magnetic field. If you're studying the magnetic field in a large area, you probably don't want to use iron filings. Instead, you

Courtesy Lou Bloomfield

Fig. 11.1.8 Supported by a liquid, this iron powder shows the magnetic flux lines around the magnet.

can hold a compass in your hand and walk in the direction its needle is pointing—the direction of the magnetic field. The path you'll follow in this compass-guided walk is a magnetic flux line. If you repeat this trip from many different starting points, you'll explore the whole magnetic field, flux line by flux line. Since a magnetic field tends to point away from north poles and toward south poles, these tours will typically take you from north poles to south poles. In fact, for our permanent magnets, every magnetic flux line begins at a north pole and ends at a south pole.

That last observation about flux lines is quite general: they never start at or end on anything other than a magnetic pole. While flux lines emerge in all directions from a north pole and converge from all directions on a south pole, that's it; flux lines never begin or end in empty space. If you're following a magnetic flux line with your compass, you will either reach a south pole or walk forever!

The possibility of that endless walk is somewhat disconcerting; if the flux line you're following doesn't end at a pole, what created its magnetic field? The answer reveals a deep connection between magnetism and electricity: some magnetic fields aren't produced by magnetic poles at all, they're produced by electricity! To see how that's possible, let's take a look at another common household magnet: the electromagnet in an ordinary doorbell.

❑ Self-educated before the French Revolution, during which his father was executed, French physicist André-Marie Ampère (1775–1836) became a science teacher in 1796. He served as a professor of physics or mathematics in several cities before settling at the University of Paris system in 1804. In 1820, only a week after learning of Oersted's experiment showing that an electric current causes a compass needle to deflect, Ampère published an extensive treatment of the subject. Evidently, he had been thinking about these ideas for a long time.

→ Check Your Understanding #5: Building Bridges
for answers, see page 384

If you sprinkle iron filings onto the magnetic stripe of a credit card, a pattern of tiny iron bridges will form. Where are the magnetic poles relative to those bridges?

Electric Doorbells and Electromagnets

A classic electric doorbell uses a magnet and a spring to drive a piece of iron into two chimes, "ding-dong." When you press the doorbell button, you close an electric circuit and the resulting electric current pushes the iron *magnetically* into the first chime, "ding." When you release the button, you open the circuit, stopping the current and its magnetism so that the spring can push the iron back into the second chime, "dong."

The big news here is that electric currents can produce magnetic forces. In fact, there is nothing optional about this connection: electric currents *are* magnetic. More specifically, moving electric charge produces a magnetic field.

First Connection between Electricity and Magnetism
Moving electric charge produces a magnetic field.

Imagine the surprise of Danish physicist Hans Christian Oersted (1777–1851) when he observed in 1820 that current in a wire caused a nearby compass needle to rotate. Until that moment, electricity and magnetism had appeared as independent phenomena. Inspired by Oersted's experiment, French physicist André-Marie Ampère ❑ undertook a 7-year study of the relationships between electricity and magnetism, and started the revolution that eventually unified them within a single overarching conceptual framework.

When we use iron powder to disclose the magnetic flux lines surrounding a long, straight current-carrying wire, we too are in for a surprise (Fig. 11.1.9). Those flux lines circle the wire like concentric rings, growing more widely separated as the distance from

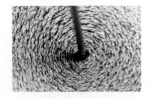

Fig. 11.1.9 Iron powder shows that flux lines around a current-carrying wire form concentric rings around that wire.

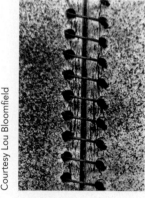

Courtesy Lou Bloomfield

Fig. 11.1.10 Iron powder shows that flux lines pass straight through a current-carrying coil and return outside it, much like the flux lines around a similarly shaped bar magnet.

(a)

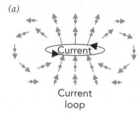

Current loop

(b)

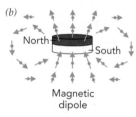

Magnetic dipole

Fig. 11.1.11 (a) The magnetic field around a loop of current-carrying wire points up through the loop and down around the outside of the loop. The magnetic field arrow passing through each green dot indicates the magnitude and direction of the force a north test pole would experience at the dot's location. (b) The field produced by a two-pole button magnet is almost identical to that of the loop.

the wire increases. The wire is an **electromagnet**—a device that becomes magnetic when it carries an electric current. But because an electromagnet has no true magnetic poles, the magnetic flux lines can't stretch from north pole to south pole. Instead, each flux line of an electromagnet is a closed loop. If you took a compass-guided walk along one of these flux lines, you'd retrace your steps over and over.

Since the flux lines are packed tightest near the surface of the current-carrying wire, that's where the magnetic field is strongest. Recalling that a piece of iron is pulled toward increasing magnetic field, we see that the wire will attract iron to it whenever it's carrying a current.

The magnetic field around a current-carrying wire is fairly weak, however, and a practical doorbell winds that wire into a coil to concentrate and strengthen its field. While the magnetic field around a current-carrying coil is complicated, we can use iron powder to make it visible (Fig. 11.1.10). Remarkably enough, the flux lines outside the coil resemble those outside a button magnet of similar dimensions (Fig. 11.1.11). It's as though the coil has a north pole at one end and a south pole at the other! But because there are no true poles present, the flux lines don't end anywhere. Instead, they continue right through the middle of the coil and form complete loops.

When current flows through the coil, nearby iron finds itself magnetized along the local magnetic field and then pulled toward increasing field—toward the tightly packed flux lines at the coil's end. But why stop there? Since the flux lines continue right into the coil and grow even more tightly packed inside, the iron will be pulled inward toward the very center of the coil!

That's how the doorbell works. When you press the doorbell button, current flows through a coil of wire and the resulting magnetic field yanks an iron rod into the center of that coil. About the time the rod reaches the center, part of it hits the first chime. When you then open the switch, stopping the current and its magnetism, a spring pushes the iron rod back out of the coil and it hits the second chime. These two chimes make the familiar ding-dong!

While current is flowing through the coil and the iron rod is inside it, the two objects act as a single, powerful electromagnet. The magnetic field surrounding the pair is the sum of the coil's modest magnetic field and the magnetized iron's much stronger field. In effect, the current in the coil magnetizes the iron and the iron then creates most of the surrounding magnetic field. Practical electromagnets, which control switches and valves in your furnace or air conditioner and can lift cars at the scrap yard, generally use iron or related materials to dramatically enhance the magnetic field produced by a current in a coil of wire (Fig. 11.1.12).

Check Your Understanding #6: Current Technology

for answers, see page **384**

In magnetic resonance imaging (MRI), a patient is immersed in an intense magnetic field. That field is created entirely without permanent magnets or even iron. How is that possible?

Fig. 11.1.12 This electric switch is controlled by an electromagnet and is called a *relay*.

Courtesy Lou Bloomfield

Courtesy Lou Bloomfield

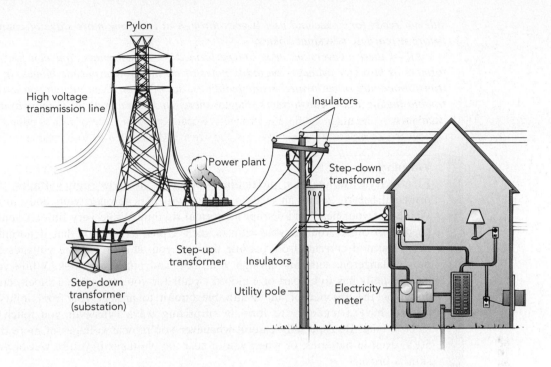

Pylon

High voltage
transmission line

Insulators

Power plant

Step-down
transformer

Step-up
transformer

Insulators

Step-down
transformer
(substation)

Utility pole

Electricity
meter

SECTION 11.2 Electric Power Distribution

Electricity is a particularly useful and convenient form of ordered energy. Because it's delivered to our homes and offices as a utility, we barely think about it except to pay the bills. The wires that bring it to us never plug up or need cleaning and work continuously except when there's a power failure, blown fuse, or tripped circuit breaker.

Just how does electricity get to our homes? In this section, we'll look at the problems associated with distributing electricity far from the power plant at which it's generated. To understand these problems, we'll examine the ways in which wires affect electricity and see how electric power is transferred and rearranged by devices called transformers.

Questions to Think About: Why do power distribution systems use alternating current? What is the purpose of high-voltage wires? Why does the power company place large electric devices on the utility poles near homes or on the ground near neighborhoods? What are the advantages and disadvantages of 120-V versus 230-V electric power?

Experiments to Do: Experiments with electric power distribution are rather dangerous, but you can observe the ways in which your local electric power is distributed. If your area is connected to a major electric power network, you'll be able to find an entire hierarchy of power conversion facilities. The power should travel from the power plant to your town on high-voltage wires, normally located overhead on tall pillars or pylons. These wires should end at a large power conversion facility, where enormous devices transfer power to the lower voltage power lines that fan out across your town. In some places, these wires are overhead; in others they're underground. However, this power is still not ready for household use. It goes through at least one more stage of conversion before it reaches individual homes.

All of these conversion steps are performed by transformers. You can find transformers as boxes or cylinders on utility poles or on the ground outside homes. In cities, transformers are often located inside buildings, out of sight. Even though you may have trouble finding these transformers, they're there. In this section, we'll see why they're necessary.

Warning

Electricity is dangerous, particularly when it involves high voltages. The principal hazard is that an electric current will pass through your body in the vicinity of your heart and disrupt its normal rhythm. While very little current is needed to cause trouble, your skin is such a poor conductor that it normally keeps harmful currents from passing through you. However, large voltages can propel dangerous currents through your skin and put you at risk. While your body usually has to be part of a closed circuit for you to receive a shock, don't count on the absence of an identifiable circuit to protect you from injury—circuits have a tendency to form in surprising ways whenever you touch an electric wire. Be especially careful whenever you're near voltages of more than 50 V, even in batteries, or when you're near any voltages if you're wet or your skin is broken.

Direct Current Power Distribution

Batteries may be fine for powering flashlights, but they're not very practical for lighting homes. Early experiments that placed batteries in basements were disappointing because the batteries ran out of energy quickly and needed service and fresh chemicals all too frequently.

A more cost-effective source for electricity was coal- or oil-powered electric generators. Like batteries, generators do work on the electric currents flowing through them and can provide the electric power needed to illuminate homes. But while generators produce electricity more cheaply than batteries, early ones were large machines that required fresh air and attention. These generators had to be built centrally, with people to tend them and chimneys to get rid of the smoke.

This was the approach taken by the American inventor Thomas Alva Edison (1847–1931) in 1882 when he began to electrify New York City. Each of the Edison Electric Light Company's generators acted like a mechanical battery, producing **direct current** that always left the generator through one wire and returned to it through another. Edison placed his generators in central locations and conducted the current to and from the homes he served through copper wires. However, the farther a building was from the generator, the thicker the copper wires had to be. That's because wires impede the flow of current and making them thicker allows them to carry current more easily.

Wire thickness is important because, like the filament of the flashlight bulb we studied in the previous chapter, wires have electrical resistance. In accordance with Ohm's law (Eq. 10.3.4), the voltage drop through a wire is equal to its electrical resistance times the current passing through it. In the case of a wire conducting current from a generating plant to a home, our primary concern is how much power the wire wastes as thermal power. We can determine this wasted power by combining Ohm's law with the equation for power consumed by a device (Eq. 10.3.1):

$$\text{power consumed} = \text{voltage drop} \cdot \text{current}$$
$$= (\text{current} \cdot \text{electrical resistance}) \cdot \text{current}$$
$$= \text{current}^2 \cdot \text{electrical resistance}.$$

The wire's wasted power is proportional to the square of the current passing through it! This relationship became all too clear to Edison when he tried to expand his power distribution systems. The more current he tried to deliver over a particular wire, the more power it lost as heat. Doubling the current in the wire quadrupled the power it wasted.

Edison tried to combat this loss by lowering the electrical resistances of the wires. He used copper because only silver is a better conductor of current. He used thick wires to increase the number of moving charges. And he kept the wires short so that they didn't have much chance to waste power. This length requirement forced Edison to build his generating plants within the cities he served. Even New York City contained many local power plants. (For an interesting tale about the early days of electric power, see ❏.)

Edison also tried to avoid waste by delivering smaller currents at higher voltages. Since the delivered power is equal to the voltage drop times the current (see Eq. 10.3.1), Edison could reduce the current flowing through the wires by raising the voltage. Although less current flowed through each home, it underwent a larger voltage drop so the power delivered was unchanged. However, high voltages are dangerous because they tend to create sparks as current jumps through the air. They also produce nasty shocks when current flows through your body. While high voltages could be handled safely outside a home, they could not be brought inside. Edison used the highest voltages that safety would allow.

Although scientists have discovered a number of materials that lose their electrical resistance at extremely low temperatures and become perfect electrical conductors or **superconductors** (Fig. 11.2.1), these superconductors are still too impractical for power distribution systems. Their use is limited to local applications such as large electromagnets and specialized electronic devices.

❏ Love Canal is the United States' most famous toxic waste dump. The dump was created in the 1920s at an abandoned section of canal constructed in 1892 by William T. Love. Love intended his canal to connect the upper and lower Niagara Rivers, so that the descending water could be used to generate DC electric power for the citizens of Niagara Falls, New York. The advent of AC power transmission systems in 1896 made the canal less useful, and it was never finished.

➤ Check Your Understanding #1: The Trouble with DC Electric Power

for answers, see page **384**

If Edison doubled the length of his delivery wires, while keeping the currents through them the same, what would happen to the power they consumed?

Courtesy Lou Bloomfield

Fig. 11.2.1 A magnetic cylinder floats above the surface of a superconductor at 78 K. Currents flowing freely in the superconductor make it magnetic and cause it to repel the magnetic cylinder.

Introducing Alternating Current

Fig. 11.2.2 This electric outlet follows the U.S. standard for 120-V AC, 15-A service. The wide slots (left) are neutral, the narrow slots (right) are hot, and the curved holes (centered) are ground. This outlet provides ground-fault circuit interruption (GFCI) protection: if any current leaving hot fails to return to neutral, or vice versa, the outlet shuts off instantly until it is reset. The test button simulates a current leak and will shut off the outlet if its protection is functioning properly.

The real problem with distributing electric power via direct current (DC) is that there's no easy way to transfer power from one DC circuit to another. Because the generator and the lightbulbs must be part of the same circuit, safety requires that that entire circuit use low voltages and large currents. DC power distribution therefore wastes much of its power in the wires connecting everything together.

However, as we'll soon see, alternating current (AC) makes it easy to transfer power from one AC circuit to another, so that different parts of the alternating current power distribution system can operate at different voltages with different currents. Most significantly, the wires that carry the power long distances are part of a high-voltage, low-current circuit and therefore waste little power.

An **alternating current** is one that periodically reverses direction—it alternates. For example, when you plug your desk lamp into an AC electrical outlet and switch it on, the current that flows through the lamp's filament reverses its direction of travel many times each second.

The power company propels this alternating current through the lamp's filament by subjecting it to an alternating voltage drop—a voltage drop that periodically reverses direction. As you may recall from Section 10.3 on flashlights, current in an ohmic filament flows down a voltage gradient from positive to negative, much as bicyclists roll down an altitude gradient from high to low. While a flashlight's battery subjects the flashlight's filament to a steady voltage drop and obtains a direct current, the power company subjects your lamp's filament to an alternating voltage drop and obtains an alternating current.

Alternating voltages are present at any AC electrical outlet. In the United States, an ordinary AC outlet offers three connections: *hot, neutral,* and *ground* (Fig. 11.2.2). In a properly installed outlet, the absolute voltage of *neutral* remains near 0 V (0 volts) while the absolute voltage of *hot* alternates above and below 0 V. *Ground,* which is an optional safety connection that we'll discuss later, also remains near 0 V absolute.

One end of your lamp's filament is connected to *hot* and the other to *neutral.* Since current always flows through the filament from higher voltage to lower voltage, it flows from *hot* to *neutral* when *hot* has a positive voltage and from *neutral* to *hot* when *hot* has a negative voltage.

In normal AC electric power, the *hot* voltage varies sinusoidally—it's proportional to the trigonometric sine function with respect to time (Fig. 11.2.3). This smoothly alternating voltage propels a smoothly alternating current through the lamp. During each reversal, the current in the filament gradually slows to a stop before gathering strength in the opposite direction. In the United States, AC voltages reverse every 120th of a second, yielding 60 full cycles of reversal (back and forth) each second (60 Hz). In Europe, the reversals occur every 100th of a second, so AC voltages complete 50 full cycles of reversal each second (50 Hz).

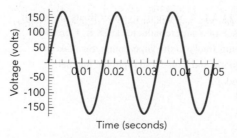

Fig. 11.2.3 The voltage of the hot wire of a U.S. 120-V AC outlet varies sinusoidally in time and completes 60 full cycles per second. While it peaks at ±170 V, its effective time average or RMS voltage is 120 V. The voltage of the neutral wire is always 0 V.

Fortunately, these reversals have little effect on many household devices. Lamps and toasters (Fig. 11.2.4) consume power because of their electrical resistances and don't care which way current passes through them. In fact, power consumption in such simple ohmic devices is used to define an effective voltage for AC electric power. An outlet's nominal AC voltage—technically, its **root mean square** or **RMS voltage**—is defined to be equal to the DC voltage that would cause the same average power consumption in an ohmic device. Thus 120-V AC power delivers the same average power to a toaster as 120-V DC power.

However, the reversals of AC power aren't without consequence. First, some electric and most electronic devices are sensitive to the direction of current flow and must handle the reversals carefully. Second, the power available from an ordinary AC outlet rises and falls with each voltage reversal and is momentarily zero at the reversal itself. Your lamp actually flickers slightly because of these power fluctuations and devices that can't tolerate even an instant without power must store energy to avoid shutting down during the reversals.

Finally, AC power's sinusoidally varying voltages peak well above their nominal values, exceeding those values by a factor of the square root of 2 (about 1.414). For example, the voltage of the *hot* connection in an ordinary 120-V AC power outlet actually swings between +170 V and −170 V. Those higher peak voltages are important for insulation and electrical safety.

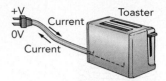
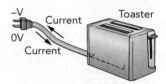

Fig. 11.2.4 The current passing through this toaster reverses directions periodically as the voltage of the toaster's hot terminal (top) reverses. The toaster's neutral terminal (bottom) remains at 0 V.

> ► Check Your Understanding #2: Timing is Everything
>
> *for answers, see page* **384**

Sticking your fingers into an electrical outlet is never a good idea, but is there a moment when you could do it without getting a shock?

Magnetic Induction

Edison was adamantly opposed to alternating current, which he viewed as dangerous and exotic. Indeed, its fluctuating voltages and moments without power don't make alternating current look attractive at all.

The champion of alternating current was Nikola Tesla (1856–1943), a Serbian-American inventor, who was backed financially by the American inventor and industrialist George Westinghouse (1846–1914). The advantage that Tesla and Westinghouse saw in alternating current was that its power could be transformed—it could be passed via electromagnetic action from one circuit to another by a device called a transformer.

A **transformer** uses two important connections between electricity and magnetism to convey power from one AC circuit to another. The first is familiar: moving electric charge creates magnetic fields. This connection allows electricity to produce magnetism. The second connection, however, is something new: magnetic fields that change with time create electric fields. Discovered in 1831 by Michael Faraday ❑, this relationship allows magnetism to produce electricity!

> **Second Connection between Electricity and Magnetism**
> Magnetic fields that change with time produce electric fields.

❑ With only a primary education, English chemist and physicist Michael Faraday (1791–1867) apprenticed with a bookbinder at 14. At 21, he became a laboratory assistant to Humphry Davy, a renowned chemist. Faraday's experiments with electrochemistry and his knowledge of work by Oersted and Ampère led him to think that, if electricity can cause magnetism, then magnetism should be able to cause electricity. Through careful experimentation, he found just such an effect. Toward the end of his career, Faraday became a popular lecturer on science and made a particular effort to reach children.

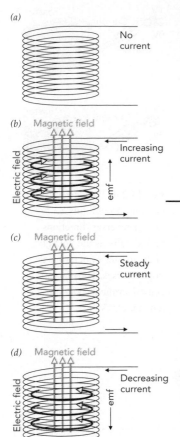

Fig. 11.2.5 (*a*) Without current, this inductor has neither electric nor magnetic fields. (*b*) An increasing current produces an increasing magnetic field in the inductor, which in turn produces an electric field. The emf resulting from that electric field opposes the current increase. (*c*) A steady current produces only a steady magnetic field. (*d*) A decreasing current produces a decreasing magnetic field in the inductor, which in turn produces another electric field. However, the resulting emf now opposes the current decrease.

Whether you wave a permanent magnet back and forth, or switch an electromagnet on and off, you are changing a magnetic field with time and thereby producing an electric field. If there are mobile electric charges around to respond to that electric field, they'll accelerate and you'll have created or altered an electric current and possibly done work on it as well. This process, whereby a time-changing magnetic field initiates or influences an electric current is called **magnetic induction.**

A transformer combines these two connections in sequence—electricity produces magnetism produces electricity. But rather than returning electric power to where it started, the transformer moves that power from the current in one coil of wire through a magnetic field to the current in a second coil of wire.

Check Your Understanding #3: Electric Phonographs

for answers, see page 384

In the days of vinyl records, a phonograph reproduced sound by sliding a diamond stylus through a record's undulating groove. A magnet attached to that stylus moved up and down with each undulation and produced a current in a nearby coil of wire. Why did the magnet's motion affect the coil?

Alternating Current and a Coil of Wire

To help us understand the power transfer that takes place in a transformer, let's start with a simpler case. What happens when you send an alternating current (AC) through a single coil of wire?

Because currents are magnetic, the coil becomes an electromagnet. However, since the current passing through it reverses periodically, so does its magnetic field. And because a magnetic field that changes with time produces an electric field, the coil's alternating magnetic field produces an alternating electric field.

This electric field has a remarkable effect: it pushes on the very alternating current that produces it! While it's not obvious how this electric field should affect that current, the result turns out to be simple (Fig. 11.2.5). As the coil's current increases, the induced electric field pushes that current backward and thereby opposes its increase (Fig 11.2.5*b*). And as the coil's current decreases, the induced electric field pushes that current forward and thereby opposes its decrease (Fig 11.2.5*d*). Amazingly, however the coil's current changes, the induced electric field always opposes that change!

This opposition to change is universal in magnetic induction, where it's known as **Lenz's law:** the effects of magnetic induction oppose the changes that produce them. In the present case, self-directed magnetic induction or "self-induction" leads our coil to oppose its own changes in current. A wire coil's natural opposition to current change makes it quite useful in electronics, where it's called an **inductor.**

Lenz's Law
The effects of magnetic induction oppose the changes that produce them.

However, magnetic induction does more than just push currents around; it can also transfer energy. Its induced electric field does work on any charge that moves with its push and negative work on any charge that moves opposite its push.

When induction does work on a charge that goes through our coil, that charge experiences a rise in voltage. The overall voltage rise, from the coil's start to its finish, is called the **induced emf** (short for *electromotive force*). But because the coil's current and induced electric field alternate, so does the induced emf—it swings between positive and negative voltages. And for related reasons, energy alternately leaves the current and returns. But where does that energy reside when it's not in the current?

The missing energy is in the coil's magnetic field! Magnetic fields contain energy. The amount of energy in a uniform magnetic field is half the square of the field strength times the volume of the field divided by the permeability of free space. We can write this relationship as a word equation:

$$\text{energy} = \frac{\text{magnetic field}^2 \cdot \text{volume}}{2 \cdot \text{permeability of free space}}, \tag{11.2.2}$$

in symbols:

$$U = \frac{B^2 \cdot V}{2 \cdot \mu_0},$$

and in everyday language:

Strong permanent magnets store so much magnetic energy that they may explode if you break them.

Common Misconceptions: Magnets as Limitless Sources of Energy
Misconception: Magnets are infinite sources of energy that could provide electric or mechanical power forever!
Resolution: While a magnet's field does contain energy, that energy is limited and was invested in it during its magnetization. To extract that energy, you'd have to demagnetize and thus destroy the magnet.

In effect, our coil is playing with the alternating current's energy, storing it briefly in the magnetic field and then returning it to the current. The coil stores energy while the magnitude of the current increases—the field strengthens and the current loses voltage. The coil returns energy while the magnitude of the current decreases—the field weakens and the current gains voltage. Because the coil's self-induced emf is responsible for bouncing this energy back to the current, it's frequently called a **back emf.**

The coil's self-induction and back emf allow it to handle alternating currents and alternating voltages with astonishing grace. You can actually plug the two ends of a properly designed coil into an AC electrical outlet without any trouble at all; the coil will rhythmically store energy and return it. That's not a stunt you'd want to try with an ordinary wire!

Unlike an ordinary wire, which can't safely receive current from the outlet at one voltage and return it to the outlet at a different voltage, the coil can use its back emf to "ride" the outlet's alternating voltage like a bottle riding ocean waves. Pushed forward or backward by the induced emf, current can enter this coil at one voltage and leave it at a completely different voltage. In fact, the coil's back emf always has just the right voltage so that current passing through the coil's *hot* end at the *hot* voltage passes through the coil's *neutral* end at the neutral voltage (0 V). For example, when *hot* is $+170$ V, the back emf is -170 V; when *hot* is -50 V, the back emf is $+50$ V; and when *hot* is 0 V, the back emf is 0 V.

→ Check Your Understanding #4: Slow Fall

for answers, see page 384

If you drop a strong magnet onto a nonmagnetic but highly conducting surface, the magnet will descend remarkably slowly. What's delaying the magnet's fall?

→ Check Your Figures #1: Field the Energy

for answers, see page 385

An MRI diagnostic unit fills about 0.1 m^3 of space with a 4-T magnetic field. How much energy is contained in that field?

Two Coils Together: A Transformer

Courtesy Lou Bloomfield

In a single coil, energy that's transferred from the current to the magnetic field must eventually return to the current. It has nowhere else to go. But what if there are two coils and two currents? In that case, energy transferred from one current to the magnetic field can move on to the second current!

That possibility is the basis for a transformer. In its simplest form, a transformer consists of two separate coils that share the same electromagnetic environment. Some or all of the energy invested in the magnetic field by current in the first coil can be withdrawn from the magnetic field by current in the second coil. Although the two currents never touch and don't exchange a single charge, power can move from one current to the other with ease.

We'll illustrate this energy transfer by examining an ordinary halogen desk lamp (Fig. 11.2.6). This device consists of a two-coil transformer and a halogen lightbulb. One coil of the transformer, its "primary" coil, is plugged directly into an electrical outlet and completes a circuit with the power company (Fig. 11.2.7). The power company pushes an alternating current through this primary circuit. The other coil of the transformer, its "secondary" coil, is connected to the lightbulb and completes another circuit—the secondary circuit. To ensure that the two coils share the same electromagnetic environment, both are wound around a ring-shaped magnetizable core. We'll discuss how that core works later in this section.

By itself, the primary coil acts as an inductor, alternately storing energy in its magnetic field and then returning that energy to the primary current by way of its back emf. Since this back emf mirrors the supply voltage, which we'll suppose is 120 volts AC, the back emf is also 120 volts AC.

Fig. 11.2.6 The bulb in this halogen desk lamp operates from low voltages provided with the help of a transformer. The lamp's two support rods also carry 12-V AC current to and from the bulb.

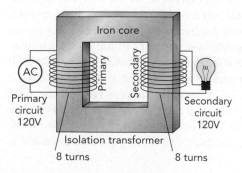

Fig. 11.2.7 This transformer conveys power from current in its primary circuit to current in its secondary circuit. The iron core guides magnetic flux so that the two coils share the same electromagnetic environment. With equal turns in its primary and secondary coils, this transformer supplies the same AC voltage from its secondary coil as it receives at its primary coil.

However, because the secondary coil shares the primary coil's electromagnetic environment, the secondary coil also experiences an induced emf and a voltage difference appears between its two ends. Since the secondary coil forms a circuit with the lamp's filament, this voltage difference imposes a voltage drop on the filament and propels a current through it. That current alternates because the emf alternates. In short, power is moving via electromagnetic action from an alternating current in the primary circuit to an alternating current in the secondary circuit and lighting up the bulb.

If the transformer is providing power to current in its secondary circuit, it must be removing the same amount of power from current in its primary circuit. Sure enough, it's using magnetic induction to do just that. However, this time the induction is reversed: the current in the secondary circuit is inducing an emf in the primary coil and that emf is removing power from the primary current!

This removal happens in an interesting way. The emf induced in the primary coil increases the primary current whenever it's investing energy in the transformer's magnetic field and decreases that current whenever it's withdrawing energy from the field. With more investment than withdrawal, the primary current is leaving energy behind in the magnetic field and the secondary current is carrying that energy away!

Remarkably, the power transfer process responds automatically to any changes in the secondary circuit's power consumption. For example, if you replace the desk lamp's bulb with one that consumes more power, more current will flow through the secondary coil so induction will transfer more power from the primary current to the secondary current. And if you remove the bulb completely, the secondary current will vanish and the primary current's energy investment and withdrawal will balance perfectly.

▶ Check Your Understanding #5: Use Only with AC Power

for answers, see page 384

If you send direct current through the primary coil of a transformer, no power will be transferred to the secondary circuit. Explain.

Changing Voltages

Transformers may be interesting, but why does the desk lamp need one? Why not just connect the bulb directly to the power outlet to form a single circuit with the power company?

In the desk lamp, the transformer's job is to provide the bulb with low-voltage AC power. Like the bulb in a flashlight, the desk lamp's bulb is designed to operate on small voltages. This low-voltage bulb derives its heating power from a large current experiencing a small voltage drop, so its filament has a small electrical resistance and is thick, short, and sturdy. In contrast, a high-voltage bulb derives its heating power from a small current experiencing a large voltage drop, so its filament needs a large electrical resistance and must be thin, long, and fragile. The shorter low-voltage filament is also a more concentrated light source, making it ideal for a desk lamp. To provide this low-voltage AC power, the transformer's secondary coil is wound differently from its primary coil.

In any transformer, the secondary coil experiences an induced emf that depends on its number of turns—the number of times its wire encircles the core. The more loops the secondary current makes around the core, the more work the transformer's electric field does on that current and the larger the induced emf. Since the amount of work done is proportional to the number of turns, so is the secondary coil's induced emf.

But what is the actual induced emf in a specific transformer? Suppose that we have a simple transformer in which the two coils, primary and secondary, are identical—they have equal numbers of turns—and that the primary coil of that transformer is plugged into a 120-V AC outlet.

Since the transformer's primary coil acts as an inductor, its back emf mirrors the AC voltage applied to it and is therefore 120 V AC. But because the two coils are identical and share the same electromagnetic environment, that same induced emf appears in the secondary coil: 120 V AC. If we connect the secondary coil to an appropriate bulb to form a complete circuit, the secondary coil will act as a source of 120-V AC electric power and light up the bulb.

This simple device is known as an *isolation transformer*. When you plug its primary coil into an AC outlet, its secondary coil acts as a source of AC power at the outlet's voltage. Though its secondary coil merely mimics the power outlet, an isolation transformer provides an important measure of electrical safety. Since its primary and secondary circuits are electrically isolated, charge can't move between those circuits and cause trouble. For example, when lightning strikes the power company's wires, the resulting burst of charge on the primary circuit can't pass to any appliances that are part of the secondary circuit. Not surprisingly, hospitals often employ isolation transformers to protect patients from shocks.

But most transformers have unequal coils and therefore different emfs in their coils. Since the secondary coil's induced emf is proportional to its number of turns, it acts as a source of AC power with a voltage equal to the voltage applied to its primary coil times the ratio of secondary turns to primary turns or

$$\text{secondary voltage} = \text{primary voltage} \cdot \frac{\text{secondary turns}}{\text{primary turns}}. \qquad \textbf{(11.2.3)}$$

An isolation transformer is simply the special case where the turn numbers are equal and their ratio is 1.

The transformer in our desk lamp is called a *step-down transformer* because it has fewer secondary turns than primary turns and provides a secondary voltage that is less than the primary voltage (Fig. 11.2.8). If we suppose that the ratio of secondary turns to primary turns is only 0.1, the secondary coil will act as a source of 12-V AC power.

Not surprisingly, there are also *step-up transformers* that have more secondary turns than primary turns and that provide secondary voltages that are greater than their primary voltages (Fig. 11.2.9). The transformer that powers a neon sign typically has 100 times

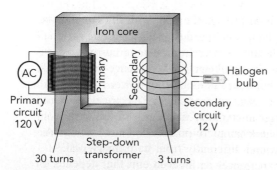

Fig. 11.2.8 With 10 times as many turns in its primary coil as in its secondary coil, this step-down transformer transforms 120-V AC power to 12-V AC power.

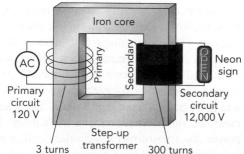

Fig. 11.2.9 With 100 times as many turns in its secondary coil as in its primary coil, this step-up transformer transforms 120-V AC power to 12,000-V AC power.

as many turns in its secondary coil as in its primary coil. When its primary coil is supplied by 120-V AC power, its secondary coil provides the 12,000-V AC power needed to illuminate the neon tube.

Even when its primary and secondary voltages are different, a transformer still manages to conserve energy. While each additional secondary coil turn increases the secondary coil's voltage, it also increases the rate at which the secondary current withdraws energy from the transformer's magnetic field. If you scale up the number of turns in the secondary coil to increase its voltage, you must scale down the current flowing through that coil in order to leave the amount of energy it withdraws from the magnetic field unchanged. As a result, the secondary current is equal to the primary current times the ratio of primary turns to secondary turns or

$$\text{secondary current} = \text{primary current} \cdot \frac{\text{primary turns}}{\text{secondary turns}}. \qquad (11.2.4)$$

In our desk lamp's step-down transformer, the primary coil has 10 times as many turns as the secondary coil. According to Eq. 11.2.3 the secondary voltage is 0.1 times the primary voltage, and according to Eq. 11.2.4 the secondary current is 10 times the primary current. If the 12-V bulb that you install in the desk lamp has been designed to consume 24 W of power, a current of 2 A will flow through the secondary circuit. To provide this power, the transformer's primary coil will carry 0.2 amperes of current supplied at 120 V AC. In all, 24 watts of power are flowing from the transformer's primary circuit to its secondary circuit.

> Check Your Understanding #6: Travel Trouble

for answers, see page 384

Your portable lava lamp operates on 120-V AC power, but you're visiting a country with 240-V AC power. You plug a travel adapter into the 240-V AC outlet and its transformer provides your lamp with the 120-V AC power it expects. Compare the numbers of turns in the transformer's two coils.

Real Transformers: Not Quite Perfect

Although we've been pretending that inductors and transformers are flawless and that their wires conduct electricity perfectly, that's not quite true. In reality, the wires used in those devices have electrical resistances and waste power in proportion to the squares of the currents they carry. To minimize this wasted power, real inductors and transformers are designed so as to minimize their resistances. To the extent practical, they employ thick wires made of highly conducting metals and those wires are kept as short as possible.

Unfortunately, inductors and transformers built only from wires can't develop the strong magnetic fields they need to store large amounts of energy unless they use long, many-turn coils. To avoid long coils, many inductors and virtually all transformers wrap their coils around magnetizable cores. Such cores respond magnetically to the alternating currents around them, boosting the resulting magnetic fields and making it easier to store large amounts of energy. Aided by those magnetizable materials—typically iron or iron alloys—cored inductors and transformers work well even with short, few-turn coils.

A core provides another crucial benefit to a transformer: it guides the transformer's magnetic flux lines so that nearly all of them pass through both coils, even when those coils are somewhat separated in space. Sharing their flux lines in that manner gives the coils a common electromagnetic environment and permits them to exchange electric power easily.

Making two separate coils share their flux lines isn't easy. Since a coil has no net magnetic pole, each flux line that emerges from it must ultimately return to it. But without a core, most flux lines leaving a coil return to it almost directly and remain nearby throughout their trip. Those unadventurous flux lines are unlikely to pass through a second, separate coil. Not surprisingly, a coreless transformer only works well when its two coils are wound so closely together that they can't help but share the same flux lines.

Winding both coils around a ring-shaped magnetic core makes it easy for the flux lines to pass through both coils because those flux lines are drawn into the core's soft magnetic material and follow it as if in a pipe. Although the flux lines leaving a coil must still return to it eventually, most of them complete that trip by way of the core—a journey that then takes them through the other coil. With nearly all the flux lines channeled by the core through both coils, power can flow easily from one coil to the other.

A core thus provides a transformer with great flexibility: its coils can be practically anywhere as long they encircle that core. However, cores aren't quite perfect pipes for flux; they leak slightly. Therefore, the most efficient transformers have coils that are wound nearby or on top of one another.

But while magnetizable cores make small, efficient transformers practical, they also introduce a few problems. First, the cores must magnetize and demagnetize easily in order to keep up with the energy investment and withdrawal processes. If they lag behind, they'll waste power as thermal power. Sadly, perfect magnetic softness is unobtainable and all cores waste at least a little power through delays in their magnetizations.

Second, because these cores are subject to the same electric fields that push currents around in the coils, they shouldn't conduct electricity. If they do, they'll develop useless internal currents known as *eddy currents* and thereby waste power heating themselves up. Since most soft magnetic materials are electrical conductors, transformer cores are frequently divided up into insulated particles or sheets so that little or no current can flow through them. But despite best efforts at minimizing resistive heating in their coils, and magnetization and eddy current losses in their cores, all transformers still waste some power. Even the best transformers are only about 99% energy efficient.

▶ Check Your Understanding #7: Winds of Change

for answers, see page 384

Large power transformers have cooling fins and often fans to blow air across them. Why does a transformer need this cooling?

Alternating Current Power Distribution

We're finally prepared to deal with the basic conflicts of power transmission. To minimize resistive heating in the power lines connecting a power plant with a distant city, electric power should travel through those lines as small currents at very high voltages. But to be practical, as well as to avoid shock and fire hazard, electric power should be delivered to homes as large currents at modest voltages.

While there is no simple way to meet both requirements simultaneously with direct current, transformers make it easy to satisfy them both with alternating current. We can use a step-up transformer to produce the very-high-voltage current suitable for cross-country

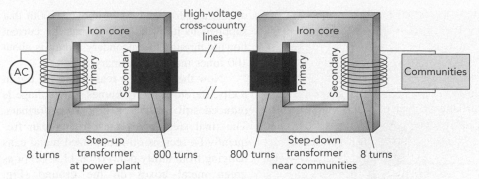

Fig. 11.2.10 Power is transmitted cross-country by stepping it up to very high voltage near the power plant, transmitting it as a small current at very high voltage, and stepping it back down to low voltage near the communities that are to be served. The secondary circuit for the step-up transformer is also the primary circuit for the step-down transformer.

transmission and a step-down transformer to produce the low-voltage current that's appropriate for delivery to communities (Fig. 11.2.10).

At the power plant, the generator pushes a huge alternating current through the primary circuit of a step-up transformer at a supply voltage of about 5000 V. The current flowing through the secondary circuit is only about 1/100th that in the primary circuit, but the voltage supplied by the secondary coil is much higher, typically about 500,000 V.

This transformer's secondary circuit is extremely long, extending all the way to the city where the power is to be used. Since the current in this circuit is modest, the power wasted in heating the wires is within tolerable limits.

Once it arrives in the city, this very-high-voltage current passes through the primary coil of a step-down transformer (Fig. 11.2.11). The voltage provided by the secondary coil

Fig. 11.2.11 This giant transformer transfers millions of watts of power from the very-high-voltage cross-country circuits above it to the medium-high-voltage neighborhood circuits to its left. Fans keep the transformer from overheating.

Courtesy Lou Bloomfield

Fig. 11.2.12 The three metal cans on this utility pole are transformers. They transfer power from the medium-high-voltage neighborhood circuits above them to the low-voltage household circuits at the lower right.

Courtesy Lou Bloomfield

of this transformer is only about 1/100th that supplied to its primary coil, but the current flowing through the secondary circuit is about 100 times that in its primary circuit.

Now the voltage is reasonable for use in a city. Before entering homes, this voltage is reduced still further by other transformers. The final step-down transformers can frequently be seen as oil-drum sized metal cans hanging from utility poles (Fig. 11.2.12) or as green metal boxes on the ground (Fig. 11.2.13). Current enters the buildings at between 110 V and 240 V, depending on the local standards. While 240-V electricity wastes less power in the home wiring, it's more dangerous than 110-V power. The United States has adopted a 120-V standard while Europe has a 230-V standard.

Courtesy Lou Bloomfield

Fig. 11.2.13 This transformer transfers power from a medium-voltage underground circuit to a low-voltage underground circuit used by nearby homes. It handles 50 kV·A or 50,000 W of power.

Check Your Understanding #8: High-Voltage Wires

for answers, see page 384

If a power utility were able to increase the voltage of its transmission line from 500,000 V to 1,000,000 V, how would that affect the power lost to heat in the wires?

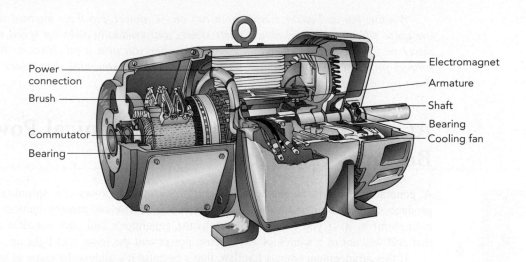

Power connection
Brush
Commutator
Bearing

Electromagnet
Armature
Shaft
Bearing
Cooling fan

SECTION 11.3 Electric Generators and Motors

We've seen what electricity is and how it's distributed. Now we'll look at how it's generated—how mechanical motion can be used to produce electric power. We'll also examine how electric power can be used to produce mechanical motion by studying electric motors. In fact, the symmetry of those two remarks foreshadows an amazing result: generators and motors are often the same devices!

Although generators are essential to electric power plants, they aren't common in your home. You encounter them mostly in cars and in emergency power equipment. However, electric motors are everywhere, spinning the parts of countless household machines. Sometimes this rotary motion is obvious, as in a fan or a mixer, but often it's disguised, as in the agitation of a washing machine or the vibration of a cell phone. Motors come in many shapes and sizes, each appropriate to its task. No matter how much torque or power a motor must provide, you can probably find one that's suitable. Some motors operate from direct current and can be used with batteries, while others require alternating current. There are even motors that work on either type of current.

Questions to Think About: *How can a moving object push electric charges through a wire and produce electricity? How does the power company determine how much power it needs to generate? If everyone turned off their lights, would the power company still have to generate the same amount of electricity? How can magnetic forces cause something to spin? Why can't a motor be built exclusively from permanent magnets? What determines which way a motor spins? Why are some motors safe near flammable chemicals while others are not?*

Experiments to Do: *Take a look at several different electric motors. You'll find them in tape recorders, fans, kitchen mixers, and car starters. A recorder motor and an automobile starter both run on DC power, but one consumes far more electric power and provides far more mechanical power than the other. What aspects of the starter motor allow it to handle so much power?*

Window fan and mixer motors both run on AC power, but their internal structures are quite different. The fan motor starts slowly and gradually picks up speed while the mixer motor reaches full speed only moments after you turn it on. How do the initial torques provided by those two motors compare? What do you think determines how fast these two motors spin?

AC Electric Generators: Mechanical Power Becomes Electric

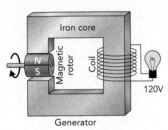

Iron core

Generator

Fig. 11.3.1 This generator resembles the transformer in Fig. 11.2.7, except that power reaches it through the motion of its spinning magnetic rotor, rather than through an electric current in a primary coil.

A generator converts mechanical power into electrical power. Its spinning magnet produces an alternating magnetic field within a coil of wire and thereby induces an alternating emf in it. If you connect a lamp to the generator's coil and complete a circuit, that coil will act as a source of AC electric power and the lamp will light up.

If this arrangement sounds familiar, that's because it's almost the same as in a transformer. In fact, the only important difference between a transformer and a generator is in what produces the alternating magnetic field!

Figure 11.3.1 shows a simple generator, one that looks strikingly like the transformer in Fig. 11.2.7. Both devices have a (secondary) coil wrapped around a magnetizable core. But in place of the transformer's primary coil, the generator has a spinning magnet or *rotor*. As that magnetic rotor spins, it produces a sinusoidally alternating magnetic field in the coil. This alternating magnetic field, in turn, produces an alternating electric field and induces an alternating emf in the coil. That emf lights the lamp.

This brief outline accurately explains how a generator provides AC electric power. But how is energy conserved in the process and what determines the frequency and voltage of the generator's alternating current? Since the generator can't create energy, whatever electric power it delivers to the current in its coil must originate as mechanical power in its rotor. Let's begin by looking at how the act of generating electric power extracts mechanical power from the rotor.

For clarity and simplicity, we're mostly going to ignore forces due to the magnetizable core itself. The one aspect of the core we won't overlook is its guidance of magnetic flux lines. The core forms a magnetic bridge between rotor and coil, conveying the rotor's magnetic field into the coil even when they're separated in space. As the rotor spins, its magnetic field sweeps through the coil as though the two objects were almost touching one another. Thanks to the core's help, the rotor and coil share the same electromagnetic environment.

With the lamp unplugged, there is no circuit and no current flows through the generator's coil. Since the generator isn't producing electric power, it shouldn't extract mechanical power from the rotor. Sure enough, the rotor turns freely; the current-free coil is nonmagnetic and the rotor's magnetic field sweeps through it effortlessly!

But with the lamp plugged into the generator, a circuit forms and current can flow. Now as the rotor's magnetic field sweeps through the coil, it induces an alternating current in the coil and lights the lamp. This electric power generation has serious consequences for the rotor; because the current-carrying coil is now magnetic, the coil interacts with the rotor and extracts mechanical power from it!

That interaction between the rotor and the coil starts the moment the rotor's magnetic field begins to sweep into the coil. The arriving magnetic field induces a current in the coil, rendering it magnetic with a field oriented *opposite* the rotor's field. Consistent with Lenz's law, the coil's induced magnetism is opposing the rotor's incoming magnetic field—the change that produced it. In effect, the coil becomes an electromagnet with its like poles turned to repel the approaching poles of the rotor. The rotor must therefore do mechanical work in order to align its poles with the coil.

The rotor keeps turning and soon begins sweeping its magnetic field back out of the coil. This withdrawal again induces current in the coil, but this time in the opposite direction. The coil becomes magnetic with a field oriented along the rotor's field. The coil's induced magnetism is again opposing the change that produced it—the rotor's *departing* magnetic field. In effect, the coil becomes an electromagnet with its opposite poles turned to attract the departing poles of the rotor. The rotor must now do mechanical work to turn itself out of alignment with the coil.

The rotor can't get a break; it must do work to align its poles with the coil and more work to undo that alignment. All of the electric power consumed by the lamp is being extracted from the generator's rotor as mechanical power.

Nothing makes this power transfer more evident than turning the rotor of a small generator by hand; you can feel the lamp draw power from you as you spin the rotor. And the more power the lamp consumes, the stronger the coil's induced current and magnetic field become and the more mechanical power you must provide to keep the rotor turning.

Of course, most generators are driven by things other than people. Industrial generators typically obtain their mechanical work from steam turbines, using steam produced by fossil fuels such as coal, oil, and natural gas or uranium. Other industrial generators use renewable resources such as water or wind to power their turbines. And smaller commercial or home generators frequently employ gas or diesel engines to keep their rotors turning.

▶ Check Your Understanding #1: Power Biking

for answers, see page 384

When you pedal a high-tech exercise bicycle, you are probably spinning the rotor of an electric generator. That generator supplies power to a heating filament with an adjustable electrical resistance. How should the bicycle alter that electrical resistance to make pedaling more difficult?

A Generator's Frequency and Output Voltage

The frequency of a generator's AC power is proportional to how fast its rotor spins. That's because the generator's alternating output voltage—the emf induced in its coil—is the result of the rhythmic sweep of its rotor's magnetic poles past its coil. This emf reverses every time a pair of poles aligns with the coil, so it takes two alignments to produce one full cycle of alternation in the generator's output voltage. For a rotor with a single pair of poles (e.g., Fig. 11.3.1), the emf alternates once per rotation of the rotor.

The generator in Fig. 11.3.1 thus provides 60-Hz AC power when its rotor spins 60 times per second. If its rotor were a more complicated magnet with two pairs of poles rather than one, it would only have to spin 30 times per second to produce 60-Hz AC power. This relationship between rotation rate and frequency explains why every generator of the 60-Hz U.S. power grid turns at either 60 rotations per second or some integer fraction of that rate. In the 50-Hz European power grid, the basic rate is 50 rotations per second. And because their output voltages must alternate together as one, all the generators of a power grid spin their rotors in perfect synchrony. Each power grid resembles a well-choreographed ballet in which every dancer is forever in step with the rest of the company.

In addition to maintaining the correct frequency, a generator must produce the right output voltage; its rotor must induce the right emf in its coil. That induced emf depends on three factors: the number of turns in the coil, the magnetic field strength, and the frequency at which that magnetic field alternates. Increasing any of those factors boosts the coil's emf and therefore the generator's output voltage.

Increasing the number of turns in a generator's coil is like increasing the number of turns in a transformer's secondary coil: the more times the wire coil encircles the generator's alternating magnetic field, the more work the resulting electric field can do on charges in that wire and the larger the coil's induced emf. As with a transformer, the generator's output voltage is proportional to the number of turns in its coil.

The magnetic field's strength and alternation frequency affect the generator's output voltage because both factors influence the electric field in the coil. Stronger or faster-changing magnetic fields produce stronger electric fields and thus larger emfs in the coil.

A good AC generator always provides power at its specified voltage and frequency. Since the rotation rate of its rotor affects both those characteristics, the generator has a control system to keep its rotor spinning steadily no matter how much electric power is being consumed. When you connect more lamps to the generator, so that the current extracts more mechanical work from its rotor, this control system boosts the flow of steam or fuel so that the rotor maintains its rotation rate. You can hear this automatic response in a gasoline- or diesel-powered generator; when you plug in more equipment, the control system powers up the engine.

With its rotor spinning steadily, a generator's output voltage will be relatively constant. However, that voltage can be further regulated by adjusting either the number of turns in the generator's coil or the strength of the magnetic field. A generator that automatically regulates its output voltage typically uses an electromagnet in its rotor, making that electromagnet stronger or weaker to raise or lower the generator's output voltage, respectively.

It's also common for a generator to allow you to use just part of its coil to obtain a smaller emf and therefore a smaller output voltage. For example, a typical household emergency generator provides 240 V AC from its full coil and 120 V AC from either half of its coil. The 120-volt outlets on this generator connect to half of the coil while the 240-volt outlet connects to the entire coil. You can even use the two voltages from the same generator at the same time!

▶ Check Your Understanding #2: Fade to Brown

for answers, see page 384

During periods of excessive power consumption, the electric grid occasionally reduces the voltage it supplies by about 5%. What can a generator do to lower its output voltage without changing its alternation frequency?

AC Electric Motors: Electric Power Becomes Mechanical

A motor converts electrical power into mechanical power. If you're wondering what motors are doing in the same section as generators, recall that a generator converts mechanical power into electrical power. As you can see, these two devices are the reverse of one another.

You might therefore expect motors and generators to be reverse in structure. Not quite. Motors and generators are actually the same in structure; what's reversed about them is their direction of current flow. Although they're usually specialized to one purpose or the other, you can often make a generator act as a motor or vice versa. Just reverse the current!

The effect of reversing the current in a generator shouldn't come as a total surprise. We have already seen that the mechanical work a generator extracts from its rotor is zero when the current is zero and rises in proportion to the current the generator propels through a lamp. What we hadn't considered was what happens if that current drops below zero—if it travels opposite its normal direction of flow through a lamp. In that case, the mechanical work extracted from the rotor should also drop below zero: the generator should *provide* mechanical work to its rotor and thereby become a motor!

However, the current that we're reversing is an *alternating* current, so how do we "reverse" it? Since a generator normally pushes current out of whichever end of its coil is momentarily at higher voltage, reversal means pushing current *into* that end of the coil. The effect of this reversal is to make the "generator" consume electric power rather than produce it. It acts as a motor.

This transition from generator to motor can also be explained in terms of a reversal of magnetic forces. Reversing the alternating current in the coil interchanges its magnetic poles so that it pushes on the rotor when it used to pull and pulls when it used to push. Now as the rotor sweeps into alignment with the coil, the coil *attracts* the approaching pole and does work on it. And as the rotor continues on, sweeping back out of alignment, the coil *repels* the departing pole and again does work on it. Overall, electric power in the coil is becoming mechanical power in the rotor—the hallmark of a motor.

It's an AC synchronous motor—a type of motor whose rotor turns in perfect synchrony with current from an AC electric power source (Fig. 11.3.2). When this type of motor is plugged into a 60-Hz electrical outlet, its rotor spins exactly 60 times per second, or at an integer fraction of that rate if the rotor has multiple pairs of poles.

Its rigid adherence to the AC power frequency makes an AC synchronous motor steady and precise, but also hard to start. If its rotor isn't turning or is turning at the wrong speed, the coil's alternating magnetic poles will push or pull on the rotor erratically and the rotor may never spin properly. That's why most practical AC synchronous motors have extra components and often multiple coils to help their rotors start turning in the desired direction and reach the proper rotation rates.

The similarity between Figs. 11.3.1 and 11.3.2 raises an important question: if a generator and a motor are the same device and you plug one into the AC power grid, what determines whether it acts as a generator or as a motor? The basic answer to that question is surprisingly simple: it acts as a generator when you do work on its rotor and as a motor when you extract work from its rotor!

The device responds to you this way in order to keep its rotor turning in synch with the AC power grid. If you attempt to speed the rotor up by twisting it forward and doing

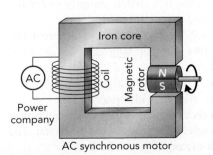

AC synchronous motor

Fig. 11.3.2 This AC synchronous motor resembles the transformer in Fig. 11.2.7, except that power leaves it through the motion of its spinning magnetic rotor, rather than through an electric current in a secondary coil.

work on it, the device acts as a generator: it extracts mechanical power from you and provides electric power to the grid. Your forward twist will have advanced the rotor's orientation to the point where its poles will be repelled as they sweep into alignment with the coil and attracted as they continue on out of alignment—the forces in a generator.

On the other hand, if you attempt to slow the rotor down by twisting it backward and having it do work on you, it acts as a motor: it extracts electric power from the grid and provides you with mechanical power. Your backward twist will have delayed the rotor's orientation to the point where its poles will be attracted as they sweep into alignment with the coil and repelled as they continue on out of alignment—the forces in a motor.

And if you simply leave the spinning rotor alone and let it coast freely, it will orient itself halfway between these two angular extremes so that the electromagnetic coil does zero average work on it.

Check Your Understanding #3: Turning the Hands of Time

for answers, see page **384**

Before the era of quartz clocks, a typical electric clock was based on an AC synchronous motor. The motor's rotor simply turned a set of gears which advanced the clock's hands. Why does the accuracy of such a clock depend on the accuracy of the power grid's alternation frequency?

DC Electric Motors

Since most portable devices are powered by batteries and direct current, they can't use AC synchronous motors to turn their components. If you send direct current through the coil of an AC synchronous motor, that coil will act like a permanent magnet and the rotor won't spin. Instead, attraction between opposite poles will snap the rotor into alignment with the coil and it will never move again. With its poles as close as possible to the coil's opposite poles, the rotor will be in a stable equilibrium.

To keep the rotor spinning, something must reverse the coil's magnetism each time the rotor reaches this stable equilibrium. That reversal will turn attraction into repulsion and the rotor will suddenly find itself in an *unstable* equilibrium. Instead of stopping, the rotor will spin onward in search of a new stable equilibrium. But when it gets there, the coil's magnetism will reverse again. The rotor thus spins forever, seeking to align its poles with the coil's opposite poles but never quite succeeding. It's the myth of Sisyphus—rolling a huge stone up a hill, only to have it roll down again as it nears the top—realized in an electromechanical device.

Flipping the coil's magnetism is a simple as reversing its current. All that's needed is a switch that interchanges the two wires connecting the coil to the battery. Whenever the rotor reaches alignment with the coil's opposite poles, this switch abruptly swaps the connections and thereby reverses the coil's current and magnetism.

Such a switch can be found in virtually every DC electric motor, but different DC motors implement it differently. There are at least two common approaches, used in at least two types of DC motors: brushless and brushed.

A brushless DC motor is just a synchronous AC motor plus a high-tech switch that reverses the current whenever the rotor's poles have aligned with opposite poles on the stationary coil (Fig. 11.3.3*a*). Although their high-tech switches make them relatively expensive, these brushless motors spin silently and safely for years without service. The fans that cool your computer are spun by brushless motors (Fig. 11.3.3*b*).

(a)

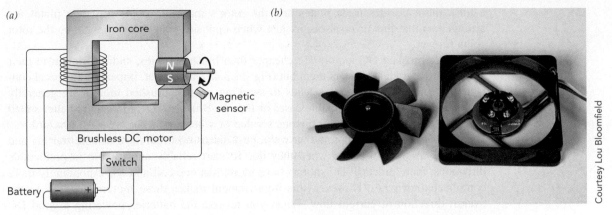

(b)

Courtesy Lou Bloomfield

Fig. 11.3.3 (*a*) A brushless DC motor uses an electromagnet to spin its magnetic rotor. A sensor monitors the orientation of the rotor and reverses the current in the electromagnet each time the rotor aligns with the magnetic field. (*b*) This computer fan uses a brushless DC motor. Its stationary four-pole electromagnet fits inside the rotor's ring-shaped permanent magnet. The black magnetic sensor is below the electromagnets.

A brushed DC motor puts its electromagnetic coil on the rotor and pushes on it with a stationary permanent magnet (Fig. 11.3.4*a*). That change is inconsequential except that now it's the rotor that interchanges its poles each time they align with opposite poles of the permanent magnet. The value of this swap is that it allows the motor to use a simple and inexpensive switch called a *commutator* to control the direction of current flow in its coil.

In its simplest form, a commutator consists of two curved plates that are fixed to the rotor and connected to opposite ends of the rotor's coil. Electric current flows into the rotor through a conducting brush that touches one of these plates and leaves the rotor through a second brush that touches the other plate. As the rotor turns, each brush makes contact first with one plate and then with the other. Every time the rotor completes half

(a)

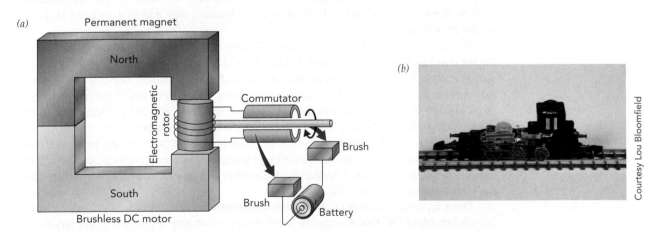

(b)

Courtesy Lou Bloomfield

Fig. 11.3.4 (*a*) A brushed DC motor spins its electromagnet rotor in the field of a permanent magnet. Each time the rotor aligns with the magnetic field, its commutator reverses the current in the electromagnet. (*b*) This toy train uses a brushed DC motor. Its electromagnetic rotor spins about a vertical axis above its commutator and brushes.

a turn, these brushes trade plates and the rotor's magnetic poles flip. The plates are arranged so that this reversal occurs just when opposite poles have aligned, so the rotor spins forever.

While brushed DC motors are cheaper than brushless ones, sliding friction in their commutators gradually wears them out (Fig. 11.3.4b). Moreover, imperfect electrical contact between brushes and pads leads to sparking, so that brushed motors are generally unsuitable for use near flammable gases or liquids. Brushless DC motors are better suited to tasks requiring long and continuous service or where sparking presents a hazard.

The single-coil DC motors that we've just discussed often have trouble starting and may spin in either direction when they do. To start reliably and to spin in predictable directions, more practical DC motors have more than one coil and correspondingly more complicated switches. However, this improvement makes these motors sensitive to the overall direction of current flow. When you reverse the batteries powering a good DC motor, the current everywhere in that motor reverses and so do all the poles of its electromagnets. Since the motor's permanent magnets are unchanged, the forces affecting the rotor reverse—attraction becomes repulsion and vice versa. As a result, the motor spins backward.

→ **Check Your Understanding #4: A Permanent Motor**

for answers, see page 385

Hoping to avoid using batteries, an inventor replaces the electromagnets of a DC motor with permanent magnets. Its rotor refuses to spin. Why can't this permanent magnet motor work?

Rotation Speed and DC Electric Generators

We've seen what makes the rotor of a DC motor spin, but not what determines how fast it spins. Since the motor effectively makes its own alternating current, won't its rotor spin faster and faster forever? The answer is no. Its rotor spins at a specific rate that's proportional to the voltage supplied to the coil. The origin of this natural rate is induction: as the rotor spins, the coil experiences an induced emf and that induced emf limits the rotor's rotation rate.

When you first connect the motor to a battery and its rotor is stationary, the battery-imposed voltage drop pushes current through its coil from higher voltage to lower voltage. But once the rotor starts spinning, an induced emf appears in the coil and opposes that current flow. That emf grows larger as the rotor spins faster and eventually becomes large enough to stop the current flow altogether. The induced emf is then equal to the battery-imposed voltage drop and the rotor is spinning at its natural rate.

The harder it is for the motor's emf to equal the battery voltage, the faster the rotor must spin to reach its natural rate. Therefore, increasing the battery voltage, weakening the motor's permanent magnet, or reducing the number of turns in its coil(s) will all increase the motor's natural rotation rate.

Once its rotor is spinning at that natural rate, the motor opposes any further change in rotation rate. If you try to slow the rotor down by twisting it backward, the motor will begin drawing current from the battery again. It will turn DC electric power into mechanical power to prevent the rotor from slowing down.

If you try to speed the rotor up by twisting it forward, however, the motor's emf will begin pushing current in the opposite direction. It will turn mechanical power into

DC electric power to prevent the rotor from speeding up. By doing work on the rotor in this manner, you will cause current to flow backward through the battery and recharge it. The DC motor will be acting as a DC generator!

After seeing that AC motors and AC generators are the same, it's not so surprising that DC motors and DC generators are often the same as well. If you do work on the rotor of a brushed DC motor, it will act as a DC generator. Edison used such DC generators in his New York City power plants. Brushless DC motors are more finicky because of their high-tech switches, but some can act as DC generators as well.

Check Your Understanding #5: Hybrid Cars

for answers, see page 385

Your new hybrid car uses a special DC motor to turn its wheels. When you accelerate at a green light, the car's battery powers the motor, which propels the wheels. But when you brake at a red light, the wheels power the motor, which recharges the battery. How is that reversal possible?

Universal Motors

If you plug a brushed DC motor into an AC electrical outlet, its rotor will hum rather than spin. That's because the rotor tries to reverse each time the AC outlet voltage reverses and it just can't make any progress either way.

But if you replace the stationary magnet in a brushed DC motor with an electromagnet and connect that electromagnet to the same power source as the rest of the motor, you'll have made a universal motor (Fig. 11.3.5a). As indicated by its name, this motor will spin properly when powered by either direct or alternating current.

To understand its flexibility, let's consider what happens when you supply a universal motor with DC power. The stationary electromagnet will magnetize in one direction

Fig. 11.3.5 (*a*) A universal motor resembles a DC motor, except that it uses only electromagnets. Since it is unaffected by the direction of current flow in its wires, it can operate on either DC or AC power. (*b*) This mixer uses a universal motor. Its electromagnetic rotor (center) spins inside a stationary electromagnet, while its commutator and brushes (right) control the direction of current flow through the rotor.

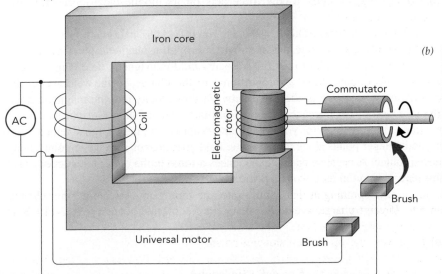

(*a*)

Iron core

AC

Coil

Electromagnetic rotor

Commutator

Brush

Universal motor

Brush

(*b*)

Courtesy Lou Bloomfield

like a permanent magnet and the rotor will spin as if it were in a DC motor. However, there's an important difference: the rotor of a universal motor won't spin backward when you reverse the supply voltage. It will continue turning in the same direction because the current reversal reverses every pole in the entire motor and therefore leaves the motor's magnetic forces unchanged. If you want to make a universal motor turn backward, you must interchange the wires of its stationary electromagnet.

Since its rotor turns in a fixed direction, regardless of which way current flows, a universal motor works fine on AC power. And like the DC motors on which it's modeled, a universal motor's speed is determined by voltages and currents rather than by the AC power frequency. Flexible and reliable, universal motors are standard in such household appliance as kitchen mixers, blenders, and vacuum cleaners (Fig. 11.3.5b).

Check Your Understanding #6: No Going Back

for answers, see page 385

You are trying to prepare whipped cream during a power failure and your electric mixer won't work. So you assemble a huge number of alkaline batteries into a 120-volt chain and supply the mixer with direct current. It works beautifully! Just for fun, you reverse the batteries but the mixer continues to turn in the same direction. Why doesn't its rotation reverse when you reverse the current flowing through it?

Induction Motors

Our final type of motor is the most conceptually sophisticated: the induction motor. Its rotor is neither a permanent magnet nor an ordinary electromagnet; it's a collection of conducting loops that are magnetized by induction alone. When the motor exposes this rotor to a rotating magnetic field, it experiences electromagnetic forces that make it spin along with that field.

To understand the forces acting on this rotor, let's start by looking at a loop of wire in a stationary magnetic field (Fig. 11.3.6). As long as that loop is motionless, it carries no current and experiences no forces. But when the loop begins to rotate through the field (Fig. 11.3.6a), a new force acts on its mobile electric charges: the Lorentz force.

Named after its discoverer, Dutch physicist Hendrik Antoon Lorentz (1853–1928), the **Lorentz force** affects a charge that is moving through a magnetic field. This force pushes the charge at right angles to both its velocity and the magnetic field (Fig. 11.3.7). The strength of the Lorentz force is proportional to the charge, to the velocity, to the magnetic field, and to the sine of the angle between the velocity and the magnetic field. Finally, the direction of the Lorentz force on a positive charge follows a right-hand rule: when the extended index finger of your right hand points along the charge's velocity and your bent middle finger points along the magnetic field, the force on the charge points along your extended thumb. A negative charge experiences a force in the opposite direction. This relationship can be written as a word equation,

$$\text{force} = \text{charge} \cdot \text{velocity} \cdot \text{magnetic field} \cdot \text{sine of angle}, \qquad \textbf{(11.3.1)}$$

in symbols:

$$F = qvB \cdot \sin(\text{angle}),$$

(a)

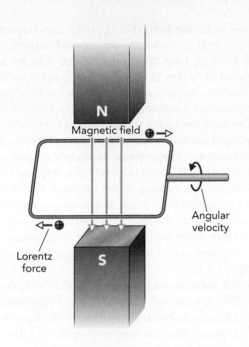

Magnetic field

Angular
velocity

Lorentz
force

Fig. 11.3.6 (*a*) When a loop of wire
rotates through a magnetic field, its
moving charges experience Lorentz
forces along the wire and begin to move
as a current around the loop. (*b*) The
circulating charges experience additional
Lorentz forces perpendicular to the wire
itself and produce a torque on the loop
opposing its rotation.

(b)

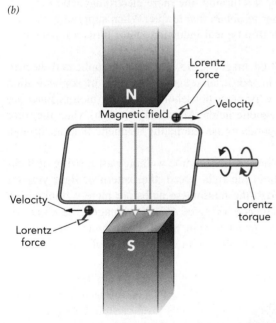

Lorentz
force

Velocity

Magnetic field

Velocity

Lorentz
force

Lorentz
torque

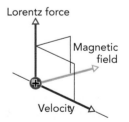

Lorentz force

Magnetic
field

Velocity

Fig. 11.3.7 A positive
charge moving through a
magnetic field experiences
a force that's perpendicular
to both its velocity and
the magnetic field. A
negatively charged particle
would experience a force
in the opposite direction.

and in everyday language:

> *When charged particles from the sun encounter the earth's magnetic field,*
> *they get pushed into spiral paths, producing the aurora borealis and the*
> *aurora australis,*

where the angle involved is between the velocity and the magnetic field, and the direction of the Lorentz force follows the right-hand rule.

When our loop of wire rotates, it carries its mobile charges through the stationary
magnetic field and they experience Lorentz forces. Since the two sides of the loop move

opposite one another, the charges in those sides are pushed in opposite directions along the wire and begin to circulate around the loop as a current—an induced current!

With charges now moving around the wire loop, the Lorentz force acts yet again (Fig. 11.3.6b). It pushes the charges toward the edges of the wire, but in opposite directions on opposite sides of the loop. Since the charges can't leave the wire, they push it along with them and the entire loop experiences a torque. That torque is directed opposite the loop's angular velocity—it acts to slow the loop's rotation through the field. This is Lenz's law again: since the loop's induced current is caused by its rotation in a magnetic field, the effect of that induction is a torque that opposes the loop's rotation.

Evidently, a loop of wire has difficulty rotating in the presence of a stationary magnetic field. Remarkably, it has equal difficulty *not* rotating in the presence of a rotating magnetic field! That's how an induction motor spins its rotor: it uses electromagnets to make a magnetic field that rotates in space and its looplike rotor spins along with that field. While a single-loop rotor is sufficient, more effective rotors contain many conducting loops. The classic rotor is the "squirrel cage," so named because of its resemblance to an animal exercise wheel (Fig. 11.3.8).

Although the rotor tries to follow the rotating magnetic field perfectly, it doesn't quite keep up. If it did, no current would flow through it and it would experience no electromagnetic torque at all. Even the friction in its bearings would slow the rotor down. Instead, the rotor always turns somewhat slower than the magnetic field. And the more torque the rotor exerts on the machinery it's turning, the more electromagnetic torque it needs and the slower it must spin in order to obtain that torque. When supplying its maximum rated mechanical power, the rotor of a typical induction motor turns a few percent slower than the motor's rotating field.

Unfortunately, creating a true rotating magnetic field requires complicated electromagnets and currents. Though routine in industrial settings, that level of sophistication is difficult to achieve at home. That's why most household induction motors make do with *alternating* rather than *rotating* magnetic fields. Amazingly enough, when the rotor is spinning at about the right rate, it responds to the alternating magnetic field as through it were a rotating one.

This effect resembles one you may have noticed while walking past a string of flickering holiday lights: if you travel at about the right speed, the pattern of light you see can appear to move with you even though the motionless bulbs are merely blinking on and off. Similarly, when the rotor spins inside a ring of AC electromagnets at about the right rate, the pattern of magnetism can appear to spin with it even though the motionless electromagnets are merely flipping their poles with each reversal of the AC power.

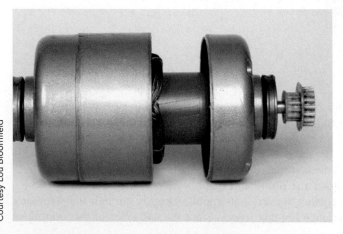

Fig. 11.3.8 This disassembled induction motor surrounds its red squirrel cage rotor with stationary electromagnetic coils. The coils' magnetic field rotates around the rotor and the rotor spins with it. To increase the motor's performance, the rotor has iron laminations within its conducting loops.

To be practical, an induction motor must add at least some rotating character to this alternating magnetic field. The rotating portion of the field gets the rotor started and up to speed, and the alternating portion takes over from there. In an induction motor that doesn't have to supply much starting torque, just a hint of rotating character is all it takes to coax the rotor gradually up to speed. That hint of rotation can be created with simple electromagnetic devices. But an induction motor that must exert a large torque while starting needs a true rotating magnetic field, at least until it reaches its proper rotation rate.

▶ Check Your Understanding #7: Lorentz Speaks

for answers, see page **385**

An ordinary audio speaker contains a wire coil immersed in a strong magnetic field. When an audio system sends currents through that coil, the coil experiences a force proportional to that current. What force is pushing on the coil?

▶ Check Your Figures #1: Lorentz Speaks with Precision

for answers, see page **385**

If the coil in an audio speaker experiences a Lorentz force of 1 N when it carries a current of 1 A, what force will it experience when it carries a current of 2 A and all the charges in it are thus traveling twice as fast as before?

Epilogue for Chapter 11

In this chapter, we studied magnetism and the ways in which magnetism relates to electricity. In *household magnets,* we looked at the concept of magnetic pole and the attractive or repulsive forces that poles exert on one another. We examined magnetic materials and saw how their magnetic properties make them useful for various purposes. And we encountered electromagnets and began to see that magnetism isn't independent of electricity. In *electric power distribution,* we saw how alternating electric currents make it possible to transfer power from one circuit to another by way of a transformer and its electromagnetic properties. We learned that transforming electric power to extremely high voltages and small currents minimizes the power wasted between power plants and cities.

And in *electric generators and motors,* we observed that when a magnet and a coil move past one another, the coil can act as a source of electric power. We also observed that when electric power is supplied to a coil, the coil and a magnet can move past one another. The symmetry of this situation, mechanical power producing electric power and electric power producing mechanical power, was impressive.

Explanation: A Nail and Wire Electromagnet

When you connect the wire from one terminal of the battery to the other, a current flows from the positive terminal to the negative terminal through the wire. (In reality, negatively charged electrons move from the battery's negative terminal, through the wire, to its positive terminal, but we've adopted a fiction that positive charges are heading the other way.) This current produces a magnetic field around the wire. Because the wire is coiled around the nail, this magnetic field passes through the nail and causes its magnetic domains to resize until most of them are aligned with the field. Without any current in the wire, the magnetic domains in the steel point in many different directions, so the nail appears nonmagnetic. However, with the current orienting the domains, they together produce a large magnetization. The nail becomes magnetic and exerts strong magnetic forces on other nearby objects.

Chapter Summary

How Household Magnets Work: Common refrigerator magnets are composed of hard magnetic materials that were permanently magnetized by their manufacturers. Simple button magnets have a single pair of magnetic poles, one north and one south, but plastic sheet magnets usually have many poles. These magnets stick to a refrigerator's surface by temporarily magnetizing that surface's soft magnetic materials and then becoming attracted to the opposite poles on that surface.

A compass is another permanent magnet, but one designed to align with the earth's magnetic field. In fact, that magnetic field can be mapped out using a compass. The magnetic fields around smaller magnets can be made visible with iron filings instead. But permanent magnets aren't the only sources of magnetic fields: we found that when current flows through the coil in a doorbell, it becomes a magnet as well—an electromagnet.

How Electric Power Distribution Works: To minimize power losses in the transmission lines between power plants and cities, power distribution systems use alternating currents and transformers. Near the power plant, relatively low-voltage, high-current electric power is transformed into very-high-voltage, low-current power for transmission on cross-country power lines. Because the power consumed by these high-voltage wires depends on the square of the currents they carry, the power losses are greatly reduced by this technique. When the power arrives at a city, it's transformed into medium-voltage, high-current power for distribution to neighborhoods. Finally, in neighborhoods, step-down transformers transform this power to low-voltage, very-high-current power for distribution to individual homes and offices.

How Electric Generators and Motors Work: Most electric power is generated electromagnetically by rotating magnets and coils. In a typical generator, a rotating permanent magnet spins inside a stationary coil of wire. As the rotating magnet's field sweeps across the coil, magnetic induction conveys power to a current flowing through that coil. That current is itself magnetic and it extracts mechanical power from the rotating magnet. Overall, the generator converts mechanical power into electrical power.

An electric motor performs the opposite task: it converts electrical power into mechanical power. It uses magnetic forces to keep a spinning rotor turning steadily. Composed of electromagnets and often permanent magnets, the motor sets the rotor on an endless, unsuccessful journey to align its magnetic poles with opposite poles on a stationary structure. Every time the rotor almost arrives at its goal, a current change in an electromagnet interchanges its poles so that the rotor is aligned with the wrong poles and must continue its travels.

Important Laws and Equations

1. Coulomb's Law for Magnetism: The magnitudes of the magnetostatic forces between two magnetic poles are equal to the permeability of free space times the product of the two magnetic poles divided by 4π times the square of the distance separating them, or

$$\text{force} = \frac{\text{permeability of free space} \cdot \text{pole}_1 \cdot \text{pole}_2}{4\pi \cdot (\text{distance between poles})^2}. \quad (11.1.1)$$

If the charges are like, then the forces are repulsive. If the charges are opposite, then the forces are attractive.

2. The Force Exerted on a Pole by a Magnetic Field: A pole experiences a force equal to the product of its pole times the magnetic field, or

$$\text{force} = \text{pole} \cdot \text{magnetic field}, \quad (11.1.2)$$

where the force points in the direction of the field.

3. Lenz's law: The effects of magnetic induction oppose the changes that produce them.

4. The Energy in a Magnetic Field: The energy of a magnetic field is equal to the square of that field times its volume, divided by twice the permeability of free space, or

$$\text{energy} = \frac{\text{magnetic field}^2 \cdot \text{volume}}{2 \cdot \text{permeability of free space}}. \quad (11.2.2)$$

5. Transformer Voltages: A transformer's secondary coil acts as a source of AC power with a voltage equal to the AC voltage applied to its primary coil times the ratio of secondary turns to primary turns or

$$\text{secondary voltage} = \text{primary voltage} \cdot \frac{\text{secondary turns}}{\text{primary turns}}. \quad (11.2.3)$$

6. Transformer Currents: The AC current in a transformer's secondary coil is equal to the AC current in its primary coil times the ratio of primary turns to secondary turns or

$$\text{secondary current} = \text{primary current} \cdot \frac{\text{primary turns}}{\text{secondary turns}}. \quad (11.2.4)$$

7. The Lorentz Force: When an electric charge moves through a magnetic field, it experiences a force equal to the product of its charge times its velocity times the magnetic field times the sine of the angle between the velocity and the magnetic field, or

$$\text{force} = \text{charge} \cdot \text{velocity} \cdot \text{magnetic field} \cdot \text{sine of angle}, \quad (11.3.1)$$

where that force is at right angles to both the velocity and the magnetic field and follows a right-hand rule.

Check Your Understanding—Answers

Section 11.1 HOUSEHOLD MAGNETS

1. The top surfaces of both halves are still north poles, and the bottom surfaces are still south poles. The two tops repel, as do the two bottoms.

Why: This puzzling phenomenon, where a shattered permanent magnet opposes attempts to reassemble it, is an illustration of the potential energy contained in a permanent magnet. The magnet is a collection of many tinier magnets, all aligned with their north poles together and their south poles together. Like poles repel one another, so the tiny magnets are difficult to hold together. Given a chance, the magnet will push apart into fragments. Very strong permanent magnets release so much potential energy when they break that they practically explode when cracked.

2. It will have a north pole.

Why: The paper clip will become magnetically polarized with its south pole nearest the permanent magnet's north pole. The other end of the paper clip will have a north pole and will be able to polarize other paper clips. These polarized clips attract one another strongly enough to cling together in a long chain.

3. Its magnetic poles will interchange.

Why: Even though the small magnet can't move, its magnetic poles can. When the repulsion between the two north poles becomes strong enough, the poles of the small magnet will interchange and it will then present its south pole to the north pole of the large magnet. You will have permanently reversed the small magnet's poles.

4. The needle will experience a force toward weak magnetic field (away from the button magnet).

Why: With its magnetic poles aligned opposite to the button's magnetic field, the compass needle experiences a force toward weaker magnetic field. Actually, if you continue to push the needle closer to the button magnet, you can accidentally remagnetize that needle; its poles will permanently interchange and it will subsequently point south rather than north! Not to worry, though, because you can simply repeat this procedure to restore the compass to normal.

5. The poles are at the ends of the bridges.
Why: The iron filings follow the magnetic flux lines, which extend from north poles to south poles. Thus one end of each bridge is a north pole and the other end is a south pole.

6. The magnetic field is created by the current in a coil of wire.
Why: MRI requires a magnetic field that is intense, uniform, and spacious enough for a patient to fit inside. The best way to create such a colossal field is with a current-carrying coil. In fact, the field is so enormous that its flux lines extend far from the magnet and can attract iron or steel objects from across the room. Understandably, magnetic materials are forbidden near MRI magnets.

Section 11.2 ELECTRIC POWER DISTRIBUTION

1. The power would roughly double.
Why: Doubling the length of a wire is like placing two identical wires one after the next. If each wire uses 1 unit of power, then two wires should use roughly 2 units of power. Electrical resistance is proportional to the length of a wire and inversely proportional to the cross-sectional area of that wire. Shortening and thickening a wire reduce its resistance.

2. Yes, at the moment the voltages are reversing.
Why: While the ground and neutral wires of the electrical outlet are normally without charge and therefore relatively safe, the hot wire is usually charged and dangerous. That hot wire's voltage alternates rapidly between high positive voltage and high negative voltage. Only when it's passing through 0 V can you touch it without risking a shock. However, that safe moment is so brief that you can't realistically avoid a shock. Don't try it!

3. The moving magnet produced an electric field, which pushed mobile charges through that wire coil.
Why: The tiny vibrating magnet affects the coil's current via magnetic induction.

4. The falling magnet is inducing currents and magnetism in the surface. In accordance with Lenz's law, that induced magnetism opposes the change that produces it: it acts to slow the magnet's descent.
Why: A strong magnet induces such powerful magnetic opposition in a good conductor that moving the magnet is difficult. This effect is most evident with a superconductor, that is, a material that conducts electricity perfectly and can sustain induced currents forever. A superconductor can slow a falling magnet to a stop and hold it suspended indefinitely (Fig. 11.2.1).

5. When direct current flows through the transformer's primary coil, it creates a constant magnetic field around the iron core. Since that field doesn't change, it doesn't create any electric fields and doesn't induce current in the transformer's secondary coil.
Why: The current through the primary coil must change so that the magnetic field in the coils will change and current will be induced in the secondary coil. Transferring power from one circuit to another is so useful that there are many DC-powered devices that switch their power on and off to mimic alternating current, so that they can use transformers.

6. The transformer's secondary coil has half as many turns as its primary coil.
Why: To step down the voltage, a transformer must have fewer turns in its secondary coil than in its primary coil. Fewer turns leads to a smaller emf in the secondary coil and a smaller output voltage for the transformer.

7. Its magnetic core converts some of the electric power into thermal power. Unless that thermal power is eliminated, the transformer will overheat.
Why: Transformers aren't perfectly energy efficient; they convert a small fraction of the electric power they handle into thermal power. Their magnetic cores contribute to that inefficiency because their limited magnetic softness and electric conductivity cause them to heat up. Fin and fans are essential to keeping large transformers cool.

8. It would reduce the amount of heat produced to only 25% of the previous value.
Why: At 1,000,000 V, the transmission line would be able to carry the same power as would a 500,000-V transmission line with only half the current. Since the power wasted by the transmission line itself is proportional to the square of the current, halving the current reduces the power waste to 25%.

Section 11.3 ELECTRIC GENERATORS AND MOTORS

1. It should reduce the heater's resistance.
Why: By lowering the heater's resistance, the bicycle increases the current flowing through the circuit. That increased current carries more power from the generator to the heater, so the generator extracts more mechanical work from the bicyclist.

2. It can weaken its rotor's magnetic field or reduce the number of turns in its coil.
Why: Since the generator can't slow its rotor without also reducing its frequency, it can only reduce its induced emf by weakening the alternating magnetic field in its coil or by using a coil with fewer turns.

3. Since an AC synchronous motor turns in synch with the alternating current, any increase or decrease in the current's frequency will make the clock run fast or slow, respectively.
Why: A clock based on an AC synchronous motor is only as accurate as the frequency of its AC power. When plugged into

an emergency generator, with its inadequately regulated frequency, the clock will not keep accurate time.

4. The rotor will quickly settle into a stable equilibrium and never turn again.
Why: Without an active system to upset the stable equilibrium each time the rotor reaches it, the motor won't continue to turn. The best it can do is coast briefly until friction slows its motion until it eventually settles into a stable equilibrium and stops completely.

5. During braking, the wheels do work on the motor, where the resulting induced emf pushes current backward through the battery and recharges it.
Why: The hybrid car's DC motor is designed to work well as a generator, too. When the car needs to accelerate forward, it draws power from its battery to propel itself forward. But when it needs to brake, it draws power from its motion to recharge its battery. Of course, the car's engine is also involved in powering the wheels and in generating electric power.

6. Its universal motor is insensitive to current direction because it contains no permanent magnets. When you reverse all the magnetic poles in the motor, the forces between those poles are unchanged.
Why: Without any permanent magnets in it, a universal motor has no reference by which to distinguish a north pole from a south pole. When you reverse the current passing through the motor, all of its poles switch but it can't detect that reversal. It keeps spinning as always.

7. The Lorentz force.
Why: The moving charges in the coil's current experience the Lorentz force as they pass through the magnetic field. This force is conveyed to the wire coil, which is attached to a moveable surface. That surface moves back and forth as the current fluctuates and produces sound. Speakers are clearly an elegant and practical application of the Lorentz force in everyday life.

Check Your Figures—Answers

Section 11.1 HOUSEHOLD MAGNETS

1. Almost 1000 N (225 lbs) in the direction of the field.
Why: According to Eq. 11.1.2, the force exerted on the wrench's north pole is equal to the product of its $1000 A \cdot m$ pole times the 1-T magnetic field. Since $1 T = 1 N/A \cdot m$, that product is 1000 N and points in the direction of the field. Such large forces are not unusual when steel or iron objects are exposed to a strong magnetic field, so be careful working near big magnets!

Section 11.2 ELECTRIC POWER DISTRIBUTION

1. About 640,000 J.
Why: Since 1 T is equivalent to $1 N/A \cdot m$, Eq. 11.2.2 gives us the energy of a 4-T field occupying $0.1 m^3$ as:

$$energy = \frac{(4 N/A \cdot m)^2 \cdot 0.1 m^3}{2 \cdot (4\pi \times 10^{-7} N/A^2)}$$

$$= 640,000 N \cdot m = 640,000 J.$$

Section 11.3 ELECTRIC GENERATORS AND MOTORS

1. 2 N.
Why: As indicated in Eq. 11.3.1, the Lorentz force is proportional to the velocity of a charge. Doubling the current in the coil doubles the velocities of its mobile charges and they experience twice the Lorentz force.

Exercises

1. Is it possible to have two permanent magnets that always attract one another, regardless of their relative orientations? Explain.

2. The magnetostatic forces between two button magnets decrease surprisingly quickly as their separation increases. Use Coulomb's law for magnetism and the dipole character of each button magnet to explain this effect.

3. If you bring two magnetic compasses nearby, they will soon begin attracting one another. Why don't they repel?

4. If you bring a button magnet near an iron pipe, they will soon begin attracting one another. Why don't they repel?

5. If you hold a permanent magnet the wrong way in an extremely strong magnetic field, its magnetization will be permanently reversed. What happens to the magnetic domains inside the permanent magnet during this process?

6. Hammering or heating a permanent magnet can demagnetize it. What happens to the magnetic domains inside it during these processes?

7. If you place a button magnet in a uniform magnetic field, what is the net force on that button magnet?

8. If you hold a magnetic compass in a uniform magnetic field pointing northward, in which direction, if any, is the net magnetic force on the compass?

9. Do more magnetic flux lines begin or end on a button magnet, or are those numbers equal?

10. Compare the number of magnetic flux lines beginning and ending on a plastic strip magnet. Explain.

11. Two plastic strip magnets differ only in how many poles they have per centimeter. One has 2 poles/cm and the other has 4 poles/cm. From which strip's surface do magnetic flux lines extend outward farther?

12. Which of the two plastic strip magnets in Exercise 11 is attracted toward a refrigerator at the greater distance?

13. How could you use iron to prevent the magnetic flux lines from a strong button magnet from extending outward into the room?

14. To keep the strong magnets in a scientific facility next door from sending flux lines through your office, should you line the office walls with aluminum or with iron?

15. Your friends are installing a loft in their room and are using thin speaker wires to provide power to an extra outlet. If they draw only a small amount of current from the outlet, the voltage drop in each of the wires will remain small. Why?

16. When your friends from Exercise 15 plug a large home entertainment system into the outlet, it doesn't work properly because the voltage rise provided by the extra outlet is only 60 V. The power company provides a voltage rise of 120 V, so where is the missing voltage?

17. A particular lightbulb is designed to consume 40 W when operating on a car's 12-V DC electric power. If you supply that bulb with 12-V AC power from a transformer, how much power will it consume?

18. Your toaster consumes 800 W when operating on 120-V AC electric power. If your rugby team is camping and all of you string together flashlight batteries to supply that toaster with 120-V DC electric power, how much power will it consume?

19. To read the magnetic strip on an ID or credit card, you must swipe it quickly past a tiny coil of wire. Why must the card be moving for the coil system to read it?

20. One type of microphone has a permanent magnet and a coil of wire that move relative to one another in response to sound waves. Why is the current in the coil related to the motion?

21. If the primary coil of a transformer has 200 turns and is supplied with 120-V AC power, how many turns must the secondary coil have to provide 12-V AC power?

22. The transformer supplying power to an artist's light sculpture provides 9600 V AC when supplied by 120 V AC. If there are 100 turns in the transformer's primary coil, how many turns are there in its secondary coil?

23. The primary coil of a transformer makes 240 turns around the iron core, and the secondary coil of that transformer makes 80 turns. If the primary voltage is 120 V AC, what is the secondary voltage?

24. The transformer in a stereo amplifier has a primary coil with 200 turns and a secondary coil with 40 turns. When the primary coil is supplied with 120 V AC, what voltage does the secondary coil provide?

25. If an average current of 3 A is passing through the primary coil of the transformer in Exercise 23, what average current is passing through the secondary coil of that transformer?

26. If the average current passing through the secondary coil of the transformer in Exercise 24 is 10 A, what average current is passing through the primary coil?

27. A magnet hanging from a spring bounces in and out of a metal ring. Although it doesn't touch the ring, the magnet's bounce diminishes faster than it would if the ring weren't there. Explain.

28. The high-voltage spark that ignites gasoline in a basic lawn mower engine is produced when a magnetic pole moves suddenly past a stationary coil of wire. From where does that spark's energy come?

29. If you have a coil of wire, a battery, a magnetic compass, and an electrical switch, how could you make the compass needle spin?

30. If you circle a permanent magnet around a magnetic compass, the compass needle will follow along. What is providing the needle with the energy it needs to continue turning despite friction in its pivot?

31. You can't make a motor using only permanent magnets. Why not?

32. You can't make a motor using direct current and electromagnets that doesn't require switches. Why not?

33. If you double the DC voltage supplied to a DC motor, what will happen?

34. If you double the AC voltage supplied to a synchronous AC motor, what will happen?

35. If you supply AC voltage to a brushed DC motor, what will happen?

36. If you supply DC voltage to a synchronous AC motor, what will happen?

37. Some decorative lightbulbs have a loop-shaped filament that jitters back and forth near a small permanent magnet. The filament wire itself isn't magnetic, so why does the filament move when alternating current flows through it?

38. If a flexible wire carrying 60-Hz alternating current runs through the gap between a north and south magnetic pole, what will happen to that wire?

39. Suppose you include an inductor in an electric circuit that includes a battery, a switch, and a lightbulb. Current leaving the battery's positive terminal must flow through the switch, the inductor, and the lightbulb before returning to the battery's negative terminal. The current in this circuit increases slowly when you close the switch, and it takes the lightbulb a few seconds to become bright. Why?

40. When you open the switch of the circuit in Exercise 39, a spark appears between its two terminals. As a result, the circuit itself doesn't open completely for about half a second, during which time the bulb gradually becomes dimmer. The bulb's behavior indicates that the current in the circuit diminishes slowly, rather than stopping abruptly when you open the switch. Why does the current diminish slowly?

41. A television picture tube produces its images using beams of electrons that move through empty space and strike phosphors on the inside of the screen. Those phosphors glow following the impact. But aiming the electron beams is a delicate task. Why is it important to avoid placing strong magnets near a television picture tube?

42. Audio speakers produce motion and ultimately sound by passing fluctuating currents through wires immersed in magnetic fields. Why would this arrangement result in motion and why is it important not to put audio speakers too close to a television picture tube (Exercise 41)?

Problems

1. If a 0.10-A·m magnetic pole is placed in an upward-pointing 1.0-T magnetic field, what force would that pole experience?

2. What is the force on a −2.0 A·m magnetic pole in a forward-pointing 0.20-T magnetic field?

3. If a −1.0-A·m magnetic pole experiences a 1.0-N force downward, what is the local magnetic field?

4. The magnetic force on a 5.0-A·m magnetic pole is 0.010 N to the right. What is the magnetic field in which that pole is immersed?

5. A magnetic pole in an upward-pointing 1.0-T magnetic field experiences a 0.10-N force upward. What is that magnetic pole?

6. A magnetic pole in an upward-pointing 0.10-T magnetic field experiences a 0.10-N force downward. What is that magnetic pole?

7. The earth's magnetic field is approximately 0.000050 T. What is the energy in 1.0 m^3 of that field?

8. The magnet in a large MRI unit may have a 1.0-T magnetic field occupying a volume of 1.0 m^3. How much magnetic field energy is in that volume?

9. What volume of the 1.0-T magnetic field in Problem 8 contains 1.0 J of energy?

10. A 0.050-T magnetic field is typical near household magnets. What volume of that magnetic field contains 1.0 J of energy?

11. What magnetic field is necessary for 1.0 m^3 of that field to contain 1.0 J of energy?

12. What magnetic field is necessary for 1.0 m^3 of that field to contain 10 J of energy?

CHAPTER 12

Electronics

Electric currents and magnetized metals can do more than simply power lightbulbs or attach notes to refrigerators. They can also represent many different things, from sounds to video images to computerized information. Moreover, they can be used to manipulate that information in remarkable ways and at astonishing speeds. In this chapter, we'll see how electricity and magnetism have given rise to the field of electronics.

Electronic devices are tools that use electric currents to perform sophisticated tasks. They first appeared in the early twentieth century with the development of vacuum tubes and have been maturing ever since. Advances in electronics have frequently followed advances in quantum and solid-state physics, so it was hardly surprising that research into the physics of semiconductors should lead to interesting electronics. But the invention of the semiconductor transistor shortly after World War II started a revolution in electronics so profound that no one could have anticipated it. Advanced jointly by physicists and engineers, electronic devices have gradually become so inexpensive and so effective that they now permeate every facet of modern society.

EXPERIMENT: Building an Electronic Kit

You've been using electronic devices all your life, so merely using them some more is unlikely to add to your insight. To get a real feel for how electronic devices work, you do best to build one!

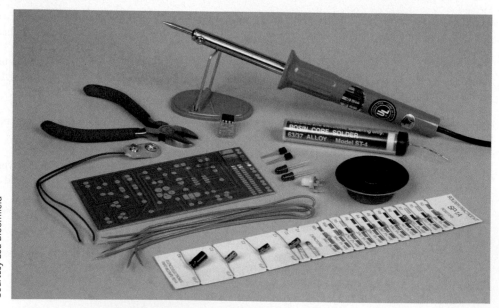

Courtesy Lou Bloomfield

There are many simple kits available over the Internet, easily found by searching for "electronic kits" or "solder practice kits." Soldering is the technique of connecting electronic components with a soft metal that melts easily when heated by a soldering iron. Snap-together experiment kits are also available, but they're more expensive and less like real electronic devices.

Look for a kit that contains some basic electronic components: resistors, capacitors, diodes, and transistors, and be sure to get or borrow a soldering iron and some solder. If you can afford it, buy a simple electronic multimeter to check the behavior of your kit. Even when you're done with the kit, that meter should come in handy. It is to the electrician or electronics technician what a ruler, square, and level are to a carpenter or surveyor.

Find a safe, well-lighted work space and read the kit instructions carefully before you begin. You'll be inserting the wire ends or "leads" of the various electronic components into holes in a printed circuit board—an insulating card bearing photographically printed metal strips. Those strips connect the kit's components, although you will have to complete those connections using solder and your soldering iron. Be careful not to burn yourself as you melt the solder into place. Make sure that the solder *wets* each metal surface—that it spreads out like water on paper towel rather than beading up like water on waxed paper.

As you assemble your kit, note the differences between the components. Some of the components must be inserted in the right orientation. Why is that? Those same directional components are usually heat-sensitive and mustn't be warmed too long with the soldering iron. What might elevated temperatures do to those components?

As you build your device, try to *predict* how current will flow through its various components and what the voltages will be at various points in the device. How might those currents and voltages change with time? From the instructions and your intuition, try to determine what role each component plays in the overall device.

When the kit is complete, turn it on and *observe* how it works (hopefully). If you have a multimeter, select the "DC Voltage" setting and *measure* the voltages at various points in the device. To do that, you'll have to give the meter a reference for 0 V by connecting its black wire to the battery's negative terminal and then touch the meter's red wire to the spot whose voltage you wish to measure. Did you *verify* your predictions?

Chapter Itinerary

To understand the remarkable properties of electronic devices, we'll examine two of them: (1) *power adapters* and (2) *audio players.* In *power adapters,* we'll look at how these simple devices use power obtained from the relatively high-voltage alternating current of electrical outlets to provide the relatively low-voltage direct current required by most electronic equipment. In *audio players,* we'll see how a small assortment of basic electronic components are brought together to build a computer and an audio amplifier, and how those two devices have been merged together into a single unit so that you can listen to thousands of songs as you lounge on the beach. For additional preview information, turn ahead to the Chapter Summary on p. 417.

While this brief survey can't go beyond the most basic issues in electronics, it should provide a good basis on which to build. Once you understand how a few important electronic components work, it's not so hard to see how to combine them in ways that no one else has ever imagined. Like building blocks that control the movements of charged particles, electronic components can be put together to do almost anything.

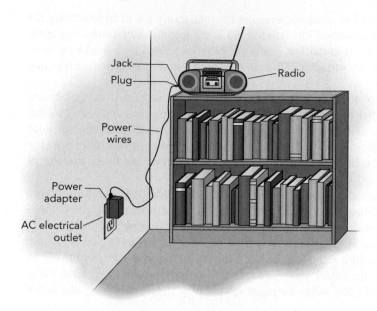

SECTION 12.1 **Power Adapters**

Virtually every electronic device you buy comes with its own power adapter, a small black cube or brick that either plugs directly into an electrical outlet or has a cord that does. These adapters obtain power from the century-old electrical grid and prepare it for this week's gadgets. Power adapters range widely in the voltages and currents they supply, the degree to which their voltages are smoothed and regulated, and even the connectors through which they deliver their power. The result is dozens of adapters that are rarely interchangeable and, consistent with Murphy's law, the adapter you need is always the one that fell behind the bookshelf.

But despite their differences, these power adapters all perform essentially the same conceptual task: they take relatively high-voltage alternating current and use its power to prepare relatively low-voltage direct current. They do this conversion efficiently, inexpensively, and with a high degree of reliability. In this section, we'll see how simple power adapters perform that conversion.

Questions to Think About: Why can't most electronic devices operate from alternating current? Why do those devices need direct current that's supplied at a certain voltage? What is the relationship between the batteries a device uses and the voltage of its power adapter? When the voltage and current of an electrical outlet reach zero during a reversal of its alternating current, what might happen to the output of the power adapter? What might the adapter do to endure that reversal?

Experiments to Do: Unless you live in a log cabin, you probably have several power adapters sitting in your closet or under your bed. Take a look at those adapters. The simplest ones will be cubes that plug directly into electrical outlets. Examine them to see that each provides a specific DC voltage with a specified maximum current. Furthermore each expects a narrow range of input voltages, such as 120 V AC in the United States or 230 V AC in Europe. What would happen if you plugged one of these adapters into the wrong electrical outlet? [Warning: Don't try it! A 120-V AC adapter will destroy both itself and the attached device when plugged into a 230-V AC outlet, while a 230-V AC adapter will survive but will fail to operate its device properly.]

Adapters that resemble bricks with power cords are usually more complicated than the wall-cube adapters. Used with computers and other more sophisticated devices, these bricks generally regulate their output voltages carefully. Furthermore, they can typically handle a wider range of input voltages, such as 100 V AC to 240 V AC. What makes accepting such a range of input voltages so challenging?

Some of your adapters may have lights to indicate when they're working. Plug these adapters in and then unplug them. Do their lights go out immediately? If not, why not?

Finally, if you have a cube adapter that you no longer need, carefully open it up. Be safe: unplug it first and don't cut yourself. Inside it, you'll find a transformer and several semiconductor devices called diodes. You may also find a storage device called a capacitor. Together, these components can prepare low-voltage DC from high-voltage AC. And that's what this section is all about.

Producing DC Power from AC Power

While electrical devices such as toasters and incandescent lightbulbs can operate from either AC or DC power, most electronic devices require DC power. There are two reasons for their pickiness. First, they contain sophisticated electronic components that are sensitive to the direction of current flow and won't operate properly if that current is reversed. Second, electronic devices typically need a continuous supply of power and can't tolerate the brief moments of powerlessness that occur when alternating current is reversing.

Addicted as they are to direct current, electronic devices operate beautifully on batteries. Of course, those batteries must be installed in the right direction or they'll act to push current the wrong way. But batteries and electronics are practically made for each other.

Unfortunately, batteries eventually run out of stored energy and must be replaced or recharged. And some electronic devices are so power-hungry that operating them on batteries would cost a fortune. Unless you're an heiress or the nephew of a battery manufacturer, you often need a cheaper, more practical source of DC power. You need a power adapter.

A power adapter's task is simple: it uses AC power from an electrical outlet to provide DC power to an electronic device. More specifically, it delivers current to the electronic device through a positive power wire and receives that current back through a negative power wire. And the adapter maintains a specified average voltage rise from the negative wire to the positive wire, as though the adapter were a battery.

The steadiness of an adapter's voltage rise depends on its sophistication. Some adapters let their voltage rises fluctuate with the AC cycle while others use complicated electronics to regulate their voltage rises precisely. Intermediate between those two extremes are adapters that use simple electronics to smooth out most of the fluctuations due to the AC cycle, but make no further attempt to regulate their voltages. In this section, we'll examine these intermediate power adapters.

To be specific, we'll examine the components in a 9-V DC power adapter (Fig. 12.1.1) that operates from 120-V AC power and supplies power to an ordinary radio. Apart from wires, this adapter contains only a few components: a transformer, four diodes, and a capacitor. To understand this power adapter, we need only examine those components and how they interact. And since we studied transformers in Section 11.2, we already have a head start.

Figure 12.1.1 These two power adapters use 120-V AC power from an electrical outlet to produce 9-V DC power for a radio. The adapter on the right has been removed from its plastic shell to reveal a transformer (top), two black diodes (lower left) and a cylindrical capacitor (lower right).

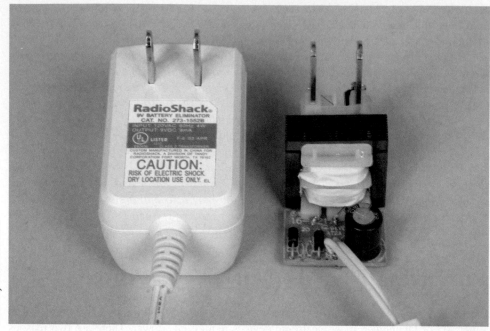

Courtesy Lou Bloomfield

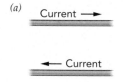

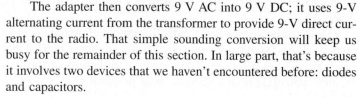

Figure 12.1.2 (*a*) A wire can carry current in either direction. (*b*) A diode can carry current only in the direction shown symbolically by its arrow. (*c*) In a schematic diagram of an electronic device, the diode is represented by an arrow and bar.

When you plug the transformer into an electrical outlet, its primary coil forms a circuit with the power company and an alternating current flows through it. The primary coil has $13\frac{1}{3}$ times as many turns as the secondary coil, so the secondary coil provides an induced emf of 9 V AC.

The adapter then converts 9 V AC into 9 V DC; it uses 9-V alternating current from the transformer to provide 9-V direct current to the radio. That simple sounding conversion will keep us busy for the remainder of this section. In large part, that's because it involves two devices that we haven't encountered before: diodes and capacitors.

Diodes are one-way conductors of electric current. Unlike a wire, which carries current equally well in either direction (Fig. 12.1.2*a*), a diode allows current to flow through it only in one direction (Fig. 12.1.2*b*). It acts as a conductor when current tries to flow the allowed direction and as an insulator when current tries to flow the forbidden direction. This switching behavior is what the power adapter uses to obtain direct current from alternating current.

Capacitors are devices that store separated electric charges. By accumulating positive charge on one side and negative charge on the other side, a capacitor stores both charge and energy. That storage helps the adapter steady its voltage rise and endure the moments when the reversing alternating current provides no power to the adapter.

Check Your Understanding #1: Safety First

for answers, see page 418

A radio operates from a power adapter that's plugged into an electrical outlet. What fraction of the current flowing through the radio also flows through the electrical outlet?

A Few Words about Quantum Physics

Since a diode is sensitive to the direction of current flow, it can't be symmetric. It must have two distinct and different ends. To understand those two ends and how they differ, we need to examine the materials from which they're both made. We need to examine semiconductors.

Semiconductors are materials with properties intermediate between those of electrical conductors and those of electrical insulators. While charge is *always* mobile in conductors and *almost never* mobile in insulators, charge is *sometimes* mobile in semiconductors. In general, a semiconductor acts like an insulator when it is cold, pure, and in the dark, and it acts like a conductor when it is hot, impure, or exposed to light. Semiconductors are so important to diodes and to most of modern electronics that we'll spend the next several pages learning how charge moves through them. And we'll begin by discussing quantum physics, which has immense influence over the tiny particles from which semiconductors are built.

As quantum physics gradually revealed itself to the scientists of the early twentieth century, they found the experience both exhilarating and disorienting. Prior to that era, the physical world seemed to divide neatly into particles and waves: scientists viewed an electron only as a particle and light only as a wave. However, one of the most basic observations of quantum physics, and the one most relevant to our present topic, is that everything has both particle and wave characteristics. Put simply, everything begins and ends as a particle, but travels as a wave.

For an electron, the quantum surprise is that it travels as a wave. For light, the quantum surprise is that it is emitted and absorbed as a particle. Called the **wave–particle duality,** this observation that everything in nature has both particle and wave characteristics has left few areas of physics unaffected. But while quantum physics is now a basic and essential part of nearly all modern physics research, its effects are subtle and often nonintuitive. They are most apparent in the microscopic world and are visible to us only indirectly. No wonder they seem so strange.

We'll encounter quantum physics and its effects several times in the next few chapters. It figures prominently in the electronic properties of semiconductors, in the light emitted by atoms and lasers, and in the radioactive decays that release nuclear energy. Our examination of semiconductors will acquaint us with the wave nature of electrons and show how the wave phenomena that we studied in Chapter 9 apply to quantum physics. Our exploration of light from atoms and lasers will acquaint us with the particle nature of light and how the collision effects we explored in Chapters 1–3 apply to quantum physics. And in our examination of radioactivity, we'll uncover particle and wave effects that we would not even have anticipated without quantum physics. With each encounter, we'll take a small bite of the quantum apple—looking at how quantum effects manifest themselves in our everyday world.

▶ Check Your Understanding #2: Wavy Atoms

for answers, see page 418

Do atoms ever exhibit wave properties?

Electrons in Solids

We learned in Chapter 10 that metals conduct electricity because they contain mobile electrons and that insulators don't conduct electricity because none of their electrons are mobile. Now it's time to see what controls electron mobility. As you might guess, the explanation lies in quantum physics.

In a nonquantum world, an electron in a solid would travel only as a particle and it would be able to move at any speed along any path. But ours is a quantum world and the electron does not travel as a particle at all; it travels as a wave. And like the waves on a violin string, drumhead, or basin of water, the electron waves in a solid have limited possibilities.

In Chapter 9, we observed that the most basic mechanical waves on a limited object are all standing waves—waves that effectively oscillate in place. This rule also applies to quantum waves. The electrons in a solid are best understood as standing waves in that solid. Each electron wave extends across part or all of the solid and has such wave characteristics as wavelength and frequency. Unlike a vibrating string or drumhead, the electron is a three-dimensional wave and its oscillation is internal. But it's still a wave.

This wave character has profound effects on the electronic structure of solids. Most significantly, it limits what electrons can do in those solids. We saw in Chapter 9 that a violin string's one-dimensional standing waves consist only of its fundamental mode (Fig. 9.2.3) and its harmonic modes (Fig 9.2.4) and that a drumhead's two-dimensional standing waves consist only of its fundamental mode and overtones (Fig. 9.2.9). Similarly, a solid's three-dimensional electron standing waves consist only of a fundamental mode and overtones. And while there are a great many overtone modes available, their possibilities are nonetheless limited.

The electron standing waves in solids are often called **levels**—a recognition that each standing wave has an amount or "level" of energy associated with it. The electron standing waves in atoms, another group of limited systems in which electrons exist as standing waves, are called orbitals—a nod to the orbiting nature of an atom's electrons. We'll see when we examine discharge lamps in Section 14.2 that each atom's limited orbital choices determine the colors of light it can emit or absorb. And we'll see in the present section that a solid's limited level choices determine its electrical conductivity.

Another remarkable observation of quantum physics is that every indistinguishable electron must have its own level or orbital, its own unique quantum wave. This law is called the **Pauli exclusion principle,** after its discoverer, Wolfgang Pauli ❑. The principle applies to a whole class of subatomic particles, the **Fermi particles,** that includes all of the basic constituents of matter: electrons, protons, and neutrons. Two indistinguishable Fermi particles can never be in the same quantum wave.

❑ Austrian physicist Wolfgang Pauli (1900–1958) rose to fame at 21 by writing an article on relativity that impressed even Einstein. He went on to discover the exclusion principle, a fundamental part of quantum theory. He was renowned for his intensely critical attitude toward new ideas, considering them all "rubbish" until convinced otherwise. Pauli was also quite interested in psychology, corresponding with Carl Gustav Jung and writing a number of articles on the subject.

The Pauli Exclusion Principle
No two indistinguishable Fermi particles ever occupy the same quantum wave.

However, a peculiar property of electrons allows two of them to share an orbital or level. Electrons have two possible internal states, usually called spin-up and spin-down. Because a spin-up electron is distinguishable from a spin-down electron, one spin-up electron and one spin-down electron can share a single orbital or level. However, two electrons is the absolute maximum allowed by quantum physics and the Pauli exclusion principle.

Despite being a wave, an electron in a level has a specific total energy—the sum of its kinetic and potential energies. That energy, which depends on the shape and structure of the electron wave, also determines the wave's oscillation frequency. According to quantum physics, the electron's total energy and the oscillation frequency of its wave are exactly proportional to one another; a low-energy electron oscillates slowly while a high-energy electron oscillates quickly. We'll return to this observation in Chapter 14, but for now we'll note that each level, each quantum standing wave in the solid, has a specific frequency and energy. The solid's fundamental standing wave has the lowest frequency and energy, while the overtone waves have progressively higher frequencies and energies.

Physicists have come to view these standing waves as abstract placeholders, independent of the electrons that may or may not exhibit them at a given moment. The levels are then analogous to seats in a theater, each of which may or may not be occupied right now. Instead of saying that the *electron* is experiencing a particular standing wave or level, we say that the *level* is occupied by an electron. In this reversed perspective, the level plays the more important role. For the rest of this section, we'll view a solid in this manner; we'll think in terms of the levels it has available and whether or not those levels are occupied by electrons.

A solid contains an enormous number of electrons and there are always plenty of levels around to accommodate those electrons. But which levels do they occupy?

At sufficiently low temperature, electrons occupy those levels which have the least energy. For thermodynamic reasons, the electrons settle into the lowest energy levels available, two electrons per level. By the time all the electrons have been accommodated, they fill the levels up to a certain maximum energy. Halfway between the highest filled level and lowest unfilled level is the **Fermi level**—a hypothetical level that defines the top of this *Fermi sea* of electrons. The energy an electron would have in that hypothetical level is called the **Fermi energy.**

Our theater analogy can again provide insight into this level-filling process. It is analogous to what happens at a popular show: the seats fill from the orchestra level on up—everyone wants to sit in the lowest (and closest) seat. When the show starts, people have filled all the seats up to a certain highest seat. Halfway between that last filled seat and the next unfilled seat is the hypothetical Fermi seat.

If we represent the levels graphically by boxes and arrange them vertically according to energy (Fig. 12.1.3), then levels (boxes) below the Fermi level contain two electrons each, while those above the Fermi level are empty. Although thermal energy

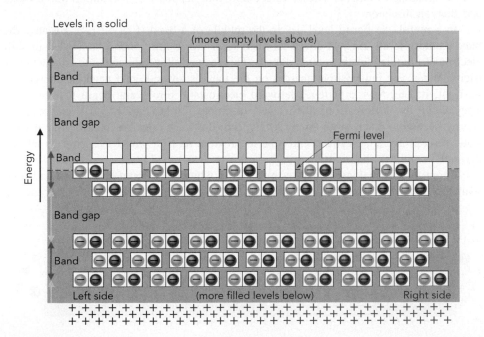

Figure 12.1.3 The levels in a solid are grouped together in bands and filled from the lowest energy level up to the Fermi level. Each level can hold at most two electrons: one spin-up electron (*blue*) and one spin-down electron (*red*). The positive charges shown at the bottom are the nuclei of atoms that make up this electrically neutral solid.

complicates this picture somewhat by shifting electrons about near the Fermi level, we can ignore that detail near room temperature or below.

Since levels are standing waves, they don't have sharply defined locations in space. But we can safely imagine that each level places its electrons near a particular location in the solid, as shown in Fig. 12.1.3. While this picture is somewhat over-simplified, it's accurate enough to illustrate much of the physics of charge motion in materials.

Of course, electrons aren't the only charged particles in a solid. The atoms also have positively charged nuclei. But those nuclei are essentially immobile and rarely participate in the flow of electricity. Instead, they form a uniform background of positive charge, shown schematically as pluses (+) at the bottom of Fig. 12.1.3, so that the object is roughly neutral throughout.

> **Check Your Understanding #3: Taking It to a Higher Level**
> *for answers, see page* **418**

If you add one extra electron to an otherwise neutral metal ball, into which level will that electron go?

Metals, Insulators, and Semiconductors

The levels in a solid occur in groups called **bands.** Each band corresponds to standing waves with a particular type of structure. Since the levels in a band involve similar waves, they also involve similar energies. Between these bands of levels there are sometimes **band gaps**—ranges of energy in which no levels exist. The solid does not and cannot contain electrons with energies that lie within a band gap.

Bands and band gaps are what distinguish metals, insulators, and semiconductors. When the Fermi level is located in a band gap, it can prevent the electrons in a solid from responding to outside forces. To see how that happens, let's examine first a metal and then an insulator.

In a **metal,** the Fermi level lies in the middle of a band (Fig. 12.1.4). Because the band's empty levels are just above its filled levels, very little energy is needed to shift electrons from filled levels to empty levels. This feature allows the metal to conduct electricity. When you put positive charges on the metal's left side and negative charges on its right, its electrons experience leftward electrostatic forces and begin to move left. They move by shifting from filled levels to empty levels (Fig. 12.1.4), obtaining the energy needed to reach those empty levels from the work done on them by the electrostatic forces. Overall, electrons enter the metal from its right and leave from its left, so the metal conducts electricity!

In our theater analogy, a metal is a theater in which only about half the ground-floor seats are filled. If you ask people in the theater to begin shifting left, those near the top

Figure 12.1.4 In a metal, the Fermi level lies in the middle of a band. When you put electrons to the metal's right and positive charges to the metal's left, electrons shift leftward through the metal with the help of the empty levels. Since there is a net flow of charges through the metal, it is conducting electricity.

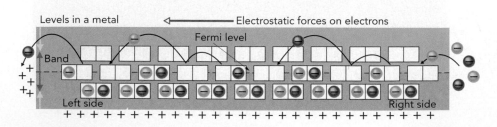

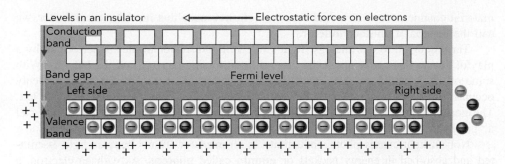

Figure 12.1.5 In an insulator, the Fermi level lies in the middle of a band gap. When you put charges near the insulator, electrons in the filled band can't shift to produce a net flow of charge through the insulator. The insulator can't conduct electricity.

of the occupied seats can shift about easily. Each finds an empty seat nearby on the left and moves over. New people are then able to enter the theater from the right while others leave the theater from the left. This "metal" theater would be "conducting" people.

Unlike the situation in a metal, an **insulator's** Fermi level lies in the middle of band gap, between the top of one band and the bottom of another band (Fig. 12.1.5). With no easily accessible empty levels available, a great deal of energy is required to shift electrons from filled levels to empty levels. When you put positive charges on the insulator's left side and negative charges on its right, its electrons experience leftward electrostatic forces but are unable to move. To shift into one of the empty levels in the upper band, an electron in the lower band would need more energy than it can get from the electrostatic forces. Since no net charge flows through the insulator, it doesn't conduct electricity!

In our theater analogy, an insulator is a theater in which all the ground-floor seats are full and in which the balcony seats are empty. When you ask people in this theater to begin shifting left, they can't do it. All of the ground-floor seats to the left are filled and they can't reach the balcony to make use of its empty seats. This "insulator" theater would be unable to conduct people.

In a metal, the band of levels containing the Fermi level is only partially filled, and electrons can easily shift from filled to unfilled levels. In an insulator, the band below the Fermi level—the **valence band**—is full and the band above the Fermi level—the **conduction band**—is empty, making such shifts extremely difficult.

But even in an insulator, an electron can shift from a **valence level** (a level in the valence band) to a **conduction level** (a level in the conduction band) if something provides the necessary energy. One such energy source is light. When an insulator is exposed to the right type of light, that light can shift electrons from the material's valence band to its conduction band (Fig. 12.1.6).

Once electrons appear in the normally empty conduction band and empty levels appear in the normally full valence band, electrons can respond to electrostatic forces. They can shift from filled levels to nearby empty levels and thus travel through the material. Electrons can then enter the material from one side and leave from the other, so the

Figure 12.1.6 When light strikes an insulator, the energy in its photons shifts some electrons from filled valence levels to empty conduction levels. Such shifts make it possible for electrons to move in response to electrostatic forces, so the insulator becomes an electrical conductor, a photoconductor.

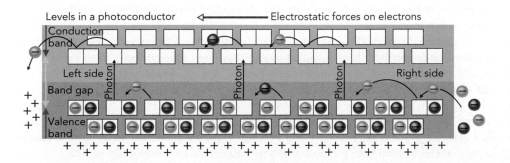

material conducts electricity. And because light has made this insulator a conductor, we call the material a **photoconductor.**

Turning again to our analogy, light's role in the insulator theater is performed by a playful gorilla that walks about the ground floor, tossing patrons into the balcony. With some of the ground-floor seats suddenly empty and some of the balcony seats suddenly occupied by dazed theatergoers, the crowd can now respond to your request to move left. The gorilla has made the insulator theater a conductor of people—what you might call a "gorillaconductor."

Not all light causes photoconductivity in an insulator. That's because light is emitted and absorbed in energy packets or **quanta** called **photons.** As with an electron, a photon's energy is proportional to its frequency; the higher the frequency of light, the more energy each of its photons contains. To shift an electron across the large band gap in a typical insulator, high-energy, high-frequency light is needed; the insulator must be exposed to violet or even ultraviolet light.

But nature also provides materials with smaller band gaps that can be crossed with the help of low-energy, low-frequency red or even infrared light. These materials are called **semiconductors** because their properties lie somewhere between those of conductors and insulators. Semiconductors have small band gaps, making it relatively easy for light, heat, or other types of energy to shift electrons between valence and conduction levels. In our analogy, a semiconductor theater is an insulator theater with a low balcony, so that even a baby gorilla can toss people into it.

For half a century, scientists and engineers have worked with semiconductors to produce an astonishing array of electronic devices. By carefully tailoring the shapes and chemical compositions of semiconducting materials such as silicon, germanium, and gallium-arsenide, they have created virtuoso instruments for electron waves in solids that are every bit as remarkable as the instruments for musical waves found in great orchestras. And of all these electronic instruments, the simplest is the semiconductor diode.

> ►Check Your Understanding #4: Stop and Go Shopping
for answers, see page 418

In many grocery checkout counters, a conveyor belt carries food to the register but stops when the food reaches the end and blocks a beam of light. How might a photoconductor be used to sense this blockage?

Diodes

A **diode** is a one-way device for current; it allows current to flow through it in one direction but not in the other. Although diodes have taken many forms over the years, the diodes in power adapters and virtually all modern electronic devices are built from semiconductors.

A semiconductor diode is made by joining together two different semiconductors. These two semiconductors have been modified so that they don't have perfectly filled valence levels and perfectly empty conduction levels. Instead, they're **doped** with atomic impurities that either create a few empty valence levels (**p-type semiconductor,** Fig. 12.1.7a) or place a few electrons in the conduction levels (**n-type semiconductor,** Fig. 12.1.7b). These empty valence levels or conduction level electrons allow p-type and n-type semiconductors to conduct electricity. The doping atoms bring with them just the right amount of positive charge in their nuclei to keep both p-type and n-type semiconductors electrically neutral.

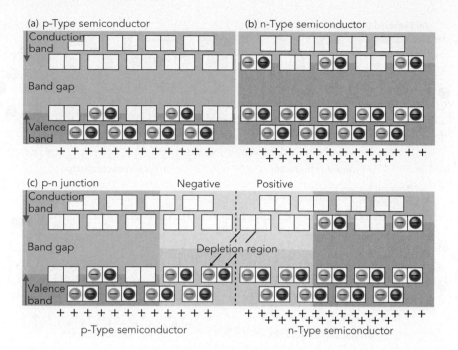

(a) p-Type semiconductor

(b) n-Type semiconductor

(c) p-n junction   Negative   Positive

p-Type semiconductor   n-Type semiconductor

Figure 12.1.7 (*a*) A p-type semiconductor has missing electrons (and missing positive atomic nuclei) and can conduct electricity through its partly filled valence band. (*b*) An n-type semiconductor has extra electrons (and extra positive atomic nuclei) and can conduct electricity through its partly filled conduction band. (*c*) When p- and n-type semiconductors touch, conduction level electrons shift from the n-type semiconductor to the p-type semiconductor, creating a thin, electrically polarized depletion region at the p-n junction.

But when a piece of p-type semiconductor touches a piece of n-type semiconductor, something remarkable happens: a **p-n junction** forms at the place where the two meet (Fig. 12.1.7*c*). To reduce their potential energies, higher energy conduction level electrons from the n-type semiconductor flow across the p-n junction and fill in empty lower energy valence levels in the p-type semiconductor. This electron flow creates separated charge. The n-type semiconductor acquires a positive net charge because it now has fewer electrons than positive charges. The p-type semiconductor acquires a negative net charge because it now has more electrons than positive charges. Electrostatic forces from this separated charge oppose the further flow of electrons across the junction and gradually bring that flow to a halt. Everything is then in equilibrium.

Near the p-n junction, there is now a **depletion region**—an area in which electron flow has emptied all of the conduction levels and filled all the valence levels. With no conduction level electrons or empty valence levels left, the depletion region can't conduct electricity and charge can't move across the p-n junction. The depletion region is an insulator, and the two pieces of semiconductor have become a diode.

In our theater analogy, the p-n junction is analogous to a theater with two halves. In the left or "p-type" half, the balcony is empty, and even the ground floor has some empty seats. In the right or "n-type" half, the ground floor is filled and there are even a few people in the balcony. Since these two halves touch, people in the right balcony notice the empty seats in the left ground floor, and a few of them near the center of the theater clamber down from the right balcony to the left ground floor to take advantage of the better seats. Near the center of the theater, the ground floor is now filled and the balcony is empty, forming a depletion region in which no one can move left or right. The theater can't conduct people!

Let's now look at what happens when we attach wires to each semiconductor half and try using a battery to push electrons across the p-n junction. If we push electrons leftward, adding them to the n-type side and removing them from the p-type side, the depletion region becomes thinner and eventually vanishes (Fig. 12.1.8*a*). We're adding electrons to the n-type conduction levels and pushing them toward the p-n junction. We're also

Figure 12.1.8 (*a*) When you add electrons to the n-type side of a p-n junction and remove them from the p-type side, the depletion region vanishes and current can flow across the junction. (*b*) When you add electrons to the p-type side and remove them from the n-type side, the depletion region thickens and no current can flow across the junction.

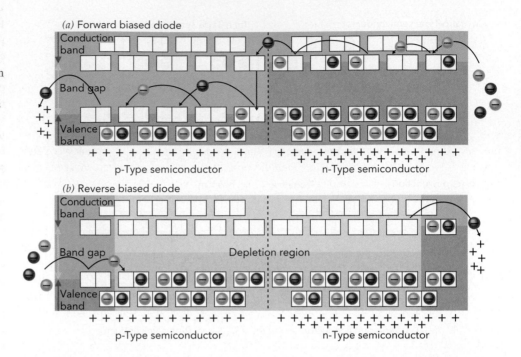

removing electrons from the p-type valence levels and pulling them away from the p-n junction.

The extra electrons on the n-type side and the missing electrons on the p-type side create a voltage difference between the diode's two halves. When the voltage of the p-type side reaches about 0.6 V above the voltage of the n-type side, the depletion region in a silicon diode disappears. Conduction level electrons in the n-type material then flow leftward across the junction and drop into empty valence levels in the p-type material. The p-n junction is conducting electric current.

In the theater analogy, we're adding people on the right to the n-type balcony and removing them on the left from the p-type ground floor. The new people in the n-type balcony can move about the empty seats and migrate toward the center of the theater. Similarly, the empty seats in the p-type ground floor allow people to shift about so that empty seats become available near the center of the theater. At that point, people in the n-type balcony can cross over to the p-type balcony and then climb down to the ground floor. There is a net leftward flow of people through the theater; the theater is conducting people from right to left.

But what happens when we try to send electrons backward through the diode, pushing them into the p-type side and removing them from the n-type side (Fig. 12.1.8*b*)? In that case, the depletion region becomes thicker as we fill in empty valence levels in the p-type side and remove conduction level electrons from the n-type side. The widening depletion region prevents charge from moving and no current flows across the p-n junction. It remains an insulator.

In the theater analogy, we're removing people on the right from the n-type balcony and adding them on the left to the p-type ground floor. Soon the n-type balcony is virtually empty and the p-type ground floor is essentially full. The entire theater is now a depletion region and behaves like the insulator theater. No one can move and the theater can't conduct people.

Since it allows current to flow in one direction but not the other, the p-n junction is a diode. For historical reasons, the diode's p-type side is called the *anode* and its n-type side is called the *cathode*. Current, which is the flow of positive charge, can only pass through a diode from its anode to its cathode. Since current naturally flows from higher voltage to lower voltage, a diode carries current only when it is **forward biased,** that is, when its anode has a higher voltage than its cathode. When it is **reverse biased**—when its anode has a lower voltage than its cathode—no current flows through the diode. This ability to control the direction of current flow is fundamental to converting alternating current into direct current.

Even when a diode is forward biased, the depletion region won't vanish until the voltage of the anode is significantly higher than the voltage of the cathode. For example, an ordinary silicon diode must have a voltage drop of about 0.6 V in order to conduct current. Since the current passing through the diode is losing voltage, each charge is losing energy in the diode. Although that missing energy is normally wasted as heat, in Chapter 14 we'll examine diodes that use this missing energy to emit light.

The production of waste heat is a problem for diodes and other semiconductor components because it can cause those components to overheat. Thermal energy, like light, can shift electrons from a semiconductor's valence levels to its conduction levels, so that it conducts a small amount of current. Though insignificant near room temperature, this thermally induced conductivity increases as the semiconductor gets hotter. Above a certain temperature, currents flowing due to thermally induced conductivity lead to even more heating and eventually to a runaway thermal catastrophe. To avoid such thermal accidents, which usually involve smoke and great unhappiness, semiconductor electronic devices must not operate too hot and are often cooled by fans.

Courtesy Lou Bloomfield

Figure 12.1.9 Capacitors store separated electric charge. Each of these capacitors contains two conducting surfaces separated by a thin insulating layer.

> ➤ Check Your Understanding #5: It's a One-Way Street
> *for answers, see page* **418**
>
> What will happen if you include a p-n junction (a diode) in the AC circuit that connects the power company to your table lamp?

Capacitors

The other new electronic component in the power adapter is its **capacitor,** a device that stores separated electric charge (Fig. 12.1.9). A capacitor consists of two conducting plates separated by a thin insulating layer (Fig. 12.1.10). When one plate is positively charged and the other is negatively charged, the opposite charges attract one another. This attraction allows the plates to store large quantities of separated charge, while leaving the capacitor as a whole electrically neutral.

You can charge a capacitor's plates by transferring charge from its negative plate to its positive one. The work you do during this transfer is stored in the capacitor as electrostatic potential energy and is released when you let the separated charge get back together. A charged capacitor acts like a battery when connected to a circuit because it pushes charge through the circuit from its positive plate and collects that charge from the circuit with its negative plate.

Since (positive) charge has more electrostatic energy on the positive plate than on the negative plate, the voltage of the positive plate is higher than the voltage of the negative plate. The voltage difference between the plates is proportional to the separated charge on them; the more separated charge the capacitor is holding, the larger the voltage difference between its plates.

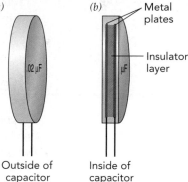

(a)   *(b)*   Metal plates

.02 μF   μF   Insulator layer

Outside of capacitor   Inside of capacitor

(c)

Symbol for capacitor

Figure 12.1.10 (*a*) A capacitor is usually a disk or cylinder with two protruding wires. Its capacitance is printed on its surface. Inside (*b*), the wires are connected to two conducting plates that are separated by a thin insulating layer. (*c*) In a schematic diagram of an electronic device, the capacitor is represented by two parallel lines.

This voltage difference also depends on the structure of the capacitor. Enlarging the plates allows the like charges on each plate to spread out, so that they repel one another less strongly. Thinning the insulating layer between the plates allows the opposite charges on the two plates to move closer together, so that they attract one another more strongly. Both of these changes lower the separated charge's electrostatic potential energy and, consequently, the voltage difference between the plates.

Because these changes allow the capacitor to store separated charge more easily, they increase its **capacitance,** that is, the separated charge it holds divided by the voltage difference between its plates. The SI unit of capacitance is the coulomb-per-volt, also called the **farad** (abbreviated F). A capacitor with a farad of capacitance stores an incredible amount of separated charge, even at a low voltage difference, while a capacitor with a billionth of a farad of capacitance is much more typical. The Greek letter μ in front of the F means millionths (μF or microfarads), the letter n in front of the F means billionths (nF or nanofarads), and the letter p in front of the F means trillionths (pF or picofarads). A capacitor's capacitance is marked on its wrapper, often in an abbreviated form.

► Check Your Understanding #6: Charging Up

for answers, see page **418**

Your stereo's amplifier obtains power from the power line. Since the power line delivers no power during the moments when its alternating current reverses direction, the stereo's power supply must store energy. How can it do this?

The Complete Power Adapter

Our power adapter uses a transformer, four diodes, and a capacitor to provide low-voltage direct current to the radio. The connections between these components are shown photographically in Fig. 12.1.11*a* and schematically in Fig. 12.1.11*b*. Engineers and scientists often use such **schematic diagrams** to represent complicated electronic devices, assigning a specific symbol to each electronic component and denoting the electrical connections between those components with lines.

Let's examine Fig. 12.1.11 from left to right. The leftmost component is the transformer. With its primary coil carrying alternating current from a 120-V AC outlet, its secondary coil develops an induced emf of 9 V AC. The secondary coil acts as a source of 9-V AC power.

If you were to connect a 9-V lightbulb to the two ends of this secondary coil, the coil's alternating induced emf would push an alternating current through the bulb's filament and the bulb would light up. But if you were to connect the 9-V radio to this secondary coil, the coil's alternating emf would try to push an alternating current through that radio and it wouldn't work. The transformer alone can't supply the steady direct current the radio requires.

The diodes in the middle of Fig. 12.1.11 solve the alternation problem by steering current so that it always flows through the positive power wire to the radio and returns from the radio through the negative power wire. Although the transformer's secondary coil continues to experience an alternating induced emf and carries an alternating current, diodes D1 and D2 guide current from the secondary coil to the radio and diodes D3 and D4 guide the returning current back to the secondary coil. Although the transformer itself is supplying AC power, the diodes allow the power adapter to supply DC power.

However, the diodes alone can't supply *steady* DC power. As the secondary coil's induced emf alternates, the voltage rise from the power adapter's negative power wire to

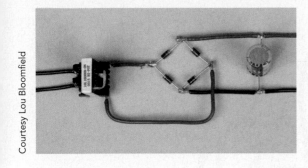

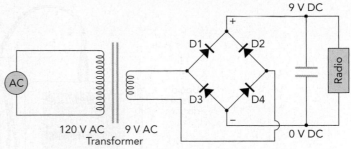

Figure 12.1.11 (*a*) A particular power adapter consists of a transformer, four diodes, and a capacitor. (*b*) The same components shown schematically.

its positive wire fluctuates up and down. That voltage rise peaks at about 12 V when the induced emf reaches its most positive or most negative value, but drops to zero while the induced emf changes sign. The transformer and diodes alone supply *pulsed* 9-V DC power—a form of direct current that offers the same average power as 9 V DC, but with severe voltage fluctuations. Since the radio will turn off between pulses, pulsed DC power won't do.

The capacitor solves the pulsing problem by storing charge and energy whenever the pulsed DC power is near its peak and releasing that charge and energy when the pulsed DC power is near its minimum. The capacitor acts like a rechargeable battery, charging whenever power is plentiful and discharging whenever its not.

When the secondary coil's induced emf is strong, one of the diodes D1 or D2 is forward biased and guides current from the secondary coil to both the radio and the capacitor's positive terminal. At the same time, one of the diodes D3 or D4 is forward biased and guides current from both the radio and the capacitor's negative terminal back to the secondary coil. The radio receives the DC power it needs and the capacitor gradually accumulates more separated charge. The voltage difference between the capacitor's terminals increases and it "charges."

When the secondary coil's induced emf is weak, all four diodes are reverse biased and current stops flowing through the coil. The capacitor takes over the job of supplying power to the radio. Current flows from the capacitor's positive terminal through the positive power wire to the radio and it returns from the radio through the negative power wire to the capacitor's negative terminal. The radio still receives the DC power it needs, but now the capacitor gradually loses its separated charge. The voltage difference between its terminals decreases and it "discharges."

By charging and discharging in this manner, the capacitor "filters" the pulsed DC provided by the transformer and diodes. The capacitor smoothes away both the voltage peaks and the voltage zeros. Although the voltage of this filtered DC power fluctuates slightly as the capacitor charges and discharges, the radio can operate from it. Like many electronic devices that accept power adapters, the radio has its own system for accommodating minor voltage fluctuations so that they don't affect its performance. As long as the radio receives a continuous current with approximately the correct voltage, it will work as though on batteries.

> **Check Your Understanding #7: Slow Waves**
>
> *for answers, see page* **418**

Power adapters in the United States operate from 60 Hz AC while those in Europe operate from 50 Hz AC. In which region do power adapters have more trouble maintaining constant output voltages?

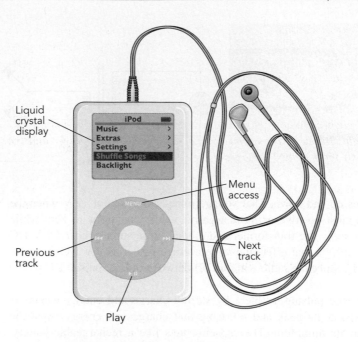

Liquid crystal display

Menu access

Previous track

Next track

Play

SECTION 12.2 Audio Players

Audio players have revolutionized portable music systems. Everywhere you look, people are sporting earpieces and listening to their favorite tunes with these little electronic marvels. Part computer and part stereo system, an audio player is a spectacular synthesis of some of the highest forms of modern electronic technology. Because they contain such a broad range of electronic components, audio players offer an excellent introduction to much of modern electronics.

To understand how audio players work, we'll need to look at how sound can be represented electronically and at how those electronic representations can be stored, retrieved, and ultimately used to recreate the sound itself. That exploration will take us all the way from the digital world of computers to the analog world of stereo amplifiers and headphones.

It will also expose us to that workhorse of modern electronics: the transistor. While early audio electronic devices were built with vacuum tubes, those relatively bulky components wasted power and aged quickly. Transistors have made audio electronics much more practical. They have also made computers so small and inexpensive that every audio player can have its own computer.

Questions to Think About: *What does it mean to store songs? How can a computer use numbers to represent music? Why does an audio player need electric power to operate? Why does an audio player become warm as it operates? Does the volume of the sound affect the battery life? What does an audio amplifier's power rating mean? How are electric power and sound volume related? How do the treble and bass controls of an audio player affect the sound?*

Experiments to Do: *Find an audio player or a related digital audio device such as CD or DVD player. Turn it on and notice that it takes a short time to "wake up." You're* *observing the boot process of a computer. But if all of the computer's information is already inside it, why does it need to do so much work as its starts? As we'll soon see, its computer uses different types of memory, some of which is wiped clean when the audio player is turned off. With that tidbit in mind, try to imagine what is happening during the boot process.*

Play some music and experiment with the sound controls. The volume control determines how much power reaches the headphones. Recall that a flashlight with a poor connection produces dim light. If you create a poor connection between the player and the earpiece, what happens to the volume? Where did the lost power go?

Now experiment with the bass and treble controls. If the player has any other audio effects, try them. Are you still hearing sound as it was originally performed? How do the player's tone settings influence the sound reproduction? Is the most perfect imitation of the original sound always what you want to hear?

Representing Sound: Analog and Digital

An audio player doesn't store sound any more than a DVD stores flickering light. Instead, the audio player stores a *representation* of sound that it can use to recreate that sound on demand. Tucked into the player's memory is enough information to reproduce some number of songs almost perfectly. But between the microphones that originally collected that information and the headphones that finally reconstruct the sound itself are a number of fascinating electronic processes. In this section, we'll explore the journey that this sound information takes between recording and playback. In doing so, we'll encounter much of the electronic basis not only for audio electronics, but for digital computers as well.

Our first task in trying to follow sound information as it moves through an audio player is to understand the two different techniques the player uses to represent sound: analog and digital. Let's begin that task by thinking about the recording process and about what information that process gathers from the original sound. In Chapter 9, we learned that sound in air is a density wave and that you hear a sound wave through its associated air pressure fluctuations. Audio recording mimics your hearing by measuring the air pressure fluctuations over time and saving those measurements in some useful format. Audio reproduction uses those measurements to reproduce a precise facsimile of the original air pressure fluctuations and thereby recreates the sound of the artists themselves.

Recording starts with microphones, which are basically sensitive electromechanical air pressure sensors. Playback ends with headphones or earpieces, which are electromechanical devices for influencing air density and pressure. Between the microphones and the headphones are some serious electronics.

The microphones and headphones connect the world of true sound with the world of electronic audio representations. What leaves the microphone is an electronic representation of the sound, not the sound itself. As the air pressure at the microphone fluctuates up and down about atmospheric pressure, the microphone produces a current that fluctuates back and forth about zero. This fluctuating current is an **analog representation** of the sound, meaning that it is a continuously variable physical quantity (current) that represents another continuously variable physical quantity (air pressure). Like any analog representation, the microphone is drawing an *analogy* between two continuously variable physical quantities: current is serving as an *analog* for air pressure. The headphones draw the reverse analogy.

But while parts of the audio player continue to use analog representations for the sound, other parts use a different representation: digital. In **digital representation,** a continuously variable physical quantity is represented by a set of physical quantities—a set of digits—each of which can have only a limited number of discrete values. In a digital representation of sound, air pressure changes are measured numerically in some units and those numerical values are then represented by sets of digits. Digital representations are widely used in computers, which represent everything imaginable as numbers and then represent those numbers as sets of digits.

For example, suppose the present air pressure at a microphone is 124 units above atmospheric pressure. We first represent the air pressure increase by the number *124* and then represent the number *124* by a set of digits. For example, *124* could be decomposed into the decimal digits 1, 2, 4 and then these decimal digits could be represented as three separate physical quantities, such as charges, currents, or voltages. Because each charge, current, or voltage only has to represent the integers from 0 to 9, its value doesn't have to be very accurate. If a voltage of 4 V is supposed to represent the digit 4, then 3.9 V or 4.1 V will still be understood to mean 4.

In everyday life, we usually break numbers into ones, tens, hundreds, thousands, and so on—the powers of 10—because we think and work in **decimal.** But we could also break numbers into ones, twos, fours, eights, sixteens, and so on. Instead of using the powers of 10, as in decimal, we would be using the powers of 2. This system for representing numbers with the powers of 2 is called **binary.**

In decimal, *124* is written as 124, meaning that *124* contains 1 hundred (10^2), 2 tens (10^1), and 4 ones (10^0). When these pieces are added together, $100 + 20 + 4$, you obtain *124*. In binary, *124* is written as 1111100, meaning that *124* contains 1 sixty-four (2^6), 1 thirty-two (2^5), 1 sixteen (2^4), 1 eight (2^3), 1 four (2^2), 0 twos (2^1), and 0 ones (2^0). When these pieces are added together, $64 + 32 + 16 + 8 + 4$, they again total *124*. This apparently complicated way to represent even a fairly small number is actually quite useful. The number has been broken into pieces that have only two possible values. There either is a thirty-two in the number being represented or there isn't. The only two symbols you need when representing a number in binary are 0 and 1.

Because *124* is 1111100 in binary, you could represent *124* by the charge, current, or voltage of seven separate objects. The first five objects would contain 1's while the last two would contain 0's. For example, if the seven objects were capacitors, the first five might hold separated charge while the last two might hold no charge. A device that needed to know what number the capacitors represented would measure their charges. Finding separated charge in the first five capacitors (11111) and no separated charge in the last two (00), it would determine that the capacitors represent 1111100 binary or *124*.

Binary is useful because fast electronic devices that operate only between two extreme values are relatively easy to build: the current is on or off, the voltage is positive or negative, the charge is present or absent. It's much harder to build fast devices that deliver the specific currents, charges, or voltages needed for decimal or even analog representation. Analog representation is also susceptible to electronic imperfections and noise because an analog device that tries to represent *124* with a voltage of 124 V might accidentally produce 123 V or 125 V. Imagine a bank computer that couldn't tell $124 from $123 or $125! Although representing *124* in binary takes at least seven separate quantities, there is no confusion about which number is being represented.

▶ Check Your Understanding #1: New Math

for answers, see page **418**

What number does binary 10000001 represent?

Resistors

After that introduction to information representations, we're ready to look at how an audio player actually works with its information. We've already hinted at the fact that this information travels about the player as currents, voltages, and a few other physical quantities. But what tools does the player use to work with those physical quantities?

The player's tools are principally electronic ones and, happily, many are familiar to us. We have already encountered capacitors, inductors, transformers, and diodes, all of which are common in audio players and most other electronic devices. However, there are at least two other electronic components that no audio player or high-tech electronic device can be without: resistors and transistors.

Resistors complete a trio of electronic components that began with inductors and capacitors. As we saw in Chapter 11, an inductor uses its induced emf to produce a voltage rise that's proportional to how quickly current is changing with time. And in the previous section, we learned that a capacitor uses its ability to store charge to produce a voltage rise that's proportional to the sum of current over time. What we're missing is a device that uses its electrical resistance to produce a voltage rise that's simply proportional to current. That missing device is called a **resistor.**

A resistor is just two wires connected by an ohmic device, an imperfect conductor of electricity (Figs. 12.2.1 and 12.2.2). Since current flows through an ohmic device only if there's an electric field pushing it forward, a current-carrying resistor must have a voltage drop across it. In accordance with Ohm's law (see Section 10.3), that voltage drop is proportional to the current in the resistor.

But the voltage drop through a resistor is also proportional to the resistor's electrical resistance—to how imperfect a conductor it is. The larger its resistance, the less current flows through the resistor at a particular voltage drop. Some are relatively good conductors ("low" resistance) while others are relatively poor conductors ("high" resistance).

As we saw in Section 10.3, a resistor's electrical resistance is defined as the voltage drop through it divided by the current flowing through it and is measured in ohms (abbreviated Ω). A resistor with a few ohms of resistance is nearly a good conductor while a resistor with a few million ohms of resistance is nearly an insulator. A k in front of the Ω means thousands (kΩ or kiloohms), and an M in front of the Ω means millions (MΩ or megaohms). Thus a 100-kΩ resistor has a resistance of 100,000 Ω and a 10-MΩ resistor has a resistance of 10,000,000 Ω. A resistor's resistance is marked on its cylindrical body, often in the form of brightly colored stripes. Ten different colors represent the digits 0 through 9, as well as various powers of ten, so that a resistor with stripes brown (1), black (0), and red ($\times$100) is a 1000-Ω resistor.

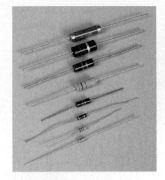

Figure 12.2.1 Resistors are ohmic devices that impede current flow and produce a drop in voltage. They convert some of that current's electric power into thermal power. The larger resistors can handle more thermal power without overheating.

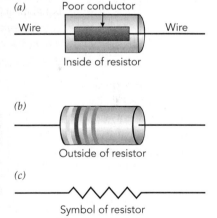

(a)

Wire — Poor conductor — Wire

Inside of resistor

(b)

Outside of resistor

(c)

Symbol of resistor

Figure 12.2.2 (a) A resistor is two wires with an imperfect conductor of electricity between them. (b) It's usually encased in a cylindrical shell, with colored stripes to indicate its resistance. (c) In a schematic diagram of an electronic device, the resistor is represented by a zigzag line.

Check Your Understanding #2: A Waste of Energy

for answers, see page **418**

Your radio produces sound when its amplifier allows current from its power supply to flow through its speaker. During a quiet moment, the amplifier sends only a small current through the speaker and that current experiences only a small voltage drop. The current thus retains most of its electric power. How can the radio use up the current's remaining power before returning it to the power supply for reuse?

Transistors

Invented in 1948 by three American physicists, William Shockley (1910–1989), John Bardeen (1908–1991), and Walter Brattain (1902–1987), **transistors** (Fig. 12.2.3) are key elements in nearly all modern electronic equipment. Like diodes, transistors are built from doped semiconductors—semiconductors such as silicon to which chemical impurities have been added. But unlike diodes, which simply prevent current from flowing backward through a single circuit, transistors allow the current in one circuit to control the current in another circuit.

While there are many types of transistors, the simplest and most important type is the *field-effect transistor*. Actually, even here there are several varieties, so we'll focus on the one that's most widely used in audio players, video equipment, and computers: the *n-channel metal-oxide-semiconductor field-effect transistor* or *n-channel MOSFET*. Despite its complicated name, the n-channel MOSFET is a relatively simple device, consisting principally of three semiconductor layers and a nearby metal surface (Fig. 12.2.4). The three layers are called the *drain,* the *channel,* and the *source,* while the metal surface is called the *gate.*

Figure 12.2.3 These MOSFETs make it possible for small electric charges to control large electric currents. The larger transistors can handle more electric power and thermal power without overheating.

The drain and the source consist of strongly doped n-type semiconductor (many conduction level electrons) and the channel between them consists of lightly doped p-type semiconductor (a few empty valence levels). When those three layers touch, they form two back-to-back p-n junctions and conduction level electrons from the drain and source flow into the channel to fill its few empty valence levels (Fig. 12.2.5a). The completed transistor is thus left with a vast depletion region—a region devoid of empty valence levels or occupied conduction levels—extending all the way from its drain to its source. With nothing to convey charge through its channel, the transistor can't conduct current between its drain and source.

However, if more electrons could be coaxed into the channel somehow, those electrons would have to go into the channel's conduction levels and the channel would then behave like an n-type semiconductor. With an n-type channel sandwiched between an n-type drain and an n-type source, the p-n junctions would vanish and so would the depletion region. The three layers would become, in effect, a single piece of n-type semiconductor and the transistor would then be able to conduct current between its drain and source.

Figure 12.2.4 (*a*) In an n-channel MOSFET, the channel is normally a depletion region that cannot carry current between the source and drain. But when positive charge on the gate attracts electrons into the channel, the channel becomes an n-type semiconductor and allows current to flow. (*b*) The symbol representing an n-channel MOSFET in a schematic diagram.

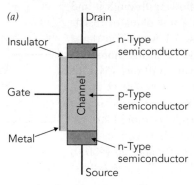

Inside of n-channel MOSFET

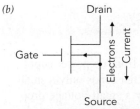

Symbol for n-channel MOSFET

Drawing extra electrons into the channel from outside the transistor is the task of the metal gate. Separated from the channel by an incredibly thin insulating layer, the gate controls the channel's ability to carry current. When a tiny positive charge is placed on the gate through a wire, it attracts electrons into the channel's conduction levels and the transistor begins to conduct current (Fig. 12.2.5b). The more positive charge there is on the gate, the more electrons enter the channel and the more current can flow through the transistor. In effect, the transistor behaves like an adjustable resistor with a resistance that decreases as the positive charge on its gate increases.

We can now understand the n-channel MOSFET's name. "n-Channel" refers to the channel's

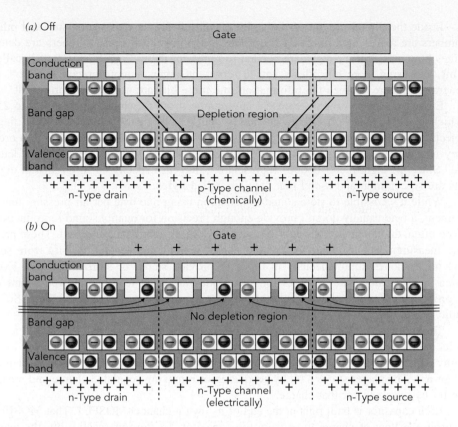

Figure 12.2.5 (*a*) When an n-channel MOSFET is formed, conduction level electrons from the n-type source and drain fill the empty valence levels of the p-type channel and form a vast, insulating depletion region. (*b*) But when positive charge is placed on the nearby gate, it attracts electrons from outside the MOSFET into the channel's conduction levels and turns the entire structure into a conducting n-type semiconductor.

n-type behavior when its gate is positively charged and the transistor can carry current. Although the channel is chemically p-type, it becomes electrically n-type when extra electrons are drawn into it and it acquires a negative net charge. "Metal-oxide-semiconductor" indicates that the metal gate is separated from the semiconductor channel by a thin insulating layer of oxide. This insulator is agonizingly easy to puncture, which is why so many electronic devices can be damaged by static electricity. "Field-effect transistor" indicates that the electric field from charge on the gate is what draws electrons into the channel and controls the current flow through the transistor.

Check Your Understanding #3: Power and Control

for answers, see page 418

Widening the channel of an n-channel MOSFET allows it to handle more current between its source and drain. However, the enlarged transistor needs more positive charge on its gate to control that current. Explain.

Storing Digital Sound Information

An audio player is half computer and half stereo system. It stores and manipulates sound information in digital form like a computer but then amplifies that information for the headphones in analog form like a stereo system. We'll mirror that sequence when examining the player's electronics: we'll start with its digital memory and processing systems and finish with its audio amplifier.

Inside the digital portion of the audio player, air pressure measurements and other numbers are represented in binary form. How large or precise those numbers are determines how many binary digits are needed to represent them. Each binary digit is called a **bit**, and using more bits allows you to represent larger or more precise numbers. In general, the more detailed the information, the more bits are needed to represent it.

Eight bits can be used to represent any number from *0* (which is 00000000) to *255* (which is 11111111). Since there are fewer than 256 of many common objects, these objects can be identified by groups of eight bits. For example, the symbols used in ordinary text have been assigned numbers between *0* and *255*, with *65* denoting the letter "A." Since the eight bits 01000001 represent *65*, they also specify an A. Groups of eight bits are so common and useful that they are called **bytes.**

While it's possible to store sound information using one byte per air pressure measurement, a byte usually doesn't provide enough precision for quality sound reproduction. More often, digital audio is saved using two bytes per pressure measurement. These pressure measurements are made tens of thousands of times per second, usually from several microphones simultaneously in order to provide stereo or surround sound. Even when sophisticated data compression techniques are used to eliminate redundant or inconsequential information, it still takes a great many bits to represent an album. So an audio player needs a lot of memory.

There are several ways in which the audio player, like any computer, stores a bit. In its main working memory (often called random access memory or RAM), each bit is a tiny capacitor that uses the presence or absence of separated electric charge to denote a 1 or a 0. The player stores a bit by producing or removing separated charge and recalls the bit by checking for that charge.

Each capacitor is built right at the end of its own n-channel MOSFET. That MOSFET controls the flow of charge to or from the capacitor. To store or recall a bit, the audio player places positive charge on the gate of the MOSFET so that the MOSFET becomes electrically conducting. The memory system can then transfer charge to or from the bit's capacitor.

Storing the bit is relatively easy; the player simply sends the appropriate charge through the MOSFET to the capacitor. But recalling the bit is harder because the charge on the capacitor is extremely small. Sensitive amplifiers in the memory system detect any charge flowing through the MOSFET from the capacitor and report what they find to the audio player. Since this reading process removes charge from the capacitor, the memory system must immediately store the bit again.

Unfortunately, these tiny capacitors can't hold separated charge forever because it leaks out to their surroundings. Memory that uses charged capacitors to store bits is called dynamic memory and must be refreshed (read and restored) hundreds of times each second to ensure that a 1 doesn't accidentally switch to a 0 or vice versa.

Dynamic memory is also volatile—its contents are lost when the audio player turns off. To conserve its batteries, the player keeps its music information in nonvolatile memory—memory which doesn't need power to retain its information. New possibilities for nonvolatile memory appear almost every year, but at present the three leading forms are flash, magnetic disk, and optical disk memories. We'll concentrate on flash and magnetic disk memory here and save optical memory for Section 15.2.

Flash memory resembles dynamic memory in that each bit is stored as the presence or absence of charge associated with a MOSFET. But in flash memory, that charge resides on the MOSFET's floating gate—a second, unattached gate located in the insulating layer between the channel and the normal gate. Since this floating gate is surrounded by insulator, it can keep its charge for decades. And as long as that charge is present, it will determine the MOSFET's conductivity and whether the bit is a 0 or a 1.

Reading bits from flash memory is easy, but storing them is a challenge. The same isolation that traps charge on the floating gate for years makes that charge difficult to change. To add or remove electrons from the floating gate, the memory system applies relatively large voltages to the MOSFET's source, drain, and normal gate and the resulting strong electric fields permit electrons to cross through the insulation separating the channel from the floating gate.

To add electrons to the floating gate, the electric fields are arranged so they accelerate channel electrons to such high speeds that those electrons simply burrow right through the insulating layer to the floating gate. To remove electrons from the floating gate, the electric fields are arranged so that the floating gate's electron standing waves are distorted into the insulator. When these distorted waves reach far enough into the insulator, electrons begin to leak through it into the channel via a process known as *quantum tunneling*. We'll return to explore quantum tunneling in Chapter 16.

Flash memory is fast to read but relatively slower to write. Moreover, the electron burrowing process causes cumulative damage to the insulating layer and limits the number of times flash memory can be written. An audio player uses a mixture of dynamic memory and flash memory: it does its computational work in dynamic memory but retains its long-term information in flash memory.

However, there's another memory concept that remains more cost-effective than flash memory for storing vast amounts of information: magnetic disk memory. Audio players that store tens of thousands of songs usually contain magnetic disks. Also called hard disks, these devices use the magnetic effects that we discussed in Chapter 11 to store music or other information.

Just as the magnetic strip on a credit card (Fig. 11.1.6) can store information in the locations of its magnetic poles, the surface of a magnetic disk can store information in the orientations of its magnetic poles. Actual hard disks are smooth aluminum platters which have been coated with hi-tech hard magnetic materials. Using microscopic electromagnets to write magnetic poles and sophisticated semiconductor magnetic sensors to read them, modern hard disks can pack almost a quarter-billion bits into a square-millimeter of surface (about 16 gigabytes per square-inch). Simply locating those microscopic bits on platters that rotate over 100 times per second is an electromechanical tour de force, yet these disks do it routinely even while you are jogging with your audio player.

> ► Check Your Understanding #4: Like Sending a Letter to Yourself
> *for answers, see page* **418**

Every few thousandths of a second, a computer's memory system stops briefly to read and then rewrite every bit in its dynamic memory. What is going on?

The Audio Player's Computer

We've seen how sound information can be represented and stored as bits, so now let's look at how an audio player's computer works with those bits. This digital processing is done by electronic devices that take groups of bits as their inputs and produce new groups of bits as their outputs. Since their output bits are related to their input bits by the rules of logic, these electronic devices are called logic elements.

The simplest logic element is the inverter, which has only one input bit and one output bit. Its output is the inverse of its input (Fig. 12.2.6). If an inverter's input bit is a 1, then its output bit is a 0 and vice versa. Inverters are used to reverse an action—

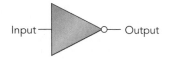

Input ─▷○─ Output

Input	Output
1	0
0	1

Figure 12.2.6 An inverter, shown here symbolically, produces one output bit that is the inverse of its one input bit.

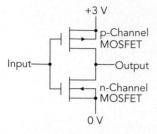

Figure 12.2.7 When negative charge arrives at the input of a CMOS inverter, its p-channel MOSFET (*top*) permits positive charge to flow to the output. When positive charge arrives at the input, the n-channel MOSFET (*bottom*) sends negative charge to the output.

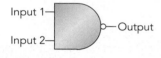

Input 1	Input 2	Output
1	1	0
1	0	1
0	1	1
0	0	1

Figure 12.2.8 The output bit of a Not-AND or NAND gate, shown here symbolically, is a 1 unless both input bits are 1's.

turning a light on rather than off or starting a song rather than stopping it. Inverters are also used as parts of more complicated logic elements.

But inverters aren't just abstract logic elements; they're real electronic devices. They act on electrical inputs and create electrical outputs. In an audio player's computer, inverters and other logic elements represent input and output bits with electric charge. Positive charge represents a 1 and negative charge represents a 0. Thus when positive charge arrives at the input of an inverter, the inverter releases negative charge from its output.

Inverters and other logic elements are usually constructed from both n-channel and p-channel MOSFETs. We've already seen that n-channel MOSFETs conduct current only when their gates are positively charged. p-Channel MOSFETs are just the reverse, conducting current only when their gates are negatively charged. The drain and source of a p-channel MOSFET are made from p-type semiconductor while the channel is made from n-type semiconductor. Since n-channel and p-channel MOSFETs are exact complements to one another, logic elements built from them are called complementary MOSFET or CMOS elements. An audio player's computer is built almost entirely from CMOS elements.

A CMOS inverter consists of one n-channel MOSFET and one p-channel MOSFET (Fig. 12.2.7). The n-channel MOSFET is connected to the negative terminal of the computer's power supply and controls the flow of negative charge to the inverter's output. The p-channel MOSFET is connected to the power supply's positive terminal and controls the flow of positive charge to the output. When negative charge arrives at the inverter's input and moves onto the gates of the MOSFETs, only the p-channel MOSFET conducts current and the output becomes positively charged. When positive charge arrives at the input, only the n-channel MOSFET conducts current and the output becomes negatively charged.

But a computer needs logic elements that are more complicated than inverters. One such element is the Not-AND or NAND gate. This logical element has two input bits and one output bit, and its output bit is 1 unless both input bits are 1's (Fig. 12.2.8). It's called a Not-AND gate because it's the inverse of an AND gate. An AND gate produces a 0 output unless both input bits are 1's. Simple memoryless logic elements are often called gates.

A CMOS NAND gate uses two n-channel MOSFETs and two p-channel MOSFETs (Fig. 12.2.9). The two n-channel MOSFETs are arranged in **series**—one after the next—so that current passing through one must also pass through the other. If either transistor has negative charge on its gate, no current can flow through the series. Components arranged in series all carry the same current, but they may experience different voltage drops.

The two p-channel MOSFETs are arranged in **parallel**—one beside the other—so that current can flow through either one of them to the output. If either transistor has negative charge on its gate, current can flow from one side of the pair to the other. Components arranged in parallel share the current they receive through one wire and deliver it together to the second wire. But while parallel components may share the current unevenly among themselves, they all experience the same voltage drop.

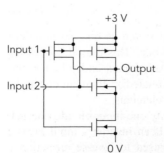

Figure 12.2.9 A CMOS NAND gate has two input bits. When negative charge arrives through either input, the chain of n-channel MOSFETs (*bottom*) stops conducting current and one of the two p-channel MOSFETs (*top*) permits positive charge to reach the output. Only if both inputs are positively charged will negative charge reach the output.

Parallel and Series Wiring

Connection in Series: Components that are wired in series, one after the other so that they form a chain, all carry the same current but can have different voltage drops. The total voltage drop from the start of the chain to its end is the sum of the individual voltage drops through the components.

Connection in Parallel: Components that are wired in parallel, one beside the other so that they all connect the same pair of wires, have the same voltage drop but can carry different currents. The total current flowing between those wires is the sum of the individual currents through the components.

If negative charge arrives at either input of the CMOS NAND gate, the series of n-channel MOSFETs will be nonconducting and one of the p-channel MOSFETs will deliver positive charge to the output. But if positive charge arrives at both inputs, both p-channel MOSFETs will be nonconducting and the series of n-channel MOSFETs will deliver negative charge to the output. Thus the CMOS NAND gate has the correct logic behavior.

These two logic elements, inverters and NAND gates, can be combined to produce any conceivable logic element. For example, they can be used to build an adder, a device that sums the numbers represented by two groups of input bits and produces a group of output bits representing that sum. These adders can themselves be used to build multipliers, and multipliers can be built into still more complicated devices. In this fashion, the simplest logic elements can be used to construct an entire computer.

Actually, a computer isn't built exclusively from NAND gates and inverters. To improve its speed and reduce its size, it uses a few other basic logic elements as well. Like the CMOS NAND gate and inverter, these elements are constructed directly from n-channel and p-channel MOSFETs.

All of these logic elements are wired together in an intricate pattern to create a complete computer (Fig. 12.2.10). In an audio player, this computer retrieves and organizes music information and prepares it for the playback electronics, which are not digital. The computer's last act is to deliver the digital music information, the air pressure measurements, to a digital-to-analog converter or *DAC*. This electronic device is the interface between the two representations of information, digital and analog. The music information leaves the DAC as a voltage that's proportional to air pressure. This voltage is the input for the audio player's main analog component: its audio amplifier. Actually, the player has two complete analog audio systems so that it can produce stereo sound. But since those systems are identical, we'll focus on only one of them.

Courtesy Intel Corporation

Figure 12.2.10 This microscope photograph shows an integrated circuit microprocessor— approximately a computer on a chip. Aluminum strips connect millions of MOSFETs and other components that have been formed by photographic techniques on the surface of a thin wafer of silicon.

Check Your Understanding #5: Getting It Together

for answers, see page 419

How could you use inverters and NAND gates to create an AND gate, a logic element that has two inputs and one output, with the output being 1 only if both of the inputs are also 1's?

The Audio Player's Audio Amplifier

The fluctuating voltage provided by the audio player's DAC is often called an *audio signal* because it represents audio information. Many analog or digital representations of information are called **signals,** including video signals, data signals, and even turn signals.

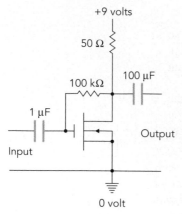

Figure 12.2.11 A simple audio amplifier can be built with one n-channel MOSFET, two resistors, and two capacitors. A 9-V battery powers the device.

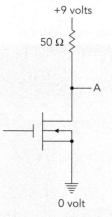

Figure 12.2.12 The voltage at A depends on the resistance of the MOSFET. The lined triangle at the bottom signifies connection to ground (often the earth itself).

But while the player's audio signal contains all the information needed to reproduce the original sound, in convenient analog format, it doesn't have the power that the headphones need to produce that sound at a reasonable volume. Something must first enlarge the audio signal; it needs to be *amplified*.

Devices that enlarge various characteristics of signals are called **amplifiers.** An audio amplifier is an amplifier that's designed to boost signals in the frequency range that we hear or feel (20 Hz to 20,000 Hz). It has two separate circuits—an input circuit and an output circuit—and it uses the small current passing through its input circuit to control a much larger current passing through its output circuit. In this manner, the amplifier provides more power to its output circuit than it receives from its input circuit.

Figure 12.2.11 shows the schematic diagram for a simple audio amplifier, built from the components we've just studied. This amplifier has only five components: an n-channel MOSFET, two resistors, and two capacitors. It draws power from a 9-V battery (or an equivalent power adapter) and amplifies a tiny alternating current in its input circuit into a large alternating current in its output circuit.

To understand how this amplifier works, let's first remove everything but the MOSFET and the 50-Ω resistor (Fig. 12.2.12). These two components are in series, so any current that passes through one must also pass through the other. When the MOSFET doesn't conduct current, no current flows through the 50-Ω resistor and it experiences no voltage drop. So the voltage at A is 9 V. But if the transistor does conduct current, a voltage drop will appear through the 50-Ω resistor and the voltage at A will decrease.

The transistor will conduct current only if positive charge is put on its gate. That can be done by connecting the gate to A with a 100-kΩ resistor (Fig. 12.2.13). Since A is at 9 V, it is positively charged and pushes charge toward anything at lower voltage. Current flows slowly through the resistor from A to the gate. But as positive charge accumulates on the gate, the transistor begins to conduct current and the voltage at A drops. When the voltage at A reaches the voltage on the gate, current stops flowing through the resistor.

The amplifier is then in a stable equilibrium; A has a voltage of approximately 5 V and the transistor's gate has a modest amount of charge on it. The 100-kΩ resistor provides the transistor with **feedback,** that is, it provides the transistor with information about the present situation at A that the transistor can use to correct or improve that situation. Although the feedback is slowed by the resistor's large electrical resistance, it perpetually acts to return the voltage at A to its equilibrium value. If the transistor conducts too little current, charge flows onto its gate and makes it conduct more. If the transistor conducts too much current, charge flows off its gate and makes it conduct less.

The amplifier is now exquisitely sensitive to small changes in the charge on the transistor's gate. If you add just a tiny bit more positive charge to the gate, down goes the voltage at A. If you remove just a tiny bit of positive charge from the gate, up goes the voltage at A. Although current in the feedback resistor tries to undo these changes, it acts too slowly to oppose short-timescale variations. The amplifier's input signal successfully adds or subtracts positive charge from the gate and the amplifier's output signal emerges from A.

The amplifier has two input wires. An analog audio signal's current flows into the amplifier through one wire and returns through the other. But the audio signal is not connected directly to the gate. Instead, it's connected to the gate through a capacitor (Fig. 12.2.14). In addition to storing charge and energy, a capacitor can transfer current between two wires that have different voltages. Such voltage flexibility is important in battery-powered audio amplifiers that must do their amplifying exclusively with positive voltages. Aided by input and output capacitors, our audio amplifier can have an average operating voltage of about +5 V while having average input and output voltages of 0 V.

To see how a capacitor passes along current, let's watch input current flow right-ward into our amplifier's input capacitor. As that current's positive charge accumulates on the capacitor's left plate, it attracts negative charge onto the capacitor's right plate and away from the gate. The capacitor remains electrically neutral but the gate becomes more positively charged. Overall, the capacitor has conveyed input current to the gate, even though no charge has actually passed through its insulating layer and its two plates remain at different voltages.

With the help of the input capacitor, the amplifier's fluctuating input current produces a fluctuating charge on the transistor's gate and the voltage at A fluctuates as a result. Even a tiny fluctuating current on the input wires creates a large fluctuating voltage at A.

This fluctuating voltage is responsible for sending a fluctuating current through the headphones. Although headphones are not truly ohmic devices, they respond to fluctu-ating voltages by carrying fluctuating currents. By applying a fluctuating voltage drop across the headphone's two wires, the amplifier can cause it to carry a fluctuating cur-rent and produce corresponding pressure fluctuations and sound.

But the voltage at A averages about 5 V, while the headphones expect an average voltage drop of 0 V. To convey the fluctuating voltages and currents from A to the head-phones, while eliminating their large voltage difference, our amplifier connects them via an output capacitor (Fig. 12.2.15). As before, the fluctuations in current and voltage on the output capacitor's left plate are mirrored by current and voltage fluctuations on its right plate. Even though the amplifier operates at a high average voltage, the output signal for the headphones has an average voltage of 0 V.

Tiny fluctuating currents in our amplifier's input circuit produce large fluc-tuating currents in its output circuit. This amplifier works remarkably well, given its simplicity. If you connect a microphone to the input wires and headphones to the output wires, the headphones will do a surprisingly good job of repro-ducing the sound in the microphone.

However, our simple amplifier isn't perfect. It distorts the sound somewhat and it doesn't handle all frequencies or amplitudes of sound equally. It also wastes a large amount of electric power heating the 50-Ω resistor. The ampli-fiers in audio players carefully correct for these problems. Most use feedback to make sure that their output signals are essentially perfect replicas of their input signals, only larger. They sense their own shortcomings and correct for them.

But perfect replication of the input signal isn't always desirable. Sometimes you want to boost the volume for part of the sound. The treble and bass controls on an audio

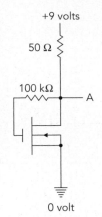

Figure 12.2.13 The 100-kΩ resistor transfers positive charge to the gate until the voltage at A drops to about 5 V.

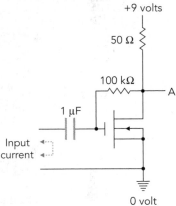

Figure 12.2.14 Because current flowing back and forth through the two input wires affects the charge on the transistor's gate, it also affects the voltage at A.

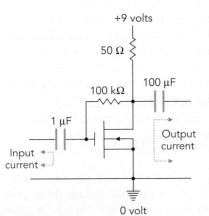

Figure 12.2.15 The amplifier causes currents to flow back and forth through the two output wires. The alternating current in the output wires is a good replica of the alternating current in the input wires, only larger.

player allow you to selectively change the volumes for the high- and low-frequency portions of the sound, respectively.

An amplifier is typically rated according to the peak power it can supply to the headphones (or speakers) and its average power should never reach that value. But the amplifier can reach its peak power during a passage that's not particularly loud. That's because sound waves often interfere with one another (see Section 9.3 to review wave interference) and when their crests and troughs coincide at the microphone, constructive interference can briefly produce enormous pressure fluctuations. To reproduce those overlapping waves properly, the audio amplifier must be able to provide several times its average power, though only for a moment. If the amplifier can't deliver that much power, the audio signal it sends to the headphones or speakers will be distorted and the sound will be unpleasant. That's why audiophiles often use powerful amplifiers even when they're playing quiet music.

As for the headphones themselves, they generally use electromagnetic effects to move a surface back and forth in sync with the amplifier's current fluctuations. In most cases, the amplifier's current is sent through a coil of wire that's immersed in a strong magnetic field and attached to a moveable surface. This current experiences a Lorentz force due to the magnetic field and that force drives the current, coil, and surface back and forth as the current fluctuates. The moving surface alternately compresses and rarefies the air, thereby reproducing the original sound.

Check Your Understanding #6: Sound Control

for answers, see page 419

When you connect a microphone to the input of an MOSFET-based amplifier, the microphone sends current back and forth through the input wires. As a result, charge moves onto or off what critical control element in the amplifier?

Epilogue for Chapter 12

In this chapter we looked at two electronic devices to see how they use electricity and magnetism to perform useful tasks. In *power adapters*, we saw how a transformer, diodes, and a capacitor are combined into a system that obtains power from high-voltage alternating current and uses it to provide low-voltage direct current power. We saw how quantum physics determines the properties of metals, insulators, and semiconductors, and we learned how semiconductor diodes limit current flow to only one direction.

In *audio players*, we studied the ways in which sound information can be represented and looked at the electronic techniques used to represent that sound in digital and analog forms. We also examined the most important modern electronic component: the transistor, a semiconductor device that allows the current in one circuit to control the current in another circuit. We saw how audio players use transistors and other electronic components in both its digital computer and its analog amplifier.

Explanation: Building an Electronic Kit

Depending on what was included in your kit, you may have worked with resistors, capacitors, diodes, and transistors. The resistors limit currents or produce voltage drops proportional to the currents they carry. The capacitors store charge and energy, or assis currents in flowing between portions of the kit that have different voltages. The diodes limit the directions of current flow, produce fixed voltage drops (typically 0.6 V) in currents, or emit light. And the transistors perform a wide variety of tasks, most often letting the current or charge in one circuit control the current or charge in another circuit. You may also have encountered integrated circuits, complex components that contain pre-assembled collections of transistors and other simpler components.

While assembling the kit, you inserted the components into the printed circuit board and completed the electrical connections using solder. It was important to install the components in the proper orientations because some of them, particularly diodes and transistors, are not symmetric. If you worked with any MOSFETs, you had to be careful about static electricity because the gate insulation is easily destroyed by errant charge. And as you melted the solder into place, it was important to avoid overheating the components, particularly those involving semiconductors, because they may have delicate electronic and electrochemical properties that can be destroyed by excessive temperatures.

Chapter Summary

How Power Adapters Work: A power adapter uses a transformer to convey power from an electrical outlet's relatively high-voltage alternating current to the relatively low-voltage alternating current in its secondary coil. The secondary coil's current flows through a set of semiconductor diodes, which act as switches to direct that current out of the adapter through its positive power wire and back to the coil through its negative power wire. The adapter uses a capacitor to store charge and energy, thereby reducing the fluctuations in voltage rise caused by the original alternating power source. The result is a unit that operates from high-voltage alternating current and provides low-voltage direct current.

Simpler adapters omit the capacitor and perhaps even some of the diodes, while more sophisticated adapters include complicated electronic systems to regulate and control their output voltages and currents, and to allow them to operate from a broader range of input voltages.

How Audio Players Work: An audio player combines a computer and an audio amplifier into a single unit. The player's computer stores and retrieves the sound information in digital form, with each air pressure measurement represented by a collection of binary bits. These bits can take only values of 0 or 1. For long-term storage, the player places its digital sound information in flash memory and/or magnetic disk memory. For temporary storage while playing a song or working on its song collection, it uses dynamic memory.

After the player's computer has manipulated that information digitally, using logic elements built primarily from MOSFETs, it passes the digital information to a digital-to-analog converter and that information then takes analog form. This analog signal represents sound as a current that's proportional to the sound's shift in air pressure away from atmospheric pressure.

The analog sound signal enters the input circuit of the player's audio amplifier, which makes use of MOSFETs. This amplifier uses power obtained from the battery to produce an output signal that represents the same sound but with increased voltage and current. This amplified output signal has enough power to produce loud sound in the headphones.

Check Your Understanding—Answers

Section 12.1 POWER ADAPTERS

1. None of it.

Why: Current from the electrical outlet flows through the primary coil of the adapter's transformer and then returns to the electrical outlet. Current from the secondary coil of that transformer flows through the radio and then returns to the secondary coil. These two circuits, the primary circuit with the power outlet and the secondary circuit with the radio, are electrically isolated: they can exchange power but not charge. The isolation between those two circuits provides a measure of electrical safety to the user of the radio and the power adapter.

2. Certainly.

Why: Like everything else in nature, atoms have both particle and wave properties. Atoms travel as waves and have recently been shown to exhibit many of the wave effects we studied in Chapter 9, including refraction, reflection, and interference.

3. The electron will go into the lowest energy empty level, the one just above the Fermi level.

Why: Since each electron goes into the lowest energy level that's available, this new electron will fill the level just above the Fermi level. Actually, if the metal ball had an odd number of electrons, then the Fermi level only contained one electron. In that case, the new electron will fill the other opening in the Fermi level.

4. The beam of light shines on the photoconductor, allowing it to carry current and turn on the conveyer belt's motor. When food blocks the light beam, the photoconductor becomes an electrical insulator and the belt stops moving.

Why: Photoconductors are commonly used in light sensors. Light allows the photoconductor to carry current and that current can be used to operate machinery, trigger a burglar alarm, or turn lights on or off. In the present case, it turns on the conveyer belt's motor.

5. Current will flow through the circuit only half the time and your lamp will be dim.

Why: If you include a diode in an AC circuit, it will prevent current from flowing in one direction. Current will flow through the circuit only during the half of each power-line cycle when the diode's cathode is negatively charged and its anode positively charged. Since the lamp will receive only about half of its normal power, its filament won't reach normal operating temperature. The lamp will glow dimly but the bulb will last an extraordinarily long time. Diodes are often used in this manner to create dim light levels in lamps or appliances.

6. It can store the energy as separated charge on a large capacitor.

Why: The stereo's power supply saves energy in large electrolytic capacitors. During the times when the power line is providing plenty of electric power, the power supply uses that power to increase the separated charge in its capacitors. When the power line isn't providing power, the power supply draws energy out of the capacitors. Unfortunately, electrolytic capacitors age quickly and often leak their liquid electrolytes. Once one leaks, it can't store energy properly and your stereo will hum.

7. They have more trouble in Europe.

Why: Since the European power grid operates at a lower frequency, it requires power adapters to endure longer periods of relative powerlessness.

Section 12.2 AUDIO PLAYERS

1. Answer: *129*

Why: Binary 10000001 contains only one hundred twenty-eight (2^7) and one (2^0). All the other powers of 2 are not present. Since 128 + 1 is *129*, that is the number represented by this binary value.

2. The amplifier can send the current through a resistor.

Why: By passing the current through a resistor, the amplifier converts most of the current's remaining electric power into thermal power. The current leaves the resistor with only enough electric power to carry it back to the power supply for reuse.

3. The larger transistor also has a larger gate. With more surface over which to spread its charge, the gate needs more positive charge in order to draw conduction level electrons into the channel.

Why: MOSFETs range in size from remarkably small (less than 0.1 μm^2) to relatively large (several square-millimeters). The smallest ones are used in computer chips, where millions of MOSFETs are created on a single wafer of silicon only a centimeter square. A tiny charge on the gate of one of these MOSFETs will allow it to conduct current. The largest MOSFETs are used in power control devices such as amplifiers and power supplies. These transistors have large gates and much more charge is needed to allow one of them to conduct current.

4. The computer is making sure that the charge stored on each capacitor in the dynamic memory adequately represents that bit's contents.

Why: Since charge leaks quickly from the capacitors in dynamic memory, the contents of each bit must be refreshed many times a second. This refreshing process, reading each bit and storing it back into memory, slows the computer down slightly.

5. You could connect an inverter to a NAND gate so that the output signal of the NAND gate is the input signal of the inverter. When both of the NAND gate's input signals are 1's, it will produce an output of 0. This 0 will arrive at the inverter, which will invert it and yield an output of 1.

Why: Connecting logic elements together one after the next is the standard method for producing more complicated logic elements. In this case, two elements produce a third.

6. The gate of a MOSFET.

Why: In all likelihood, the input current adds or removes charges from the gate of a MOSFET in the amplifier's first stage of amplification.

Exercises

1. If a very small piece of material contains only 10,000 electrons and those electrons have as little energy as possible, how many different levels do they occupy in that material?

2. If electrons had four different internal states that could be distinguished from one another, how many electrons could occupy the same level without violating the Pauli exclusion principle?

3. If electrons were not Fermi particles, any number of them could occupy a particular level. How would these electrons tend to arrange themselves among the levels in an object?

4. If electrons were not Fermi particles (Exercise 3), would there still be a distinction between metals and insulators? Explain your answer.

5. Thermal energy can shift some of the electrons in a hot semiconductor from valence levels to conduction levels. What effect do these shifts have on the semiconductor's ability to conduct electricity?

6. Why do semiconductor devices often self-destruct when they are overheated?

7. Is the p-type half of a p-n junction electrically neutral, positive, or negative?

8. In which direction does the electric field point in the middle of a p-n junction?

9. Two capacitors are identical except that one has a thinner insulating layer than the other. If the two capacitors are storing the same amount of separated electric charge, which one will have the larger voltage difference between its plates?

10. Dynamic memory stores bits as the presence or absence of separated charge on tiny capacitors. It takes energy to produce separated charge, and a computer that minimizes this energy will use less electric power. Why does making the insulating

layers of the memory capacitors very thin reduce the energy it takes to store each bit in them?

11. Suppose a battery is transferring positive charges from one plate of a capacitor to the other. Why does the work that the battery does in transferring a charge increase slightly with every transfer?

12. A charged capacitor resembles a battery in that both can supply electric power. However, as the capacitor delivers its power, the voltage of the current it provides decreases. Why?

13. We combine the three decimal digits 6, 3, and 1 to form 631 in order to represent the number *631*. What does the 6 in 631 mean? What are there 6 of?

14. We combine the three binary bits 1, 0, and 1 to form 101 in order to represent the number *5*. What does the leftmost 1 in 101 mean? What is there 1 of?

15. What numbers do the two binary bytes 11011011 and 01010101 represent?

16. How is the number *165* represented in binary?

17. Why are there no 2's in the binary representation of a number? (In other words, why isn't 1101121 a valid binary representation?)

18. For convenience, hexadecimal (powers of 16) is often used in place of binary. The traditional symbols used to represent hexadecimal digits are 0–9 and A–F. In hexadecimal, 10 represents the number *16*. Show that one hexadecimal digit can substitute perfectly for four binary digits.

19. The gate of a MOSFET is separated from the channel by a fantastically thin insulating layer. This layer is easily punctured by static electricity, yet the manufacturers continue to use thin layers. Why would thickening the insulating layer spoil the MOSFET's ability to respond to charge on its gate?

20. In an n-channel MOSFET, the source and drain are connected by a thin strip of p-type semiconductor. Why is this device labeled as having an n-channel rather than a p-channel?

21. A MOSFET doesn't change instantly from a perfect insulator to a perfect conductor as you vary the charge on its gate. With intermediate amounts of charge on its gate, the MOSFET acts as a resistor with a moderate electrical resistance. This flexibility allows the MOSFET to control the amount of current flowing in a circuit. Explain why a MOSFET becomes warm as it controls that current.

22. The tiny MOSFETs that are used to move charge onto and off the capacitors in dynamic memory are so small that they're never very good conductors. Why do their modest electrical resistances lengthen the time it takes to store or retrieve charge from the memory capacitors?

23. Why does the effect described in Exercise 22 limit the speed with which a computer can store or retrieve bits from its dynamic memory?

24. If you connect the output of one inverter to the input of a second inverter, how will the output of the second inverter be related to the input of the first inverter?

25. Suppose you connected a microphone directly to a large unamplified speaker. Why wouldn't the speaker reproduce your voice loudly when you talked into the microphone?

26. Why can't an audio amplifier operate without batteries or a power supply?

27. You like to listen to old phonograph records but your new stereo amplifier has no input for a phonograph. You connect the phonograph to the stereo's CD player input but find that the volume is extremely low. Why?

28. To correct the volume problem in Exercise 27, you buy a small preamplifier and connect it between the phonograph and the CD player input of the stereo. The volume problem is gone. What is the preamplifier doing to fix the problem?

CHAPTER 13

Electromagnetic Waves

Electric and magnetic fields are so intimately related that each can create the other even in empty space. In fact, the two fields can form electromagnetic waves, in which they recreate one another endlessly and head off across space at an enormous speed. These electromagnetic waves are all around us and are the basis for much of our communications technology, for radiative heat transfer, and for our ability to see the universe in which we live.

EXPERIMENT: Boiling Water in an Ice Cup

You can experiment with electromagnetic waves using a microwave oven. As we'll see in Section 13.2, microwaves are a type of electromagnetic wave that falls between radio waves and light. For reasons that we'll discuss in that section, microwaves transfer energy easily to water molecules in a liquid but not to water molecules in an ice crystal. That difference allows for some interesting cooking tricks.

Take a block or cube of ice from the freezer and melt a shallow bowl-shaped depression in its top surface. Then put the block on a microwave-safe ceramic plate and return it to the freezer to cool. Once the block and plate are cold and frozen, take them out of the freezer and put them quickly into the microwave. Fill the block's bowl-shaped depression with water and immediately start the microwave.

If you've been quick enough, the water in the depression will still be liquid when the oven starts producing microwaves. The liquid water will absorb power from the microwaves that fill the cooking chamber but the ice will not. The water will become extremely hot and will begin to melt into the block. Why was it important to chill the plate before putting the block into the microwave oven?

If you're lucky and the block is large enough, the water in the depression will reach a full boil before it melts its way through the block. Try to *predict* how the shape of the liquid region will change as the microwaves heat it up. *Observe* what happens and try to *verify* your prediction. *Measure* how quickly the ice melts and think about how you

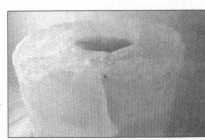

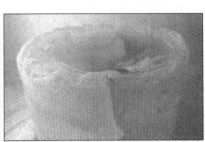

Courtesy Lou Bloomfield

could use that measurement to calculate how much power the microwave oven is delivering to the water.

Chapter Itinerary

In this chapter, we'll discuss how electromagnetic waves are formed and detected in two common situations: (1) *radio* and (2) *microwave ovens*. In *radio,* we'll examine the ways in which charge moving in an antenna can emit or respond to electromagnetic waves and how those waves can be controlled in order to send an audio signal from a radio transmitter to a radio receiver. In *microwave ovens,* we'll see how microwaves affect water molecules and metals, and also how they are produced in the oven's magnetron tube. While we can't see the electromagnetic waves that these two devices use, they clearly play important roles in our world. For more about what we'll study, turn to the Chapter Summary on p. 439.

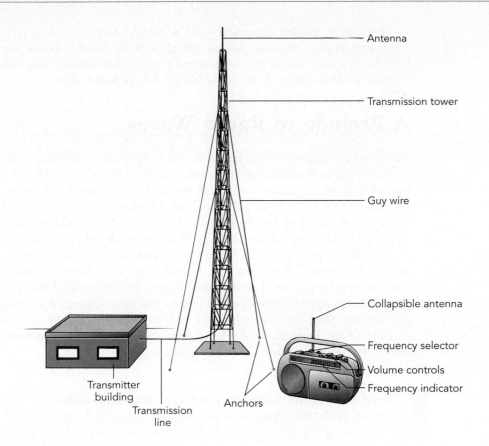

Antenna

Transmission tower

Guy wire

Collapsible antenna

Frequency selector

Volume controls

Frequency indicator

Transmitter
building

Transmission
line

Anchors

SECTION 13.1 **Radio**

We've seen that electric currents can represent sound and carry speech and music any-where wires will reach. But how can we send sound to someone who is moving? We need a way to represent sound that doesn't involve wires. We need radio.

This section describes how radio works. We'll look at how radio waves are trans-mitted and how they're received. We'll also examine the common ways in which sound is represented by radio waves so that it can travel through space to a receiver far away.

Questions to Think About: *How might the movement of electric charge in one metal antenna affect electric charge in a second antenna nearby? What about when the sec-ond antenna is far away from the first antenna? What does it mean when a radio sta-tion claims to transmit 50,000 watts? How does your radio select one channel from among all the possibilities?.*

Experiments to Do: *Listen to a small AM radio and notice that the volume of the sound depends on the radio's orientation or location. Radio waves are pushing electric charges back and forth along the radio's hidden internal antenna. You can sometimes find an ori-entation in which the radio is silent because in that orientation the radio waves are unable to move charges along the antenna. If you put the radio inside a metal box, it will also become silent. Can you explain why?*

You can try similar experiments with a cordless telephone—actually a radio transmitter and receiver. See how far you can go with the handset before you lose contact with the base unit. Notice that the antenna's size and orientation affect its range. What happens to the reception if you stand behind a large metal object?

A Prelude to Radio Waves

Before we can examine radio and radio waves, let's take a moment to finish the introduction to electrodynamics that we began in Chapters 10 and 11. Although we've already learned most of the fundamental relationships between electricity and magnetism, the remaining one is about to become important. To refresh your memory, we have observed so far that electric fields can be produced by electric charges and by changing magnetic fields, and that magnetic fields can be produced by moving electric charges (Table 13.1.1).

In 1865, Scottish physicist James Clerk Maxwell (1831–1879) discovered a second source of magnetic fields: *changing electric fields.* That effect is subtle and scientists overlooked it for most of the nineteenth century. It wasn't until Maxwell was trying to formulate a complete electromagnetic theory that he uncovered this additional connection between electricity and magnetism. This final relationship completed the set shown in Table 13.1.1. Together, these relationships allowed Maxwell to understand one of the most remarkable phenomena in nature: electromagnetic waves!

Third Connection between Electricity and Magnetism
Electric fields that change with time produce magnetic fields.

Since electric fields can create magnetic fields and magnetic fields contain energy, it's clear that electric fields must contain energy, too. The amount of energy in a uniform electric field is the square of the field strength times the volume of the field divided by 8π times the Coulomb constant. We can write this relationship as a word equation:

$$\text{energy} = \frac{\text{electric field}^2 \cdot \text{volume}}{8\pi \cdot \text{Coulomb constant}}, \tag{13.1.1}$$

in symbols:

$$U = \frac{E^2 \cdot V}{8\pi \cdot k},$$

and in everyday language:

When you charge a large capacitor, it stores a great deal of energy in the electric field between its plates.

Table 13.1.1 Sources of Electric and Magnetic Fields

SOURCES OF ELECTRIC FIELDS	SOURCES OF MAGNETIC FIELDS
Electric charge	Moving electric charge
Changing magnetic fields	Changing electric fields

With these observations, we have finished the prelude and are ready to see how radio works.

►Check Your Understanding #1: A Real Flux Capacitor

for answers, see page **440**

When a capacitor has separated charge on its plates, there is a strong electric field between those plates. Connecting the plates with a wire will discharge the capacitor and its electric field will suddenly vanish. As the electric field disappears, what other field is present between the plates?

►Check Your Figures #1: Lightning in the Fields

for answers, see page **441**

During a thunderstorm, the charged clouds produce an electric field of about 10,000 V/m near the ground. How much energy is contained in 1.0 cubic meter of that electric field?

Antennas and Tank Circuits

A radio transmitter communicates with a receiver via radio waves. These waves are produced by electric charge as it moves up and down the transmitter's antenna and are detected when they push electric charge up and down the receiver's antenna. But what exactly are radio waves and how does charge on the antenna produce them?

We've already seen that electric charge produces electric fields and that moving charge produces magnetic fields. However, something new happens when charge *accelerates*. Accelerating charge produces a mixture of changing electric and magnetic fields that can reproduce one another endlessly and travel long distances through empty space. These interwoven electric and magnetic fields are known generally as **electromagnetic waves.** In the case of radio, the electromagnetic waves have low frequencies and long wavelengths and are known as **radio waves.**

But before we look at the structure of a radio wave and at how it travels through space, let's start with a much simpler situation. We'll look at how two nearby metal antennas affect one another. Figure 13.1.1 shows a radio transmitter and a radio receiver, side by side. Because of their proximity, electric charge on the transmitter's antenna is sure to affect charge on the receiver's antenna.

To communicate with the nearby receiver, the transmitter sends charge up and down its antenna. This charge's electric field surrounds the transmitting antenna and extends all the way to the receiving antenna, where it pushes charge down and up. Unfortunately, the resulting charge motion in the receiving antenna is weak and the receiver may have difficulty distinguishing it from random thermal motion or from motion caused by other electric fields in the environment. Therefore the transmitter adopts a clever strategy—it moves charge up and down its antenna rhythmically at a particular frequency. Since the resulting motion on the receiving antenna is rhythmic at that same frequency, it's much easier for the receiver to distinguish from unrelated motion.

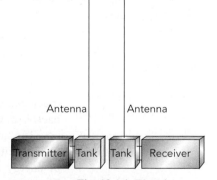

Fig. 13.1.1 Electric charge rushing on and off the transmitting antenna causes a similar motion of electric charge in the receiving antenna.

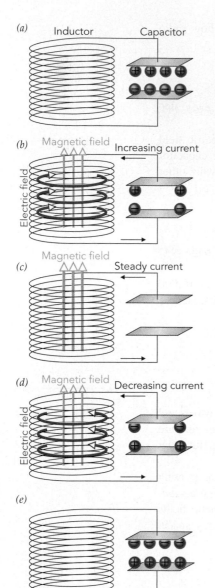

Fig. 13.1.2 A tank circuit consists of a capacitor and an inductor. Energy sloshes rhythmically back and forth between the two components.

Using this rhythmic motion has another advantage: it allows the transmitter and receiver to each use a **tank circuit**—a resonant electronic circuit consisting only of a capacitor and an inductor (Fig. 13.1.2). Charge "sloshes" back and forth through the tank circuit at a particular frequency, much as a child swings back and forth on a swing. And just as you can get a child swinging strongly at the playground by giving her a gentle push every swing, so the transmitter can make charge slosh strongly through its tank circuit by giving that charge a gentle push every cycle. By helping the transmitter move larger amounts of charge up and down the antenna, the tank circuit dramatically strengthens the transmission.

A second tank circuit attached to the receiving antenna helps the receiver detect this transmission. Gentle, rhythmic pushes by fields from the transmitting antenna cause more and more charge to move through the receiving antenna and its attached tank circuit. While the motion of charge on this antenna alone may be difficult to detect, the much larger charge sloshing in the tank circuit is unmistakable.

We can understand how a tank circuit works by looking at how charge moves between its capacitor and its inductor. Let's imagine that the tank circuit starts out with separated charge on the plates of its capacitor (Fig. 13.1.2a). Since the inductor conducts electricity, current begins to flow from the positively charged plate, through the inductor, to the negatively charged plate. But the current through the inductor must rise slowly and, as it does, it creates a magnetic field in the inductor (Fig. 13.1.2b).

Soon the capacitor's separated charge is gone and all of the tank circuit's energy is stored in the inductor's magnetic field (Fig. 13.1.2c). But the current keeps flowing, driven forward by the inductor's opposition to current changes. The inductor uses the energy in its magnetic field to keep the current flowing, and separated charge reappears in the capacitor (Fig. 13.1.2d). Eventually, the inductor's magnetic field decreases to zero and everything is back to its original state—almost. While all of the tank circuit's energy has returned to the capacitor, the separated charge in that capacitor is now upside-down (Fig. 13.1.2e).

This whole process now repeats in reverse. The current flows backward through the inductor, magnetizing it upside down, and the tank circuit soon returns to its original state. This cycle repeats over and over again, with charge sloshing from one side of the capacitor to the other and back again.

A tank circuit is an electronic harmonic oscillator, equivalent to the mechanical harmonic oscillators we examined in Chapter 9. Like all harmonic oscillators, its period (the time per cycle) doesn't depend on the amplitude of its oscillation. Thus no matter how much charge is sloshing in the tank circuit, the time it takes that charge to flow over and back is always the same.

The tank circuit's period depends only on its capacitor and its inductor. The larger the capacitor's capacitance, the more separated charge it can hold with a given amount of energy and the longer it takes that charge to move through the circuit as current. The larger the inductor's **inductance**—its opposition to current changes—the longer that current takes to start and stop. A tank circuit with a large capacitor and a large inductor may have a period of a thousandth of a second or more, while one with a small capacitor and a small inductor may have a period of a billionth of a second or less.

Inductance is defined as the voltage drop across the inductor divided by the rate at which current through the inductor changes with time. This division gives inductance the units of voltage divided by current per time. The SI unit of inductance is the volt-second-per-ampere, also called the **henry** (abbreviated H). While large electromagnets have inductances of hundreds of henries, a 1-μH (0.000001-H) inductor is more common in radio.

Its resonant behavior makes the tank circuit useful in radio. That's because small, rhythmic pushes on the current in a tank circuit can lead to enormous charge oscillations in that circuit. In radio, these rhythmic pushes begin when the transmitter sends an alternating current through a coil of wire. Fields from this coil push current back and forth through the nearby transmitting tank circuit, causing enormous amounts of charge to slosh back and forth in it and travel up and down the transmitting antenna. That charge's electric field then pushes rhythmically on charge in the receiving antenna, causing substantial amounts of charge to travel down and up it and slosh back and forth in the receiving tank circuit (Fig 13.1.3). The receiver can easily detect this sloshing charge.

Energy flows from the transmitter to the receiver via resonant energy transfer—from the transmitter, to the transmitting tank circuit and antenna, to the receiving antenna and tank circuit, and finally to the receiver. This sequence of transfers can work efficiently only if all the parts have resonances at the same frequency. Tuning a radio receiver to a particular station is largely a matter of adjusting its capacitor and inductor so that its tank circuit has the right resonant frequency.

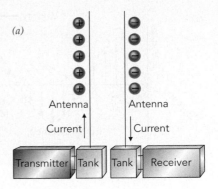

(a)

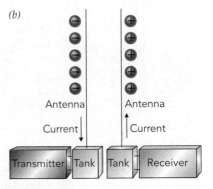

(b)

Fig. 13.1.3 (*a*) As current flows up the transmitting antenna, it causes current to flow down the receiving antenna and (*b*) vice versa.

► Check Your Understanding #2: No Tanks

for answers, see page 440

Why doesn't the radio transmitter simply push electric charge directly on and off the antenna, without using a tank circuit?

Radio Waves

When the two antennas are close together, charge in the transmitting antenna simply exerts electrostatic force on charge in the receiving antenna. But when the antennas are far apart, the interactions between them are more complicated. Charge in the transmitting antenna must then emit a radio wave in order to push on charge in the receiving antenna. Like a water wave, a radio wave is a disturbance that carries energy from one place to another. But unlike a water wave, which must travel in a fluid, a radio wave can travel through otherwise empty space, from one side of the universe to the other.

Like all electromagnetic waves, a radio wave consists only of a changing electric field and a changing magnetic field. These fields recreate one another over and over again as the wave travels through empty space at the speed of light—exactly 299,792,458 m/s (approximately 186,282 miles-per-second).

The radio wave is created when electric charge in the antenna accelerates. While stationary charge or a steady current produces constant electric or magnetic fields, accelerating charge produces fields that change with time. As charge flows up and down

(a)

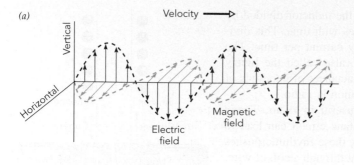

(b)

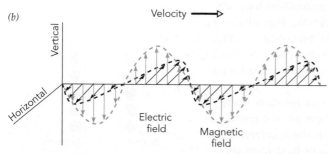

Fig. 13.1.4 (*a*) In a vertically polarized electromagnetic wave, the electric field along the wave's level path points alternately up and down while the magnetic field along that path points alternately right and left. (*b*) In a horizontally polarized wave, the electric field points alternately left and right while the magnetic field points alternately up and down.

the antenna, its electric field points alternately up and down and its magnetic field points alternately left and right. These changing fields then recreate one another again and again, and sail off through space as an electromagnetic wave.

The wave emitted by a vertical transmitting antenna has a **vertical polarization,** that is, its electric field points alternately up and down (Fig. 13.1.4*a*). We identify those "ups" as crests and the distance between adjacent crests is its wavelength. For radio waves, that wavelength is usually 1 m (3.3 ft) or more. The wave's magnetic field is perpendicular to its electric field and thus points alternately left and right.

Had the transmitting antenna been tipped on its side, the wave's electric field would have pointed alternately left and right and the wave would have had a **horizontal polarization** (Fig. 13.1.4*b*). The wave's magnetic field would then point alternately up and down. Whatever the polarization, the electric and magnetic fields move forward together as a traveling wave, so the pattern of fields moves smoothly through space at the speed of light.

Common Misconceptions: Electromagnetic Waves and Undulations

Misconception: Since the fields of an electromagnetic wave appear wavy (Fig. 13.1.4), the light wave itself undulates; it actually undulates up and down or back and forth as it heads rightward!

Resolution: The arrows drawn to represent the fields in an electromagnetic wave are associated with points along the straight axis of the wave. Each wave in Fig. 13.1.4 is heading directly rightward along the axis line and the arrows indicate field values at points along that line.

If you stood in one place and watched this wave pass, you'd notice its electric field fluctuating up and down at the same frequency as the charge that created it. When the wave passes a distant receiving antenna, it pushes charge up and down that antenna at this frequency. If the receiving tank circuit is resonant at this frequency, the amount of charge sloshing in it should become large enough for the receiver to detect.

A radio station can optimize its transmission by using a transmitting antenna of the proper length. When that length is exactly a quarter of the wavelength of the radio wave it's transmitting, charge sloshes vigorously up and down the antenna in a natural resonance. Surprisingly enough, the antenna is another electronic harmonic oscillator, with a period that depends only on its length. (In fact, the antenna is the top half of a tank circuit, with its tip acting as one plate of a capacitor and its length acting as the top half of an inductor. Objects at the base of the antenna complete the tank circuit.) When the transmitting tank circuit and antenna have resonances at the same frequency, there's a resonant energy transfer from one to the other. As you might expect, these resonant effects help to produce a powerful radio wave.

The transmitting antenna sends the strongest portion of its radio wave out perpendicular to its length. That's not unexpected because the motion of charge on the antenna is most obvious when viewed from a line perpendicular to its length. Thus a vertical antenna sends most of its wave out horizontally, where people are likely to receive it. No wave emerges from the ends of an antenna.

Both electric and magnetic fields contain energy, so as the electromagnetic wave travels through space, it carries energy away from the transmitter. When a radio station advertises that it "transmits 50,000 watts of music," it's claiming that its antenna emits 50,000 J of energy per second or 50,000 W of power in its electromagnetic wave. The receiving antenna must absorb enough of this power to detect the wave. But the farther the wave gets from the transmitting antenna, the more spread out and weak it becomes. Trees and mountains also absorb or reflect some of the wave and hinder reception.

For the best reception, a listener should be located where the radio wave is strong and where there's an unobstructed path from the transmitting antenna to the receiving antenna. The receiving antenna should be a quarter-wavelength long and it should be oriented along the radio wave's polarization—vertical for a vertically polarized radio wave or horizontal for a horizontally polarized radio wave. Aligning the receiving antenna with the wave's polarization makes certain that the wave's electric field pushes charge *along* the antenna, not *across* it.

To ensure good reception regardless of receiving antenna orientation, many radio stations transmit a complicated *circularly polarized* wave that combines both vertical and horizontal polarizations. To form this wave, they need several quarter-wavelength antennas. For wavelengths under a few meters, these antennas can all be attached inexpensively to a single mast. That's why commercial FM and TV broadcasts, which use short-wavelength radio waves, are usually transmitted with circular polarization. However, commercial AM broadcasts, which use long-wavelength radio waves, are transmitted only with vertical polarization.

Because commercial FM radio waves usually include both polarizations, FM receiving antennas can be vertical or horizontal. Portable FM receivers often use vertical telescoping antennas while home receivers frequently use horizontal wire antennas. All of these antennas are roughly a quarter-wavelength long.

A quarter-wavelength antenna for commercial AM radio would have to be about 100 m (330 ft) long, so straight AM antennas (such as those on cars) are much shorter than optimal. That's why many AM antennas are designed to respond to the radio wave's horizontal magnetic field rather than to its vertical electric field. These magnetic antennas are horizontal coils of wire that experience induced currents when exposed to fluctuating magnetic fields.

→ Check Your Understanding #3: There's No Place Like Home
for answers, see page **440**

Why do cordless telephones only work when they're close to their base units?

Representing Sound: AM and FM Radio

A radio transmitter does more than simply emit a radio wave. It uses that radio wave to *represent sound.* Because sound waves are fluctuations in air density and radio waves are fluctuations in electric and magnetic fields, a radio wave can't literally "carry" a sound wave. However, a radio wave can carry sound information and instruct the receiver how to reproduce the sound.

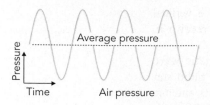

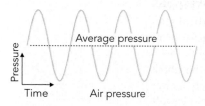

Fig. 13.1.5 When sound is transmitted using amplitude modulation, air pressure is represented by the strength of the radio wave. A compression is represented by strengthening the radio wave and a rarefaction is represented by weakening it.

To convey sound information, the radio station alters its radio wave to represent compressions and rarefactions of the air. The receiver then recreates those compressions and rarefactions. There are two common techniques by which a radio wave can represent those density fluctuations. One is called amplitude modulation and involves changing the overall strength of the radio wave. The other is called frequency modulation and involves small changes in the frequency of the radio wave.

In the **amplitude modulation** (AM) technique, air density is represented by the strength of the transmitted wave (Fig. 13.1.5). To represent a compression of the air, the transmitter is turned up so that more charge moves up and down the transmitting antenna. To represent a rarefaction, the transmitter is turned down so that less charge moves up and down the antenna. The frequency at which charge moves up and down the antenna remains steady, so only the amplitude of the radio wave changes. The receiver measures the strength of the radio wave and uses this measurement to recreate the sound. When it detects a strong radio wave, it pushes its speaker toward the listener and compresses the air. When it detects a weak radio wave, it pulls its speaker away from the listener and rarefies the air.

In the **frequency modulation** (FM) technique, air density is represented by the frequency of the transmitted wave (Fig. 13.1.6). To represent a compression of the air, the transmitter's frequency is increased slightly so that charge moves up and down the transmitting antenna a little *more* often than normal. To represent a rarefaction, the transmitter's frequency is decreased slightly so that the charge moves up and down a little *less* often than normal. These changes in frequency are extremely small—so small that charge continues to slosh strongly in all the resonant components and reception is unaffected. The receiver measures the radio wave's frequency and uses this measurement to recreate the sound. When it detects an increased frequency it compresses the air, and when it detects a decreased frequency it rarefies the air.

Fig. 13.1.6 When sound is transmitted by frequency modulation, air pressure is represented by changing the frequency of the radio transmitter slightly. A compression is represented by increasing that frequency and a rarefaction by decreasing it.

Although the AM and FM techniques for representing sound can be used with a radio wave at any frequency, the most common commercial bands in the United States are the AM band between 550 kHz and 1600 kHz (550,000 Hz and 1,600,000 Hz) and the FM band between 88 MHz and 108 MHz (88,000,000 Hz and 108,000,000 Hz). Elsewhere in the spectrum of radio frequencies are many other commercial, military, and public transmissions, including TV, short wave, amateur radio, telephone, police, and aircraft bands. These other transmissions use AM, FM, and a few other techniques to represent sound and information with radio waves.

> **Check Your Understanding #4: Another Volume Control**
>
> for answers, see page 440

When you are listening to the AM radio in a car and drive through a tunnel, the volume becomes very low. Explain.

Bandwidth and Cable

An audio signal on a wire has a range of frequencies present in it, from zero frequency up to the highest pitch sound it is representing. Similarly, an audio signal traveling via a radio wave has a range of radio frequencies present in it, stretching from somewhat below the official frequency of the radio wave, the *carrier frequency,* to somewhat above that frequency. The wider the audio frequency range of the sound, the more sound information must be sent each second and the wider the range of radio frequencies needed to represent that sound. The range of frequencies needed to transmit such a stream of information is known as the transmission's **bandwidth.**

By international agreement, an AM radio station may use 10 kHz of bandwidth, 5 kHz above and below its carrier frequency. To stay within that bandwidth, the audio signal can't contain frequencies above 5 kHz. While this restricted frequency range is bad for music, it allows competing stations to function with carrier frequencies only 10 kHz apart, so that 106 different stations can operate between 550 kHz and 1600 kHz.

An FM radio station may use 200 kHz of bandwidth, 100 kHz on each side of its carrier frequency. This luxurious allocation permits FM radio to represent a very broad range of audio frequencies, in stereo, which is why an FM radio station can do a much better job of sending music to your radio than an AM station can.

However, the spectrum of electromagnetic waves is a limited resource and if it could only be used once, it would quickly run out of bandwidth. Fortunately, distance and enclosures make it possible to reuse the spectrum many times. Cell phones that are far from one another can share the same carrier frequencies and bandwidth because their radio waves weaken with distance and essentially don't overlap. But even nearby radio transmissions can use the same carrier frequencies by enclosing their electromagnetic waves inside cables.

Cable radio, television, and data networks are similar to broadcast networks except that they send electromagnetic waves through cables rather than through empty space. A typical radio or television cable consists of an insulated metal wire inside a tube of metal foil or woven metal mesh. This wire-inside-a-tube arrangement is called *coaxial cable* because its two metal components share the same centerline or axis. In contrast, a typical computer-data cable consists of a number of insulated metal wires that are twisted into several pairs.

Electromagnetic waves can propagate easily through a coaxial or twisted-pair cable, following its twists and turns from the transmitter that produces the waves to the receiver that uses them. The fact that wires are assisting these waves in their travels makes them more complicated than waves in empty space. However, they still involve electric and magnetic fields and still propagate forward at nearly the speed of light.

Because the electromagnetic waves inside a cable don't interact with those outside it, the transmitter and receiver can use whatever parts of the spectrum they chose, without concerns about sharing. A typical coaxial cable can handle frequencies up to about 1000 MHz and typical twisted-pair cable can reach 350 MHz, so either one can carry a great deal of information each second.

However, coaxial cables must now compete with optical fiber cables that guide light from one place to another. We'll examine optical fibers in Section 15.2. Like radio waves, light is an electromagnetic wave and can be amplitude or frequency modulated to represent information. But light's frequency is extremely high; the frequencies of visible light range from 4.5×10^{14} Hz to 7.5×10^{14} Hz. If we were to allocate FM radio channels 200 kHz apart throughout the visible spectrum, there would be about 1.5 billion channels available!

Check Your Understanding #5: Beaten by the Bandwidth
for answers, see page 440

You're playing piano for an AM radio station and you strike the highest key on the piano. The pitch of the resulting sound is 4186 Hz. Can the listeners hear that note from their radios?

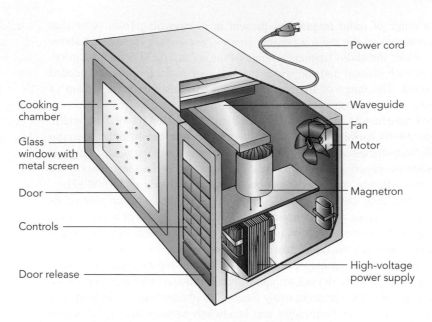

Microwave Ovens

In addition to carrying sounds from one place to another, electromagnetic waves can carry power. One interesting example of such power transfer is a microwave oven. It uses relatively high-frequency electromagnetic waves to transfer power directly to the water molecules in food, so that the food cooks from the inside out. This section discusses both how those waves are created and why they heat food.

Questions to Think About: *Why do microwave ovens tend to cook food unevenly if you don't move the food during cooking? How can part of a frozen meal become boiling hot while another part remains frozen? Why must you be careful with metal objects placed inside the oven? Why do some objects remain cool in the microwave oven while other objects become extremely hot? How does microwave popcorn work?*

Experiments to Do: *A microwave oven transfers power primarily to the water in food. You can see this effect by placing completely water-free food ingredients such as salt, baking power, sugar, or salad oil on a microwave-safe ceramic dish in a microwave oven. Cook the ingredients briefly. You will find that the ingredients and dish remain relatively cool. Add just a little water to the collection. What happens when you cook them this time?*

Now try cooking a very cold ice cube. The cube should come directly from the freezer on an ice-cold plate, so that its surface is solid and dry. What happens? If ice contains water and water is what absorbs power in a microwave oven, why doesn't the ice absorb power and melt?

Microwaves and Food

When studying incandescent lightbulbs in Section 7.3, we discussed the *wavelengths* of electromagnetic waves. While examining radio, we concentrated on the *frequencies* of

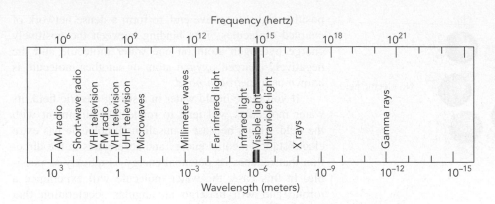

Fig. 13.2.1 The electromagnetic spectrum. Microwaves have wavelengths between about 1 m and 1 mm, corresponding to frequencies from 300 MHz up to 300 GHz.

electromagnetic waves. However, we know from Eq. 9.2.1 that wavelength and frequency of a wave aren't independent. A basic electromagnetic wave in empty space has both a wavelength and a frequency, and their product is the speed of light. That relationship can be written as a word equation:

$$\text{speed of light} = \text{wavelength} \cdot \text{frequency} \qquad \textbf{(13.2.1)}$$

in symbols:

$$c = \lambda \cdot \nu,$$

and in everyday language:

The higher the frequency of an electromagnetic wave, the shorter its wavelength becomes.

Like Fig. 7.3.2, Fig. 13.2.1 shows the approximate wavelengths of many types of electromagnetic waves but it also shows their frequencies.

Radio broadcasts use the low-frequency, long-wavelength portion of the electromagnetic spectrum. Commercial AM radio is at frequencies of 550 kHz to 1600 kHz (wavelengths of 545 m to 187 m) and commercial FM radio is at frequencies of 88 MHz to 108 MHz (wavelengths of 3.4 m to 2.8 m). Because these waves have wavelengths longer than 1 m, they are called radio waves. But the electromagnetic waves used in microwave ovens have wavelengths shorter than 1 m and are called microwaves. **Microwaves** extend from wavelengths of 1 m (3.3 ft) down to 1 mm (0.04 inches).

To explain how a microwave oven heats food (❑), let's begin by looking at water molecules. Water molecules are electrically polarized—that is, they have positively charged ends and negatively charged ends. This polarization comes about because of quantum physics and the tendency of oxygen atoms to pull electrons away from hydrogen atoms. The water molecule is bent, with its two hydrogen atoms sticking up from its oxygen atom like Mickey Mouse's ears. When the oxygen atom pulls the electrons partly away from the hydrogen atoms, its side of the molecule becomes negatively charged while the hydrogen atoms' side becomes positively charged. Water is thus a polar molecule.

In ice, these polar water molecules are arranged in an orderly fashion with fixed positions and orientations. But in liquid water, the molecules are more randomly oriented (Fig. 13.2.2). Their arrangements are constrained only by their tendency to bind together,

❑ Though he was orphaned as child and never completed grade school, American Percy Lebaron Spencer (1894–1970) had a brilliant career as a scientist and microwave engineer. In 1945, while visiting a magnetron testing laboratory, he leaned over an operating magnetron and the candy bar in his shirt pocket melted. Immediately recognizing what had happened, he soon had popcorn popping about the lab and even cooked an egg until it exploded. Cooking has never been the same since.

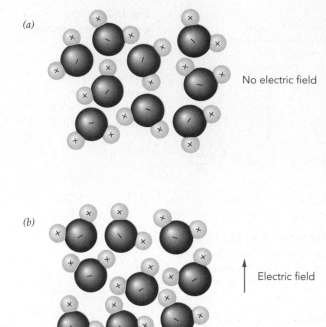

No electric field

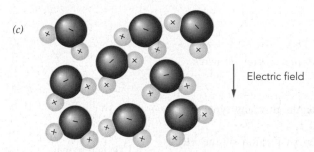

Electric field

Electric field

Fig. 13.2.2 (*a*) The water molecules in liquid water are randomly oriented when there's no electric field. (*b, c*) But an electric field tends to orient them with their positive ends in the direction of the field.

positive end to negative end, to form a dense network of coupled molecules. This binding between the positively charge hydrogen atom on one water molecule and the negatively charged oxygen atom on another molecule is known as a *hydrogen bond*.

If you place liquid water in a strong electric field, its water molecules will tend to rotate into alignment with the field. That's because a misaligned molecule has extra electrostatic potential energy and accelerates in the direction that reduces its potential energy as quickly as possible. In this case, the water molecule will experience a torque and will undergo an angular acceleration that makes it rotate into alignment. As it rotates, the molecule will bump into other molecules and convert some of its electrostatic potential energy into thermal energy.

A similar effect occurs at a crowded party when everyone is suddenly told to face the front of the room. People brush against one another as they turn and sliding friction converts some of their energy into thermal energy. If the people were told to turn back and forth repeatedly, they would become quite warm. The same holds true for water. If the electric field reverses its direction many times, the water molecules will turn back and forth and become hotter and hotter.

A microwave's fluctuating electric field is well suited to heating water. A microwave oven uses 2.45-GHz (2.45-gigahertz or 2,450,000,000-Hz) microwaves to twist food's water molecules back and forth, billions of times per second. As the water molecules turn, they bump into one another and heat up. The water absorbs the microwaves and converts their energy into thermal energy. This particular microwave frequency was chosen not because of any resonant effect but because it was not in use for communications and because it cooks food uniformly. If the frequency were higher, the microwaves would be absorbed too strongly by food and wouldn't penetrate deep into large items. If the frequency were lower, the microwaves would pass through foods too easily and wouldn't cook efficiently.

This twisting effect explains why only foods or objects containing water or other polar molecules cook well in a microwave oven. Ceramic plates, glass cups, and plastic containers are water-free and usually remain cool. Even ice has trouble absorbing microwave power because its crystal structure constrains the water molecules so they can't turn easily.

But while ice melts slowly in a microwave oven, the liquid water it produces heats quickly. This peculiar heating behavior explains why it's so easy to burn yourself on frozen food heated in a microwave oven. Portions of the food that defrost first absorb most of the microwave power and overheat while the rest of the food remains frozen solid. You never know whether your next bite will break your teeth or sear the roof of your mouth. To address this problem, many microwave ovens have defrost cycles in which microwave heating is interrupted periodically to let heat flow naturally through the food to melt the ice. Once the frozen parts have melted, all of the food can absorb microwaves.

► Check Your Understanding #1: Microwave Popcorn

for answers, see page **440**

A popcorn kernel contains moist starch trapped inside a hard, dry hull. You can scorch this hull by cooking the corn in hot oil but not when you cook it in a microwave oven. How can the microwave oven pop the corn without risk of overheating the hull?

► Check Your Figures #1: Shopping for Food

for answers, see page **441**

The red light used by many grocery store checkout stations to scan product codes is produced by a helium-neon laser. This light is an electromagnetic wave with a wavelength of approximately 633 nm. What is its frequency?

Metal in a Microwave Oven

Contrary to popular lore, metal objects and microwave ovens aren't always incompatible. In fact, the walls of the oven's cooking chamber are metal, yet they cause no trouble when exposed to microwaves during cooking. Like most metal surfaces, the walls reflect microwaves. They do this by acting as both receiving and transmitting antennas. Electric fields in the microwaves cause mobile charges in the metal surfaces to accelerate and absorb the original microwaves. But as these charges accelerate, they emit new microwaves. The emitted microwaves have the same frequencies as the original ones, but they travel in new directions. The original microwaves have been reflected by the surface.

The cooking chamber walls reflect the oven's microwaves and keep them bouncing around inside. Even the metal grid covering the window reflects microwaves. That's because charge has enough time during a microwave cycle to flow around each hole in the grid and compensate for the hole's presence. As long as the wavelength of an electromagnetic wave is much larger than the holes in a metal grid, the wave reflects perfectly from that grid. In fact, if there's nothing inside the oven to absorb the microwaves, they'll bounce around inside it until they return to their source, a vacuum tube called a magnetron (Fig. 13.2.3), and eventually cause it to overheat.

While metal surfaces help confine the microwaves inside the oven, cooking your food and not you, extra metal inside the microwave can cause trouble. If you wrap food in aluminum foil, the foil will reflect the microwaves and the food won't cook. However, food placed in a shallow metal dish cooks reasonably well because microwaves enter the open top, pass through the food, reflect, and pass through the food again.

Sometimes metal's mobile charges do more than just reflect microwaves. If enough charge is pushed onto the sharp point of a metal twist-tie or scrap of aluminum foil, some of it will jump right into the air as a spark. This spark can start a fire, particularly when the twist-tie is attached to something flammable, like a plastic or paper bag. As a rule of thumb, never put a sharp metal object in the microwave oven.

Some metal objects heat up in a microwave oven. When microwaves push charge back and forth in a metal, the metal experiences an alternating current. If the metal has

Fig. 13.2.3 This oven's magnetron microwave source is located in the middle of the picture, just to the left of its cooling fan. Microwaves travel to the cooking chamber through the rectangular metal duct on top of the oven. The high-voltage transformer at the bottom right provides power to the magnetron.

Courtesy Lou Bloomfield

a substantial electrical resistance, this alternating current will experience a voltage drop and heat up the metal. While thick oven walls and cookware have low resistances and remain cool, thin metal strips quickly overheat. Metallic decorations on porcelain dinnerware are particularly susceptible to damage in a microwave oven, so that warming up coffee in Grandma's gold-rimmed teacup is sure disaster. When putting metal into a microwave oven, make sure that it is thick enough to conduct electricity well and that it has no sharp points.

Resistive heating in conducting objects can actually be useful at times. Since microwave ovens cook food inside and out at the same time, the food's surface never gets particularly hot and the food doesn't brown or become crisp. To improve their textures and appearances, some foods come with special wrappers that conduct just enough current to become very hot in a microwave oven. These wrappers provide the high surface temperatures needed to brown the foods.

Another peculiar feature of microwave ovens is that they don't always cook evenly. That's because the amplitude of the microwave electric field isn't uniform throughout the oven. As the microwaves bounce around the cooking chamber, they pass through the same spot from several different directions at once. When they do, they exhibit interference effects (see Section 9.3). At one location, the individual electric fields may point in the same direction and experience constructive interference, so that food there heats up quickly. But at another location, those fields may point in opposite directions and experience destructive interference, so that food there doesn't cook well at all.

If nothing is moving in the microwave oven, the pattern of microwaves inside it doesn't move either. There are then regions in which the electric field has very large amplitudes and regions in which the amplitudes are very small. The larger the amplitude of the electric field, the faster it cooks food.

To heat food uniformly in such a microwave oven, you must move the food around as it's cooking. Many ovens have turntables inside that move the food automatically. Another solution to this problem is to stir the microwaves around the oven with a rotating metal paddle. The pattern of microwaves inside the chamber changes as the paddle turns and the food cooks more evenly. Still other microwave ovens use two separate microwave frequencies to cook the food. Because these two frequencies cook independently, it's unlikely that a portion of the food will be missed by both waves at once.

▶Check Your Understanding #2: Half and Half

for answers, see page 440

You place a thick metal divider into your microwave oven, so that it divides the cooking chamber exactly in half. The oven sends its microwaves into the right half of the chamber. If you put food in the left half of the chamber, will it cook?

Creating Microwaves with a Magnetron

Clearly, changing electric fields cook the food as microwaves bounce around the inside of an oven. But how are these microwaves created? From the previous section on radio, you might guess that the oven creates an alternating current at 2.45 GHz and that this current causes charge to slosh in a tank circuit and move up and down an antenna. That's pretty much what actually happens inside a magnetron tube.

A magnetron is a special vacuum tube—a hollow chamber from which all the air has been removed. Composed primarily of metal and ceramic parts, the magnetron uses beams of electrons to make charge slosh in a number of microwave tank circuits. These tank circuits have resonant frequencies of 2.45 GHz, the operating frequency of the oven. With the help of a tiny antenna, the magnetron emits the microwaves that cook the food.

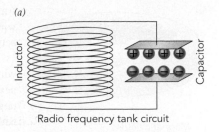

(a)

Inductor

Capacitor

Radio frequency tank circuit

The microwave tank circuits are arranged in a ring around the magnetron's evacuated chamber. For one of these tank circuits to oscillate naturally at 2.45 GHz, its capacitor must have an extremely small capacitance and its inductor must have an extremely small inductance. These requirements can be met by a C-shaped strip of metal (Fig. 13.2.4). Its curve is the inductor and its tips are the capacitor.

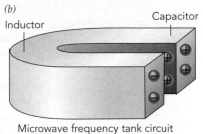

(b)

Inductor   Capacitor

Microwave frequency tank circuit

Electric charge sloshes back and forth on the C-shaped strip just as it does in a conventional tank circuit (Fig. 13.1.2). Known as a **resonant cavity,** this strip is another electronic harmonic oscillator and therefore has a period that doesn't depend on the amount of charge that's sloshing.

The magnetron of a microwave oven typically contains eight of these resonant cavities, each carefully adjusted in size and shape so that its natural resonance occurs exactly at 2.45 GHz. Because these cavities are arranged in a ring and each one shares its tips with those of its neighboring cavities, they tend to oscillate alternately (Fig. 13.2.5). At the start of an oscillatory cycle, half the metal tips are positively charged and half are negatively charged (Fig. 13.2.5a). Currents begin to flow through the ring and produce magnetic fields in the resonant cavities (Fig. 13.2.5b). These magnetic fields propel the currents around the ring even after the charge separations have vanished. Soon the charge separations reappear but with the positive and negative tips interchanged (Fig. 13.2.5c).

Fig. 13.2.4 (a) At radio frequencies, a tank circuit's inductor is a coil of wire and its capacitor is a pair of separated plates. (b) At microwave frequencies, a tank circuit's inductor is merely the curve of a C-shaped strip and its capacitor is the tips of that strip.

These currents oscillate back and forth around the cavities at 2.45 GHz, filling the magnetron with alternating electric and magnetic fields. But as the energy in these fields is extracted to cook the food or is lost to the imperfect conductivities of the cavities themselves, something must continuously replenish it. That replacement power is supplied to the cavities by four streams of energetic electrons.

At the center of the magnetron tube, surrounded only by empty space, is an electrically heated cathode that tends to emit electrons (Fig. 13.2.6a). A high-voltage power supply pumps negative charge onto this cathode so that a strong electric field points toward it from the positively charged cavity tips. If there were no other fields present in the magnetron, negatively charged electrons would emerge from the hot cathode and accelerate toward the positively charged tips as four beams of electrons (Fig. 13.2.6b).

However, the magnetron also includes a large permanent magnet. Why else would it be called a magnetron? This magnet creates a strong, steady magnetic field that points upward along the axis of the magnetron, parallel to the cathode itself (Fig. 13.2.6c). If there were no other fields inside the magnetron, electrons would experience only Lorentz forces perpendicular to their velocities and would circle around the magnetic flux lines in counterclockwise loops—a behavior known as **cyclotron motion.** The circling electrons would remain near the cathode and would never go near the cavities.

But in a real magnetron, the electric field of Fig. 13.2.6b and the magnetic field of Fig. 13.2.6c are present simultaneously. Because both of these fields exert forces on moving electrons, the paths the electrons follow are extremely complicated (Fig. 13.2.6d). The outward directed and circulating motions merge together into four electron beams which arc outward and rotate counterclockwise, like the spokes of a spinning bicycle wheel.

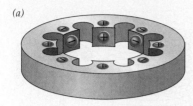

(a)

Magnetic fields

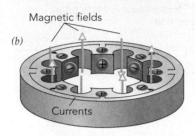

(b)

Currents

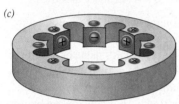

(c)

Magnetron resonators

Fig. 13.2.5 A typical magnetron has eight C-shaped resonant cavities arranged in a ring. (*a*) Separated charge on the tips of the cavities (*b*) flows as currents through the ring and (*c*) becomes reversed. As the currents flow, magnetic fields appear in the eight cavities, pointing alternately up and down.

An electron beam reaches each cavity not at its positively charge tip, as it would without the magnetic field, but at its negatively charged tip. The electron beams actually add to the charge separations in the cavities!

The electron beams sweep around the cathode in perfect synchronization with the oscillating charge on the cavities. The beams sweep from one tip to the next in the same amount of time it takes for the charge separation on the tips to reverse. As a result, the beams always arrive on the negatively charged tip. By adding to the charge separations, the electron beams provide power to the oscillations in the cavities, keeping them going and allowing them to transfer power to the food. The electron beams actually initiate the oscillation in the cavities by adding energy to tiny random oscillations that are always present in electric systems.

But how does the oscillating charge inside the magnetron create microwaves inside the oven's cooking chamber? There are many ways to extract microwaves from the ring of cavities. One extraction method is to insert a single-turn wire coil into one of the magnetron's cavities. As the magnetic field in that cavity changes, it induces a 2.45-GHz alternating current in the coil. One end of this coil is attached to the ring but the other end passes out of the magnetron through an insulated, air-tight hole in the ring and connects to a quarter-wavelength antenna. This 3-cm (1.2-in) antenna emits microwaves into a metal pipe attached to the cooking chamber. These microwaves reflect their way through the pipe and into the cooking chamber, where they cook the food.

Check Your Understanding #3: Large Economy Size

for answers, see page 440

If a manufacturer made a magnetron that was slightly larger than normal in every dimension, how would that magnetron behave?

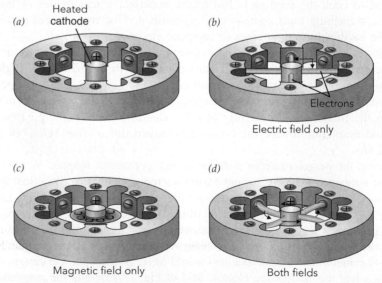

Fig. 13.2.6 (*a*) Electrons are emitted by the hot cathode in the center of a magnetron's ring of resonant cavities. (*b*) Electric fields alone would accelerate the electrons toward the positively charged cavity tips. (*c*) A magnetic field alone (pointing upward) would make the electrons orbit the cathode in counterclockwise loops. (*d*) Together, these fields create spokelike electron beams that circle the cathode counterclockwise and always strike negatively charged tips of the cavities.

Epilogue for Chapter 13

This chapter examined two common devices that are based on electromagnetic waves. In *radio*, we saw that electromagnetic waves can be created by accelerating electric charge and that these waves can be detected by looking for their effects on other electric charges. We also examined the techniques that are used to send audio signals through space by having them control either the amplitude or the frequency of electromagnetic waves.

In *microwave ovens*, we explored the ways in which electromagnetic waves can interact directly with polar water molecules and can transfer energy to those molecules. We also saw how interactions between microwaves and a metal object can lead to reflection, sparking, or heating. And we examined the technique used in ovens to create powerful microwave radiation.

Explanation: Boiling Water in an Ice Cup

The water molecules in ice are held rigidly in place by ice's crystalline structure. Unable to move or turn, these water molecules can't follow the microwave radiation's rapidly changing electric field. As a result, ice is unable to absorb much energy from the microwave radiation.

In contrast, molecules in liquid water can move about relatively freely and turn back and forth with the changing electric field of the microwaves. These water molecules absorb energy from the microwave radiation and become hotter. The liquid water's temperature rises.

While heat tends to flow out of the hot water and into the ice, water is not a good conductor of heat. Convection stops working once the water temperature rises above 4 °C and the hotter water begins to float on the colder water beneath it. In favorable circumstances, the water can begin to boil before it has melted the ice cup that conta

Chapter Summary

How Radio Works: A radio transmitter creates a radio wave when electric charge accelerates up and down its antenna. To get as much charge moving as possible, the transmitter attaches a tank circuit to the antenna and slowly adds energy to that tank circuit until an enormous amount of charge is flowing up and down the antenna. If the antenna is a quarter-wavelength long, it's also resonant at the transmission frequency and boosts the amount of charge sloshing up and down.

The radio receiver detects this radio wave when it causes charge to accelerate up and down the receiving antenna. If both the receiving antenna and the receiver's tank circuit are resonant at the transmission frequency, large amounts of charge will slosh back and forth in the receiver's tank circuit and the receiver will detect the transmission.

This radio wave can represent sound using either the AM or FM technique. In the AM technique, the strength of the wave is increased or decreased to represent compressions and rarefactions of the air, respectively. In the FM technique, the precise frequency of the transmission is increased or decreased to represent those compressions or rarefactions.

How Microwave Ovens Work: A microwave oven uses microwaves to cook food. These microwaves bounce around the cooking chamber, where they transfer energy to water molecules in the food. Because a water molecule is polar, having a positive end and a negative end, it tends to align with an electric field. The microwave's fluctuating electric field causes the tightly packed water molecules to twist back and forth rapidly, and the ensuing collisions heat the water and cook the food.

The oven's microwaves are produced by a magnetron, a vacuum tube containing resonant cavities and a heated cathode. By combining strong electric and magnetic fields, the magnetron produces powerful beams of electrons that add energy to the charge oscillating in the cavities. A loop of wire and a short antenna extract power from the resonant cavities and emit the microwaves that then cook the food.

Important Laws and Equations

1. The Energy in an Electric Field: The energy of an electric field is equal to the square of that field times its volume, divided by 8π times the Coulomb constant, or

$$\text{energy} = \frac{\text{electric field}^2 \cdot \text{volume}}{8\pi \cdot \text{Coulomb constant}}. \qquad (13.1.1)$$

2. Relationship between Wavelength and Frequency: The product of the frequency of an electromagnetic wave times its wavelength is the speed of light, or

$$\text{speed of light} = \text{wavelength} \cdot \text{frequency}. \qquad (13.2.1)$$

Check Your Understanding—Answers

Section 13.1 RADIO

1. A magnetic field.
Why: Since a changing electric field produces a magnetic field, the plates have a magnetic field between them while their electric field is disappearing.

2. The amount of charge that the transmitter can move directly on and off the antenna is too small to create a strong radio signal.
Why: The tank circuit is useful because it allows the transmitter to move much more charge. Just as a tuning fork is inefficient at emitting sound waves by itself, so a radio antenna is inefficient at emitting radio waves by itself. You can make the tuning fork much louder by coupling it to an object that resonates at its frequency. Similarly, you can make the radio antenna emit a much stronger radio wave by coupling it to a tank circuit that resonates at its frequency.

3. When the cordless telephone is too far from its base unit, their electromagnetic waves become so spread out that they have trouble communicating.
Why: The powers emitted by the base unit and the handset are small, so that their waves are relatively difficult to detect. As long as the handset and base unit are nearby, they are able to detect each other's waves. But when the distance between them becomes too great, the waves become too spread out to detect and the handset and base unit lose contact with one another.

4. The tunnel blocks most of the radio wave. Since only a small fluctuating wave reaches your radio, the radio produces only small fluctuations in air density with its speaker.
Why: An AM radio has trouble distinguishing between a distant transmission representing loud music and a nearby

transmission representing soft music. In both cases, the receiver detects only small variations in the current moving up and down its antenna. That's why you must turn up the volume of an AM radio as you move farther from the transmitting antenna or as you enter a tunnel.

5. Yes, at least in principle.
Why: The bandwidth of an AM radio station extends to 5000 Hz above and below its carrier frequency, so the station is officially permitted to represent sound frequencies as high as 5000 Hz. However, in practice the station probably begins to filter out sounds well below that frequency to avoid accidentally violating their license.

Section 13.2 MICROWAVE OVENS

1. The microwave oven transfers heat to water molecules in the starch so that the hull never becomes hotter than the material inside it.
Why: A corn kernel cooked in oil is heated by contact with the hot oil and pot. You can easily overheat the outer hull and burn it. But microwaves transfer heat to the water molecules inside the kernel. The hull can't overheat because the hottest thing it touches is the starchy insides of the kernel. When the pressure of steam inside the kernel becomes high enough, the hull breaks and the kernel "pops."

2. No.
Why: The metal divider would reflect microwaves and keep them from entering the left half of the oven.

3. It would operate at a frequency below 2.45 GHz.
Why: The frequency of the microwaves emitted by a magnetron is determined exclusively by the natural resonances of

its cavities. If those cavities are enlarged, both the inductances of their curves and the capacitances of their tips will increase.

Their resonant frequencies will decrease and the magnetron will emit lower frequency microwaves.

Check Your Figures—Answers

Section 13.1 RADIO

1. About 0.00045 J.

Why: Since 10,000 V/m is equivalent to 10,000 J/C·m, Eq. 13.1.1 gives us the energy of a 10,000 V/m field occupying 1.0 m^3 as:

$$\text{energy} = \frac{(10,000 \text{ J/C} \cdot \text{m})^2 \cdot 1.0 \text{ m}^3}{8\pi \cdot 8.988 \times 10^9 \text{N} \cdot \text{m}^2/\text{C}^2}$$

$$= \frac{0.00045 \text{ J}^2}{\text{N} \cdot \text{m}} = 0.00045 \text{ J}.$$

While that isn't much energy per cubic meter, a single lightning strike may release the electric field energy in almost a billion cubic meters. No wonder it produces such a bang!

Section 13.2 MICROWAVE OVENS

1. About 4.74×10^{14} Hz.

Why: Since the product of frequency and wavelength for an electromagnetic wave is equal to the speed of light, its frequency is equal to the speed of light divided by its wavelength:

$$\frac{299,792,458 \text{ m/s}}{0.000000633 \text{ m}} = 4.74 \times 10^{14} \text{ Hz}.$$

Exercises

1. If you pull a permanent magnet rapidly away from a tank circuit, what is likely to happen in that circuit?

2. Will the speed with which you pull the magnet away from the tank circuit (Exercise 1) affect the period of its charge oscillation?

3. A tank circuit consists of an inductor and a capacitor. Give a simple explanation for why the magnetic field in the inductor is strongest at the moment the separated charge in the capacitor reaches zero.

4. The metal wires from which most tank circuits are made have electrical resistances. Why do these resistances prevent charge from oscillating forever in a tank circuit, and what happens to the tank circuit's energy as time passes?

5. To add energy to the charge oscillation in a tank circuit with an antenna, at which time during the oscillation cycle should you bring a positively charged wand close to the antenna?

6. Two identical tank circuits with antennae are next to one another. Explain why charge oscillating in one tank circuit can continue to do work on charge oscillating in the other tank circuit.

7. The ignition system of an automobile produces sparks to ignite the fuel in the engine. During each spark process, charges suddenly accelerate through a spark plug wire and across a spark plug's narrow gap. Sometimes this process introduces noise into your radio reception. Why?

8. To diminish the radio noise in a car (see Exercise 7), the ignition system uses wires that are poor conductors of electricity. These wires prevent charges from accelerating rapidly. Why does this change improve your radio reception?

9. The electronic components inside a computer transfer charge to and from wires, often in synchrony with the computer's internal clock. Without packaging to block electromagnetic waves, the computer will act as a radio transmitter. Why?

10. To save power in a computer, its thousands of wires usually avoid sharp bends. Why do sharp bends in current-carrying wires waste power?

11. The sun emits a stream of energetic electrons and protons called the solar wind. These particles frequently get caught up in the earth's magnetic field, traveling in spiral paths that take them toward the north or south magnetic poles. When they head northward and collide with atoms in the earth's upper atmosphere, those atoms emit light we know as the aurora borealis, or northern lights. These particles also interfere with radio reception. Why do they emit radio waves?

12. When a radio signal travels through a coaxial cable, charge moves back and forth on both the central wire and the surrounding tube. Show that both electric and magnetic fields are present in the coaxial cable.

13. If you wave a positively charged wand up and down vertically, the electromagnetic wave it emits has which polarization?

14. If you set a magnetic compass on the table and spin its magnetic needle horizontally, its accelerating poles will emit an electromagnetic wave with which polarization?

15. While a particular AM radio station claims to transmit 50,000 W of music power, that's actually its average power. There are times when it transmits more power than that and times when it transmits less. Explain.

16. When your receiver is too far from an AM radio station, you can only hear the loud parts of the transmission. When it's too far from an FM station, you lose the whole sound all at once. Explain the reasons for this difference.

17. When an AM radio station announces that it's transmitting at 950 kHz, that statement isn't quite accurate. Explain why it may also be transmitting at 948 kHz and 954 kHz.

18. The Empire State Building has several FM antennas on top, added in part to increase its overall height. These antennas aren't very tall. Why do short antennas, located high in the air, do such a good job of transmitting FM radio?

19. Porous, unglazed ceramics can absorb water and moisture. Why are they unsuitable for use in a microwave oven?

20. Why are most microwave TV dinners packaged in plastic rather than aluminum trays?

21. Why is it so important that a microwave oven turn off when you open the door?

22. Compare how a potato cooks in a microwave oven with how it cooks in an ordinary oven.

23. When you're listening to FM radio near buildings, reflections of the radio wave can make the reception particularly bad in certain locations. Compare this effect to the problem of uneven cooking in a microwave oven.

24. Dish-shaped reflectors are used to steer microwaves in order to establish communications links between nearby buildings. Those reflectors are often made from metal mesh. Why don't they have to be made from solid metal sheets?

25. Why is the thin metal handle of a Chinese food container dangerous when placed in a microwave oven?

26. Is a thick, smooth-edged stainless steel bowl dangerous in a microwave oven?

27. A cyclotron is a particle accelerator invented in 1929 by American physicist Ernest O. Lawrence. It uses electric fields to do work on charged particles as they follow circular paths in a strong magnetic field. Lawrence's great insight was that all the particles take the same amount of time to complete one circle, regardless of their speed or energy. That fact allows the cyclotron to do work on all the particles at once as they circle together. How can a faster moving electron take the same time to circle as a slower moving electron?

28. An extremely fast-moving charged particle traveling in a magnetic field can radiate X-rays, a phenomenon known as synchrotron radiation. Why is the magnetic field essential to this emission?

Problems

1. How much energy is contained in 1.0 m^3 of a 1.0-V/m (or 1.0-N/C) electric field?

2. How much energy is contained in 1.0 m^3 of a 10,000-V/cm electric field?

3. What volume of a 1000-N/C electric field contains 1.0 J of energy?

4. How much volume of a 500-V/m electric field contains 0.0010 J of energy?

5. What electric field is needed for 1.0 m^3 to contain 1.0 J of energy?

6. What electric field contains 0.05 J in 10 m^3?

7. The frequency of the radio wave emitted by a cordless telephone is 900 MHz. What is the wavelength of that wave?

8. Citizens band (CB) radio uses radio waves with frequencies near 27 MHz. What are the wavelengths of these waves and how long should a quarter-wavelength CB antenna be?

9. The electromagnetic waves in blue light have frequencies near 6.5×10^{14} Hz. What are their wavelengths?

10. Amateur radio operators often refer to their radio waves by wavelength. What are the approximate frequencies of the 160-m, 15-m, and 2-m wavelength amateur radio bands?

11. The radio waves used by cellular telephones have wavelengths of approximately 0.36 m. What are their frequencies?

14

Light

While radio waves and microwaves are useful for communications and energy transfer, there's another portion of the electromagnetic spectrum that we find far more important: light. Light consists of very-high-frequency, very-short-wavelength electromagnetic waves. Light's frequencies are so high that normal antennas can't handle it. Instead, it's absorbed and emitted by the individual charged particles in atoms, molecules, or materials. Because of its special relationship with the charged particles in matter, light is important to physics, chemistry, and materials science. Moreover, it's one of the principal ways by which we interact with the world around us.

EXPERIMENT: Splitting the Colors of Sunlight

We see light because it stimulates cells in our eyes. This stimulation is an example of light's ability to influence chemistry. And because our eyes are able to distinguish between

different wavelengths of light, we perceive colors. While sunlight normally appears uncolored, that's because it contains a rich mixture of wavelengths that our eyes interpret as whiteness. But there are situations in which sunlight becomes separated into its constituent colors.

You can observe this separation of colors by looking at sunlight passing through a cut crystal glass or bowl, or by reflecting sunlight from a CD or DVD. Hold the object in direct sunlight and observe the light that it redirects toward your eyes or projects onto a white sheet of paper nearby. While some of the light you see will still be white, you should see colors as well.

Turn the object slowly in your hand and observe how the colors change. You will see gradual progressions from one color to the next. *Predict* the order of colors you'll see and then *observe* the actual sequence. Were you able to *verify* your predictions? Is the sequence of colors always the same? How does this sequence relate to the colors of the rainbow? What is the relationship between this sequence and the wavelengths of light? Can you *measure* the relative spacings of the classic rainbow colors: red, orange, yellow, green, blue, indigo, and violet?

Chapter Itinerary

In this chapter, we'll examine three sources of light: (1) *sunlight*, (2) *discharge lamps*, and (3) *lasers and LEDs*. In *sunlight*, we'll see how sunlight travels to our eyes and how its passage through the atmosphere, raindrops, and soap bubbles can separate it into its constituent colors. In *discharge lamps*, we'll explore the ways in which atoms and molecules emit and absorb light, and how different atoms and molecules can be used to produce light of different colors. In *lasers and LEDs*, we'll look at how atoms, molecules, and solids can duplicate or amplify the light passing through them and thus produce intense beams of highly ordered light. In the process of studying these three light sources, we'll also learn about three different types of light: thermal light, atomic resonance light, and coherent light. For additional preview information, flip to the Chapter Summary on p. 471.

Sunlight

For thousands of years, people have marked the passage of time by the rising and setting of the sun over the horizon. The sun first appears as a red disk in the east every morning, rises white in the blue sky, and then sets once again as a red disk in the west. The sunlight that we see takes about 8 minutes to travel the 150,000,000 km (93,000,000 miles) from the sun to our eyes and provides most of the energy and heat that make life on earth possible. While the light in sunlight is really just another electromagnetic wave, and could be considered part of the previous chapter, it's so important to everyday life that it deserves special attention. And so we'll begin by looking at how sunlight interacts with our world.

Questions to Think About: Why is the sun red at sunrise and sunset? Why is the sky blue during the day? Why can you occasionally see sunbeams when sunlight enters an otherwise dark room through a small window? Why do we see colors when sunlight passes through cut crystal or a soap bubble?

Experiments to Do: Sunlight is actually composed of many different electromagnetic waves. These waves differ in frequency and wavelength like the radio waves from your two favorite stations. But you don't need a machine to help you distinguish between various wavelengths of light; you can use your eyes. Take a look at a soap bubble on a bright, sunny day. You see the bubble because it reflects light. In fact, the clear bubble appears colored, even though the sunlight hitting it is white. That's because the bubble separates sunlight according to wavelength and sends only certain wavelengths toward your eyes.

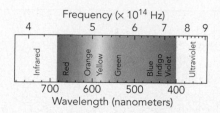

Fig. 14.1.1 The visible portion of the spectrum of sunlight. Each wavelength of visible light has a particular frequency and is associated with a particular color. At the ends of the visible spectrum are invisible infrared and ultraviolet lights.

Sunlight and Electromagnetic Waves

Electromagnetic waves can have any wavelength, from thousands of kilometers to a fraction of the width of an atomic nucleus. The radio waves and microwaves that we examined in the previous chapter have wavelengths longer than 1 mm. In this chapter, we'll turn our attention to shorter wavelength radiation. In particular, we'll study electromagnetic waves with wavelengths between 400 nm and 750 nm (recall that 1 nm or 1 nanometer is 10^{-9} m). These are the electromagnetic waves that we perceive as **visible light** and the principal components of sunlight.

Because the electromagnetic waves in sunlight have such short wavelengths, their frequencies lie between 10^{14} Hz and 10^{15} Hz (Fig. 14.1.1). As one of these waves of sunlight passes by, its electric field fluctuates back and forth almost 1,000,000,000,000,000 times each second. Since producing microwaves, which have much longer wavelengths and much lower frequencies, already requires specialized components and tiny antennas, what can possibly emit or absorb light waves? The answer is the individual charged particles in atoms, molecules, and materials. These tiny particles can move extremely rapidly, often vibrating about at frequencies of 10^{14} Hz, 10^{15} Hz, or even more. As these charged particles accelerate back and forth, they emit light waves. Similarly, passing light waves cause individual charged particles in atoms, molecules, and materials to accelerate back and forth, thereby absorbing the light waves as well.

Sunlight originates at the outer surface of the sun, in a region called the photosphere. There, atoms and other tiny charged systems (mostly atomic ions and electrons) jostle about at 5800 °C. Since these charged particles accelerate as they bounce around, they emit electromagnetic waves.

Because the sun's surface emits light through the random, thermal motions of its charged particles, the distribution of wavelengths it emits is determined only by its temperature. It emits a blackbody spectrum, like the incandescent lightbulbs we discussed in Section 7.3. Because the photosphere's temperature is 5800 °C, the jostling motions are extremely rapid and most of the sunlight falls in the visible portion of the electromagnetic spectrum (Fig. 14.1.2).

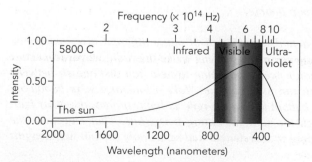

Fig. 14.1.2 Sunlight comes from the sun's photosphere, where the temperature is 5800 °C. This light has a blackbody distribution of wavelengths, with much of its intensity concentrated in the visible portion of the overall electromagnetic spectrum.

However, not all sunlight is visible. On the long-wavelength, low-frequency side of visible light is **infrared light.** We can't see infrared light with our eyes but we feel it when we stand in front of a hot object. In sunlight, infrared light is produced by charges that are accelerating back and forth more slowly than average.

On the short-wavelength, high-frequency side of visible light is **ultraviolet light.** We can't see ultraviolet light either, but we're aware of its presence because it induces chemical damage in molecules. It causes sunburns and encourages skin to tan. In sunlight, ultraviolet light is produced by charges that are accelerating back and forth more rapidly than average.

Check Your Understanding #1: The Rosy Glow of Candlelight

for answers, see page 472

Why does a burning candle emit reddish or yellowish light?

Sunlight's Passage to the Earth

Sunlight travels from the sun to the earth at the speed of light. But what sets the speed of light? Actually, as we learned in Section 4.2, it's one of the fundamental constants of nature, with a defined value of 299,792,458 m/s in empty space. While one could argue that the speed of light is set by the relationships between the electric and magnetic fields, that observation simply passes the buck. If you were then to ask what sets the relationships between the electric and magnetic fields, the answer would be the speed of light.

Rather than justifying why sunlight travels as fast as it does in empty space, let's look at what happens to it when it enters a region that's not empty. After all, sunlight eventually reaches the earth's atmosphere and, when it does, several interesting things happen.

First, the sunlight slows down as its electric and magnetic fields begin to interact with the electric charges in the atmosphere. Light polarizes the molecules it encounters, a process that delays its passage and reduces its speed. Since most transparent materials respond much more strongly to light's electric field than to its magnetic field, we'll concentrate on only electric effects.

The factor by which light slows down in a material is known as the material's **index of refraction.** Light travels particularly slowly through materials that are easy to polarize, and some of them have indices of refraction of 2 or even 3. Because air near sea level is only slightly polarizable, however, its index of refraction is just 1.0003. While this reduction in light's speed is too small to notice, we do notice the polarized air particles that cause it. These polarized air particles are what makes the sky blue (Fig. 14.1.3).

The particles in air consist of individual atoms and molecules, small collections of atoms and molecules, water droplets, and dust. As a wave of sunlight passes through one of these particles, the particle becomes polarized. Its electric charges accelerate back and forth as the sunlight's electric field pushes them around and they reemit a new electromagnetic wave of their own.

This new wave draws its energy from the original wave. In effect, the particle acts as a tiny antenna, temporarily receiving part of the electromagnetic wave and immediately retransmitting it in a new direction. This process, whereby a tiny particle redirects the path of a passing light wave, is called **Rayleigh scattering,** after the English physicist Lord Rayleigh (John William Strutt, 1842–1919) who first understood it in some detail.

While most sunlight travels directly to our eyes, some of it undergoes Rayleigh scattering and reaches us by more complicated paths. We see the direct light as coming from

Fig. 14.1.3 During the day, the sky above Monument Valley appears blue because the earth's atmosphere scatters mostly blue sunlight toward us. At night, the scattered sunlight is gone and the atmosphere is a clear window through which to observe the stars. The time-lapse photograph on the right shows these stars as nearly vertical streaks in the night sky.

the brilliant disk of the sun, but the scattered light gives the entire sky a fairly uniform blue glow (Fig. 14.1.4). But why is this glow blue?

The sky's blue color comes about because the tiny air particles that Rayleigh scatter sunlight are too small to make good antennas for that light. We observed in Section 13.1 that an antenna works best when it is one-quarter as long as the wavelength of its electromagnetic wave. The air particles make particularly bad antennas for long-wavelength red light, so that very little red sunlight undergoes Rayleigh scattering on its way through the atmosphere. But the air particles are not such bad antennas for short-wavelength blue light. Some of the blue sunlight does Rayleigh scatter and reaches our eyes from all directions. We see this Rayleigh scattered light as the blue glow of the sky.

Rayleigh scattering not only makes the sky blue; it also makes the sunrises and sunsets red. As the sun rises or sets, its light must travel long distances through the earth's atmosphere in order to reach your eyes. Its path is so long that most of the blue light Rayleigh scatters away miles to your east or west and all you see is the remaining red light. Sometimes the whole local sky appears reddish because there simply isn't any blue light left to scatter toward you. Sunrises and sunsets are particularly colorful when extra dust or ash is present in the atmosphere to enhance the Rayleigh scattering. Air pollution, forest fires, and volcanic eruptions tend to create unusually red sunrises and sunsets.

In contrast, clouds and fog appear white because they're composed of relatively large water droplets. These droplets are larger than the wavelengths of visible light and scatter all of sunlight's wavelengths equally well. Although this scattering is often so effective that you can't see the sun's disk through a cloud, it doesn't give the cloud any color. The cloud simply looks white.

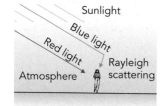

Fig. 14.1.4 As sunlight passes through the atmosphere, some of its blue light undergoes Rayleigh scattering from particles in the air. We see this redirected blue light as the diffuse blue sky. The remaining light reaches our eyes directly from the sun and tends to be reddish, particularly at sunrise and sunset.

Check Your Understanding #2: Seeing the Blues

for answers, see page 472

The air in a dark, smoky room often looks bluish when illuminated by white light. What creates this bluish appearance?

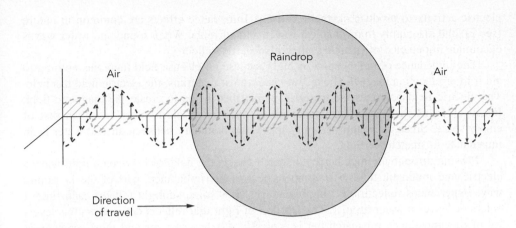

Fig. 14.1.5 As an electromagnetic wave enters a material, its speed decreases and the waves bunch up together. Its wavelength decreases.

Rainbows

Sometimes water droplets do separate the colors of sunlight. When sunlight shines on clear, round raindrops as they fall during a storm, these raindrops can create a rainbow. To understand how clear spheres of water can bend sunlight's path and separate it according to wavelength, we must understand three important optical effects: refraction, reflection, and dispersion. We encountered those same wave phenomena while studying water surface waves in Section 9.3, but now they appear in a new context: light waves!

Let's begin by looking at what happens when a wave of sunlight passes directly through a raindrop. Because water is more polarizable than air, the wave slows down inside the raindrop and its cycles bunch together (Fig. 14.1.5). While this bunching effect reduces the light's wavelength inside the drop, the light's frequency remains unchanged. The cycles don't disappear as they go through the raindrop, they just move more slowly.

If a narrow wave of sunlight is aimed directly through the center of the raindrop, it will follow a straight path and emerge essentially unaffected from the other side (Fig. 14.1.6a). But if that wave strikes the raindrop near the top, it will bend as it enters the water (Fig. 14.1.6b). Because the lower edge of the wave will reach the water first and slow down, the upper edge will overtake it and the wave will bend downward. The wave will head more directly into the water.

As the wave in Fig. 14.1.6b leaves the raindrop, its upper edge emerges first and speeds up while the lower edge lags behind. The wave bends downward even further and heads less directly into the air and away from the water.

This bending of sunlight at the boundaries between materials is *refraction*. It occurs whenever sunlight changes speeds as it passes through a boundary at an angle. If sunlight slows down at a boundary, it bends to head more directly into the new material. If sunlight speeds up at a boundary, it bends to head less directly into the new material. The amount of the bend increases as the speed change increases.

However, part of the sunlight striking a boundary doesn't pass through the boundary at all. Instead, it *reflects*. In Section 9.3, we attributed wave reflection specifically to changes in wave speed at a boundary. However, the more general cause of wave reflections is an **impedance mismatch**—an abrupt change in how the wave moves through its environment. In general, **impedance** is the measure of a system's opposition to the passage of a current or a wave. For an electromagnetic system, impedance measures how much voltage or electric field is needed to produce a particular current or magnetic field. In other words, electrical impedance measures how hard it is for

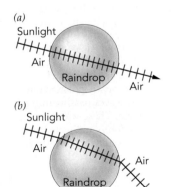

Fig. 14.1.6 A side view of two narrow waves of sunlight entering and leaving raindrops. The lines drawn across each light wave represent upward electric field maxima and are bunched together as light slows down in water.

❑ When electromagnetic waves travel through wires, they reflect from impedance mismatches. These mismatches occur whenever the relationship between electric and magnetic fields changes and must carefully be avoided in television wiring. If you don't provide an impedance matching device when connecting an antenna wire (300 V) to a video cable input (75 V), you'll have reflections in the wires and ghost images on the screen.

❑ Sand appears white because it redirects sunlight in all directions. One explanation for this effect is that the sand grains act as tiny antennas that respond to and reemit light's electromagnetic waves. A second explanation is that the sand grains present the sunlight with thousands of air–sand boundaries from which to reflect. However, both explanations are descriptions of exactly the same physics—the charged particles in the sand grains are electrically polarized by the waves passing through them. These waves are randomly redirected without being absorbed, and they give the sand its white appearance.

electric activity to produce magnetic activity. Impedance effects are common in nature (see ❑) and also apply to mechanical waves and currents. When sound and water waves encounter impedance mismatches, they also partly reflect.

The impedance of empty space is high because an electric field there has nothing to aid it in producing a magnetic field. But inside most materials, the electric field has help. The electric field polarizes the material, which then helps to create the magnetic field. Because of this assistance, the impedance of most materials is much less than that of empty space. Since air is almost empty space, the boundary between air and water is an impedance mismatch for light.

Passing through an impedance mismatch upsets the balance between a light wave's electric and magnetic fields. To compensate for this imbalance, part of the incoming wave experiences reflection off the boundary. Thus some sunlight reflects each time it enters or leaves a water droplet. The fraction of light that reflects depends on the severity of the impedance mismatch but is typically 4% between air and most transparent materials, including water (for reflection from sand, see ❑). In contrast, metals polarize so easily that their impedances are essentially zero and they reflect light almost perfectly.

There is one more important point about sunlight's passage through water: red light travels about 1% faster through water than violet light does. That's because higher-frequency violet light polarizes the water molecules a little more easily than lower-frequency red light, and that increased polarization slows down the violet light. This frequency dependence of light's speed in a material is *dispersion*. Dispersion affects refraction. The more light slows as it enters a raindrop, the more it bends at the boundary. Since violet light slows more than red light, violet light also bends more and the different colors of sunlight follow somewhat different paths through the raindrop.

A rainbow is created when raindrops separate sunlight according to color (Fig. 14.1.7). To see the rainbow, you stand with the sun at your back and look up at the sky. When sunlight hits the raindrops, they redirect some of that light back toward you. Since each raindrop redirects light only in a narrow range of angles, you can't see light from every raindrop. Only the raindrops in a narrow arc of the sky redirect visible light toward you. This arc appears brightly colored because raindrops at the inner edge of the arc send violet light toward you while raindrops at the outer edge of the arc send red light toward you. In between, you see all the colors of the rainbow.

Figure 14.1.8 shows how a raindrop redirects different colors of light in different directions. While there are many possible paths light can take through the raindrop, this path is the one that produces rainbows. Sunlight enters near the top of the raindrop and bends inward. Violet light bends more than red light, so the sunlight begins to separate

Steve Satushek/The Image Bank/Getty Images

Fig. 14.1.7 A rainbow forms when water droplets reflect sunlight back toward your eyes. Because the different wavelengths of light follow slightly different paths, we see the different colors coming from slightly different directions and observe bands of color.

according to color. Some sunlight is also reflected from the raindrop but doesn't contribute to rainbows.

When the light inside the raindrop strikes the back surface, most of it leaves the drop and is lost. A small fraction of the light reflects from that surface, however, and continues to travel through the raindrop. When this light reaches the raindrop's front surface, most of it leaves the drop. Violet light bends more strongly than red light as they reenter the air, so the different colors of light leave the drop heading in different directions. Since violet light is redirected more upward than red light, you see violet light coming toward you from the lower raindrops. Red light is redirected more downward so you see it coming toward you from the upper raindrops. Thus the upper arc of the rainbow is red while the lower arc is violet.

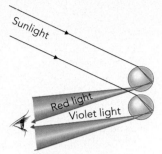

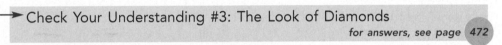

Check Your Understanding #3: The Look of Diamonds

for answers, see page 472

A diamond pendant sparkles with color when you look at it in sunlight. From where do the colors come?

Soap Bubbles

Soap bubbles also separate sunlight into its various colors (Fig. 14.1.9), but they use another wave phenomenon: interference. We encountered interference of mechanical waves in Section 9.3 when we studied wave beat at the seashore and interference of electromagnetic waves in Section 13.2 when we considered the unevenness of microwave cooking. Now we'll look at interference in another type of electromagnetic wave: light.

Light's interference effects stem from the summing or *superposition* of its electromagnetic waves. When several light waves overlap at a particular location, their electric fields sum together and so do their magnetic fields. If their individual fields all point in the same direction, the waves experience *constructive interference*—they sum together in a mutually assisting way and the light intensity at that location is enhanced (Fig. 14.1.10*a*). But when their individual fields point in opposing directions, the waves experience *destructive interference*—they sum together in a canceling way and the light intensity at that location is reduced (Fig. 14.1.10*b*).

Both forms of interference occur when sunlight reflects from the outer skin of a soap bubble. As each wave of sunlight hits that thin film of soapy water, the film's front surface reflects about 4% of the wave and the film's back surface reflects another 4%

Fig. 14.1.8 As sunlight passes through spherical raindrops, its colors separate. Violet light bends more at each air/water boundary than does red light, and the two emerge from the raindrops heading in different directions. You see red light coming toward you from the upper raindrops and violet light from the lower raindrops.

Fig. 14.1.9 Light reflected by the front and back surfaces of this soap film interferes with itself and gives the film its colorful appearance. Since the colors are determined by film thickness and since the film's thickness increases in the downward direction, the film displays horizontal bands of color.

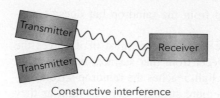

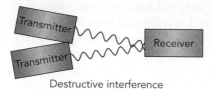

Constructive interference

Destructive interference

Fig. 14.1.10 (*a*) When waves from two separate paths arrive in phase at the receiver, constructive interference produces a particularly large effect on the receiver. (*b*) When waves arrive out of phase, destructive interference produces a particularly weak effect on the receiver.

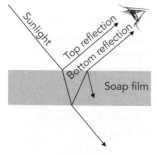

Fig. 14.1.11 Sunlight reflects from both the front and back surfaces of a soap film. The back surface reflection is delayed relative to the front surface reflection because it must pass twice through the soap film itself. If the two reflected waves arrive in phase, you see a bright reflection. If they are out of phase, you see a dim reflection.

(Fig. 14.1.11). Since both reflections travel in the same direction, the reflected light that you see reaches your eyes via two different paths, one from each reflection. If these two waves arrive **in phase,** that is, with their electric fields synchronized and assisting one another—you see the particularly bright reflection of constructive interference. If the two waves arrive **out of phase**—with their electric fields canceling one another—you see the particularly dim reflection of destructive interference.

Whether you see constructive or destructive interference depends on the wavelength of the sunlight. The back surface reflection has to travel twice through the soap film, so it's delayed relative to the front surface reflection. If the delay is just long enough for the wave to complete an integral number of cycles, then the two reflected waves are in phase with one another as they head toward your eye and you see a bright reflection. If the delay allows the back surface reflection to complete an extra half cycle, then the two reflected waves are out of phase with one another and you see a dim reflection.

Sunlight contains many different wavelengths of light, and these wavelengths behave differently during the reflection process. You see a colored reflection, consisting mainly of those wavelengths of light that experience constructive interference. Because the delay experienced by the back surface reflection depends on the thickness of the soap film, you can actually determine the film's thickness by studying its colors.

Check Your Understanding #4: A Slick-Looking Oil Slick

for answers, see page 472

A thin layer of oil or gasoline floating on water appears brightly colored in sunlight. From where do these colors come?

Sunlight and Polarizing Sunglasses

All sunglasses absorb some of the sunlight passing through them, but the best ones absorb horizontally polarized light much more strongly than vertically polarized light. These polarizing sunglasses dramatically reduce glare by eliminating most of the light reflected from horizontal surfaces.

When light strikes a transparent surface at right angles, about 4% of that light is reflected, regardless of its polarization. But when light strikes a horizontal surface at a shallow angle, horizontally polarized light reflects much more strongly than vertically

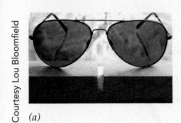

(a)

(b)

Fig. 14.1.12 (*a*) Ordinary sunglasses simply darken the scene. (*b*) Polarizing sunglasses block horizontally polarized light, the main component of glare.

polarized light. That's because horizontally polarized light's horizontal electric field pushes electric charges back and forth along the surface. Charges shift relatively easily in that direction and the surface becomes even easier to polarize as the angle becomes shallower. As a result, the surface reflects more horizontally polarized light at shallow angles than it does at steeper angles.

But vertically polarized light's vertical electric field pushes electric charges up and down vertically. At shallow angles, this field acts to lift the charges in and out of a horizontal surface. Since the charges can't leave the surface, the surface becomes harder to polarize as the angle becomes shallower. A horizontal surface doesn't reflect much vertically polarized light, and there is even a special angle, **Brewster's angle,** at which no vertically polarized light reflects at all.

Sunlight is an even mixture of vertically and horizontally polarized waves. Since horizontally polarized waves reflect most strongly from horizontal surfaces, sunglasses that absorb horizontally polarized light will prevent you from seeing most of this reflected light (Fig. 14.1.12). That's why polarizing sunglasses are so effective at reducing glare.

Check Your Understanding #5: Looking into the Reflecting Pool

for answers, see page 472

When you look into a pool of water, you see mostly a reflection of the sky. But when you wear polarizing sunglasses, you see into the water clearly. Explain.

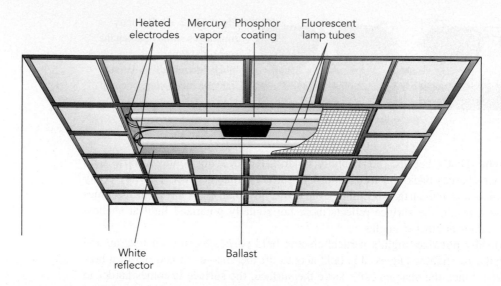

Energy efficiency is crucial to modern lighting. While incandescent lamps provide pleasant, warm illumination, most of the power they consume is wasted as invisible infrared light. Fluorescent and other gas discharge lamps produce far more visible light with the same amount of electric power and now dominate office, industrial, and street lighting. In this section we'll explore several types of discharge lamps—fluorescent, mercury vapor, sodium vapor, and metal–halide lamps—that all share a common theme: current passing through a gas.

Questions to Think About: Why do different fluorescent lamps often have slightly different colors? Why is the light from a neon sign red? Are neon atoms red? Why do most fluorescent lamps take a few seconds to turn on? Why are streetlights so dim when they first turn on? Why are some highway lights orange?

Experiments to Do: Examine the discharge lamps around you, particularly the white fluorescent tube lamps. Where do their lights originate? In a fluorescent lamp, it comes from the white phosphor coating on the tube's inner surface. What about in a neon lamp or a mercury vapor streetlight? Compare the colors of various lamps, including several different fluorescent lamps. Are their lights identical?

Both fluorescent and neon lamps start almost immediately. But watch how long it takes a streetlight to start. While fluorescent and neon lamps remain cool during operation, the mercury, sodium, and metal–halide lamps used in street lighting get hot. Watch a streetlight warm up. Does its color change? If the streetlight loses power even for a moment, it must wait about 5 minutes before it can start again. Why must it wait?

How We See Light and Color

Before examining discharge lamps, let's look at how our eyes recognize color. While it might seem that they actually measure the wavelengths of light, that's not the case. Instead, our retinas contain three groups of light-sensing *cone cells* that respond to three

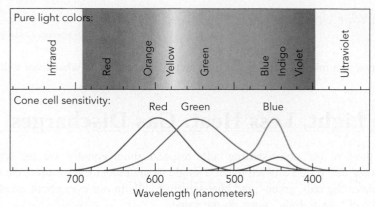

Fig. 14.2.1 The red-sensitive cells in our retina detect light near 600 nm, the green-sensitive cells near 550 nm, and the blue-sensitive cells near 450 nm. The red-sensitive cells also respond near 440 nm, so that we see violet.

different ranges of wavelengths. One group of cone cells responds to light near 600 nm and lets us see red, another responds to light near 550 nm and lets us see green, and a third responds to light near 450 nm and lets us see blue (Fig. 14.2.1). These cone cells are most abundant at the center of our vision. While our retinas also contain *rod cells,* which are more light-sensitive than cone cells, rod cells can't distinguish color. They are most abundant in our peripheral vision and provide us with night vision.

Having only three types of color-sensing cells doesn't limit us to seeing just three colors. We perceive other colors whenever two or more types of cone cells are stimulated at once. Each type of cell reports the amount of light it detects, and our brains interpret the overall response as a particular color.

While light of a certain wavelength will stimulate all three types of cone cells simultaneously, the cells don't respond equally. If the wavelength is 680 nm, the red cone cells will respond much more strongly than the green or blue cells. Because of this strong red response, we see the light as red.

Other wavelengths of light stimulate the three types of cells somewhat more evenly. Light with a wavelength of 580 nm is in between red and green light. Both the red-sensitive and the green-sensitive cone cells respond about equally to this light and we see it as being yellow.

But we also see yellow when looking at an equal mixture of 640-nm light (red) and 525-nm light (green). The 640-nm light stimulates the red-sensitive cone cells and the 525-nm light stimulates the green-sensitive ones. Even though there is no 580-nm light entering our eyes, we see the same yellow color as before.

In fact, mixtures of red, green, and blue light can make us see virtually any color. For that reason, these three are called the **primary colors of light** or the *primary additive colors* (Fig. 14.2.2*a*). Color televisions and computer screens use tiny sources of red, green, and blue light to produce their full-color images.

While the idea of mixing primary colors also applies to paints, inks, and pigments, the palette is different. The **primary colors of pigment** or the *primary subtractive colors* are cyan, magenta, and yellow (Fig. 14.2.2*b*). When you apply one of these primary pigments to a white surface, it absorbs or subtracts one of the primary colors of light from the surface's reflection. Cyan subtracts the reflection of red, magenta subtracts the reflection of green, and yellow subtracts the reflection of blue. Color printers, photographs, magazines, and books use tiny patches of cyan, magenta, and yellow pigments to produce their full-color images.

(a) Additive colors (light)

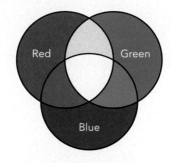

(b) Subtractive colors (pigment)

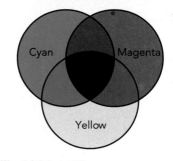

Fig. 14.2.2 (*a*) The primary colors of light or additive colors, red, green, and blue, can be combined to form any colors of light. (*b*) The primary colors of pigment or subtractive colors, cyan, magenta, and yellow, subtract red, green, and blue, respectively, from reflected or transmitted light and can be combined to form any colors of pigment.

► Check Your Understanding #1: Mixed Up Light

for answers, see page 472

If you look at a mixture of 70% red light and 30% green light, what color will you see?

More Light, Less Heat: Gas Discharges

When all three of our color-sensing cells respond about equally, we see white light. That's because our vision evolved under a single incandescent light source: the sun. Sunlight stimulates the red-, green-, and blue-sensitive cells in our eyes about evenly, so any other source of "white light" must do the same.

While an incandescent lightbulb makes a good attempt at producing white light, it suffers from two serious drawbacks. First, because its filament can't reach the temperature of the sun's surface (5800 °C), its blackbody spectrum contains too little blue light and appears redder than sunlight. Second, because most of its electromagnetic radiation is invisible infrared light, it doesn't use energy efficiently to produce visible light.

Fortunately, modern science has given us some alternative sources of light, sources that don't use heat and thermal radiation to produce their light. Among these sources are gas discharge lamps, which emit light when electric currents pass through their gases. While some discharge lamps are colored, others do excellent jobs of producing white light. And many of them are far more energy efficient at producing visible light than are incandescent lightbulbs.

To get an understanding of gas discharges and the lights they emit, let's start with one of the simplest examples: a neon sign tube. Though neon's rich red glow isn't suitable for lighting and isn't particularly energy efficient, it's great for signs and also relatively easy to understand.

A neon sign tube is a sealed glass tube that contains pure neon gas at a density less than 1% that of the atmosphere outside the tube. It has metal electrodes at each end so that electric current can enter the gas through one electrode and leave through the other. Of course, gases are normally electrical insulators and neon is no exception. To transform the tube's neon into a conductor, a large voltage difference is applied between its two electrodes. As we saw in Section 10.2, a gas breaks down when exposed to large voltage gradients; its few naturally occurring ions initiate a cascade of ionizing collisions that quickly fill the gas with charged particles and render it conducting.

While ordinary air breaks down at a voltage gradient of about 30,000 V/cm, low-density neon breaks down at a much lower voltage gradient. That's because in a low-density gas, charged particles can travel farther and accumulate more kinetic energy before each collision. When about 10,000 volts is applied between the electrodes of a neon sign tube, the gas breaks down and begins to emit its familiar red glow.

The lamp is then experiencing a **discharge;** that is, current is flowing through the neon gas. This current consists mostly of electrons, flowing from the tube's negatively charged electrode to its positively charged electrode (Fig. 14.2.3). While these electrons collide frequently with neon atoms, they have so little mass that they usually just bounce off the atoms without losing much energy. Like a Ping-Pong ball rebounding from an elephant, the electron does most of the bouncing and then continues on its way.

Every so often, however, an electron will collide with a neon atom and something different will happen; the neon atom will rearrange internally and absorb part of the electron's kinetic energy. The electron will rebound with less energy than it had before and the neon atom will go on to emit light, probably red light. To understand why that light is probably red, however, we'll have to look at quantum physics and the structure of the neon atom.

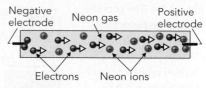

Negative electrode   Neon gas   Positive electrode
Electrons   Neon ions

Check Your Understanding #2: Making Whiter Light

for answers, see page **472**

Some photographic lamps simulate sunlight by placing a blue filter in front of an incandescent bulb. This filter absorbs some of the red light so that the lamp appears whiter. Does this filter increase or decrease the lamp's energy efficiency?

Atoms, Light, and Quantum Physics

According to the wave–particle duality we encountered in Section 12.1, electrons have both particle and wave characteristics, and it is as waves that an atom's electrons are best understood. Like all objects in our universe, electrons travel as *waves* when they move from place to place, and it's only when you go looking for them that you find them as *particles* at particular locations (see ❏).

In an atom, each electron exists in a standing wave known as an **orbital,** with a specific energy and quantum frequency. Although these orbitals are three-dimensional standing waves, they resemble the two-dimensional standing waves of the drumhead shown in Fig. 9.2.9. And like the drumhead modes, the orbitals are distinguished from one another by their patterns of nodes—surfaces along which the electron wave has no amplitude.

Like *levels* in a solid, *orbitals* in an atom can be viewed as placeholders that may or may not be occupied by electrons. From that perspective, an atom's orbitals fill with electrons from lowest energy on up. In accordance with the Pauli exclusion principle, each orbital accommodates two electrons, one spin-up and one spin-down, until all of the atom's electrons have been placed. The chemical nature of a particular atom, and its location in the periodic table of the elements (Fig. 14.2.4), is determined by how many electrons it has and how those electrons fill the available orbitals. Atoms are electrically neutral, so the number of negatively charged electrons in an atom is the same as the number of positively charged protons in its nucleus, its **atomic number.**

Just as there are patterns to levels, giving rise to the bands and band gaps that distinguish different solids, so there are patterns to orbitals, giving rise to the shells and energy gaps that distinguish different atoms. **Shells** are groups of orbitals that have similar energies and tend to fill with electrons at about the same time. Although the orbital patterns and shells are complicated by the fact that charged electrons influence one another and distort one another's standing waves, many atomic properties are determined simply by which orbitals are occupied by an atom's electrons.

The atomic orbitals are identified primarily by an integer and a letter, both of which relate to their node patterns. The integer (1, 2, 3, ...) is one more than the number of node surfaces in the orbital (0, 1, 2, ...) while the letter (*s, p, d, f, g, h,* ...) indicates how many of those node surfaces pass through the atom's center (0, 1, 2, 3, 4, 5, ...). Although *s* orbitals appear one at a time, *p* orbitals appear in groups of three, *d* orbitals in groups of five, *f* orbitals in groups of seven, etc.

A neon atom has 10 electrons, so it takes 5 orbitals to accommodate them. The first two electrons go into the 1*s* orbital, the fundamental mode with zero nodes. Filling the

Fig. 14.2.3 A neon sign tube sends electrons through low-pressure neon gas. These electrons collide with neon atoms, transferring energy to them and causing them to emit primarily red light. Positively charged neon ions, created by particularly energetic collisions, keep the electrons from repelling one another to the walls of the tube.

❏ In 1927, American physicists Clinton Joseph Davisson (1881–1958) and Lester H. Germer (1896–1972) showed that electrons travel as waves by observing interference effects when electrons reflected from different atomic layers in a crystal of nickel metal. When the various electron waves arrived at a detector in phase, the detector found many electrons. When the waves arrived out of phase, the detector found few electrons. Their work was aided by a fortuitous accident in which air entered the experiment's glass vacuum tube. While carefully eliminating oxygen from their nickel sample after that leak, they managed to perfect its crystalline structure, making it possible to observe the interferences.

Fig. 14.2.4 The periodic table of the elements is organized by the filling of electron orbitals. Major shells of orbitals fill with electrons from left to right. The number of electrons in a neutral atom is equal to its atomic number.

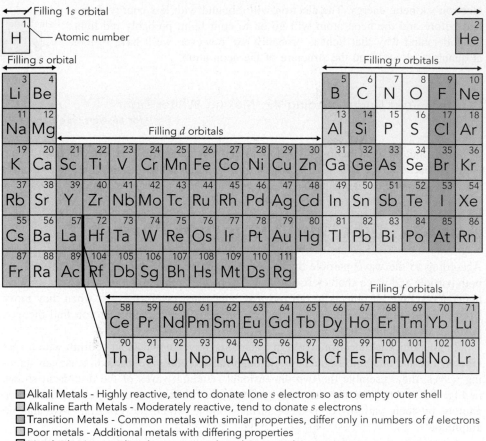

Alkali Metals - Highly reactive, tend to donate lone *s* electron so as to empty outer shell
Alkaline Earth Metals - Moderately reactive, tend to donate *s* electrons
Transition Metals - Common metals with similar properties, differ only in numbers of *d* electrons
Poor metals - Additional metals with differing properties
Metaloids - Intermediate between metals and non-metals
Non-metals - Semiconductors and insulators
Halogens - Highly reactive, tend to steal one *p* electron so as to complete outer shell
Noble gases - Non-reactive gases with completed outer shells
Lanthanides - Moderately reactive metals, differ only in numbers of 4*f* electrons
Actinides - Moderately reactive metals, differ only in numbers of 5*f* electrons

1*s* orbital completes the first major shell. The next two electrons go into the 2*s* orbital, which has one node surface that doesn't pass through the atom's center. Finally, neon's last six electrons go into the three 2*p* orbitals, each of which has one node surface passing through the atom's center. Since filling the 2*s* and 2*p* orbitals completes the second major shell, the neon atom (Ne in Fig. 14.2.4) is chemically inert. That is why neon exists as a gas of individual atoms!

An arrangement of occupied orbitals is called a **state** and the state we have just described is neon's **ground state**—the state with the lowest possible total energy. The neon atom has other states available, but they all involve additional energy. That's where the discharge enters the picture. When a charged particle collides with a ground state neon atom, there's a chance that the collision will knock an electron out of its usual orbital into one of the empty orbitals. The neon atom will then be in an **excited state**— it will have extra energy.

Suppose, for example, that a collision has just shifted one of the neon atom's electrons from a 2*p* orbital to a 3*p* orbital. The atom will quickly undergo a series of state-to-state transitions that eventually return it to its ground state. With each transition, the atom will drop to a lower energy state and emit a photon. The photon carries away the energy released in transitioning from the higher energy state to the lower energy state. In all likelihood, one of those transitions will shift an electron from a 3p orbital to the 3s orbital and produce a photon of red light, the red of a neon sign!

As long as the electron remains in the 3p orbital—a particular standing wave—it can't emit an electromagnetic wave. That standing wave has a quantum oscillation, however, the oscillation is internal to the electron and neither the electron nor its charge has any overall motion. Since charge must accelerate to emit electromagnetic waves (see Chapter 13), without such overall motion there can be no emission of electromagnetic waves.

But when the 3p electron begins its transition to the empty 3s orbital, its quantum wave begins to change with time. The wave moves rhythmically back and forth during the transition and its charge accelerates with it. The atom begins to emit an electromagnetic wave. By the time the transition is complete, the atom has emitted a single *particle* of light, a photon of red light. Like an electron, light travels as a *wave* but behaves as a particle when you try to locate it. While it's being emitted or absorbed by an atom, light exhibits its particle nature.

We've seen that hot objects emit light, a process called **incandescence,** but here the neon atom emits light without heat, a process called **luminescence.** That luminescence is the result of a **radiative transition**—a transition between states in which a photon of light is emitted or absorbed. In this case, the radiative transition emits a photon that carries away the energy released when the excited neon atom's 3p electron shifts into the empty 3s orbital. The difference in energy between those two states determines the photon energy, which in turn determines the frequency and color of the photon's light wave.

According to quantum physics, a photon's frequency is proportional to its energy. Specifically, the photon's energy is equal to the frequency of its electromagnetic wave times a fundamental constant of nature known as the Planck constant. This relationship between energy and frequency can be written as a word equation:

$$\text{energy} = \text{Planck constant} \cdot \text{frequency}, \qquad \textbf{(14.2.1)}$$

in symbols:

$$E = h \cdot v,$$

and in everyday language:

> *Ultraviolet light and X-rays can injure your skin and tissues because each*
> *particle of high-frequency light carries a great deal of energy.*

First used by German physicist Max Planck (1858–1947) in 1900 to explain the light spectrum of a hot object, the **Planck constant** has a measured value of 6.626×10^{-34} J · s. And while we have just encountered the Planck constant and Eq. 14.2.1 in the context of light waves, it actually applies to all quantum waves. For example, the quantum frequency of an electron is related to its energy by Eq. 14.2.1.

The Planck constant is so tiny that even a photon of 10^{15}-Hz ultraviolet light has an energy of only 6.626×10^{-19} J. A typical beam of light thus contains so many photons that you can't see that they're arriving as particles. However, the energy in a single ultraviolet photon is substantial on a molecular scale. It can damage a molecule in your skin, contributing to a sunburn and inducing your skin to tan as a defensive response. X-rays have even higher frequencies, and their energetic photons can cause more severe molecular damage.

A neon atom can only emit photons corresponding to energy differences between two of its states, a constraint that severely limits its light spectrum. Neglecting nuclear issues that we'll discuss in Chapter 16, all neon atoms are identical and emit the same characteristic spectrum of light. The visible part of that spectrum is dominated by the warm red glow of photons emitted when electrons shift from $3p$ to $3s$ orbitals, which is why a neon sign glows red.

Since atoms of different elements have different numbers of electrons and different states, each emits its own unique spectrum of light after being excited. Copper atoms emit a blue-green spectrum, strontium atoms a deep red, and sodium a bright yellow-orange. Chemists, astronomers, and manufacturers rely on those emission spectra for information and applications. And, as we shall soon see, so do scientists and engineers working on illumination.

► Check Your Understanding #3: Colorful Atoms

for answers, see page 472

Many fireworks involve brilliantly colored lights. How do the atoms in burning chemicals produce particular colors of light?

► Check Your Figures #1: The Particles in a Radio Wave

for answers, see page 473

How much energy is carried by a photon from a 1000-kHz AM radio station?

Fluorescent Lamps

If neon tubes appeal to you for illumination, you probably march to your own drummer. Most people opt for a somewhat better simulation of sunlight in their discharge lamps. As an energy-efficient source of artificial sunlight, it's hard to beat fluorescent lamps.

At the heart of a fluorescent lamp is a narrow glass tube filled with argon, neon, and/or krypton gases at about 0.3% of atmospheric density and pressure. The tube also contains a few drops of liquid mercury metal, some of which evaporates to form mercury vapor. About one in every thousand gas atoms inside the tube is a mercury atom, and it's these mercury atoms that are responsible for the light.

Like a neon sign tube, a fluorescent lamp uses a discharge in its gas to produce light. While fluorescent lamps occasionally use high voltages to initiate their discharges, most operate their discharges at household voltages and must rely on alternative techniques to render their gases conducting. These low-voltage lamps normally heat their electrodes so that thermal energy ejects electrons from their surfaces into the gas. But regardless of how the discharge is started, the result is current flowing through the gas and the emission of light.

But the fluorescent lamp has a problem. While its mercury atoms emit most of the light in its discharge, that light is almost entirely ultraviolet. The final radiative transition that returns each mercury atom to its ground state ($6p \rightarrow 6s$) releases a large amount of energy and produces a photon with a wavelength of 254 nm. This light can't go through the glass walls of the tube, and you couldn't see it if it did. So the fluorescent lamp converts it into visible light with the help of phosphor powder on the inside of the glass tube.

Phosphors are solids that *luminesce*—they emit light—when something transfers energy to them. Their behavior is similar to that of an atom: an energy transfer shifts the phosphor from its ground state to an excited state and it then undergoes a series of transitions that return it to its ground state. Some of those transitions are radiative ones and emit light.

In a fluorescent lamp, the phosphor is excited by ultraviolet light. This excitation or energy transfer is actually a radiative transition, but one in which the photon is absorbed by the phosphor; one of its electrons makes a transition from a lower energy level to a higher energy level. During this absorption, the light's electric field pushes the electron's wave back and forth rhythmically until it shifts to the new level. The photon disappears and the phosphor receives its energy.

Once the phosphor is in an excited state, its electrons begin to make transitions back to their ground state levels. Most of those transitions radiate visible light, the light that you see when you look at the lamp. However, some of the transitions radiate invisible infrared light or cause useless vibrations in the phosphor itself. Despite this wasted energy, phosphors are relatively efficient at turning ultraviolet light into visible light, a process called **fluorescence.**

The phosphors in a fluorescent lamp are carefully selected and blended to fluoresce over a broad range of visible wavelengths. While this light doesn't have the same spectrum as sunlight, it appears white because it stimulates the red-, green-, and blue-sensitive cells in our eyes about equally. The blending of phosphors is necessary because, like atoms, each phosphor fluoresces with a characteristic spectrum of light that's determined by the energy differences between its levels. Several different phosphors are needed to create the right balance between red, green, and blue lights. While some advertising and novelty lamps use brightly colored, unblended phosphors, fluorescent lighting tubes come in six standard color blends: cool white, deluxe cool white, warm white, deluxe warm white, white, and daylight.

When they were first introduced, fluorescent tubes used the daylight phosphor. This phosphor emits too much blue light and makes everything look cold and medicinal. Phosphors have improved over the years so that the most common phosphors today, cool white and warm white, look much more pleasant. Light from the "cool" phosphors resembles daylight, while that from the "warm" phosphors resembles incandescent lighting. Household fluorescent lamps often use deluxe cool white and deluxe warm white, which are even more pleasant versions of white but which are slightly less energy efficient.

▶ Check Your Understanding #4: Sorry, We Forgot the Coating

for answers, see page 472

If there were no phosphor coating on the inside of a fluorescent tube, what would you see when the lamp operated?

A Few Practical Issues

A fluorescent tube needs a substantial electric field inside it to keep its electrons moving forward. Since this electric field is proportional to the voltage drop through the tube, longer tubes need higher voltages. Power line voltages (110 V to 240 V) are appropriate for tubes up to 3 m (10 ft) in length, but the longer and often colored fluorescent tubes used in artwork or advertising require much higher voltages.

To keep electrons from pushing one another into its walls, a fluorescent tube must also contain positively charged mercury ions. The discharge naturally produces these ions

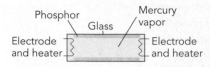

Fig. 14.2.5 In a hot-electrode fluorescent tube, the electrodes are actually filaments and are heated by running currents through them.

during particularly energetic collisions. The result, a gaslike mixture of positively charged ions and negatively charged electrons, is called a **plasma.** A plasma is distinct from a gas because its charged particles exert forces on one another over considerable distances. All operating discharge lamps contain plasmas, including the neon sign that we discussed earlier.

Since a typical fluorescent lamp must heat its electrodes red-hot in order to form its plasma, it runs current through filaments at each end of its tubes (Fig. 14.2.5). Once the discharge is operating, the heating current can usually be turned off because the electrodes will be kept hot by electrons that hit them from the discharge. However, some fluorescent fixtures can be dimmed, in which case electron heating alone won't sustain the plasma. Dimmable fluorescent fixtures must continue to pass heating currents through their filament/electrodes.

Unfortunately, the filament/electrodes are fragile. They're damaged by a process called **sputtering,** in which positive mercury ions from the plasma collide with the filament/electrodes and chip away their tungsten atoms. Because sputtering is particularly severe during start-up, a typical filament/electrode breaks after a few thousand starts. That's why you shouldn't turn a fluorescent lamp on and off more than once every few minutes.

> Check Your Understanding #5: Getting Red-y
>
> for answers, see page 472
>
> Why do the ends of some fluorescent lamps glow red during starting?

Mercury, Metal–Halide, and Sodium Lamps

While a low-pressure mercury discharge emits mostly ultraviolet light, a high-pressure mercury discharge emits more visible light than ultraviolet light. This change occurs because ultraviolet light becomes trapped in densely packed mercury atoms and only visible light is able to escape from the discharge.

Known as **radiation trapping,** this effect occurs because mercury atoms absorb 254-nm photons just as well as they emit them. The same radiative transition that causes a mercury atom to emit a 254-nm photon ($6p \rightarrow 6s$) can also run backward to absorb that photon ($6s \rightarrow 6p$). In a dense gas of mercury, whenever one mercury atom emits a 254-nm photon, another mercury atom snaps that photon up. So while the discharge keeps pouring energy into the mercury atoms, they can't get rid of it as 254-nm photons. Instead, they emit most of their energy through radiative transitions between other excited states. Since this light is much less likely to be captured by other mercury atoms, it emerges from the lamp as bluish visible light.

When you first turn on a high-pressure mercury lamp, most of the mercury is liquid and the pressure is low. The lamp starts like a small fluorescent tube without phosphor, so you see very little visible light. But the tube is designed to heat up during operation so that the liquid mercury evaporates to form a dense gas. As the gas pressure rises, the tube's color charges until it emits brilliant, blue-white light.

To make a lamp that's a little less bluish, some high-pressure mercury lamps contain additional metal atoms. These atoms are introduced into the lamps as metal–iodide compounds, making them metal–halide lamps. Sodium, thallium, indium, and scandium iodides all contribute their own emission spectra to the outgoing light and help to strengthen the red end of its spectrum. They give metal–halide lamps a warmer color than pure mercury lamps.

Courtesy Lou Bloomfield

Fig. 14.2.6 The active component of a high-pressure sodium vapor lamp is a small translucent tube. As the lamp warms up, sodium metal in the tube evaporates to form a brilliant yellow discharge. The dense vapor of sodium atoms in the tube traps the 590-nm light so that the lamp emits a richer spectrum of wavelengths and a less monochromatic glow than a low-pressure sodium lamp.

Pure sodium lamps resemble mercury lamps, except that they use sodium atoms. Sodium is a solid at room temperature, so both low- and high-pressure sodium lamps must heat up before they begin to operate properly. A low-pressure sodium lamp is extremely energy efficient because its 590-nm light comes directly from a sodium atom's strongest radiative transition. That transition is from sodium's lowest excited state to its ground state ($3p \rightarrow 3s$). Many highways are illuminated by the yellow-orange glow of low-pressure sodium lamps.

But this monochromatic illumination is unpleasant and permits no color vision at all. While it may be acceptable on a highway, you wouldn't want it near your home. That's why people buy high-pressure sodium lamps for home use (Fig. 14.2.6).

Remarkably enough, the 590-nm emission itself smears out at high pressure to cover a wide range of wavelengths, from yellow-green to orange-red. This spreading occurs because of the many collisions suffered by the densely packed sodium atoms as they try to emit 590-nm light. These collisions distort the atomic orbitals so that the photons emerge with somewhat shifted energies. Overall, a high-pressure sodium lamp emits remarkably little light exactly at 590 nm because the ground-state sodium atoms trap that light; they run the ($3p \rightarrow 3s$) transition backward and absorb the photons ($3p \rightarrow 3s$). This trapping is so effective that there is actually a hole in the lamp's spectrum right at 590 nm.

High-pressure discharge lamps suffer from a problem not found in low-pressure lamps: they're difficult to start when hot. It's much harder to initiate a discharge in a high-pressure gas than in a low-pressure gas, so they all start at low pressure and then evolve to high pressure. If the discharge in a high-pressure mercury, sodium vapor, or metal–halide lamp is interrupted and loses its plasma, the lamp must cool down before it can be restarted.

Check Your Understanding #6: Slow Glowing

for answers, see page **473**

At dusk, a mercury streetlight turns on. It glows dimly at first and gradually increases in brightness. What is happening during this warm-up period?

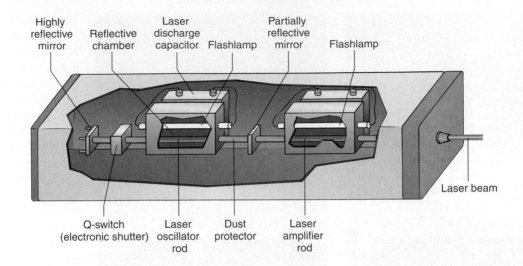

Highly reflective mirror   Reflective chamber   Laser discharge capacitor   Flashlamp   Partially reflective mirror   Flashlamp

Laser beam

Q-switch (electronic shutter)   Laser oscillator rod   Dust protector   Laser amplifier rod

SECTION 14.3   # Lasers and LEDs

Few devices have inspired our imaginations more than lasers. Since their invention in the late 1950s, lasers have found countless uses, from cutting metal and clearing human arteries to surveying land and playing CDs. But lasers are more than just novel applications of old ideas. Instead, they bring together quantum and optical physics to produce a new type of light. This light is radically different from that produced by incandescent and fluorescent lamps, and its properties make it particularly useful for many applications. In this section we'll examine the nature of this new light and the ways in which lasers produce it.

Questions to Think About: *Why is laser light usually brightly colored? Why does laser light often appear as a narrow beam? In movies, "lasers" are often shown to emit bright streaks of light that can be dodged if one jumps quickly enough; is that view realistic?*

Experiments to Do: *While lasers are household objects, the ones in CD or DVD players and laser printers are relatively inaccessible. If you don't own a laser pointer, look at a store barcode scanner. The scanning system contains a gas or solid-state laser that emits a very narrow beam of bright red light. This system also contains a rotating mirror or holographic disk that directs the beam as a pattern of thin stripes onto anything that passes through the scanner. A light sensor inside the window watches for this moving beam of light to travel across a label. If you look down at the laser light emerging from the scanner's window or observe the spot of a laser pointer on the wall, you'll see both the purity of its color and its strange speckled character; the light appears to consist of tiny light and dark speckles. These speckles are caused by interference effects, which are extremely pronounced in the ordered laser light. This light also appears unusually bright and, as when looking at the sun, you should keep your gaze brief to avoid eye injury.*

Lasers and Laser Light

To understand lasers, you must understand how laser light differs from the normal light emitted by hot objects or by individual atoms in an electrical discharge. Each particle of normal light, each photon, is emitted willy-nilly without any relationship to the other light particles being emitted nearby. Because of this light's independent and unpredictable character, it's called *spontaneous light* and its creation is called **spontaneous emission of radiation.**

But theoretical work by Albert Einstein and others in the 1920s and 1930s predicted the existence of a second type of light, *stimulated light,* that can be created when an excited atom or atomlike system duplicates a passing photon. While this **stimulated emission of radiation** can occur only when the excited atom is capable of emitting the duplicate photon spontaneously, the copy that it produces is so perfect that the two photons are absolutely indistinguishable. Together, these two photons form a single electromagnetic wave.

To get a slightly better picture for how such stimulation occurs, think about an isolated atom in an excited state. That atom will eventually return to its ground state, but it must emit one or more photons in order for this to happen. That atom waits around in the excited state until it begins a spontaneous radiative transition. During the transition, one of the atom's electrons accelerates back and forth and the atom emits a photon.

But if a similar photon passes through the atom as it's waiting in the excited state, that photon's electric field can stimulate the radiative transition process through sympathetic vibration. The field pushes and pulls on the atom's electrons and makes them accelerate back and forth. While this effect is small, it may be enough to trigger the emission of light. If the atom does emit light, the photon it will produce will be a perfect copy of the stimulating photon.

When this stimulated emission process was first discovered, people immediately recognized that it made light amplification possible. If enough excited systems could be assembled together, a single passing photon could be duplicated exactly over and over again. Instead of a single particle of light, you would soon have thousands, or millions, or even trillions of identical light particles.

Implementing this idea had to wait until the late 1950s, however, when the technical details for how to actually achieve light amplification were worked out. In 1960, the first laser oscillators were constructed. These were devices that emitted intense beams of light, in which each particle of light was identical to every other particle of light. A single particle of light had been duplicated by the stimulation process into countless copies.

When individual excited atoms or atomlike systems emit light through spontaneous emission, the particles of light head off separately as many independent electromagnetic waves (Fig. 14.3.1a). Light consisting of many independent electromagnetic waves is called **incoherent light.**

(a)

Incoherent radiation
from excited atoms

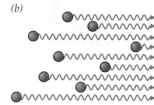

(b)

Coherent radiation
from excited atoms

Fig. 14.3.1 (a) Photons of incoherent light are created independently and have somewhat different wavelengths and directions of travel. (b) Photons of coherent light are produced by stimulated emission and are identical to one another in every way.

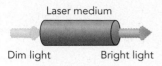

Fig. 14.3.2 A laser amplifier uses excited atoms or atomlike systems to increase the number of light particles leaving the laser medium. The incoming light is duplicated by stimulated emission.

But when that same collection of excited atoms or atomlike systems emit light by stimulated emission, all of the light particles are *absolutely* identical and form a single electromagnetic wave (Fig. 14.3.1*b*). Light consisting of many identical photons and a single electromagnetic wave is called **coherent light.** Because of its single wave nature, coherent light exhibits remarkable interference effects. These effects are easily seen in the coherent light emitted by lasers.

Check Your Understanding #1: A Light Dusting of Photons
for answers, see page 473

If you were to measure the electric field in the light from a flashlight, you would find that it fluctuates about randomly. Why is the light's electric field so disorderly?

Light Amplifiers and Oscillation

Producing coherent light requires amplification. You must start with only one particle of light and duplicate it many times. The basic tool for this duplication of light is a **laser amplifier** (Fig. 14.3.2). When weak light enters an appropriate collection of excited atoms or atomlike systems—the **laser medium**—that light is amplified and becomes brighter. The new light has exactly the same characteristics as the original light, but it contains more photons.

When we think of lasers, we rarely imagine a device that duplicates photons from somewhere else. We usually picture one that creates light entirely on its own. To do that, the laser must produce the initial particle of light that it then duplicates to produce others. A **laser oscillator** is a device that uses the laser medium itself to provide the seed photon, which it then duplicates many times (Fig. 14.3.3). If a laser medium is enclosed in a pair of carefully designed mirrors, it's possible for the stimulation process to become self-initiating and self-sustaining. However, the mirrors must be curved properly and must have the correct reflectivities. One mirror must normally be extremely reflective, while the other must transmit a small fraction of the light that strikes its surface.

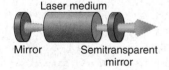

Fig. 14.3.3 A laser oscillator is a laser amplifier enclosed in mirrors. Oscillation occurs when the laser medium spontaneously emits one photon in just the right direction. This photon bounces back and forth between the two mirrors and is duplicated many times. Some of the light is extracted from this laser by making one of the mirrors semitransparent.

When the laser medium is placed between the two mirrors, there is a chance that a photon, emitted spontaneously by one of the excited systems, will bounce off a mirror and return toward the laser medium. As that returning photon passes through the laser medium, it's amplified. Because the photon was emitted by one of the excited systems, it has the right wavelength to be amplified by other excited systems. (For a discussion of a photon's properties, see ❏.)

By the time the original photon leaves the laser medium, it has already been duplicated many times. This group of identical photons then bounces off the second mirror and returns for another pass through the laser medium. It continues to bounce back and forth between the mirrors until the number of identical photons in the collection is astronomical.

Eventually there are so many identical photons that the laser medium is no longer able to amplify them. The laser medium has only so much stored energy and only so many excited systems in it. If the laser medium continues to receive additional energy, it may continue to amplify the light somewhat. But if it doesn't receive more energy, light amplification will eventually cease.

To let the light out of this laser oscillator, one of its mirrors is normally *semitransparent*—that is, some of the photons that strike the surface of the mirror travel through it rather than reflecting. This transmission creates a beam of outgoing light, a *laser beam.*

The laser beam continues to emerge from the mirror as long as the amplification process can support it. Because this laser beam consists of duplicates of one original photon, it's coherent light. For technical reasons, many lasers duplicate more than one original photon simultaneously, so that their laser beams are a little less coherent than they might be. However, with suitable fine-tuning, one original photon can usually be made to dominate the laser beam.

When you focus a flashlight's beam with a lens, its independent photons won't end up exactly together at the focus of the lens. That's because the photons leave the flashlight heading in somewhat different directions and because their broad range of wavelengths leads to dispersion problems in the lens. But since virtually all of the photons in a laser beam are identical, they can all focus together to an extremely small spot. That's why a laser printer employs a laser; a laser beam can illuminate a very small spot on the photoconductor drum that is used in the xerographic process to produce a printed image.

❑ While it might seem that a photon should have an exact wavelength and frequency, and travel in only one direction, that's not the case. Photons travel as electromagnetic waves and spread in more than a single direction. And because each photon has a beginning and an end, its wave contains more than a single wavelength or a single frequency. Thus while lasers can produce some of the most perfect electromagnetic waves imaginable, those waves still spread outward slightly and still have a range of wavelengths and frequencies.

➤ Check Your Understanding #2: More of a Good Thing
for answers, see page 473

If you take the laser beam from a particular laser oscillator and send it through a similar laser amplifier, what will happen to the laser beam?

How a Laser Medium Works

Obtaining the excited systems needed to amplify light is a critical issue for lasers. Ideally, a laser involves four different states of an atom or atomlike system: the ground state, an excited state, the upper laser state, and the lower laser state. The reason for having four separate states should become clear in a moment.

Let's consider an atom that acts as an ideal laser amplifier (Fig. 14.3.4). The atom starts in its ground state. A collision or the absorption of a photon shifts it to the excited state, giving it the energy it needs to amplify light. The atom then shifts to the upper laser state, either by emitting a photon or as the result of a collision. This preliminary shift is important because it prevents the excited atom from returning directly back to the ground state and avoiding the amplification process. Once it has shifted to the upper laser state, the atom is stuck there and will wait around long enough to amplify light.

When a suitable photon passes through the atom, that photon stimulates the emission of a duplicate photon and the atom undergoes a radiative transition to the lower laser state. So far, so good. However, if the atom remains in the lower laser state, it might absorb a photon of the laser light and return to the upper laser state. To prevent this sort of radiation trapping, the atom must quickly shift to the ground state, either by emitting a photon or as the result of another collision. The atom is then ready to begin the cycle all over again.

This four-state cycle, or something close to it, is found in nearly all lasers. The cycle helps the laser develop a **population inversion** between its upper and lower laser states—a situation in which there are more atoms in the upper laser state prepared to *emit* the laser light than there are atoms in the lower laser state prepared to *absorb* that light. Developing a population inversion is critical to laser amplification because without it, the laser medium is more absorbing than amplifying and there can't be a buildup of light intensity.

In each laser, something provides the energy needed to shift atoms or atomlike systems in the laser medium from their ground states to their excited states in order to

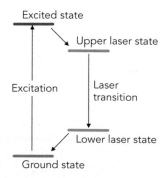

Fig. 14.3.4 An ideal laser system passes through four different states during the laser's operation.

Fig. 14.3.6 This flashlamp-pumped laser amplifier contains a purple neodymium:YAG rod. The rod is in the bottom half of the opened, gold-lined amplifier box and is protected by a glass tube. Light from a long flashlamp in the top half of the box excites the neodymium ions so that they can amplify infrared light passing horizontally through the rod.

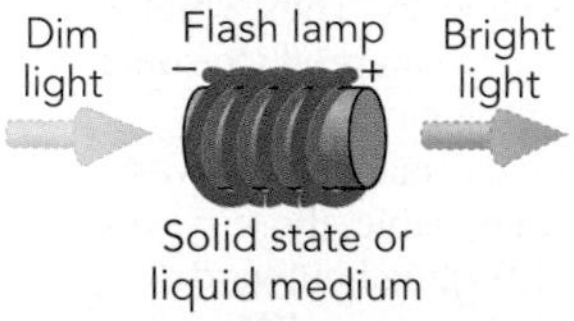

Fig. 14.3.5 In an optically pumped laser, intense light from a flashlamp, an arc lamp, or even another laser transfers energy to the laser medium. Atoms or atomlike systems inside the medium store this energy and use it to amplify light.

develop a population inversion. This transfer of energy into the laser medium to prepare it for amplifying light is called *pumping*. How a particular laser medium is pumped depends on the laser.

The most common pumping mechanisms are electronic and optical. In electronic pumping, currents of charged particles use their kinetic or electrostatic energies to excite the medium's atoms or atomlike systems from their ground states to their excited states. In optical pumping, intense light is shone on the laser medium, causing a similar excitation.

The most important examples of optical pumping are ion-doped solid-state lasers. These lasers are based on atomic ions embedded in transparent solids. Common ions are titanium (Ti), neodymium (Nd), and erbium (Er), and they are often embedded in sapphire, yttrium aluminum garnet (YAG), or glass. Ti:sapphire, Nd:YAG, and Er:glass lasers are important to modern research, technology, and optical communications systems. When these laser media are exposed to extremely bright light, their ions become excited, and they can act as laser oscillators or amplifiers (Figs. 14.3.5 and 14.3.6).

> Check Your Understanding #3: Not So Fast, 007

for answers, see page **473**

A secret agent in a movie uses a tiny, hand-held laser to burn a hole through a thick metal plate. From the point of view of power, why is such a laser essentially impossible to make?

Light-Emitting Diodes and Laser Diodes

The most common electronically pumped lasers are diode lasers. Found in pointers, barcode readers, printers, and CD and DVD players, these lasers are closely related to light-emitting diodes, which themselves resemble the ordinary diodes we discussed in Section 12.1. But whereas ordinary diodes are designed merely to control currents while wasting as little power as possible, laser and light-emitting diodes are optimized to produce light.

When a diode is forward biased and current flows from its anode to its cathode, conduction level electrons in the n-type cathode travel across the p-n junction to become conduction level electrons in the p-type anode. In effect, the anode is then in an excited state: it has conduction level electrons and empty valence levels.

What happens next depends on the characteristics of the diode. In a normal silicon diode, those conduction level electrons shift to empty valence levels without producing significant light. Silicon's band structure has characteristics that discourage light emission, so most of these electron transitions produce internal vibrations and heat the diode instead of producing light.

But in specialized diodes made from more exotic semiconductors, conduction level electrons in the p-type anode frequently undergo radiative transitions to empty valence levels and thereby emit light. Composed primarily from combinations of gallium, indium, aluminum, arsenic, phosphorus, and nitrogen, they are known as light-emitting diodes or LEDs. LEDs now come in just about any color of the rainbow, including infrared, red, orange, yellow, green, blue, violet, and ultraviolet (Fig. 14.3.7). Although white LEDs also exist, they're actually violet or ultraviolet LEDs with built-in phosphors that fluoresce white.

Fig. 14.3.7 These LEDs are connected in series, so that the same current flows sequentially through each of them. However, their different band gaps cause them to emit different colors of light.

Courtesy Lou Bloomfield

An LED's color is directly related to the energy released when an electron in its p-type anode shifts from a conduction level to a valence level. The most convenient unit in which to measure that energy is the **electron volt** (abbreviated eV)—the energy released when 1 elementary unit of electric charge experiences a 1 V decrease in voltage (1 eV is equal to 1.6021×10^{-19} J). In a typical red LED, an electron releases 1.9 eV as it shifts from a conduction level to a valence level and can produce a photon with an energy of 1.9 eV. Since energy and frequency are related by Eq. 14.2.1 and frequency and wavelength are related by Eq. 9.2.1, that 1.9-eV photon has a frequency of 4.6×10^{14} Hz and a wavelength of 650 nm.

To operate and produce these 1.9-eV photons, the red LED must be forward biased with a voltage drop of at least 1.9 V. The current-carrying diode uses that voltage drop to inject electrons into the anode's conduction band, where they have energies 1.9 eV above the valence band. Many of those electrons subsequently release their excess energies as 1.9-eV photons of light.

The shorter the wavelength of the light an LED emits, the more energy each electron must release as it shifts from a conduction level to a valence level and the larger the semiconductor's band gap must be. A violet LED that emits 400-nm light requires a band gap of about 3.1 eV to produce its 3.1-eV photons. That LED also needs a forward-bias voltage in excess of 3.1 V. The larger voltage drops required by LEDs near the violet end of the spectrum explain why those LEDs need higher voltage power sources.

Unfortunately, less than a quarter of the electrons sent across an LED's p-n junction succeed in lighting the room. Although a substantial fraction of those electrons emit photons, most of the photons are reabsorbed by the semiconductor before they leave the LED; the same radiative transitions that emit this light (conduction level → valence level) can also absorb it (valence level → conduction level). But despite those difficulties, modern LEDs can produce visible light with energy efficiencies comparable to those of fluorescent lamps. LED efficiencies continue to rise along with their operating lifetimes and it's only a matter of time before they become a primary form of illumination.

A laser diode is quite similar to an LED, except that a laser diode uses its radiative transitions to amplify light. Since that amplification can only occur when light emission exceeds light absorption, the laser diode must produce a population inversion between an upper laser state and a lower laser state.

The laser diode achieves such an inversion by concentrating current into a very narrow p-n junction made from heavily doped semiconductors. The intense current injects an enormous density of electrons into the anode's conduction band, where they quickly settle into the lowest energy conduction levels—the upper laser state. The heavy doping empties most of the anode's highest energy valence levels—the lower laser state. With many electrons in the upper laser state and few in the lower laser state, the diode has a population inversion and can amplify light.

Fig. 14.3.8 This tiny semiconductor chip is a diode laser that emits an intense beam of coherent light when a current flows through it.

Most laser diodes act as laser oscillators (Fig. 14.3.8), amplifying their own spontaneously emitted light until it forms an intense coherent beam. The ends of the anode itself are usually reflective enough to act as mirrors and form a complete laser oscillator. However, to concentrate the laser light in one direction and to control its beam characteristics, many laser diodes have complicated structures and coatings.

As in an LED, a laser diode's wavelength and color depend primarily on the band gap of its anode. Infrared and red laser diodes were developed first and quickly incorporated into a variety of household products. Developing laser diodes with larger band gaps and shorter wavelengths has been a long and arduous process. However, the spectrum of laser diode colors now extends into the ultraviolet and as these short-wavelength lasers become less expensive and more reliable, they begin to appear in everyday products.

Check Your Understanding #4: Lighten Up

for answers, see page 473

When you increase the current flowing through an incandescent lightbulb, it brightens and its color shifts toward the blue end of the spectrum. If you increase the current flowing through an LED or laser diode, what happens to its brightness and color?

Epilogue for Chapter 14

In this chapter we explored the creation and movement of light. In *sunlight,* we looked at how light is scattered during its passage through the atmosphere and at how it bends and reflects when it moves from one material to another. We also examined interference effects that occur when a light wave follows more than one path to a particular destination and learned how polarizing sunglasses can diminish glare.

In *discharge lamps,* we examined electrical discharges in gases. We saw that atoms excited by collisions with charged particles can subsequently emit light through radiative transitions. We studied the primary colors of light to see how the phosphors on the inside surface of a fluorescent tube are able to produce a reasonable facsimile of white sunlight. And in *lasers and LEDs,* we looked at the difference between common incoherent light and the unusual coherent light emitted by a laser. We saw that lasers use stimulated emission to duplicate photons, so that a small number of initial photons can be amplified into an enormous number.

Explanation: Splitting the Colors of Sunlight

Light bends as it passes through the cut facets of the crystal glass or bowl. Because of dispersion, the angle of each bend depends slightly on the wavelength of the light involved, so that the different colors of sunlight follow slightly different paths through the crystal. When the light emerges from the crystal, its various wavelengths head in somewhat different directions; you see colors. Color sequences progress from long wavelength to short wavelength, or vice versa. You see red, orange, yellow, green, blue, indigo, and violet, or the reverse—the colors of the rainbow.

Chapter Summary

How Sunlight Works: Sunlight originates at the 5800 °C outer surface of the sun, when electrically charged particles there accelerate back and forth rapidly and emit electromagnetic waves. This sunlight travels at the speed of light through empty space until it reaches the earth's atmosphere. There it slows slightly and some of it Rayleigh scatters. Short-wavelength light is Rayleigh scattered more strongly than long-wavelength light, so the sky appears blue.

As the sunlight passes through various objects, it slows down and its colors separate. When sunlight passes through falling raindrops, its different wavelengths follow different paths and create rainbows. As sunlight reflects from thin films such as soap bubbles, its waves are divided and may follow several different paths to the same destination. These light waves then interfere with one another so that some waves appear strong and bright while others are weak and dim. Interference depends on the wavelengths of light so thin films appear brightly colored.

How Discharge Lamps Work: Discharge lamps produce light by passing electric currents through gases. Those gases are turned into electrical conducting plasmas by filling them with charged particles, either by exposing them to strong voltage gradients or by injecting electrons into them from heated electrodes. Once a plasma is formed, current can pass through it and collisions within that current-carrying plasma cause the gas particles to emit light.

A fluorescent lamp emits visible light when the phosphor coating on the inside of its tube is exposed to 254-nm ultraviolet light produced inside the tube by a low-pressure mercury vapor discharge. This ultraviolet light excites the phosphor coating and causes it to emit visible light. In contrast, mercury, metal–halide, and sodium lamps use discharges to produce visible light directly and don't employ phosphors. By operating at high pressures, those discharge lamps produce relatively broad light spectrums and provide energy-efficient illumination.

How Lasers and LEDs Work: Lasers amplify light through the process of stimulated emission. In this process, energy is transferred to atoms or atomlike systems contained in a laser medium. While these excited systems might emit light spontaneously, they can be stimulated into emitting duplicates of a passing photon. When a photon with just the right wavelength passes through an excited system, that system is likely to give up its stored energy by emitting an exact copy of the initial photon. In a laser oscillator, two mirrors cause light to bounce back and forth through the laser medium. An initial photon is duplicated endlessly to produce coherent light. One of the mirrors is semitransparent so part of the light emerges from the laser as a laser beam.

In an LED, light is emitted when electrons that have crossed the diode's p-n junction into the anode's conduction levels undergo radiative transitions to the empty valence levels. The color of light produced by the LED is determined primarily by the anode's band gap. A diode laser uses the same concepts as an LED, except that a population inversion is developed between the relatively empty valence levels and the relatively full conduction levels. With more electrons prepared to emit light than absorb it, the diode laser can then amplify light.

Important Laws and Equations

1. Relationship between Energy and Frequency: The energy in a photon of light is equal to the product of the Planck constant times the frequency of the light wave, or

energy = Planck constant · frequency.   (14.2.1)

Check Your Understanding—Answers

Section 14.1 SUNLIGHT

1. Charged particles in the hot flame accelerate back and forth and emit electromagnetic waves that include the low-frequency end of the visible spectrum.
Why: The accelerating charged particles in hot objects emit light. The hotter the object, the faster the charged particles move and accelerate and the higher the frequencies of light they emit. A candle isn't hot enough to emit the whitish light of the sun, so it emits mostly reddish or yellowish light.

2. Rayleigh scattering from the microscopic smoke particles scatters more blue light than red light and gives the air a bluish glow.
Why: When smoky air in a dark room is illuminated by a bright spotlight or another white light source, the tiny particles in the air Rayleigh scatter some of the light. When you look at the air against a dark background you can see this glow. Since Rayleigh scattering affects blue light most strongly, the air appears bluish.

3. A diamond exhibits dispersion so that different frequencies of sunlight follow somewhat different paths through the diamond's polished facets. The different colors of sunlight emerge from the diamond traveling in slightly different directions, so that you can see them individually.
Why: One of the delightful aspects of a diamond is its strong dispersion. It bends violet light much more than red light so that sunlight is separated into its different colors as it passes through the stone. Cleverly cut facets help to isolate the individual colors.

4. From interferences between light reflected by the top and bottom surfaces of the floating layer of oil.
Why: A thin film of almost anything on water will appear colored because of interference. Part of each light wave striking the film reflects from the top surface, and part reflects from the bottom surface. These two reflected waves interfere with one another in a wavelength-dependent manner and make the layer appear brightly colored. Different colors correspond to different thicknesses of the thin films.

5. Most of the light that reflects from the water is horizontally polarized. The sunglasses block horizontally polarized light so that you see mostly light from within the pool of water.
Why: Sunlight reflecting from a horizontal surface is mostly horizontally polarized waves. By blocking horizontally polarized light, polarizing sunglasses virtually eliminate this reflected light, and you see mostly light from below the surface. If you turn the sunglasses sideways, they will block the wrong polarization of light and you will see mostly the reflected light. Try it.

Section 14.2 DISCHARGE LAMPS

1. Orange.
Why: Your red-, green-, and blue-sensitive cone cells have the same response to this light mixture as they would have to pure 600-nm light. Such light appears orange, so you see orange when viewing the mixture.

2. It decreases the lamp's energy efficiency.
Why: Although the bluish filter changes the spectrum of wavelengths so that the amount of blue light leaving the lamp increases relative to the red light, it does this by absorbing some of the red light. The filter gets hot, and less electric power used by the bulb leaves as visible light.

3. When fire adds energy to an atom and shifts it to an excited state, it emits photons in order to return to its ground state. Each photon's energy, frequency, and color are determined by energy differences between the atom's states.
Why: Each color in fireworks is produced by a particular type of atom. Some atoms, such as strontium and lithium, emit mostly red light as they return to their ground states. Barium atoms emit green light, copper atoms emit blue-green light, and sodium atoms emit yellow-orange light.

4. You would see only a dim (blue-white) glow from the lamp because most of the light produced by the mercury discharge is invisible ultraviolet light.
Why: Without its phosphor coating, a fluorescent tube produces very little visible light. The phosphor coating is needed to convert the discharge's ultraviolet light into visible light. But even if you could see ultraviolet light, you would see very little of it leaving the tube. The tube is made of glass, which absorbs virtually all ultraviolet light with wavelengths shorter than 350 nm. Even in a normal fluorescent tube, any ultraviolet not converted to visible light by the phosphors is absorbed by the glass tube.

5. To start many lamps, their electrodes must be heated red-hot. You can often see light emitted by these hot-filament/electrodes.
Why: In most fluorescent fixtures, the filaments are heated to high temperatures just before the discharge starts. Thermal energy then ejects electrons from the filament so that they can carry current through the gas.

6. The small high-pressure discharge tube is heating up, vaporizing more mercury.

Why: When the streetlight first turns on, the pressure of mercury atoms inside it is low and it emits very little visible light. As the discharge heats the tube, more mercury atoms evaporate until eventually all of the mercury inside the tube is gaseous.

Section 14.3 LASERS AND LEDS

1. The flashlight produces incoherent light, with many independent light waves contributing their individual electric fields to the overall electric field. If you were to measure that overall field, you would find it very disorderly.

Why: Because the photons of incoherent light are independent, their individual electric fields don't fluctuate together. At any place and time, the individual electric fields will add in a complicated, random fashion. As time passes, this overall field will fluctuate randomly. But coherent light, in which each photon is the same as all the others, has a much more orderly electric field because all of the photons contribute equally.

2. It will become even brighter.

Why: A laser oscillator normally emits as intense a beam of light as it can, based on the amount of stored energy in the laser medium. This beam of light can be amplified further by sending it through a separate laser amplifier. Most high-powered lasers use a laser oscillator and one or more laser amplifiers to create particularly bright beams of light.

3. The light in such a laser beam would carry an enormous amount of power. Something must transfer that power to the laser medium, an impossible task in a tiny, hand-held unit.

Why: The power in a laser beam comes from the laser medium. The laser medium must have gotten it from somewhere else. Since batteries aren't up to delivering thousands of watts of electric power, it's unlikely that powerful hand-held lasers will ever be developed. Even if a suitable power source were available, these lasers would tend to overheat. The conversion of power to light is not perfectly efficient, and most of the energy ends up as thermal energy in the laser's components. Lasers require cooling to remove this wasted energy.

4. The diode becomes brighter, but its color doesn't change significantly.

Why: The diode's brightness depends on how many electrons cross its p-n junction each second and increasing that number increases the brightness. However, the color of light the diode emits depends mostly on its band gap and therefore doesn't change much when you increase the current in the diode.

Check Your Figures—Answers

Section 14.2 DISCHARGE LAMPS

1. 6.626×10^{-28} J.

Why: Since the Planck constant is 6.626×10^{-34} J · s and the frequency of the radio wave is 10^6 Hz or 10^6 cycles/second, the energy per photon is given by Eq. 14.2.1 as

$$energy = (6.626 \times 10^{-34} \text{ J} \cdot \text{s}) \cdot (10^6 \text{ cycles/s})$$
$$= 6.626 \times 10^{-28} \text{ J.}$$

This energy is so small that it's virtually impossible to observe the particulate character of radio waves.

Exercises

1. How long should an antenna be to receive or transmit red light well?

2. How long should an antenna be to receive or transmit violet light well?

3. When astronauts aboard the space shuttle look down at the earth, its atmosphere appears blue. Why?

4. When astronauts walked on the surface of the moon, they could see the stars even though the sun was overhead. Why can't we see the stars while the sun is overhead?

5. If you shine a flashlight horizontally at a glass full of water, the glass will redirect the light beam. How?

6. A laser light show uses extremely intense beams of light. When one of these beams remains steady, you can see the path it takes through the air. What makes it possible for you to see this beam even though it isn't directed toward you?

7. To make the beams at a laser light show (see Exercise 6) even more visible, they're often directed into mist or smoke. Why do such particles make the beams particularly visible?

8. Why can you see your reflection in a calm pool of water?

9. Use the concepts of refraction, reflection, and dispersion to explain why a diamond emits a spray of colored lights when sunlight passes through its cut facets.

10. Diamond has an index of refraction of 2.42. If you put a diamond in water, you see reflections from its surfaces. But if you put it in a liquid with an index of refraction of 2.42, the diamond is invisible. Why is it invisible, and how is this effect useful to a jeweler or gemologist?

11. Why is a pile of granulated sugar white while a single large piece of rock candy (solid sugar) is clear?

12. Basic paper consists of many transparent fibers of cellulose, the main chemical in wood and cotton. Why does paper appear white, and why does it become relatively clear when you get it wet?

13. On a rainy day you can often see oil films on the surfaces of puddles. Why do these films appear brightly colored?

14. When two sheets of glass lie on top of one another, you can often see colored rings of reflected light. How do the nearby glass surfaces cause these colored rings?

15. If you're wearing polarizing sunglasses and want to see who else is wearing polarizing sunglasses, you only have to turn your head sideways and look to see which people now have sunglasses that appear completely opaque. Why does this test work?

16. Why is it easier to see into water when you look directly down into it than when you look into it at a shallow angle?

17. Light near 480 nm has a color called cyan. What mixture of the primary colors of light makes you perceive cyan?

18. What is different about the two mixtures of red and green lights that make you see yellow and orange, respectively?

19. What colors of light does red paint absorb?

20. What colors of light does yellow paint absorb?

21. If you illuminate red paint with pure blue light, what color will that paint appear?

22. Fancy makeup mirrors allow users to choose either fluorescent or incandescent illumination to match the lighting in which they'll be seen. Why does the type of illumination affect their appearances?

23. While a sodium atom is in its ground state, it cannot emit light. Why not?

24. When a sodium atom is in its lowest energy excited state, it can emit light. Why?

25. You expose a gas of argon atoms to light with photon energies that don't correspond to the energy difference between any pair of states in the argon atom. Explain what happens to the light.

26. A discharge in a mixture of gases is more likely to emit a full white spectrum of light than a discharge in a single gas. Why?

27. If the low-pressure neon vapor in a neon sign were replaced by low-pressure mercury vapor, the sign would emit almost no visible light. Why not?

28. Increasing the power to an incandescent bulb makes its filament hotter and its light whiter. Why doesn't increasing the power to a neon sign change its color?

29. While many disposable products no longer contain mercury, a potential pollutant, fluorescent tubes still do. Why can't the manufacturers eliminate mercury from their tubes?

30. When white fabric ages it begins to absorb blue light. Why does this give the fabric a yellow appearance?

31. To hide yellowing (see Exercise 30), fabric is often coated with fluorescent "brighteners" that absorb ultraviolet light and emit blue light. In sunlight, this coated fabric appears white, despite absorbing some blue sunlight. Explain.

32. Camera flashes use discharges in high-pressure xenon and krypton gases to produce brief, intense white light. Why is it important that they use these complicated atoms?

33. A CD player uses a beam of laser light to read the disc, focusing that light to a spot less than 1 μm (10^{-6} m) in diameter. Why can't the player use a cheap incandescent lightbulb for this task, rather than a more expensive laser?

34. Why can't a CD player (Exercise 33) use a light-emitting diode (LED) in place of its diode laser?

35. Explain why the electromagnetic wave emitted by a radio station is coherent—a low-frequency equivalent of coherent light.

36. One of the most accurate atomic clocks is the hydrogen maser. This device uses excited hydrogen molecules to duplicate 1.420-GHz microwave photons. In the maser, the molecules have only two states: the upper maser state and the lower maser state (which is actually the ground state). To keep the maser operating, an electromagnetic system constantly adds excited state hydrogen molecules to the maser and a pump constantly removes ground state hydrogen molecules from the maser. Why does the maser require a steady supply of new excited state molecules?

37. Why must ground state molecules be pumped out of a hydrogen maser (see Exercise 36) as quickly as possible to keep it operating properly?

38. While some laser media quickly lose energy via the spontaneous emission of light, others can store energy for a long time. Why is a long storage time essential in lasers that produce extremely intense pulses of light?

39. One of the first lasers used synthetic ruby as its laser medium. However, a ruby laser is a three-state laser; its lower laser state is its ground state. Why does that arrangement make the ruby laser relatively inefficient?

40. If most of the highest energy valence levels in a diode laser's p-type anode weren't empty, it would become relatively inefficient and probably wouldn't emit laser light at all. Why not?

41. Why doesn't increasing the current passing through an LED affect the color of its light?

42. Why does increasing the current passing through an LED affect the brightness of its light?

Problems

1. The yellow light from a sodium vapor lamp has a frequency of 5.08×10^{14} Hz. How much energy does each photon of that light carry?

2. If a low-pressure sodium vapor lamp emits 50 W of yellow light (see Problem 1), how many photons does it emit each second?

3. A particular X-ray has a frequency of 1.2×10^{19} Hz. How much energy does its photon carry?

4. A particular light photon carries an energy of 3.8×10^{-19} J. What are the frequency, wavelength, and color of this light?

5. If an AM radio station is emitting 50,000 W in its 880-kHz radio wave, how many photons is it emitting each second?

6. If an FM radio station is emitting 100,000 W in its 88.5-MHz radio wave, how many photons is it emitting each second?

CHAPTER 15

Optics

Many of the devices around us perform useful tasks by manipulating light. Cameras record images of the objects in front of them while magnifying glasses allow us to see details we'd miss with our eyes alone. Still other gadgets, such as CD or DVD players, use light in ways that have nothing to do with vision. But all of these objects manipulate light using similar techniques—the techniques of optics.

While optical tools such as lenses and prisms have been around for hundreds of years, advances of modern technology have accelerated the development of optics. Just as the invention of transistors has sped the growth of the electronics industry, so the invention of lasers has sped the growth of the optics industry. The two fields are not far apart in many ways, and there is hope that one day computers will be as much optical devices as they are electronic.

EXPERIMENT: Focusing Sunlight

Courtesy Lou Bloomfield

There are many household devices that manipulate light, and one of the most familiar is a magnifying glass. A magnifying glass bends light rays toward one another as they pass through it. In this chapter, we'll see how a simple converging lens of this sort can magnify an object or cast its image onto a light-sensitive surface. For the moment, we'll simply use it to bring light rays together.

On a clear day with the sun roughly overhead, take a magnifying glass and a dark piece of paper outdoors to a place where there is nothing flammable around. You will use light from the sun to heat the paper until it smokes, so be prepared to extinguish a fire if necessary. Hold the magnifying glass just above the paper and slowly lift the lens upward toward the sun. Make sure that the lens is turned so that its surface faces the sun. You should see a bright circle appear on the dark paper, and that circle should become smaller and smaller as you lift the lens. The sun's light rays are bending inward as they pass through the lens and are converging together as they head toward the paper. Be careful not to let this bright circle of light touch your skin because it will burn you. If the spot isn't round, try tilting the lens differently.

When the lens is just the right height above the paper, the circle of light will reach a minimum size. It will appear as a brilliant white disk that is actually an image of the sun itself. If the magnifying glass is wide enough and the sun is bright enough, the paper will begin to smoke. The image of the sun is then so bright that it is able to heat the paper to its ignition temperature. If the lens is quite large, the paper may begin to burn with a flame.

You can project images of other brightly illuminated objects on a sheet of white paper. Try to *predict* the sizes and orientations of those images. *Observe* what happens. Did you *verify* your predictions? Try to *measure* the distance to an object by seeing how far the lens must be from the paper to get a sharp image.

Chapter Itinerary

This process of bringing light together to form a small spot or an image of a distant object is a common theme in optics. In this chapter, we'll examine several systems that are based on this sort of manipulation of light: (1) *cameras* and (2) *optical recording and communication*. In *cameras*, we'll see how lenses bend light to form images and how those images are used to create photographs. In *optical recording and communication*, we'll explore the roles of lasers in optics while investigating several novel optical effects. For additional preview information, turn to the Chapter Summary on p. 498.

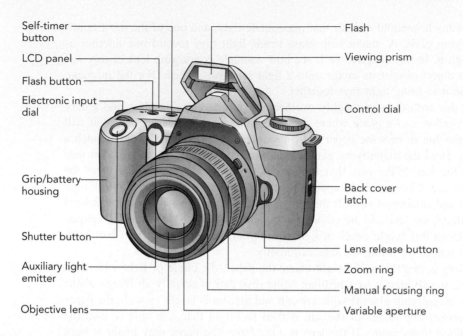

Self-timer button
LCD panel
Flash button
Electronic input dial
Grip/battery housing
Shutter button
Auxiliary light emitter
Objective lens

Flash
Viewing prism
Control dial
Back cover latch
Lens release button
Zoom ring
Manual focusing ring
Variable aperture

SECTION 15.1 Cameras

In the two centuries since their invention, cameras have become extremely easy to use. What started as a hobby for a few dedicated enthusiasts has evolved into an everyday activity. But despite all of the technological improvements, photography still employs many of the same principles it did in the 1800s. Cameras still use lenses to project images onto light-sensitive surfaces, and photographers still have to worry about getting the exposure right, focusing properly, and avoiding the blur of rapid motion. In this section, we'll explore some of the principles that make cameras work.

Questions to Think About: *Why are expensive camera lenses so complicated, with so many separate pieces of glass? Why does a longer lens seem to bring the objects nearer to you? What does a camera's aperture do? How does the camera's shutter speed affect the photograph?*

Experiments to Do: *The basic activity of a camera—projecting an image of the scene in front of you onto its image sensor—requires nothing more than a simple magnifying glass. In fact, you can use almost any simple lens that's bowed outward in the middle, including the eyeglasses of a farsighted person with no astigmatism or drugstore reading glasses. Stand in a darkened room, opposite a window. Hold a sheet of white paper so that it's parallel to the window and move the lens back and forth in front of the paper. Make sure that the lens itself is parallel to the window.*

You should find a distance at which the lens casts a clear image of the window onto the sheet of paper. You'll find that the scene appears upside down and backwards and that the image becomes fuzzy if you move the lens toward or away from the paper. You'll also find that, when the window's image is sharp, the outside scene's image is fuzzy and vice versa. Which way must you move the lens to bring more distant objects into focus?

Cut a hole in a piece of cardboard and use it to cover all but the center portion of the lens. The image will now be substantially darker, but it will also be sharper over a wider range of distances from the paper. If the hole is narrow enough, virtually everything focuses at once. Evidently, the diameter of the lens and its ability to focus on several things at once are closely related. Why?

Lenses and Real Images

When you take a picture of the scene in front of you, the lens of your camera bends light from that scene into a real image on a light-sensitive surface. A **real image** is a pattern of light, projected in space or on a surface, that exactly reproduces the pattern of light in the original scene. Since the real image that's projected looks just like the scene you're photographing, recording the light in that image is equivalent to recording the appearance of the scene itself.

While that light-sensitive surface was once always photographic film, digital cameras are gradually replacing film with electronic image sensors. Fortunately, the two light-sensing surfaces are essentially interchangeable, so we can refer to them both as *image sensors*: one is electronic and one is photochemical.

Real images don't occur without help. When light from a candle falls directly on an image sensor, it produces only diffuse illumination (Figs. 15.1.1*a* and 15.1.2*a*). You can't tell by looking at the sensor what the candle looks like because the light that leaves the candle travels in all directions and is as likely to hit the top of the sensor as it is to hit the bottom.

That's why a camera needs a **lens,** a transparent object that uses refraction to form images. The light passing through a lens bends twice, once as it enters the glass or plastic and again as it leaves. In a camera lens, this bending process brings much of the light from one point on the candle back together at one point on the sensor. As you can see in Figs. 15.1.1*b* and 15.1.2*b*, the real image that forms is upside-down and backward. This inversion of the real image relative to the object always happens when a single lens creates a real image.

The curved shape of the camera lens allows it to form a real image. Light passing through the upper half of the lens is bent downward while light passing through the lower half is bent upward. Because the camera lens bends light rays toward one another, it's a **converging lens.** You can see how it forms an image in Fig. 15.1.1*b* by following some of the rays of light leaving one point on the candle.

The upper ray from the candle flame travels horizontally toward the top of the lens. As it enters the lens and slows down, this ray of light bends downward. It bends downward again as it leaves the lens and travels downward toward the bottom of the image sensor.

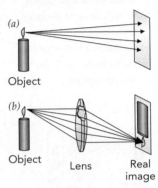

Fig. 15.1.1 (*a*) Without a lens, light from a candle uniformly illuminates an image sensor. (*b*) When a lens is introduced between the candle and the sensor, it brings light from each point on the candle back together on the sensor's surface, forming an upside-down and backward real image of the candle. The distance between the lens and the sensor must be chosen correctly, or the image will be blurry.

Fig. 15.1.2 (*a*) When candlelight falls directly on a sheet of paper, it produces no image. (*b*) But a lens inserted between the candle and paper forms an inverted real image of the flame on the paper.

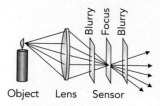

Fig. 15.1.3 The real image is in focus only when the image sensor is just the right distance from the lens. If the sensor is too near or too far from the lens, the image is blurry.

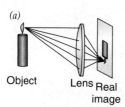

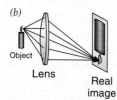

Fig. 15.1.4 (*a*) The light from a distant candle is traveling in almost the same direction and the lens focuses it easily. The real image forms close to the lens. (*b*) The light from a nearby candle diverges, and the lens has more difficulty bending it back together. The real image forms far from the lens. If the candle is too close to the lens, no real image forms at all.

The lower ray from the candle flame travels downward toward the bottom of the lens and bends upward as it enters the lens. It bends upward again as it leaves the lens and travels horizontally toward the bottom of the image sensor.

These two rays of light reach the image sensor at the same point. They are joined there by many other rays from the same part of the candle flame, so that a bright spot forms on the sensor. Overall, each part of the candle illuminates a particular spot on the image sensor, so the lens creates a complete image of the candle on the sensor.

However, the lens can bring the light back together to form a sharp image on the sensor only if the lens and sensor are separated by just the right distance (Fig. 15.1.3). If the sensor is too close to the lens, then the light doesn't have room to come together. If the sensor is too far from the lens, then the light begins to come apart again before reaching the sensor. In either case, the image on the sensor is blurry. The candle's real image is only *in focus* at one distance from the lens.

If the candle moves toward or away from the camera lens, the distance between the lens and the image sensor must also change (Fig. 15.1.4). When the candle is far away, all of its light rays that pass through the lens arrive traveling almost parallel to one another and the inward bend caused by the lens makes those rays converge together quickly. The rays come into focus relatively near the lens and that's where the sensor must be (Fig. 15.1.4*a*). The candle's image on the sensor is much smaller than the candle itself because the light rays have only a short distance over which to move up or down after leaving the lens.

When the candle is nearby, its light rays that pass through the lens are diverging rapidly and the inward bend caused by the lens is just barely enough to make those rays converge at all. As a result, the rays come into focus relatively far from the lens (Fig. 15.1.4*b*). The candle's image on the sensor is quite large because the light rays have considerable distance over which to move up or down after leaving the lens.

Because distant and nearby objects form real images at different distances from the camera lens, they can't both be in focus on the same image sensor. When you take a picture of a person standing in front of a mountain, only one of them can be in sharp focus. However, if you're willing to compromise a little bit on sharpness, a lens can sometimes form acceptable images of both objects.

► Check Your Understanding #1: Seeing the Lights

for answers, see page **498**

If you hold a magnifying glass at the proper distance above a sheet of white paper, you will see an image of the overhead room lights on the paper. Explain.

Focusing and Lens Diameter

A disposable camera is little more than a box with a lens. The lens projects a real image of the scene in front of it onto the camera's image sensor. Light in the real image exposes the image sensor, which records the image permanently. While it may also have a shutter that starts and stops the exposure, a flash to provide extra light, and a mechanism that prepares for the next photograph, there's little else to this simple camera.

However, there are limitations to the disposable camera design. One of the most severe limitations is that you can't focus it—the camera has a fixed distance between the lens and image sensor. Nonetheless, it manages to form relatively sharp real images on the sensor, even when there are objects at various distances from the camera. These simple cameras work because they use narrow (small-diameter) lenses. A narrow lens gathers less light than a wide lens but doesn't require *focusing*—adjusting the distance between the lens and the image sensor.

Because a wide (large-diameter) lens brings rays together from many different directions, you must focus it (Fig. 15.1.5*a*); if the image sensor is even slightly too near or too far from the lens, the recorded image will be blurry. But a narrow lens forms a reasonably clear image even without focusing. Any rays from one part of the scene that succeed in passing through the narrow lens must already be fairly close together. Their initial closeness means that these converging rays illuminate only a small part of the image sensor even when the sensor isn't exactly the right distance from the lens (Fig. 15.1.5*b*). Since the image sensor can't record every minute detail anyway, the image that forms on it doesn't have to be in absolutely perfect focus. As a result, a camera with a narrow lens and no focus adjustment manages to take pretty good pictures.

Unfortunately, these simple cameras collect very little light and need extremely light-sensitive image sensors. These high-speed sensors can't record as sharp photographs as low-speed sensors. Furthermore, the pictures produced by simple cameras lack fine details; while everything is almost in focus, most things are a bit fuzzy if you look carefully or make enlargements.

More sophisticated cameras use wider lenses that gather more light and expose the image sensors much more rapidly. They also automatically adjust the distance between the lens and the sensor. They identify the object you are photographing and position the camera lens so that it projects a sharp image on the sensor. As the camera focuses, you can usually see components in the lens moving backward or forward to arrive at the correct distance from the sensor.

Even a camera with a wide lens can take advantage of the narrow lens trick for focusing. Its lens contains an internal diaphragm that reduces its **aperture** or effective diameter. The *diaphragm* is a ring of metal strips with a central opening. These strips can swing in or out, changing the diameter of the diaphragm's opening and with it the aperture of the lens (Fig. 15.1.6).

When its lens aperture is narrow, the sophisticated camera imitates a simple camera. Nearly everything is essentially in focus simultaneously. In such a situation, the camera has a large *depth of focus*. Actually, the sophisticated camera can bring the most important object into perfect focus, so it produces pictures that are superior to those from simple cameras. However, narrowing the aperture of the lens also reduces the amount of light reaching the image sensor. The scene in front of the camera must either be very bright or the exposure must be relatively long. You don't get something for nothing.

While widening the aperture of a large lens makes full use of its light-gathering capacity, focusing then becomes crucial. Even a small error in the lens-to-sensor distance produces a blurry picture, so the depth of focus is very small. This trade-off, between light gathering and depth of focus, is a continual struggle for photographers. However, photographers sometimes take advantage of the small depth of focus in wide lenses to blur the background or foreground of a photograph deliberately. A camera's portrait setting adopts this strategy to produce sharp images of people against blurred backgrounds.

At other times photographers choose long exposures at narrow apertures in order to bring an entire scene into sharp focus. A camera's landscape setting takes this route, so that everything in the photograph shows full detail. And to capture fast motion while retaining a large depth of focus, photographers use a flash to brighten the scene and shorten the exposure. Unfortunately, a camera flash is ineffective at brightening a distant

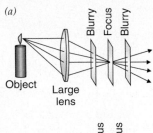

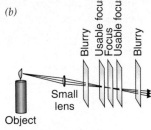

Fig. 15.1.5 (*a*) A large lens collects lots of light but its focus is critical. The image is blurry except at the actual focus. (*b*) A small lens collects less light but its image is relatively sharp anywhere near the focus.

Courtesy Lou Bloomfield

Fig. 15.1.6 The aperture of this lens can be reduced by closing its internal diaphragm, dimming its image but increasing its depth of focus.

scene and it can produce unpleasant reflections from windows and eyes. A camera's sports setting emphasizes brief exposures to avoid speed blur, even though that may require a wide aperture and small depth of focus.

➤ Check Your Understanding #2: Portrait Photos

for answers, see page **498**

While taking a photograph of your friend, with the diaphragm of your large camera lens wide open, you notice that the background is blurry. Explain.

Focal Lengths and f-Numbers

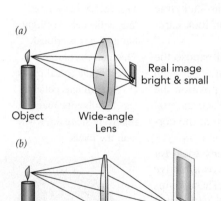

(a)

Object   Wide-angle
Lens

Real image
bright & small

(b)

Object   Telephoto   Real image
Lens   dim & large

Fig. 15.1.7 (*a*) A wide-angle lens has a short focal length and forms a small, bright real image near the lens. (*b*) A telephoto lens has a long focal length and forms a large, dim real image far from the lens.

Lenses are characterized by two quantities: focal length and f-number. The **focal length** of a lens is the distance between the lens and the real image it forms *of a very distant object*. For example, if a real image of the moon forms 100 mm (4 inches) behind a particular lens, then that lens has a focal length of 100 mm. The focal lengths of camera lenses range from less than 10 mm (0.4 inches) in many compact cameras to about 2 m (7 feet) in cameras used for nature photography.

When light from a scene passes through a short focal length lens, it comes to a focus near that lens and produces a relatively small image on the image sensor. Because a long focal length lens permits the light passing through it to spread out more before coming to a focus, it produces a larger real image on the sensor.

The "normal" lens for a particular camera has a focal length that allows all of the objects in your central field of vision to fit onto the image sensor. When you hold the finished photograph about 30 cm (1 foot) from your eyes, the objects in the picture appear about the same size they did when the photograph was taken. The focal length of a camera's normal lens is about 1.5 times the horizontal width of its image sensor.

A wide-angle lens has a shorter focal length than the normal lens (Fig. 15.1.7*a*). The image it projects onto the image sensor is smaller but brighter, and most of the objects in your entire field of vision appear in the photograph. A telephoto lens has a longer focal length than the normal lens (Fig. 15.1.7*b*). The image it projects onto the sensor is larger but dimmer, with only objects at the center of the scene appearing in the photograph.

In addition to indicating where the image of a distant object forms, the focal length of the camera lens relates the object distance to the image distance. The **object**

Table 15.1.1 Several Cameras, the Widths of the Image Sensor They Use, and Their Normal Lenses

TYPE OF CAMERA	SENSOR WIDTH	NORMAL LENS
Typical digital camera	8 mm	12 mm
35-mm Camera	36 mm	50 mm
2¼-inch Medium format camera	2¼ inches	80 mm
5-inch Portrait camera	5 inches	180 mm

distance is the distance between the lens and the object you're photographing. The **image distance** is the distance between the lens and the real image it forms (Fig. 15.1.8). The relationship is called the **lens equation** and can be written in a word equation:

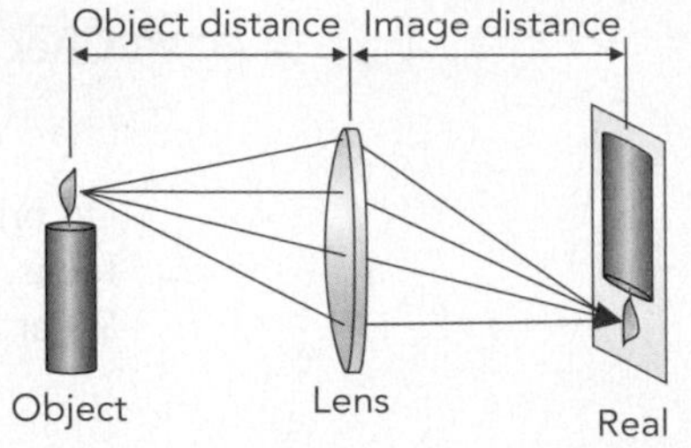

$$\frac{1}{\text{focal length}} = \frac{1}{\text{object distance}} + \frac{1}{\text{image distance}}, \qquad \textbf{(15.1.1)}$$

in symbols:

$$\frac{1}{f} = \frac{1}{o} + \frac{1}{i},$$

and in everyday language:

The farther away an object is, the closer to the lens its image forms.

Fig. 15.1.8 The relationship between the object distance, the image distance, and the focal length of the lens is given by the lens equation.

> **The Lens Equation**
> One divided by the focal length of a lens is equal to the sum of one divided by the object distance and one divided by the image distance.

According to the lens equation, the image distance for a distant object is equal to the focal length of the lens. That agrees with our earlier discussion of focal length. But when the object is nearby, the image distance becomes larger than the focal length. That's why a camera lens moves away from the image sensor as you focus closer. And when the object distance becomes less than the focal length, the image distance becomes negative and no real image forms at all. That's why you can't focus on an object that's too close to the lens.

A lens's **f-number** characterizes the brightness of the real image it forms on the image sensor, with smaller f-numbers indicating brighter images. The f-number is calculated by dividing the lens's focal length by its diameter. Since long focal length lenses naturally produce larger and dimmer images on the image sensor, the f-number takes into account both the light-gathering capacity of the lens and its focal length. Increasing a lens's diameter increases its light-gathering capacity and decreases its f-number. Increasing a lens's focal length decreases the brightness of its real image and increases the f-number. Doing both at once, increasing the lens diameter and focal length equally, leaves the brightness and f-number unchanged.

Most sophisticated cameras use large-diameter lenses so that their f-numbers are generally less than 4. Since it's difficult to fabricate a lens that's larger in diameter than its focal length, the smallest practical f-number is about 1. And because long focal length lenses need large apertures to keep their f-numbers small, some telephoto lenses are huge.

The diaphragm inside a lens allows you to decrease the lens's aperture and thus increase its f-number. A factor of 2 increase in f-number corresponds to a factor of 2 decrease in the lens's effective diameter and a factor of 4 decrease in the lens's light-gathering area. Thus when you double the f-number of the lens, you must compensate by quadrupling the exposure time. Although closing the aperture increases the lens's depth of focus, it requires a longer exposure.

▶ Check Your Understanding #3: Bright and Sharp Photographs

for answers, see page 499

On bright, sunny days, your automatic camera takes photographs with large depths of focus, while on dark, overcast days its pictures have much smaller depths of focus. What causes this difference?

▶ Check Your Figures #1: The Image of an Apple

for answers, see page 500

If the distance from an apple to a converging lens is twice the focal length of that lens, where will the real image of the apple form?

Improving the Quality of a Camera Lens

A high-quality camera lens isn't a single piece of glass or plastic. Instead, it's composed of many separate elements that function together as a single lens. This complexity improves the quality of the real image. To begin with, dispersion in a single element lens causes different colors of light to bend differently and focus at different distances behind that lens. Known as *chromatic aberration*, this problem can be fixed by using several lens elements made of different types of glass, with different amounts of dispersion. These elements compensate for one another so that the overall lens, known as an *achromat*, has very little dispersion and almost no color focusing problems.

After correcting for color and other technical image problems, a sophisticated camera lens may contain more than ten individual elements. For the purposes of the lens equation, this complicated lens has an effective center from which to calculate object and image distances. But having so many separate elements creates reflection problems; each time light passes from air to glass or vice versa, some of it reflects. To avoid fogging the photographs with this bouncing stray light, the individual elements are *antireflection coated* with thin layers of transparent materials. The best coatings use interference effects to cancel out the reflected light waves and give the lens only a weak violet reflection.

Many modern cameras are equipped with zoom lenses. A zoom lens is a complicated lens that can change the size of the real image it projects onto the image sensor. By carefully moving its lens elements relative to one another, the zoom lens can adjust its effective focal length.

A common type of zoom lens contains three separate groups of lens elements and produces a sequence of three images (Fig. 15.1.9). The first lens group forms a first image of the scene in front of the camera. The second lens group forms a second image of that first image. And the third lens group projects a third, real image of the second image onto the image sensor. Zooming the lens, that is, changing its focal length, involves altering the spacings between the lens groups in order to vary the second lens group's object and image distances and thus the relative sizes of the first and second images.

As the zoom lens changes from short focal length to long focal length, the image it projects on the sensor becomes larger. This effect allows you to compose the picture so that the scene fills the photograph completely. A lens that can change its focal length while retaining the same f-number and still keep the real image in focus on the sensor is a truly remarkable achievement.

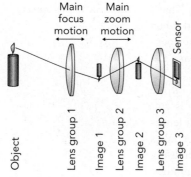

Fig. 15.1.9 A common type of zoom lens uses three lens groups to project a variable size real image on the image sensor. Zooming is done mostly by moving the second lens group to change its image and object distances. The first lens group is responsible for focusing and the third lens group projects the real image on the sensor.

► Check Your Understanding #4: Not Such a Good Picture
for answers, see page **499**

If you use a large magnifying glass to form a real image of an overhead fluorescent light fixture, you will find that the corners of the light fixture are blurry and have rainbow colors in them. Why is the image quality so poor?

Courtesy Lou Bloomfield

The Viewfinder and Virtual Images

SLR or <u>s</u>ingle <u>l</u>ens <u>r</u>eflex cameras permit you to change their lenses so that you can choose a lens that's optimized for the task at hand. When you peer through the viewfinder of an SLR camera (Fig. 15.1.10), you're looking at the same real image that will be projected onto the image sensor during the exposure. The light you see travels through the camera's main lens, reflects off a mirror, and projects onto a translucent screen inside the top of the camera. You're simply inspecting this screen and the real image through a magnifying lens in the eyepiece. During the exposure, the mirror flips out of the way and the real image projects briefly onto the image sensor.

Since the screen and real image are only an inch or two from your eye, you can't focus on them without the help of the eyepiece lens. The eyepiece lens is converging, but in this case it doesn't form a real image. Instead, it forms a **virtual image**—an image located at a negative image distance, that is, on the wrong side of the lens!

The screen displaying the scene that you're photographing is so close to the eyepiece lens that the object distance is less than that lens's focal length. According to the lens equation, the image distance should be negative and it is; the image is located on the screen side of the eyepiece lens (Fig. 15.1.11). You can't put your fingers in the light and project this image on your skin because the image is virtual rather than real.

You can, however, see this image through the eyepiece. It's located farther away than the screen itself, so your eye can comfortably focus on it. And the image is magnified— the eyepiece lens is acting as a magnifying glass (Fig. 15.1.12). This lens provides magnification because when you look at the screen through it, the screen image covers a wider portion of your field of vision. This magnification increases as the eyepiece lens's focal length decreases. That's because a shorter focal length eyepiece lens must be quite close to the screen, where it can bend light rays coming from a smaller region so that they fill your field of vision. The eyepiece lens in a typical camera has been chosen so that the screen fills a comfortable portion of your visual field, allowing you to examine the virtual image in great detail and adjust the lens and camera settings until you have just the right picture in your view. Then all you have to do is take the photograph.

Fig. 15.1.10 The mirror in the center of this reflex camera directs light from the lens (removed for this photograph) onto the focusing screen above it. During the exposure, the mirror swings upward to permit light from the lens to strike the image sensor at the back of the camera.

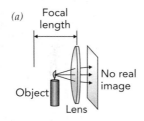

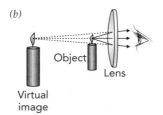

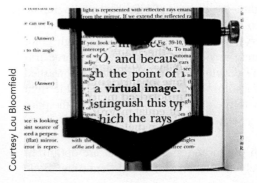

Courtesy Lou Bloomfield

Fig. 15.1.12 This magnifying glass creates an enlarged virtual image located far behind the printed text. You can't touch the image or put your fingers in its light, but you can see it clearly with your eyes.

Fig. 15.1.11 (*a*) Light from an object very near a converging lens diverges after passing through the lens and no real image forms. (*b*) Your eye sees a virtual image that is large and far away.

Cameras with fixed lenses often have two separate viewfinder systems. A typical digital camera has a small electronic viewfinder, which displays the real image being projected onto its image sensor. But many digital and all film cameras also have optical viewfinders. While optical viewfinders vary in style and sophistication, the best combine real and virtual images. In a *real-image optical viewfinder*, a system of lenses, mirrors, and/or prisms produces an erect real image of the scene and you then examine that real image through an eyepiece magnifying glass. The lenses projecting the real image zoom along with the camera's main lens so that what you see through the viewfinder is similar to what the camera's image sensor will record.

➤ Check Your Understanding #5: Real and Virtual Images

for answers, see page 499

As you move a magnifying glass slowly toward the photograph in front of you, you see an inverted image that grows larger and nearer to your eye. This image eventually becomes blurry, and then a new upright and enlarged image appears on the photograph's side of the lens. What's happening?

Image Sensors

Once the lens has projected its real image onto the image sensor, it's the image sensor's job to record that pattern of light. Interestingly enough, both film and electronic image sensors use semiconductors and both detect light when its photons shift electrons from valence levels to conduction levels. But how those two image sensors act on the electron transitions is quite different.

Photographic film detects light photochemically. Embedded in the film are tiny crystals of silver salts. Composed primarily of silver and halogen atoms, these semiconductor crystals are extremely sensitive to light. When a silver halide crystal absorbs a photon of visible light, it can undergo a radiative transition that shifts an electron from a valence level to a conduction level and eventually frees one silver atom from a silver halide molecule. After several nearby silver atoms have been freed by light, they can form a tiny particle of silver metal. When the film is developed, this silver particle transforms the entire silver halide crystal into metallic silver. The microscopically rough structure of that silver makes it appear black rather than shiny.

In black-and-white photography, the silver particles themselves form a negative image on the developed film. Wherever the film was struck by light, it acquires a dense, black pattern of silver particles. Wherever light was absent, the film becomes clear once the unexposed silver salts are washed away. Although the image on the developed film itself is negative—light is dark and dark is light—the process of preparing photographic prints reverses light and dark a second time so that the image on the prints is positive.

In color photography, the silver halide crystals are exposed to light through color filters and sensitizers, so that the film separately records its exposure to the three primary colors of light (see Section 14.2). During development, the silver itself is washed away, but a negative color image remains in the film. For example, wherever blue light struck the film, it acquires a yellow tint and therefore absorbs blue light. Again, the photographic printing process reverses the colors a second time so that the prints have positive images.

Electronic image sensors detect light with **photodiodes**—diodes that are optimized to detect light. A photodiode is approximately the reverse of a light-emitting diode

(Section 14.3). Recall that a light-emitting diode *emits* light when conduction level electrons cross from the cathode to the anode and undergo radiative transitions to the valence levels. The LED is using an electric current to produce light. In contrast, a photodiode *absorbs* light when valence level electrons undergo radiative transitions to conduction levels and cross from the anode to the cathode. The photodiode is using light to produce an electric current.

An electronic image sensor uses a vast array of microscopic photodiodes to record the pattern of light in the real image. As it's exposed to light, each photodiode accumulates electrons on its cathode and the camera subsequently measures the accumulated charge on each of its photodiodes. To obtain color information, the image sensor's photodiodes are covered with a pattern of red, green, and blue filters so that each photodiode measures the intensity of only one primary color of light.

Check Your Understanding #6: Foggy Photos

for answers, see page 499

Airport screening devices often use X-rays to search for hidden items. These high-energy photons can damage film. How?

Eyes and Eyeglasses

Not all cameras involve modern technology. Most people are born with two of them: their eyes. Like the cameras we discussed above, each eye consists primarily of a converging lens and an image sensor (Fig. 15.1.13). In this case, the lens is a combination of the front surface of the eyeball, its *cornea*, and the internal lens just beneath the cornea. The image sensor is the retina, a vast pattern of light-sensitive cells and nerves at the back of the eyeball.

When you look at the scene in front of you, the cornea and lens of your eye project a real image of that scene onto your retina and your retina reports the resulting pattern of light to your brain. As usual, the real image is inverted and reversed left to right, but your brain compensates for that effect.

Since your eyeball can't alter the distance between the lens and the image sensor, it focuses the real image by adjusting the focal length of the lens. When you look at nearer objects, the lens in your eye becomes more highly curved and its focal length decreases. The light rays from that nearer object thus converge more sharply and form a real image on your retina. When you view a more distant object, the lens becomes less curved and its focal length increases.

Like a sophisticated camera, your eye has an iris within its lens system. When you view a bright scene, that iris shrinks to limit the amount of light striking your retina. As a side effect, your depth of focus increases and everything appears sharper. It's easier to focus when you read or work in a well-lighted environment.

But not all eyes are perfect and many need help forming sharp real images on their retina. Although modern laser surgical techniques can reshape corneas to improve image sharpness, the classic approach to better vision is to wear eyeglasses or contact lenses. An eye's lens system already consists of two components, the cornea and the lens, so adding a third component, eyeglasses, is no big deal.

A person who is farsighted can't see nearby objects sharply because her lens system has too long a focal length (Fig. 15.1.14*a*). While it can project real images of distant

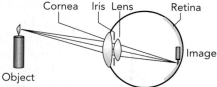

Fig. 15.1.13 An eye is a camera, with its cornea and lens forming a real image on the retina. The eye focuses by changing the curvature of its lens. The iris changes the eye's f-number.

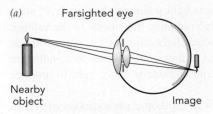

(a) Farsighted eye

Nearby
object   Image

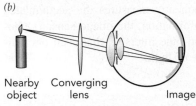

(b)

Nearby   Converging
object   lens   Image

Fig. 15.1.14 (*a*) A farsighted eye bends
light too weakly to focus on a nearby
object. The real image forms beyond
the retina. (*b*) A converging lens shifts
the real image forward so that it
focuses on the retina.

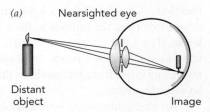

(a) Nearsighted eye

Distant
object   Image

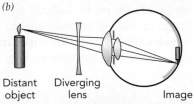

(b)

Distant   Diverging
object   lens   Image

Fig. 15.1.15 (*a*) A nearsighted eye
bends light too strongly to focus on a
distant object. The real image forms
before the retina. (*b*) A diverging lens
shifts the real image backward so that it
focuses on the retina.

objects on her retina, nearby objects focus too far away from the front of her eye and
the light strikes her retina before it forms a real image.

To compensate for farsightedness, she wears eyeglasses with converging lenses
(Fig. 15.1.14*b*). These lenses begin the task of bending light rays together even before
they enter her eyes. Her own lens system completes the bending and the real images
form closer to the front of her eyes. She is thus able to see distant objects clearly.

In contrast, a person who is nearsighted is unable to focus on distant objects because
his lens system has too short a focal length (Fig. 15.1.15*a*). Real images of those distant objects form too close to the front of his eye and the light has already begun to
spread apart by the time it reaches his retina.

To compensate for nearsightedness, he wears eyeglasses with diverging lenses
(Fig. 15.1.15*b*). A **diverging lens** is one which bends light rays apart and therefore has a
negative focal length. Typically thinner at its middle than at its edge, a diverging lens
bends the nearly parallel rays of light from a distant object so that they diverge more
rapidly. Those rays then appear to come from a much nearer object, actually a nearby
virtual image, and his eyes are able to focus them properly on his retina.

▶ Check Your Understanding #7: Keeping Up Your Image

for answers, see page **499**

The eyeglasses of a farsighted person can project a real image of a distant scene on
a white wall. However, the eyeglasses of a nearsighted person produce no real image.
Why not?

Cover
Previous track
Next
Stop
Play/pause
Compact disc
LCD display
Cover latch
Line output
Volume
Headphone jack
Tone control

SECTION 15.2   Optical Recording and Communication

Using light to convey information is as old as signal fires and as natural as sight itself. But advances in light sources, optical materials, and electronics have radically increased the possibilities for optical information systems. Optics and information go so well together that they're partly responsible for the current information revolution. The introduction of compact disc players in the early 1980s transformed the music industry virtually overnight, and optical fibers are knitting our world together at an astonishing pace. In this section, we'll look at how optical devices use light to manipulate information.

Questions to Think About: *Why are CDs and DVDs so colorful? Where is their information stored? How can CD or DVD players ignore fingerprints, dust, and other surface contamination during playback? Why is a laser involved in recording and playing back a CD or DVD? What limits the duration of a CD's audio or a DVD's video? Why are CDs and DVDs so free of noise? How can a glass fiber direct light in a curving path? How can light travel through kilometers of glass fiber without becoming dim?*

Experiments to Do: *Hold a prerecorded CD or DVD by its edges and look at its unlabeled surface. Beneath the clear plastic face is a smooth, shiny layer that reflects a rainbow of colors. Tiny pits in this layer cause this coloration. These pits form a spiral track around the center of the disc, and adjacent arcs of this track are so closely spaced that light waves reflecting from them interfere, as from a soap film. The reflective layer is so thin that you can see through it if you hold it in front of a bright light. Can you see the reflections of dust particles on the unlabeled surface? How deep below the plastic surface is the shiny layer?*

Digital Recording

Although analog techniques dominated audio recording for about a century, they were replace by digital techniques in less than a generation. It was mostly a matter of sound quality. Analog recording, in which a continuous physical quantity represents the density

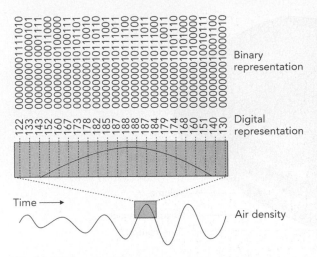

Binary representation

Digital representation

Time ⟶

Air density

Fig. 15.2.1 Sound can be represented as a series of numbers. Each number corresponds to the air density at a particular moment in time.

fluctuations that are sound in air, is susceptible to noise, wear, and a myriad of other imperfections. Phonograph records and analog tape recordings simply can't provide perfect sound.

In contrast, digital recordings can approach perfection. By representing sound in digital form, using many physical quantities with discrete values to represent density measurements numerically, digital recording can provide exact sound information that's free of noise and wear. As long as those density fluctuations are measured accurately, digital recording can recreate them with uncanny precision.

Like the audio players we examined in Section 12.2, audio CDs and DVDs represent sound's compressions and rarefactions of the air as numbers (Fig. 15.2.1). These numbers are essentially air density measurements taken over and over again during the recording process, and the player uses them to reproduce the recorded sound.

In the conventional CD format, density measurements are made 44,100 times each second for two independent audio channels and these measurements are recorded on the disc in binary form, using 16 bits for each measurement. Since the air density can go down as well as up, these bits represent the positive and negative integers from $-32,768$ to 32,767, which in turn represent how much the air density is above or below the average density. Density measurements with 16 bits of precision are sufficient to reproduce both loud and soft music with almost perfect fidelity.

Like CDs, DVDs record information digitally. But DVDs are newer technology and therefore more sophisticated. Audio DVDs can choose from several measurement rates, bits per measurement, and numbers of channels. A typical DVD might have five audio channels: left-front, center-front, right-front, left-rear, and right-rear. The three front channels might have 96,000 density measurements per second at 24 bits per measurement, and the two rear channels might have 48,000 measurements per second at 20 bits per measurement. While all of these samples, bits, and channels involve far more information than is stored on a CD, a DVD compresses that information before storing it. In contrast, a CD's information is uncompressed.

In either case, air density measurements aren't simply recorded one after the next on a disc's surface. Instead, these numbers are extensively reorganized before they're stored. This reorganization allows the player to reproduce the sound perfectly even if the disc can't be read completely. As we'll see shortly, reading these discs is a technological tour de force and susceptible to various failures. To be sure that the sound (and video) can be reproduced completely and without interruption, the numbers are recorded in an encoded manner. They appear redundantly so that even if one copy of a number is illegible, there is still enough legible information along the same arc of the disc's spiral track to completely recreate that missing number. This duplication of information reduces the playing time of both CDs and DVDs but is essential for reliability.

Its encoding scheme leaves a CD or DVD almost completely immune to all but the most severe playback problems. In principle, you can damage or obscure a 2-mm-wide swath of the disc, from its center to its edge, and the player will still be able to reproduce the sound (and video) perfectly. But damage along an arc of the spiral track is far more threatening to the data. If the player can't read a long stretch of a single arc, it won't be able to recover the information. That's why you should always clean a CD or DVD from its center outward to its edge.

► Check Your Understanding #1: Cordless Clarity Away from Home

for answers, see page 499

Some cell phones transmit your voice in digital form. In general terms, how does this digital transmission work?

The Structure of CDs and DVDs

Standard CDs and DVDs are 120 mm (4.72 inches) in diameter and 1.2 mm (0.05 inches) thick. One side of a CD is clear and smooth but the other side contains a sandwich of layers: a thin film of aluminum, a protective lacquer, and a printed label (Fig. 15.2.2a). In contrast, a DVD is laminated from two 0.6-mm-thick clear plastic discs, with one, two, or four reflective layers of aluminum, gold, or silicon stacked up in between them (Fig. 15.2.2c). The more layers in the DVD, the more information it holds.

The reflective layers are the recording surfaces. These layers are so thin that they actually transmit a small amount of light. In the gold or silicon DVD layers, this semi-transparency is essential because it allows the optical system that reads information to send light through the semitransparent layer to the aluminum layer beyond it. But even the aluminum layers transmit some light. While aluminum's electrons accelerate in

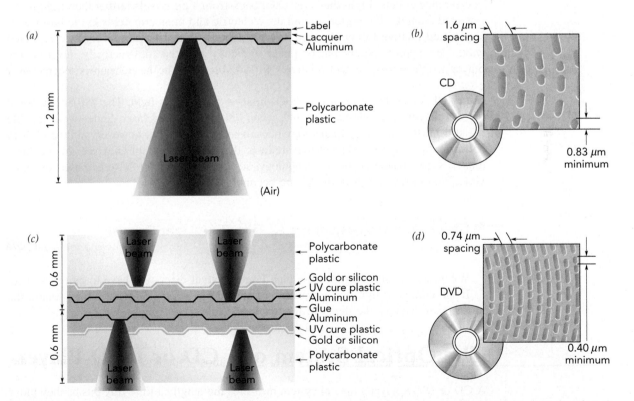

Fig. 15.2.2 (*a*) A CD contains a thin layer of aluminum on one side of a clear plastic disc. (*b*) The aluminum layer has tiny pits that are detected by a 780-nm laser beam. (*c*) A DVD contains up to four aluminum, gold, or silicon layers sandwiched between two clear plastic discs. The gold or silicon layers are semireflective. (*d*) Pits in those layers are detected by a 650-nm laser beam. The beam can focus either on the semireflective layer or, by passing through that layer, on the aluminum layer beyond it.

response to the light's electric field and normally reflect that light completely, there aren't enough electrons in these 50- to 100-nm-thick layers to do the job and some light gets through.

The reflective layers aren't perfectly smooth. Instead, each has a narrow spiral track formed in its surface (Fig. 15.2.2b,d). This track is a series of microscopic pits, as little as 0.83 μm long on a CD or 0.40 μm long on a DVD. Adjacent arcs in the spiral track are only 1.6 μm apart on a CD and just 0.74 μm apart on a DVD. The lengths of the pits and the flat "lands" that separate them represent numbers. The player examines these pits and lands as the disc turns and converts their lengths into numbers, sound, and video.

The pit lengths and the spacings between arcs weren't chosen arbitrarily. Since electromagnetic waves are unable to detect structures much smaller than their wavelengths (see Section 13.2), the laser beam's wavelength limits the smallest features on a disc. In a CD player, that beam's wavelength is 780 nm in air and 503 nm in polycarbonate plastic—short enough to detect the pits of a CD easily. In a DVD player, the laser beam's wavelength is between 635 and 650 nm in air and between 410 and 420 nm in plastic— just short enough to detect the pits in a DVD. The wavelength reduction inside the disc occurs because polycarbonate plastic has an index of refraction of 1.55, meaning that the light's speed in that plastic is reduced from its vacuum speed by a factor of 1.55. Its wavelength is reduced by the same factor.

The player detects a pit by bouncing light from the disc and determining how much of it reflects. As the focused laser beam passes over a pit, the reflection becomes dim, in part because the curved pit scatters light in all directions and in part because of interference effects. Light that's reflected back from a pit travels farther than light that's reflected from the flat region around it, so electric and magnetic fields in the two waves are shifted relative to one another. The pit depth was chosen so that the two reflected waves are approximately out of phase and they interfere destructively. Overall, the player's light sensors detect relatively little light when the laser beam is located over a pit.

A CD or DVD player uses a laser diode to produce its light. The 780-nm standard for CD players was adopted in 1980, when 780-nm infrared laser diodes were reliable but still fairly expensive. Technology has advanced since then, however, and the 635- to 650-nm standard for DVD players reflects the development of inexpensive red laser diodes. New standards follow technology, so optical recording systems based on blue, violet, and ultraviolet laser diodes are beginning to appear.

➤ Check Your Understanding #2: The Blue Light Special

for answers, see page **499**

When disc players begin to use blue lasers, optical discs will have to be redesigned. If the new laser light has a wavelength of 400 nm in air or vacuum, how should the designers change the depth of the pits in the reflective layer?

The Optical System of a CD or DVD Player

A CD or DVD player's optical system measures the lengths of the tiny pits as they move by on a spinning disc. That reading process requires incredible precision. Not only must the player focus its spot of laser light exactly on the reflective layer, but it must also follow the spiral track as it moves by. The disc itself is neither perfectly flat nor perfectly round, so the player must continuously adjust its reading unit during playback.

The optical system must keep its laser beam focused on the reflective layer (autofocusing) and must follow the track as it passes (autotracking). These two automatic processes are beautiful examples of the use of feedback.

The basic structure of a typical CD or DVD player is shown in Fig. 15.2.3. Light from a laser diode passes through several optical elements on its way to the disc's reflective layer. It comes to a tight focus on that layer, where it illuminates only a single track. Some light reflects from the layer and returns through the optical elements. Finally, the reflected light turns 90° at a special mirror called a polarization beam splitter and focuses on an array of light detectors. The player measures the electric currents flowing through the detectors and uses those measurements both to obtain data from the disc and to control the focusing and tracking systems.

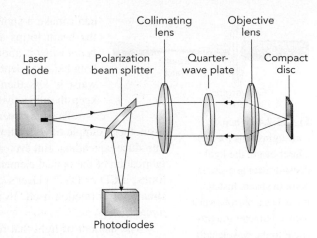

Fig. 15.2.3 In the optical system of a CD or DVD player, light from a laser diode passes through a polarization beam splitter, a collimating lens, a quarter-wave plate, and an objective lens before focusing on the reflective layer inside the disc. Reflected light turns 90° at the polarization beam splitter and focuses on an array of photodiodes.

Let's examine this optical system one element at a time. After leaving the laser diode, light passes through a *polarization beam splitter*. This device analyzes the light's polarization. As we saw in Section 14.1, different polarizations of light reflect differently when they strike a transparent surface at an angle. In this case, polarized light from the laser passes through the 45° surface, but light of the other polarization reflects. The beam splitter is specially coated to separate the two polarizations almost perfectly.

Light from the laser diode diverges rapidly as it passes through the beam splitter. It's not that the laser diode is broken or poorly designed; it's that a light wave emerging from a small opening naturally spreads outward, like ripples on a pond. This spreading is known as **diffraction** and occurs whenever a light wave is truncated by passage through an opening. The smaller the opening, the worse is the spreading. Because the emitting surface of the laser diode is essentially a very small opening, the laser beam spreads rapidly as it heads away from the diode. The player uses a converging lens located after the beam splitter to stop this spreading. At that point, the light beam is already wide enough that diffraction resulting from the passage of the light through the aperture of the lens causes little additional spreading. The beam leaves the lens *collimated*, meaning that it maintains a nearly constant diameter after passing through the lens.

The laser light then passes through a *quarter-wave plate*. This remarkable device performs half the task of converting horizontally polarized light into vertically polarized light or vice versa. Horizontally and vertically polarized lights are said to be **plane polarized** because their electric fields always oscillate back and forth in one plane as they move through space. But the quarter-wave plate turns plane polarized light into circularly polarized light. We encountered circular polarization before in the radio transmissions from FM stations. In **circularly polarized** light, the electric field actually rotates about the direction in which the light is traveling.

Now the light passes through an objective lens that focuses it onto the reflective layer of the disc. On its way to the reflective layer, the light enters the plastic surface of the disc. At its entry point, the beam is still more than 0.5 mm in diameter, which explains why dust or fingerprints on the disc's surface don't cause much trouble. While contamination may block some of the laser light, most of it continues onward to the reflective layer.

The light comes to a tight focus just as it arrives at the reflective layer. Although it might seem that all of the light should converge together to a single point on that surface, it actually forms a spot roughly one wavelength in diameter. This spot size is limited by the wave nature of light. No matter how perfectly you try to focus light, you

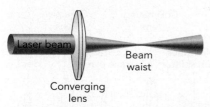

Fig. 15.2.4 When a converging lens focuses a laser beam, the light doesn't meet at a single point in space. Instead, it reaches a narrow waist with a diameter roughly equal to the wavelength of the light.

can't make a spot that's much smaller than the light's wavelength. Instead, the beam forms a narrow waist and then spreads apart (Fig. 15.2.4). This *beam waist* is about one wavelength of the light in diameter and a few wavelengths long, depending on the f-number of the converging lens. Since that waist is less than 2 microns long, the player's autofocusing system must keep the objective lens just the right distance from the reflective layer.

This fundamental limitation on how tightly a beam of light can be focused is another example of diffraction; the focusing lens truncates the light wave and inevitably introduces spreading. But even reaching this ideal focusing limit requires careful design and fabrication of the optical elements. While most other optical systems fall short of their ideal limits, a CD or DVD player's optical system does as well as can be done within the constraints of diffraction itself. Its optics are essentially perfect and are said to be *diffraction limited*.

The amount of light that reflects from the layer depends on whether or not the laser spot hits a pit. This reflected light follows the optical path in reverse. It's collimated by the objective lens and then passes through the quarter-wave plate again. The plate now finishes the job it started earlier; the light ends up plane polarized, but with the opposite polarization from what it had when it left the laser. Horizontally polarized light is now vertically polarized and vice versa.

The reflected light then passes through the collimating lens, which makes the light converge, and then strikes the polarization beam splitter. Because the light's polarization has changed, the beam splitter no longer allows the beam to pass directly through. Instead, it turns this reflected beam 90° and directs it toward the detector array. This clever redirection scheme is important for two reasons. First, it conserves laser light by allowing most of it to travel from the laser diode to the detector. Second, it prevents reflected light from returning to the laser diode, where that light would be amplified and cause the laser diode to misbehave.

The light comes to a focus on an array of photodiodes. This array allows the player to detect pits via the reflected light intensity and also to determine whether the objective lens is properly positioned relative to those pits. For generality, Fig. 15.2.3 omits optical elements involved in autofocusing and autotracking. However, because of those elements, the pattern of light hitting the detector array indicates which way the objective lens should move, if necessary. That lens is attached to coils of wire that are suspended near permanent magnets. By varying the currents flowing through those coils, the player uses Lorentz forces to move its objective lens about rapidly and keep it in the right place over the disc.

Check Your Understanding #3: That's Why It's Called a Laser Disc

for answers, see page **499**

Why must the CD or DVD player use a laser diode rather than a more conventional light source such as an incandescent bulb?

Optical Fibers

Optical playback of prerecorded discs is fine for music and movies, but it isn't of much use when you want up-to-the-minute information. The Internet and World Wide Web require communication links that operate at lightning speed. Yet even here, optics and light have an important role to play. The fastest way to send enormous amounts of information is to use optical fibers.

An optical fiber is a glass conduit that guides light from one place to another. Nearly every photon that enters the fiber at one end emerges from the other end moments later. In its simplest form, the fiber is made from two different glasses: a solid core of one glass surrounded by a cladding of the other glass. Both glasses are so incredibly transparent that light can travel through them for kilometers with little loss. For comparison, look through the edge of an ordinary piece of window glass and you'll see how dark the glass looks. It absorbs far too much light to be suitable for optical fibers. They're made of the purest glasses known.

If both glasses are almost perfectly transparent, what keeps the light from leaking out of the sides of the fiber? The answer is a phenomenon known as total internal reflection. As light tries to move from the inner glass core to the outer glass cladding, it's reflected perfectly and thus can't escape.

Total internal reflection is an extreme case of refraction. When light encounters the boundary between two materials with different indices of refraction, refraction causes that light to bend (Fig. 15.2.5). If the material it enters has a smaller index of refraction than the one it leaves, the light bends away from a line perpendicular to the boundary. The amount of this bend depends on the two indices of refraction and on the angle at which the light approaches the boundary. As long as the approach angle is steep enough, light will succeed in entering the second material. But if the approach angle is too shallow, the light won't enter the second material at all. Instead, it will reflect perfectly from the boundary. In fact, total internal reflection is far more efficient at reflecting light than a conventional metal mirror.

To keep the light inside the fiber's core, the core glass must have a higher index of refraction than the cladding glass. As light in the high-index core encounters the boundary with the low-index cladding, it experiences total internal reflection and bounces back into the core (Fig. 15.2.6a). As long as the fiber doesn't bend too sharply, the light bounces back and forth inside the core and can't escape. As a result, light entering the fiber core through one of its cut ends follows the fiber all the way to its other cut end.

But a large-diameter fiber (typically 50 μm or more) has a problem. Light rays bouncing through it at slightly different angles travel different distances during their passage through the fiber. Light heading almost straight down the center of the fiber rarely bounces and takes less time to complete its trip than light that bounces many times. Because this wide fiber has many bouncing paths or "modes" in which light can travel through it, a short pulse of light going through the fiber gets stretched out in time (Fig. 15.2.6a). This pulse broadening severely limits the rate at which information can be sent through a multimode fiber.

To reduce stretching problems, the core of a better performance optical fiber has a graded refractive index. The core glass is specially treated so that its index of refraction decreases smoothly away from its center toward the cladding. Instead of bouncing abruptly when it reaches the boundary between core and cladding, light in this graded index environment turns smoothly back toward the core (Fig. 15.2.6b). The path differences between the modes in a graded index multimode fiber aren't so different and a short pulse of light isn't stretched very much in a fiber of moderate length.

But in very long fibers, even small path differences add up and short pulses become blurred in multimode fibers. Therefore, the highest performance optical conduits are single-mode fibers. These fibers have very narrow graded index cores that permit light to travel only in one mode—effectively right down the center of the fiber (Fig. 15.2.6c). The core is typically only 9 μm in diameter. A pulse of light entering this narrow core broadens very little in time during its passage.

What little broadening occurs in a single-mode fiber isn't caused by the light taking different paths; it's caused by ordinary dispersion in the glass. To carry information,

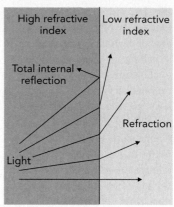

Fig. 15.2.5 When light traveling through one material enters a second material with a lower refractive index, it bends toward the boundary between those materials. If its approach angle is too shallow, the light will bend so much that it will simply reflect from the boundary. This effect is called total internal reflection.

Fig. 15.2.6 Light trying to leave the high-index core of an optical fiber at a shallow angle undergoes total internal reflection at the boundary with the low-index cladding. (*a*) In an ordinary multimode fiber, a pulse of light can follow many paths through the core and becomes spread out in time. (*b*) A core with a graded refractive index exhibits less temporal spreading because the reflection process is more gradual. (*c*) The least spreading, however, occurs in a single-mode fiber. The small core diameter of this fiber provides light with only one mode of travel so that the only spreading occurs because of ordinary dispersion.

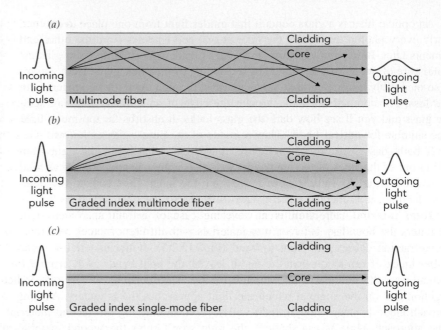

the light wave must change with time and thus must have a range of frequencies and wavelengths. As usual, the shorter wavelengths travel slower than the longer wavelengths and the pulses get stretched out in time. To minimize these dispersion effects, the highest speed optical fibers operate at wavelengths that minimize dispersion. They also operate at wavelengths that minimize loss of light through absorption in the glass. These two wavelengths coincide at 1550 nm in so-called dispersion shifted fibers, so this infrared wavelength is commonly used in long-haul optical communication.

> **Check Your Understanding #4: If It Looks Like a Mirror . . .**
> *for answers, see page* 499

When you look into an aquarium from the front, the sides of the aquarium often look like mirrors. Those sides are actually clear glass, so why do they reflect light so well?

Optical Communication

A typical optical communication transmitter uses a 1550-nm laser diode to generate short pulses of light. These pulses carry information from the transmitter to a receiver somewhere far away. The transmitter produces its pulses by varying the current passing through the laser diode. Light emerging from the laser diode is focused into the exposed core of a single-mode optical fiber, and it follows the core all the way to the other end of the fiber. When the light emerges, it's gathered by a lens and focused onto the receiver's photodiode. Each pulse of light causes a pulse of current to flow through the photodiode, allowing the receiver to begin processing the information.

A laser diode and a single-mode optical fiber can send billions of bits of data per second for 50 km or 100 km without significant errors. At longer distances, the gradual absorption of light in the glass makes it difficult to receive the information reliably. The easiest solution to this problem is to receive the data before the light becomes too weak and then to retransmit it with another laser diode.

But instead of interrupting the optical transmission with a receiver and retransmitter, some long-haul communication systems employ erbium-doped fiber amplifiers (EDFAs). An EDFA is a piece of optical fiber that has about 0.01% erbium ions in its glass core. When the EDFA is exposed to 980-nm or 1480-nm light, it becomes a laser amplifier for 1550-nm light. As the weakened pulses of light from a long fiber pass through the fiber amplifier, the amplifier duplicates photons and brightens the pulses. These amplified pulses then continue through ordinary fiber before being amplified again. Undersea optical cables often splice fiber amplifiers into the fibers every 50 km or so. These amplifiers allow light to travel thousands of kilometers through a continuous path, from one side of an ocean to the other.

To get the most out of a single optical fiber, many communication systems use several laser diodes operating at somewhat different wavelength ranges around 1550 nm. Light from these diodes is merged together and focused into the fiber. When the light emerges at the far end of the fiber, its different wavelength ranges are split apart and directed onto individual receivers. The different wavelength ranges are like different channels, so that this wavelength-division multiplexing allows one fiber to carry far more information than it could with light from a single laser. Remarkably enough, an EDFA can amplify all of these different channels at once because erbium ions can copy a wide range of wavelengths.

▶Check Your Understanding #5: Making Something Out of Nothing

for answers, see page **499**

If light weakens gradually as it passes through a long optical fiber, why can't you simply wait until the very end of the fiber to amplify what little light remains and then send the amplified light into a receiver?

Epilogue for Chapter 15

In this chapter we looked at several common objects and studied the ways in which they use optics. In *cameras*, we saw how a converging lens can form a real image of an object and how that real image can be used to record a scene on a piece of film or an image sensor chip. We learned about the roles of focal length and f-number in determining the size and brightness of the real image and explored some of the complications that must be considered when designing lenses.

In *optical recording and communication*, we looked at how laser light is affected by various optical devices and looked at what happens to it when it's focused to a small spot. We learned about total internal reflection and how this effect makes it possible to send light hundreds or thousands of kilometers through a fiber of ultraclear glass.

Explanation: Focusing Sunlight

The magnifying glass forms a real image of the sun on the sheet of paper. Because the sun is so far away, the image distance between the lens and the paper is equal to the focal length of the lens. The real image is relatively small, yet it contains all of the sunlight that strikes the entire surface of the lens. With so much light focused onto a small region of the paper, the radiative heat transfer from the sun to the paper becomes quite large and the paper's temperature rises until it begins to smoke or burn. The lens's image also contains light from other parts of the sky, and this light can be seen surrounding the sun's disk. Because of the way light is bent during the formation of the real image, the clouds in the sky appear upside down and backward on the sheet of paper.

Chapter Summary

How Cameras Work: A camera lens projects a real image of the scene in front of the camera onto the image sensor inside the camera. This real image is formed when the lens bends all of the light reaching it from one part of the scene onto one part of the sensor. For this imaging process to work well, the distance between the lens and the image sensor must be adjusted so that the light converges together (focuses) just as it reaches the sensor. If the sensor is too close to or too far from the lens, the image is blurry. The depth of focus depends on the effective diameter of the lens—its aperture. The smaller the lens's aperture, the less critical the focus but the less light the lens gathers. A long focal length lens brings the light to a focus far behind it, forming a relatively large but dim real image on the image sensor. To brighten this image, the long focal length lens must have a large aperture so that it gathers lots of light. The f-number, the ratio of a lens's focal length to its aperture, characterizes the brightness of the lens's image.

How Optical Recording and Communication Work: A CD or DVD player uses a beam of laser light to read the pits on a thin reflective layer inside the disc. As the disc rotates, the pits pass through the focused laser beam and the amount of reflected light fluctuates up and down. The player monitors the reflected light and uses it both to recreate the music and to keep its optical system properly aligned. This continual realignment allows the player to follow the spiral track of pits as the disc turns and to keep the laser beam tightly focused on the reflective layer. The optical system, which includes a laser diode, several photodiodes, and a variety of lenses and other optical elements, is so well designed and executed that the spot it creates on the reflective layer is limited in smallness only by diffraction effects due to the wave nature of light.

Laser light can also carry information long distances through an optical fiber. Made from a core of extremely transparent glass, clad in a second glass, this fiber confines light via total internal reflection. As light attempts to leave the core at a shallow angle, it is perfectly reflected from the boundary with the cladding and continues through the fiber for great distances.

Important Laws and Equations

1. The Lens Equation: One divided by the focal length of a lens is equal to the sum of one divided by the object distance and one divided by the image distance, or

$$\frac{1}{\text{focal length}} = \frac{1}{\text{object distance}} + \frac{1}{\text{image distance}}. \quad (15.1.1)$$

Check Your Understanding—Answers

Section 15.1 CAMERAS

1. It is a real image of the room lights, created by the converging magnifying glass.
Why: A magnifying glass is a converging lens and can form a real image. It brings light spreading outward from the room lights together onto the sheet of paper as a real image.

2. Light from the more distant background focuses closer to the lens than light from your friend. Since the camera is focusing on your friend, the background appears out of focus.
Why: When the aperture of your camera lens is wide open, focusing becomes critical. Objects in front of or behind your friend are out of focus on the image sensor and appear blurry.

3. On a bright day, your camera needs only a small aperture to gather enough light for an exposure, so the depth of focus is large. On a dark day, it uses the largest aperture available to gather light and has a small depth of focus.

Why: Light gathering and depth of focus go hand in hand. If you must open the aperture of your camera's lens to gather enough light for an exposure, focusing will become critical and your photographs will have small depths of focus.

4. The magnifying glass has only one glass element and suffers from chromatic aberration (and many other image imperfections).

Why: A single converging lens can't form a high-quality image on a flat surface because it has chromatic aberration, spherical aberration, coma, and astigmatism. If the lens is small, these failings are often invisible. But in a large magnifying glass, they are all readily apparent and you can't form a sharp image of the entire light fixture.

5. The lens is initially creating a real image near your eye. This real image moves past you as the lens approaches the photograph. Finally, the lens is close enough to create an enlarged virtual image of the photograph.

Why: A magnifying glass forms either a real or virtual image, depending on object distance and the lens's focal length. If the lens and photograph are separated by more than the focal length, the image is real and you see it near your eye. You can touch this inverted image or project it on a piece of paper. If the lens and picture are separated by less than the focal length, the upright image is virtual and you see it on the far side of the lens.

6. X-rays cause radiative transitions in the silver salt crystals.

Why: X-rays can penetrate through normally opaque objects and reach the film. If a silver salt crystal in the film absorbs an X-ray, it will respond as though it was exposed to visible light. Although most modern screening devices use such weak X-ray sources that the effects are minimal, repeated screenings of high-speed film will gradually ruin it.

7. Nearsighted eyeglasses use diverging lenses, which bend light rays apart and don't focus them together into a real image.

Why: To form a real image of a distant scene, you need to bring light rays together with a converging lens. Diverging lenses spread light rays apart and can't form real images on their own.

Section 15.2 OPTICAL RECORDING AND COMMUNICATION

1. Your cell phone's microphone converts the sound of your voice into a fluctuating current. This current is measured periodically and represented as a sequence of numbers. These numbers are then represented by radio waves and transmitted through space. The numbers eventually work their way to a receiving cell phone, where the radio waves are converted back into numbers and the numbers are converted back into a fluctuating current. Finally, the current is amplified and sent through a speaker to create sound.

Why: Many modern communication devices use a digital representation of sound. In radio wave communication, digital representations are particularly useful because they are immune to the noise that plagues analog radio transmissions and also permit more efficient use of the radio bandwidth.

2. The pits should become about half as deep as they are now.

Why: The pits in the reflective layer should be about a quarter of a wavelength deep so that light reflected from the bottom of a pit interferes destructively with light reflected from an adjacent flat region. If the light wavelength changes to 400 nm, about half the present value, then the pits should change to about half their present depth.

3. The light in a CD or DVD player must be coherent so that it can be focused to a single diffraction-limited spot on the reflective layer and experience destructive interference when it encounters a pit.

Why: You can't focus light from an incandescent bulb to a diffraction-limited spot. Each photon leaving the bulb is independent and will focus at a somewhat different point in space. The bulb's incoherent light also doesn't experience the strong interference effects of laser light.

4. The light you see is experiencing total internal reflection as it tries to leave the glass at a shallow angle. You see perfect reflections of the fish inside the aquarium.

Why: Light has trouble passing from water or glass into the air at a shallow angle. When it tries, it experiences total internal reflection and bounces off the surface without losing any intensity at all.

5. As the light is absorbed, the number of photons in a pulse gradually diminishes. When that number gets to zero, there's nothing left to amplify.

Why: Each pulse of light in the fiber contains a limited number of photons. The glass absorbs many of these photons during their long passage through a fiber, so they must be amplified before their numbers become so small that there's a chance that none of them makes it at all.

Check Your Figures—Answers

Section 15.1   CAMERAS

1. The image will form at a distance twice the focal length behind the lens.

Why: Since the object distance is twice the focal length, we can use Eq. 15.1.1 to find the image distance:

$$\frac{1}{\text{image distance}} = \frac{1}{\text{focal length}} - \frac{1}{\text{object distance}}$$

$$= \frac{1}{\text{focal length}} - \frac{1}{2 \cdot \text{focal length}}$$

$$= \frac{1}{2 \cdot \text{focal length}}$$

The image distance is twice the focal length of the lens.

Exercises

1. On a bright, sunny day you can use a magnifying glass to burn wood by focusing sunlight onto it. The focused sunlight forms a small circular spot of light that heats the wood until it burns. Why is the spot of light circular?

2. Light passing through a curved water glass forms an image of the candle flame on the wall beside the table. Why would moving the candle toward or away from the glass spoil this effect?

3. Simple reading glasses are converging lenses that come in a variety of strengths, ranging from about 0.25 diopter (almost flat glass) to 3.00 diopters (highly curved). Which lens has the shorter focal length: 1.0 diopter or 2.0 diopters?

4. In a movie, an actor's eyeglasses send a brief undistorted reflection of the sun at the camera. Why does that simple reflection tell you that the glasses are props and that the actor doesn't need eyeglasses?

5. One part of a fiber optic communication system uses a lens to focus light from a semiconductor laser onto the end of an optical fiber. The tiny light source and fiber are on opposite sides of the lens, and each is 1.0 cm away from the lens. Why must the lens have a focal length of 0.5 cm?

6. Light from a distant object approaches the objective lens of a telescope as a group of parallel light rays. Draw a picture of the light rays from three stars passing through a converging lens to show that they focus at three separate locations.

7. Sports photographers often use large aperture, long focal length lenses. What limitations do these lenses impose on the photographs?

8. If you're taking a photographic portrait of a friend and want objects in the foreground and background to appear blurry, how should you adjust the camera's aperture and shutter speed?

9. Your new 35-mm camera comes with two lenses, a 50-mm focal length "normal" lens and a 200-mm focal length "telephoto" lens. The minimum f-number for the 50-mm lens is 1.8. Although the 200-mm lens has much larger glass elements in it, its lowest f-number is 4. Why is the telephoto lens's f-number so much larger?

10. If you hold your camera up to a small hole in a fence, you will be able to take a picture of the scene on the other side. However, the exposure time will have to be quite long, and the depth of focus will be surprisingly large. Explain.

11. Why does squinting increase your depth of focus?

12. Why is it easiest for the lens of your eye to form sharp images on your retina when the scene in front of you is bright and the iris of your eye is very small?

13. People who need glasses find it hardest to see clearly without them in dimly lit situations when the irises of their eyes are wide open. Why?

14. Photographic telescopes are simply enormous cameras. With such small f-numbers, they have small depths of focus. Why isn't that a problem for astronomical work?

15. Two similar looking magnifying glasses have different magnifications. One is labeled 2× (two times) and the other 4× (four times). Which lens has the longer focal length?

16. Which magnifying glass from Exercise 15 must you hold closer to the object you're looking at in order to see a virtual image of that object far in the distance?

17. The objective lens in a DVD player must move quickly to keep the laser beam focused on the disc's reflective layer. Why must this lens have a very small mass?

18. Why don't portable CD or DVD players play properly if you shake them back and forth too quickly?

19. What happens to a ray of light entering plastic from the air at an angle to the interface?

20. Why is a laser beam's focus delayed by its entry into the plastic of a CD or DVD? (Draw a picture.)

21. Why does a laser beam spread out quickly after it passes through a tiny pinhole?

22. Scientists measure the distance to the moon by bouncing laser light from reflectors left on the moon by the Apollo astronauts. This light is sent backward through a telescope so that it begins its trip to the moon from an enormous opening. This procedure reduces the size of the laser beam when it reaches the moon. Explain.

23. As it passes through the lens that focuses it onto a CD's aluminum layer, the player's laser beam is more than 1 mm in diameter. Why does its large size as it leaves the lens allow the beam to focus to a smaller spot?

24. Why can light from a blue laser form a narrower beam waist than light from an infrared laser?

25. Sometimes the surfaces of a glass of water look mirrored when you observe them through the water. Explain.

26. When you look into the front of a square glass vase filled with water, its sides appear to be mirrored. Why do the sides appear so shiny?

27. Some of the laser light striking a DVD's reflective layer hits the flat region around a pit and reflects back toward the photodiode. How does this reflected wave actually reduce the amount of light detected by the photodiode?

28. Why does the surface of a DVD look so colorful in white light?

Problems

1. Your camera has a 35-mm focal length lens. When you take a photograph of a distant mountain, how far from that lens is the real image of the mountain?

2. If you use a 35-mm focal length lens to take a photograph of flowers 2 m from the lens, how far from that lens does the real image of the flowers form?

3. You're trying to take a photograph of two small statues with a 200-mm telephoto lens. One statue is 4 m from the lens, and the other statue is 5 m from the lens. The real image of which statue forms closer to the lens? How much closer?

4. When you place a saltshaker 30 cm from your magnifying glass, a real image forms 30 cm from the lens on the opposite side. You can see this image dimly on a sheet of paper. What is the focal length of the magnifying glass?

5. When light from your desk lamp passes through a 50.0-mm focal length lens, it forms a sharp real image on a sheet of paper located 50.5 mm from the lens. How far is the desk lamp from the lens?

Modern Physics

In recent years, scientists have been looking deeper into the atom to see how it's made, farther into space to see how the universe works, and more carefully into objects to see how complicated things can be understood in terms of simple laws. Among the most important tools that these scientists have to work with are quantum theory and the theory of relativity. This chapter will look at some of the ways in which modern physics contributes to our lives.

EXPERIMENT: Radiation Damaged Paper

One path of modern physics involves the control and use of high-energy radiation. While our access to most forms of high-energy radiation is restricted, there's one source that anyone can use: the sun. Because the sun's ultraviolet light is energetic enough to damage chemical bonds and rearrange molecules, it can provide us with a glimpse of the effects that occur with X-rays and beyond.

To see sun damage for yourself, expose some sheets of colored construction paper to direct sunlight for a few days. Cover the paper with some opaque objects such as coins and place it outdoors. Don't cover the paper with glass because glass absorbs enough ultraviolet light to slow the damage process. After a day or two, you should find that the exposed portions of the paper have lightened; the sun's ultraviolet radiation has destroyed some of the dye molecules in the paper. If you find that nothing happens, the dye is evidently robust enough to tolerate ultraviolet light for a while.

Try the experiment again with different papers and different colors. Can you *predict* which papers will fade fastest? *Observe* the results and see if they *verify* your predictions. Try to *measure* the rates at which dye molecules are damaged. Do they seem to have a half-life? How could you tell?

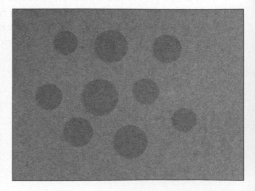

Courtesy Lou Bloomfield

This same optical bleaching appears on items that are displayed in shop windows and on outdoor furniture. It was once the only method people had for whitening fabrics. The sun's ultraviolet light also damages your skin when you sit in the sun: a sunburn isn't thermal damage, it's radiation damage.

Chapter Itinerary

Fortunately ultraviolet light can't penetrate far into your body. In *nuclear weapons,* we'll look at more penetrating forms of radiation. We'll also explore the structures of atomic nuclei and see how taking them apart or joining them together can release enormous amounts of energy. In *medical imaging and radiation,* we'll examine high-energy radiation and see how it's used to help rather than hurt. We'll study the ways in which X-rays and gamma rays are produced and how they interact with the atoms and molecules in a patient. We'll also look at the particle accelerators that produce high-energy particles for radiation therapy. Finally, we'll discuss the bases for CT imaging and MRI, which make it possible to prepare detailed maps of patients' insides without ever touching their bodies. For additional preview information, see the Chapter Summary on p. 528.

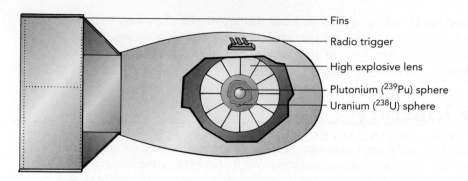

Fins
Radio trigger
High explosive lens
Plutonium (^{239}Pu) sphere
Uranium (^{238}U) sphere

SECTION 16.1   Nuclear Weapons

The atomic bomb is one of the most remarkable and infamous inventions of the twentieth century. It followed close on the heels of various developments in the understanding of nature, developments that in many ways made the invention of nuclear weapons inevitable. By the late 1930s, scientists had discovered most of the principles behind nuclear energy and were well aware of how those principles might be applied. The onset of World War II prompted concern that Germany would choose to follow the military path of nuclear energy. Propelled by fear, curiosity, and temptation, the scientists, engineers, and politicians of that time brought nuclear weapons into existence. The world has lived in the shadow of these terrible devices ever since.

Questions to Think About: Where is nuclear energy stored in the atoms? From where did this nuclear energy come? How do nuclear weapons release nuclear energy? Why are nuclear weapons so difficult to build? Why do we associate uranium and plutonium with nuclear weapons? How much uranium or plutonium does it take to build a bomb?

Experiments to Do: Since uranium and plutonium aren't sold in hardware stores, you won't be able to build your own bomb. However, you can get a feel for the way a chain reaction works by playing with a box of dominoes. If you stand the dominoes on end on a level table, each of them will have extra gravitational potential energy that it can release by tipping over. If you spread the dominoes widely about the table and then give the table a gentle shake, they'll tip over one by one.

However, if you pack the dominoes tightly together, so that one falling domino can knock over others, they'll no longer be independent. As you jiggle the table, nothing will happen until the first domino falls, but then many or even all of the dominoes will tip over in quick succession. You will have created a chain reaction, where a single event triggers an ever-increasing number of subsequent events. What characteristics of the dominoes and their arrangement determine whether or not such a chain reaction occurs? Can you envision a scenario in which a single tipping domino could trigger the release of an enormous amount of stored energy? Another chain reaction, this time in the decay of atomic nuclei, is what makes nuclear weapons possible.

Background

At the end of the nineteenth century, "classical physics" reigned supreme. Here classical physics means the rules of motion and gravitation identified by such people as Galileo, Newton, and Kepler, and the rules of electricity and magnetism developed by others including Ampère, Coulomb, Faraday, and Maxwell. It was generally felt that most of physics was well understood: physicists knew all of the laws governing the behavior of objects in our universe, and what was left was to apply those laws to more and more complicated examples. It was a time when physicists didn't know what they didn't know.

However, a few nagging problems remained—specific difficulties that couldn't be explained by the rules of classical physics. Among these were the spectrum of light emitted by a blackbody, the photoelectric effect in which electrons are ejected from metals by light, and the apparent absence of an ether or medium in which light traveled. At the beginning of the twentieth century, the whole of classical physics collapsed under the weight of these seemingly trivial difficulties, and a largely new understanding of the universe emerged. The major advances took 25 years, from 1901 to 1926, and the time since has largely been spent applying those new laws to more and more complicated examples.

The two main developments, both essential to the making of the atomic bomb, were the discoveries of quantum physics and relativity. Often these are called *quantum theory* and the *theory of relativity*. But while the word *theory* might imply that they're somehow on shaky ground, they're not theories in the sense of hypotheses waiting to be tested. In fact, they've been confirmed countless times since they were developed and have been shown to have enormous predictive power. Rather, they're theories in the sense of being carefully constructed and codified rules that model the behavior of the physical universe in which we live. Between them, these two theories made the discovery of nuclear forces and nuclear energy inevitable. Finally, given peoples' love for gadgets and power, they also made the development of nuclear weapons inevitable.

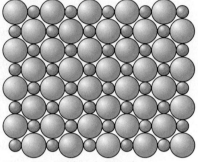

Table salt crystal
(Sodium chloride)

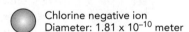
Chlorine negative ion
Diameter: 1.81×10^{-10} meter

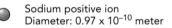
Sodium positive ion
Diameter: 0.97×10^{-10} meter

Fig. 16.1.1 A salt crystal is an orderly array of positively charged sodium ions and negatively charged chloride ions. These ions are held together by the attractive forces between oppositely charged ions. The ions are so small that there are about 7.2 million ions on the edge of a 1-mm-wide salt crystal.

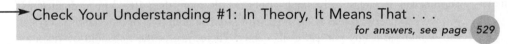

Check Your Understanding #1: In Theory, It Means That . . .

for answers, see page 529

If something is referred to as a "theory," how likely is it to be true?

The Nucleus and Radioactive Decay

Though the name "atomic" bomb has stuck for more than half a century, the more correct name would be "nuclear" bomb. The items that are responsible for the energy released by nuclear weapons are not atoms, but tiny pieces of atoms—their nuclei (plural of nucleus). But before we can discuss nuclei, let's put them into context. Let's start by looking at atoms.

To get an idea of just how tiny atoms are, imagine magnifying a grain of table salt, 1 mm (0.04 inch) on a side, until it was the size of the state of Colorado. That grain would then appear as an orderly arrangement of spherical particles, each about the size of a grapefruit (Fig. 16.1.1). These spherical particles would be single atoms, and there would be about 7.2 million of them along each edge of the grain.

Like most solids, table salt is a crystal and its atoms are bound to one another by their outermost components: their electrons. Electrons dominate the chemistry of atoms and molecules. Sodium is a reactive metal because of its electrons, and chlorine is a reactive gas because of its electrons. When mixed, these two chemicals react violently to form table salt and release a considerable amount of light and heat. This, then, is a true "atomic bomb."

Obviously, something is missing here. If a crazy person could buy a kilogram or two of sodium and a tank of chlorine from a chemical company and destroy an entire city, rural living would be a whole lot more popular. Fortunately the energy released by chemical reactions is fairly limited. A kilogram of chemical explosives just can't do that much damage. But nuclear bombs tap an entirely different store of energy deep within the atoms.

While all of nuclear weaponry is often attributed to Einstein's famous equation, $E = mc^2$, that notion is vastly oversimplified. Nonetheless, this equation is quite significant. As we noted in Section 4.2, one of Einstein's discoveries at the beginning of the twentieth century was that matter and energy are in some respects equivalent. In certain circumstances mass can become energy or energy can become mass. This equivalence is part of the theory of relativity and has some interesting consequences. It implies that an object can reduce its mass by transferring energy to its surroundings. Thus, if you weigh an object before and after it undergoes some internal transformation, you can use any weight loss to determine how much energy was released from the object by that transformation.

Because of this equivalence, mass and changes in mass can be used to locate energy that's hidden within normal matter. This technique is important in nuclear physics, but it also applies to chemistry. When sodium and chlorine react to form table salt, their combined mass decreases by a tiny amount. What's missing is some chemical potential energy, which becomes light and heat and escapes from the mixture. In leaving, this chemical potential energy reduces the mass of the sodium and chlorine mixture by about 1 part in 10 billion. That tiny change in mass is too small to detect with present measuring devices, although scientists are working on techniques that will soon make it possible to measure mass changes due to chemical bonds.

Electrons are, however, by far the lightest part of an atom and thus have relatively little mass to release as energy. Most of an atom's mass is located in its **nucleus.** The nucleus is fantastically small—only a little more than 10^{-15} m in diameter. If you were to peer into one of the grapefruit-sized sodium ions of our giant salt crystal, you would see a tiny particle at its center. There, just at the threshold of visibility, would be the ion's nucleus, only 1 μm (0.00004 inch) in diameter. The remaining 99.9999999999999% of the ion is occupied only by its 10 electrons in their orbitals.

The sodium nucleus contains 11 protons and 12 neutrons (Fig. 16.1.2). Each of these nuclear particles or **nucleons** has about 2000 times as much mass as an electron, so that 99.975% of the sodium ion's mass is in this nucleus. Thus, while the electrons are certainly important to chemistry and matter as we know it, their contribution to the ion's mass is insignificant. The ion is mostly empty space, lightly filled with fluffy electrons and having a tiny nuclear lump at its center.

The nucleons that make up this nucleus experience two competing forces. The first of these forces is the familiar electrostatic repulsion between like electric charges. Because each proton in the nucleus has a single positive charge, they're constantly trying to push one another out of the nucleus. However, the second of these forces is attractive and holds the nucleus together. This new force is called the **nuclear force**, and at short distances it dominates the weaker electrostatic repulsion. However, the nuclear force only attracts the nucleons toward one another when they're touching. As soon as they're separated, they're on their own.

Sodium nucleus
(11 protons, 12 neutrons)

Neutron
Diameter: 1 x 10⁻¹⁵ meter

Proton
Diameter: 1 x 10⁻¹⁵ meter

Fig. 16.1.2 At the center of a sodium ion is a tiny nucleus containing about 99.975% of the ion's mass. It consists of 11 positively charged protons and 12 uncharged neutrons. The protons repel one another at any distance, but the protons and neutrons are bound together by the highly attractive nuclear force as long as they touch one another.

The competition between these two forces, the repulsive electrostatic force between like charges and the attractive nuclear force between nucleons, is analogous to what happens in a familiar toy (Figs. 16.1.3 and 16.1.4). This hopping toy has a suction cup attached to a spring, so that the spring tries to separate the toy's top from its base while the suction cup tries to keep the two parts together. When the two parts are well separated, only the spring exerts a force. But when the two parts touch, the suction cup begins to act and holds the two parts together.

What makes the hopping toy exciting is that its suction cup leaks. Eventually, the suction cup lets go and allows the spring to toss the toy into the air. But suppose that the suction cup didn't leak. Once pushed together, the pieces would never separate and the spring would retain its stored energy indefinitely. To get the suction cup to let go, you would have to pull it away from the base. Only then could the spring release its stored energy.

In effect, an energy barrier would be preventing the leak-free toy from hopping. Until you did a little work on it by pulling the suction cup off the base, it wouldn't be able to release its stored energy. Another example of a system that needs energy to release energy is a bottle of champagne, where you must push on the cork to help it out of the neck. After that initial investment of energy, a great deal of energy is released as gas blasts the cork across the room.

The nucleus is in a similar situation. The attractive nuclear force prevents the nucleus from coming apart, despite the enormous amount of electrostatic potential energy it contains. The nuclear force creates an energy barrier that prevents the nucleons from separating. Unless something adds energy to the nucleus to help the nucleons break free of the nuclear force, the nucleus will remain together forever. At least that's the prediction of classical physics.

Quantum physics, however, has an important influence on the behavior of the nucleus. One of the many peculiar effects of quantum physics is that you can never really tell exactly where an object is located, or at least not for long. That fuzziness is a manifestation of the **Heisenberg uncertainty principle,** which observes that some pairs of physical quantities, such as position and momentum or energy and time, are not entirely independent and cannot be determined simultaneously beyond a certain accuracy. This principle is a result of the partly wave and partly particle nature of objects in our universe. Since waves are normally broad things that occupy a region of space rather than a single point, objects in our universe normally don't have exact locations.

The smaller an object's mass, the fuzzier it is and the more uncertain its location. While the fuzzy nucleons in a nucleus will normally stay in contact with one another for an extremely long time, there's always a tiny chance that they'll find themselves temporarily separated by a distance that's beyond the reach of the nuclear force. The nucleons will then suddenly be free of one another, and electrostatic repulsion will push them apart in a process called **radioactive decay.** The quantum process that allows the nucleons to escape from the nuclear force without first obtaining the energy needed to surmount the energy barrier is called **tunneling** because the nucleons effectively tunnel through the barrier. We first encountered quantum tunneling in Section 12.2, when we saw that erasing flash memory requires that electrons tunnel through an insulating barrier.

Natural radioactive decay is a perfectly random process. Although half of a large population of identical radioactive nuclei will decay in a certain amount of time, you absolutely cannot predict in advance which of the original nuclei will have survived. Because of this randomness, radioactive decay is characterized simply by a **half-life—** the time required for half of the nuclei to decay. If you wait a second half-life, only a quarter of the original nuclei will remain (half of a half). After a third half-life, only an eighth will remain (half of a half of a half). And so on.

Fig. 16.1.3 This hopping toy stores energy as its spring is compressed and retains that energy while its suction cup grips its base. When the suction cup lets go, the toy leaps into the air.

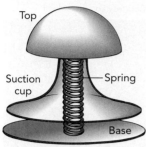

Hopping toy with
suction cup released

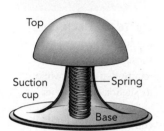

Hopping toy with
suction cup holding

Fig. 16.1.4 A spring and
a suction cup combine to
form a toy that hops
suddenly after a long
wait. The spring tries to
separate the top from the
base, while the suction
cup tries to hold the two
parts together. Energy you
store in the spring is
released when the leaking
suction cup eventually
allows the spring to
expand.

This halving of the population with each additional half-life is a type of *exponential decay*. In general, the fraction of nuclei remaining after a given amount of time is one-half raised to the power of the time divided by the half-life. Written as a word equation, that relationship is:

$$\text{fraction remaining} = \left(\frac{1}{2}\right)^{\text{elapsed time}/\text{half-life}} \qquad \text{(16.1.1)}$$

in symbols:

$$\frac{N}{N_0} = \left(\frac{1}{2}\right)^{t/T_{1/2}}$$

and in everyday language:

Radioactivity goes away, but you have to wait it out.

While most radioactive nuclei have short half-lives and don't linger long in our environment, there are a few with half-lives of billions of years. It is those long-lived radioactive nuclei, particularly uranium and thorium nuclei, that have survived since the formation of the earth, remain abundant in nature, and gave rise to nuclear weapons.

▶ Check Your Understanding #2: Who Hid the Energy?

for answers, see page 529

When a large nucleus breaks apart into fragments, it releases a great deal of energy. In what form was that energy stored in the intact nucleus?

▶ Check Your Figures #1: A Real Hand-Me-Down

for answers, see page 530

A small fraction of the carbon found in the earth's atmosphere is carbon 14, a rare, radioactive form that's synthesized by cosmic rays and has a half-life of 5730 years. While plants and animals are alive, they incorporate carbon 14 into their tissues along with ordinary carbon. But once they die, the fraction of carbon 14 in their tissues begins to decrease as the carbon 14 nuclei decay. If you are a museum curator and someone donates a garment that is supposed to be 1001 years old, what fraction of the original carbon 14 should you expect to find in that garment when you examine its carbon content?

Fission and Fusion

The more protons there are in a nucleus, the more they repel one another and the more likely they are to cause radioactive decay. Adding additional neutrons to the nucleus reduces this proton–proton repulsion by increasing the size of the nucleus without adding to its positive charge. However, adding too many neutrons also destabilizes the nucleus for reasons that we'll discuss in the next section. So constructing a stable nucleus is a delicate balancing act.

In nuclei with only a few protons, the attractive nuclear force wins big over the repulsive electrostatic force and the nucleons stick like crazy. These nuclei resemble hopping toys with weak springs and big suction cups: once brought together, the pieces never come apart. In fact, the average binding energy of the nucleons (the energy required to separate them from one another divided by the number of nucleons) would increase if these nuclei had even more protons and neutrons.

In nuclei with many protons, the electrostatic repulsion is so severe that the nuclear force can't hold the nucleons together for long. These nuclei decay rapidly. They resemble hopping toys with strong springs and small suction cups. The average binding energy of the nucleons would increase if these nuclei had fewer protons and neutrons.

In nuclei with roughly 26 protons, in between the two extremes we've just considered, the attractive nuclear force and repulsive electrostatic force are nicely balanced. These nuclei are extremely stable, and you can't increase the average binding energy of their nucleons by adding or subtracting nucleons. Smaller nuclei can release potential energy by growing to reach this intermediate size, while larger nuclei can release potential energy by shrinking toward the same goal.

For a small nucleus to grow, something must push more nucleons toward it. Electrostatic repulsion will initially oppose this growth, but once everything touches, the nuclear force will bind the particles together and release a large amount of potential energy. This coalescence process is called **nuclear fusion.**

For a large nucleus to shrink, something must separate its pieces beyond the reach of the nuclear force. Electrostatic repulsion will then push the fragments apart and release a large amount of potential energy. This fragmentation process is called **nuclear fission.**

The energies released when small nuclei undergo fusion or when large nuclei undergo fission are enormous compared to chemical energies. Uranium, a large nucleus, converts about 0.1% of its mass into energy when it breaks apart. Hydrogen, a tiny nucleus, converts about 0.3% of its mass into energy when it fuses with other hydrogen nuclei. Kilogram for kilogram, nuclear reactions release about 10 million times more energy than chemical reactions. Fortunately, they're much harder to start.

With that scientific background, let's follow a sequence of discoveries near the start of the twentieth century. Natural radioactive decay was discovered accidentally by French physicist Antoine-Henri Becquerel (1852–1908) in 1896. Intrigued by the recent discovery of X-rays, he began looking for materials that might emit X-rays after exposure to light. To his surprise, he found that uranium fogged photographic plates, even through an opaque shield and even without exposure to light. His discovery was soon confirmed and elaborated on by Polish-born French physicist Marie Curie (1867–1934) and French chemist Pierre Curie (1859–1906). This wife and husband team discovered several new radioactive elements, including polonium (named after Marie's homeland) and radium.

In 1911, British physicist Ernest Rutherford (1871–1937) discovered that atoms have nuclei. He subsequently found that nuclei sometimes shatter when struck by energetic helium nuclei. And in 1932, British physicist James Chadwick (1891–1974) discovered a fragment of the nucleus, the neutron, which has no electric charge and can thus approach a nucleus without any electrostatic repulsion. It was soon discovered that neutrons stick to the nuclei of many atoms.

But the crucial discovery that made the atomic bomb possible was neutron-induced fission of nuclei. In 1934, Italian physicist Enrico Fermi (1901–1954) and his colleagues were trying to solve a particular riddle about the nucleus, a radioactive decay process called *beta decay*. They were adding neutrons to the nuclei of every atom they could get hold of. When they added neutrons to uranium, with its huge nucleus, they observed the production

❑ Austrian-born physicist Lise Meitner moved to Berlin in 1907 and soon began a 30-year collaboration with chemist Otto Hahn. In 1934, she convinced Hahn to join her in studying nuclear processes and they made great progress. Unfortunately, Meitner's Jewish ancestry made her a target of Nazi academics restriction and she fled to Sweden in 1938. Meitner continued to guide their collaboration through letters. Only months after she left, Hahn and his assistant Fritz Strassmann found that neutron irradiation of heavy elements was creating smaller rather than larger nuclei. Meitner and her nephew Otto Frisch soon developed a model of nuclear fission based on these measurements. Hahn, however, published the results without Meitner's name on the paper, ostensibly to avoid Nazi interference. As a result of this omission, the 1944 Nobel Prize in Chemistry was awarded to Hahn alone. Hahn went on to claim that Meitner was simply his assistant, rather than the leader of their joint effort. In recognition of this gross injustice, element 109 was named meitnerium in 1994.

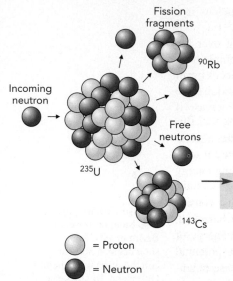

= Proton

= Neutron

Fig. 16.1.5 When a neutron strikes a uranium nucleus, there's a good chance that the nucleus will fall apart into fragments. This process is called induced fission. Among the fragments of induced fission are other neutrons.

of some very short-lived radioactive systems. They thought that they had formed ultraheavy nuclei and even went so far as to give these new elements tentative names.

Four years later, however, Austrian physicists Lise Meitner (1878–1968) and Otto Frisch (1904–1979) and German chemists Otto Hahn (1879–1968) and Fritz Strassmann (1902–1980) collectively showed ❏ that what Fermi's group had actually done was to fragment uranium into lighter nuclei (Fig. 16.1.5). Many of the fragments created by this **induced fission** were neutrons, which could themselves cause the destruction of other uranium nuclei.

Check Your Understanding #3: A Sticky Nuclear Problem
for answers, see page **529**

If you take two intermediate-sized nuclei and combine them to make a single uranium nucleus, will the process release or consume energy?

Chain Reactions and the Fission Bomb

Physicists quickly realized that a **chain reaction** was possible, a reaction in which the fission of one uranium nucleus would induce fission in two nearby uranium nuclei, which would in turn induce fission in four other uranium nuclei, and so on (Fig. 16.1.6). The result would be a catastrophic nuclear process in which many or even most of the nuclei in a piece of uranium would shatter and release fantastic amounts of energy.

In a sense, the remaining work toward both the atomic bomb and the hydrogen bomb was a matter of technical details. Only four conditions had to be satisfied in order for an atomic or *fission bomb* to be possible:

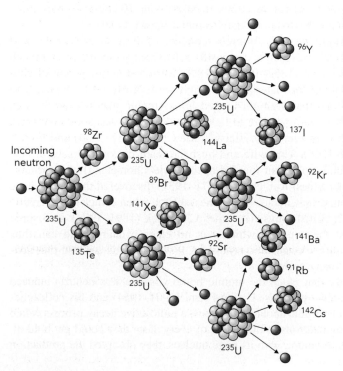

Fig. 16.1.6 A chain reaction occurs when the fragments of one fissioning nucleus induce fission in at least one additional nucleus. Such a chain reaction is particularly easy in ^{235}U, the light isotope of uranium, where each fissioning nucleus releases an average of 2.5 neutrons.

1. A source of neutrons had to exist in the bomb to trigger the explosion.
2. The nuclei making up the bomb had to be **fissionable,** that is, they had to fission when hit by a neutron.
3. Each induced fission had to produce more neutrons than it consumed.
4. The bomb had to use the released neutrons efficiently so that each fission induced an average of more than one subsequent fission.

Meeting the first condition was easy. Many radioactive elements emit neutrons. But meeting the second and third conditions was more difficult. Here is where uranium fit into the picture. It was known to be fissionable, and it was known to release more neutrons than it consumed.

But not all uranium nuclei are the same. While a uranium nucleus must contain 92 protons, so that it forms a neutral atom with 92 electrons and has all of the chemical characteristics of uranium (U in Fig. 14.2.4), the number of neutrons in that nucleus is somewhat flexible. Nuclei that differ only in the numbers of neutrons they contain are called **isotopes,** and natural uranium nuclei come in two isotopes: ^{235}U and ^{238}U, where the number specifies how many nucleons are in each nucleus. The ^{235}U nucleus contains 92 protons and 143 neutrons, for a total of 235 nucleons. In contrast, the ^{238}U nucleus contains 238 nucleons—92 protons and 146 neutrons.

It turns out that only ^{235}U is suitable for a bomb. It's marginally stable, with too many protons for the nuclear force to keep together, even with the diluting effects of 143 neutrons. Like many proton-rich nuclei, ^{235}U eventually undergoes **alpha decay**— it emits a helium nucleus (^{4}He) and thereby loses two protons and two neutrons. ^{235}U has a radioactive half-life of 710 million years. But when struck by a neutron, ^{235}U shatters immediately into fragments and this induced fission releases about 2.5 neutrons.

^{238}U is slightly more stable than the lighter isotope, and its half-life is 4.51 billion years. But this nucleus absorbs most neutrons without undergoing fission. Instead, it undergoes a series of complicated nuclear changes that eventually convert it into plutonium, an element not found in nature. It becomes ^{239}Pu, a nucleus with 94 protons and 145 neutrons. As we'll see later on, plutonium itself is useful for making nuclear weapons.

So ^{238}U actually slows a chain reaction rather than encouraging it. Since only ^{235}U can support a chain reaction, natural uranium had to be separated before it could be used in a bomb. But ^{235}U is quite rare. The earth's store of uranium nuclei was created long ago, in the explosion of a dying star. That *supernova* heated the nuclei of smaller atoms so hot that they collided together and stuck. Uranium nuclei were formed, with the supernova's energy trapped inside them. They were incorporated in the earth during its formation about 4 or 5 billion years ago and have been decaying ever since. The only uranium isotopes that remain in any quantity are ^{235}U and ^{238}U. Since ^{235}U is less stable, it has dwindled to only 0.72% of the uranium nuclei. The remaining 99+% of the uranium is ^{238}U.

Separating ^{235}U from ^{238}U is extremely difficult. Since atoms containing these two nuclei differ only in mass, not in chemistry, they can only be separated by methods that compare their masses. Because the mass difference is relatively small, heroic measures are needed to extract ^{235}U from natural uranium. During World War II and the Cold War era, the U.S. government developed enormous facilities for separating the two uranium isotopes. The need for such installations is one of the major obstacles to the proliferation of nuclear weapons.

The last condition for sustaining a chain reaction is that the bomb must use neutrons efficiently, so that each fission induces an average of more than one subsequent fission. That means that the bomb's contents can't absorb neutrons wastefully and that it can't let too many of them escape without causing fission. A lump of relatively pure ^{235}U wouldn't absorb neutrons wastefully, but it might allow too many of them to escape through its surface. For a chain reaction to occur, the lump must be large enough that each neutron has a good chance of hitting another nucleus before it

Courtesy U.S. Dept. of Energy

Courtesy Defense Nuclear Agency

Fig. 16.1.7 The Little Boy (*top*) used a cannon to fire a cylinder of uranium into an incomplete sphere of uranium. The Fat Man (*bottom*) used high explosives to crush a sphere of plutonium to extreme density. Little Boy destroyed Hiroshima, while Fat Man destroyed Nagasaki.

leaves the lump. The lump also should have a minimal amount of surface. It should be a sphere.

But how large must that sphere be? Since atoms are mostly empty space, a neutron can travel several centimeters through a lump of uranium without hitting a nucleus. Thus a golf ball-sized sphere of uranium would allow too many neutrons to escape. For a bare sphere of ^{235}U, the **critical mass** required to initiate a chain reaction is about 52 kg (115 lbm), a ball about 17 cm (7 inches) in diameter. At that point, each fission will induce an average of one subsequent fission. But for an **explosive chain reaction,** in which each fission induces an average of much more than one fission, additional ^{235}U is needed: a **supercritical mass.** About 60 kg (132 lbm) will do it.

By 1945, the scientists and engineers of the Manhattan Project had found ways to meet these four conditions and were prepared to initiate an explosive chain reaction. They had accumulated enough ^{235}U to construct a supercritical mass. Carefully machined pieces of ^{235}U would be put into the bomb so that they would join together at the moment the bomb was to explode. When the critical mass was reached, a few initial neutrons would start the chain reaction. When the uranium became supercritical, catastrophic fission would quickly turn it into a tremendous fireball.

However, assembly was tricky. The supercritical mass had to be completely assembled before the chain reaction was too far along; otherwise the bomb would begin to overheat and explode before enough of its nuclei had time to fission. In pure ^{235}U, the time it takes for one fission to induce the next fission is only about 10 ns (10 nanoseconds or 10 billionths of a second). In a supercritical mass, each generation of fissions is much larger than the previous generation, so it takes only a few dozen generations to shatter a significant fraction of the uranium nuclei. The whole explosive chain reaction is over in less than a millionth of a second, with most of the energy released in the last few generations (about 30 ns).

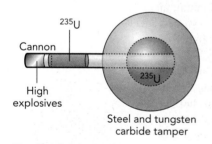

Fig. 16.1.8 The concept behind Little Boy was simple: a sphere of ^{235}U was divided into two parts so that neither was a critical mass on its own. One part was a hollow sphere, and the other part was a cylinder that would complete that sphere. When the bomb was detonated, a cannon fired the cylinder into the hollow sphere, creating a supercritical mass and initiating an explosive chain reaction.

To make sure that the assembly was complete before the bomb exploded, it had to be done extraordinarily quickly. In the ^{235}U bomb called "Little Boy" (Fig. 16.1.7), exploded over Hiroshima on August 6, 1945, at 8:15 a.m. and responsible for the deaths of about 200,000 Japanese citizens, the supercritical mass was assembled when a cannon fired a cylinder of ^{235}U through a hole in a sphere of ^{235}U (Fig. 16.1.8). When the cylinder was centered in the hole, it completed a 60-kg sphere of uranium, housed in a tungsten carbide and steel container. This container confined the uranium, holding it together with its inertia as the explosion began. An explosive chain reaction started immediately, and by the time the uranium blew itself apart, 1.3% of the ^{235}U nuclei had fissioned. The energy released in that event was equal to the explosion of about 15,000 tons of TNT.

But Little Boy was actually the second nuclear explosion. Its concept was so foolproof and its ^{235}U so precious that Little Boy was dropped without ever being tested. However, the Manhattan Project had also developed a plutonium-based bomb that involved a much more sophisticated concept. This bomb was much less certain to work, so it was tested once before it was used.

"The Gadget," as the first atomic bomb was called, didn't use ^{235}U. Instead, it used plutonium that had been synthesized from ^{238}U in nuclear reactors. A nuclear reactor carries out a controlled chain reaction in uranium, and neutrons from this chain reaction can convert ^{238}U into ^{239}Pu.

Like ^{235}U, ^{239}Pu meets the conditions for a bomb. The ^{239}Pu nucleus is relatively unstable, with a half-life of only 24,400 years. It fissions easily when struck by a neutron and releases an average of 3 neutrons when it does. Thus ^{239}Pu can be used in a chain reaction. For a bare sphere of ^{239}Pu, the critical mass is about 10 kg (22 lbm)—a ball about 10 cm (4 inches) in diameter.

But ^{239}Pu has a problem. It's so radioactive and releases so many neutrons when it fissions that a chain reaction develops almost instantly. There is much less time to assemble a supercritical mass of plutonium than there is with uranium. The cannon assembly method won't work because the plutonium will overheat and blow itself apart before the cylinder can fully enter the sphere.

Thus a much more sophisticated assembly scheme was employed. When The Gadget (Fig. 16.1.9) was detonated at Alamogordo on June 16, 1945, at 5:29 a.m., over 2000 kg (4400 lbm) of carefully designed high explosives crushed or *imploded* a sphere of plutonium (Fig. 16.1.10). By itself, the 6.1-kg (13.4-lbm) sphere wasn't large enough to be a critical mass; it was a **subcritical mass.** But it was surrounded by a *tamper* of ^{238}U whose massive nuclei reflected many neutrons back into the plutonium like marbles bouncing off bowling balls. And the implosion process compressed the plutonium well beyond its normal density. With the plutonium nuclei packed more tightly together, they were more likely to be struck by neutrons and undergo fission.

The scheme worked. The chain reaction that followed caused 17% of the plutonium nuclei to fission and released energy equivalent to the explosion of about 22,000 tons of TNT. The tower and equipment at the Trinity test site disappeared into vapor, and the desert sands turned to glass for hundreds of meters in all directions. A nearly identical device named "Fat Man" (Fig. 16.1.7) was dropped over Nagasaki on August 9, 1945, at 11:02 a.m., where it ultimately killed about 140,000 people.

In the years following the first fission bombs, development focused on how best to bring the fissionable material together. The longer a supercritical mass could be held together before it overheated and exploded, the larger the fraction of its nuclei that would fission and the greater the explosive yield. The crushing technique of The Gadget and Fat Man became the standard, and bombs grew smaller and more efficient at using their nuclear fuel. The implosion process reduced the amount of plutonium needed to reach a supercritical mass, so that very small fission bombs were possible. The smallest atomic bomb, the "Davy Crockett," weighed only about 220 N (50 lbf).

Courtesy Los Alamos Scientific Laboratory, University of California

Fig. 16.1.9 The first atomic bomb, nicknamed "The Gadget," was detonated on a tower at a remote desert site near Alamogordo, New Mexico, on June 16, 1945. Here employees of the top secret Manhattan Project are seen hoisting parts of the plutonium bomb onto the tower before the explosion.

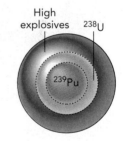

Fig. 16.1.10 The concept used in The Gadget and Fat Man was relatively sophisticated. Carefully shaped high explosives crushed a baseball-sized sphere of ^{239}Pu inside a neutron-reflecting shell of ^{238}U. The plutonium was compressed to high density and quickly reached supercritical mass, initiating an explosive chain reaction.

▶ Check Your Understanding #4: Tickling the Dragon's Tail

for answers, see page 529

What would happen if you slowly moved two 30-kg (66-lbm) hemispheres of ^{235}U together to form a single sphere?

The Fusion or Hydrogen Bomb

Since fissionable material begins to explode as soon as it exceeds critical mass, this critical mass limits the size and potential explosive yield of a fission bomb. In search of a way around this limit, bomb scientists took a look back at small nuclei and figured out how to extract energy by sticking them together.

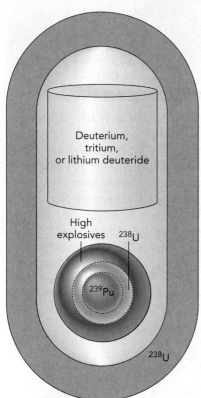

The fission bomb brought to the earth, for the first time, temperatures that had previously only been observed in stars. Stars obtain most of their energy by fusing hydrogen nuclei together to form helium nuclei, a process that releases a great deal of energy. Because hydrogen nuclei are protons and repel one another with tremendous forces, hydrogen doesn't normally undergo fusion here on earth. To cause fusion, something must bring those protons close enough for the nuclear force to stick them together. The only practical way we know of to bring the nuclei together is to heat them so hot that they crash into each other. That is what happens in a *fusion bomb,* also called a *thermonuclear* or *hydrogen bomb.*

In a fusion bomb, an exploding fission bomb heats a quantity of hydrogen to about 100 million degrees Celsius (Fig. 16.1.11). At that temperature, hydrogen nuclei begin to collide with one another. To ease the nuclear fusion processes, heavy isotopes of hydrogen are used: *deuterium* (^{2}H) and *tritium* (^{3}H). While the normal hydrogen nucleus (^{1}H) contains only a proton, the deuterium nucleus contains a proton and a neutron. The tritium nucleus contains a proton and two neutrons. When a deuterium nucleus collides with a tritium nucleus, they stick to form a helium nucleus (^{4}He) and a free neutron. Because this process converts about 0.3% of the original mass into energy, the helium nucleus and the neutron fly away from one another at enormous speeds.

Since hydrogen won't explode spontaneously, even in huge quantities, a hydrogen bomb can be extremely large. A fission bomb is used to set it off, but after that, the sky's the limit. Some enormous fusion bombs were constructed and tested during the early days of the Cold War. These bombs usually consisted of a fission starter and a hydrogen follower, all wrapped up in a tamper of ^{238}U. The tamper confined the hydrogen as the fusion began. Once fusion was underway, converting deuterium and tritium into helium and neutrons, the neutrons collided with ^{238}U nuclei in the tamper. These fusion neutrons were so energetic that they were able to induce fissions even in ^{238}U nuclei and release still more energy. Overall, this structure is sometimes called a fission– fusion–fission bomb.

A variation on this bomb is the so-called neutron or enhanced radiation bomb. That bomb has no ^{238}U tamper so that the energetic neutrons from the fusion process travel out of the explosion and irradiate everything in the vicinity. This bomb is lethal to humans but not particularly destructive to property.

Tritium is a radioactive isotope created in nuclear reactors. It has too many neutrons to be a stable nucleus and slowly decays into a light isotope of helium (^{3}He). Because tritium has a half-life of 12.3 years, fusion bombs containing tritium require periodic maintenance to replenish their tritium.

Many fusion bombs use solid lithium deuteride instead of deuterium and tritium gases. Lithium deuteride is a salt containing lithium (^{6}Li) and deuterium (^{2}H). When a neutron from the fission starter collides with a ^{6}Li nucleus, the two fragment into a helium nucleus (^{4}He) and a tritium nucleus (^{3}H). In the bomb, lithium deuteride is

Fig. 16.1.11 A fusion bomb releases energy by fusing deuterium (^{2}H) and tritium (^{3}H) nuclei together to form helium (^{4}He) and neutrons. This fusion is initiated by heating the hydrogen to more than 100 million degrees Celsius with a fission bomb. High-energy neutrons released by the fusion process then induce fission in the ^{238}U tamper, releasing still more energy. Lithium (^{6}Li) produces tritium when exposed to neutrons.

quickly converted into a mixture of deuterium, tritium, and helium, which then undergoes fusion.

► Check Your Understanding #5: It's Hard to Get Together

for answers, see page 529

Why must hydrogen be heated to extremely high temperature to initiate fusion?

Heat, Radiation, and Fallout

Once a nuclear weapon has exploded—after its fissionable material has fissioned and its fusible material has fused—what then? First, a vast number of nuclei and subatomic particles emerge from the explosion at enormous speeds, many at nearly the speed of light. These particles crash into nearby atoms and molecules, heating them to fantastic temperatures and producing a local fireball around the bomb itself. They also cause extensive radiation damage in the surrounding area.

Second, a flash of light emerges from the explosion, caused partly by the fission and fusion processes themselves and partly by the ultrahot fireball that follows. This light is not only visible light, but also every portion of the electromagnetic spectrum from infrared to visible to ultraviolet to X-rays to gamma rays. It burns things nearby, inside and out.

Third, the explosion creates a huge pressure surge in the air around the fireball. A shock wave propagates outward from the fireball at the speed of sound, knocking over everything in its path for a considerable distance. Fourth, the rarefied and superheated air rushes upward, lifted by buoyant forces, to create a towering mushroom cloud.

But the most insidious aftereffect of a nuclear explosion is *fallout,* the creation and release of radioactive nuclei. Fission converts uranium and plutonium nuclei into smaller nuclei. Each new nucleus has several dozen protons and its share of neutrons from the nucleus that fissioned. These new nuclei attract electrons and become seemingly normal atoms like iodine or cobalt. But while large nuclei such as uranium need extra neutrons to dilute their protons and reduce their electrostatic repulsions, intermediate and small nuclei like those of iodine and cobalt don't need as many neutrons. The new nuclei wind up with too many neutrons and are radioactive. They have half-lives that are anywhere from thousandths of a second to thousands of years.

Until they decay, the atoms that contain these nuclei are almost indistinguishable from normal atoms. They are radioactive isotopes of common atoms, and our bodies naively incorporate them into our tissues. There they sit, performing whatever chemical tasks our bodies require of them. But eventually these radioactive atoms fall apart and release nuclear energy. Because each radioactive decay that occurs near us or inside us releases perhaps a million times more energy than is present in a chemical bond, these decays cause chemical changes in our cells. They can kill cells or damage the cells' genetic information, potentially causing cancer.

This **transmutation of elements**—the restructuring of nuclei to transform atoms of one element into another—occurs in an uncontrolled fashion in nuclear weapons and produces a lethal mix of unstable isotopes. Those isotopes take years to decay out of the environment and all anyone can do is wait. Even nuclear weapons with poor explosive yields, so-called dirty bombs, can litter the surrounding landscape with radioactive debris. Nuclear reactors produce similar mixtures of radioactive isotopes in their fuel assembles and core structures, which is why disposing of spent nuclear fuel remains so problematic.

On the other hand, radioactive isotopes have been a boon to medicine and biochemistry, where many of them have found valuable and life-saving applications. Moreover, in controlled circumstances, elements can be transmuted systematically to generate primarily desirable isotopes rather than a random assortment. But such transmutation is difficult and expensive, and it involves nuclear rather than chemical processes. Although the alchemists' dream of transmuting lead into gold is finally possible, it's not a path to riches.

Common Misconceptions: Radiation and Radioactivity

Misconception: When a material such as food is exposed to microwave, radio wave, infrared, or ultraviolet radiation, it may become radioactive.

Resolution: To render a material radioactive, something must alter its nuclei so that they are no longer stable. Such an alteration requires vastly more energy than one of those low-energy photons can provide. The only forms of electromagnetic radiation with enough energy per photon to affect nuclei are gamma rays and, occasionally, X-rays.

Check Your Understanding #6: Not So Good to Eat

for answers, see page 529

Normal iodine ^{127}I is stable forever. But the fission product ^{131}I is radioactive and has a half-life of about 8 days. What are the consequences of eating ^{131}I?

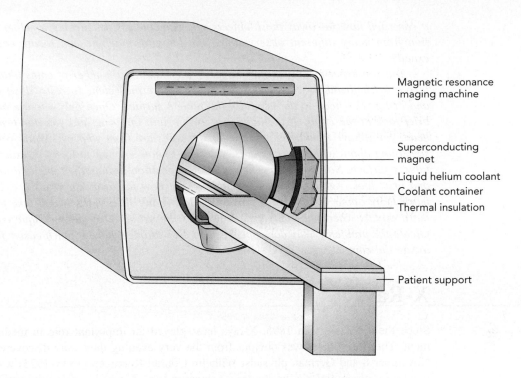

Magnetic resonance
imaging machine

Superconducting
magnet

Liquid helium coolant

Coolant container

Thermal insulation

Patient support

SECTION 16.2 Medical Imaging and Radiation

Some of the most important recent advances in health care have occurred at the border between medicine and physics. As scientists have refined their understanding of atomic and molecular structure and learned to control various forms of radiation, they have invented tools that are enormously valuable for diagnosing and treating illness and injury. The developments continue, with new applications of physics appearing in clinical settings almost every time you turn around. In this section, we'll examine two of the most significant examples of medical physics: the imaging techniques that are used to detect problems and the radiation therapies that are used to treat them.

Questions to Think About: What's different about bones and tissue that bones appear light in an X-ray image while tissue appears dark? How can a CT scan or an MRI image show a cross-sectional view of a living person without touching that person? If X-rays are a form of electromagnetic radiation, why don't opaque materials absorb them? Why does a CT scan show primarily bone while an MRI image shows primarily tissue? The subatomic particles used in radiation therapy often have enormous energies. From where do these energies come?

Experiments to Do: One triumph of medical imaging is its ability to locate objects inside a person without actually entering the body. This is often done using X-rays to look through the person from several different angles. From each vantage point, the imaging machine determines which objects are to the left or right of one another, but it can't tell how far away those objects are. Nonetheless, by piecing together information from many different observations, the imaging machine can locate each object exactly.

You can experiment with this process by sprinkling a number of coins onto the surface of a small table, without looking to see where they come to rest. Close your eyes and bring your head to the height of the table's surface. Open only one eye and take a brief look at the coins. If the lighting is bright and uniform, and you don't move your head, you'll have trouble telling how far the coins are from your eye. While you'll know something about where each coin is, you won't know enough to locate it exactly on the table's surface. Now move your head to a new position around the table and take a second brief look. Again, you won't be able to tell how far away the coins are, but you'll learn more about their relative positions. How many such views will it take for you to learn exactly where the coins are? How does the presence of many opaque coins complicate the problem? Why does opening two eyes at once make it much easier for you to locate the coins?

X-Rays

Since their discovery in 1895, X-rays have played an important role in medical treatment. Their usefulness was obvious from the very evening they were discovered. It was November 8 and German physicist Wilhelm Conrad Roentgen (1845–1923) was experimenting with an electric discharge in a vacuum tube. He had covered the entire tube in black cardboard and was working in a darkened room. Some distance from the tube a phosphored screen began to glow. Some kind of radiation was being released by the tube, passing through the cardboard and the air, and causing the screen to fluoresce. Roentgen put various objects in the way of the radiation, but they didn't block the flow. Finally, he put his hand in front of the screen and saw a shadowed image of his bones. He had discovered X-rays and their most famous application at the same time.

The first clinical use of X-rays was on January 13, 1896, when two British doctors used them to find a needle in a woman's hand. In no time, X-ray systems became common in hospitals as a marvelous new technique for diagnosis. But this imaging capability was not without its side effects. Although the exposure itself was painless, overexposure to X-rays caused deep burns and wounds that took some time to appear. Evidently the X-rays were doing something much more subtle to the tissue than simply heating it.

X-rays are a form of electromagnetic radiation, as are radio waves, microwaves, and light. These different forms of electromagnetic radiation are distinguished from one another by their frequencies and wavelengths—while radio waves have low frequencies and long wavelengths, X-rays have extremely high frequencies and short wavelengths. But they're also distinguished by their photon energies. Because of its low frequency, a radio wave photon carries little energy. A medium-frequency photon of blue or ultraviolet light carries enough energy to rearrange one bond in a molecule. But a high-frequency X-ray photon carries so much energy that it can break many bonds and rip molecules apart.

In a microwave oven, the microwave photons work together to heat and cook food. The amount of energy in each microwave photon is unimportant because they don't act alone. But in radiation therapy, the X-ray photons are independent. Each one carries enough energy to damage any molecule that absorbs it. That's why X-ray burns involve little heat and appear long after the exposure—the molecular damage caused by X-rays takes time to kill cells.

► Check Your Understanding #1: Forms of Radiation

for answers, see page 529

Which is more closely related to X-rays: the beam of electrons traveling through a microwave oven's magnetron tube or the infrared light from the hot filament of a toaster?

Making X-Rays

Medical X-ray sources work by crashing fast-moving electrons into heavy atoms. These collisions create X-rays via two different physical mechanisms: bremsstrahlung and X-ray fluorescence.

Bremsstrahlung occurs whenever a charged particle accelerates. This process is nothing really new, since we know that radio waves are emitted when a charged particle accelerates on an antenna. But in a radio antenna, the electrons accelerate slowly and emit low-energy photons. Bremsstrahlung usually refers to cases in which a charged particle accelerates extremely rapidly and emits a very high-energy photon. In X-ray tube bremsstrahlung, a fast-moving electron arcs around a massive nucleus and accelerates so abruptly that it emits an X-ray photon (Fig. 16.2.1). This photon carries away a substantial fraction of the electron's kinetic energy. The closer the electron comes to the nucleus, the more it accelerates and the more energy it gives to the X-ray photon. However, the electron is more likely to miss the nucleus by a large distance than to almost hit it, so bremsstrahlung is more likely to produce a lower energy X-ray photon than a higher energy one.

In **X-ray fluorescence,** the fast-moving electron collides with an inner electron in a heavy atom and knocks that electron completely out of the atom (Fig. 16.2.2). This collision leaves the atom as a positive ion, with a vacant orbital close to its nucleus. An electron in that ion then undergoes a radiative transition, shifting from an outer orbital to this empty inner one and releasing an enormous amount of energy in the process. This energy emerges from the atom as an X-ray photon. Because this photon has an energy that's determined by the ion's orbital structure, it's called a **characteristic X-ray.**

To discuss the energies carried by X-ray photons, we'll use the energy unit we encountered in Section 14.2: the electron volt or eV. Photons of visible light carry energies of between 1.6 eV (red light) and 3.0 eV (violet light). Because the ultraviolet photons in sunlight have energies of up to 7 eV, they are able to break chemical bonds and cause sunburns. But X-ray photons have much larger energies than even ultraviolet photons.

In a typical medical X-ray tube, electrons are emitted by a hot cathode and accelerate through vacuum toward a positively charged metal anode (Fig. 16.2.3). The anode is a tungsten or molybdenum disk, spinning rapidly to keep it from melting. The energy of the electrons as they hit the anode is determined by the voltage difference across the tube. In a medical X-ray machine, that voltage difference is typically about 87,000 V, so each electron has about 87,000 eV of energy. Since an electron gives a good fraction of its energy to the X-ray photon it produces, the photons leaving the tube can carry up to 87,000 eV of energy. No wonder X-rays can damage tissue!

When the electrons collide with a target of heavy atoms, they emit both bremsstrahlung and characteristic X-rays (Fig. 16.2.4). The characteristic X-rays have specific energies so they appear as peaks in the overall X-ray spectrum. The bremsstrahlung X-rays have different energies but are most intense at lower energies. Because lower energy X-ray photons injure skin and aren't useful for imaging or

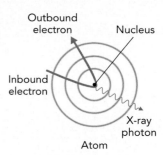

Fig. 16.2.1 When a fast-moving electron arcs around a massive nucleus, it accelerates rapidly. This sudden acceleration creates a bremsstrahlung X-ray photon, which carries off some of the electron's energy.

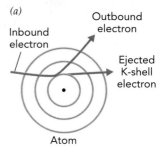

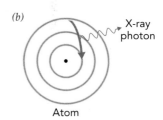

Fig. 16.2.2 (*a*) When a fast-moving electron collides with an electron in one of the inner orbitals of a heavy atom, it can knock that electron out of the atom. (*b*) An electron from one of the atom's outer orbitals soon drops into the empty orbital in a radiative transition that creates a characteristic X-ray.

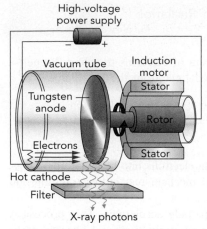

Fig. 16.2.3 In a medical X-ray machine, electrons from a hot filament accelerate toward a positively charged metal disk. They emit X-rays when they collide with the disk's atoms. A motor spins the disk to keep it from melting. The filter absorbs useless low-energy X-rays.

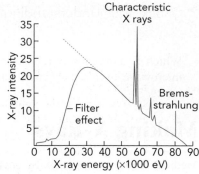

Fig. 16.2.4 When electrons with 87,000 eV of energy collide with tungsten metal, they emit X-rays via bremsstrahlung and X-ray fluorescence. While the bremsstrahlung X-rays have a broad range of energies, an absorbing filter blocks the low-energy ones. X-ray fluorescence produces characteristic X-rays with specific energies.

radiation therapy, medical X-ray machines use absorbing materials, such as aluminum, to filter them out.

> ➤ Check Your Understanding #2: The Origins of Synchrotron Radiation
> *for answers, see page* 529

The giant particle accelerators used in high-energy physics are often built as rings so that they can use the same electrically charged particles over and over again. As these particles travel in circles around the rings, they emit X-rays. Explain.

Using X-Rays for Imaging

X-rays have two important uses in medicine: imaging and radiation therapy. In X-ray imaging, X-rays are sent through a patient's body to a sheet of film or an X-ray detector. While some of the X-rays manage to pass through tissue, most of them are blocked by bone. The patient's bones form shadow images on the film behind them. In X-ray radiation therapy, the X-rays are again sent through a patient's body, but now their lethal interaction with diseased tissue is what's important.

X-ray photons interact with tissue and bone through four major processes: *elastic scattering*, the *photoelectric effect*, *Compton scattering*, and *electron–positron pair production*. **Elastic scattering** is already familiar to us as the cause of the blue sky: an atom acts as an antenna for the passing electromagnetic wave, absorbing and reemitting it without keeping any of its energy (Fig. 16.2.5). Because this process has almost no effect on the atom, elastic scattering isn't important in radiation therapy. However, it's a nuisance in X-ray imaging because it produces a hazy background: some of the X-rays passing through a patient bounce around like pinballs and arrive at the film from odd angles. To eliminate these bouncing X-ray photons, X-ray machines use filters to block X-rays that don't approach the film from the direction of the X-ray source.

The **photoelectric effect** is what makes X-ray imaging possible. In this effect, a passing photon induces a radiative transition in an atom: one of the atom's electrons absorbs the photon and is tossed completely out of the atom (Fig. 16.2.6). If the atom were using the X-ray photon to shift an electron from one orbital to another, that photon would have to have just the right amount of energy. But because a free electron can have any amount of energy, the atom can absorb any X-ray photon that has enough energy to eject one of its electrons. Part of the photon's energy is used to remove the electron from the atom, and the rest is given to the emitted electron as kinetic energy.

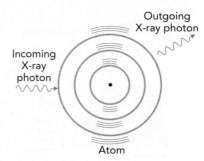

Fig. 16.2.5 When an X-ray photon scatters elastically from an atom, the whole atom acts as an antenna. The passing photon jiggles all of the charges in the atom, and these charges absorb the photon and reemit it in a new direction.

However, the likelihood of such a *photoemission* event decreases as the ejected electron's energy increases. This decreasing likelihood makes it difficult for a small atom to absorb an X-ray photon. All of its electrons are relatively weakly bound and the X-ray photon would give the ejected electron a large kinetic energy. Rather than emit a high-energy electron, a small atom usually just ignores the passing X-ray photon.

In contrast, some of the electrons in a large atom are quite tightly bound and require most of the X-ray photon's energy to remove them. These electrons would depart with relatively little kinetic energy. Because the photoemission process is most likely when low-energy electrons are produced, a large atom is likely to absorb a passing X-ray. Thus, while the small atoms found in tissue (carbon, hydrogen, oxygen, and nitrogen) rarely absorb medical X-rays, the large atoms found in bone (calcium and phosphorus) absorb X-rays frequently. That's why bones cast clear shadows onto X-ray film. Tissue shadows are also visible, but they're less obvious.

Although one shadow image of a patient's insides may help to diagnose a broken bone, more subtle problems may not be visible in a single X-ray image. For a better picture of what's going on inside the patient, the radiologist needs to see shadows from several different angles. Better yet, the radiologist can turn to a *computed tomography (CT) scanner*. This computerized device automatically forms X-ray shadow images from hundreds of different angles and positions and produces a detailed three-dimensional X-ray map of the patient's body.

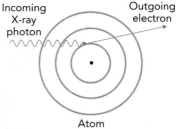

The CT scanner works one "slice" of the patient's body at a time. It sends X-rays through this narrow slice from every possible angle, including the two shown in Fig. 16.2.7, and determines where the bones and tissues are in that slice (Fig. 16.2.8). The scanner then shifts the patient's body to work on the next slice.

Fig. 16.2.6 In the photoelectric effect, an absorbed photon ejects an electron from an atom. Part of the photon's energy is used to remove the electron from the atom, and the rest becomes kinetic energy in the electron.

> Check Your Understanding #3: Aluminum X-Ray Windows
> for answers, see page 529

Aluminum atoms are much smaller than calcium atoms. While aluminum metal blocks visible light, it's relatively transparent to high-energy X-rays. Explain.

Using X-Rays for Therapy

Radiation therapy also uses X-rays, but not the ones used for medical imaging. Even though tissue absorbs fewer imaging photons than bone, most imaging photons are absorbed before they can pass through thick tissue. For example, only about 10% of the

(a)

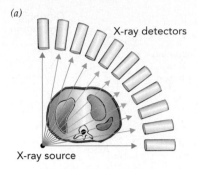

X-ray detectors

X-ray source

(b)

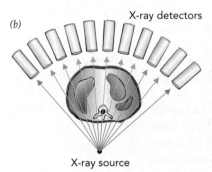

X-ray detectors

X-ray source

Fig. 16.2.7 A computed tomography or CT scan image is formed by analyzing X-ray shadow images taken from many different angles and positions. An X-ray source and an array of electronic X-ray detectors form a ring that rotates around the patient as the patient slowly moves through the ring.

Fig. 16.2.8 The computed tomography (CT) scanner on the left uses X-rays to image layer after layer of the patient's body. With the help of a computer, it produces a three-dimensional (3D) map of heavy elements in the patient. Part of that map, showing the patient's head, appears in the image on the right.

imaging photons make it through a patient's leg even when they miss the bone. That percentage is good enough for making an image, but it won't do for radiation therapy because most imaging X-rays would be absorbed long before they reached a deep-seated tumor. Instead of killing the tumor, intense exposure to these X-rays would kill tissue near the patient's skin.

To attack malignant tissue deep beneath the skin, radiation therapy uses extremely high-energy photons. At photon energies near 1,000,000 eV, the photoelectric effect becomes rare in tissue and bone, and the photons are much more likely to reach the tumor. Photons still deposit lethal energy in the tissue and tumor, but they do this through a new effect: Compton scattering.

Compton scattering occurs when an X-ray photon collides with a single electron so that the two particles bounce off one another (Fig. 16.2.9). The X-ray photon knocks the electron right out of the atom. This process is different from the photoelectric effect because Compton scattering doesn't involve the atom as a whole and the photon is scattered (bounced) rather than absorbed. The physics behind this effect resembles that of two billiard balls colliding, although it's complicated by the theory of relativity. The fact that it occurs at all is proof that a photon carries both energy and momentum and that these quantities are conserved when a particle of light collides with a particle of matter.

Compton scattering is crucial to radiation therapy. When a patient is exposed to 1,000,000-eV photons, most of the photons pass right through them, but a small fraction undergo Compton scattering and leave some of their energy behind. This energy kills cells and can be used to destroy a tumor. By approaching a tumor from many different angles through the patient's body, the treatment can minimize the injury to healthy tissue around the tumor while giving the tumor itself a fatal dose of radiation.

But Compton scattering isn't the only effect that occurs when high-energy photons encounter matter. X-rays with slightly more than 1,022,000 eV can do something remarkable when they pass through an atom: they can cause **electron–positron pair production.** A **positron** is the **antimatter** equivalent of an electron. Our universe is symmetrical in many ways, and one of its nearly perfect symmetries is the existence of antimatter. Almost every particle in nature has an antiparticle with the same mass but opposite characteristics. A positron or antielectron has the same mass as an electron, but it's positively charged. There are also antiprotons and antineutrons.

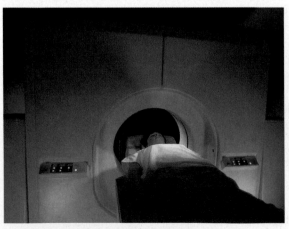

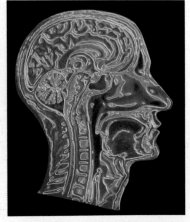

Deep Light Productions/Photo Researchers, Inc.

Alfred Pasieka/Photo Researchers, Inc.

Antimatter doesn't occur naturally on the earth, but it can be created in high-energy collisions. When an energetic photon collides with the electric field of an atom, the photon can become an electron and positron. In the previous section, we discussed matter becoming energy; pair production is an example of energy becoming matter. It takes about 511,000 eV of energy to form an electron or a positron, so the photon must have at least 1,022,000 eV to create one of each. Any extra energy goes into kinetic energy in the two particles.

The positron doesn't last long in a patient. It soon collides with an electron and the two annihilate one another—the electron and positron disappear and their mass becomes energy. They turn into photons with a total of at least 1,022,000 eV. So energy became matter briefly and then turned back into energy. This exotic process is present in high-energy radiation therapy and becomes quite significant at photon energies above about 10,000,000 eV. Not surprisingly, it also helps to kill tumors.

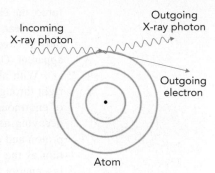

Fig. 16.2.9 In Compton scattering, an X-ray photon collides with a single electron and the two bounce off one another. The electron is knocked out of the atom.

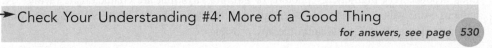

Check Your Understanding #4: More of a Good Thing

for answers, see page **530**

About how much energy would a photon need to create a proton–antiproton pair?

Gamma Rays

Producing very high-energy photons isn't quite as easy as producing those used in X-ray imaging. In principle, a power supply could create a huge voltage difference through an X-ray tube so that very high-energy electrons would crash into metal atoms and produce very high-energy photons. But million-volt power supplies are complicated and dangerous, so other schemes are used instead.

One of the easiest ways to obtain very high-energy photons is through the decay of radioactive isotopes. The isotope most commonly used in radiation therapy is cobalt 60 (^{60}Co). The nucleus of ^{60}Co has too many neutrons and, like many neutron-rich nuclei, ^{60}Co undergoes **beta decay**—one of its neutrons breaks up into a proton, an electron, and a *neutrino* (or more precisely, an *antineutrino*). Beginning with that beta decay, ^{60}Co undergoes a series of transformations that produce two high-energy photons: one with 1,170,000 eV and one with 1,330,000 eV. These photons penetrate tissue well and are quite effective at killing tumors.

Although the process by which ^{60}Co produces those two high-energy photons is complicated, beta decay itself shows that protons, electrons, and neutrons are not immutable and that there are other subatomic particles in our universe. Neutrons that are by themselves or in nuclei with too many neutrons are radioactive and experience beta decay. When that beta decay process occurs in a ^{60}Co nucleus, the negatively charged electron and neutral neutrino quickly escape from the nucleus but the newly formed proton remains. The nucleus thus becomes nickel 60 (^{60}Ni).

The **neutrino** is a subatomic particle with no charge and little mass. Neutrinos aren't found in normal atoms. Though important in nuclear and particle physics, neutrinos are difficult to observe directly because they travel near the speed of light and hardly ever collide with anything. Without charge, they don't participate in electromagnetic forces and, unlike the electrically neutral neutron, they don't experience the nuclear force. They experience only gravity and the **weak force,** the last of the four **fundamental forces** known to exist in our universe. (The other three fundamental forces are the gravitational force, the electromagnetic force, and the **strong force**—that's a more complete version of the nuclear force that we discussed in the previous section.) Because it's weak and occurs only between particles that are very close together, the weak force rarely makes itself apparent. One of the few occasions where it plays an important role is in beta decay.

With almost no way to push or pull on another particle, a neutrino can easily pass right through the entire earth. Neutrinos are detected occasionally, but only with the help of enormous detectors. That's why physicists first showed that neutrinos are emitted from decaying neutrons by measuring energy and momentum before and after the decay. The proton and electron produced by the decay don't have the same total energy and momentum as the neutron had before the decay. Something must have carried away the missing energy and momentum, and that something is the neutrino.

Once ^{60}Co has turned into ^{60}Ni, the decay isn't quite over. The ^{60}Ni nucleus that forms still has extra energy in it. Nuclei are complicated quantum physical systems just as atoms are, and they have excited states, too. The ^{60}Ni nucleus is in an excited state, and it must undergo two radiative transitions before it reaches the ground state. These radiative transitions produce very high-energy photons or **gamma rays** that are characteristic of the ^{60}Ni nucleus—one with 1,170,000 eV of energy and the other with 1,330,000 eV. These gamma rays are what make ^{60}Co radiation therapy possible.

Fig. 16.2.10 In a linear accelerator, moving charged particles are pushed forward by electric fields that change with time. (*a*) While the moving positive charge is passing through the first of a series of microwave cavities, the field there pushes it forward. (*b*) By the time the moving charge has entered the second cavity, the fields have reversed and the field there pushes it forward again.

> ➤ Check Your Understanding #5: A Visit from the Snake Oil Salesman
>
> for answers, see page 530

If someone offered to sell you a bottle of neutrinos, you'd be foolish to buy it. What's wrong with the idea of a bottle of neutrinos?

Particle Accelerators

Electromagnetic radiation isn't the only form of radiation used to treat patients. Energetic particles such as electrons and protons are also used. Like tiny billiard balls, these fast-moving particles collide with the atoms inside tumors and knock them apart. As usual, this atomic and molecular damage tends to kill cells and destroy tumors.

However, obtaining extremely energetic subatomic particles isn't easy. High-voltage power supplies can be used to accelerate an electron or proton to about 500,000 eV, but that isn't enough. When a charged particle enters tissue, it experiences strong electric forces and is easily deflected from its path. To make sure that it travels straight and true, all the way to a tumor, the particle must have an enormous energy. Giving each charged particle the millions or even billions of electron volts it needs for radiation therapy takes a particle accelerator.

Particle accelerators use metal cavities that behave like the tank circuits and antennas we discussed in Section 13.1. Almost any metal structure can act simultaneously as a capacitor and an inductor and thus have a natural resonance for sloshing charge. In the resonant cavities of a particle accelerator, this sloshing charge creates huge electric fields that change with time. Those electric fields push charged particles through space until they reach incredible energies.

One important type of particle accelerator is the *linear accelerator*. In this device, the electric fields in a series of resonant cavities push charged particles forward in a straight line (Fig. 16.2.10). Each of these cavities has

(*a*)

Microwave cavities

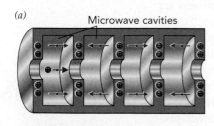

(*b*)

Electric fields

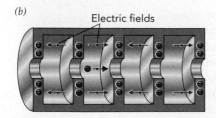

charge sloshing back and forth rhythmically on its wall. When a small packet of charged particles enters the first cavity through a hole, it's suddenly pushed forward by the strong electric field inside that cavity (Fig. 16.2.10a). The packet accelerates forward and leaves the first resonant cavity with more kinetic energy than it had when it arrived: the electric field in that cavity has done work on the packet.

If the fields in the cavities were constant, the electric field in the second cavity would slow the packet down. In Fig. 16.2.10a, you can see that the electric field in the second cavity points in the wrong direction. But by the time the packet reaches the second cavity, the charge sloshing in its walls has reversed and so has the electric field (Fig. 16.2.10b). The packet is again pushed forward, and it emerges from the second cavity with still more kinetic energy.

Each resonant cavity in this series adds energy to the packet, so that a long string of cavities can give each of the packet's charged particles millions or even billions of electron volts. This energy comes from microwave generators that cause charge to slosh in the accelerator's resonant cavities. The linear accelerator then only has to inject charged particles into the first cavity, using equipment resembling the insides of a television picture tube, and those charged particles will come flying out of the last cavity with incredible energies (Fig. 16.2.11).

This acceleration technique, however, has a few complications. Most importantly, each cavity must reverse its electric field at just the right moment to keep the packet accelerating forward. For simplicity of operation, all the cavities have the same resonant frequency and reverse their electric fields simultaneously. Since the packet spends the same amount of time in each cavity and since it speeds up as it goes from one cavity to the next, each cavity must be longer than the previous one.

But as the packet approaches the speed of light, something strange happens. The packet's energy continues to increase as it goes through the cavities, but its speed stops increasing very much. This effect is a consequence of special relativity, the rules governing motion at speeds comparable to the speed of light. As we saw in Section 4.2, the simple relationship between kinetic energy and speed given in Eq. 2.2.1 isn't valid for objects moving at almost the speed of light; we must use Eq. 4.2.4 instead. As a further consequence of relativity, the packet can approach the speed of light but can't actually reach it. Though each charged particle's kinetic energy can become extraordinarily large, its speed is limited by the speed of light.

Because the packet's speed stops increasing significantly after it has gone through the first few cavities of the linear accelerator, the lengths of the remaining cavities can be constant. Only the first few cavities have to be specially designed to account for the packet's increasing speed inside them. The charged particles emerge from the accelerator traveling at almost the speed of light. They pass through a thin metal window that keeps air out of the accelerator and enter the patient's body. They have so much energy that they can penetrate deep into tissue before coming to a stop.

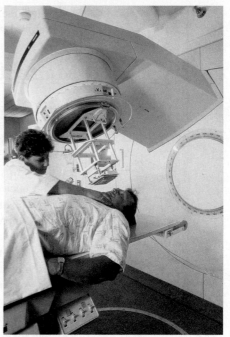

Doug Martin/Photo Researchers, Inc.

Fig. 16.2.11 This radiation therapy unit uses a linear accelerator to produce extremely high-energy subatomic particles. These particles penetrate deep inside the patient to destroy a cancerous tumor. The linear accelerator itself is hidden from view in the room behind this one. Its beam is steered by magnets in the rotatable arm toward a particular spot in the patient. The arm moves periodically during treatment so that its beam intersects the tumor from many different directions and causes less damage to healthy tissue nearby. That multidirectional strategy is also widely used in X-ray and gamma ray therapies.

Check Your Understanding #6: Particle Recycling

for answers, see page 530

Many research accelerators send each packet of electrons through the same series of resonant cavities several times. After the packet leaves the last cavity, magnets steer it around in a circle and send it back through the cavities again. With each pass through the cavities, the packet acquires more energy, so how can it possibly stay synchronized with the reversing electric fields in the cavities?

Magnetic Resonance Imaging

While X-rays do an excellent job of imaging bones, they aren't as good for imaging tissue. A better technique for studying tissue is *magnetic resonance imaging* or *MRI*. This technique locates hydrogen atoms by interacting with their magnetic nuclei. Since hydrogen atoms are common in both water and organic molecules, finding hydrogen atoms is a good way to study biological tissue.

The nucleus of an ordinary hydrogen atom, ^{1}H, is a proton. Protons, like electrons, have two possible internal quantum states, usually called spin-up and spin-down. Calling it *spin* is appropriate because spin-up and spin-down protons have equal but oppositely directed angular momentum. And when electric charge and rotation are both present, it's not surprising that magnetism is, too; electric currents are magnetic, after all. Sure enough, protons have magnetic dipoles—equal north and south poles at a distance from one another. A spin-up proton acts as though it has its north pole on top, while a spin-down proton acts as though it has its south pole on top.

When a proton is immersed in a magnetic field, it tends to align its magnetic dipole with that field. Doing so minimizes its magnetic potential energy. But while protons would align perfectly with the field at absolute zero, they are less successful near room temperature. Thermal energy agitates the protons so that, even in a strong, upward-pointing magnetic field, spin-up protons only slightly outnumber spin-down protons.

In that upward-pointing magnetic field, each proton has two possible quantum states: alignment with the field (spin-up) or anti-alignment (spin-down). Because alignment reduces the proton's magnetic potential energy, it's the ground state—the lower energy of the two possible states. The anti-aligned state is the excited state.

With its two possible states, ground and excited, a proton in a magnetic field can exhibit many of the behaviors we explored when looking at atoms in Section 14.2. Most importantly, the proton can experience radiative transitions between its two states. A ground state proton can *absorb* a photon while making a radiative transition to its excited state and an excited state proton can *emit* a photon while transiting to its ground state.

In Section 14.2, we saw that a given atom can absorb or emit only certain photons, photons carrying exactly the right amount of energy to shift the atom from one quantum state to another. For example, neon signs are red because neon atoms have states that are separated in energy by the energy of red photons. Similarly, a proton in a magnetic field can absorb or emit only certain photons, photons carrying exactly the right amount of energy to shift the proton from one quantum state to the other.

However, unlike a neon atom, which always interacts with red photons, a proton in a magnetic field interacts with photons that vary in "color" according to the strength of the magnetic field. That's because the energy separating the proton's two states is proportional to the magnetic field in which it resides. As a result, the photon energy needed to cause radiative transitions between the proton's two states is also proportional to the magnetic field. If the field changes, so does the photon energy.

When a patient enters the strong magnetic field of an MRI machine (Figs. 16.2.12 and 16.2.13), the protons in the patient's body respond to the field and a small excess of aligned protons develops. Only these excess aligned protons matter to the MRI machine because effects due to the remaining protons, which are equally aligned and anti-aligned, cancel completely. The excess aligned protons are in their ground state and they are what the MRI machine studies.

Magnetic
field

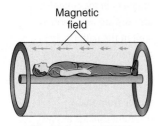

Fig. 16.2.12 An MRI machine places the patient in a strong magnetic field. This field varies spatially, so that protons at different places in the patient's body experience different fields and absorb different radio wave photons.

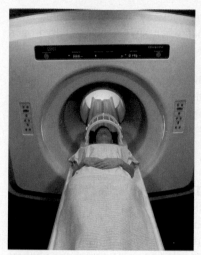

Charles Thatcher/Stone/Getty Images

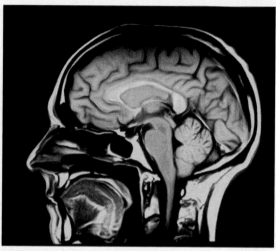

Med. Illus. SBHA/Stone/Getty Images

Fig. 16.2.13 The patient on the left is entering the intense magnetic field of a magnetic resonance imaging (MRI) system. Using the interactions between electromagnetic waves and protons that take place in magnetic fields, it produces a three-dimensional map of hydrogen atoms in the patient's body. A portion of this map, showing the patient's head, is displayed in the image on the right.

The MRI machine interacts with these ground state protons using radio wave photons—photons with energies equal to the energy difference between their ground and excited states. The protons can absorb and subsequently emit those radio wave photons and they can also exhibit a variety of fascinating and useful quantum interference effects.

If the protons in the patient's body were all experiencing exactly the same magnetic field, they would all interact with the same radio wave photons. But the protons don't all experience the same field. The MRI machine introduces a slight spatial variation to its magnetic field. Because the magnetic field is different for different protons, only some of them can interact with radio wave photons of a particular energy. This selective interaction is how the MRI imager locates protons within a patient.

In its simplest form, an MRI machine applies a spatially varying magnetic field to the patient's body. It then sends radio wave pulses through the patient and looks for those waves to be reemitted by the protons, often using the same antennae to emit the waves as to receive their echos. Since only a proton that is experiencing the right magnetic field can interact with a particular radio wave photon, the MRI machine can determine where each proton is by the photons with which it interacts. By changing the spatial variations in the magnetic field and adjusting the energies of the radio wave photons, the MRI machine gradually locates the protons in the patient's body. It builds a detailed three-dimensional map of the hydrogen atoms. A computer manages this map and can display cross-sectional images of the patient from any angle or position.

Check Your Understanding #7: Major Magnets

for answers, see page **530**

Some of the most modern MRI machines use extremely strong magnetic fields. As a machine's magnetic field gets stronger, what must happen to the radio waves that are used to interact with protons in a patient?

Epilogue for Chapter 16

In this chapter we examined some of the applications of modern physics. In *nuclear weapons*, we examined nuclear fission to see how nuclear chain reactions can be used to release electrostatic potential energy stored in the large nuclei of uranium and plutonium atoms. We also studied nuclear fusion and found that when small nuclei of hydrogen bind together, they release a potential energy associated with the nuclear force. We learned about radioactive isotopes and fallout.

In *medical imaging and radiation*, we learned how X-rays are produced and why those X-rays pass more easily through tissue than through bone. We saw how X-rays can be used to make images and explored the uses of gamma rays for radiation therapy. We looked at particle accelerators and concluded by examining magnetic resonance imaging.

Explanation: Radiation Damaged Paper

The paper fades because the ultraviolet light photons in sunlight have enough energy to shift electrons out of the orbitals that bond the dye molecules together. These dye molecules then fall apart and leave the paper without color. In some cases, the photons of light completely remove the electrons from the dye molecules. That process is the same photoelectric effect that makes it possible to distinguish tissue from bone in X-ray imaging.

Chapter Summary

How Nuclear Weapons Work: A fission bomb releases nuclear energy through a chain reaction in the fissionable isotopes of uranium or plutonium. In a chain reaction, each fission induces, on average, at least one subsequent fission. Assembling a supercritical mass must be done rapidly so that the bomb can't blow itself apart before most of the nuclei have undergone fission.

A fusion bomb uses heat from a fission bomb to initiate fusion in the heavy isotopes of hydrogen—deuterium and tritium. Because tritium has a short half-life, it must be replaced frequently. In some fusion bombs, the tritium is created during the explosion by neutron collisions with lithium.

How Medical Imaging and Radiation Work: Because X-rays pass more easily through tissue than they do through bone, X-rays form shadow images of a patient's bones. These X-rays are produced when energetic electrons accelerate near metal nuclei and when they knock other electrons out of the metal atoms.

Radiation therapy is done with very high-energy X-rays and gamma rays because they pass more easily through tissue. These electromagnetic waves kill tumor cells by depositing energy in the atoms and molecules of those cells. Gamma rays are usually obtained from radioactive nuclei. Some radiation therapy is done with energetic particles that are given enormous energies by particle accelerators.

Magnetic resonance imaging uses the magnetic nature of hydrogen nuclei (protons) to locate hydrogen atoms in a patient's body. The patient is put in a strong magnetic field, and the protons tend to align with this field. The imaging machine then uses radio waves to reverse the alignments of those protons. By making the magnetic field vary slightly from place to place, the machine is able to locate the protons in the patient's body. A computer records and analyzes the results so that it can present cross-sectional images of the hydrogen atoms in a patient's body.

Important Laws and Equations

1. The Exponential Decay of Radioactive Nuclei and Particles: The fraction of a large original population of identical radioactive systems remaining is equal to ½ raised to the power of the elapsed time divided by the half-life of those systems, or

$$\text{fraction remaining} = \left(\frac{1}{2}\right)^{\text{elapsed time}/\text{half-life}}. \tag{16.1.1}$$

Check Your Understanding—Answers

Section 16.1 NUCLEAR WEAPONS

1. That depends completely on the theory and the extent to which it has been compared to the real world.

Why: A great many theories have been formulated to explain the world around us. Each of these theories is an attempt at describing some particular behavior of our universe in terms of various rules or mechanisms. But until a theory has been tested by comparing it carefully to the system it's trying to explain, you can't tell i it's true or not. Some theories are eventually proved true, others false, and many remain uncertain. The theories of relativity and quantum physics have long since been proved true, although there is always the possibility that they may be only parts of a more complete theory.

2. It was stored as electrostatic potential energy in the repulsion between protons.

Why: Although it is routinely called nuclear energy, much of the energy stored in radioactive nuclei is actually electrostatic potential energy. Assembling a giant nucleus out of positively charged particles requires a considerable amount of work against electrostatic forces and it is that stored work that's released when the nucleus decays.

3. It will consume energy.

Why: To merge two smaller nuclei and form a uranium nucleus, you will have to push the two together with considerable force because they contain many protons. You will have to do considerable work to bring these nuclei close enough for the nuclear force to bind them. The energy you invested in this new nucleus is the same energy that is released when it undergoes fission.

4. Before they actually touched, a chain reaction would occur.

Why: As soon as the hemispheres became close enough that each fission began to induce an average of one subsequent fission, a chain reaction would occur. You would have assembled a critical mass. The hemispheres would begin to get hotter and hotter and would eventually melt or explode. Because they would blow apart before most of the nuclei could fission, any explosion would be rather small. But you would suffer severe radiation injury. Just such an accident killed Louis Slotin during the Manhattan Project.

5. The protons in hydrogen nuclei repel one another quite strongly at short distances. They must be moving very quickly with lots of thermal energy for them to touch.

Why: In a gas, thermal energy takes the form of kinetic energy. The hotter the gas, the faster the particles are moving. At 100 million degrees Celsius, the hydrogen nuclei are moving so fast that they can overcome their electrostatic repulsion and touch. When they do, the nuclear force pulls them together and they fuse.

6. The ^{131}I is incorporated into your body and exposes you to radiation, particularly during the first few weeks before most of it has decayed away.

Why: Your body can't distinguish ^{127}I from ^{131}I because they're chemically identical. Since your body uses iodine in its functions, any ^{131}I that you ingest is likely to become part of your body's iodine supply. Over the next 8 days, about half of this iodine will decay and expose you to radiation. The remaining ^{131}I is also radioactive, and half of this amount will decay in the next 8 days. Thus after 16 days, only one-quarter of the original amount will remain. After 24 days, only one-eighth will remain. And so on.

Section 16.2 MEDICAL IMAGING AND RADIATION

1. The infrared light from the hot filament.

Why: Infrared light and X-rays are both forms of electromagnetic radiation. All that distinguishes them is their frequencies and wavelengths, and the energy of their photons. The beam of electrons in a magnetron tube is also a form of radiation, but it involves particles of matter, not electromagnetic waves.

2. Because a particle traveling in a circle is accelerating, it emits electromagnetic waves. In this case, those waves are X-rays.

Why: A rapidly accelerating charged particle will emit an X-ray, whether it's accelerating around the nucleus of a heavy atom or around the ring of a particle accelerator. In an accelerator, these X-rays are called synchrotron radiation. Synchrotron radiation is useful in research and industry and is often deliberately enhanced by adding special magnets to the ring of the accelerator.

3. The electrons in an aluminum atom are so weakly bound that they are unlikely to absorb high-energy X-ray photons through the photoelectric effect.

Why: Like the small atoms in biological tissue, aluminum atoms rarely use the photoelectric effect to absorb high-energy

X-rays. This result makes it possible to use thin films of aluminum as windows and filters for X-ray sources.

4. About 2,000,000,000 eV.
Why: A proton has about 2000 times the mass of an electron, so producing a proton–antiproton pair should take about 2000 times the 1,000,000-eV energy required to produce an electron–positron pair.

5. The bottle couldn't confine neutrinos because it barely interacts with them.
Why: Because neutrinos experience only gravity and the weak force, the bottle couldn't trap them for long. Actually, you're being bathed in neutrinos from the sun all the time without even noticing it. They're as common as dirt and you can't do anything with them anyway.

6. The packet is traveling at almost the speed of light, so its speed barely changes as its energy increases.

Why: If the packet were speeding up with each trip through the cavities, it wouldn't stay synchronized with the reversing electric fields inside them. But the packet's speed is so nearly constant once it nears the speed of light that it can travel through the cavities over and over again without any problem.

7. Each radio wave photon must carry more energy (the radio waves must have higher frequencies).
Why: The stronger the magnetic field in which it's immersed, the more energy separates a proton's two states. The MRI machine must use higher frequency, more energetic radio waves to cause radiative transitions between those two states. There are many advantages to using extremely strong magnetic fields, including lower noise and better spatial resolution in the image. However, the magnetic fields of these advanced MRI machines are so strong that they can erase credit cards from across the room and rip steel objects right out of your pockets.

Check Your Figures—Answers

Section 16.1 NUCLEAR WEAPONS

1. The fraction of the original carbon 14 that remains should be about 0.886.
Why: According to Eq. 16.1.1, the fraction of carbon 14 nuclei remaining after 1001 years should be

$$\text{fraction remaining} = \left(\frac{1}{2}\right)^{1001\ \text{years} / 5730\ \text{years}}$$

$$= \left(\frac{1}{2}\right)^{0.1747} = 0.886$$

Finding that fraction indicates that the fibers in the garment come from plants or animals that died approximately 1001 years ago, although you can't tell exactly when the fabric itself was woven. Finding a larger fraction would prove that the garment is less than 1001 years old.

Exercises

1. Naturally occurring copper has two isotopes, ^{63}Cu and ^{65}Cu. What is different between atoms of these two isotopes?

2. Why is it extremely difficult to separate the two isotopes of copper, ^{63}Cu and ^{65}Cu?

3. If a nuclear reaction adds an extra neutron to the nucleus of ^{57}Fe (a stable isotope of iron), it produces ^{58}Fe (another stable isotope of iron). Will this change in the nucleus affect the number and arrangement of the electrons in the atom that's built around this nucleus? Why or why not?

4. If a nuclear reaction adds an extra proton to the nucleus of ^{58}Fe (a stable isotope of iron), it produces ^{59}Co (a stable isotope of cobalt). Will this change in the nucleus affect the num-

ber and arrangement of the electrons in the atom that's built around this nucleus? Why or why not?

5. When two medium-sized nuclei are stuck together during an experiment at a nuclear physics lab, the result is usually a large nucleus with too few neutrons to be stable. The nucleus soon falls apart. Why could more neutrons make it stable?

6. The large nucleus in Exercise 5 is likely to undergo alpha decay. What emerges from the nucleus during alpha decay and why does this decay reduce the nucleus's total potential energy?

7. When a large nucleus is split in half during an experiment at a nuclear physics lab, the result is usually two medium-sized

nuclei with too many neutrons to be stable. These nuclei eventually fall apart. Why don't these smaller nuclei need as many neutrons as they received from the original nucleus?

8. One of the medium-sized nuclei in Exercise 9 is likely to undergo beta decay. What emerges from that nucleus during beta decay and what happens to the nucleus as a result?

9. Light can't penetrate even a millimeter into plutonium, so why is a neutron able to travel centimeters into plutonium?

10. Why is it difficult or impossible to make very small atomic bombs?

11. Explain the strategy of putting highly radioactive materials in a storage place for many years as a way to make them less hazardous. That wouldn't work for chemical poisons, so why does it work for radioactive materials?

12. Once fallout from a nuclear blast distributes radioactive isotopes over a region of land, why is it virtually impossible to separate many of those radioactive isotopes from the soil?

13. Burning chemical poisons in a gas flame often renders those poisons harmless. Why won't that strategy make radioactive materials less dangerous?

14. Sunscreen absorbs ultraviolet light while permitting visible light to pass. Why does this coating reduce the risk of chemical and genetic damage to the cells of your skin?

15. Why is it important to keep foods and drugs out of direct sunlight, even when they're not in danger of overheating?

16. Why are many drugs packaged in amber-colored containers that block ultraviolet light?

17. Museums often display priceless antique manuscripts under dim yellow light. Why not use white light?

18. The most troubling radioactive isotopes in fallout are those with half-lives between a few days and a few thousand years. Why are those with much shorter or much longer half-lives less of a problem?

19. The most troubling radioactive isotopes in nuclear waste are those with half-lives of between a few years and a few hundred thousand years. Explain.

20. An X-ray technician can adjust the energy of the X-ray photons produced by a machine by changing the voltage drop between the X-ray tube's cathode and anode. Explain.

21. Lead, with 82 electrons per atom, is an excellent absorber of X-rays. Why?

22. Electric charge is strictly conserved in our universe, meaning that the net charge of an isolated system can't change. Why doesn't the production of an electron–positron pair in a patient cause a change in the patient's net charge?

23. Which has more mass: a positron or an antiproton?

24. Magnetic resonance imaging (MRI) differs from computed tomography imaging in that it involves no "ionizing radiation." What electromagnetic radiation is used in MRI and why aren't the photons of this radiation able to remove electrons from atoms and convert those atoms into ions?

25. Magnetic resonance imaging isn't good at detecting bone. Why not?

26. No magnetic metals such as iron or steel are permitted near a magnetic resonance imaging machine. In part, this rule is a safety precaution since those magnetic metals would be attracted toward the machine. But the magnetic fields from these magnetic metals would also spoil the imaging process. Why would having additional magnetic fields inside the imaging machine spoil its ability to locate specific protons inside a patient's body?

27. Why does the strength of the magnetic field used in MRI affect the frequency of the radio waves used to detect the protons?

28. The stronger the magnetic field used in MRI, the larger the fraction of protons that align their spins with the field. Why does this increased alignment make it easier for the MRI to study the protons?

Problems

1. Gallium 67 (^{67}Ga) is a radioactive isotope with a half-life of 3.26 days. It's used in nuclear medicine to locate inflammations and tumors. Accumulations of ^{67}Ga in a patient's tissue can be detected by looking for the gamma rays it emits when it decays. A radiologist usually begins looking for the ^{67}Ga about 48 hours after administering it to a patient. What fraction of the original ^{67}Ga nuclei remain after 48 hours?

2. Two weeks after ^{67}Ga was administered to the patient (Problem 1), what fraction of the ^{67}Ga nuclei remain?

3. Technetium 99m (^{99m}Tc) is a radioactive nucleus with a half-life of 6.03 hours. It's used in nuclear medicine to trace biological pathways. The ^{99m}Tc nucleus is actually an excited state of the ^{99}Tc nucleus and the radiative transition ^{99m}Tc $\rightarrow$ ^{99}Tc emits a gamma ray. That gamma ray shows where the nucleus was located when it decayed. If the radiologist begins looking for ^{99m}Tc 4.00 hours after administering it, what fraction of the ^{99m}Tc nuclei remain?

4. After a ^{99m}Tc nucleus emits a gamma ray (Problem 3), it becomes a ^{99}Tc nucleus. ^{99}Tc is also radioactive, with a half-life of 213,000 years, and it decays into ^{99}Ru. Two weeks after

^{99m}Tc was administered to the patient, what fraction of it is ^{99m}Tc? What fraction of it is ^{99}Tc? What fraction of it is ^{99}Ru?

5. While most natural potassium nuclei are stable ^{39}K (93.3%) or ^{41}K (6.7%) nuclei, about 0.0117% are radioactive ^{40}K nuclei, which have a half-life of 1.26 billion years. What fraction of all the potassium nuclei in your body will undergo radioactive decay in the next 1.00 year period?

6. If you're worried about ^{40}K radioactivity (Problem 5) and wanted to wait for 99% of the ^{40}K in the environment to decay away, how long would you have to wait?

Appendix A

Vectors

Many of the quantities of physics are vectors, meaning that they have both magnitudes (amounts) and directions. Among these vector quantities are position, velocity, acceleration, force, torque, momentum, angular momentum, and electric and magnetic fields. Of these vector quantities, position is probably the easiest to visualize: you specify an object's position by giving its position vector—its distance and direction from a reference point. For example, you can specify the library's position with respect to your home by giving both its distance from your home (say, 3.162 km or 1.965 miles) and its direction from your home (18.43° east of due north). That position vector is all the information someone would need to travel from your home to the library.

In illustrations, such as Fig. A.1, vector quantities are drawn as arrows. The length of each arrow indicates the vector's magnitude, while the direction of the arrow indicates which way the vector points. Suppose that you live in a city with major east–west and north–south streets spaced 1 km apart. Figure A.1 shows four aerial views of your city. The vector **A** in Fig. A.1a shows the position of the library with respect to your home. It begins at your home and ends at the library, thus indicating both the magnitude and direction of the library's position.

Let's look at another position vector. The vector **B** in Fig. A.1b begins at the library and ends at your friend's home. This position vector shows the position of your friend's home with respect to the library and is 2.828 km long and points 45° east of south. If you happen to be at the library, you could use this vector to find your friend's home.

But how can you go from your home to your friend's home? To make this trip, you must add two vectors: you first follow vector **A** from your home to the library and then follow vector **B** from the library to your friend's home. This combined trip is shown as the upper path in Fig. A.1c. But you could also go directly from your home to your friend's home by following a new vector in Fig. A.1c—vector **C**. This vector from your home to your friend's home is the sum of vectors **A** and **B** and is 3.162 km long and points 18.43° north of east. Using bold letters to indicate that **A**, **B**, and **C** are vectors, we can write **A** + **B** = **C**, meaning that vector **C** is the sum of vectors **A** and **B**.

Another interesting path from your home to your friend's home is shown in Fig. A.1d: you first travel along vector **B** and then along vector **A**. On this path, you would arrive at your friend's home without visiting the library. The first leg of this journey would take you into new territory, but the

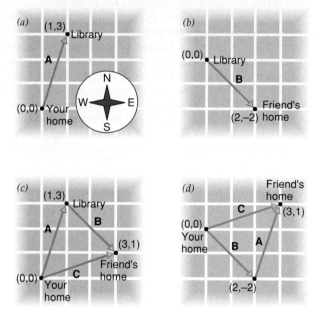

Fig. A.1. Four aerial views of your city, showing the major north–south and east–west streets that are spaced 1 km apart. (a) To go from your home to the library, you must travel 3.162 km at a direction 18.43° east of north—vector **A**. (b) To go from the library to your friend's home, you must travel 2.828 km at a direction 45° east of south—vector **B**. (c) You can go from your home to your friend's home either by going first to the library along vector **A** and then going to your friend's home along vector **B**, or you can travel directly along vector **C**, which is the sum of vectors **A** and **B**. (d) You can also reach your friend's home by going first along vector **B** and then along vector **A**. The sum of these two vectors is still vector **C**. However, you will not visit the library on that trip.

second leg would leave you at your friend's home. The sum of vectors **B** and **A** is still vector **C,** or **B + A = C.** Thus vectors added in any order yield the same sum.

While you can estimate the sum of two vectors by drawing arrows on a sheet of paper, obtaining an accurate sum requires some thought. Adding their magnitudes is unlikely to give you the magnitude of the sum vector, and adding their directions doesn't even make sense. To add two vectors, it helps to specify them in another form: as their *components* along two or three directions that are at right angles with respect to one another. In the present example of travel in a city, two right-angle directions are all we need. If height were also important, we'd need three right-angle directions.

For the two right-angle directions we need in the city, let's choose east and north. We can then specify vector **A** as its component toward the east and its component toward the north. Its east component is 1 km and its north component is 3 km, so vector **A,** the position of the library with respect to your home, is 1 km to the east and 3 km to the north.

This new form for vector **A,** a pair of distances, is often more convenient than the old form, a distance and a direction. If you go to the library by walking 3.162 km in the direction 18.43° east of north, you'll have to pass through a lot of other buildings and backyards. Walking to the library by heading 1 km east and 3 km north allows you to travel on the side walks.

If we designate the position of your home as 0 km east and 0 km north, then the position of the library is 1 km east and 3 km north. These positions are labeled in Fig. A.1*a* as (0, 0) and (1, 3), respectively. The first number in the paren-

theses is the distance east, measured in kilometers, and the second number is the distance north, also in kilometers.

To go from the library to your friend's home, you must go 2 km east and 2 km south. That's the new form of vector **B.** Because a southward position has a negative component along the northward direction, the position of your friend's home with respect to the library is 2 km east and −2 km north. In Fig. A.1*b*, the library is at (0, 0) and your friend's home is at (2, −2).

Now adding these vectors **A** and **B** is relatively easy. To go from your home to your friend's home by way of the library, you must move east first 1 km and then 2 km for a total of 3 km, and you must move north 3 km and then −2 km for a total of 1 km. Thus the position of your friend's home with respect to your home, vector **C,** is 3 km east and 1 km north. In Fig. A.1*c*, your home is at (0, 0) and your friend's home is at (3, 1).

Similarly, you could go from your home to your friend's home by following first vector **B** and then vector **A,** as shown in Fig. A.1*d*. This trip would take you through unknown regions, but you'd still arrive at the right place. You would move east first 2 km and then 1 km, and you'd move north first −2 km and then 3 km. In the end, you'd be 3 km east and 1 km north of your home, the position of your friend's home. Thus the sum of vectors **B** and **A** is still vector **C.**

As you can see, vectors are useful for specifying physical quantities in our three-dimensional world. When you encounter vectors, remember that their directions are just as important as their magnitudes and that these directions must be taken into account when you add two vectors together.

Appendix   B

Units, Conversion of Units

When you return from a camping trip and begin to describe it to your friends, there are a number of physical quantities that you may find useful in your conversation. Distance, weight, temperature, and time are as important in everyday life as they are in a laboratory. And when you explain to your friends how far you hiked, how much your backpack weighed, how cold the weather was, and how long the trip took, you'll have to relate those quantities to standard units or your friends won't appreciate just how difficult the trip was.

Most physical quantities aren't simple numbers like 7 or 2.9. Instead, they have units like length or time and are specified in multiples of widely accepted standard units such as meters or seconds. When you say a door is 3.0 meters tall, you're comparing the door's height to the meter, a widely accepted standard unit of length. With this comparison, most people can determine just how tall the door is, even though they've never seen it.

But the meter isn't familiar to everyone; many people prefer to measure length in multiples of another standard unit—the foot. These people might be more comfortable hearing that the door is 9.8 feet tall. These quantities, 3.0 meters and 9.8 feet, are the same length.

Determining the door's height in feet doesn't require a new measurement because we can convert the one in meters to one that's in feet. To perform this conversion, we need to know how to express one particular length in both units. Any length will do. For example, Table B.1 states that 1 foot is the same length as 0.30480 meter. Because of that equality, we know that the following equation is true:

$$\frac{1 \text{ foot}}{0.30480 \text{ meter}} = 1.$$

We can multiply 3.0 meters, the height of the door, by this version of 1 and obtain the door's height in feet:

$$3.0 \text{meters} \cdot \frac{1 \text{ foot}}{0.30480 \text{ meter}} = 9.8425 \text{ feet}.$$

Notice that the original units, meters, cancel and are replaced by the new units, feet. Since we only knew the door's height to two digits of precision in meters, we can't report the door's height to any more than two digits of precision in feet. So we round the result to 9.8 feet.

You can change the units of almost any physical quantity by multiplying that quantity by a version of 1. You should form this 1 by dividing new units by old units, where the number of new units in the numerator is equivalent to the number of old units in the denominator. You can obtain these pairs of equivalent quantities from Table B.1. When you do this multiplication, the old units will cancel and you'll be left with the physical quantity expressed in the new units.

One last note about units: when you use physical quantities in a calculation, make sure that you keep the units throughout the process. They're as important in the calculation as they are anywhere else. Some of the units may cancel, but in all likelihood the result of the calculation will have some units left and these units must be appropriate to the type of result you expect. If you're expecting a length, the units of your result should be meters or feet or another standard unit of length. If the units you obtain are seconds or kilograms, you've made a mistake in the calculation.

| Table B.1 | Conversion of Units[a]

1. Acceleration (SI unit: 1 meter/second^2 or 1 m/s^2)

$$1 \text{ foot/second}^2 = 0.30480 \text{ m/s}^2$$

2. Angle (SI unit: 1 radian)

$$1 \text{ degree } (1°) = 0.017453 \text{ radian}$$

3. Area (SI unit: 1 meter^2 or 1 m^2)

$$1 \text{ foot}^2 = 0.092903 \text{ m}^2$$
$$1 \text{ inch}^2 = 6.4516 \times 10^{-4} \text{ m}^2$$

4. Density (SI unit: 1 kilogram/meter^3 or 1 kg/m^3)

$$1 \text{ pound/foot}^3 = 16.018 \text{ kg/m}^3$$

5. Energy (SI unit: 1 joule or 1 J)

$$1 \text{ Btu} = 1054.7 \text{ J}$$
$$1 \text{ calorie, thermochemical} = 4.1840 \text{ J}$$
$$1 \text{ electron volt } (1 \text{ eV}) = 1.6022 \times 10^{-19} \text{ J}$$
$$1 \text{ foot-pound} = 1.3558 \text{ J}$$
$$1 \text{ kilocalorie (food Calorie)} = 4186.8 \text{ J}$$
$$1 \text{ kilowatt-hour} = 3,600,000 \text{ J}$$

6. Force (SI unit: 1 newton or 1 N)

$$1 \text{ pound} = 4.4482 \text{ N}$$

7. Length (SI unit: 1 meter or 1 m)

$$1 \text{ angstrom } (1 \text{ Å}) = 10^{-10} \text{ m}$$
$$1 \text{ centimeter } (1 \text{ cm}) = 0.01 \text{ m}$$
$$1 \text{ fermi } (1 \text{ fm}) = 10^{-15} \text{ m}$$
$$1 \text{ foot} = 0.30480 \text{ m}$$
$$1 \text{ inch} = 0.02540 \text{ m}$$
$$1 \text{ kilometer } (1 \text{ km}) = 1000 \text{ m}$$
$$1 \text{ light-year} = 9.4606 \times 10^{15} \text{ m}$$
$$1 \text{ micron } (1 \text{ } \mu\text{m}) = 10^{-6} \text{ m}$$
$$1 \text{ mil} = 2.5400 \times 10^{-5} \text{ m}$$
$$1 \text{ mile} = 1609.3 \text{ m}$$
$$1 \text{ millimeter } (1 \text{ mm}) = 0.001 \text{ m}$$
$$1 \text{ nanometer } (1 \text{ nm}) = 10^{-9} \text{ m}$$
$$1 \text{ picometer } (1 \text{ pm}) = 10^{-12} \text{ m}$$

8. Mass (SI unit: 1 kilogram or 1 kg)

$$1 \text{ gram } (1 \text{ g}) = 0.001 \text{ kg}$$
$$1 \text{ metric ton} = 1000 \text{ kg}$$
$$1 \text{ pound (at the earth's surface)} = 0.45359 \text{ kg}$$
$$1 \text{ slug} = 14.594 \text{ kg}$$

9. Power (SI unit: watt or W)

$$1 \text{ Btu/hour} = 0.29307 \text{ W}$$
$$1 \text{ horsepower} = 745.70 \text{ W}$$

10. Pressure (SI unit: 1 pascal or 1 Pa)

$$1 \text{ atmosphere} = 101,325 \text{ Pa}$$
$$1 \text{ millimeter of mercury } (1 \text{ torr}) = 133.32 \text{ Pa}$$
$$1 \text{ pound/inch}^2 (1 \text{ psi}) = 6894.8 \text{ Pa}$$

11. Temperature (SI units: degrees Celsius or °C; kelvins or K)

Because temperature in the three common units, °C, K, and °F, aren't multiples of one another, you must use special formulas to convert between them:

$$\text{Temperature in } °C = \tfrac{5}{9} \cdot (\text{temperature in } °F - 32)$$
$$\text{Temperature in } °C = \text{temperature in } K - 273.15$$
$$\text{Temperature in } K = \text{temperature in } °C + 273.15$$

12. Time (SI unit: 1 second or 1 s)

$$1 \text{ day} = 86,400 \text{ s}$$
$$1 \text{ femtosecond } (1 \text{ fs}) = 10^{-15} \text{ s}$$
$$1 \text{ hour} = 3600 \text{ s}$$
$$1 \text{ microsecond } (1 \text{ } \mu\text{s}) = 10^{-6} \text{ s}$$
$$1 \text{ millisecond } (1 \text{ ms}) = 0.001 \text{ s}$$
$$1 \text{ minute} = 60 \text{ s}$$
$$1 \text{ nanosecond } (1 \text{ ns}) = 10^{-9} \text{ s}$$
$$1 \text{ picosecond } (1 \text{ ps}) = 10^{-12} \text{ s}$$

13. Torque (SI unit: 1 newton-meter or 1 N·m)

$$1 \text{ inch-pound} = 0.11298 \text{ N·m}$$
$$1 \text{ foot-pound} = 1.3558 \text{ N·m}$$

14. Velocity (SI unit: 1 meter/second or 1 m/s)

$$1 \text{ foot/second} = 0.30480 \text{ m/s}$$
$$1 \text{ kilometer/hour } (1 \text{ km/h}) = 0.27778 \text{ m/s}$$
$$1 \text{ knot} = 0.51444 \text{ m/s}$$
$$1 \text{ mile/hour } (1 \text{ mph}) = 0.44704 \text{ m/s}$$
$$1 \text{ mile/hour } (1 \text{ mph}) = 1.6093 \text{ km/h}$$

15. Volume (SI unit: 1 meter^3 or 1 m^3)

$$1 \text{ cup} = 2.3659 \times 10^{-4} \text{ m}^3$$
$$1 \text{ fluid ounce} = 2.9574 \times 10^{-5} \text{ m}^3$$
$$1 \text{ foot}^3 = 0.028317 \text{ m}^3$$
$$1 \text{ gallon} = 0.0037854 \text{ m}^3$$
$$1 \text{ liter } (1 \text{ L}) = 0.001 \text{ m}^3$$
$$1 \text{ milliliter } (1 \text{ mL}) = 10^{-6} \text{ m}^3$$
$$1 \text{ quart} = 0.00094635 \text{ m}^3$$

[a] This table lists pairs of equivalent quantities, one in SI units and one in other units. Each pair can be used to convert measurements expressed as multiples of one unit into measurements expressed as multiples of the other unit. These pairs are grouped according to physical quantity and are given to a precision of five digits.

Glossary

absolute temperature scale A scale for measuring temperature in which 0 K corresponds to absolute zero.

absolute zero The temperature at which all thermal energy has been removed from an object or system of objects. Because it's impossible to find and remove all the thermal energy from an object, absolute zero can be approached but is not actually attainable.

acceleration A vector quantity that measures how quickly an object's velocity is changing: the greater the acceleration, the more the object's velocity changes each second. It consists of both the amount of acceleration and the direction in which the object is accelerating. This direction is identical to the direction of the force causing the acceleration. The SI unit of acceleration is the meter-per-second^2.

acceleration due to gravity A physical constant that specifies how quickly a freely falling object accelerates and also relates an object's weight to its mass. At the earth's surface, the acceleration due to gravity is 9.8 m/s^2 (or 9.8 N/kg).

activation energy The energy required to initiate a chemical reaction. This energy serves to break or weaken the bonds in the starting chemicals so that the reaction can proceed to form the reaction products.

adverse pressure gradient A region of fluid flow in which the fluid must flow toward higher pressure. The fluid's momentum and kinetic energy carry it through this situation, although the fluid does slow down.

aerodynamic forces The forces exerted on an object by the motion of the air surrounding it. The two types of aerodynamic forces are lift and drag.

aerodynamics The study of the dynamic (moving) interactions of air with objects.

airfoil An aerodynamically engineered surface, designed to obtain particular lift and drag forces from the air flowing around it.

airspeed An object's speed relative to the air through which it moves.

alpha decay A radioactive decay in which a helium nucleus (two protons and two neutrons) escapes from a larger proton-rich nucleus via quantum tunneling.

alternating current An electric current that periodically reverses its direction of flow. Abbreviated as AC.

ampere (A) The SI unit of electric current (synonymous with the coulomb-per-second). One ampere is defined as the passage of 6.25×10^{18} charged particles per second and is roughly the current flowing through a 100-W lightbulb operating on household electric power.

ampere-meter (A·m) The SI unit of magnetic pole.

amplifier A device that replicates an input signal as a larger output signal.

amplitude The maximal displacement of an oscillator away from its equilibrium position.

amplitude modulation A technique for representing sound or data by changing the amplitude (strength) of a wave.

analog representation The representation of numbers directly as continuous values of physical quantities such as voltage, charge, or pressure.

angle of attack The angle at which an airfoil is tilted relative to the airflow around it.

angular acceleration A vector quantity that measures how quickly an object's angular velocity is changing: the greater the angular acceleration, the more the object's angular velocity changes each second. It consists of both the amount of angular acceleration and the direction about which the angular acceleration occurs. This direction is identical to the direction of the torque causing the angular acceleration. The SI unit of angular acceleration is the radian-per-second^2.

angular impulse The mechanical means for transferring angular momentum. One object gives an angular impulse to a second object by exerting a certain torque on the second object for a certain amount of time. In return, the second object gives an equal but oppositely directed angular impulse to the first object.

angular momentum A conserved vector quantity that measures an object's rotational motion. It is the product of that object's rotational mass times its angular velocity. The SI unit of angular momentum is the kilogram-meter^2-per-second.

angular position A quantity that describes an object's orientation relative to some reference orientation.

angular speed A measure of the angle through which an object rotates in a certain amount of time.

angular velocity A vector quantity that measures how rapidly an object's angular position is changing: the greater the angular velocity, the farther the object turns each second. It consists of both the object's angular speed and the direction about which the object is rotating. This direction points along the axis of rotation in the direction established by the right-hand rule. The SI unit of angular velocity is the radian-per-second.

anharmonic oscillator An oscillator in which the restoring force on an object is not proportional to its displacement from a stable equilibrium. The period of an anharmonic oscillator depends on the amplitude of its motion.

antimatter Matter resembling normal matter, but with many of its characteristics such as electric charge reversed.

aperture The diameter or effective diameter of a lens or opening.

apparent weight The sum of a person's weight plus their feeling of acceleration. All three quantities are vectors, so that apparent weight can be quite large if the weight and feeling point in the same direction or quite small if they point in opposite directions.

Archimedes' principle   The observation that an object partially or wholly immersed in a fluid is acted on by an upward buoyant force equal to the weight of the fluid it's displacing.

atmospheric pressure   The pressure of air in the earth's atmosphere. Atmospheric pressure reaches a maximum of about 100,000 Pa near sea level and diminishes with increasing altitude.

atom   The smallest portion of a chemical element that retains the chemical properties of that element.

atomic number   The number of protons present in an atomic nucleus and equal to the number of electrons in a neutral atom.

axis of rotation   The straight line in space about which an object or group of objects rotates. More specifically, the axis of rotation points in a particular direction along that line to reflect the sense of rotation according to the right-hand rule.

back emf   The self-induced electromotive force that develops in an inductor when its current changes or in the coil of an electromechanical system such as a motor when its current causes magnets to move.

band   A group of levels in a solid that involve similar standing waves and thus have similar energies.

band gap   A range of energies over which a solid has no levels available.

bandwidth   The range of frequencies involved in a group of electromagnetic waves.

base of support   A surface outlined on the ground by the points at which an object is supported.

Bernoulli's equation   An equation relating the total energy of an incompressible fluid in steady-state flow to the sum of its pressure potential energy, kinetic energy, and gravitational potential energy.

beta decay   A radioactive decay in which the weak force allows a neutron in a neutron-rich nucleus to disintegrate into an electron, a proton, and an antineutrino. The proton remains in the nucleus, but the electron and antineutron escape.

binary   The digital representation of numbers in terms of the powers of two. The number 6 is represented in binary as 110, meaning 1 four (2^2), 1 two (2^1), and 0 ones (2^0).

bit   A single binary value, either a 0 or a 1.

black hole   A region of space, normally spherical, within which the gravitational distortions of space and time are so severe that not even light can escape.

blackbody spectrum   The distribution of thermal electromagnetic radiation emitted by a black object. This distribution is the amount of radiation emitted at each wavelength and depends only on the temperature of the black object.

blunt   Not streamlined so that the fluid flowing around it stalls and experiences flow separation and pressure drag.

boiling   Accelerated evaporation that occurs when stable gas-phase bubbles form and grow inside a material's liquid phase.

boiling temperature   The threshold temperature at which gas-phase bubbles first become stable within a material's liquid phase.

Boltzmann constant   The constant of proportionality relating a gas's pressure to its particle density and temperature. It has a measured value of 1.381×10^{-23} Pa·m^3/(particle·K).

boundary layer   A thin region of fluid near a surface that, because of viscous drag, is not moving at the full speed of the surrounding airflow.

bremsstrahlung   The process in which a rapidly accelerating charge emits electromagnetic radiation, usually an X-ray or gamma-ray photon.

Brewster's angle   The angle at which no vertically polarized light reflects from a transparent surface that is oriented horizontally. The precise angle depends on the surface's index of refraction.

buoyant force   The upward force exerted by a fluid on an object immersed in that fluid. The buoyant force is actually caused by pressure from the fluid. That pressure is highest below the object so the force exerted upward on the object's bottom is greater than the force exerted downward on the object's top.

byte   Eight binary bits that collectively can represent a number from 0 to 255. Bytes are often used to represent letters and other characters, where a convention associates each character with a specific number.

calibration   The process of comparing a local reference object to a generally accepted standard.

capacitance   The amount of separated charge on the plates of a capacitor divided by the voltage difference across those plates. The SI unit of capacitance is the farad.

capacitor   An electronic component that stores separated electric charge on a pair of plates that are separated by an insulating layer.

Celsius   A temperature scale in which 0 °C is defined as the melting point of water and 100 °C is defined as the boiling point of water at sea level. Absolute zero is −273.15 °C.

center of gravity   The unique point about which all of an object's weight is evenly distributed and therefore balanced. Because weight is proportional to mass, the center of mass is identical to the center of gravity for objects that are much smaller than the earth. For larger objects, the centers of mass and gravity differ slightly. An object suspended from its center of gravity will balance and will experience no net torque due to gravity. In many situations, you can accurately predict an object's behavior by assuming that all of the object's weight acts at its center of gravity.

center of mass   The unique point about which all of the object's mass is balanced. The center of mass is the natural pivot point for a free object. In the absence of outside forces or torques, a rigid object's center of mass travels at constant velocity while the object rotates at constant angular velocity about this center of mass.

center of percussion   The special spot on a bat or racket where a collision with a second object will not cause any acceleration of the bat's handle.

centripetal acceleration   An acceleration that is always directed toward the center of a circular trajectory.

centripetal force   A centrally directed force on an object. A centripetal force is not an independent force but rather the sum of other forces, such as gravity, acting on the object.

chain reaction   A process in which one event triggers an average of one or more similar events so that the process becomes self-sustaining.

chaos   Unpredictable behavior in which minute changes in a system's initial arrangement lead to very different final arrangements. These differences grow more dramatic with each passing second.

chaotic system   A dynamic system that is exquisitely sensitive to initial conditions. Minute changes in how you set up a chaotic system can lead to wildly different final configurations.

characteristic X-ray An X-ray emitted by X-ray fluorescence from an atom. The energy of the characteristic X-ray is determined by the atom's orbital structure.

charges Objects, particularly small particles, that carry electric charge.

chemical bond An energy deficit that holds two or more atoms together to form a molecule and that must be repaid in order to separate the atoms. Chemical bonds form when chemical potential energy is released during a molecule's assembly.

chemical potential energy Energy stored in the chemical forces between atoms. Those chemical forces are electromagnetic in origin.

chemical reaction An encounter between two or more atoms and molecules that results in a rearrangement of the atoms to form different atoms and molecules.

circularly polarized A light wave in which the electric and magnetic fields rotate about the direction in which that wave is heading.

closed circuit A complete electric circuit through which electric current can flow continuously.

coefficient of restitution The measure of a ball's liveliness, determined by bouncing the ball from a rigid, immovable surface. It's the ratio of the ball's rebound speed to its collision speed.

coefficient of volume expansion The fractional change in an object's volume caused by a temperature increase of 1 °C.

coherent light Light consisting of identical photons that together form a single electromagnetic wave.

collision energy The amount of kinetic energy removed from two objects as they collide.

color temperature The temperature at which a black object will emit thermal electromagnetic radiation with this particular distribution of wavelengths.

components The portions of a vector quantity that lie along particular directions.

compressible A substance that changes density significantly as its pressure changes. A gas is compressible since its density is proportional to its pressure.

Compton scattering The process in which a photon bounces off a charged particle, usually an electron. The photon and charged particle exchange energy and momentum during the collision.

condensation The phase transformation whereby a gas becomes a liquid.

conduction The transmission of heat through a material by a transfer of energy from one atom or molecule to the next. The atoms themselves don't move with the heat. In metals, mobile electrons also contribute to heat conduction.

conduction band The group of quantum levels in an insulator that lies above the Fermi level.

conduction level A quantum level in an insulator that requires more energy than the Fermi level and that is normally unoccupied by electrons.

conserved quantity A physical quantity, such as energy, that is neither created nor destroyed within an isolated system when that system undergoes changes. A conserved quantity may pass among the objects within an isolated system, but its total amount remains constant.

constructive interference Interference in which two or more waves arrive at a location in space and time in phase with one another and produce a particularly strong effect.

convection The transmission of heat by the movement of a fluid. Convection normally entails the natural circulation of the fluid that accompanies differences in temperatures and densities.

convection cell A loop of fluid flow that is propelled by convection. Fluid in a convection cell normally rises in a hotter region and descends in a colder region.

convection current A fluid flow propelled by convection.

converging lens A lens that bends the individual light rays passing through it toward one another so that they either converge more rapidly than before or at least diverge less rapidly from one another. A converging lens has a positive focal length and often produces real images.

corona discharge A faintly glowing discharge that surrounds a small, highly charged object in the presence of a gas. In the discharge, electric charge is transferred from the object to the gas molecules.

coulomb (C) The SI unit of electric charge. One coulomb is about 1 million times the charge you acquire by rubbing your feet across a carpet in winter.

Coulomb constant The fundamental constant of nature that determines the electrostatic forces two charges exert on one another. Its measured value is 8.988×10^9 $N \cdot m^2/C^2$.

Coulomb's law The magnitudes of the electrostatic forces between two objects are equal to the Coulomb constant times the product of their two electric charges divided by the square of the distance separating them. If the charges are like, then the forces are repulsive. If the charges are opposite, then the forces are attractive.

crest A peak positive excursion of an extended system that is experiencing a wave.

critical mass A portion of a fissionable material that is able to sustain a fission chain reaction. The amount of material required depends on its mass, shape, and density.

crystalline Having its atoms arranged in an orderly pattern that extends for many atomic spacings in all directions.

current The amount of electric charge flowing past a point or through a surface per unit of time. The SI unit of current is the ampere.

cycle-per-second (1/s) The SI unit of frequency (synonymous with hertz).

cyclotron motion The circular or spiral motion of a charged particle in a magnetic field. The charged particle tends to loop around the magnetic flux lines.

decimal The digital representation of numbers in terms of the powers of ten. The number *124* is represented in decimal as 124, meaning 1 hundred (10^2), 2 tens (10^1), and 4 ones (10^0).

demagnetization The disappearance of magnetic polarization in a material.

density The mass of an object divided by its volume. The SI unit of density is the kilogram-per-meter^3.

depletion region The nonconducting region around a p-n junction in which all of the valence levels are filled with electrons and there are no conduction level electrons.

deposition The phase transformation whereby a gas becomes a solid.

destructive interference Interference in which two or more waves arrive at a location in space and time out of phase with one another and produce a particularly weak effect.

diffraction A wave phenomenon that limits the focusability of light and alters the way in which it travels after passing through an opening.

digital representation The representation of numbers by decomposition into digits that are then individually represented by discrete values of physical quantities such as voltage, charge, or pressure.

diode A semiconductor device that allows charge to flow through it only in one direction. Diodes are commonly formed by bringing n-type silicon into contact with p-type silicon.

direct current An electric current that always flows in one direction. Abbreviated as DC.

direction The line or course on which something is moving, is aimed to move, or along which something is pointing or facing.

discharge A flow of electric current through a gas.

dispersion The dependence of light's speed through a material on the frequency of that light.

distance The length between two positions in space. The SI unit of distance is the meter.

diverging lens A lens that bends the individual light rays passing through it away from one another so that they either converge less rapidly than before or don't converge at all. A diverging lens has a negative focal length and often produces virtual images.

domain wall A boundary surface between magnetic domains having different directions of magnetic orientation.

doped Modified by adding chemical impurities that change its physical properties.

Doppler effect A difference between the frequency at which a wave is sent and the frequency at which that wave is received caused by relative motion between the sender and receiver.

drag forces The frictionlike forces exerted by a fluid and a solid on one another as the solid moves through the fluid. These forces act to reduce the relative velocity between the two.

dynamic stability An object's stability when it's in motion.

dynamic variations Changes in a physical quantity such as pressure that are caused by motion.

elastic collision A collision in which all of the kinetic energy present before the impact is again present as kinetic energy after the impact.

elastic limit The most extreme distortion of an object from which it can return to its original size and shape without permanent deformation.

elastic potential energy The energy stored by the forces within a distorted elastic object.

elastic scattering The process in which two particles bounce off one another without losing any of their kinetic energies.

electric charge An intrinsic property of matter that gives rise to electrostatic forces between charged particles. Electric charge is a conserved physical quantity. A specific charge can have a positive amount of electric charge (a positive charge) or a negative amount (a negative charge). The SI unit of electric charge is the coulomb.

electric charges Objects, particularly small particles, that carry electric charge.

electric circuit A complete loop of conductors, loads, and power sources through which an electric current can flow continuously.

electrical conductor A material that permits electric charges to flow through it.

electric current The movement or flow of electric charge.

electric field An attribute of each point in space that exerts forces on electrically charged particles. An electric field has a magnitude and direction proportional to the force it would exert on a unit of positive charge at that location. While electric fields are often created by nearby charges, they can also be created by other electromagnetic phenomena. The SI unit of electric field is the volt-per-meter or, equivalently, the newton-per-coulomb.

electric polarization A distribution of electric charge that is nonuniform so that the object has a region of positive charge and a region of negative charge.

electrical insulator A material that prevents any net movement of electric charge through it.

electrical resistance The measure of how much an object impedes the flow of electric current. The SI unit of electrical resistance is the ohm.

electromagnet A coil of wire, with or without an iron core, that becomes a magnet when an electric current flows through the coil.

electromagnetic waves Waves consisting of electric and magnetic fields that travel through empty space at the speed of light. These waves carry energy and momentum and are emitted and absorbed as particles called photons. Radio waves, microwaves, infrared, visible, and ultraviolet light, X-rays, and gamma rays are examples of electromagnetic waves.

electron–positron pair production The formation of an electron and a positron during an energetic collision.

electrons The tiny negatively charged particles that make up the outer portions of atoms and that are the main carriers of electricity and heat in metals.

electron volt (eV) A unit of energy equal to the energy obtained by an elementary charge (electron or proton) as it moves through a voltage difference of 1 V. One electron volt is equal to about 1.602×10^{-19} J.

electrostatic force The force experienced by a charged particle in the presence of other charged particles.

electrostatic potential energy Energy stored in the forces between electric charges.

elementary unit of electric charge The basic quantum of electric charge, equal to about 1.6×10^{-19} C.

emissivity A surface's capacity to emit or absorb thermal radiation, relative to that of a perfectly black object at the same temperature.

energy The capacity to do work. Each object has a precise quantity of energy, which determines exactly how much work that object could do in an ideal situation. The SI unit of energy is the joule.

English system of units An assortment of antiquated units that were used throughout the English colonies and remain in common use in the United States today. Units in this system include feet, ounces, and miles-per-hour.

entropy The physical quantity measuring the amount of disorder in a system. The system's entropy would be zero at absolute zero.

equilibrium The state of an object in which zero net force (or zero net torque) acts on it. An object that is stationary or in uniform motion is in equilibrium.

equilibrium position The point at which an object experiences zero net force and doesn't accelerate.

escape velocity The speed a spacecraft needs in order to follow a parabolic orbital path and escape forever from a particular celestial object.

evaporation The phase transformation whereby a liquid becomes a gas.

exhaust velocity The velocity of exhaust gas relative to the rocket engine from which it emerged.

excited state A configuration of a system having excess energy; its electrons (or other particles) are in an arrangement of quantum waves (e.g., orbitals or levels) that has more than the least possible energy.

explosive chain reaction A chain reaction in which each fission induces an average of much more than one subsequent fission and the fission rate skyrockets.

Fahrenheit A temperature scale in which 32 °F is defined as the melting point of water and 212 °F is defined as the boiling point of water at sea level. Absolute zero is −459.67 °F.

farad (F) The SI unit of electric capacitance. A 1-farad capacitor will have a voltage difference between its plates of 1 volt when storing 1 coulomb each of separated positive and negative charge.

feedback The process of using information about a system's current situation to control changes you are making in that system.

feeling of acceleration A person undergoing acceleration experiences a weightlike sensation in the direction exactly opposite the direction of acceleration. The amount of this feeling of acceleration is proportional to the amount of the acceleration.

Fermi energy The energy of an electron in the Fermi level.

Fermi level A hypothetical level located halfway between the highest occupied level and the lowest unoccupied level in a solid.

Fermi particles A class of fundamental particles that includes electrons, protons, and neutrons and that obeys the Pauli exclusion principle.

ferromagnetic Composed of magnetic atoms that all have the same magnetic orientation within a magnetic domain.

firm Having a large spring constant and thus experiencing large restoring forces in response to small distortions.

first law of thermodynamics The change in a stationary object's internal energy is equal to the heat transferred into that object minus the work that object does on its surroundings. This law restates the conservation of energy.

fissionable Able to undergo induced fission.

fluid A substance that has mass but no fixed shape. A fluid can flow to match its container. Gases and liquids are both fluids.

fluorescence An emission of light that immediately follows an absorption of light.

f-number The ratio of a lens's focal length to its effective aperture.

focal length The distance after a converging lens at which the real image of a distant object forms. The focal length of a diverging lens is negative and is the distance before the lens at which the virtual image of a distant object forms.

force An influence that if exerted on a free body results chiefly in an acceleration of the body and sometimes in deformation and other effects. A force is a vector quantity, consisting of both the amount of force and its direction. The SI unit of force is the newton.

forward biased A p-n junction in which the voltage of the p-type semiconductor has been raised relative to the voltage of the n-type semiconductor.

freezing The phase transformation whereby a liquid becomes a solid.

frequency The number of cycles completed by an oscillating system in a certain amount of time. The SI unit of frequency is the hertz.

frequency modulation A technique for representing sound or data by changing the exact frequency of a wave.

friction The force that resists relative motion between two surfaces in contact. Frictional forces are exerted parallel to the surfaces in the directions opposing their relative motion.

fundamental forces The four basic forces that act between objects in the universe—the gravitational force, the electromagnetic force, the strong force, and the weak force.

fundamental vibrational mode The slowest and often broadest vibration that an extended object can support.

gamma rays Extremely high-energy photons of electromagnetic radiation, often produced during radioactive decays.

gas or gaseous A form of matter consisting of tiny, individual particles (atoms or molecules) that travel around independently. A gas takes on the shape and volume of its container.

general theory of relativity The physical rules governing all motion, even motion involving speeds comparable to the speed of light and occurring in the presence of massive objects.

gravitational constant The fundamental constant of nature that determines the gravitational forces two masses exert on one another. Its value is 6.6720×10^{-11} N·m^2/kg^2.

gravitational mass The mass associated with the gravitational attraction between objects.

gravitational potential energy Potential energy stored in the gravitational forces between objects.

gravity The gravitational attraction of the mass of the earth, the moon, or a planet for bodies at or relatively near its surface. All objects exert gravitational forces on all other objects.

ground state The lowest energy configuration of a system; its electrons (or other particles) are in the arrangement of quantum waves (e.g., orbitals or levels) that has the least possible energy.

half-life The time needed for half the nuclei of a particular radioactive isotope to undergo radioactive decay.

hard magnetic material A material that is relatively difficult to magnetize and that retains its magnetization once the magnetizing field is removed. Hard magnetic materials are suitable for permanent magnets.

harmonic An integer multiple of the fundamental frequency of oscillation for a system. The second harmonic is twice the frequency of the fundamental, and the third harmonic is three times the frequency of the fundamental. In principle, harmonics can continue forever.

harmonic oscillator An oscillator in which the restoring force on an object is proportional to its displacement from a stable equilibrium. The period of a harmonic oscillator doesn't depend on the amplitude of its motion.

heat The energy that flows from one object to another as a result of a difference in temperature between those two objects.

heat capacity The amount of heat that must be added to an object to cause its temperature to rise by 1 unit.

heat engine A device that converts thermal energy into work as heat flows from a hot object to a cold object.

heat exchanger A device that allows heat to flow naturally from a hotter material to a colder material without any actual exchange of those materials.

heat pump A device that pumps heat against its natural direction of flow, transferring it from a cold object to a hot object. To satisfy the second law of thermodynamics, a heat pump normally converts some ordered energy into thermal energy.

Heisenberg uncertainty principle A quantum physical law that states that an object's position (a particle characteristic) and momentum (a wave characteristic) can't be sharply defined at the same time. This principle gives objects with small masses a fuzzy character.

henry (H) The SI unit of inductance. A 1-henry inductor will experience a 1-ampere change in the current flowing through it each second when subjected to a 1-volt voltage drop.

hertz (Hz)   The SI unit of frequency (synonymous with cycles-per-second).

higher-order vibrational mode   A vibrational mode that is more complicated than the fundamental mode and in which different parts of the extended system move in opposite directions.

Hooke's law   The general law covering spring and elastic behavior. Hooke's law states that a spring exerts a restoring force that is proportional to the distance the spring is distorted from its equilibrium length.

horizontal polarization   An electromagnetic wave in which the electric field always points left or right (horizontally). The magnetic field always points vertically.

ideal gas law   The law relating the pressure, temperature, and particle density of an ideal gas. An ideal gas is one that is composed of perfectly independent particles. The particles don't stick and bounce perfectly from one another.

image distance   The distance between the lens and the image that the lens creates. Real images form at positive image distances while virtual images form at negative image distances.

impedance   A measure of a system's opposition to the passage of a current or a wave.

impedance mismatch   An abrupt change in the opposition to a wave's passage, typically accompanied by reflections.

impulse   The mechanical means for transferring momentum. One object gives an impulse to a second object by exerting a certain force on the second object for a certain amount of time. In return, the second object gives an equal but oppositely directed impulse to the first object.

incandescence   The emission of thermal radiation from a hot object.

incoherent light   Light consisting of individual photons, each its own independent electromagnetic wave.

incompressible   A substance that doesn't change density significantly as its pressure changes. Liquids and solids are incompressible since their densities change very little as their pressures change dramatically.

index of refraction   The factor by which the speed of light in a material is reduced from its speed in empty space, equal to the speed of light in empty space divided by light's speed in the material.

induced drag   The drag force that occurs when a wing deflects the stream of air passing across it to obtain lift.

induced emf   An overall voltage difference between the ends of a coil produced by a changing magnetic field in that coil and the resulting electric field.

induced fission   A fission event that's caused by a collision, usually with a neutron.

inductance   The voltage drop across an inductor divided by the rate at which current through that inductor is changing with time. The SI unit of inductance is the henry.

inductor   An electronic component that stores magnetic energy in a coil of wire and opposes changes in current in that wire.

inelastic collision   A collision in which some of the kinetic energy present before the impact is no longer present as kinetic energy after the impact.

inert gas   A gas consisting of atoms that are chemically inactive and rarely bond permanently with other atoms or molecules. Inert gases include helium, neon, argon, krypton, and xenon.

inertia   A property of matter by which it remains at rest or in uniform motion in the same straight line unless acted on by some outside force.

inertial   Moving because of inertia alone and therefore not accelerating.

inertial frame of reference   A frame of reference that is not accelerating and is thus either stationary or traveling at constant velocity. The laws of motion accurately describe any situation that is observed from an inertial frame of reference.

inertial mass   The mass associated with an object's inertia, its resistance to acceleration.

infrared light   Invisible light having wavelengths longer than about 750 nanometers.

in phase   The relationship between two waves in which they complete the same portions of their oscillatory cycles at the same time and place.

insulator   A solid in which the Fermi level falls within a band gap.

interference   A wave phenomenon in which waves passing through the same location from different directions reinforce or oppose one another.

interference pattern   A pattern of intensity variations in time and space that occurs when two or more waves are superposed and experience constructive and destructive interferences.

internal energy   The sum of an object's thermal energy and any additional potential energy stored entirely within the object.

internal kinetic energy   The portion of an object's kinetic energy that involves only the relative motion of particles within the object and that excludes the object's overall translation or rotation.

internal potential energy   The portion of an object's potential energy that involves only forces between particles within the object and that excludes the object's interactions with its surroundings.

ion   An atom or molecule with a net electric charge.

isotopes   Chemically indistinguishable atoms containing nuclei that differ only in their numbers of neutrons.

joule (J)   The SI unit of energy and work (synonymous with newton-meter). Lifting 1 liter of water upward 10 centimeters near the earth's surface requires about 1 joule of work.

joule-per-kilogram-kelvin (J/kg·K)   The SI unit of specific heat.

joule-per-second (J/s)   The SI unit of power (synonymous with watt).

Kelvin   The SI scale of absolute temperature, in which 0 K is defined as absolute zero. The spacing between units is the same as that used in the Celsius scale.

Kepler's first law   All planets move in elliptical orbits, with the sun at one focus of the ellipse.

Kepler's second law   A line stretching from the sun to a planet sweeps out equal areas in equal times.

Kepler's third law   The square of a planet's orbital period is proportional to the cube of that planet's mean distance from the sun.

kilogram (kg)   The SI unit of mass. (The standard kilogram is a platinum–iridium cylinder kept at the International Bureau of Weights and Measures near Paris.) A liter of water has a mass of about 1 kilogram.

kilogram-meter^2 (kg·m^2)   The SI unit of rotational mass. One kilogram-meter^2 is roughly the rotational mass of your forearm as it pivots about your elbow.

kilogram-meter-per-second (kg·m/s)   The SI unit of momentum. One kilogram-meter-per-second is about the momentum in a baseball traveling 25 km/h (16 mph).

kilogram-meter-per-second^2 (kg·m/s^2)   The SI unit of force (synonymous with newton).

kilogram-meter^2-per-second (kg·m^2/s)   The SI unit of angular momentum. One kilogram-meter^2-per-second is about the angular momentum of a 7.3 kg (16 lbm) bowling ball spinning 34 times/second as it rolls down the lane.

kilogram-per-meter^3 (kg/m^3) The SI unit of density. One kilogram-per-meter^3 is about the density of air at 2000 m (about 1 mile) above sea level.

kinetic energy The form of energy contained in an object's translational and rotational motion.

laminar flow Smooth, predictable fluid flow in which nearby portions of the fluid remain nearby as they travel along.

laser amplifier A device that amplifies weak incoming light to produce brighter outgoing light. The outgoing light is a brighter copy of the incoming light.

laser medium An assembly of excited atoms or other quantum systems that is capable of amplifying passing light through stimulated emission.

laser oscillator A laser amplifier that is surrounded by mirrors so that it can amplify one or more spontaneously emitted photons to form an intense beam of coherent light.

latent heat of evaporation The heat required to transform a unit mass of material from liquid to gas without changing its temperature.

latent heat of fusion Latent heat of melting.

latent heat of melting The heat required to transform a unit mass of material from solid to liquid without changing its temperature.

latent heat of vaporization Latent heat of evaporation.

law of universal gravitation Every object in the universe attracts every other object in the universe with a force equal to the gravitational constant times the product of the two masses, divided by the square of the distance separating the two objects.

laws of thermodynamics The four laws that govern the movement of heat between objects.

lens A transparent optical device that uses refraction to bend light, often to form images.

lens equation The equation relating a lens's focal length to the object and image distances.

Lenz's law When a changing magnetic field induces a current in a conductor, the magnetic field from that current always opposes the change that induced it.

level An electron standing wave in a solid, one of the basic electron wave modes allowed in a solid by quantum physics.

lever arm The directed distance from the pivot or axis of rotation to the point at which the force is exerted.

lift forces Forces exerted by a fluid on a solid that are at right angles to the fluid flow around that solid.

light See visible, infrared, and ultraviolet light.

linear momentum A conserved vector quantity that measures an object's motion. It is the product of that object's mass times its velocity. The SI unit of linear momentum is the kilogram-meter-per-second.

liquid A form of matter consisting of particles (atoms or molecules) that are touching one another but that are free to move relative to one another. A liquid has a fixed volume but takes the shape of its container.

longitudinal wave A wave in which the underlying oscillation is parallel to the wave itself.

Lorentz force The force experienced by a charged particle when it moves through a magnetic field.

lumen A common unit of total radiated light as perceived by a human eye.

luminescence The emission of light by any means other than thermal radiation.

magnetic dipole A pair of equal but opposite poles separated by a distance.

magnetic domains Regions of uniform alignment within a magnetic material.

magnetic field An attribute of each point in space that exerts forces on magnetic poles. A magnetic field has a magnitude and direction proportional to the force it would exert on a unit of north magnetic pole at that location. The SI unit of magnetic field is the tesla.

magnetic flux lines Abstract strands following along the local magnetic field direction and having a density proportional to that local field. Flux lines can only begin at north poles and end at south poles.

magnetic induction The process whereby a time-changing magnetic field initiates or influences an electric current.

magnetic monopole An isolated magnetic pole, either north or south. None has ever been observed.

magnetic polarization A distribution of magnetic poles that is nonuniform so that the object has a region of north pole and a region of south pole.

magnetic pole A property of nature that gives rise to magnetostatic forces between

magnetic poles. A specific pole can have a positive amount of magnetic pole (a north pole) or a negative amount (a south pole). The SI unit of magnetic pole is the ampere-meter.

magnetic poles Objects that carry magnetic pole.

magnetization The development of magnetic polarization in a material.

magnetostatic force The force experienced by a magnetic pole in the presence of other magnetic poles.

magnitude The amount of some physical quantity.

Magnus force A lift force experienced by a spinning object as it moves through a fluid. The Magnus force points toward the side of the ball moving away from the onrushing airstream.

mass The property of a body that is a measure of its inertia or resistance to acceleration, that is commonly taken as a measure of the amount of material it contains, and that causes it to have weight in a gravitational field. The SI unit of mass is the kilogram.

mechanical advantage The process whereby a mechanical device redistributes the amounts of force and distance that go into performing a particular amount of mechanical work.

mechanical wave A natural and often rhythmic motion of an extended object about its stable equilibrium shape or situation.

melting The phase transformation whereby a solid becomes a liquid.

melting temperature The temperature at which a material's solid and liquid phases can coexist in stable equilibrium.

metal A solid in which the Fermi level falls within a band of levels.

meter (m) The SI unit of length or distance. (One meter is formally defined as the distance light travels through empty space in 1/299,792,458th of a second.) One meter is about the length of a long stride or about 3.28 feet.

meter^2 (m^2) The SI unit of area. One square meter is about twice the area of an opened newspaper.

meter^3 (m^3) The SI unit of volume. One cubic meter is about the volume of a four-drawer file cabinet.

meter-per-second (m/s) The SI unit of velocity or speed. One meter-per-second is a typical walking pace or about 2.2 mph.

meter-per-second² (m/s²)   The SI unit of acceleration. One meter-per-second² is about the acceleration of an elevator as it first begins to move upward.

microwaves   Electromagnetic waves with wavelengths between about 1 meter and 1 millimeter.

mode   A basic pattern of distortion or oscillation.

molecule   A particle formed out of two or more atoms. A molecule is the smallest portion of a chemical compound that retains the chemical properties of that compound.

moment of inertia   Rotational mass.

momentum   Linear momentum.

natural resonance   A mechanical process in which an isolated object's energy causes it to perform a certain motion over and over again. The rate at which this motion occurs is determined by the physical characteristics of the object.

net electric charge   The sum of all charges on an object, both positive and negative. Positive charges increase the net charge while negative charges decrease it. Net charge can be negative.

net force   The sum of all forces acting on an object, considering both the magnitude of each individual force and its direction. The magnitude of the net force is often less than the sum of the magnitudes of the individual forces, since they often oppose one another in direction.

net magnetic pole   The sum of all poles on an object, both north and south. Since there are no isolated magnetic poles, an object's net magnetic pole is always zero.

net torque   The sum of all torques acting on an object, considering both the magnitude of each individual torque and its direction. The magnitude of the net torque is less than the sum of the magnitudes of the individual torques, since they often oppose one another in direction.

neutral   Having zero net electric charge.

neutrinos   Chargeless and nearly massless particles created during radioactive decays and other nuclear events. They rarely interact with matter.

neutrons   The electrically neutral subatomic particles that, together with protons, make up atomic nuclei.

newton (N)   The SI unit of force (synonymous with the kilogram-meter-per-second²). Eighteen U.S. quarters have a weight equal to about 1 newton. The common English unit of force, the pound, is about 4.45 newtons.

newton-meter (N·m)   The SI unit of energy and work (synonymous with the joule). Also the SI unit of torque, exerted by a 1-newton force located 1 meter from the axis of rotation. One newton-meter is about the torque exerted on your shoulder by the weight of a baseball held in your outstretched arm.

newton-per-ampere-meter (N/A·m)   The SI unit of magnetic field (synonymous with tesla).

newton-per-coulomb (N/C)   The SI unit of electrical field (synonymous with volt-per-meter).

newton-per-meter²   The SI unit of pressure (synonymous with pascal).

Newton's first law of motion   An object that is free from all outside forces travels at a constant velocity, covering equal distances in equal times along a straight-line path.

Newton's first law of rotational motion   An object that is not wobbling and is free from all outside torques rotates with constant angular velocity, spinning steadily about a fixed axis.

Newton's second law of motion   An object's acceleration is equal to the force exerted on that object divided by the object's mass. This equality can be manipulated algebraically to state that the force on the object is equal to the product of the object's mass times its acceleration (Eq. 1.1.2).

Newton's second law of rotational motion   An object's angular acceleration is equal to the torque exerted on that object divided by the object's rotational mass. This equality can be manipulated algebraically to state that the torque on the object is equal to the product of the object's rotational mass times its angular acceleration (Eq. 2.1.2). The law doesn't apply to wobbling objects.

Newton's third law of motion   For every force that one object exerts on a second object, there is an equal but oppositely directed force that the second object exerts on the first object.

Newton's third law of rotational motion   For every torque that one object exerts on a second object, there is an equal but oppositely directed torque that the second object exerts on the first object.

normal   Directed exactly away from (perpendicular to) a surface. A line that is normal to a surface meets that surface at a right angle.

normal force   Support force.

n-type semiconductor   A semiconductor such as silicon that contains impurity atoms such as phosphorus, arsenic, antimony, or bismuth that place electrons in the semiconductor's conduction level.

nuclear fission   The shattering of a heavy nucleus into smaller fragments. During fission, the positively charged fragments repel one another and release energy.

nuclear force   An attractive force that binds nucleons together once they touch one another.

nuclear fusion   The merging of two small nuclei to form a larger nucleus. During fusion, the nuclear force binds the nucleons together and releases energy.

nucleation   Forming an initial seed of one material phase in the midst of another material phase.

nucleon   A general name given to the particles that make up atomic nuclei: protons and neutrons.

nucleus   The positively charged central component of an atom, containing most of the atom's mass and about which the electrons are arranged. Plural is nuclei.

object distance   The distance between the lens and the object that it is imaging.

ohm (Ω)   The SI unit of electrical resistance. A 1-ohm resistor exhibits a voltage drop of 1 volt when 1 ampere of current flows through it.

ohmic   Exhibiting a voltage drop that's proportional to current, consistent with Ohm's law.

Ohm's law   The observation that the voltage drop across an ordinary electrical conductor is proportional to both the electric current passing through it and to its electrical resistance.

open circuit   An incomplete electric circuit where a gap in the electrical conductors stops electric current from flowing.

orbit   The path an object takes as it moves in the presence of a centripetal force.

orbital   An electron standing wave in an atom, one of the basic electron wave modes allowed in an atom by quantum physics.

orbital period   The time required to complete one full orbit.

ordered energy   Energy that can easily be used to do work.

oscillation   A repetitive and rhythmic movement or process that usually takes place about an equilibrium situation.

out of phase   The relationship between two waves in which they complete opposite portions of their oscillatory cycles at the same time and place.

parallel (wiring arrangement) An arrangement in which the current reaching two or more electric devices divides into separate parts to flow through those devices and then joins back together as it leaves them. Current experiences the same change in voltage in each device.

particle density The number of particles in an object divided by its volume. The particle density of water is about 3.35×10^{28} molecules-per-meter^3. The particle density of air at sea level is about 2.687×10^{25} molecules-per-meter^3.

pascal (Pa) The SI unit of pressure (synonymous with newton-per-meter^2). Atmospheric pressure at sea level is about 100,000 pascals. A 1-millimeter-high water droplet exerts a pressure of about 10 pascals on your hand.

Pascal's principle A change in the pressure of an enclosed incompressible fluid is conveyed undiminished to every part of the fluid and to the surfaces of its container.

Pauli exclusion principle An observed property of nature that indistinguishable Fermi particles must each have their own unique quantum wave.

period The time required to complete one full cycle of a repetitive motion.

permanent magnet An object that can be magnetized and that retains that magnetization for a long time.

permeability of free space The defined constant that relates two poles and the magnetostatic forces they exert on one another. Its value is $4\pi \times 10^{-7}$ N/A^2.

phase A form of matter, notably solid, liquid, gas, and plasma.

phase equilibrium A situation in which two material phases coexist stably, neither one growing at the expense of the other.

phase transition A transformation from one material phase to another.

phosphor A solid that luminesces (emits light) when energy is transferred to it by light or by a collision with a particle.

photoconductor A solid that is an electrical insulator in the dark but that becomes an electrical conductor when exposed to light of the correct wavelength.

photodiode A diode that permits current to flow backward across the p-n junction when exposed to light. Light provides the energy needed to move charges across the junction's depletion region in the wrong

direction. The current flowing in the reverse direction through a photodiode is proportional to the light intensity.

photoelectric effect The process in which an atom absorbs a photon in a radiative transition that ejects one of the electrons out of the atom.

photon A particle or quantum of light, having energy and momentum but no mass.

pitch The frequency of a sound.

Planck constant The fundamental constant of quantum physics, equal to the energy of an object divided by the frequency of its quantum wave. It is about 6.626×10^{-34} J·s.

plane polarized A light wave in which the electric field (and the magnetic field) fluctuates back and forth in a plane as the light travels through space.

plasma A gaslike phase of matter consisting of electrically charged particles, such as ions and electrons. The strong electromagnetic interactions between its particles distinguish a plasma from a gas.

p-n junction The interface between an n-type semiconductor and a p-type semiconductor that gives the diode its unidirectional characteristic for electrons.

Poiseuille's law The volume of fluid flowing through a pipe each second is equal to ($\pi/128$) times the pressure difference across that pipe times the pipe's diameter to the fourth power, divided by the pipe's length times the fluid's viscosity.

poles Objects that carry a magnetic pole.

population inversion A nonequilibrium population of quantum systems in which more are in a higher energy state than in a lower energy state.

position A vector quantity that specifies the location of an object relative to some reference point. It consists of both the length and the direction from the reference point to the object.

positron The antimatter counterpart of the electron. The positron is positively charged.

potential energy The stored form of energy that can produce motion. Potential energy is stored in the forces between or within objects.

power The measure of how quickly work is done on an object. The SI unit of power is the watt.

precession The change in orientation of a spinning object's rotational axis that occurs when it's subject to an outside torque.

pressure The average amount of force a fluid exerts on a certain region of surface area. Pressure is reported as the amount of force divided by the surface area over which that force is exerted. The SI unit of pressure is the pascal.

pressure drag The drag force that results from higher pressures at the front of an object than at its rear.

pressure gradient A distribution of pressures that varies continuously with position.

pressure potential energy The product of a fluid's volume times its pressure. However, this energy isn't really stored in the fluid. Instead, it's energy that's provided by a pump (or other source) when the fluid is delivered.

primary colors of light The three colors of light (red, green, and blue) that are sensed by the three types of color-sensitive cone cells in our eyes. Mixtures of these three colors of light can make our eyes perceive any possible color.

primary colors of pigment The three colors of pigment (cyan, magenta, and yellow) that absorb the three primary colors of light (red, green, and blue, respectively). Mixtures of these three pigments can be applied to a white surface to make it reflect any possible mixture of the three primary colors of light and thus to make our eyes perceive any possible color.

principle of equivalence The principle that gravitational mass and inertial mass are truly identical and therefore that no experiment you can perform in a small region of space can distinguish between free fall and the absence of gravity.

protons The positively charged subatomic particles found in atomic nuclei.

p-type semiconductor A semiconductor such as silicon that contains impurity atoms such as boron, aluminum, gallium, indium, or tellurium that remove electrons from the semiconductor's valence levels.

quanta The fundament, discrete units in which an item is emitted, absorbed, or otherwise observed, reflecting the particulate character of that item.

quantized Existing only in discrete units or quanta. Quantized physical quantities are only observed in integer multiples of the elementary quantum.

radian The natural unit in which angles are measured. There are 2π radians in a full circle, so 1 radian is $180/\pi$ degrees or approximately 57.3°.

radian-per-second (1/s) The SI unit of angular velocity or angular speed. An object turning at 1 radian-per-second completes a full revolution in just less than 6.3 seconds.

radian-per-second^2 (1/s^2) The SI unit of angular acceleration.

radiation The transmission of heat through the passage of electromagnetic radiation between objects.

radiation trapping The phenomenon in which a particular wavelength of light has trouble propagating through a material that eagerly absorbs and emits it. The light passes from one atom or atomlike system to the next and makes little headway.

radiative transition The shift of an atom or atomlike system from one state to another through the emission or absorption of an electromagnetic wave.

radio waves Electromagnetic waves, usually with wavelengths longer than about 1 m.

radioactive decay The spontaneous decay of a nucleus into fragments.

ramp An inclined plane that allows work to be done over a longer distance, thereby requiring less force.

Rayleigh scattering The redirection of light due to its interaction with small particles of matter.

real image A pattern of light, projected in space, that exactly reproduces the pattern of light at the surface of the original object. A real image forms after the lens that creates it and can be projected onto a surface.

rebound energy The amount of kinetic energy returned to two objects as they push apart following a collision.

reflection The redirection of all or part of a wave so that it returns from a boundary between media.

refraction The bending of a wave's path that occurs when the wave crosses a boundary between media and experiences a change in speed.

relative humidity The actual humidity as a percentage of the humidity required to achieve phase equilibrium between liquid and gaseous water.

relative motion The movement of one object from the perspective of another object. Two objects that are moving relative to one another have different velocities.

relativistic energy An object's energy according to the relativistic laws of motion and including its rest energy.

relativistic laws of motion The laws of motion in the special theory of relativity. They correct deficiencies in the Newtonian laws of motion that appear primarily at speeds comparable to the speed of light.

relativistic momentum An object's momentum according to the relativistic laws of motion.

resistor An electronic component that impedes the flow of electric current, converting some of its energy into heat.

resonant cavity A simple resonant circuit consisting of a carefully shaped conducting strip or shell and equivalent to a capacitor and an inductor. Energy flows back and forth between the cavity's electric and magnetic fields.

resonant energy transfer The gradual transfer of energy to or from a natural resonance caused by small forces timed to coincide with a particular part of each oscillatory cycle.

restoring force A force that acts to return an object to its equilibrium shape. A restoring force is directed toward the position the object occupies when it's in its equilibrium shape.

reverse biased A p-n junction in which the voltage of the p-type semiconductor has been lowered relative to the voltage of the n-type semiconductor.

Reynolds number A dimensionless number that characterizes fluid flow through a system. At low Reynolds numbers a fluid's viscosity dominates the flow, while at high Reynolds numbers a fluid's inertia dominates.

right-hand rule The convention whereby the specific direction of an object's angular velocity is established. According to this rule, if the fingers of your right hand are curled to point in the direction of the object's rotation, your thumb will point in the direction of the angular velocity.

root mean square (RMS) voltage A measure of AC voltage defined as the DC voltage that would cause the same average power consumption in an ohmic device.

rotational equilibrium The state of an object in which zero net torque acts on it. An object that has constant angular momentum is in rotational equilibrium.

rotational inertia A property of matter by which it remains at rest or in steady rotation about the same rotational axis unless acted on by some outside torque.

rotational mass The property of a body that is a measure of its rotational inertia. An object's rotational mass is determined by its mass and by how far that mass is from the axis of rotation. The SI unit of rotational mass is the kilogram-meter^2.

rotational motion Motion in which an object rotates about an axis. The orientation of an object undergoing only rotational motion will change, but its position will remain unchanged.

saturated In phase equilibrium with another material phase. The gaseous phase of a material is saturated when it is in phase equilibrium with that material's liquid and/or solid phase.

schematic diagram A symbolic picture of the conceptual structure of an electronic device.

second (s or sec) The SI unit of time. (One second is formally defined as the duration of 9,192,631,770 periods of the radiation corresponding to the transition between two hyperfine levels of the ground state of the cesium 133 atom.)

second law of thermodynamics The entropy of a thermally isolated system of objects never decreases. This law recognizes that creating disorder is easy; restoring order is hard.

semiconductor An insulator with a small band gap, so that only a modest amount of energy is needed to shift an electron from an occupied valence level to an unoccupied conduction level.

series (wiring arrangement) An arrangement in which the current reaching two or more electric devices flows sequentially through one device after the next before leaving them. Current may experience different changes in voltage in the different devices.

shell A group of atomic orbitals having similar energies.

shock wave A narrow region of high pressure and temperature that forms when the speed of an object through a medium exceeds the speed at which sound, waves, or other vibrations travel in that medium.

short circuit A defect in a circuit that allows current to bypass the load it's supposed to operate.

SI Units A system of units (Système Internationale d'Unités) that carefully defines related units according to powers of 10. SI units are now used almost exclusively throughout most of the world, with the notable exception of the United States.

signal An electrical or optical representation of information.

simple harmonic motion The regular, repetitive motion of a harmonic oscillator. The period of simple harmonic motion doesn't depend on the amplitude of oscillation.

sliding friction The forces that resist relative motion as two touching surfaces slide across one another.

soft Having a small spring constant and thus experiencing small restoring forces in response to large distortions.

soft magnetic material A material that is relatively easy to magnetize and that loses its magnetization once the magnetizing field is removed. Soft magnetic materials are suitable for electromagnets.

solid A form of matter consisting of particles (atoms or molecules) that touch and that are not free to move relative to one another. A solid has a fixed volume and shape.

sound In air, sound consists of density waves, patterns of compressions and rarefactions that travel outward from their source at the speed of sound.

special theory of relativity The physical rules governing all motion, even motion involving speeds comparable to the speed of light.

specific heat The amount of heat that must be added to a unit mass of a material to cause a unit rise in its temperature. The SI unit of specfic heat is the joule-per-kilogram-kelvin.

speed A measure of the distance an object travels in a certain amount of time. The SI unit of speed is the meter-per-second.

speed of light The speed with which an electromagnetic wave travels through space. In empty space, a vacuum, the speed of light is exactly 299,792,458 m/s.

speed of sound The speed at which sound's compressions and rarefactions travel in a medium such as air or water.

spontaneous emission of radiation Light emission that occurs when an excited atom or atomlike system releases stored energy randomly through a radiative transition. The photon that results is independent and unique.

spring constant As a measure of the stiffness of an elastic object, the spring constant relates the object's distortion to the restoring force it exerts. The larger the spring constant, the stiffer the spring.

springlike force A force that is proportional to displacement, consistent with Hooke's law.

sputtering Ejection of atoms from a surface caused by the impact of energetic ions, atoms, or other tiny projectiles.

stable equilibrium A state of equilibrium to which an object will return if it's disturbed. At equilibrium, the object is free of net force or torque. If an object is moved away from that equilibrium state, however, the net force or torque that will then act on it will tend to return it to equilibrium.

stall When a fluid flow stops and spoils steady-state flow. In the aerodynamic flow around an airfoil, stalling refers to airflow separation triggered by a stall in the flow near the airfoil's surface.

standard units Agreed on amounts of various physical quantities, which define a system in which those quantities are subsequently measured.

standing wave A wave in which all the nodes and antinodes remain in place.

state A possible arrangement of electrons (or other particles) in a quantum system.

static friction The forces that resist relative motion as outside forces try to make two touching surfaces begin to slide across one another.

static stability An object's stability when it's not in motion.

static variations Changes in a physical quantity such as pressure that are not caused by motion.

steady-state flow A situation in a fluid where the characteristics of the fluid at any fixed point in space don't change with time.

Stefan–Boltzmann constant The constant of proportionality relating a surface's radiated power to its emissivity, temperature, and surface area. It has a measured value of 5.67×10^{-8} J/(s·m^2·K^4).

Stefan–Boltzmann law The equation relating a surface's radiated power to its emissivity, temperature, and surface area.

stiffness A measure of how rapidly a restoring force increases as the system exerting that force is distorted.

stimulated emission of radiation Light emission that occurs when an excited atom or atomlike system releases stored energy through a radiative transition by duplicating a photon passing through that system.

streamline The path followed by a particular portion of a flowing fluid.

streamlined Carefully tapered so that the fluid flowing around it doesn't stall and doesn't experience flow separation of pressure drag.

strong force The fundamental force that gives structure to nuclei and nucleons and is the basis for the nuclear force.

subatomic particles The fundamental building blocks of the universe, from among which atoms and matter are constructed.

subcritical mass A portion of fissionable material that is too small to sustain a chain reaction.

sublimation The process by which atoms or molecules go directly from a solid to a gas.

superconductor An electrical conductor that permits electrons to flow without losing any of their kinetic energy to thermal energy. Electrons will continue to flow in a superconductor indefinitely. Materials only become superconducting at extremely low temperatures.

supercritical mass A portion of fissionable material that is well in excess of a critical mass so that it undergoes an explosive chain reaction.

superheated Above the temperature at which a phase transition should have occurred. Superheating results from a failure to nucleate the new phase.

superposition The overlapping of two or more waves so that their amplitudes add together and they form a combined wave.

support force A force that is exerted when two objects come into contact. Each object exerts a force on the other object to keep the two from passing through one another. Support forces are always normal, or perpendicular, to the surfaces of objects.

surface area The extent of a two-dimensional surface bounded by a particular border. The SI unit of surface area is the meter^2.

surface waves Disturbances in the stable equilibrium shape of a surface.

sympathetic vibration The transfer of energy between two natural resonances that share a common frequency of oscillation.

tank circuit A simple resonant circuit consisting of a capacitor and an inductor. Energy flows back and forth between these two devices repetitively.

temperature A measure of the average internal kinetic energy per particle in a material. In a gas, temperature measures the average kinetic energy of each atom or molecule.

tension Outward forces on an object that tend to stretch it.

terminal velocity The velocity at which an object moving through a fluid experiences enough drag force to balance the other forces on it and keep it from accelerating.

tesla (T)   The SI unit of magnetic field (synonymous with newton-per-ampere-meter).

thermal conductivity   The measure of a material's capacity to transport heat by conduction from its hotter end to its colder end.

thermal energy   A disordered form of energy contained in the kinetic and potential energies of the individual atoms and molecules that make up a substance. Because of its random distribution, this disordered energy can't be converted directly into useful work. Other names for thermal energy include internal energy and heat.

thermal equilibrium   A situation in which no heat flows in a system because all of the objects in the system are at the same temperature.

thermal motion   The random motions of individual particles in a material due to the internal or thermal energy of that material.

third law of thermodynamics   As an object's temperature approaches absolute zero, its entropy approaches zero. This law points out that absolute zero is the unattainable state in which an object has no disorder.

thrust   A forward, propulsive force.

tidal forces   The differences between one celestial object's gravity at particular locations on the surface of a second object and the average of that gravity for the entire second object. Tidal forces tend to stretch the second object into an egg shape.

timbre   The mixture of tones in an instrument's sound that are characteristic of that instrument.

torque   An influence that if exerted on a free body results chiefly in an angular acceleration of the body. A torque is a vector quantity, consisting of both the amount of torque and its direction. The SI unit of torque is the newton-meter.

total internal reflection   Complete reflection of a light wave that occurs when that wave tries unsuccessfully to leave a material with a large refractive index for a material with a small refractive index at too shallow an angle.

traction   The largest frictional force that an object can obtain in its present situation.

trajectory   The path taken by an object as it moves.

transformer   A device that uses magnetic fields to transfer electric power from one

circuit to another circuit. The two circuits are electrically isolated since no charges actually travel between the two circuits.

transistor   An electronic component that allows a tiny amount of electric charge, either moving or stationary, to control the flow of a large electric current.

translational motion   Motion in which an object moves as a whole along a straight or curved line.

transmutation of elements   Changing the atoms of one element into another via nuclear processes that alter the numbers of protons in their nuclei.

transverse wave   A wave in which the underlying oscillation is perpendicular to the wave itself.

traveling wave   A wave that moves steadily through space in a particular direction.

trough   The peak negative excursion of an extended system that is experiencing a wave.

tunneling   Because of the Heisenberg uncertainty principle, small objects have somewhat ill-defined positions and occasionally move through energy barriers to places they can't reach classically. That quantum process is tunneling.

turbulence   The unpredictable swirls and eddies of turbulent fluid flow.

turbulent flow   Irregular, fluctuating, unpredictable fluid flow in which nearby portions of the fluid quickly become widely separated.

ultraviolet light   Invisible light having wavelengths shorter than about 400 nanometers.

uniform circular motion   Motion at a constant speed around a circular trajectory. An object undergoing uniform circular motion is accelerating toward the center of the circle.

unstable equilibrium   An equilibrium situation to which the object will not return if it's disturbed. At equilibrium, the object is free of net force or torque. However, if the object is moved away from that equilibrium situation, the net force or torque that will then act on it will tend to accelerate it further away from the equilibrium situation.

valence band   The group of quantum levels in an insulator that lies below the Fermi level.

valence level   A quantum level in an insulator that requires less energy than the Fermi level and that is normally occupied by electrons.

vector quantity   A quantity, characterizing some aspect of a physical system, that consists of both a magnitude and a direction in space.

velocity   A vector quantity that measures how quickly an object's position is changing: the greater the velocity, the farther the object travels each second. It consists of both the object's speed and the direction in which the object is traveling. The SI unit of velocity is the meter-per-second.

vertical polarization   An electromagnetic wave in which the electric field always points up or down (vertically). The magnetic field always points horizontally.

vibration   A spontaneous repetitive and rhythmic movement about an equilibrium position.

vibrational antinode   A region of a vibrating object that is experiencing maximal motion.

vibrational node   A region of a vibrating object that is not moving at all.

virtual image   A pattern of light that appears to come from a particular region of space and reproduces the pattern of light at the surface of the original object. A virtual image forms before the lens that creates it and can't be projected onto a surface.

viscosity   The measure of a fluid's resistance to relative motion within that fluid.

viscous drag   A drag force that results from viscous forces on a moving surface immersed in a fluid.

viscous forces   The forces exerted within a fluid that oppose relative motion. Layers of fluid that are moving across one another exert viscous forces on each other.

visible light   Light having wavelengths between about 400 nanometers (violet) and 750 nanometers (red). This small portion of the electromagnetic spectrum is all that we are able to detect with our eyes.

volt (V)   The SI unit of voltage (synonymous with joule-per-coulomb). The voltage on the positive terminal of a common battery is about 1.5 volts above that on its negative terminal.

volt-per-meter (V/m)   The SI unit of electric field (synonymous with newton-per-coulomb).

voltage   The electrostatic potential energy of each unit of positive electric charge at a particular location. The SI unit of voltage is the volt.

voltage drop   The amount of electrostatic potential energy that each coulomb of positive charge loses in passing through a device. It's equal to the voltage of the charges entering the device minus the voltage of the charges leaving that device.

voltage gradient   A gradual slope in the voltage across a region of space. A voltage gradient is an electric field.

voltage rise   The amount of electrostatic potential energy that each coulomb of positive charge receives in passing through a device. It's equal to the voltage of the charges leaving the device minus the voltage of the charges entering that device.

volume   The extent of a three-dimensional region of space bounded by a particular enclosure. The SI unit of volume is the meter^3.

vortex   A whirling region of fluid that is moving in a circle above a central cavity.

wake   The trail left behind by an object as it moves through a fluid.

wake deflection force   A lift force experienced by a spinning ball when it deflects its turbulent wake to one side. The wake deflection force points toward the side of the ball moving away from the onrushing airstream.

water hammer   The impact of a moving mass of water that is suddenly stopped.

watt (W)   The SI unit of power, equal to the transfer of 1 joule-per-second. One watt is the power used by the bulb of a typical flashlight.

wave velocity   The speed and direction of the moving crests of a wave.

wave–particle duality   The observation that everything in nature has both particle and wave characteristics. An item is primarily particlelike when it is emitted, absorbed, or otherwise observed and primarily wavelike as it travels through time and space.

wavelength   A structural characteristic of a wave, corresponding to the distance separating adjacent crests or troughs.

weak force   The fundamental force that allows electrons and neutrinos to interact and that's responsible for beta decay.

weight   (near the earth's surface) The downward force exerted on an object due to its gravitational interaction with the earth. An object's weight is equal to the product of that object's mass times the acceleration due to gravity. The direction of the weight is always toward the center of the earth.

work   The mechanical means of transferring energy. Work is defined as the force exerted on an object times the distance that object travels in the direction of the force. A large force exerted for a short distance or a small force exerted for a long distance can perform the same amount of work. The SI unit of work is the joule.

X-rays   Very high-energy photons of electromagnetic radiation.

X-ray fluorescence   The process in which an electron in one of the outer orbitals of an atom undergoes a radiative transition to an empty inner orbital, emitting an X-ray photon.

zeroth law of thermodynamics   Two objects that are each in thermal equilibrium with a third object are also in thermal equilibrium with one another. This law is the basis for a meaningful system of temperatures.

Solutions to Selected Exercises and Problems

Chapter 1

E.1 The dolphin's inertia carries it upward, even though its weight makes it accelerate downward and gradually stop rising.

E.3 Your feet accelerate upward rapidly when they hit the ground and the snow continues downward, leaving your feet behind.

E.5 Any collision in which the car accelerates forward, such as when the car is hit from behind by a faster moving car.

E.7 As it turns left, the car accelerates left. The loose objects remain behind and end up on the right side of the dashboard.

E.9 Backward, in the direction opposite your forward velocity.

E.11 The anvil's large mass slows its acceleration, so the hot metal is squeezed between the moving hammer and the stationary anvil.

E.13 The pad's inertia tends to keep it in place. If you pull the paper away too quickly, the pad won't be able to accelerate with the paper.

E.15 Regardless of their horizontal components of velocity, all objects fall at the same rate. The ball and bullet descend together.

E.17 The forward component of his velocity remains constant as he falls, and he follows an arc that carries him forward over the rocks.

E.19 In the absence of air effects, a ball hit at 45° above horizontal will travel farthest. A ball hit higher or lower won't travel as far.

E.21 Its magnitude must equal the suitcase's weight.

E.23 Zero net force.

E.25 The astronaut exerts an upward force of 850 N on the earth.

E.27 Both forces have exactly the same magnitude.

E.29 The wall exerts a support force to accelerate you backward.

E.31 A person does more work. Movements that appear large compared to the ant's height still involve small distances and little work.

E.33 Yes, it pushed the wall inward and the wall dented inward.

E.35 As it rolls on the surface, it always accelerates downhill.

E.37 Its greatest acceleration is at the steepest point; its greatest speed is at the bottom of the slide.

E.39 The kinetic energy becomes gravitational potential energy.

P.1 3200 N.　　　　**P.3** 11.13 m/s.

P.5 Your Mars weight would be about 38% of your earth weight.

P.7 About 0.64 s (0.32 s on the way up and 0.32 s on the way down).

P.9 48 N　　　　**P.11** About 0.20 s.

P.13 4800 N　　　**P.15** 9800 N.

P.17 14,700,000 J.　**P.19** 12.5 J.

P.21 8000 N.

Chapter 2

E.1 The angle by which the front, center seat would have to be rotated, as viewed from above, to have each seat's orientation.

E.3 It would rotate about its center of mass, not its geometric center. It would appear to wobble as it turned.

E.5 The wheel has rotational inertia, as measured by its rotational mass, making it hard to start and stop spinning.

E.7 A force exerted at the hinges produces no torque about them.

E.9 The farther the water is from the water wheel's pivot, the more torque its weight produces on the wheel.

E.11 Your force far from the hinges produces a large torque. To oppose this torque, the nut must exert a huge force near the hinges.

E.13 The weights of your chest and your feet exert torques in opposite directions about your knees. They partially balance one another.

E.15 It reduces the car's rotational mass so that the car can undergo rapid angular accelerations and change directions quickly.

E.17 By pushing far from the pivot, you exert more torque on the lid.

E.19 Your small effort exerted on the crowbar far from its pivot produces a large force on the box, located near the crowbar's pivot.

E.21 Skidding sideways does work against sliding friction, converting some of the skier's kinetic energy into thermal energy.

E.23 The pin's surface turns with the crust and doesn't slide across it.

E.25 The nearer the frictional force is to the pivot, the less torque it produces to slow the yo-yo's rotation. Slipperiness reduces friction.

E.27 A static frictional force from the pavement pushes you forward.

E.29 Pressing the wheels more tightly against the pavement increases the maximum force that static friction can exert on the wheels.

E.31 The chalk experiences sliding friction as you write and leaves visible wear chips on the blackboard.

E.33 To avoid accelerating when pushed on with a force, the villain would have to have infinite mass. That's impossible.

E.35 Angular momentum.

E.37 The collapsing star's angular momentum can't change. Since its rotational mass decreases, its angular velocity must increase.

E.39 As you descend, you land hard and your knees and legs must convert your kinetic energy into thermal energy. Injuries can occur.

E.41 Sliding friction converts some into heat, but a slippery pole converts a considerable fraction into kinetic energy.

P.1 About 122.5 N·m to the left.

P.3 Three times as much torque.

P.5 12.5 N·m, slowing the blade.

P.7 Its energy would be only 0.2 times as large as before.

P.9 2400 kg·m/s forward. **P.11** 3600 J.

P.13 450 kg·m/s to the right.

Chapter 3

E.1 As you draw the string away from its equilibrium shape, it experiences a restoring force proportional to its displacement.

E.3 The top of the curl supports more weight than the bottom.

E.5 15 N. **E.7** Correct.

E.9 You support your clothes and the scale supports you.

E.11 A 30% loss of rebound height is a 30% loss of gravitational potential energy—equal to the energy that became thermal energy.

E.13 You do work on the sand as you step on it, but the sand doesn't return this energy to you as you lift your foot back up again.

E.15 An increase in the balls' coefficients of restitution.

E.17 Your relative velocity is small—in your frame of reference the car in front of you is barely moving, so the impact is very gentle.

E.19 Because the trains' relative velocity is zero, a person jumping between them views them both as essentially stationary.

E.21 During a bounce, the work done on a RIF ball—to store energy in it—involves a smaller force exerted for a longer distance.

E.23 At high pressure, the shoes are stiffer and exert larger forces when distorted. They accelerate more rapidly and bounce faster.

E.25 The force that the hammer exerts on the nail during their collision would diminish if the hammer's surface distorted easily.

E.27 While you can feel accelerations, you can't feel velocity.

E.29 When the rattle accelerates, the beads inside it continue on and hit the walls of the rattle. The rattle then makes noise.

E.31 The sharper the curve, the more centripetal force the train needs to accelerate around the curve. If the track can't supply it . . . disaster.

E.33 The nail exerts an enormous force on the hammer to slow it down. The hammer pushes back, driving the nail into the wood.

E.35 At the bottom of each swing.

E.37 As the salad undergoes rapid centripetal acceleration, the water travels in straight lines and runs off the salad.

E.39 After going over a bump, the car begins to accelerate downward and your apparent weight is briefly less than your real weight.

E.41 As the car accelerates upward, you must pull upward on the briefcase extra hard to make it accelerate upward too.

P.1 15,000 N/m. **P.3** 10 mm.

P.5 It must turn at about 31 m/s.

P.7 20 m/s².

Chapter 4

E.1 As a sprinter accelerates, the ground must push toward the sprinter's center of mass to avoid exerting a torque on the sprinter.

E.3 The leftward frictional force on your feet keeps you balanced.

E.5 As long as the inward force from the U-shaped surface points toward the skateboard's center of mass, the skateboarder doesn't begin rotating.

E.7 The torque you exert on the crank is much larger than the torque the blades exert on the batter.

E.9 Each second, the blade pushes the dough a short distance. To do significant work, the force it exerts on the dough must be large.

E.11 As the blower pushes air toward the leaves, that air pushes the blower away from the leaves.

E.13 They are equivalent. **E.15** Yes.

E.17 The ball transfers momentum between you so that you acquire momentum in the opposite direction from your friend.

E.19 Directly toward the earth.

E.21 A rocket must do much less work against the moon's gravity.

E.23 Since a line from the sun to the comet must sweep out area at a steady rate, the closer the comet gets to the sun, the faster it must arc around the sun.

P.1 About 0.076 m/s.

P.3 About 0.00028 times your earth weight.

P.5 About 4.7×10^{14} N.

P.7 1.73×10^{11} kg·m/s. **P.9** 5.56×10^{-10}.

P.11 1.1×10^{-14} kg· **P.13** 1.26.

Chapter 5

E.1 It will sink. Its average density is larger than that of helium.

E.3 As water enters the car, the car's average density will increase.

E.5 The upper liquid is less dense than the lower liquid. It floats.

E.7 One kilogram of gasoline takes up more space than 1 kg of water.

E.9 An upward force equal in magnitude to the fish's weight.

E.11 Air pressure decreases steadily with altitude.

E.13 Air pressure holds the dimple down when the jar has a vacuum inside, but the pressure difference vanishes when the jar is opened.

E.15 Cooling the air in the container reduced its pressure. The resulting pressure imbalance across the lid pushes the lid inward.

E.17 As it cooled, the air became more dense.

E.19 Higher pressure in the can produces greater distance of spray. As for direction, the spray won't travel as far upward because some of its kinetic energy becomes gravitational potential energy on the way up.

E.21 The pressure imbalance on the base of the dam is larger.

E.23 The pressure outside your lungs is above atmospheric pressure.

E.25 Gas only accelerates toward lower pressure.

E.27 As wind slows in the bag, its pressure rises and inflates the bag.

P.1 101,400 Pa.

P.3 Three times as much as before.

P.5 0.94 times as much as before.

P.7 122.4 kg or 0.1224 m³. **P.9** 28.4 N.

P.11 About 250,000 Pa above atmospheric pressure.

P.13 141 m/s or 510 km/h.

P.15 About 3,100,000 Pa.

Chapter 6

E.1 The increased flow in the cold water pipe requires a larger pressure difference in it, leaving too little pressure to operate the shower.

E.3 Flow through an artery scales with the fourth power of radius.

E.5 In warm weather, the molecules in a viscous liquid have more thermal energy and are able to move past one another more easily.

E.7 Squeezing highly viscous frosting through a narrow "pipe" requires an enormous pressure difference across that pipe.

E.9 Air near the stationary bridge surface is slowed by viscous forces, creating a slower moving boundary layer.

E.11 The pressure at the can bottom rises suddenly due to water hammer. The pressure at the can top falls or remains the same.

E.13 In turbulent flow, adjacent regions of water become separated.

E.15 The bucket stops the water, so its pressure rises. The higher pressure above the bucket pushes it downward hard.

E.17 The water slows down just upstream and downstream of the posts and speeds up on the sides of the posts.

E.19 The dimpled ball will hit first.

E.21 The front runner drags the air forward so that you experience less drag while running through forward-moving air.

E.23 It balances the backward force due to air drag.

E.25 Both objects are slowed by similar pressure drags. But the spear's larger mass and momentum keep it from slowing as quickly.

E.27 It increases pressure drag and reduces your terminal velocity.

E.29 An airfoil, round in front and tapered behind, would be better.

E.31 Lift supports the skier and drag pulls the skier backward.

E.33 Without backspin, the ball can't obtain lift and won't go as far.

E.35 Small disturbances in the airflow around the sides of the nonspinning volleyball produce lift forces that push the ball to the side.

E.37 The water speeds up around the curve and its pressure drops. The resulting pressure imbalance pushes the spoon into the stream.

E.39 Air travels faster over your hand than under it, so the pressure above your hand is less than the pressure under your hand.

E.41 Turbulent air pockets form behind its moving blades, so swirling vortices and eddies flow out of the fan.

P.1 For a fish with an obstacle length (width) of about 2 cm, turbulence will appear if it moves faster than about 0.1 m/s.

P.3 About 84 times as long.

P.5 About 400,000 Pa. **P.7** About 18.3 m/s.

Chapter 7

E.1 Your body temperature would decrease.

E.3 The work you do in pushing the bread in the direction it moves.

E.5 Foods are poor conductors of heat. The metal skewer carries heat better.

E.7 The grate lets convection carry air upward past the wood.

E.9 Heat rises with convection and ignites the rest of the stick.

E.11 Black is the best emitter of thermal radiation, whereas white and silver are poor emitters.

E.13 The solid material sublimes and its molecules become gaseous.

E.15 The ice is colder than its melting temperature and must warm up before it starts melting.

E.17 The bag traps steam so that the humidity soon rises to 100% so that net evaporation ceases.

E.19 It goes into the ice's latent heat of melting.

E.21 The boiling water will remove heat quickly enough to protect the pot.

E.23 It prevents evaporation from cooling the water.

E.25 Water boils at a lower temperature in Denver.

E.27 Anything dissolved in water stabilizes the water so that it has trouble freezing or evaporating.

E.29 Different chemicals leave the liquid perfume at different rates, so some evaporate away earlier than others. Higher skin temperature promotes evaporation, even in the slower leaving chemicals.

E.31 From the color of its light; whiter is hotter.

E.33 The cooler the object, the longer the wavelengths of its blackbody spectrum. The spectrum of a very cold object peaks in the microwave portion of the electromagnetic spectrum.

E.35 In a three-way bulb, any filament that's on is at full temperature. A dimmed bulb's filament operates below full temperature.

E.37 Cooler skin emits dimmer infrared with longer wavelengths.

E.39 The sidewalk has a different coefficient of volume expansion than the ground and would break when the temperature changed.

E.41 The metal lid has a larger coefficient of volume expansion than the glass jar, and it pulls away from the jar when you heat them both.

P.1 18,800 W   **P.3** 73,500,000 W.

P.5 The amount of heat radiated away is proportional to the area of the radiating surface. Double that surface area and the radiated heat will double.

Chapter 8

E.1 They are heat pumps and transfer heat to the surrounding air.

E.3 Without the fan, the air conditioner won't be able to transfer its waste heat to the surrounding air. It will stop pumping heat.

E.5 Its temperature increased. Gravity did work on the gas, and this work appeared in the gas as thermal energy. The gas became hot.

E.7 The gas still in the container does work pushing the other gas into the siphon and uses up some of its thermal energy. It cools.

E.9 Though not forbidden by the laws of motion, it's extraordinarily unlikely for the vase fragments to reassemble themselves.

E.11 This uneven distribution of thermal energies is extraordinarily unlikely. It would violate the second law of thermodynamics.

E.13 Though not forbidden by the laws of motion, such uneven distributions of snow are remarkably unlikely.

E.15 Heat flowing into or out of the pavement during weather changes allows the pavement to do work as it tears itself apart.

E.17 As heat flows from the hot spots to the cold spots, by way of huge convection cells, some heat is becoming ordered kinetic energy.

E.19 Without temperature differences, heat won't naturally flow. Such heat flow is what powers a heat engine such as the winds.

E.21 The hotter the burned gas, the larger the fraction of heat that can be converted to work as it flows to the outdoor air.

E.23 Converting burned fuel entirely into work would violate the second law of thermodynamics.

P.1 4190 J.   **P.3** 0.77 °C.

P.5 1000 J.   **P.7** 1000 J.

P.9 40 J.   **P.11** 400 J.

P.13 95% of the heat can become work.

Chapter 9

E.1 It would run slow.

E.3 The period will decrease as the pendulum gets shorter.

E.5 Push on it as it flexes away from you (so you do work on it).

E.7 Multiplying sound's roundtrip time by its speed gives twice the distance to the wall.

E.9 Less mass, more tension, or less length.

E.11 The copper wrap adds mass to the string to lower its pitch.

E.13 The longer rods have more mass vibrating and are less stiff.

E.15 A piccolo's air columns are half the length of those in a flute, so they vibrate at twice the pitch or one octave higher.

E.17 Ocean water can't enter or leave the Mediterranean Sea quickly enough to allow its high tide to be very different from its low tide. The tide simply rearranges the water levels within the sea itself.

E.19 Resonant energy transfer occurs when your steps are synchronized with the rhythmic motion of the coffee sloshing in the cup

E.21 The wave's energy is proportional to its circumference and its height. As its circumference grows, its height must diminish.

E.23 Crests and troughs don't travel along the string. Instead, the center of the string becomes alternately a crest and a trough.

E.25 There isn't enough water in front of the largest waves to complete their crests as they pass over the sand bar.

E.27 Longitudinal.

E.29 A surface can't vibrate as half- or third-surfaces, so its overtones are complicated and don't occur at harmonic frequencies.

E.31 The surface projects sound waves far better than strings alone.

E.33 No.

E.35 The sound waves reflect from the stone surface.

E.37 Some of the wave reflects but the rest of its energy becomes thermal energy.

E.39 Refraction occurs as a wave slows down over the shallow coral.

E.41 Interference between the two sound waves causes a beating effect.

P.1 3.01 m. **P.3** 264 m/s.

P.5 5.2 m/s.

Chapter 10

E.1 You'd have to compare the objects with a reference. The object that repelled the reference would have its charge.

E.3 You can't keep dividing the charge in half because charge comes in discrete units, the elementary unit of electric charge.

E.5 The charged paint particles electrically polarize the surface being coated, so that the paint particles are attracted to it and stick.

E.7 The wall's attractive and repulsive forces would be equal but oppositely directed, summing to zero net force on the balloon.

E.9 The two pieces are equivalent, so they acquire like charges while separating from the tape dispenser. Like charges repel.

E.11 The positive balloon has a higher voltage than the negative one.

E.13 Their voltages change. The positive balloon's voltage increases; the negative balloon's voltage decreases.

E.15 Conductive materials allow charge to flow and minimize its energy. Everything will tend toward electrical neutrality.

E.17 For you to accumulate a great deal of charge, you must not be able to lose it easily through conducting paths to the ground.

E.19 The forces between a positive charge and the hairbrush decrease with their separation. Thus the electric field that acts on that positive charge also decreases with distance from the hairbrush.

E.21 The 9-V battery has the larger voltage gradient and therefore the stronger electric field because its two terminals have a greater voltage difference and a shorter distance between them.

E.23 The like charges repel and flow through the conducting car to its outside surfaces.

E.25 Downward, away from the cloud.

E.27 The voltage falls or rises rapidly toward zero in the vicinity of the sharp point and that large voltage gradient is a strong electric field.

E.29 The charged cloud overhead would have induced a large opposite charge to flow up from the ground onto your hair.

E.31 While charge can accumulate on the bird, it can't flow as a current because it has nowhere else to go.

E.33 Current arrives through one wire and leaves through the other.

E.35 Yes, the battery will have lost some chemical potential energy.

E.37 The socket's central pin has the higher voltage.

E.39 Batteries, because something is supplying power to the current.

E.41 The power deposited in a metal is proportional to its electric resistance, so high-resistance metals heat more.

P.1 3.3×10^{-9} C. **P.3** 8.3×10^{19} N.

P.5 94,800 m. **P.7** 0.05 N upward.

P.9 20 N/C (20 V/m) pointing forward.

P.11 0.18 N toward the negative terminal.

P.13 1 A. **P.15** 0.15 W.

P.17 Each battery in this problem supplies 3 W.

P.19 About 13,333 s, or 3.7 hours.

P.21 0.05 Ω. **P.23** 0.6 V.

P.25 240,000,000 W. **P.27** 0.05 A.

P.29 0.1 Ω. **P.31** 5 V.

Chapter 11

E.1 No. Magnetic monopoles have never been found and the forces between magnetic dipoles depend on their relative orientations.

E.3 The two compass needles will pivot so as to minimize their total potential energies and will soon have opposite poles pointing toward one another. So aligned, they will then attract.

E.5 The domains aligned with the new applied field grow while those that are anti-aligned with that field shrink.

E.7 Zero. **E.9** They are equal.

E.11 The flux lines extend farther from the 2-poles/cm strip.

E.13 Encase the magnet in an iron box. The iron will then guide the flux lines.

E.15 The voltage drop in each wire is proportional to the current.

E.17 It will consume 40 W.

E.19 The card's magnetic poles must move in order to induce currents in the coil of wire.

E.21 20 turns. **E.23** 40 V AC.

E.25 9 A.

E.27 The bouncing magnet heats the ring by inducing current in it. That current extracts this heating energy from the magnet by exerting magnetic forces on it and doing negative work on it.

E.29 Form a circuit from the coil, battery, and switch and close the circuit briefly each time the compass needle reaches anti-alignment with the coil's field.

E.31 The magnets will all eventually orient themselves in the minimum energy configuration and never move again.

E.33 The rotor's rotation speed will double.

E.35 The rotor will vibrate but will not spin.

E.37 Current in a magnetic field experiences the Lorentz force and bends the filament back and forth.

E.39 The inductor opposes changes in current and slows the rise in current that occurs when you close the switch.

E.41 The electron beams will be deflected by Lorentz forces as they pass through the stray magnetic fields from those magnets.

P.1 0.10 N upward. **P.3** 1.0 T upward.

P.5 0.10 A·m. **P.7** 0.0010 J.

P.9 2.5×10^{-6} m^3.

P.11 0.0016 T.

Chapter 12

E.1 5000 levels.

E.3 All in lowest energy level.

E.5 They will increase the semiconductor's electrical conductivity.

E.7 The p-type half has a negative net charge.

E.9 The one with the thicker insulating layer.

E.11 The voltage difference between the plates increases with each transfer, so the energy required to complete the subsequent transfer is greater.

E.13 There are 6 hundreds.

E.15 *219* and *85*.

E.17 Instead of reporting 2 twos, the binary representation should report it as 1 four, the next higher power of two.

E.19 The gate's charge must be very near the channel so that it can attract opposite charge and draw that charge into the channel.

E.21 When current experiences a voltage drop as it flows through a MOSFET, some of its energy is converted into thermal energy.

E.23 It takes time for a MOSFET to change the charge on a wire in order to store or retrieve a bit.

E.25 The microphone can't provide enough power.

E.27 The phonograph produces a much smaller voltage rise than a CD player would provide. The amplifier expects a larger voltage.

Chapter 13

E.1 Charge will oscillate in the tank's capacitor and inductor.

E.3 The inductor's magnetic field contains energy and it peaks when the capacitor's energy is zero.

E.5 Each time the antenna reaches its peak positive charge, push the positive wand close to it. You'll then be doing work on the oscillating charge.

E.7 As charges accelerate in the wires, they emit radio waves.

E.9 As charges accelerate in the computer, they emit radio waves.

E.11 The spiraling charges are accelerating and thus emit electromagnetic waves.

E.13 Vertical polarization.

E.15 The AM station changes the power of its transmission in order to represent air pressure fluctuations with the radio wave.

E.17 Amplitude modulation introduces additional frequencies that extend as much as 5 kHz above and below the carrier wave.

E.19 Water trapped in the ceramic would absorb microwaves, and the ceramic would become extremely hot. It might even shatter.

E.21 Releasing the microwaves into the room wouldn't be healthy.

E.23 Both involve destructive interference in electromagnetic waves.

E.25 It is thin enough to be heated by the resulting currents and its sharp ends may spark.

E.27 The path of a faster moving electron bends more gradually, so it travels in a larger circle. It returns to its starting point at the same time the slower moving electron returns to its starting point.

P.1 4.4×10^{-12} J.   **P.3** 230,000 m^3.

P.5 480,000 V/m.   **P.7** 0.333 m.

P.9 0.461×10^{-6} m (461 nm).

P.11 8.33×10^8 Hz.

Chapter 14

E.1 About 160 nm.

E.3 Rayleigh scattering deflects blue light in all directions.

E.5 The light refracts as it enters and leaves the glass.

E.7 The large particles scatter light better than air molecules.

E.9 Light bends and disperses on entry, reflects from the back surface, and bends and disperses more on exit from the diamond.

E.11 Each surface in the granulated sugar reflects some of the light passing through it. These random reflections make the sugar white.

E.13 Light reflects from the top and bottom surfaces of an oil film, and the two reflections interfere with one another. The type of interference

depends on the film's thickness and the light's wavelength.

E.15 Their polarizing sunglasses transmit only vertically polarized light, which your rotated polarizing sunglasses now block. You see no light.

E.17 Green and blue.

E.19 The green, blue, and violet end of the spectrum.

E.21 Black.

E.23 There is no lower energy state to which it can make a transition and, while it remains in the ground state, its electrons are in standing waves and cannot emit electromagnetic waves.

E.25 Nothing happens because the atoms have no radiative transitions that can absorb photons of that light.

E.27 Excited mercury atoms emit primarily invisible ultraviolet light.

E.29 Mercury atoms themselves produce the tubes' ultraviolet light.

E.31 Fluorescence from brighteners replaces the missing blue light.

E.33 The lightbulb's photons are all different and won't focus to the same tiny spot.

E.35 The radio station's photons are identical—part of a single wave.

E.37 They'll absorb the microwaves the maser is trying to produce.

E.39 The ground state systems absorb much of the amplified light before it can leave the ruby.

E.41 The color of light emitted by an LED depends primarily on the band gap in the LED's semiconductor.

P.1 3.37×10^{-19} J.

P.3 7.95×10^{-15} J.

P.5 8.58×10^{31} photons/s.

Chapter 15

E.1 It is a real image of the circular sun itself.

E.3 The 2.0-diopter lens has the shorter focal length.

E.5 The lens equation gives 0.5 cm as the focal length of the lens when the image and object distances are both 1.0 cm.

E.7 Their depths of focus are small.

E.9 To have an f-number of 1.8, the elements in the 200-mm lens must be four times the diameter of the elements in the 50-mm lens.

E.11 Reducing the size of your eye's aperture increases its depth of focus.

E.13 The depth of focus is smallest when the whole lens is used.

E.15 The lower magnification 2 × glass has the longer focal length.

E.17 The low mass lens can be accelerated rapidly by modest forces.

E.19 The ray bends to travel more nearly perpendicular to the interface.

E.21 Diffraction makes the narrowed light wave spread severely.

E.23 The lens's large opening reduces diffraction effects.

E.25 Light is experiencing total internal reflection inside the glass.

E.27 Waves reflected by the pits and flats interfere destructively.

P.1 35 mm.

P.3 The real image of the more distant statue forms 2.2 mm closer.

P.5 5 m.

Chapter 16

E.1 Nuclei of ^{65}Cu atoms contain two more neutrons than those of ^{63}Cu.

E.3 No, the number of electrons is set by the number of protons.

E.5 Neutrons would experience the attractive nuclear force while diluting the repulsion between protons.

E.7 With fewer protons, the diluting effect of neutrons matters less.

E.9 A neutron has no charge and only interacts with nuclei. The nuclei are so small that they're hard to hit.

E.11 With time, radioactive nuclei decay spontaneously and the materials become less and less hazardous.

E.13 Chemical reactions have no effect on nuclei.

E.15 Photons in sunlight can cause chemical damage.

E.17 Photons in blue or ultraviolet light can cause chemical damage to the molecules in the manuscripts. Yellow light generally can't.

E.19 Isotopes with shorter half-lives decay quickly and can be waited out, while those with much longer half-lives are relatively less likely to decay during a human lifetime.

E.21 Almost any X-ray matches the energy of one of lead's many electrons and thus can cause efficient photoelectron emission.

E.23 An antiproton has more mass than a positron.

E.25 MRI detects hydrogen. Bone contains little hydrogen.

E.27 The stronger the magnetic field, the more energy a proton needs to change from aligned to anti-aligned. The radio wave photons must have more energy, so the frequency must be higher.

P.1 65% of the ^{67}Ga nuclei remain.

P.3 63% of the ^{99m}Tc nuclei remain.

P.5 6.44×10^{-12} percent of the nuclei will decay.

Index